Botany for Degree Students

GYMNOSPERMS

BOTANY

FOR DEGREE STUDENTS

GYMNOSPERMS

Dr. P. C. VASHISHTA
M.Sc. (Hons. Sch.) ; Ph.D.; P.E.S. (I)
Retd. Ex Principal, Government College for Boys Chandigarh and D.A.V. College, Hoshiarpur

Revised by

Dr. A.K. SINHA
M.Sc., Ph.D.; F.P.S.I.
Reader & Head (Retd.)
Department of Botany
Feroze Gandhi Postgraduate College
(Kanpur University), Raebareli, U.P.

Dr. ANIL KUMAR
M.Sc., Ph.D.
Reader
Department of Botany
Feroze Gandhi Postgraduate College
(Kanpur University), Raebareli, U.P.

S. CHAND
PUBLISHING

S Chand And Company Limited
(ISO 9001 Certified Company)

S. CHAND PUBLISHING

S Chand And Company Limited

(ISO 9001 Certified Company)

Head Office: D-92, Sector–2, Noida – 201301, U.P. (India), Ph. 91-120-4682700

Registered Office: A-27, 2nd Floor, Mohan Co-operative Industrial Estate, New Delhi – 110 044, Phone: 011-49731800

www.**schandpublishing.com;** e-mail: **info@schandpublishing.com**

Marketing Offices:

Chennai	:	Ph: 23632120; chennai@schandpublishing.com
Guwahati	:	Ph: 2738811, 2735640; guwahati@schandpublishing.com
Hyderabad	:	Ph: 40186018; hyderabad@schandpublishing.com
Jalandhar	:	Ph: 4645630; jalandhar@schandpublishing.com
Kolkata	:	Ph: 23357458, 23353914; kolkata@schandpublishing.com
Lucknow	:	Ph: 4003633; lucknow@schandpublishing.com
Mumbai	:	Ph: 25000297; mumbai@schandpublishing.com
Patna	:	Ph: 4011400; patna@schandpublishing.com

First Edition 1976
Subsequent Editions and Reprints 1976, 78, 80, 83 (Twice), 85, 86, 87, 89, 90, 91, 92, 94, 95, 96, 98, 99, 2001, 2004, 2005, First Multicolour Illustrative Revised Edition 2006, Reprints 2007, 2009, 2011, 2016 (Twice), 2018 (Twice), 2020, 2021 (Twice), 2023 (Twice)

Reprint 2026

ISBN: 978-81-219-2618-8 **Product Code:** H6BGY68BOTN10ENZX06O

PRINTED IN INDIA

By Shri Shyamlal Printing Press Private Limited, Plot 20/4, Site-IV, Industrial Area Sahibabad, Ghaziabad – 201 010 and Published by S. Chand And Company Limited, A-27, 2nd Floor, Mohan Co-operative Industrial Estate, New Delhi – 110 044.

Preface to the Revised Edition

This thoroughly revised, improved and multicoloured edition of this popular book **"Botany for Degree Students—Gymnosperms"** has been brought out for the benefit of undergraduate and post-graduate students. While the original format remains unchanged, lot of matter has been added. A new chapter dealing with recent topics like Experimental Embryology and Biotechnology and application of biotechnology, world distribution of gymnosperms, impact of gymnosperms on human society and areas to be explored in studies of gymnosperms is an added attraction. In addition, each and every chapter in the book has been thoroughly revised and wherever needed, recent matter has been added. All the diagrams have been redrawn and several new diagrams have been added. As per the new trend of question papers, the three types of questions essay type, short answer type and objective type questions have been added. We have taken materials from the research journals and recent textbooks. The editors wish to express their sincere gratitude to the authors of the research papers and books and to their publishers.

We have been guided and assisted by a number of university and college professors and friends. We express our grateful thanks to each one of them.

Our heartiest thanks are due to Mr. R.K. Gupta, Managing Director, Shri R.S. Saxena, Shri Navin Joshi, General Manager (Sales and Marketing), Ms. Malini Kothiyal Asstt. Editor, Mr. Santosh Kumar Singh Asstt. Editor & Mr. Kaliram, Sr. D.T.P. Operator for their help, encouragement and cooperation in bringing out the book in this improved form. It is hoped that this multicolour, revised and improved edition of the book will prove more useful to all its users.

Suggestions for further improvement of this book will be gratefully accepted and acknowledged.

A.K. Sinha
Anil Kumar

Preface to the First Edition

This text has been written in a manner to suit the needs of degree students of Indian Universities. It is fifth in the series Botany for Degree Students that was started by my father in the year 1960, and describes the morphology and reproduction of both living and fossil gymnosperms. Special care has been taken to present the subject matter in a simple and lucid style and to include maximum of recent information available uptodate. Large number of research papers have appeared and many new horizons of both living and fossil forms have been unravelled. Much work has been done on the fossil gymnosperms and several details, hitherto unknown, have been brought to our knowledge. The works of Arnold, Andrews, Florin, Delevoryas, Beck, Taylor, Eggert, Miller, Rothwell, Phillips, Crepet, Banks, and Jennings needs special mention. In India Awasthi, Bharadwaja, Bose, Chowdhury, Ganju, Ghosh, Gupta, Jain, Mehta, Pant, Prakash, Tripathi, Sahni, M.R. Srivastava, Sharma, and Vishnu Mittre have done much work on the fossil gymnosperms and have kept up the tradition started by the Indian stalwarts of the calibre of Birbal Sahni and M.B. Raizada. Lot of work has been done on the living gymnosperms and in this context names of Prof. P.N. Mehra, Prof. P. Maheshwari, Prof. B.M. Johri, Prof. T.N. Khoshoo, Prof. D.D. Pant, Dr. L.N. Rao, Prof. R.N. Konar, Dr. R. Khan, Dr. Hardev Singh and their students will be remembered for ever. The works of Coulter and Chamberlain, Chamberlain, Pearson, Thomas, Thompson, Sporne, Martens, Waterkeyen, Namboodari, C.B. Beck, T.A. Steeves, Alosi and Alfieri and large number of other workers have added a great deal to our knowledge of extant gymnosperms. In writing the present text the author has tried to consult all the available literature and especially by the above mentioned botanists of international fame. The author takes this opportunity to express his deep sense of gratitude to these famous botanists and the publishers of the books and journals that contain the valuable information.

The illustrations, wherever possible, have been drawn from original specimens and slides. A number of photographs have been given. All of them have been taken by the author from various places in India. They clearly illustrate the morphological features of plants and give an actual and a clear picture of the habit and external features of the sporophytes.

I am indebted to my father Prof. B.R. Vasishta who has introduced me in this series and from whom, besides inheriting my interest in the subject of botany, I have inherited the capacity to write. I express my deep sense of gratitude to Prof. Y.P. Oberoi for sending me reprints of his research papers and some other valuable papers on various aspects of gymnosperms. In this context I am also grateful to Prof. Tej Nath Lakhanpaul, Prof. P.S. Gill, Prof. V.P. Chadha, Prof. K.P. Sood, Prof. S.P. Kanwal, Prof. S.S. Malik, Prof. M.S. Sethi, and many other friends in India and abroad. I also thank Mr. Muni Lal Verma, the artist for taking keen interest in drawing illustrations for this book. I shall be failing in my duty if I do not thank my publishers M/s. S. Chand & Company Ltd., who have extended me full co-operation during the process of printing this book and in bringing it out in the present form.

I request my colleagues all over India to send me their healthy criticism and their useful suggestions to improve the book.

AUTHOR

CONTENTS

GNETUM MALE STROBILUS

CHAPTER 1

Introduction

- Characteristic features
- Reproductive Organs
- Embryogeny
- Origin of Gymnospermous Structures
- Classification of Gymnosperms
- Gymnosperm in India, their distribution and Economic Importance
- Comparisions with Pteridophytes and Angiosperms
- Indian contribution on Gymnosperm
- Progymnospermopsida: Archaeopteris

The word '**Gymnosperm**' was used in 300 B.C. by Theophrastus, a pupil of Aristotle, in his book "Enquiry into Plants". Theophrastus used this term to embrace all those plants whose seeds are unprotected (*Gymnos* = naked; *sperma* = seeds). The seed plants (Spermatophyta) are grouped into two major groups on the basis of protection afforded to the ovule before and after fertilisation. These groups are the **gymnosperms** and the **angiosperms**. The gymnosperms have their **ovules** freely exposed before and after fertilisation. They are not enclosed by any ovary wall, whereas in the angiosperms (*angios* = vessel; *sperma* = seed) the ovule or the ovules are completely enclosed within a structure called the **ovary**, which forms the lower distended part of the **carpel**. Due to the protection afforded to the ovule and the seed, the angiosperms are considered to be advanced than the gymnosperms, which are, therefore, regarded as intermediate between the **pteridophytes** and the **angiosperms**. The evolutionary high status of angiosperms and their conquest over other groups of the plant kingdom is due to their

- diversified habit and vegetative forms;
- higher degree of perfection of the vascular system; the xylem in addition to the tracheids, contains wood vessels, and the phloem possesses companion cells;
- successful invasion of all habitats;
- adaptation of flower to insect pollination;
- bisexual flower, the bisexuality ensures self-pollination if cross-pollination fails, hence the seeds are formed in any case;
- development of ovules within the ovary which ensures proper protection to the developing ovules and seeds;
- efficient and effective dispersal by insects, birds, other animals, wind, water and by many other specialised mechanisms;
- efficient and varied means of vegetative propagation, which result in rapid multiplication;
- their utmost economic importance; and
- besides the above listed reasons, there must be some hereditary causes such as better equipment of gene potential and useful gene mutations, which enabled them to encounter variations in temperature and other environmental changes and led to their conquest over other plant groups.

The gymnosperms are not in such an advantageous position because :

- they lack vegetative means of reproduction by means of cuttings, layering, etc., and are slow growers;
- limited means of dispersal (only wind and by man) and their failure to grow under varied habitats (*e.g.*, water);
- absence of vessels (a few exceptions are there) in xylem and of companion cells in the phloem;
- absence of bisexuality which reduces the chances of self-fertilisation and lot of pollen is wasted as wind pollination is the main source of carriage of pollen grains;
- unprotected ovules and seeds and
- their decline may also be due to the running out of the gene potential for environmental adaptation and appearance of certain harmful mutations and chromosomal alterations. Due to their beautiful foliage and stately appearance, man has attempted and is attempting to grow them in his gardens, but with little success, because they are slow growers and have little means to adapt to varied habitats. Economically too they are less important as compared to angiosperms.

In both the gymnosperms and the angiosperms, the sporophyte, in the course of evolution has become perfected to adapt to the various environmental conditions, composed of widely different land habitats, and climates of plains, of mountains and valleys of oceans and rivers and their bank and shores, of deserts and of a great variety of micro-environments. Through millions of years of evolution, the sporophyte of the seed plants was organised, capable of bracing the terrestrial environments. That the evolutionary experimentation leading to death was still at work at the stage of origin of the seed plants, is supported by the fact that the early gymnosperms became extinct in the same way as the early pteridophytes. The origin of angiospermous plant-body represents, perhaps the last or say latest outcome of sporophyte evolution in the plant kingdom. The sporophyte of the seed plants is the only visible individual bearing the '*gametophytes*', both male and female. The male gametophyte or the '*pollen grain*' is mobile whereas the female gametophyte or '*megaspore*' is '*quiescent*' and enclosed in a sporophytic vegetative envelope, the pollen and the seed are the two structures that are shed from the sporophyte during the life cycle.

The gymnosperms are the most ancient seed plants that originated during the late Palaeozoic (ca 350 million years ago from the upper Devonian to the present) era, but flourished well during the Mesozoic. Their long evolutionary history is full of diverse lines of evolution within the group, during which many examples of organisms flourished and then became extinct. One of these lines of evolution included the early fern-like seed bearing group of plants known as the *Cycadofilicales* or the *Pteridospermae*. *Neuropteris decipiens*, *Medullosa noei*, *Lygenopteris*, are some of the examples of such plants. They had fern-like general appearance and foliage but possessed the primitive type of seed. The seeds were found attached to the pinnatifid fronds (Fig. 1.1). These fossil seed-ferns or Cycadofilicales are an evidence of the origin of gymnosperms from an ancient stock of fern-like individuals. The *Cordaitales* and the *Bennettitales* are the other two extinct lines of the gymnosperms. The former existed during the late Paleozoic and early Mesozoic eras and have been regarded as an ancestral stock from which the modern conifers and *Gnetales* might have originated. The *Bennettitales* resembled the living cycads and flourished well during the early Mesozoic era (185 million years ago) along with the extinct cycads. The striking bisporangiate cones of the *Bennettitales* have been proposed by some botanists as possible precursors of the angiospermous flower and this now extinct group may have fathered a line leading to the earliest angiosperms.

The extinct gymnosperms are included under four orders in this text. Their period of existence and life span in million years is given in the following table:

TABLE I

	Extinct Groups	Period of Existence	Life Span in Million years
1a.	Palaeozoic Pteridospermales (Medullosaceae, Glossopteridaceae, Lygonopteridaceae, Calamopilyaceae)	Lower Carboniferous to Permian *i.e.* 345 million years to 230 million years.	115 million years
1b.	Mesozoic Pteridospermales (Caytoniaceae, Petaspermaceae and Corplospermaceae)	Upper Triassic (190 m) to Lower Cretaceous (100 m)	90 million years
2.	Pentoxylales	Lower Jurassic (175 m) to Middle Jurassic (150 m)	25 million years
3.	Bennettitales	Triassic (220 m) to Upper Cretaceous (65 m)	155 million years
4.	Cordaitales	Lower Carboniferous (345 m) to Permian (230 m)	115 million years

The modern gymnosperms are commonly grouped under four orders: The *Cycadales*, the *Ginkgoales*, the *Coniferales* and the *Gnetales* (*see box*). The *Ginkgoales* and the *Cycadales* include living members that have a long fossil history and can be regarded as 'Living fossils'. *Ginkgoales* in the past (early Mesozoic) were represented by widely distributed group of plants, but now the order is represented by a single species *Ginkgo biloba*. Similarly Cycads flourished well during the Mesozoic and are now represented by nine well-defined genera that are confined to limited areas in the tropical and subtropical countries of the world. The *Coniferales* form the most conspicuous order of the living gymnosperms and include the most familiar and economically important plants like *Pinus*, *Cedrus*, *Abies*, *Juniperus*, *Cupressus*, *Thuja*, etc. Some of them are the world's tallest and long-lived plants,

LIVING "GYMNOSPERM" GROUPS - COMPARISON

	Cycadales	Ginkgoales	Gnetales	"Conifers" (incl. Taxales)
(1) History	group 200 M y old	*Ginkgo*-like leaves 200 M y old	fossils extremely sparse	group more than 200 M y old
(2) Distribution	Nine genera, Central America, S. Africa, E. Asia	sole, monotypic genus *Ginkgo* cultivated in China; introduced elsewhere	three genera, with varying distributions	57 genera distributed variously, world-wide
(3) Stem	woody, unbranched large, typically	branching trees with dimorphic shoots (long, short)	woody trees, shrubs, lianes, hemicryptophytes	branching woody plants, often with long and short shoots
(4) Leaves	pinnate compound	leathery, fan-shaped	leaves simple, but diverse in size and form	needle- or scale-like, arranged in spirals, or whorled
(5) Reproduction	dioecious	dioecious (XX females, XY males)	mostly dioecious	mostly monoecious
(6) Male structures	Microsporophylls scale-like or peltate; sperm with spiral band of flagella	microsporophylls on catkin-like axes; sperm with spiral band of flagella	Microsporophylls in flower-like structures on compound strobili; sperm reduced to two nuclei in pollen tube	Microsporophylls on simple strobili; sperm reduced to two nuclei in pollen tube
(7) Female structures	Megasporophylls leaflike, in cones (except *Cycas*); seeds large	ovules in pairs, seed large	ovules in compound flower-like, strobili; double fertilization	ovules on ovuliferous scales in compound seed cones, or solitary and terminal (Taxaceae)

e.g., *Sequoia gigantea* lives for 4,000 years and grows very tall. The conifers are cosmopolitan and widely distributed throughout the northern and southern hemispheres of the globe and form extensive forests. The *Gnetales* are represented by three living genera, *e.g.*, *Gnetum*, *Ephedra* and *Welwitschia*. The last named genus is monotypic and is represented by *Welwitschia mirabilis* in some deserts in South-West Africa. *Gnetum* (40 species) and *Ephedra* (30-40 species) include many species that are distributed in tropical and temperate regions of Asia, Africa and South America. The *Gnetales* have an obscure past (probably in the Upper Permian 230 million years ago. Pollen grains of *Ephedra* (*Gnetaciopollenites*) and *Welwitschia* (*Welwitschiasites*) have been recorded from Raniganj series in India by Bharadwaja in 1960 and due to certain peculiar reproductive structures that are not yet completely understood, they are an isolated and little understood order of gymnosperms. The living gymnosperms include 22 genera that are monotypic. The following table (Table II) gives the details:

TABLE II

MONOTYPIC GENERA OF LIVING GYMNOSPERMS (AFTER P.N. MEHRA, 1988 P. 242)

Sl. No.	Species	Family	Order	Distribution
1.	*Stangeria paradoxa*	Cycadaceae	Cycadales	Natal
2.	*Microcycas of calocoma*	,,	,,	Cuba
3.	*Ginkgo biloba*	Ginkgaceae	Ginkgoales	S. China
4.	*Welwitschia mirabilis*	Welwitschiaceae	Gnetales	S. W. Africa
5.	*Metasequoia glyptostroboides*	Taxodiaceae	Coniferales	Central China

6. *Sequiodendron gigantenum*	,,	,,	California
7. *Sequoia sempervirens*	,,	,,	,,
8. *Cryptomeria Japonica*	,,	,,	Japan-*China*
9. *Glyptostrobus lineatus*	,,	,,	S. E. China
10. *Tawania cryptomerioides*	,,	,,	S. W. China Formosa
11. *Sciadopitys verticillata*	,,	,,	Japan
12. *Austrocedrus chilensis*	Cupressaceae	,,	Chile, Argentina
13. *Diselma archeri*	,,	,,	Tasmania
14. *Fitzroya cuperessoides*	,,	,,	Chile, Argentina
15. *Neocallitropsis araucarioides*	,,	,,	New Caledonia
16. *Pilgerodendron uviferum*	,,	,,	S. Chile
17. *Tetraclinus articulata*	,,	,,	Algeria, Morocco, Malta
18. *Thujopsis dolobrata*	,,	,,	Japan
19. *Biota orientalis*	,,	,,	N. & W. China
20. *Microcachrys tetragona*	Podocarpaceae	,,	Tasmania
21. *Saxegothaea conspicua*	,,	,,	S. America
22. *Austrotaxus spicata*	Taxaceae	,,	New Caledonia

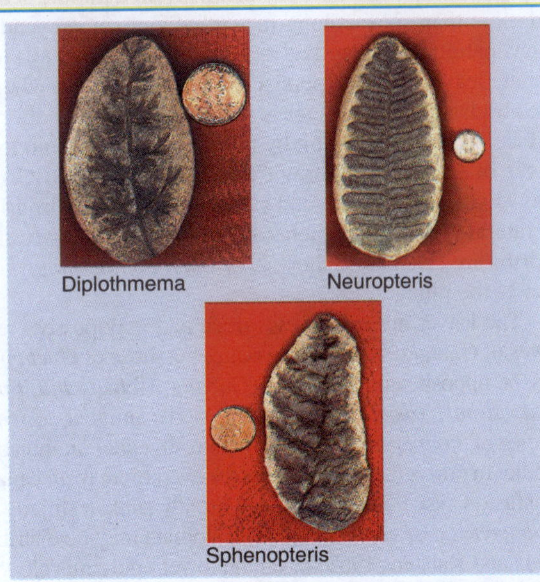

Seed fermin (fossil gymnosperms) foliage reseulaeled that of modern forn, bd reproduce by means of seed.

Fig. 1.1. *Medullosa noei.* A. reconstruction of the entire plant. It is fossil gymnosperm belonging to the order Pteridospermales. Note the fern-like leaves. *B.* Frond of *Emplectopteris triangularis* bearing seeds. *C.* Frond genus *Sphenopteris tenuis bearing seeds.* (*A, after Andrews; B, and C after Halle*).

Cycadaceae has *ten* living genera out of which *two* are monotypic; Ginkgoaceae 1 living genus which is monotypic and the same is the case with family Welwitschiaceae; Taxodiaceae has *ten* genera out of which *seven* are monotypic; Cupressaceae has *nineteen* to *twenty-two genera* out of which *eight* are monotypic; Podocarpaceae has *seven* genera out of which *two* are monotypic; Taxaceae has *five* genera out of which *one* is monotypic; Taxodiaceae and Cupressaceae have maximum number of monotypic genera. Pinaceae, Arucariaceae and Cephalotaxaceae have no monotypic genera. The reason for these genera being monotypic is difficult to assess. There may be one of the following three causes:

They did not possess the genetic capability to diversify.

They did not have enough time to do so after their origin.

They may be the only survivors of the once flourishing multispecific flora.

CHARACTERISTIC FEATURES (Extinct or Living Gymnosperms only) (See Table III)

Morphology of Vegetative Organs

The living gymnosperms include middle-sized or tall trees and shrubs and number approximately 70 genera and 725 species. There are no herbs and climbers. They do not reproduce vegetatively by means of cuttings, layering or budding, but vegetatively propagating structures like bulbils are known in some groups, *e.g.*, Cycads. The roots are generally tap-roots, but mycorrhizic and coralloid roots are known in some genera (*Pinus*, *Cycas*). The stems are aerial, erect, underground (*Welwitschia*), unbranched (*Cycas*, *Zamia*, *Welwitschia*) or branched (*Pinus*, *Cedrus* etc.).
In some genera (*Pinus*, *Ginkgo*, *Cedrus*, *Larix*, *Pseudolarix*) the branches may be of two types: (*i*) the long shoots; and (*ii*) the dwarf shoots that bear at their apices bundles or clusters of green leaves and are collectively known as spur. Both the *microphyllous* and the *megaphyllous* leaves are known among the gymnosperms. The former are small and scaly, whereas the latter are large and well-developed and their vascular supply always leaves a leaf gap in the stem stele. The leaves (Fig. 1.2) may be simple or compound and vary in size from a minute scale leaf (some species of *Juniperus*) to a large and more than six feet long megaphylls of some cycads. They show a great variation in shape also. The venation may be reticulate (*Gnetum*), parallel (*Welwitschia*) or even dichotomous (*Ginkgo*). The pinnate leaves in cycas persist for 3-10 years, and when they fall, the leaf bases persist even longer. The abscission of the leaves is smooth, but usually abscission takes place in progressive stages as the surface periderm is lost bit by bit until the surface no longer shows signs of leaf attachment. Pinnate leaves of *Cycas* alone show *Circinate venation*. In *Ceratozamia* alone the entire leaves show circinate venation. The pinnae in *Cycas* have a single midrib and there are no lateral veins. In *Stangeria* the lateral veins branch dichotomously after arising from the midrib. In other genera of Cycadales the midribs are absent and they show *open dichotomous* venation with maximum dichotomies near the base of the pinnae.

The leaves are mostly evergreen and may possess resin passages (*Pinus*, *Cedrus*, *Abies*, etc.). The leaves of *Gnetales* lack resin passages but those of *Gnetum* possess latex tubes. The arrangement of leaves may be opposite and decussate (*Gnetum*, *Welwitschia*, *Ephedra*), whorled (*Cedrus*) or spirally arranged (*Podocarpus*, *Taxus*, *Taxodium*, etc.). The number of resin canals varies in various species. In all the species of *Juniperus* there is a single resin canal on the abaxial side between the epidermis and vascular bundle. In others there may be two or even more resin canals. The leaves of conifers and cycads possess a transfusion tissue. The stomata are usually sunken slightly (*Ginkgo*) or deeply (*Cycas*, *Pinus*) and may be *syndetocheilic* or *haplocheilic*. The stomata may be distributed on both the leaf surfaces (*Ginkgo biloba*, Kanis and Karstens 1963) or on the lower epidermis alone (*Cycas*, *Taxus*). The stomata have guard cells usually surrounded by subsidiary cells whose number varies from 2-4 or even more. The subsidiary cells in many genera overlap the guard cells (*Abies*, *Larix*, *Juniperus*, *Tsuga*). In *Tsuga dumosa* there are 4 subsidiary cells that are unequal in size. In *Abies forrestii* there are two subsidiary cells that are parallel to the guard cells. Their arrangement and number varies in various species. The mesophyll may (*Cycas*, *Taxus*, *Gnetum*) or may not (*Pinus*, leaves on spurs of *Ginkgo*) be distinguished into palisade tissue and spongy parenchyma. In *Pinus* the walls of mesophyll cells are thrown into folds. The mesophyll is distinguished into palisade and spongy parenchyma in *Abies*, *Tsuga*, *Larix* and many other coniferales. The leaves may be triangular (*Pinus roxburghii*), semi-circular (*Pinus sylvestris*), circular (*Pinus monophylla*) and bifacial (leaflets of *Cycas*, *Zamia*, leaves of *Gnetum*, *Ginkgo*). The leaflets in some (*Cycas revoluta*) have revolute margins. The leaves in *Pinus kempferi* are flattened. It is found in highlands of South Vietnam and is considered to be a possible link between *Pinus* and *Pseudolarix* and *Keteleeria*.

TABLE III

Some Morphological Characteristics of Living Gymnosperms (Based on Burhorst, 1997)

Family	Genus	Number of Species	Distribution	Leaf Form	Leaf Arrange-ment	Vernation	Sporangia per Micro-sporangiophere	Number of ovules per Megasporophyll	Number of Cotyledons
Cycadaceae	*Cycas*	Twenty	Japan to Queenland Madagascar, China, India	Dimoorphic, pinnate, pinnae, linear or broad and lanceolate with a single midrib. No veins.	Spiral	Circinate in pinnae	Many, arranged in sori	Two to eight	Two
Stangeriaceae	*Stangeria*	One	Natal (S.E. Africa)	Pinnate, pinnae broad with pinnate venation	Spiral	Sub-circinate	Many	Two	Two
Zamiaceae	*Zamia*	Thirty to Forty	From Florida to Chile, West-Indies, Tropical America	Pinnate, broad, without a mid-rib	Spiral	Straight	Many	Two	Two
	Ceratozamia	Four	Mexico	Pinnate, broad without a mid-rab	Spiral	Whole-leaf in circinate	Many	Two	One
	Macrozamia	Fourteen	Australia	Pinnate, pinnae broad without a mid-rib	Spiral	Sub-circinate	Many	Two	Two
	Encephalartos	Thirty	South Africa	Pinnate, pinnae broad; no mid-rib.	Spiral	Straight	Many	Two	Two
	Dioon	Three to Five	Mexico and Central America	Pinnate, pinnae broad without a mid-rib Leaves dimorphic like Cycas	Spiral	Straight	Many	Two	Two
	Bowenia	Two	North America	Bi-pinnate, pinnae broad, no mid-rib	Spiral	Sub-circinate	Many	Two	Two
	Microcycas	One	Cuba	Pinnate, pinnae broad, no mid-rib	Spiral	Straight	Many	Two	Two
	Lepidozamia	Two	Australia	Pinnate, pinnae broad, no mid-rib	Spiral	Sub-circinate	Many	Two	Two
Ginkgoaceae	*Ginkgo*	One	South China, now cultivated throughout the world	Simple, fan-shaped with dichotomous venation. Those on long shoots bilobed, on short shoots entire margin wavy	Spiral	Staight	Two, rarely three to seven	Two on ovuliferous stalk	Two
Pinaceae	*Pinus*	90-105 Mehra, (1988) 70-100 (Bierhorst 1971)	Cosmopolitan Five species in India	Needle-shaped and scale leaves. Needles borne only on dwarf shoots (1-5 or 7) with one or two veins. Leaves flattened in P. *kempferi*	Spiral	Straight	Two	Two	Four to Fifteen
	Cedrus	Four	Middle East; Algeria Cyprus, India (one Species)	Needle-shaped on dwarf shoots	Spiral	Straight	Two	Two	Nine to Ten

Family	Genus	Number of Species	Distribution	Leaf Form	Leaf Arrange-ment	Vernation	Sporangia per Micro-sporangiophore	Number of ovules per Megasporophyll	Number of Cotyledons
	Pseudolarix	Two	China	Needle-like borne on indeterminate spur shoots.	Spiral	Straight	Two	Two	Five to Seven
	Larix	Ten to Twelve	Europe, N. Asia, N. America, India (Eastern Himalayas) one Species only	Needle-like,on dwarf shoots.	Spiral	Straight	Two	Two	Six
	Abies	Forty six to Fifty species Seventy according to Gaussen (1964) Two Species in India	Central America North Temperate India, N. Africa, Mexico, South Europe, Western N. America.	Needle-like, borne on branches. No distinction into dwarf and long shoots	Spiral	Straight	Two	Two	Five to Seven
	Cathya	Two	China	Linear, no distinction into dwarf and long shoots.	Spiral	Straight	Two	Two	Three to Four
	Keteleeria	Four to Eight	Indo-China, East Asia	Linear as well as broad shoots of one type.	Spiral	Straight	Two	Two	Two
	Picea	Fifty	North-temperate Two Species in India	Needle-like, shoots of one type	Spiral	Straight	Two	Two	Four to Fifteen
	Pseudotsuga	Seven	East Asia, W.N. America	Linear, shoots of one type	Spiral	Straight	Two	Two	Six to Twelve
	Tsuga	Ten to Fifteen	East Asia, India (I), N. America	Linear, shoots of one type	Spiral	Straight	Two	Two	Three to Six
Taxaceae	*Taxus*	Ten	North temperate, one in India	Linear	Spiral	Straight	Three to Nine	There are no female cones and Megasporophylls. The ovule is orthotropus and is borne terminally on a secondary axis.	Two

Family	Genus	Number of Species	Distribution	Leaf Form	Leaf Arrange-ment	Vernation	Sporan-gia per Micro sporangiophere	Number of ovules per Megasporophyll	Number of Cotyledons
	Torreya	Six	East Aisa, California, Florida	Linear	Spiral	Straight	Three or more	There are no female cones and Megaspo-rophylls	Two
	Amentotaxus	One to Four	India (Assam), West China	Linear	Spiral	Straight	Three or more	The reproductive structures are not organised into strobili or cones.	Two
	Austrotaxus	One	North Caledonia,	Broad	Spiral	Straight	Three or more	Do	Two
	Pseudotaxus	One	China	Linear	Spiral	Straight	Three or more	Do	Two
Cephalotax-aceae	*Cephalotaxus*	Four to Seven	East Asia	Linear	Opposite	Straight	Three or Eight	Two	Two
Podocarpaceae	*Podocarpus*	One hundred	Japan, India (2) Southern Hemisphere	Scale like, Linear broad	Spiral	Straight	Two	One	Two
	Acmopyle	Three	North Caledonia, Fiji	Linear	Spiral	Straight	Two	One	Two
	Dacridium	Twenty To Twenty Five	North Newzealand, Indomalaysia, Chile	Linear, Scale-like	Spiral	Straight	Two	One	Two
	Microcachrys	One	Tasmania	Scale-like	Spiral	Straight	Two	One	Two
	Microstrobus	Two	Australia, Tasmania	Scale-like	Spiral	Straight	Two	One	Two
	Phyllacladus	Seven	North, Newzealand Tasmania, Philippines	Linear, Scale-like	Spiral	Straight	Two	One	Two
	Saxegotheca	One	Andes, Argentina	Linear	Spiral	Straight	Two	Two	Two
Araucariaceae	*Araucaria*	Eighteen	North Newzealand, Australia, S. America, North Calle-donia, South Brazil, Chile, New Guinea, India	Broad	Spiral	Straight	Five to Fifteen	One to Two	Two to Four
	Agathis	Twenty	Indo-China, W. Malayasia, Philippines, N. Newzealand, Fiji	Broad	Spiral	Straight	Five to Fifteen	One	Two
Cupressaceae	*Cupressus*	Sixteen to Twenty	Mediterranean, Sahara, Asia, N. America, India (I)	Scale-like	Spiral	Straight	Two to Six	Six to Twenty	Two to Five
	Actinostrobus	Two	South-West Australia	Scale-like	Spiral	Straight	—	One to Two	—

Family	Genus	Number of Species	Distribution	Leaf Form	Leaf Arrange-ment	Vernation	Sporangia per Micro-sporangiophere	Number of ovules per Megasporophyll	Number of Cotyledons
	Austrocedrus	One	South America	Scale-like	Opposite	Straight	—	One to Two	—
	Callitris	Sixteen	Australia, N. Caledonia	Scale-like	Opposite	Straight	—	Many	—
	Calocedrus	Three	Pacific N. America Formosa, South-east Asia	Scale-like	Opposite	Straight	Two to Four	Two	—
	Chamaecyperus	Seven	Japan, N. America, Formosa	Scale-like	Opposite	Straight	Two to Five	Two to Five	Two
	Diselma	One	Tasmania	Scale-like	Opposite, Whorled	Staight		Two	—
	Fitzroya	One	Chile	Scale-like	Whorled	Straight		One to Six	Two
	Fokienia	One to three	Indochina, China	Scale-like	Whorled	Strairght	Two to Five	Two	—
	Juniperus	Sixty to Seventy	Northern Hemisphere, India (8 species) occ-urring wild in the Himalayas	Needle-like in *j. Communis* and *J. oxycedrus*. In others scale-like	Spiral	Straight	Three to Six	One to Three	Two to Six
	Libocedrus	Five	North Caledonia North Newzealand	Scale-like	Opposite	Straight	Three to Five	One to Two	—
	Neocallitropsis	One	North Caledonia	Linear as well as scale-like	Whorled	Straight	Numerous	Two to Five	Two
	Papuacedrus	Three	Moluccaas, North Guinea	Scale-like	Opposite	Straight	Numerous	Two	—
	Pilgerodendron	One	South Chile	Scale-like	Opposite	Straight	Four to Eight	One to Two	—
	Tetreclinis	One	Malta, Spain, Tunis	Scale-like	Opposite	Straight	—	—	—
	Thuja	Five	China, Japan, N. America, India (cultivated)	Scale-like	Opposite	Straight	Two to Five	Two to Three	—
	Thujopsis	One	Japan	Scale-like	Opposite	Straight	Two to Five	Three to Five	—
	Widdringtonia	Five	Tropia, South America	Scale-like	Oppoosite	Straight	Three to Five	Five or more	Two
Taxodiaceae	*Taxodium*	Three	Southern U.S.A, Texas to Guatemala, Mexico	Linear	Spiral	Straight	Four to Seven	Two	—
	Metasequoia	One	Central China	Linear	Opposite	Straight	—	Five to Eight	—
	Sequoia	One	California	Linear, Short Lanceolate	Spiral	Straight	Three or Four	Three to Seven	Two
	Sequoiaden-dron	One	California	Needle-like, Scalex lanceolate	Spiral	Straight	Three to Six	Five to Eight	Three to Five
	Cryptomeria	One	Japan and China	Oval-shaped	Spiral	Straight	Five to Seven	Three to Six	—

Family	Genus	Number of Species	Distribution	Leaf Form	Leaf Arrangement	Vernation	Sporangia per Micro-sporangiophore	Number of ovules per Megasporophyll	Number of Cotyledons
	Sciadopitys	One	Japan	Scale-like, Linear, resemble two pine needles fused face to face, 12 cm. long	Spiral	Straight	Two	Five to Nine	—
	Athrortaxis	Three	Australia, Tasmania	Leaves small, scale-like crowded, overlapping	Spiral	Straight	Two	Three to six	—
	Cunninghamia	Three	China, Formosa	Broad	Spiral	Straight	Two to Four	Two to Three	Two
	Glyptostrobus	One	China	Linear as well as Broad	Spiral	Straight	Four to Seven	Two	—
	Taiwania	Three	China, Formosa	Scale	Spiral	Straight	—	Two	—
Gnetaceae	*Gnetum*	Forty	Amazon basin, West Africa, India, South China, Malaysia	Broad, simple with reticulate venation, Elliptic, Scale-like	Opposite	Straight	Two	The single ovule constitutes the entire strobilus	Two
Ephedraceae	*Ephedra*	Forty	North and South America, From Mediterranean to China, India, Jeursalem	Scale-like	Opposite or whorled, fused into a basal sheath, non-green at maturity	Straight	Usually Four Septate Sporongia	Megasporangiate strobilus bears one or two terminal ovules	Two
Welwitschiaceae	*Welwitschia*	One	South-West Africa	Large and woody with a basal meristem. Strap shaped	Opposite	Straight	Each micro-sporangiophore bears a sporangium of three fused sporangia	Each female flower contains a single ovule. The flowers are arranged in a distinct strobilus or female cone.	Two

Fig. 1.2. Vegetative organs of gymnosperms. A. Dwarfshoot of *Pinus roxburghii* bearing three needle-like leaves. B.Vegetative shoot of *Picea smithiana* bearing needle like leaves. C.A single leaf of *Picea* D. Leaves of *Araucaria bidwillii.* E. Leaves of *Ginkgo* borne on a dwarf or short shoot. F. Leaves of *Taxus* baccata. G. Leaves of *Gnetum*. H. Leaves of *Sequoia sempervirens* on a branch. I. A scale-like leaves of *Cupressus sempervirens* closely arranged on a branch J.Same as I, enlarged. K. Needle-like leaves of *Cedrus deodara*

The vasculature of the petiole in living gymnosperms shows greater amount of variations. In the cycads the petiole of *Cycas* shows omega shaped disposition of the vascular bundles. There is variation in the structure of the vascular bundle as it traverses from the bases of the petiole towards the apex. Many vascular strands enter the petiole. The bundles are open. A characteristic feature of the vascular bundles of petiole and lamina in *Cycas* is the *diploxylic* xylem. In this case there is a patch of centripetally developed metaxylem and two patches of centrifugally developed xylem. The leaf traces in *Cycas* are *endarch* at their attachment in the pith region of the stem. They became *mesarch* during their course through the cortex. During their course from the pith and the cortex their *centrifugal* xylem is progressively reduced and their *centripetal* metaxylem is enlarged. In the petiole the trace xylem is weakly mesarch and nearly *exarch*. In the *pinnae* many traces become exarch. In *Ginkgo* only two vascular bundles enter the petiole. They show secondary growth. On entering the leaf blade these strands diverge right and left and then divide dichotomously. In *Taxus* the leaf is enervated by a single strand that is surrounded by a sclerenchymatous sheath. In *Ephedra* the unilacunar nodes receive two leaf traces that have quite distinct points of origin. They lack cambium. In *Welwitschia* large number of traces depart into the leaf from a saucer-shaped vascular mass of the old stem. In a young stem there is a ring of discrete vascular bundles. The base of the leaf remains permanently

Pinus - needle

Picea - needle

Cedrus deodara - needle

Araucaria - leaves

Gnetum - leaves

Gingo biloba - leaves

Taxus - leaves

Sequoia sempervirens - leaves

Cupressus sempervirens - scale leaves

meristematic, while the tip keeps wearing and dying off. In *Gnetum* the vascular bundles are arranged in an arc in the midrib. In this case three to seven traces supply each leaf. Half the departing number goes to each of the two opposite leaves.

The secondary wood in the gymnosperms may be: (*i*) **manoxylic** or (*ii*) **pycnoxylic**. The manoxylic wood is found in **Cycadophyta** and is porous, soft and more parenchymatous in nature. It has wide medullary rays and is useless commercially. The pycnoxylic wood is characteristic of **Coniferophyta** and is compact and has narrow medullary rays. It is of great commercial use. Seward (1919) suggested to divide gymnosperms into two groups which he named as manoxylic and pycnoxylic groups. The xylem lacks wood vessels (except in *Ephedra*, *Welwitschia* and *Gnetum*), whereas the phloem is devoid of companion cells. The xylem is usually *mesarch* or *endarch* (exarch in roots). The secondary vasculature may be **monoxylic** (when there is a persistent cambium as in *Pinus*) or **Polyxylic** when different layers of cambium appear and result in distinct concentric rings of secondary xylem and phloem (*Cycas*, *Gnetum*). The secondary wood tracheids usually have bordered pits, sometimes scalariform thickenings are also present (*Zamia* and *Stangeria*). The medullary rays may be uniseriate or multiseriate and may be **homogenous** (*Ginkgo*) or **heterogenous** (*Pinus*). The resin canals may also be present in the secondary wood of many gymnosperms. In some cases, alongwith some other characters of xylem they help in distinguishing the various species. For example in the wood of *Larix griffithii* both resin canals and ray tracheids are present. In *Tsuga dumosa* the wood lacks resin canals but possesses ray tracheids. In *Abies forrestii* both the resin canals and ray tracheids are absent. All these genera belong to the family *Pinaceae*. In other members of this family *e.g.*, *Pseudolarix* lacks both ray tracheids and resin ducts. It is considered to be the simplest. In *Cedrus* the wood possesses only ray tracheids and no resin ducts. In *Pinus* both ray tracheids and resin ducts are present. *Pinus* differs from *Larix* in that the epithelial cells are thin-walled whereas in *Larix* the epithelial cells lining the resin ducts have thick walls. These genera possess both the longshoots and dwarf shoots (*Pseudolarix*, *Cedrus*, *Larix* and *Pinus*). In genera with only long shoots, *Abies* and *Keteeleria* lack both resin ducts and ray tracheids, *Tsuga* has only ray tracheids, *Picea* has both ray tracheids and resin ducts. *Pseudotsuga* possesses both resin ducts (with thick walled epithelial cells) and ray tracheids and in addition the tracheids have tertiary spiral thickenings on their walls. The genera that lack both the resin ducts and ray tracheids are considered to be primitive, next come those with either of them and next to them are those which possess both of them. Last of all the most advanced ones possess both of them with some other peculiarities as in the case with *Pinus*, *Larix* and *Pseudotsuga*. This conclusion is also supported by evidence obtained from fossil data. Some fossil gymnospermous woods identified as *Dadoxylon* and *Araucarioxylon* were found in the Carboniferous, Permian and Triassic strata. In both of them resin ducts have not been observed. *Araucarioxylon lindeli* described from mid-Jurassic in the coniferous wood (Seward, 1904) in which the resin ducts appeared in tangential series. The same is the case in *Cupressinoxylon cedrioides* (Holden 1913) described from the same strata. In *Protopiceoxylon* found from Upper Jurassic and Lower Cretaceous strata the resin canals were found in a scattered condition in addition to tangential series (Seward, 1919). The fossil data clearly indicate that woods without resin ducts or with restricted distribution of ducts are primitive as compared to those with scattered and widely dispersed resin ducts. Greguss (1955) made extensive studies on the secondary wood of gymnosperms and concluded that forms with homogenous rays are primitive than those with ray tracheids. So taxa with homogenous rays and without resin ducts are primitive as compared to those with ray tracheids and resin ducts. The protoxylem in the seedling stages has spiral thickenings but later on it consists of scalariform pattern. The phloem has sieve tubes on lateral walls rather than on end walls.

In the seedling stage the stem has a ring of discrete vascular bundles with narrow or broad medullary rays. The bundles are conjoint, collateral, endarch and open.

Reproductive Organs

The gymnosperms are heterosporous, there may not be any difference in the spore size or the so-called **microspores**, may be larger than the **megaspores** but the two kinds of spores produce two kinds of gametophytes, *i.e.*, the microspores or the pollen grains produce male gametophytes, whereas the single megaspore enclosed within the **megasporangium** (nucellus) develops into a female gametophyte that bears two or more **archegonia** or female sex organs. The two kinds of spores are produced within sporangia that are borne on sporophylls which are arranged spirally along an axis to form lax or compact strobili or cones. The strobili bearing microsporophylls and microsporangia are called microsporangiate or male strobili, whereas those bearing megasporophylls with ovules or megasporangia are called macrosporangiate or female strobili. The two types of cones or strobili may be borne on the same tree (*Pinus*) or on different trees (*Ginkgo*, *Cycas*). The microsporangium contains numerous microspores, whereas the megasporangium contains only one megaspore. The microspores and the megaspores are haploid and developed as a result of meiosis in the respective spore mother cells. They are the pioneer structures of the male and female gametophytes, respectively. The gametophytes are **endosporic**, *i.e.*, they develop within the respective spore wall. The strobili or the cones

Fig. 1.3. Male cones and microsporophylls of gymnosperms. A—C. Male strobilus and microsporophylls of *Ginkgo biloba*. D— F. Male flowers and microsporophylls of *Ephedra foliata*. G— H. Male cone and a male flower of *Gnetum ula*. I— J. Male cone and a male flower of *Welwitschia*. K—L.Male cone and microsporophyll of *Pinus roxburghii*. M— N. Male strobili of *Taxus baccata*. O— P. Male cone and a microsporophyll of *Cycas circinalis*.

are of varying sizes and shapes (Figs. 1.3 and 1.4). Their position on the plant also varies. In *Cycas* the male or microsporangiate cones are the largest among gymnosperms and arise singly at the apex of the male plant. The sporangia are arranged in sori on the ventral surfaces of microsporophylls. All the microporangia in serus receive a common vascular supply. They are compact. The female or macrosporangiate cones in *Cycas* consist of spirally arranged megasporophylls that do not form a compact strobilus but are loosely arranged in between successive crowns of foliage leaves. In all other genera of cycads the megasporophylls are compactly arranged to form distinct female cones or strobili. In *Dcoon spinulosum* the mature female cone weighs upto sixty pounds and is about two feet long. In some species of *Macrozamia* the female cones are axillary. In all others except *Cycas* they are terminal in position. In *Ginkgo* the microsporangiate or catkinlike staminate strobili are about an inch in length and are borne singly in leaf axils or on the upper sides of the bases of petioles. They are borne at the ends of spurs, about 6 on each spur. The axis bears lateral microsporangiophores with two pendant microsporangia. The macrosporangiate cones consist of two sessile ovules borne at the tip of the slender stalk. Such stalks occur in groups. Each ovule has a fleshy cup or collar at its base. In the *Coniferales* there is great variation. In *Taxus* the microsporangiate cones arise singly in the axils of leaves and bear 6 to 14 peltate microsporophylls bearing 5 to 7 pendant microsporangia each; the macrosporangiate cone is represented by a single ovule. In *Pinus* the oblong staminate cones ($\frac{1}{2}''$ to 1″ long) arise in compact clusters at the ends of long shoots; and each cone arises in the axil of a scale leaf, each microsporophyll bears two oblong microsporangia on their ventral side; the female cones are large and woody at maturity, each megasporophyll bears two ovules on its adaxial side. In *Podocarpaceae* the staminate cones are small and arranged in spikes; the macrosporangiate strobilus is represented by two ovules with several sterile bracts at base. In *Cephalotaxaceae* the macrosporangiate cones are solitary axillary and have several pairs of megasporophylls that are arranged spirally on a short axis. They occur in spikes or globose heads. In other coniferous families the macrosporangiate cones are large and woody and possess many spirally arranged megasporophylls with two (*Pinus*, *Cedrus*, *Abies*, etc.), one (*Araucaria*), 2-9 (*Taxodiaceae*), or 1-many (*Cupressaceae*) erect ovules at the bases. In *Ephedra* the small male and female strobili arise singly in the axils of leaves. In *Gnetum* the male strobili are large and conspicuous and possess a long central axis with numerous bracts arranged in decussate pairs; these paired bracts are fused along their margin to form a cup-shaped structure in whose axil are borne whorls of microsporophylls (3-5 in each whorl). The macrosporangiate cones are also large and possess a long axis with collar-like rings arranged one above the other, each collar has 4-10 ovules arranged in a ring. (Biosporangiate cones are also known in some species of *Gnetum*). In *Welwitschia* the dioecious plants bear dichasially branched inflorescences that arise from a series of transverse ridges parallel to leaf bases and terminate in beautiful male or the female cones, with opposite and decussate cone scales or bracts that are usually crimson or scarlet coloured at maturity. Each cone scale or a bract subtends a male or a female flower.

Fig. 1.4. Female strobili of the gymnosperms. A. Megasporophyll of *Cycas revoluta*. B. Twig bearing female strobilus of *Taxu baccata*. C. Female strobilus of *Taxus baccata* enlarged. Note the scales. D. Female strobilus of *Ginkgo biloba*. E — F. Female strobili of *Ephedra foliata*. G. Female cone of *Gnetum ula*.H. Female strobilus of *Welwitschia*. I. Ovule of *Welwitschia*. J — K. Female cone and megasporophyll of *Pinus roxburghii*.

Ovule and Female Gametophyte (Figs. 1.5 and 1.6)

The ovules are naked and are borne on megasporophylls that are usually spirally arranged around a central axis. The ovules are sessile and those of *Cycas* (Fig. 1.5 A) are the largest in the plant kingdom. The ovule consists of a parenchymatous mass of cells called the **nucellus** or the **megasporangium**. The nucellus encloses a single diploid **megaspore mother cell** that undergoes meiosis to form a linear tetrad of 4 haploid **megaspores**, out of which only one (usually the lower) remains functional, whereas the rest degenerate. The functional megaspore enlarges and undergoes free nuclear division to form large number of free nuclei. This is the young female gametophyte (that has developed within the megaspore). By now the nucellus becomes enveloped by a single massive integument that grows around leaving a small pore or the **micropyle** at one end. This is the micropylar end of the ovule. This integumented megasporangium is called the **ovule**. Now centripetal wall formation starts around the free nuclei of the female gametophyte. It continues till the whole female gametophyte becomes cellular. The single integument in gymnospermous ovule consists of three layers : (*i*) **Sarcotesta** or the outer fleshy layer; (*ii*) **Sclerotesta** or the middle stony layer; and (*iii*) the inner **sarcotesta** or the inner fleshy layer. The integument may be free from the nucellus right upto its base as in all *Cupressaceae*, *Taxus*, *Agathis*, *Araucaria*, *Cryptomesia* etc. In *Pinus*, *Ephedra* and *Gnetum* the integument is partly fused and partly free from the nucellus. In *Cycas* the integument is fused with the nucellus except in the apical region. In *Ginkgo* the integument is free from the nucellus in young ovules but is fused with it in the older ovules. In *Podocarpus* the ovuliferous scale is folded around the ovule and encloses it completely or partially. It forms an extra envelope called *epimatium* which may completely be fused with the integument as in *P. macrophylla* and *P. spicatus* or it may be free from the integument as in *P. ferruginins*. In *Phyllocladus* the integument is enveloped by a true *aril* that is free from the integument and encloses it only partially. The nucellus is usually *crassinucellate*. In some cases as in *Gnetum*, *Welwitschia* and *Ephedra*, the inner integument grows into a long micropylar tube. (Fig. 1.5 F & 21.6 C — E). In *Podocarpus gracilior* the entire integument grows into a micropylar tube in mature ovule. The apical region of the nucellus degenerates to form a *pollen chamber* in which the semi-germinated pollen grains or microspores remain till further growth. Towards the micropylar end the female prothallus develops two or more *archegonia* or the female sex organs (Fig. 1.7). The archegonia have a short or long neck made up of two (*Cycas*, *Zamia*, *Ginkgo*), 4 (*Taxus*), 8 (*Biota*), and 8 tiers of 4 cells each (*Ephedra*). The *ventral* canal cell is present in *Ginkgo* and *Pinus* whereas it is absent in *Cycas* and *Ephedra*, which have only *ventral canal nucleus*. The venter has a large egg or the *oosphere*. In *Biota orientalis*, *Cryptomeria* and some other conifers the archegonia occur in groups and form an *archegonial complex*. Such groups may also develop at the chalazal end of the female gametophyte, *e.g.*, *Biota orientalis* and *Cryptomeria japonica*.

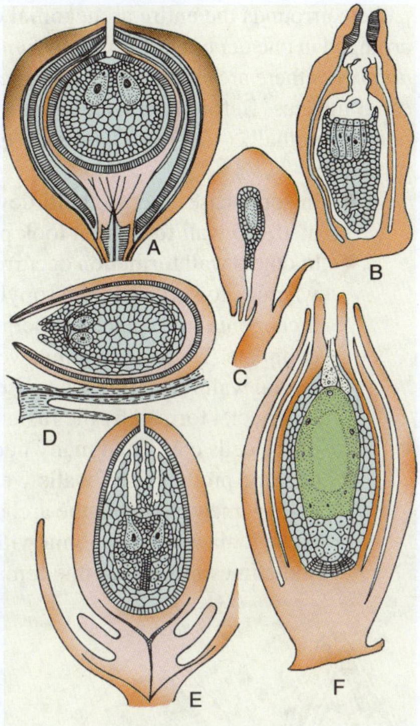

Fig. 1.5. Ovules of gymnosperms. A. Ovule of *Cycas revoluta*. B. Ovule of *Cryptomeria japonica*. C. V.S. Ovule of *Podocarpus*. D. Ovule of *Pinus roxburghii*. E. Ovule of *Taxus baccata*. F. Ovule of *Gnetum ula*.

The female gametophyte in the Cupressaceae (Dogra, 1984) presents certain interesting observations. In this family the archegonia occur in groups or complexes towards the micropylar end of the female gametophyte or even on the lateral sides of the female gametophyte. In one prothallus there is usually one group or complex of archegonia, but variations are there in some genera like *Callitris* and *Widdringtonia*. In *Callitris* and *Widdringtonia* the archegonial complexes have been observed to develop only at places where the pollen tubes touch the female gametophyte. Looby and Dayle (1940) and Baird (1953) observed that in *Callitris* the number of archegonial complexes formed within one female gametophyte depends upon the number of pollen tubes growing along the side of female gametophyte. In *Widdringtonia cupressoides* Moseley (1943) reported lateral archegonial complex with 25–100 archegonia. In addition there were also several archegonial groups near the micropylar end. These archegonia were deep seated and were devoid of neck cells. Sonton (1934) reported 200 archegonia extending along the entire length of the female gametophyte. In the various species of *Cupressus* and some other genera there is only one archegonial complex present towards the micropylar end of the female gametophyte. The number of archegonia in each complex varies with species *e.g.*, it is 8–14 in *Cupressus funiberis*, 12–24 in. *C. glabra*, 10-18 in *C. torulora* and 10–16 in *C. arizonica*. Martin (1950) reported 5–7 in *Thuja occidentalis*, 7-9 in *Thujopsis dolabrata* (Sugihara, 1939), 15–28 in *Biota orientalis* (Singh and Oberoi, 1962), 5–17 in *Chamaecyparis pisifera* (Sugihara, 1938) and 5–19 in *C. nootkatensis* (Owens and Molder, 1975). A one to three layered archegonial jacket surrounds the entire archegonial complex. Normally the archegonia have a neck of four cells arranged in one tier but in *Cupressus funiberis* there may be two tiers of four to eight cells. In *Thujo*psis *dolabrata* there are eight cells in a single tier. The archegonia have an elongated egg cell and a ventral canal nucleus. Before fertilization the ventral canal nucleus degenerates. Dogra observed some abnormal prothalli (40%) in *Cupressus glabra*. He recognised three types of developments in such prothalli:—

- In some cases the prothalli developed upto advanced free nuclear stage and then degenerated. No wall formation took place.
- In others wall formation occurred in the lower part (chalazal end) of the female gametophyte whereas towards the micropylar end only free nuclei were formed. The development of archegonia took place in the lower cellular portion but most of these archegonia were incomplete. These archegonia were either scattered or formed scattered groups. In some cases partial wall formation was observed around the free nuclei in the micropylar region. Polyploid nuclei formed by the fusion of some free nuclei were also observed in this region. Some of the cells contained many nuclei.
- In some prothalli cell walls were laid down around all the nuclei resulting in a cellular female gametophyte but the archegonia failed to develop or developed only partially. Inverted archegonia and archegonia with joined eggs were also observed in some prothalli.

Sometimes more than one female gametophytes developed one above the other. The upper one developed micropylar archegonia, whereas in the lower prothallus the archegonia developed partially.

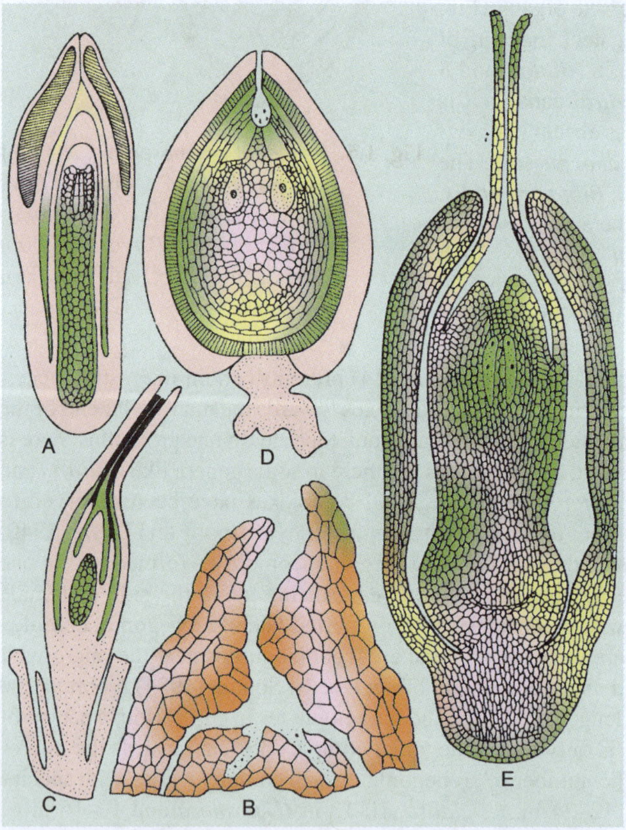

Fig. 1.6. Ovules of gymnosperms. A. V.S. ovule of *Biota*. B. V. S. ovule of *Biota* showing upper part. C. V.S. ovule of *Welwitschia*. D. V.S. ovule of *Ginkgo biloba*. E. V.S. ovule of *Ephedra foliata*.

In Podocarpaceae (Fig. 1.5) and Taxodiaceae also the archegonia occur in complexes or groups. Their number in each complex varies with species (For details see chapter on Coniferales).

There is no archegonium in *Gnetum*, a few free nuclei lying at the micropylar end of the female prothallus act as eggs. In *Gnetum* there are 8-15 megaspore mother cells and all of them undergo meiosis without wall formation. This leads to the formation of 8–15 **tetrasporic female gametophytes**. They start developing by free nuclear division but ultimately most of them abort and only one matures. The female gametophyte has cellular tissue towards its lower end whereas towards its apex a few nuclei remain free and act as eggs. It, however, becomes cellular after fertilisation. In *Welwitschia* the megagametophyte is **tetrasporic** and develops into a cellular tissue before fertilisation. It does not bear archegonia, but a few of the cells of the gametophyte grow upwards into the nucellus as **prothallial tubes** which fuse with pollen tubes.

Regarding vascular supply of the ovules, the *cycads* have vascular strands traversing both the outer and inner fleshy layers of the integument. In *Ginkgo* only the inner fleshy layer is enervated whereas in *Pinaceae* there are no vascular strands in the integuments.

Microspores (Pollen grains)

In the gymnosperms the sporophyte is the only visible individual, bearing the gametophytes, both male and the female. The male gametophyte or the pollen grain is mobile while the female gametophyte or the megaspore is quiescent and enclosed in a sporophytic vegetative envelope. It should be noted that the organisation of the functionally different male and female gametophytes, characterised by mobility and quiescence respectively, is accomplished within the same sporophyte body. Further the haploid pollen grains and the diploid seeds (both free life units) are the two units that are shed from the mother plant body during the life cycle. The pollen grain obtains its wall material and complete protective compliment from the sporophyte. They complete their **differentiation stage** within the sporophyte and are released from the parent plant when they are fully mature. It is therefore imperative that the whole range of activity in the male strobilus development is designed to provide an ideal environment for microsporogenesis and to ensure the life potential of pollen for leading a free life when left alone. The isolation of the pollen from the diploid mother plant as a result of dehiscence of the microsporangium is an important event because after that the pollen lives independently in the air or on the bodies of insects that carry it and later deposit it on the micropyle of the ovule. This delicate haploid structure, enclosing within its walls the genetic compliment of the future plant, is open to the environmental hazards and is faced with the danger of loss to its life at any stage during the course of its movement. This free period is usually short and during this period its metabolic activities are at a slow pace so as to economise its energy reserve. The structure of the pollen in various gymnosperms is attempted in following pages.

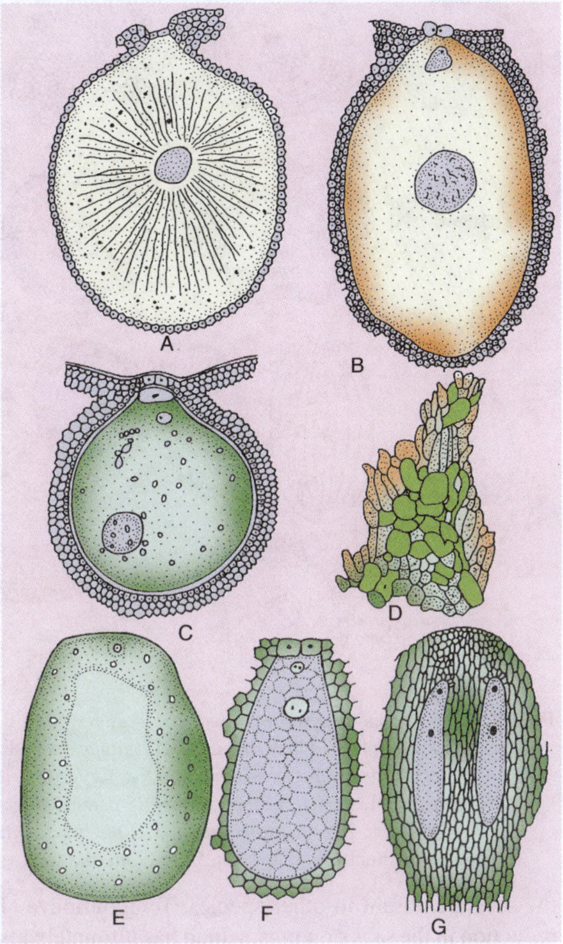

Fig. 1.7. Archegonia of gymnosperms. A. *Pinus roxburghii.* B. *Cycas revoluta.* C. *Ginkgo biloba.* D. Prothallial tubes in *Welwitschia.* E. Embryo sac of *Gnetum* showing eggs. F. *Taxus baccata.* G. *Ephedra foliata.*

The microspores or the pollen grains (Fig. 1.8) are unicellular and haploid structure produced within microsporangia. In *Cycas* the microspores are tetrahedral in shape and show a distinct polarity because the exine is thicker towards the base. They are uniaperturate. In *Ginkgo* the microspores are almost spheroidal and uniaperturate. In *Taxus baccata* the uniaperturate and spheriodal microspores have psilate exine (without any pattern). In *Abies* the microspores are winged (saccate), uniaperturate

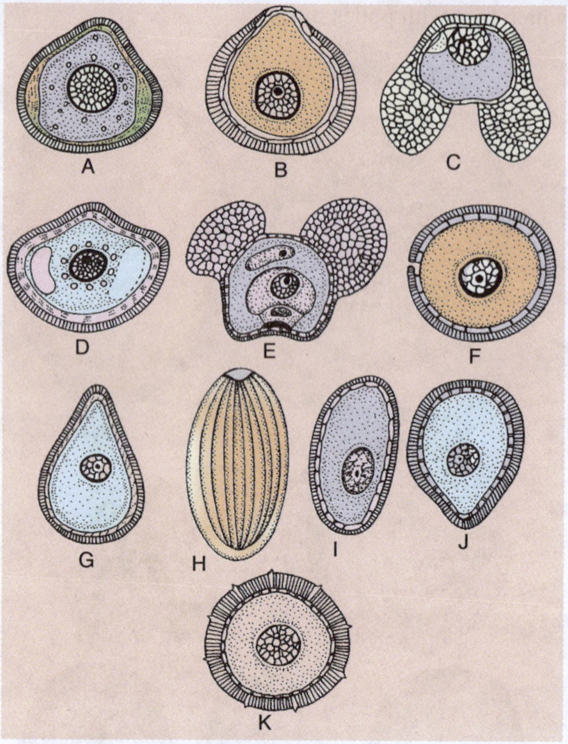

Fig.1.8. Microspores of gymnosperms. A. *Cycas revoluta*. B. *Ginkgo biloba* C. *Pinus roxburghii*. D. *Araucaria sp*. E. *Abies pindrow*. F. *Taxus baccata*. G. *Cryptomeria japonica*. H. *Ephedra foliata* (showing parallel ridges and two sac-like protuberances). I. *Ephedra foliata* (showing nucleus). J. *Welwitschia*. K. *Gnetum* sp.

and the exine is reticulate. The same is the case in *Cedrus*, *Picea*, *Pinus*, *Pseudolarix* and *Ketleeria*. In other members of Pinaceae, such as, *Tsuga*, *Pseudostuga* and *Larix* the grains are *non-saccate*. *Tsuga mertensiana Carr.* is an exception. It has winged pollen. In the families *Taxodiaceae*, *Cupressaceae*, *Arucaliaceae*, *Cephalotaxaceae* and *Taxaceae* the wings are absent (non-saccate). In *Podocarpaceae* except *Saxegotheca*, all other genera possess bisaccate (two winged) pollen grains. Three or more sacci (wings) have been reported in *Microcachrys*, *Phaerosphaera* and *Sacryearpus*. In *Podocarpus*, *Dacrydium*, *Acmopyle* and *Phyllocladus* the pollen grains are bisaccate. In *Cryptomeria griffithiana* the spherical microspores have a well marked papilla on one side. It indicates the aperturate area. In *Cupressaceae* the pollen grains are inaperturate and spheroidal with psilate (*Cupressus funiberus*), faintly granulose (*C. torulosa*) or granulose (*Juniperus communis*) exine. In *Ephedra* the grains are inaperturate elongate with plicate exine surface showing parallel ridges along the long axis of the grain. In *E. foliata* the grains possess two tiny sac-like structures at the extremities; whereas the sacs are absent in other species. Tchigourioeva (1954) suggested that there has been a gradual reduction in the sacs or wings, which has ultimately resulted in non-saccate pollen grains like those of *Ephedra*. Presence of minute sacs in *E. foliata* indicates a relationship with the *Coniferales* and *Cordailates*. In *Gnetum* and *Welwitschia* too the microspores are spheroidal and have thick stratified exine. The pollen grains are uni-nucleate.

The development of wings in some taxa of the gymnosperms (Pinaceae and Podocarpaceae) is considered to be a significant development during the course of pollen evolution. Woodhouse (1935) was of the opinion that the wings aid in pollen flotation in the air. Dogra (1964) is of the opinion that the wings help in fertilization process. In the dry pollen when the pollen is in flight or in natural condition, the wings touch each other and keep the apertural slit lightly closed. Under wet conditions the apertural area expands. These observations suggest that the wings are protective in function. A question arises as to whether the wings contain air and if so from where the air enters the wings. The development of the wings takes place after meiosis when the outer layer of exine (*i.e.*, ectine or sesine) falls loose from the inner layer of exine (endine or nesine) and expands to produce bladder like structures. It is clear that the wings are lined on the outside by the *ectine* with its hanging *collumellae*, and on the inside by the more homogenous and close-grained layer called the *endine*. These structural peculiarities do not provide any clue for the entrance of air into the bladders. Nair (1974, p. 51) suggests that the wings may be vacuumatic providing additional protection to the protoplasm contained in the body of the grain. The vacuum does not obstruct the expansion and contraction of the wings, during dry and wet conditions, respectively. Another important development which took place during the course of spore evolution is the shifting of the apertural mask from *proximal* position *in* the bryophytes and pteridophytes to the *distal* position in the gymnosperms and the angiosperms.

Abnormal Microspores: Lots of work has been done in the occurrence of abnormal pollen grains in the Pinaceae (Puri, 1945; Vishnu Mittra, 1957; Lakhanpal and Nair, 1956; Srivastva, 1961; Mehra and Dogra, 1965; Mesa, Stanley and Kirby, 1983 and Mehra, 1966). Mehra (1988) reported the occurrence of abnormal pollen grains in the conifers. Mithilesh Chaturvedi (1979) reported some morphotypes in the pollen grains of *Juniperus pseudozabina*. A brief account of the abnormalities is given below:—

Pinaceae: Mehra (1988) reported abnormal pollen grains in three species of Indian pines. These include *Pinus griffithii*, *P. roxburghii* and *P. kesiya*. In *P. griffithii* tetrasaceate (with four wings) pollen grains were reported. In *Pinus roxburghii* pollen grains with three and four wings were recorded. In some grains there was a single *saccus* or *wing* encircling the grain all over except the distal end. In others the saccus was poorly developed approaching nearly the *nonsaccate* condition. Boraccate *micropollen* were also noticed in some cases. In *P. kesiya trisaceate* pollen and pollen with poorly developed saccus on one side only were reported. The percentage of such abnormal grains is very little. It is 5.87% in *P. griffithii*, 7.59% in *P. roxburghii* and 3.26% in *P. kesiya*. Mesa, Stanley and Kirby (1983) reported *tetrasaccate* pollen grains in *P. elliothi* collected from Central Florida. Their occurrence was only 0.3%. They were able to promote germination in such abnormal grains which produced pollen tubes similar to those of normal grains upto two nucleate stage. Chira (1967) reported oversized pollen grains in *P. edulis*. The cause of production of such abnormal pollen grains is attributed to low temperature at the time of meiosis. Ho and Owens (1974) reported the occurrence of multiple sacci in *P. contorta*.

The other members of the family that show such abnormalities include *Cedrus deodara*, *Abies pindrow*, *Abies spectabilis* and *Picea smithiana*. Micropollen with two wings or sacs have been observed in *Cedrus deodara* and *Abies pindrow*. Exceptionally large diploid pollen grains have been observed in *Abies spectabilis*. Trisaccate grains are present in *Cedrus deodara*, *Abies spectabilis*, *Abies pindrow* and *Picea smithiana*. In the latter species tetrasaccate grains have also been reported. Saccus almost all around the grain except at the distal end has been reported in *Abies pindrow*, *A. spectabilis* and *Cedrus deodara*. Such a condition had been a normal feature in the *Cordaitales*, *Lebachia* and *Ernestiodendron* (Fossil conifers).

Pinus, *Cedrus*, *Abies*, *Picea*, *Pseudolarix* and *Ketleeria* are normally bisaccate genera, whereas *Psudostuga*, *Larix* and *Tsuga* are nonsaccate genera. Vishnu Mittre (1957) reported the presence of two rudimentary wings in *T. patoniana* and *T. brunomiana*. The fossil representatives of this genus from the Tertiary strata, *T. viridifleminipites* (Woodhouse, 1933) and *T. sp* (Kircheimer, 1934) possessed a rudimentary bladder encircling the grain all over. It appears that bisaccate condition of the pollen grains has not been firmly established in its constituent genera. The genes and gene block determining this character are absent in *Larix* and *Pseudotsuga*. The variations in the number of sacci in other genera is explainable on the basis of **multiallelic** or **muligenic** nature of the gene block determining the development of wings or sacci. The occurrence of micro- and macro-pollen (large diploid grains) is perhaps due to aberrations in the meiotic divisions in some of the pollen mother cells.

In the *Taxodiaceae*, *Cupressaceae*, *Arancariaceae*, *Cephalotaxaceae* and *Taxaceae* all the genera are **nonsaccate** and no abnormal grains with wings or bladders have even been reported. However, micropollen and exceptionally large or diploid pollen have been observed in *Crytomeria japonica*, *Cunninghamia lanceolata*, *Taxodium mucronatum* (all Taxodiaceae), *Cupressus funiberis*, *C. macrocarpa*, *Juniperus communis*, *J. squamata* and *J. bermudiana* (all cupressaceae). Micropollen have also been reported in *Agathus robuslus* and *Araucaria cunninghami* (Arucariaceae). Abnormal pollen grains with small size (micropollen) or with a papillate outgrowth surrounded by intine have been reported in *Cephalotaxus* and *Taxus*. Papillate pollen grains have also been reported in *Cryptomeria japonica* and *Cunninghamia lanceolata*. Chaturvedi (1979) reported pollen grains of different sizes in one population of *Juniperus pseudosabina*. The pollen grains ranged in size from 22-32 m in diameter. Under the scanning electron microscope, he observed the presence of *orbicules* of various sizes and shapes on the surface of the *exine*, which is 1.11 m thick. These orbicules are usually nodular and uniform. In some cases these orbicules are arranged in the form of islands of different sizes and shapes. He also observed variability in the same pollen mass in *Ephedra intermedia var pessica*. He recorded circular grains with internal thickenings and elongate grains with external thickenings. Formation of diploid pollen grains, due to irregularities in Cytokinesis, has been reported in *Gnetum ula* and *Ephedra saxatilis*.

Male Gametophyte

In order to describe the various cells, formed during the development of pollen grain into male gametophyte, various terms have been used by different authors. The terminology adopted here is according to Sterling (1963) with a little change. The pollen grain enclosses a uninucleate cell which is called *embryonal cell* by Sterling but this has been objected to by some authors. Singh and Johri (1972) and Singh (1978) have pointed that the usage of term embryonal cell is not proper because this term is used in the descriptions of the development of embryo. Dogra (1984) has rightly suggested the use of term *androspore* as proposed by Doyle (1953). The androspore cuts off two *primary prothallial cells* usually by periclinal divisions and the large cell is called the *antheridial initial* or *cell*. The primary prothallial cells may divide further to form *secondary prothallial cells*. The antheridial initial divides to form a *tube cell* and a *generative cell*. The generative cell may divide periclinally to form a *sterile* (stalk cell) and a *spermatogenous cell* (body cell). The spermatogenous cell divides ultimately into two *male gametes (or male cells). In some cases where prothallial cells are not formed the androspore cell acts directly as an antheridial initial.*

The male gametophytes in gymnosperms show variations regarding their release from sporangia, the number of male prothallus cells, size and motility of male gametes and their time of formation and discharge. They are **endosporic** and complete their development partly in the microsporangium and partly in the pollen chamber of the ovule. A brief account of their comparative morphology is presented here. In *Cycadaceous microgamet* ophytes there is one male prothallus cell that protrudes into a large **sterile cell** or the **stalk cell** followed by a **body cell** or a **spermatogenous** cell and a **tube cell**. They are arranged in a linear row. The body cell divides into two multicilliate **male gametes**. The pollen tube is formed in all Cycads and is more haustorial in nature than a sperm carrier. In *Microcycas* the **stalk cell** divides into 10 or 11 **spermatogenous** (body cell) **cells** all of which divide to produce 20 or 22 spermatozoids. In *Ceratozamia* there are 4 spermatozoids. The division of **generative cell** into **stalk cell** and **body cell** is periclinal except in *Cycas revoluta* where it is anticlinal.

In *Ginkgo biloba* there are two male **prothallus cells**, a **stalk cell**, a **body cell** and a **tube cell**. The generative cell divides anticlinally into **stalk cell** and a **body cell**. At a later stage the stalk cell surrounds the second male prothallus cell. The two male gametes are multiflagellate.
The sequence of microgametophyte development shows slight variations in the different families of *Coniferales*. In the *Pinaceae* the microspore nucleus divides twice by periclinal walls to cut off two **male prothallus cells** and an **antheridial cell**. The latter divides periclinally to form a **generative cell** and a **tube cell**.

At this four-celled stage the grains are liberated from the sporangium. The generative cell divides periclinally to form a stalk cell and a body cell (spermatogenous cell). All these cells lie in an axial row. After pollination the semigerminated pollen grains may germinate immediately or a month later. On entering the nucellus the pollen tube may branched. The tube nucleus migrates into one of the branches and moves to its tip. The generative cell remains within the spore wall and divides sometime between the fall of the year of pollination and the following spring into stalk cell and the body cell. These two move into the pollen tube. About a week before fertilisation the body cell divides into two non-motile male gametes that are unequal in size (Sethi, 1928; Konar, 1960). Some workers regard the two male gametes as two nuclei of a binucleate sperm cell; whereas others regard them as two sperm cells. In many genera, *e.g.*, *Abies, Picea, Larix, Cedrus*, etc., the generative cell divides before pollination. In *Abies* the body cell sometimes divides into two male gametes before pollination or shortly after pollination. Equal male gametes have been reported in *Cedrus* by Chowdhury (1960). The period between pollination and fertilisation varies in various members of *Pinaceae*. It is longest in *Pinus* (up to one year), up to nine months in *Cedrus*, 3 months in *Keteleeria*, one month in *Larix*, 3 weeks in *Pseudotsuga*, 4-5 weeks in *Abies*, a few days in *Picea*. In *Abies balsamea*, Hutchinson (1915) reported the formation of 3 or 4 prothallial cells (Fig. 1.9, A-C). In *Taxodiaceae* the microgametophytes lack prothallial cells and the microspore functions directly as antheridial cell and divides into a generative cell and a tube cell before pollination. The generative cell divides into stalk cell and body cell after pollination. The male gametes are equal and non-motile.

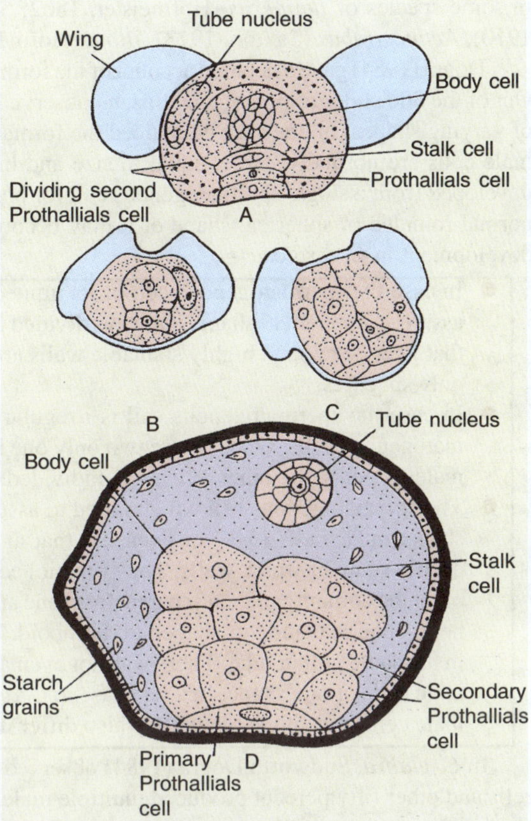

Fig. 1.9. Germination of pollen grain in *Abies* and *Araucaria*. A. Semi-germinated pollen grain of *Abies balsamea*, B.Abnormal male gametophyte of *A. balsamea* showing division of second prothallial cell. C. Abnormal micro-gametophyte of *A. balsamea* showing bi-antheridial gametophyte. D. Semi-germinated pollen gain of *Araucaria* cunninghamii with several derivatives of prothallial cell. Sterile cell, spermatogenous cell and tube nucleus are also shown.

In the nonsaccate pollen of *Cupressaceae* too the prothallial cells are absent and the androspore acts as an antheridial initial and may divide into generative cell and tube cell before or after pollination. In most of the Indian species of *Cupressus* the pollen is shed in a unicellular condition but in some species it is shed in a bi-celled condition (Dogra, 1984) *i.e.*, after the formation of generative and tube cell. The pollen tube is formed when the grain is on the nucellus. It is narrow at first but it broadens as it enters the nucellus. In the nucellus it takes an irregular course and sometimes may even branch. The generative cell divides to form body cell (or spermatogenous cell) and a stalk cell (or the sterile cell). On reaching the *archegonial complex* the tip of the pollen tube becomes broad and bulbous and is pressed against the neck cells of the archegonia. At this stage the body cell or the spermatogenous cell divides into two large and equal male gametes that are hemispherical in shape. This is the normal development in most of the *Cupressaceae e.g*., *Libocedrus decurrens*, *Cupressus funileris*, *C. arizonica, Thuja occidentalis*, *Thuja* orientalis (*Biota*), *Chamaecyparis pisifera, C. obtusa, Thujopsis dolabrata, Juniperus procera, J. chinensis, Juniperus communis, J. virgioniana, Widdringlonia, Cupressoides, Callitris sp., Actinostrobus pyramidalis*, and *Tetraclinis articulata*. In some species of *Cupressus* formation of more than two male gametes has been reported (Juel 1904, Doak, 1932, Mehra and Malhotra 1947, Sughiara 1956, Dogra, 1984). More than two male gametes have also been reported in some species of *Iuniperus* (Hofmeister, 1862; Strassburger, 1877, 1879; Noren, 1907; Nichols, 1910), *Actinostrobus* (Saxton, 1913), *Biota* (Hofmeister, 1862) and *Callitris* (1910).

Dogra (1984) gave a detailed account of the formation of multiple male cells in *Cupressus torulosa*. Out of the 300 slides of pollen tube tips, he observed that 90% contained or produced 3-12 male cells of varying shapes. In one case he noticed the formation of two unequal male cells (in the family the male cells are normally, large, equal in size and hemispherical in shape). The multiple male cells developed from a single spermatogenous cell (or body cell). The spermatogenous cell may assume a normal rounded or spherical shape or it may become irregular in shape. It exhibited three types of development in *C. torulosa*:—

- In case the spermatogenous cell or its immediate derivate (secondary spermatogenous cell) assumed a spherical shape, it always divided to form two large and equal sized male gametes, that later developed highly stainable walls around them. Such a development was noticed in only sp. cases.
- In case the spermatogenous cell is irregular in shape it is divided into two irregular spermatogenous cells. Out of these two only one irregular cell developed further into two normal male gametes, the other remained undivided. This happened only in two percent cases.
- The spermatogenous cell which failed to assume a round shape divided into many irregular-shaped male cells (3-12 or even 28) that at first developed thin walls but later developed deeply stainable thick walls. They functioned as male gametes. The mature male cells round off at the corners and then separate from one another assuming various shapes such as rounded, hemi-spherical and top-shaped to rhomboid. These varying shaped male cells were all found in the same pollen tube. There was not even a single case where a pollen tube contained only male cells of one shape. In a rare case in *C. arizonica*, Doak (1932) observed all pear-shaped male cells in a single tube. They also differ in sizes.

In *C. glabra*, Sudworth Dogra (1984) observed that sp. of the pollen tubes showed normal male cells and other fifty percent produced multiple male gametes that numbered from 13-28. In this species he observed that if the spermatogenous cell assumes a normal rounded shape with a well defined wall around it, it divided into two large and hemispherical functional male gametes that developed thick walls around them at maturity. In case the spermatogenous cell fails to develop a wall around it and becomes irregular in shape, it divided repeatedly to form 13-28 male gametes. Some of these may undergo degeneration. In one abnormal case he discovered a pollen tube containing two normal male gametes and a large number of free nuclei in the cytoplasm. In such tubes the sterile stalk nucleus and pollen tube nucleus were not found. It appears that these two nuclei divided to form several free nuclei. Dogra (1984) considers this multigamete formation as an advanced condition among the *Cupressaceae*. Incidentally the archegonia in the archegonial complex of cuprassaceous ovules are also known to contain multiple female gametes. This condition which is found in both male and female gametophytes leads to polyembryony due to multiple fertilization.

The *Araucariaceae* are unique in producing many prothallial cells (Fig. 1.9, D). In *Araucaria* as many as 13 to 40 **secondary prothallial** cells have been recorded (Lopriore, 1905; Thomson, 1905; Burlingame, 1913; Chamberlain, 1935). In *Agathis* 6-10 secondary prothallial cells have been recorded (Eames, 1913). In both these genera the generative cell divides anticlinally and the spermatogenous cell (body cell) divides to produce two male gametes that may be equal (Ghose, 1924) or unequal in size (Eames, 1913). The archegonia in the family are solitary. The male gametes are non-motile. The large number of prothallial cells is due to continued division of the two prothallial cells formed by the periclinal division of the andropore cell. The generative and tube cells are formed by the third periclinal division of the antheridial initial. The walls of the secondary prothallial cells and of the sterile cell (stalk cell) break down before pollination. Their nuclei and the spermatogenous cell (body cell) along with tube nucleus float freely in the tube cell.

In *Podocarpaceae* (Fig. 1.10 A-C) the number of male prothallus cells may be 1–8 (*Podocarpus*), 3–6 (*Dacrydium*), 1–3 (*Phyllocladus*), 2–4 in *Microcachyrys* and 3–4 in *Saxegotheca*. The pollen grain nucleus divides into two particular primary prothallial cells by periclinal divisions. The larger cell functions as an antheridial initial and divides periclinally to form the generative and tube cells. The primary prothallial cells may divide later by anticlinal divisions to form secondary prothallial cells whose number varies in various genera as described above. The generative cell, as in *Araucariaceae*, divides by an anticlinal division into a large spermatogenous (body cell) cell and a smaller sterilecell (stalk cell) that lie side by side. The walls of the prothallial cells and the sterile cell (stalk cell) break down before pollination so that their free nuclei along with the tube nucleus and the spermatogenous cell lie free in the tube cell. According to Boyle and Doyle (1953) this break down occurs after pollen germination in *Podocarpus nivalis*. After pollination the free nuclei move to the pollen tube followed by the spermatogenous cell which divides just before fertilization into two unequal male gametes in *Podocarpus*, *Dacrydium* and *Phaerosphaera* and *Phyllocladus*. In *Phyllocladus* the male gametes of equal size have been reported by Kildahl (1908). In *Microcachrys* the male gametes are usually equal in size (Lawson, 1923). In *Saxagotheca* the male gametes show lots of variations in size (Looby and Doyle, 1939). In *Pherosphaera* there are no male prothallus cells.

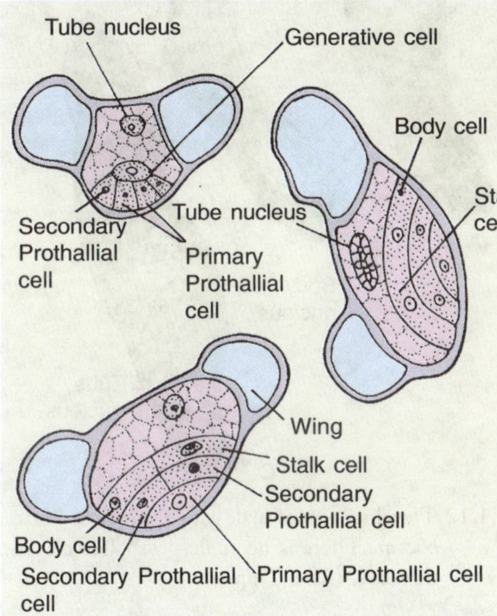

Fig. 1.10. (A—C). Stages in the development of male gametophyte in *Podocarpus totarra*. Primary prothallial cells divide to form secondary prothallial cells (After Burlinagme, 1908).

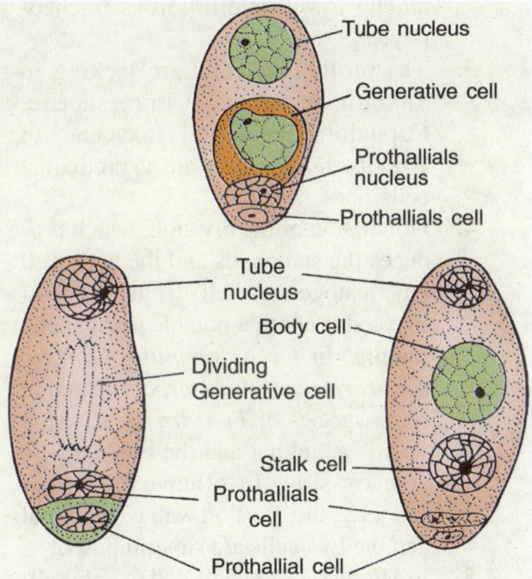

Fig. 1.11. Male gametophyte in *Ephedra foliata* (After Land).

There are no prothallial cells in any species of *Cephalotaxus* (*Cephalotaxaceae*). The male gametes may be unequal (Sughihara, 1947; Kaur, 1958) or equal (Singh, 1961) in size. There are no male prothallus cells in *Taxus* (Fig. 1.12, A — C) also. The male gametes are unequal. In *Ephedra* there are two male prothallus cells, a stalk cell, a body cell and a tube cell that are arranged in a single axial row. Body cell and stalk cells are formed by the division of the generative cell (Fig. 1.11 B, C). Out of the prothallial cells one is without a cell wall and is thus called prothallial nucleus (Fig. 1.11). The body cell divides into two male gametes after pollination. In *Gnetum* there is only one male prothallial cell, a tube nucleus and a generative nucleus. At the time of pollination the pollen grain contains only three nuclei. The generative nucleus divides into two male gametes. In *Welwitschia* there is no prothallial cell and the released grain contains a tube nucleus and a generative nucleus.

The above discussion on the general characteristics of male gametophyte in gymnosperms highlights the following points:—

1. Except for the formation of prothallial cells the *Cycadales*, *Ginkgoales*, *Coniferales*, and *Ephedra* among *Gnetales* all have the same cellular constitution of the microgametophytes including the tube cell, the stalk cell and the male gametes.
2. The very large number of prothallial cells of *Araucariaceae* and *Podocarpaceae*, developed secondarily at the base of the microspore, may well be a later evolutionary specialisation. This condition is not found elsewhere among living or fossil gymnosperms or among lower heterosporous tracheophytes.
3. The prothallial cells are lacking in Taxodiaceae, Cupressaceae, Caphalotaxaceae and Taxaceae. In *Welwitschia* also there are no prothallial cells.
4. Orientation of the division, which produces the stalk cells and the body cell (spermatogenous cell) from a generative cell is another notable ontogenetic feature. In *Cycas revoluta*, *Ginkgo*, *Araucariaceae, Podocarpaceae* and in some species of *Ephedra* the division wall is anticlinal and the two cells lie side by side. In other cycads and *Pinaceae* the division wall is periclinal and the two cells are superimposed.

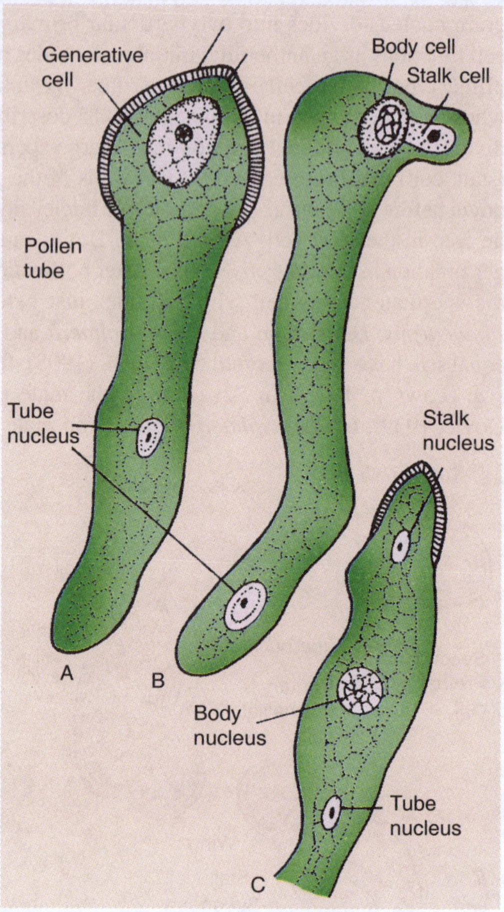

Fig. 1.12. Development of male gametophyte in *Taxus baccata*.There is no male prothallus cell in this case. (After Dupler).

5. In *Microcycas* the stalk cell (sterile cell) divides and gives rise to additional spermatogenous or body cells.
6. In *Cupressus* and sometimes in *Juniperus* more than two male gametes are produced. This condition appears to be a derived state since the gametes are formed from the sub-division of a single body or spermatogeneous cell. This feature has been related to the presence of the large number of archegonia in the archegonial complex. Such a condition has been regarded by Juel (1904) as a case of atavism. The male cells are of variable shapes and sizes depending upon the mutual pressure exerted upon each other during their formation and maturation. In *Cupressus glabra* and *C. tarulora* some of the male cells may degenerate whereas the rest of them remained fertile.
7. The development of the archegonia in Podocarpaceae, Cupressaceae and Taxodiaceae is markedly influenced by the presence of the pollen tube. In *Callitris*' and *Windringtonia* the archegonial complex is lateral because pollen tube travels along the sides of female gametophyte. In *Cupressus* as well as in *Podocarpus* and *Tavaodium* the pollen tube rests at the micropylar end of the female gametophyte and hence archegonial complex is terminal. Konar

and Oberoi (1969) have observed, "Usually the archegonia lying near the body cell show more stage of advancement than those away from it. Some archegonial initials start enlarging but soon become arrested. This is probably due to the fact that the pollen tube stimulates the superficial cells to grow as archegonial initials and by the time these cells start enlarging the tube advances on to some other cells. Such immature archegonia can be seen on the gametophyte along the path of the pollen tube (*Podocarpus gracilior*)".

8. Pollen sterility (5%) has also been reported in some species of *Cupressus* (Dogra, 1984) *e.g.*, *C. arizonia* and *C. glabra*. In these species due to the formation of chromosomal bridges and fragmentation during microsporogenesis, some pollen grains become shrivelled. Some of them contained micro and macro nuclei.

Pollination

The pollination in gymnosperms takes place by means of wind and results in the transference of semigerminated pollen grains on the micropyle of the ovule. These pollen grains, in most of the gymnosperms, are caught into a pollination drop secreted by the micropylar end of the ovule. The semi-germinated microspores are drawn into the ovule by the drying of the pollination drop. The micropyle closes after the act of drawing in of the microgametophyte.

In the Pinaceae the **pollination mechanism** has been studied in *P. sylvestris, P. tabulae formis, P. elliottii, P. nigra, P. griffithii, Tsuga, Cedrus, Pseudostuga, Abies* and *Picea* by different workers all over the world. In all these genera and species the pollination drop is secreted by the nucellus and fills the micropylar tube upto the brim. In *P. sylvestris* the secretion of the drop begins at night and is completed by 2 A.M. The drop completely dries by 7 A.M. The drop is usually secreted at night and not during day time. It contains sucrose, glucose and fructose. In *Pinus* the winged microspores are so oriented in pollination drop so that their germ fusion is closely adpressed to the nucleus. In *Larix* and *Pseudolarix* the pollen does not lie on the nucellus but laterally placed. The pollen reaches the nucellus either by **flotation mechanism** or by a **guttation phenomenon**. In *Abies* no pollination drop is secreted and the pollen grains fall directly on a funnel-like **integumentary expansion** around the micropyle and nucellus grows up to come in contact with the grains. Formation of three or four prothallial cells and biantheridial gametophytes have been reported in *Abies bekamea.* Division of the tube nucleus into two and that of sterile cell also have been reported as abnormalities in this family. The pollen grains do not immediately start germinating on the nucellus but remain dormant for variable periods *e.g.*, 9 months in *Cedrus*, 3 months in *Keteleeria evelyniana*, 3 weeks in *Pseudostuga* and for a few days in *Picea*. In *Pinus* they rest for about a year after the migration of the tube nucleus into the pollen tube (soon after pollination). It is during the next spring that the generative cell divides into sterile (stalk) cell and spermatogenous (body) cell, which then migrate into the tube. The spermatogenous cell divides only a week before fertilization into two unequal male gametes (*Abies, Keteleeria, Larix, Picea, Pinus* and *Pseudostuga*). In *Cedrus deodara* the male gametes have been reported to be equal (Roy Chowdhury, 1961). Konar (1960) reported that the male cells are equal in *Pinus roxburghii* but later become unequal due to enlargement of one of them.

In Taxodiaceae the pollen grains are discharged from the microsporangium at a two celled stage. In this family primary prothallial cells are not formed and the androspore acts directly as antheridial cell and divides into generative cell and tube cell before pollination. The drop mechanism is there and the two celled pollen grains reach the nucellus where a pollen tube is formed and during its growth through the nucellus, the generative cell divides into spermatogenous cell (body cell), sterile cell (stalk cell). The former enlarges in size and divides into two equal male cells shortly before fertilization. Tahara (1937, 1940) reported two unequal male gametes in *Sciodopitys*. This is the only genus in Taxodiaceae where the male gametophyte overwinters on the nucellus. In majority of the genera the pollen tube grows straight into the nucelleus but in *Athrotaxis selaginoides* and *Sequoia sempervirens* it grows laterally between the integument and the nucellus for some distance and then penetrates into the nucellus and establishes contact with the megaspore membrane, which in the former genus is still in the free nuclear stage.

In Cupressaceae the pollination mechanism is similar to that of the Taxodiaceae. The primary prothallial cells are not formed in any member of this family and the androspore cell functions directly as the antheridial initial. The pollen grain is liberated in a unicellular condition in some genera (*Juniperus, Actinostrobus, Callitris, Widdringtonia, Tetraclinis*) and in a bicellular condition in other (*Biota, Chamaecyparis, Libocedrus, Thuja, Jhujopsis*). In *Cupressus* it may be released in either of the two conditions depending upon the species. The pollen are carried away by wind and fall on the pollinative drop that exudes from the micropyles of ovules. The pollen grains reach the nucellus after the contraction of the drop. Open self-pollination has been recorded by Dogra (1984) in *Jhujopsis dolabrata* and *Chamaeajparis obtusa*. Controlled self-pollination was also carried out by Dogra and Tandon in *Biota orientalis* achieving 75% seed fertility. In pruned and reshaped garden trees of *Biota* natural self-pollination takes place (Dogra, 1984). Normally in wild populations open pollination is prevalent.

In Podocarpaceae which is also wind pollinated, Doyle O'Leary (1935) reported that in *Saxegotheca* the pollen grains fall directly on the nucellar flare or even on the scale from where it gives out a long pollen tube that traverses through the nucellus and reaches the embryo sac. It has been reported that all the prothallial nuclei move into pollen tube alongwith the tube nucleus, body cell and stalk cell and persist till fertilization (*Saxgotheca, Podocarpus*). On coming in contact with the gametophyte the tip of pollen tube spreads out like a thin disc, covering the open of archegonial complex. The body cell divides shortly before fertilization into two male gametes that are usually unequal except in *Saxegotheca* in *Microcachyrus*.

In *Cephalotaxus drupacea* the pollination takes place in March when the pollen grain is binucleate (prothallial cells are absent). Wind pollinated and semi-germinated grains are entangled in the pollination drop and settle on the nucellus after being drawn in by the drying of the drop. On the nucellus, in this bicelled condition (generative cell and tube cell), the pollen grains overwinter. Later the generative cell divides into spermatogenous cell and stalk cell. The spermatogenous cell divides into two equal or unequal male gametes after the nucleus of sterile cell is set free.

In the Araucariaceae (*Araucaria* and *Agathis*) the pollen grains with many prothallial cells (13-40 or 6-10) are carried by wind to the cone scales and not to the micropyles of ovules because in both the genera the ovules are immature at that time or are even in early stages of development or even not developed (*A. anguslifolia*). In *Araucaria bidwillii* the ovules are mature at the time of pollination but the pollen falls on the cone scale. The germinating pollen grains develop branched pollen tubes that grow towards the micropyles alongwith the prothallial nuclei, spermatogenous nucleus, stalk nucleus and tube nucleus. Pollination takes place in May followed by a long resting period with fertilization taking place in next April.

In the Taxacae the unwinged pollen grains are usually released in a unicellular condition. The androspore nucleus functions directly as antheridial cell as no primary prothallial cells are formed. The unicellular grain is carried by wind and is caught in the pollination drop secreted by the ovule. The drop fills the micropyle and exudes out of it. The drop later dries and the pollen is sucked in on the nucellar tip. The pollen grain starts germinating without undergoing any period of rest. The antheridial cell divides into tube cell and generative cell. The pollen tube emerges out of the single aperture and is covered by intine. The tube nucleus moves to its tip and the tube grows through the nucellus. The generative cell remains in the grain end and divides into spermatogenous cell and stalk cell. Later the stalk nucellus comes out of the stalk cell and moves to the pollen tube tip alongwith the tube nucleus. Later the spermatogenous cell divides into two unequal male gametes. Further growth through the nucellus lodges the pollen tube at the tip of the female gametophyte. All the nuclei move to the tip.

Fertilization. This process embraces all the changes that take place from the time of liberation of the male gametes from the pollen tube to their fusion with the female gametes. It shows variations in the different groups of gymnosperms. In *Cycas* the **grain end** of the pollen tube becomes extremely turgid, distended and ultimately bursts to release its contents into the archegonial chamber which is dry or slightly humid. The fluid of the pollen tube fills the archegonial chamber. In this fluid the multicilliate sperms start swimming and find their way towards the archegonial necks. The fluid discharged by the pollen tubes into the archegonial chamber is believed to be of high osmotic concentration. It is believed that it plasmolyses the turgid neck cells of the archegonia and permits the sperms to enter into the archegonia. The exact manner of entrance of the sperms into the archegonia is not exactly known. Chamberlian (1935) is of the opinion that after the act of plasmolysis of the neck cells some contents of the upper portion of the egg cell of the archegonia move into the archegonial chamber. In doing so a large vacuole appears at the upper part of the egg. This leads to the drawing in of the sperm into the egg with such a force that its cytoplasm and cilia are torn as under and only the nucleus is drawn into the lower portion to meet the female nucleus. De Silva and Tambiah (1952) suggested that in *C. rumphii* the neck cells are pushed apart, after they loose their turgidity, by the spermatozoids which are attracted by the egg nucleus by a chaemotactic response. The sperms move by amoeboid movements. Swamy (1948) also observed amoeboid movements of sperm in a species of *Cycas* from Mysore. He observed that in this species the base of the pollen tube ruptures so little that the sperm has to squeeze its way through it by amoeboid movements. Ikeno (1896) was the first to report this phenomenon of fertilization by ciliated and mobile sperms in *Cycas*. Such a fertilization is called **zoodiogamy**. In *Cycas* as well as in *Ginkgo* (where also the sperms are ciliated and are discharged in the same way into the archegonial chamber that is not dry, but is already filled with a fluid for the sperms to swim). Zoodiogamy is accompained by **siphonigamy** *i.e.*, the formation of pollen tubes whose function is only haustorial and not deliverance of male gametes. The male nucleus has been reported to shrink in size before penetrating into the larger female nucleus. After the entry of the male gamete or the sperm the archegonium is sealed perhaps by a substance released by the degenerating spermatozoid. The process of fertilization is almost similar in all cycads except for minor variations as for example, in *Encephalartos villosus* the ventral canal nucleus does not disintegrate but moves towards the egg nucleus and enlarges (Sedgwick, 1924) and may probably effect fertilization if male nucleus fails to do so. Such a phenomenon has been reported to occur in *Ginkgo, Pinus*, and *Picea*. In *Bowenia serrulata* Lawson (1926) reported the unique phenomenon of secretion of droplets of a liquid by archegonia into the archegonial chamber. The spermatozoids of *Zamia floridona* are reported to be the largest in the plant kingdom (332 m long). In *Microcycas* as many as 200 archegonia have been observed to develop in the female gametophyte (Caldwell, 1907) but only those towards the micropylar end develop fully (Raynolds, 1924). The archegonia have a reduced neck and may possess two or three free nuclei in the egg cell. This multi-nucleate condition may be due to the division of ventral canal nucleus or by the fusion of two or more archegonia. Land (1907) has suggested that these multinucleate eggs can be likened to multi-nucleate archegonial tubes of *Welwitschia*.

In the conifers the pollen tube acts as the carrier of male gametes. In most cases the male gametes, the stalk or sterile nucleus and the tube nucleus move to the tip of the pollen tube, which moves through the nucleus to the archegonial neck. It pierces through the neck and bursts to liberate its contents in the egg cell. This type of fertilization is called **siphonogamous**. One of the male gametes fuses with the egg to form the diploid fusion nucleus or the *zygote*. This process is called syngamy. In *Pinus* and *Cedrus*, after entering through the archegonial neck, the pollen tube releases all its nuclei, starch and cytoplasm into the egg cell. In *Keteleeria* Sugihara (1943) and Wang (1948) reported that the pollen tube travels through the egg cytoplasm to a considerable distance before releasing its contents. It has been reported that in some species of *Pinus* (*P. lambertiana* and *P. cembroides var. monophylla*) the egg cytoplasm forms a receptive vacuole in its upper part before the contents of pollen tube are released into it. The functional male nucleus travels to the female nucleus or the egg and sinks into it. A double membrane separates the male and the female nuclei. Later this membrane dissolves and the chromatin material in each nucleus condenses and later assumes the form of distinct chromosomes.

Singh and Chatterjee (1963) have described fertilization in *Cryptomeria japonica* whereas Buchholz (1939) studied it in *Sequoia* and Brennan and Doyle (1956) in *Alhrotaxis*. The process of fertilization in other families of coniferales has been discussed in chapter dealing with the general account of coniferales. It is, however, thought proper to discuss it in the Cupressaceae in which family Dogra (1984) has described it in detail. In this family the archegonia occur in groups called archegonial complexes which are micropylar in position in *Cupressus, Jhuja, Biota, Juniperus*, etc., whereas in *Callitris, Widdringtonia* and *Actinotrobus* they are lateral in position. In *Biota* chalazal complexes are also recorded. In *Cupressus* and other genera with micropylar complexes, the pollen tube travels to the micropylar end only, whereas in genera with lateral complexes, the pollen tube travels along the side of the female gametophyte upto half of its total length where the archegonial complex is situated. The tip of the pollen tube swells considerably and presses against and covers the necks of most of the archegonia of the complex. The neck cells are displaced to permit the pollen tube to discharge its contents into the egg cell. In most cases two and in species of *Cupressus* with multiple male cells enter the same or different archegonia of the complex. Some portion of male cytoplasm of the male gamete mixes with the female cytoplasm near the upper part of the egg whereas the rest of the male cytoplasm travels down along with male nucleus and mixes with female cytoplasm around the female nucleus. The male nucleus comes in contact with the female nucleus in a crescent-shaped depression. The male nucleus is smaller in size than the female nucleus and only one male nucleus fuses with it. Both of them may completely fuse *in situ* or they may travel to the base of the egg cell and then fuse. In *Cupressus* one to four pollen tubes have been observed to reach the archegonial complex. Chesnoy and Thomas (1971) studied the process of fertilization in *Biota orientalis* under the electron microscope. They observed that the entire archegonial complex is surrounded by a common chamber. The pollen tube discharges its contents into this chamber. This enables the gametes from a single pollen tube to fertilize two neighbouring eggs. The male gamete penetrates the egg cytoplasm. The peripheral or the outer zone of male cytoplasm mixes with the female cytoplasm. The male nucleus surrounded by its perinuclear and inner zones of cytoplasm moves down into the maternal cytoplasm. During this movement the male nucleus frees itself from inner cytoplasmic zone and reaches near the female nucleus. The female nucleus develops a depression in its upper part. The male nucleus settles in this depression. The membranes of the two nuclei join together and the nuclei take a turn of 180 degree. The inner layers of cytoplasm of male nucleus surround the two fusing nuclei thus isolating them from the maternal cytoplasm. This surrounding **male nucleus cytoplasm** later, after the first **zygotic mitosis**, penetrates the original nucleoplasm and surrounds the two **proembryonal nuclei**.

The pollen tubes with multiple male cells in *Cupressus glabra* and *C. torulosa* discharge all the male cells near the necks of the archegonial complex. One or many male cells may enter the archegonium but only one of them is able to effect fertilization. Dogra (1984) also observed fertilization in *Juniperus procera*, a species which forms only two equal male gametes in a pollen tube. In this species also the entire archegonial complex is surrounded by a common chamber or an archegonial jacket which is one cell thick. The bulbous pollen tube tip presses against the archegonial necks and displaces them. It enters the archegonium and releases the two large and equal sized male gametes. One or both of them may enter the egg cell but only one of them effects fertilization. Its outer male cytoplasm is discarded outside the female cytoplasm and the male nucleus slips into the egg and moves towards the female nucleus, which develops a crescent-shaped depression. The male nucleus comes in contact with the female nucleus at this depression. He could not observe the role of male cytoplasm during fertilization but observed that the first zygotic mitosis to form two free proembryonal nuclei takes place *in situ*.

In *Ephedra* the pollen tube grows through the archegonial neck and discharges its contents into the cytoplasm of the egg by the bursting of its swollen tip. The larger male nucleus moves towards the female nucleus which is surrounded by a dense cytoplasmic sheath. The male nucleus moves through the sheath and fuses with the female nucleus to form a fusion nucleus which along with its surrounding cytoplasm is called **zygote**. Khan (1940, 41) and Mulay (1941) observed the fusion of the second male gamete with the ventral canal nucleus in *E. foliata. The fusion nucleus thus formed does not germinate to form embryo. In Gnetum* there are no archegonia. The pollen tube tip bursts to release the male cells, one of which enters the **egg cell**. It leaves its sheath outside the egg cell. The male and the female nuclei lie as a pair for some time and then fuse to form the diploid zygote. Both the male cells are functional and can fertilize two eggs lying side by side. In *Welwitschia* also there are no archegonia. In this case the cells at the micropylar end of the female gametophyte grow upwards into small tubes. These are the **embryo sac tubes** and grow through the **nucellar cap** and meet the downward growing pollen tubes. Their tips meet, one female nucleus moves to the tip of each embryosac tube. The wall of contact between the pollen tube and embryosac tube dissolves and the female nucleus moves into the pollen tube to fuse with one of the male nuclei to form a zygote. After fertilization the zygote moves down into the female gametophyte, where further development takes place.

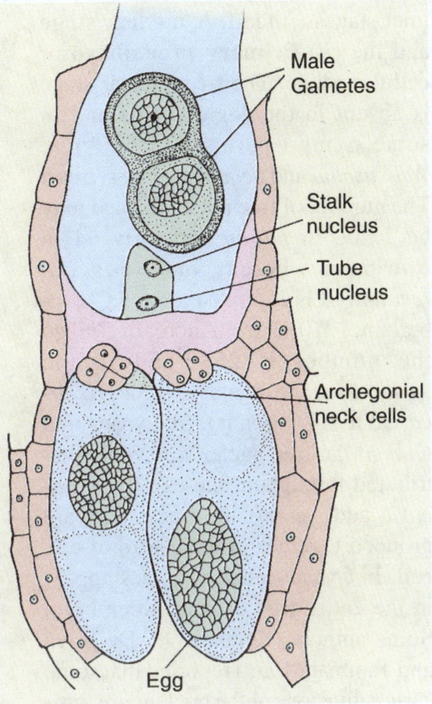

Fig. 1.13. *Tetraclinis articulata* (*Cupressacceae*) showing passage of pollen tube containing two equal male gametes. Also note the archegonia. (After *Saxton*).

The extinct orders of gymnosperms, namely *Cycadofilicales*, *Bennettilales* and *Cordaitales* did not produce pollen tubes and the motile sperms were liberated directly into the pollen chambers. This process is called **zooidogamous**.

In some gymnosperms (*Cycas, Ephedra, Ginkgo*) a distinct **pollen chamber** is formed at the apex of the nucellus and it receives the microgametophytes on pollination (For details see the concerned chapters). In the extinct gymnosperms of the palaeozoic the pollen chambers contained liquid containing cavities into which the motile sperms were liberated as a result of dehiscence of the microspore wall. In all the living gymnosperms the microspores produce tubular outgrowths or pollen tubes that grow through the nucellar tissue. In *Ginkgo* and the *cycads* the pollen tubes act mainly as the haustorial organs. In both these genera pollen drop mechanism effects pollination. They grow for several months into the nucellar tissue (Fig. 1.13) and absorb food and convey it to the microgametophyte at the lower end or **grain end** of the tube.

EMBRYOGENY (Figs. 1.14 — 1.16)

In living gymnosperms (except in *Gnetum, Welwitschia* and *Sequoia sempervirens*) the basic gymnosperm plan of embryo development can be divided into three stages:—

- **Free Nuclear Proembryo:** In the primitive gymnosperms like the cycads, *Ginkgo* and some families like the *Araucariaceae* the nucleus of the zygote divides by free nuclear divisions that may range from 10 mitoses in some cycads, 8 in *Ginkgo* and six (*Bowenia*) or five in other primitive members. In advanced gymnosperms there may be two to three mitoses. In Cycads the zygote moves towards the base of the archegonium and without undergoing any rest starts further development. Its nucleus starts dividing. The whole fertilized egg does not develop into an embryo. The divisions of zygote do not directly lead to the formation of an embryo (as is the case in the angiosperms) but produce a structure called the **proembryo**, which can be defined as an "*early embryonic stage beginning after the first division of the zygote and ending when a basal embryonic region begins to differentiate the distinctive organs of the embryo.*" Proembryo can be clearly distinguished into two distinct stages (*i*) the **free nuclear stage** and the (*ii*) **Primary proembryo** or **cellular stage.** The free nuclear stage is absent in the angiosperms and in some gymnosperms like *Gnetum, Welwitschia* and *Sequoia sempervirens*. The number of free nuclei formed may be 1,000 in *Dioon* and only 64 in *Bowenia*, whereas in *Cycas* (*C. revoluta*) it is 256 (Coutter and Chamberlain, 1910) or even more. In *Ginkgo* the number is 258, and in the *Araucariaceae* it may be 32 (*Araucaria*) to 64 (*Agathis*). In Cycads and *Ginkgo* the free nuclei are evenly distributed throughout the egg. In *Araucaria* and *Agathis* the free nuclei are grouped together in the centre of egg cell. In Cycads small vacuoles appear in the entire free nuclear proembryo. Some authors (Keno, 1898, De Silva and Tambiah, 1952) reported that some earlier divisions of the nucleus are **amitotic**. Swamy (1948) reported a few intranuclear divisions during first few divisions of the zygote nucleus in his species of *Cycas* from Mysore. Such divisions are, however, absent during late stage of free nuclear proembryo.

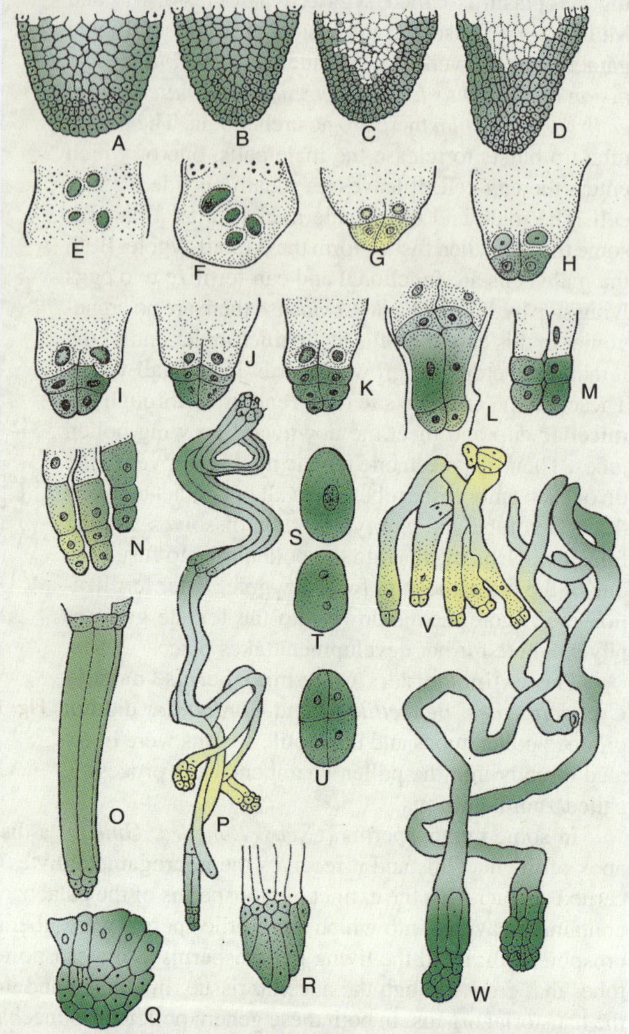

Fig. 1.14. Embryology of gymnosperms. A — D. Stages of embryo development in *Zamia* (cycadales) (After Lawson). E — P. Embryo development in *Pinus* (After Buchhloz). Q — R. Pro-embryos of *Araucaria species* (*After Johansen*). S — W. Embryo development in *Sequoia sempervirens.* In this case there is no free nuclear division.

- **Primary Proembryo or The Cellular Stage:** Beginning of the stage is marked by the appearance of evanescent segmentation of the cytoplasm and later formation of walls around the nuclei. The primitive proembryo formed as a result of complete segmentation and wall formation consists of a homogenous mass of cells which completely fills the entire archegonium. This cellular primary proembryo shows no differentiation except a *crude physiological differentiation* into upper (or micropylar) and lower regions made up of less active larger cells and more active smaller cells respectively. The former is designated as the **primary upper region** (pUr) and the latter is called the **primary embryonal region** (pEr). This kind of differentiation is also found in *Ginkgo*. There is no differentiation into upper, suspensor and embryonal tiers (U.S.E. pattern). The morphological suspensor or suspensor system found in the coniferales is entirely absent. During late embryogeny a crude and nonfunctional pre-suspensor system is developed from some cells of the primary embryonal region. There is no cleavage and the terminal embryonal mass invades the central region of the female prothallus as a blunt cylinder.

During the development of the primary proembryo, the processes of segmentation and wall formation start from the basal region and gradually extend towards the upper end. This results in the formation of small celled primary embryonal region and the large celled primary upper region. In the cycads formation of a central vacuole has been reported either during the free nuclear stage (Swamy, 1948 and De Silva and Tambiah, 1952) or after the cellular stage (Treub, 1884). The cellular proembryo in *Cycas*, therefore consists of a basal small celled compact mass of several cells (pEr) and a layer or two of comparatively larger cells around the sides and upper end of the vacuole (pUr). In some species of *Cycas, Macrozamia*, and *Encephalartos* the central vacuole is later filled up by a cellular tissue whereas in *Bowenia* it is retained as such and the upper region remains noncellular.

- **Direct Development of Primary Proembryo into Late Embryo:** The primary proembryo directly develops into the late embryo. The terminal or the apical region of the lower pEr region develops into the **embryonal mass** of mature embryo. The cells in the proximal portion of the pEr elongate slightly to form a columella like or a columnar structure called E_t region (E_t r). This region gives rise to the **pre-suspensor system** which can be regarded as the precursor of the suspensor-system of conifers. The larger cells of pUr region become flabby and drawn out and top the E_t r region. Later they show dissolution and degeneration. The E_tr region elongates very rapidly and pushes the embryonal mass deep into the food laden female prothallus. The suspensor like E_t r region later becomes much elongated and coiled and twisted. The neighbouring archegonia may also develop into such embryos which grow rapidly but only one of them reaches maturity. Their E_tr regions may twist together and persist and remain attached to the micropylar end of the egg membrane. The embryonic mass divides separately to form several cells in the embryonic region. The development of the various parts of the embryo takes more than a year, after the act of fertilization. In some cases they become morphologically distinct after the seed has fallen (*Cycas circinalis*, *C. rumphii*).

Similar stages have also been observed in *Ginkgo biloba* (Dogra, 1980) and to some extent in Araucariaceae. In the Araucariaceae (*Araucaria* and *Agathis*) the cell walls around the free nuclei (32—64) are laid down in a manner that a central group of two tiers of cells is surrounded by an outer layer or a jacket on all sides. The upper cells of the jacket (those lying towards the archegonial neck) elongate to form the suspensor, whereas cells towards the chalazal end act as **cap cells** whose function is not known. The cells acting as suspensor may elongate (*Agathis*) and grow out of archegonial necks. The upper cells of central two tiers may elongate and divide periclinally to form secondary suspensor cells. The lower tier of central group of cells divides to form embryonal cells that give rise to the embryo proper. There is no cleavage. Formation of cap cells has also been recorded in *Zamia umbrosa* (Bryan, 1952). The U.S.E. system is, therefore, also absent in the Araucariaceae.

The basal plan of embryology in the coniferales and other advanced gymnosperms can be discussed under the following four headings:—

1. Free Nuclear Proembryo :

It begins with the first division of the zygote. In the conifers the number of mitotic divisions varies from 2—5 or more resulting in the formation of four to sp. free nuclei in the archegonium. The first one or two divisions occur *in situ* whereas the rest occur after the free nuclei move to the base of archegonium. The free nuclei arrange themselves in two distinct groups after the last free nuclear division. The number of free nuclei in each group is variable. The nuclei in the **primary upper group** (pU) arrange themselves in a single horizontal tier. The free nuclei in the lower group are called **primary embryonal group** (pE). They are arranged either in tiers or in irregular storied arrangement.

2. Primary Proembryo :

The free nuclei in both the groups are surrounded by walls to form a primary proembryo, consisting of two groups of cells pU and pE. This wall formation determines the number of cells in the primary embryonal (pE)group. The structure of the primary proembryo determines the general conifer plan and its differentiation into two separate morphological parts pU and pE.

3. Internal Division :

The internal division of the proembryo gives rise to distinct upper, suspensor and embryonal tiers. This arrangement is called U.S.E. pattern, which is characteristic of the conifers. The cells of the primary upper group (pU) divide by transverse walls to form a distinct upper tier (U) and suspensor tier (S). The number of cells in each tier is usually the same. The internal division is, therefore, a very important step in the conifer embryology because it establishes a distinct suspensor tier (S) for the first time in the gymnosperm embryology. This highly specialised and a regular division results in the formation of distinct upper (U), suspensor (S) and embryonal (E) tiers (U.S.E.). It occurs only in the *pU* tier. The cells in the lower pE group remain undivided and form the embryonal mass (E).

4. Secondary Proembryo :

The internal division leads to the formation of secondary proembryo with a clear cut differentiation of the suspensor tier from the pU. The cells of the suspensor tier elongate and form the upper segment of the **suspensor system** in majority of the conifers. This U.S.E. pattern is characteristic of the conifers and is absent in the primitive gymnosperms like Cycadaceae, Ginkgoaceae and Araucariaceae. It is due to the introduction of internal division in the embryonal plan after the formation of cellular primary proembryo. The cells in the embryonal mass may undergo one division in a transverse plane to form upper tier of cells that act as multicellular **secondary suspensor** and the lower tier of cells that develop into embryo. The number of cells in the embryonal tier determines the number of embryonal units in the embryo system. Since the embryonal tier is away from the archegonial neck, the embryogeny is said to be **endoscopic**.

It will be worthwhile to discuss the conifer plan in the various families of order coniferales, because there are slight variations among them. In *Pinus* the zygote nucleus divides twice *in situ* to form four nuclei which migrate to the chalazal end of the archegonium. Here these four nuclei divide again to form eight nuclei that arrange themselves in two tiers of four nuclei each. This is the free nuclear proembryo which later gives rise to the primary proembryo as a result of wall formation. The lower four nuclei are surrounded by walls on all the sides whereas the upper ones are incompletely surrounded by walls. There is no wall formation on the upper side. This determines the formation of primary upper (pU) tier that takes no further part in the development of embryo. The lower four cells or the primary embryonal tier (pE) develops further. Due to internal divisions the four cells of the pE divide transversely to form eight cells arranged in two tiers of four cells each. The cells of the lower tier divide again so that three tiers of four cells each are formed. These are from below upwards (*i*) embryonal tier, (*ii*) suspensor tier, and (*iii*) rosette tier. The fourth one or the upper tier takes no part in embryonal development. The same sequence of development has been reported in *Cedrus, Larix, Keteleeria, Abies* and *Picea*. In *Pseudostuga* the rosette tier is absent. So in Pinaceae, the secondary proembryo consists of upper (U), rosette (R), suspensor (S) and embryonal (E) tiers.

In the Taxodiaceae four kinds of pro-embryogeny is recognised: (*a*) Taxodian type found in species of *Taxodium, Cryptomeria japonica, Cunninghamia, Glyptostrobus* and *Sequoia dendron*. (*b*) *Sequoia sempervirens* type; (*c*) *Athrotaxis* type; and (*d*) *Sciodopitys* type. In the typical taxodian type three free nuclear mitoses give rise to eight celled primary proembryo with pU and pE which on internal division result into the usual U.S.E. pattern of embryo development. In *Taiwania* the upper tier is surrounded by walls on all sides. The suspensor cells elongate considerably and directly followed by the formation of secondary suspensor (E_t). In *Cryptomeria japonica* the zygote shifts to the base of archegonium and divides thrice to form eight free nuclei that soon arrange into pU and pE tiers. The former has 5 to 6 nuclei whereas the latter has 2 or 3 nuclei. Walls are laid around them so that the two or three cells of the primary embryo tier (pE) are completely surrounded by walls whereas the primary upper (pU) cells are open towards the top. The pU cells divide once to form the upper tier and a middle suspensor tier. The upper tier degenerates after some time. The secondary proembryo then contains thirteen or fourteen cells arranged in three distinct tiers (U.S.E.) with 5 or 6 cells in upper, 5 or 6 in suspensor and 2 or 3 in the embryonal tier. The proembryo of *Sequoia* is unique in having no free nuclear stage and no tier formation in its development. The first division of the zygote is followed by a transverse wall forming two cells. These two cells divide further to form variously arranged four cells. These cells divide to form a primary suspensor cell and an embryonal cell. In *Sciadopitys* thirty-two or rarely fifty to sixty free nuclei. All these nuclei arrange themselves in three or four tiers with twelve to twenty cells in the embryonal tier, seven to thirteen in suspensor tier, two to nine cells in rosette tier and twenty to thirty cells in the upper tier. In *Athrotaxus selaginoides* the zygote forms four free nuclei that arrange themselves in three tiers. Wall formation takes place and four cells are formed. There is one cell in embryonal, two in suspensor and one in upper tier (1 : 2 : 1). All the four cells divide vertically resulting in eight cells (2 : 4 : 2) arranged in three tiers.

The proembryogeny of Cupressaceae resembles very closely with that of the Taxodiaceae. Eckenwalder (1976) has proposed the merger of Taxodiaceae and Cupressaceae into a larger family Cupressaceae on the basis of resemblances in morphological and phytochemical features. The basal chromosome member in both the families is X = 11 (Khoshoo, 1961). The fossil record in both of them also supports such a merger. Miller (1977) also suggested a merger of the two families. Dogra (1984) has reviewed the embryogeny of these families and is strongly in favour of merging them into a larger family Cupressaceae. The family Cupressaceae is sub-divided into two sub-families depending upon various morphological and embryological data. These are (*i*) the Cupressoideae of northern hemisphere and (*ii*) the southern Callitroideae. Among the Cupressoideae the embryology of *Cupressus arizonica, C. glabra, C. funeberis, C. torulosa, Chamaecyparis pitifera, C. obtusa, C. lawsoniana, Thuja occidentalis, Biota orientalis, Thujopsis dolabrata, Juniperus communis, J. virginiana, J. chinensis, J. squamata, J. procera* and *Calocedrus* (*Libocedrus*) *decurrens*, has been studied in detail. Among the Callitroideae the embryogeny of *Actinostrobus*, *Callitris* and 'Widdringtonia'has been studied.

The first proembryonal division among the Cupressoideae may take place *in situ* (near the middle of the egg where fusion takes place) or at the base of the archegonium. In some advanced genera like *Juniperus* the division usually takes place at the base of the archegonium. The two free nuclei may thus lie in the middle or at the base of the archegonium. Formation of a distinct spherical **proembryonal sheath** has been recorded by Dogra in Taxodiaceae, by Sugihara (1939) in *Thujopsis dolabrata*, by Nichols (1910) in *Juniperus communis var depressa* and by Mathews (1939) in *Juniperus virginiana*. Such a sheath has not been observed in other genera and species. Three successive free nuclear divisions take place and form eight free nuclei that lie at the base of the archegonium. These nuclei arrange themselves in two primary groups. The four free nuclei that form the primary upper tier (pU) lie in one plane, whereas those forming the primary embryonal group (pE) get variously arranged. They may form one or two tiers or irregular storied group. Normally these two groups (pU and pE) are composed of eight nuclei but in some rare cases in several genera of Taxodiaceae (Dogra, 1966) and Cupressaceae (Dogra, 1984), there is extrusion of some relict nuclei so that the proembryo contains less than eight nuclei. Wall formation commences only after the third mitosis and results in the formation of **primary proembryo** which consists of two distinct groups of cells called the **primary upper** (pU) and primary **embryonal** (pE) groups of cells. The pU is called the **primary upper tier** because it consists of cells that are open towards the archegonial neck and are arranged in a single tier. The arrangement of cells in the pE group is variable. The cells may be arranged in a single tier or in two tiers or even in irregular storied arrangement. Since the cells in this group are not always arranged in definite tiers, it is termed as primary embryonal group of cells (pE). The number of cells composing pU and pE (pU/pE) varies. It ranges from 6/2, 5/3, 4/4, 3/5 to a rare 7/1 (seven in pU and one in pE). The 4/4 arrangement is common and constant in the Pinaceae but it is rare and at random in Cupressaceae and Taxodiaceae. In *Thujopsis dolabrata* (Sugihara, 1939), *Juniperus communis* (Nichols, 1910), *Cupressus arizonica* and *Juniperus procera* (Dogra, 1984) four free nuclear mitoses occur resulting in the formation of sixteen free nuclei. The primary proembryo consists of sixteen cells. This condition is regarded as primitive as compared to the eight celled condition. The *internal division* leads to the formation of *secondary proembryo* and is a regular step that occurs simultaneously in all the cells of the primary proembryo or it may take place first in the pU and then in the pE or *vice versa*. In the cells of the pU tier this division usually results in the formation of **secondary upper tier** (U) and the **suspensor tier** (S). These two tiers belong to the same morphological part in the secondary proembryo. The internal division in the pE cells is not always simultaneous in all the cells. There may not be regular tier formation except in rare cases. As a result of internal division each pE becomes a two celled unit either as a result of simultaneous division or as a result of independent division in each pE cell of the proembryo. In rare cases the internal division in the pE cell may be delayed to a later stage in late embryogeny.

In the secondary proembryo the cells of the U tier are usually open from above whereas those of S tier are enclosed by walls from all sides. Both these tiers are regularly formed by the internal divisions of the pU cells. In most of the conifers this S tier is the only true morphological and functional suspensor (Doyle, 1957, 1963; Dogra, 1966, 1967, 1978; Mehra and Dogra, 1975) but it is only in Pinaceae that this tier, though formed, remains non-functional. Dogra and Tandon (1984) and Dogra (1984) have reported that S tier may not elongate in some cases in *Juniperus procera* and *J. chinensis*. As already stated each pE cell divides simultaneously or independently to form two cell units of E cells. There is no regular tier formation of E cells. This is the usual mode of internal division of pE group of cells to form E cells in most of the conifers except for the Pinaceae. In Pinaceae all the internal divisions occur simultaneously in all the cells of pU and pE tiers to form a four tiered secondary proembryo. This formation of four tiered secondary proembryo is extremely rare in the Cupressaceae and occurs only at random. The usual U.S.E. pattern as proposed by Doyle (1957, 1963) is not applicable to the Cupressaceae.

The development of the secondary proembryo in the Cupressaceae, as discussed above, is characteristic of most of the species of Cupressaceae except for specialised types such as Callitroids. The development of the proembryo is uniform and follows similar stages at the species as well as generic levels.

Among the Callitroideae, embryogeny has been studied in *Callitris*, *Actinostrobus* and *Widdringtonia*. In *Widdringtonia* the details of proembryo development are not known. The details are known only for *Actinostrobus pyramidalis* (Saxton, 1913) and *Callitris verrucosa* (Saxton, 1910), *C. rhomboidea* (Looby and Doyle, 1940), *C. robusta*, *C. drumondii, C. roei* and *C. g Canca* (Baird, 1953). In *Actinostrobus pyramidalis* and *Callitris* a distinct proembryonal sheath with starch is present around the zyotic nucleus. The first division takes place at the base of egg and the two free daughter nuclei are enclosed by a proembryonal sheath made up of granular cytoplasm. Only four free nuclei are formed. Wall formation takes place after the second division. In *A. pyramidalis* (Fig. 1.15, F—I) wall formation results in the formation of four cells. The upper two cells are commonly arranged in a transverse plane and the lower two one above the other (Fig. 1.15, F, G). In *Callitris rhomboidea* there may be two cells in a transverse plane flanked on upper and lower side by a single cell or all the four cells lie in a single linear row. The third division takes place only in the lower two cells, the upper ones do not divide (Fig. 1.15, G) and take no further part in the embryo development. The division in the lower cells cuts off four embryonal initials. These are polarised in a transverse plane and each of them divides to form a small distal initial cell and a large suspensor cell so that there are four small embryonal cells and four large suspensor cells (Fig. 1.15, H). The suspensor cells elongate considerably. There is cleavage polyembryony.

Among the Podocarpaceae, which includes seven genera (*Podocarpus, Dacrydium, Phyllocladus, Pherosphaera, Saxegotheca, Acmopyle* and *Microcachyrus*), the embryo development has been thoroughly studied in *Podocarpus* (106 species). The embryogeny of *Dacrydium* has been studied in detail by Quinn (1964, 1965, 1966). The embryogeny of other genera is incompletely known. The embryogeny of *Acmopyle* has yet to be studied. Due to variations in the structure and development of various species of *Podocarpus*, the genus has been variously split into sections and subsections. Pilger (1926) divided it into two sub-genera: *Stachycarpus* and *Protopodocarpus*. The later is further split into four sections namely *Nageia, Microcarpus, Dacrycarpus* and *Eupodocarpus*. Buchholz and Gray (1948-51) and Gray (1953-62) divided the genus into eight sections. Buchholz (1941) gave an account of the embryogeny of different species of *Podocarpus* belonging to various sections and concluded that the genus *Podocarpus* is exceptional among conifers in showing several distinct types of embryogeny. Brownlie, G. (1953) studied the embryogeny of New Zealand species of *Podocarpus* belonging to the section *Eupodocarpus*. Buchholz (1936) studied it in three species of sub-genus *Stachcarpus*. Quinn (1964) studied it in some species belonging to the section *Dacrycarpus*. Looby and Doyle (1944a) studied it in *Podocarpus andinus*. In this species, which is considered to be the simplest in the genus and in *P. amarus, P. usambarensis, P. gracilor, P. nagi, P. falcatus* and in other species of *Stachycarpus* section, five free mitoses take place and result in the formation of thirty-two free nuclei. In *P. nivalis* (Boyle and Doyle, 1954) and other species of the section *Eupodocarpus* (Buchh, 1941, Brownlie, 1953) only four free mitoses take place and form sixteen free nuclei. In *P. andina* and in some other species with 32 free nuclei six to ten free nuclei may escape. In *P. gracilior* some free nuclei move to the upper part of the archegonium and are called **relict nuclei**. They may persist as such or may degenerate. Wall formation results in the formation of two tiers. The lower tier consists of 9-12 embryonal cells whereas the upper open tier consists of 15-22 cells. The embryonal cells are arranged in a single tier. The cells in the upper open tier divide transversely followed by wall formation. This results in the formation of an open upper tier and a middle **suspensor tier**. The embryonal cells undergo internal nuclear division not followed by wall formation so that each embryo cell is binucleate. This condition is also found in *Saxegothaca*, all species of *Podocarpus*, *Phyllocladus* and *Dacrydium*. In *Pherosphaera* (Elliot, 1950) binucleate cells were observed but later wall formation takes place to produce a bi-celled stage. The suspensor cells elongate and push the embryonal cells into the gametophyte. Later **cap like** structures appear at the tips of embryonal cells. The binuclear embryonal cells undergo another nuclear division followed by wall formation. This results in the formation of **tetrads of cells**. This is called an **embryo tetrad stage**. There is simple polyembryony in *P. andinus, P. spicatus, P. ferrugineus, P. falcatus, Saxegothaea* and *Phyllocladus alpinns*. Cleavage polyembryony is known as *P. gracilior* and some other species.

In Cephalotaxaceae the zygote nucleus divides into sixteen free nuclei (the first division of zygote nucleus is intra-nuclear). They arrange themselves in four tiers named as (*i*) rosette tier; (*ii*) suspensor tier, (*iii*) embryonal tier; and (*iv*) cap cells (Favra-Ducharte, 1957). According to Singh (1961) the cap cells do not form a separate tier but are lowermost cells of the embryonal tier. He did not observe even the rosette tier in *C. drupacea* so that there are only two tiers *i.e.*, the upper suspensor tier, and the embryonal tier. Later the upper or the suspensor tier divides into an upper open tier and a prosuspensor tier. The former degenerates after some time and the latter grows and pushes the embryonal tier out of the archegonial venter. The embryonal tiers are pushed into a central cavity in the centre of female gametophyte. Later the cap cells degenerate and embryonal cells enlarge and may undergo another division to form secondary suspensor cells above them. These cells elongate further and push the embryonal cells still further. The embryonal cells divide further to form a massive embryonal tissue. The prosuspensor cells degenerate after the formation of secondary suspensor cells. There is no cleavage polyembryony. Singh (1961) reported three such cases in *C. drupacea*.

In Araucariaceae the zygote nucleus undergoes free nuclear divisions to form thirty-two (*Araucaria*) or sixty-four (*Agathis*) free nuclei that are grouped together in the centre of the egg cells. Walls are laid down in a manner that a central group of two tiers of cells is surrounded by an outer layer or the jacket on all sides. The upper cells of the jacket (those facing the archegonial neck) elongate to form the suspensor, whereas the cells towards the chalazal end act as cap cells whose function is not known. The suspensor cells may elongate and divide and grow out of the archegonial necks in *Agathis*. The upper cells of the central two tiers divides periclinally to form an upper tier of **secondary sus-**

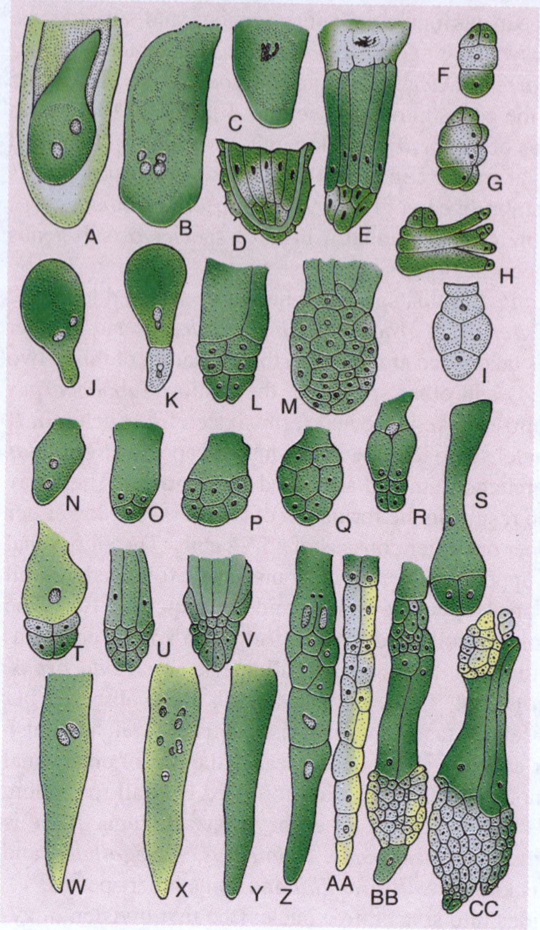

Fig. 1.15. Embryology in Gymnosperms. A—E. Stages in the development of embryo in *Podocarpus gracilior* (After Konar and Oberoi). F—I. Early embryogeny in *Actinostrobus pyramidalis* (After Saxton). J—M. Early embryogeny in *Ephedra trifurca* (After Johansen). N—R. Early embryogeny in *Torreya nucifera* (After Buchholz). S—V. Early Embryogeny in *Welwitschia mirabilis* (After Johansen). W—CC. Embryogeny in *Cephalataxus sp.* (After Hardev Singh).

pensor cells and the lower tier of **embryonal cells**. The latter divide and develop into the embryo proper that has two cotyledons. There is no cleavage polyembryony. The embryo is surrounded by endosperm, nucellus and seed coat (that has three layers, outer exocarp, middle mesocarp and inner endocarp).

Polyembryony is of common occurrence among the gymnosperms. This is possible because more than one archegonia are fertilised and more than one zygotes are formed. These develop into embryos, but only one of them succeeds in developing into a complete embryo. In the conifers there is a **cleavage polyembryony**. In this case all the four cells of the young embryo separate and develop into 4 embryos, but only one completes development and others abort. In *Thuja* there is no cleavage polyembryony and only one embryonal initial develops into an embryo. In *Sequoia sempervirens* there is no free nuclear division (Fig. 1.14 S—W) and the zygote divides first by a transverse wall forming two cells which divide further by longitudinal walls to form four cells (Fig. 1.14 T, U). Each of these four cells may function as embryo initial and give rise to a filamentous embryo. There is an indication of formation of an apical cell in rare cases but generally the embryo initial divides by a longitudinal wall. Cleavage polyembryony is present. The embryo has two cotyledons. In majority of conifers the zygote develops four free nuclei and later further division and wall formation results in forming a **proembryo** with a distal embryonal **tier** (4 cells), middle **suspensor tier** and upper **rosette tier** (*Pinus, Cedrus, Tsuga, Pseudolarix*). In *Abies, Picea* and *Larix* the rosette tier disappears at a later stage. The lower embryo tier develops into all the organs of the embryo. The suspensor tier develops into long and coiled suspensor. The rosette tier cells may often develop into additional embryos. The uppermost tier of cells, of which distal ends are in open contact with the egg cytoplasm is apparently active in transmission of nutrients to the growing embryo below. The number of cotyledons in *Pinus* embryo varies in different species. It is 10 in *P. roxburghii*, 3–6 in *P. banksiana*, 2–8 in *P. contorta*, 12–18 in *P. lambertiana* and 7–18 in *P. sabiniana*. In *Araucariaceae* 32–45 free nuclei are formed in *Araucaria angustifolia* (Fig. 1.14, Q-R). Wall formation is accompanied by polar elongation. The cells at the distal and develop into enlarged **cap cells** (Fig. 1.14, R), those in the centre develop into embryo (Fig. 1.14, R) whereas those towards the micropylar end give rise to the suspensors. There is no **cleavage polyembryony**.

Among *Cupressaceae* the genus *Actinostrobus* (Fig. 1.15, F—I) shows some curious developmental stages. In this case 4 free nuclei are formed. Wall formation results in forming 4 cells, two vertically placed cells below archegonial necks and two transversally placed cells below. The upper cells do not divide further. The lower ones divide once to form four embryonic initials. These are polarised in a transverse plane and each of them forms a small distal initial cell and a large suspensor cell so that there are 4 small embryonal cells and four large suspensor cells that elongate considerably. There is cleavage polyembryony. There are no rosette tier and upper tier in this case. The embryo has two cotyledons. In the *Podocarpaceae* the embryonal cells are binucleate (Fig. 1.15 A—E). The number of binucleate embryonal cells and the **pro-suspensor cells** varies with species and genera. The least number of embryonal cells is 1-2 in *Podocarpus totarra*, *P. hallii*, *P. nivalis* and *P. macrophyllus* and the maximum number is 9-12 in *P. spicatus*. The number of pro-suspensor cells is 6-9 in *Podocarpus spicatus* and 14-15 in *P. microphyllus*. The embryo has two cotyledons.

The details of embryo development in the *Gnetales* are interesting. In *Ephedra trifurca* the zygote nucleus divides into eight free nuclei of unequal size that are unevenly distributed in the protoplasm. Three to five of these nuclei become enclosed individually by irregular walls which later become globular. Each of these globular cells is pro-embryo, *i.e.*, there is polyembryony. Further the globular **pro-embryo** develops into an embryo with the (Fig. 1.15, J—M) massive **embroyonal** mass at the distal end and a number of suspensors. In *E. foliata* there is only one suspensor tube.

In *Gnetum* (Fig. 1.16, A—T) there is no free nuclear division, the zygote divides and develops into a two-celled body that gives out elongated and tube-like suspensors. These suspensor cells may branch and all have a distal, densely granulose embryonal cell. These terminal cells develop into embryos, out of which only one reaches maturity. Embryo development is completed after the detachment of the seed.

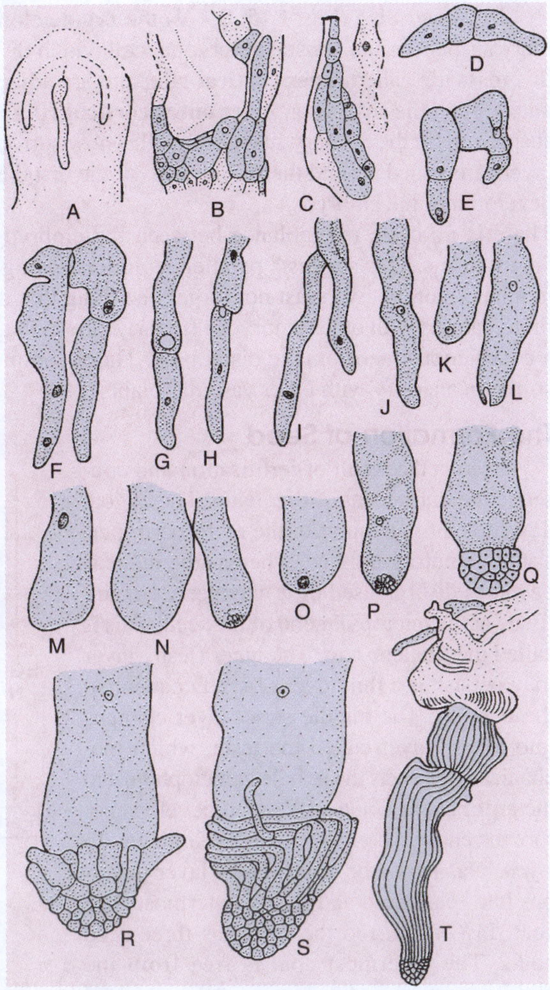

Fig. 1.16 (A—T). Stages in the development of Embryo in *Gnetum ula* (After Vasil). A. L.S. ovule showing *zygote*. B. Part of female gametophyte showing zygote with two primary suspensor tubes. C. Showing a gametophyte with three primary suspensor tubes. D. Two primary suspensor tubes, out of which one has further divided to form another cell. E—F. Each of the primary suspensor tubes has divided to form a daughter cell. G—I. show formation of additional primary suspensor tubes. J. Part of a primary suspensor tube. K. Tip of primary suspensor tube with two nuclei. L—M. Peculiar cells on the tip of the tube. N—P. Stages in the formation of eight cells from the peculiar cell. Q. Mass of cells is formed from the peculiar cell. R—S. Upper cells of the mass enlarged to form secondary suspensor cells. T. Embryo with enlarged and coiled secondary suspensor and embryonal mass.

In *Welwitschia* (Fig. 1.15, S—V) the zygote elongates and divides into an upper elongated primary suspensor and a distal embryonal cell which divides to form an apical mass. The outer cells of this mass are called **inner cortical ring** and are adjacent to the primary suspensor. These also elongate around the suspensor. Later **outer cortical ring** is formed and its cells also elongate around the suspensor. So the primary suspensor cell is now surrounded by two layers of 8 to 16 cells. Later more layers are added around these. The cells at the distal end form cap cells whereas those in the middle develop into the embryo.

There is no close resemblance between the embryogeny of the gymnosperms, pteridophytes and angiosperms. This suggests parallel evolution among these groups rather than descent from a common ancestor, but we must not forget one thing that the gymnospermous embryos share some common features with other embryos : (*i*) early determination of polarity; (*ii*) axial development and (*iii*) a conspicuous meristematic distal pole. These common features do not rule out the possibility of a common ancestry with other vascular plants.

The Formation of Seed

Seed is the result of fertilisation and consequent enlargement of the ovule. The zygote develops into an embryo, the endosperm persists as a nutritive tissue, whereas the nucleus becomes disorganised or it may persist as dry tissue at the micropylar end of the seed. This is called the **nuclear cap**. The inner fleshy layer may persist as a thin layer of seed coat called the **tegmen**. The middle stony layer changes into a hard layer called the testa, which mechanically protects the female gametophyte and the **embryo**. In *Gnetum* the seed develops before the embryo completes its development. In *Cycas* and *Taxus* the outer fleshy layer develops into scarlet red and fleshy outermost seed coat. In *Gnetum* too the seed has three seed coats. The outermost coat is free from the middle stony coat, from base to the apex and possesses stomata, scattered fibres and sclereids. The innermost layer is thin and fused with nucleus upto $\frac{2}{3}$ of its length. In *Taxus* a fleshy **aril** develops from the basal cup-shaped structure. The seeds of all gymnosperms except those of *Cycas* and *Ginkgo* remain dormant for some time. In these two genera the seeds germinate immediately on falling on a moist substratum or else they loose their viability. The seed is a remarkable combination of two sporophytic generations, and one gametophytic generation :

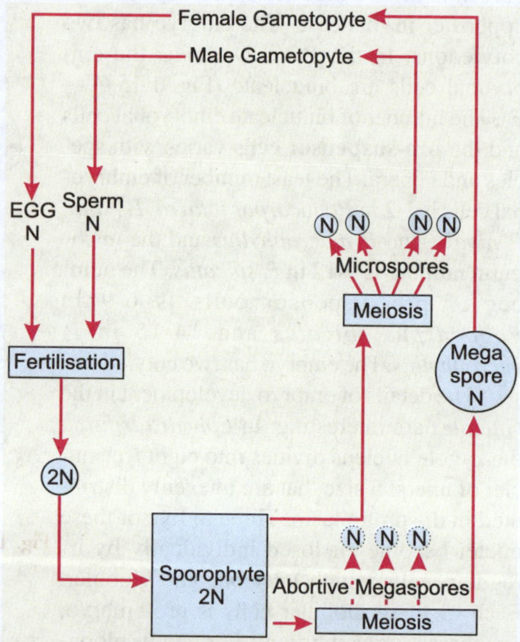

Fig. 1.17. It illustrates the life cycle pattern among gymnosperms.

- The seed coat represents the old sporophytic generation.
- The young embryo represents the new sporophytic generation.
- The endosperm represents the gametophytic generation.

The germination of seed is epigeal, *i.e.*, the cotyledons come above ground, in most of the genera except in *Ginkgo* where the cotyledons remain embedded in the endosperm. Vivipary has been reported in *Ephedra trifurcata*.

The alternation of generations is heterologous in all the gymnosperms (Fig. 1.17).

ORIGIN OF GYMNOSPERMOUS STRUCTURES

C.B. Beck (1960, 1962) included among the group, '**progymnosperms**' all those fossil plants that he regarded as ancestral to the gymnosperms. These peculiar plants possessed certain gymnospermous and some pteridophytic characters and existed during the Devonian era. By the Upper Devonian times these plants had already acquired many gymnospermous characters. They throw much light on the origin of vegetative and reproductive structures of gymnosperms. The fossil genera relegated to this group include *Archaeopteris, Aneurophyton, Tetraxylopteris, Ginkgophyton* and *Callixylon*. The light thrown by these genera on the origin of gymnospermous structures is discussed below :

Origin of Secondary Wood

The progymnosperm genus *Archaeopteris* was discovered from Upper Devonian of north temperate and arctic latitudes about 360 million years ago. It has characteristic fern-like fronds about 1 metre long. The fronds appear to be bipinnate and bear wedge-shaped pinnules. The rachis of pinnules in the basal region of the frond bore numerous stalked sporangia of two kinds (smaller and larger). They bore two kinds of spores (heterosporous). *Callixylon* is a stem genus of *Archaeopteris* and was found in petrified fossils from the same age and locality. Transverse sections of the petrified stem of *Callixylon* that varied in diameter from 3-5 feet and in length upto 28 feet, showed gymnosperm-like wood. The primary vascular bundles in this genus are arranged in a ring around a central pith and are mesarch. The secondary wood is of coniferous type, consisting of tracheids with circular border pits aligned in radial rows, the pits are crowded and multiseriate but lacked torus; the secondary medullary rays are narrow and only two cells broad; the tracheids appear rectangular in a cross-section. Beck (1960) reported *Archaeopteris* fronds in organic connection with *Callixylon* like stem, so this important discovery gave us a plant with fern-like foliage and sporangia and a gymnospermous wood. It combines the characteristics of pteridophytes and gymnosperms and hence throws significant light on the gymnospermous ancestry in at least regarding the origin of secondary xylem. The secondary wood of another progymnosperm named *Tetraxylopteris* also shows gymnospermous characters outlined above. This genus has pteridosperm affinities: (*i*) the presence of secondary wood; (*ii*) round bordered pits of tracheids; and (*iii*) sclerotic outer cortex. Its pteridophytic characteristics include : (*i*) three dimensional branching fronds; and (*ii*) large clustered sporangia. Uniform pitting of radial and tangential walls is another interesting gymnospermous characteristic of this plant. Another coniferous characteristic in the *Callixylon* wood is that the narrow rays have ray tracheids with paired bordered pits on their radial walls. In *C. eria num* there is a helical wall layer superimposed on that containing the bordered pits, thus forming any tracheids similar to those of *Sequoia*. The arrangement of bordered pits in various species of *Callixylon* (now *Archaeopteris*) may be either multiseriate (*C. newberryi*) or uniseriate (*C. arnoldii*). Groups of pits are separated by bands of thickened primary wall that are equivalent to **crassulae** of living conifers. The pit borders may be angular or circular in outline and have elliptical to narrow slit-like apertures that never extend beyond the pit border. The pit membrane is homogenous, *i.e.*, there is no torus. The torus, among the living gymnosperms, is characteristic of the Pinaceae alone, and is lacking in most of the gymnosperms.

The body of facts stated above clearly reveals that the secondary xylem of the living gymnosperms had evolved in the progymnosperms of the Upper Devonian.

Origin of Primary Vasculature or Evolution of Stelar System

The living gymnosperms have a eustele consisting of discrete strands. The origin of such a stele from protostelic vasculature is discussed below. Namboodri and Beck (1968) have studied the primary vasculature of large number of living conifers and have come to the conclusion that they consist of a cylindrical system of discrete sympodia, each sympodium consisting of an axial vascular bundle from which leaf traces diverge. These authors have also studied the primary vasculature of a number of fossil genera belonging to the progymnosperms and the ancient pteridosperms and have revealed that they also possess sympodial vascu lar pattern. They also suggest that two trace node is an advanced condition as compared to the single trace node, which condition is prevalent in most of the living and fossil forms. They further state that in gymnosperms the leaf traces do not leave any leaf gaps.

Two views are prevalent regarding the origin of the eustele from the primitive protostele :

1. Jeffrey (1899, 1902, 1917) suggested that from the basic protostelic condition the first step in stelar evolution was the appearance of a pith in the centre. This led to the siphonostelic condition. He suggested that the pith appeared by the enclosure of cortical tissue within the stele with the correlated appearance of leaf gaps. These leaf gaps are interruptions in the wall of the tubular stem stele above the points of departure of the leaf traces. He believed in extrastelar origin of the pith. Some authors are in favour of intrastelar origin of the pith by the appearance of parenchyma in the central xylem core. The siphonostele became dissected by overlapping leaf gaps leading to discrete bundles or meristeles which constitutes a dictyostele. In some cases loss of internal endodermis, internal pericycle and internal pith lead to discrete vascular bundles of collateral type. This is the eustele, characteristic of gymnosperms and angiosperms. In Jeffrey's own words, "It is now clear... that the lower gymnosperms have come from the *Filicales* as a result of the simplification and reduction of the primary structures of the stele of the stem on the one hand, accompanied by the marked development of secondary fibrovascular tissues on the other". Many workers including Bailey, Eames, Smith and Zimmermann are of the same view as Jeffrey, regarding the origin of eustele from protostele by the appearance of central pith and leaf gaps. This view is illustrated in Fig. 1.18.

Majority of the members of *Lyginopteridaceae* do not show the above features and Scot (1923, p. 14) regarded the appearance of internal tracheary elements to be the result of a response to injury.

2. Namboodiri and Beck (1968) and Beck (1970) are of the opinion that there is no good evidence on which to interpret the vascular system of living gymnosperms in terms of a siphonostele having become dissected by overlapping of leaf gaps as proposed by Jeffrey. They postulate that the gymnosperm eustele has no leaf gaps and has evolved through longitudinal dissection of the photostele. This view was put forth by Geyler in 1867 and was later supported by D.H. Scott (1923). Beck (1960, 1962, 1970) has suggested that gymnosperms have evolved from the pteridophytic progymnosperms which, in turn, may have evolved directly from some psilophytic group. Namboodiri and Beck (1968) and Beck (1960, 1962, 1970) have suggested that evolution of

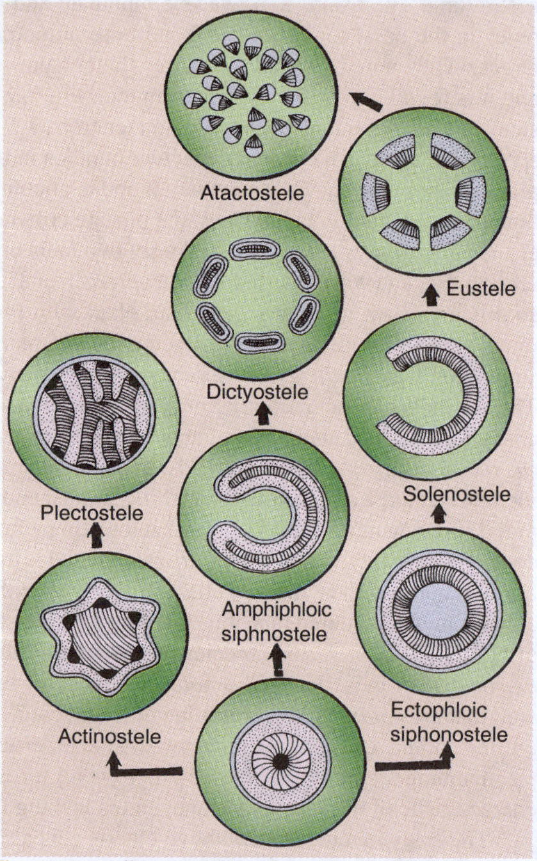

Fig. 1.18. It illustrates the older view of origin of eustele from the protostele. (After Zimmermann). Jeffery based his conclusions on the anatomical observations : (i) a single instance in a specimen of Lyginopteris of some indications of internal phloem; (ii) the occurrence of internal secondary wood in Cycadoxylon robustum (= Lyginodendron robustum), and (iii) the occurrence of internal secondary xylem and phloem in Cycadoxylon fremyii, which he claimed to be closely related to Lyginodendron.

gymnospermous eustele by the longitudinal dissection of protostelic condition might have followed the following successive stages (Fig. 1.19) :

- The primitive protostelic condition with a massive three-ribbed stele is exemplified by some progymnosperms like *Aneurophyton, Tristichia, Tetrastichia*, etc. Leaf traces to appendages diverge radically from the outer surface of the stelar column (Fig. 1.19, A).
- The next stage is represented by the Upper Devonian *Stenomyelon muratum, S. tripartitum, Archaeopteris, Calamopitys* and *Eristophyton*. In these there is an evident tendency for gradual dissection of the stele (protostele) into three or more longitudinal columns (Fig. 1.19, B) with consequent appearance of pith (medullation). The traces diverge from the outer margins of the strands by tangential division, without forming leaf gaps.
- The next step is the stele with larger pith and dis crete vascular strand near the periphery. At the time of leaf trace formation, a stelar bundle divides into an outer and an inner bundle, the former diverging as a **leaf trace** and the latter continuing as the **reparatory strand** without leaf gaps, *e.g.*, *Archaeopteris, Calamopitys* (Fig. 1.19 C).
- The next step is the change in the nature of the division of the sympodium from a division resulting in radially arranged bundles (*i.e.*, leaf trace and reparatory strand oriented along the same radius) to a division of the sympodium to form two bundles initially placed tangentially (*i.e.*, oriented along the same circumference). This condition is characteristic of *Lyginopteris, Pitys, Mesoxylon, Poroxylon* and *Caliostophyton* and also the living conifers (Fig. 1.19, D).
- In addition to the initial tangential orientation of the leaf trace and the sympodial segment (reparatory trace), the sympodial segment takes an undulating upward path (Fig. 1.19, E). This feature is clearly evident in the primary vasculature of the living *Coniferales*.

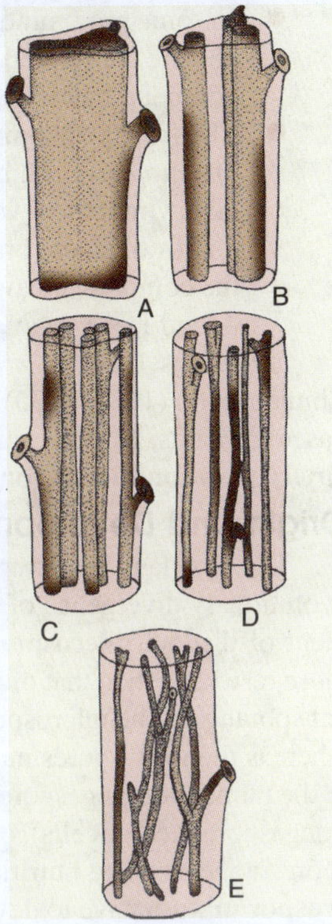

Fig. 1.19. It illustrates Namboodiri and Beck's view of origin of eustele from the protostele.

An increase in the number of traces supplying a leaf from one to two and change in the direction of divergence of the traces may lead to the vascular pattern of *Ginkgo, Ephedra*, and reticulate patterns of those conifers with opposite leaves. Beck (1967) suggested that there are two lines of evolution that have led to the eustelic organisation of living gymnosperms. These lines diverge from the Aneurophytalean progymnosperms. One of them diverges towards cycadophytic gymnosperms by way of *Calamopityeae* and the other towards the coniferophytic gymnosperms through *Archaeopteris*. The first line includes such genera as *Stenomyleon, Calamopitys, Heterangium, Lyginopteris*, etc. The second line includes *Callixylon* (*Archaeopteris*), *Archaeopitys*, *Pitys, Mesoxylon* (*Cordaites*), and living conifers. Both these lines follow the pattern of stelar evolution as discussed above *i.e.*, the sympodial stelar system, evolved similarly and independently in both evolutionary lines.

Origin of Leaf

The following steps, according to Beck (1970), lead to the origin of leaves in gymnosperms :

- The progymnosperms of the Middle and Upper Devonian such as *Aneurophyton* and *Tetraxylopteris* possess a three dimensional and dichotomously branched ultimate branch system. In *Svalbardia* and *Actinoxylon* (Middle Devonian) leaf-like ultimate appendages appeared.

- In the Upper Devonian *Archaeopteris* simple wedge-shaped and laminated leaves appeared first. In *A. fissilis* the leaves were made up of slender dichotomous divisions.
- The simple leaf is considered to have evolved in the progymnosperms through the process of plantation and lamination of an original three-dimensional ultimate branch system.
- In some progymnosperms like *Siderella* and *Archaeopteris* the lateral branches are flattened and are considered to represent an intermediate stage in the evolution of the gymnospermous compound leaf.
- The fronds of some lower Carboniferous pteridosperms such as *Tristichia, Tetrastichia* and *Sphenopteris* exhibit cauline characteristics such as bifurcation of the main axis and a radially symmetrical vascular system. This feature reflects the origin of gymnosperm compound leaf from lateral branch system. It cannot be said with certainty as to whether the pinnules of the compound leaves evolved before or after plantation of lateral branch systems. The compound leaves may also have originated by the dissection of the lamina of the large simple leaves.

Zimmermann (1930, 1935) illustrated the origin of megaphyllous and microphyllous leaves, among vascular plants, while discussing his **Telome concept**. Bower (1935) explained the origin of microphylls in his **ennation theory**. These theories have been discussed in detail in the book on Pteridophyta.

Origin and Development of Heterospory

The first land plants were homosporous, but by the Early Devonian period there had been an evolutionary divergence of spore sizes in some plants. Continuation of this trend led to the development of distinct **microspores** and **megaspores**, each with a developmental expression which was more restricted than that of the original homospore. The two kinds of spores were produced in different sporangia. The microspores were the more numerous and remained approximately the same size, whereas the megaspores numbered less and ultimately to one per **megasporangium**. The reduction in the number of megaspores has been achieved either through decrease in the number of **sporocytes** (megaspore mother cells), or by abortion of developing spores. This difference in size and number of spores appears to be nutritional. The small microspores that are produced in large number per **microsporangium** have to develop into microgametophytes that are of temporary duration and contain one or two antheridia with a limited number of male gametes. After the liberation of male gametes they shrivel and are required no more. On the other hand, the large megaspores take more food because they have to reserve food, produce female sex organs that harbour egg and have to retain permanently after fertilisation or at least upto complete development of young embryo. So megaspores need more food, because they have to nourish the young embryo, after developing into a food laden **macrogametophyte.**

Another important phenomenon that is linked with heterospory is the sex determining process of the plant. With the development of heterospory there is a gradual shifting of the sex determination process from the gametophytic generation to the sporophytic phase. In all the homosporous vascular plants sex determination exerts its influence during the differentiation of sex organs (antheridia and archegonia), that are the first sexually differentiatied structures. In some species of *Equisetum* sex determination occurs somewhat earlier, during the vegetative growth of the prothallus. The male prothalli are smaller in size and reddish in colour, whereas the female prothalli are comparatively robust and dark green in colour. In the heterosporous lycopsids *Selagi nella* and *Isoetes* sex becomes evident during the differentiation of spores *e.g.*, microspores are fashioned in sporangia where all or most of the mother cells remain functional whereas megaspores are fashioned in sporangia where only one or rarely two or more mother cells are functional. The microspores are small and are destined to give rise to the male prothalli whereas the megaspores are large and germinate into female prothalli. This is an example where sex determinants exert their influence still earlier. In some plants sex determination occurs in the sporophytic phase *e.g.* in *Salvinia* and *Azolla* the micro- and megasporangia are structurally different and are borne on the same sporophylls. These plants are **heterangic**, and sex determination evidently occurs during the development of each sporangia prior to meiosis. In the gymnosperms and angiosperms the distinction is evident during the differentiation of **microsporophylls** and **megasporophylls**.

In many gymnosperms and monoecious angiosperms reproductive structures are produced on specialised branches which may be identified prior to the development of the reproductive structures, by their distinctive morphology or position (Wareing, 1958). In these, sex determination occurs during the vegetative growth of the branch. Finally there are examples in both gymnosperms and angiosperms of the entire sporophyte becoming persistently microsporangiate or megasporangiate in its reproduction. It is evident from these examples that with the development of heterospory there is a significant shifting of the time of sex determination from the gametophyte to the sporophyte.

Palaeobotanical evidence of heterospory

The origin of heterospory can be properly understood only after examining the fossil record. Fossil record clearly suggests that there was an early and widespread occurrence of heterospory in almost all the major groups of plants. By the Late Devonian or Early Carboniferous periods heterosporous genera were known from the *Lycopsida* (*Lepidostrobus, Sigillariostrobus, Pleuromia, Lepidocarpon, Miadesmia*), and the *Pteropsida* (*Archaeopteris, Lagenstoma, Trigonocarpus*). Some of them illustrate difference in spore size and spore number, whereas others show complete attainment of heterospory and still others have almost approached the seed condition. Some examples are given below to illustrate the sequence.

Williamson and Scott (1894) described *Calamostachys binneyana* and *C. casheana*. The former is homosporous but in some sporangia the spores of some tetrads are of unequal size. In the latter species there are two types of sporangia containing smaller (microspores) and larger spores (megaspores). The megaspores were about three times the diameter of the microspores. The megasporangia also contained some abortive spores. In *Stauropteris burntislandica* there is evidence of more extensive spore abortion leading to a reduction in spore number. Chaloner (1958) described megasporangia in which two spores out of a tetrad were large and two were abortive. In *Lepidocarpon* the reduction is still greater where three spores of the tetrad are abortive and one functional. In *Sphenophyllum dawsonii* in some sporangia there is an indication of abortion of whole tetrads leaving behind a few large spores. In seed-like megasporangia of *Miadesmia* there is no indication of abortion of spores. Such cases (*Miadesmia*) may be due to extreme reduction in the number of sporocytes in the megasporangia. Among the progymnosperms two kinds of sporangia were found in *Archaeopteris latifolia* from the Upper Devo nian of Pennsylvania. Beck (1957) also found heterosporous condi tion in *Tetraxylopteris schmidtii* in which case sporangia of different sizes were found (they were of course without spores). There are many other examples of fossil plants exhibiting heterospory. In a primitive fern member *Enigmophyton superbum* the sporangia are borne (one per sporophyll) on sporophylls arranged in distinct bifurcating spikes. Some of them contain smaller spores (60-85 m diam.) and other contain larger spores (up to 250 m diam.) In *Discinites major* (Halle, 1954), a fossil member belonging to the order *Noeggerathiales*, there is no distinction between sporangia. Every sporangium has one large megaspore and many small microspores. Walton (1957) discovered two types of spores in the sporangia of *Protopitys scotica*. Isolated spore members have been discovered from various horizons of the earth, that show heterospory. Chaloner and Pettitt (1963, 1964) isolated a spore tetrad from the Upper Devonian rocks of Quebec, which they named as *Cystosporites*. Each tetrad is composed of one large megaspore and three abortive spores. It is considered to belong to *Lycopsids*. A number of other examples also exist, *e.g.*, *Ocksisporites* has a large triangular megaspore (560 m broad) and *Nikitinosporites* (610 m diameter) has a spherical megaspore with spinous exine. These fossil records only illustrate the existence of two kinds of spores in the past, but no evidence is available as to whether they germinated to produce two kinds of gametophytes. In the absence of such an evidence no concrete conclusions can be drawn out.

Developmental evidence

The living representation of the major divisions of the vascular cryptogams (*Lycopsida, Sphenopsida, Pteropsida*) include a number of heterosporous genera (*Selaginella, Isoetes, Marsilea, Pilularia, Regnellidium, Salvinia, Azolla* and *Stylites*). They afford us the possibilities of extensive morphological, developmental and comparative studies, that can throw much light on the origin of heterosporous condition in the plants. The development of the sporangium in all the living plants follows a common developmental pattern. In all cases archesporium is formed by the lower cell or cells cut off from the sporangial initial or a group of initials. The outer cells differentiate into the jacket initials. The archesporial cells by further division become sporogenous cells or spore mother cells, which undergo reduction division to form the spores. This is the common basic pattern of sporangial development.

Differences are, however, there in the details of the development in different groups of plants. These are :

- Whether the sporangium arises from a single initial or a group of initials.
- The number of cells making up the primary archesporial tissue.
- Whether the archesporial cells act directly as spore mother cells or they divide mitotically and then act as sporocytes (spore mother cells).
- Number of functional sporocytes.
- Number of spores which mature.
- Origin of the tapetal layer and its time of disorganisation.

A few of these differences have played a role in the development of heterospory. These comparative and developmental studies illustrated the various pathways that have led to heterospory. The time at which the sex determinants activate and exert their influence to segregate sex seems to play a major role in illustrating the different developmental pathways leading to heterospory. The following examples will illustrate the statement.

In the homosporous vascular cryptogams *e.g.*, *Lycopodium, Dryopteris, Pteridium* etc., the sex determinants exert their influence during the development of the sex organs.

In *Equisetum*, Hauke (1963) observed that the gametophytes that are all alike in early stages of development show distinction into two types during further vegetative growth and it is easy to determine which will bear antheridia and which archegonia. In this case the sex determinants exert their influence earlier, *i.e.*, before the formation of sex organs. In these cases sex determinants exert their influence in the gametophytes, their time, however, differs.

In *Selaginella*, which is a heterosporous vascular cryptogam, the sex determinants exert their influence during the differentiation of the sporocytes (spore mother cells). This is a significant drift in the time of activation of sex determinants. It has shifted from the gametophyte to the sporophyte. It is a small shift, but very significant, because it has brought along with it the condition of heterospory in the vascular cryptogams. In *Selaginella* we have read that there is no difference in the sporangial development up to the stage of development of the sporocytes. Thereafter the differences begin. In the microsprorangia almost all the sporocytes produce microspores, whereas in the megasporangia usually one sporocyte (rarely 2, 4, 10) is functional and others abort or disintegrate. This functional sporocyte produces 4 megaspores. At this stage we can differentiate between the micro- and megasporangia. The former produce microspores and the latter megaspores which produce male and female gametophytes, respectively.

The same is the case in *Marsilea*. In the microsporangia all the sporocytes develop into 64 microsopres. In megasporangia only one sporocyte undergoes meiosis and produces 4 megaspores out of which only one matures.

In the angiosperms, the stamens (microsporophylls) and carpels (megasporophylls) appear quite different at maturity so that we can tell in advance that the stamens will bear microsporangia and the carpels will bear ovules that contain megasporangia. Here sex determinants exert their influence before the maturation of the sporophylls. In many gymnosperms and some angiosperms the reproductive bodies are borne on special branches which can be identi fied before they bear the reproductive bodies. This can be ascertained by their position and morphology. In these cases the sex determination occurs duing the vegetative growth of the branch. So in the seed plants the sex determinants exert their influence much earlier in the heterosporous pteridophytes.

The above discussion is based on the fact that the spermatophytes are heterosporous and that microspores and megaspores are sexually differentiated structures, *i.e.*, the former give rise to male gametophytes that bear male sex organs, whereas the latter germinate to produce female gametophytes that bear female sex organs. Thompson (1927, 1934) showed that in some spermatophytes the microspores were larger than the megaspores and concluded that the spermatophytes may have originated from the homosporous ancestors. This conclusion does not seem to be sound because the difference in size is there, whether the microspores are larger than megaspores or *vice-versa*. There is no question of homosporous origin, moreover there are some other structural and cytological differences between the two types of spores of the same species.

The question now arises is as to what was the first step towards heterospory, because all the examples cited above had developed heterospory in an advanced state. There are some living pteridophytes that provide some clue towards this direction. The genus *Equisetum* although homosporous has frequently been cited as showing stages prior to heterospory in that the gametophytes of many species appear to be functionally heterothallic (Schratz, 1928). In these species the gametophytes are of one morphological type during their early growth, but can subsequently be recognised as two types which occur in approximately equal numbers. Each type usually produces only one type of sex organs. Although female plants are stated to be able to produce antheridia late in their development. A careful further examination can, however yield some significant results. Tryon (1964) has shown that in a fern *Platyzoma microphylla* two kinds of spores are produced. They vary in size in the ratio of 1 : 2. The sporangia producing these two types of spores occur intermingled on the sporophylls. The larger spores occur 16 per sporangium where as the smaller number 32 per sporangium. It is not known whether this difference is due to abortion or from reduction in the number of megaspore mother cells. In pot cultures the smaller spores germinated into filamentous male gametophytes that produced antheridia at a later stage if fertilisation failed to take place. These examples illustrate restriction of sexual expression in the gametophyte, and it is quite probable that heterothallism may have preceded heterospory in evolution.

Experimental Evidence

Some experiments have been carried out in this direction on the basis that spore abortion is responsible for heterospory as it leaves greater amount of nutrition for the surviving spores that enlarge in size. Shattuck (1910) conducted experiments on *Marsilea* by manipulating environmental factors such as light, temperature and moisture. He found that under favourable nutritional conditions some of the spores in microsporangia aborted, and if it affected over half of the spores the remaining enlarged and increased about 16 times in diameter. In extreme cases only a single spore survived and developed all the structural characteristics of a megaspore but he was not able to germinate these spores. Similarly he found that under unfavourable circumstances the megasporangia developed large number of mature microspores. He could not induce them to germinate and hence no probable conclusion could be arrived Stow (1930) subjected the flowers of *Hyacinth* to sudden temperature fluctuations and found that many anthers contained cells similar to the embryo sacs and many of the spores,had aborted. Stow postulated that such affected cells produced necrohormones which altered the sexual expression of the developing spores. Naithani (1937) considered the spore abortion to be the result and not the cause of the hypertrophy of others. Geitler (1941) regarded it to be the result of genetic peculiarity of some strains and varieties.

The spore differentiation, besides nutrient supply and environmental factors, is controlled by a more deeply seated component of the cellular control system. The genetic system of the cell controls the spore type in a plant. The genetic control of spore differentiation may be exercised at the level of protein synthesis or during the event of gene transcription. The differences between microspores and megaspores may be : (*i*) in the quantity and type of food reserves; (*ii*) the number and appearance of organelles; (*iii*) nuclear shape and position; and (*iv*) thickness and sculpturing of wall. All these differences are regarded to be gene controlled and it is suggested that many genes are involved in the determination of spore type. These genes must be regulated as a single functional unit because in heterosporous plants the spore will either have a characteristic of a microspore or of a megaspore, and are never bisexual. No cytological proof is available for the time of action of particular genes or groups of genes for systems comparable to spore differentiation. Such an evidence for other systems is available. Researchers in this direction will certainly throw a great light on the genetic phenomenon of spore differentiation.

It may be mentioned that heterospory originated in many groups of pteridophytes and was prevalent in the progymnosperms which are believed to be the ancestors of gymnosperms. This phenomenon was accompanied by partial or complete development of the gametophytes within the spore wall and consequent reduction in the gametophytic tissue.

Evolution of the Male Gametophyte

The male gametophytes of modern gymnosperms differ appreciably from those of the homosporous pteridophytes in being endosporous, much reduced and in possessing a siphonogamous outgrowth called the pollen tube. The male gametophyte is an outcome of the germination of the microspore or the pollen grains. The germination is partially **precocious** and partially it takes place on the nucellus of the ovule. It is an extremely reduced structure as compared to the massive gametophytes of the homosporous bryophytes and pteridophytes. It is devoid of any orderly arrangement of cells as compared to the antheridia of mosses and pteriodophytes. Some sort of comparison can be made with the male gametophytes of heterosporous pteridophytes like *Selaginella* and *Isoetes*. Both these pteridophytes have completely endosporic male gametophytes with one prothallial cell and an antheridium with 256 antheriozoids in *Selaginella* and only four in *Isoetes*. Liebig (1931) pointed out a sort of similarity between the male gametophytes of *Isoetes* and some fossil gametophytes like *Cordainthus*. In *Cordainthus*, Florin (1936) described a row-like disposition of spermatozoids (Fig. 1.20, A) and a tendency towards a similar disposition is evident in *Isoetes*. Sewards and Ford (1906) also pointed out towards a resemblance between the organisation of male gametophytes of such heterosporous pteridophytes and some living and fossil *Araucariaceae*.

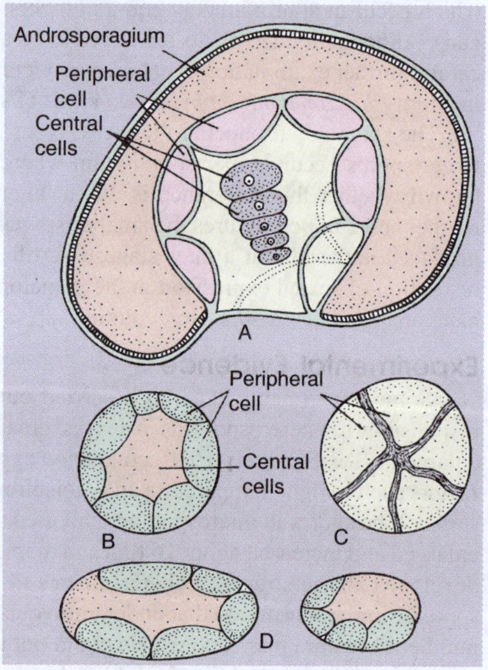

Fig. 1.20. Germinated pollen grains of some gymnosperms. A. Male gametophyte of *Cordainthus* (After Florin). B. Male gametophyte of *Physostoma elegans* (After Oliver). C. Male gametophyte of *Stephanosperma akenoides* (After Oliver). D. Male gametophyte of *Cy- cadeoidea dacotensis* (After Wieland).

Structure of male gametophytes of fossil gymnosperms is not completely known for all fossil groups. No data are available regarding the fossil remains of fossil microgametophytes of **Coniferophytes** of Carboniferous and Permian periods. Well preserved male gametophytes of

Cycadofilicales and *Cordaitales* are available and all such specimens show multicellular nature of the microgametophytes.

Germinated pollen grains have been described in the nucellus of ovules of many genera of *Cycadofilicales*. Oliver (1904) and Renault (1885) described the pollen grains and microgametophytes of the fossil genus *Stephanospermum*. In *S. akenoides*, Oliver described about 20 large and parenchymatous cells filling the microspore cavity (Fig. 1.20, C). In some specimens these cells have been seen to form a parietal layer enclosing a central cavity. In *S. caryoides* he reported two cells lying in this cavity and a single peripheral cell enclosing them partially. The peripheral cell was regarded by Oliver to be comparable to a tube cell of the modern gymnosperms. The central two cells perhaps divided further to produce antherozoid mother cells.

Hoskins and Cross (1946) and Stewart (1951, 1954, 1958) studied the male gametophytes of *Pachytesta vera*, *P. hexangulata*, *P. illionensis* and *P. composita*. The male gametophytes have many peripheral cells enclosing a number of internal cells lying in the central cavity of the microspore. Renault (1902) described a small protuberance on one side of the microspore of *Aetheotesta* and *Stephanospermum*. He interpreted it as the base of the pollen tube. In *Physostoma elegans* (*Cycadofilicales*) Oliver described the presence of sterile peripheral cells (Fig 1.20, B).

Florin (1936) described the germinated pollen grains of *Cordainthus* (*Cordaitales*) and reported the presence of definite peripheral cells enclosing a central space containing a row of about five nuclei along the polar axis of the microspore (Fig. 1.20, A). Florin compared the peripheral cells to the male prothallial cells of living gymnosperms and the five nuclei to the gamete nuclei. These germinated pollen grains were found in the microsporangia.

In *Cycadeoidea* (Fig. 1.20, D) Wieland (1906) found germinated pollen grains in microsporangia, they also contained 1-5 peripheral cells that were interpreted as prothallial cells.

The fossil genera completely lacked a pollen tube.

A careful study of the microgametophytes of modern gymnosperms (see under general characteristics) reveals that there has been a gradual reduction in the size and content of the male gametophyte. In *Ginkgo* and *Welwitschia* the male prothallial cells are altogether wanting.

Origin and Evolution of the Ovule

The fundamental characteristic of all the spermatophytes is "*sensulato*" to produce seeds and that the seed or more exactly the ovule that will develop into a seed on maturation, *is a tegumented megasporangium that the plants disperse together with their contents*. The nucellus of an ovule contains a megaspore that develops into the female gametophyte (prothallus, archegonia) and is hence comparable to megasporangium. It is surrounded by one or two protective structure called the *integuments*. In order to elucidate the origin and evolution of ovule the botanists have to analyse ontogenetically and phylogenetically the various elements of this body (integuments, nucellus, megaspore, prothallus, embryo). The research in this direction has almost been exclusively restricted to the integument, and it is still a subject of dispute.

Regarding its morphological nature, three main theories have been advanced :—

- The **axial theory**, which held that nucellus is of the nature of a bud bearing integuments that are regarded as lateral foliar appendages.
- The **sui-generis theory** that regarded the ovule to be an independent structure borne on either foliar or axial organs.
- The **foliar theory** regarded the ovule to be homologous to, with three lobed leaflet of the female sporophyll, the two lateral lobes of the leaflet are fused to form the outer integument, the nucellus is an emergence borne on the cup-shaped terminal lobe. Eames (1961) supported this interpretation in case of angiosperm ovule.

After the discovery of *Lagenostoma lomaxii*, a pteridosperm ovule, by Oliver and Scott in 1904, a number of theories have been advanced to explain the origin of ovule and seed. A brief description of Palaeozoic ovules will be essential before discussing these theories.

Structure of Palaeozoic Ovules

Oliver (1903, 1904) classified Palaeozoic ovules into two main types : (*i*) **Platyspermeae** with bilaterally symmetrical ovules; and (*ii*) **Rediospermeae** with radially symmetrical ovules. Seward (1917) classified them into three types:—

1. Cardiocarpales. These are equivalent to Oliver's Platyspermeae and have Cordaitean affinities. The integument is free from the nucellus and consists of two or three distinct layers (*a*) outer usually fleshy **sarcotesta**; (*b*) middle stony **sclerotesta**; and (*c*) inner **endotesta**, which is also fleshy. Roth (1955) described the *Cardiocarpus spinatus*. Besides the three layered single integument the nucellus appeared as a thin band of tissue surrounding the female gametophyte which consists of parenchymatous mass of cells with two archegonia towards the micropylar end. The nucellus is free from the integument except at the base. There is a distinct **pollen chamber** at the nucellus tip. The vascular supply in some ovules consists of a single **central strand** that enters the sarcotesta (Fig. 1.21). The main bundle then passes through the **sclerotesta** and enters the base of the **nucellus** where it forms a **tracheidal disc** from which arise numerous discrete bundles which run up the peripheral part of the nucellus. In *Cardiocarpus spinatus, Taxocarpum undulatum*, and *Lyrasperma scotica* (Fig. 1.21, A) there are no **nucellar strands**. The integument in the last named ovule is bilobed and is fused with the nucellus. Other ovules of this type include *Rhabdospermum, Diplotesta, Cardainthus williamsonii, Cycadiocarpus angustudunensis, Kamarospermum leeanum*, and *Cardiocarpus magnicellularis* (Brotzman and Schabilion, 1972).

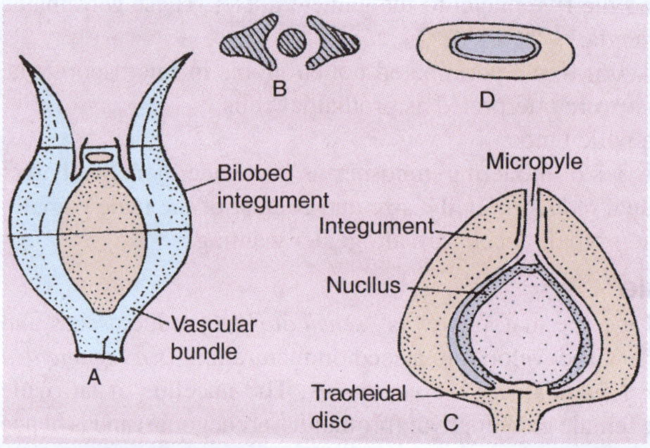

Fig. 1.21. Caridocarpalean ovules. A. L.S. ovule of *Lyrasperma scotica*. Long, with two transverse sections at levels indicated (After Long. 1960). B. L.S. Cardiocarpalean ovule; dotted lines represenet vascular strands.

It has been suggested (Smith, 1964) that the cardiocarpalean ovules have originated from a common ancestor having ovules that were radially symmetrical and had an integument derived from several telomes.

2. Lagenostomales. In these ovules the nucellus and the integuments are united and are radially symmetrical. A protective structure called the **cupule** is present and may enclose one or more ovules (1 in *Lagenostoma*, 4 in *Stamnostoma* and about sixty in *Calathospermum*). Only the integument is vascularised. **Cupule** is found external to

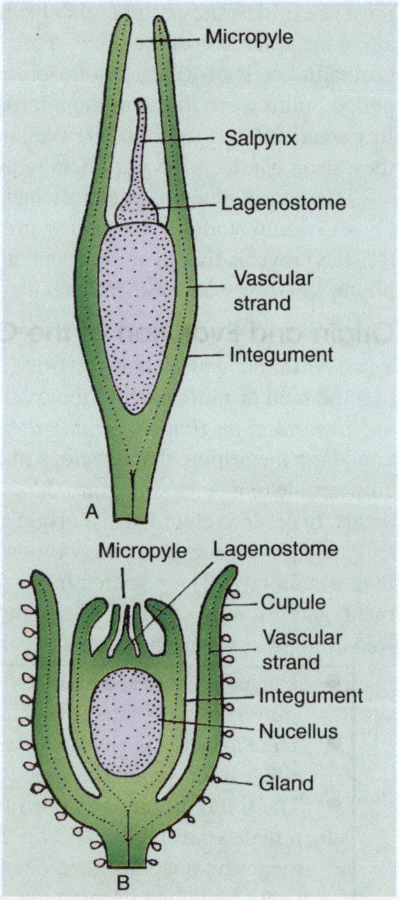

Fig. 1.22. Lagenostomalean ovules, A. *Salpingostoma dasu*. B. *Lagenostoma lomaxi* (After Gordon).

the integument. The common examples (Fig. 1.22) are *Lagenostoma lomaxi* (= *Lygenopteris oldhamia*), *Salpingostoma dasu* (Fig. 1.22, A) *Sphaerostoma ovale* (1.23, B) and *Tyliosperma orbiculum* (1.23, A).

3. Trigonocarpales. They are characterised by : (*i*) the single integument is free from the nucellus; (*ii*) both the nucellus and integument are vascularised; (*iii*) cupules have not so far been discovered; (*iv*) they are radially symmetrical; and (*v*) the nucellar apex is complex and has a distinct **pollen chamber** with a small projection. *Pachytesta illionense* (Fig. 1.24, A) and *Stephanospermum elongatum* are good examples of this type.

There are some ovules (*Eosperma oxroadensis, Deltasperma fouldense, Camptosperma burmiciense, etc.*) that resemble *Lagenostomales* in one respect and *Cardiocarpales* in being bilaterally symmetrical. All of them possess bilobed integument with a single vascular strand in each lobe.

Some lower carboniferous ovules (Fig. 1.25, A, B) do not fit into any of the above three categories because they resemble one or the other type in one or the other character, *e.g.*, they resemble Cardicarpales and Trigenocarpales in possessing integument and nucellus free from each other but they also resemble Lagenostomales in nucellus having no vascular strands. Common examples include *Genosperma kidstonii* (Fig 1.25A), *G. latens*, and *Geminitheca scotica* (Fig. 1.25, B). In the last type cupule is present enclosing two ovules.

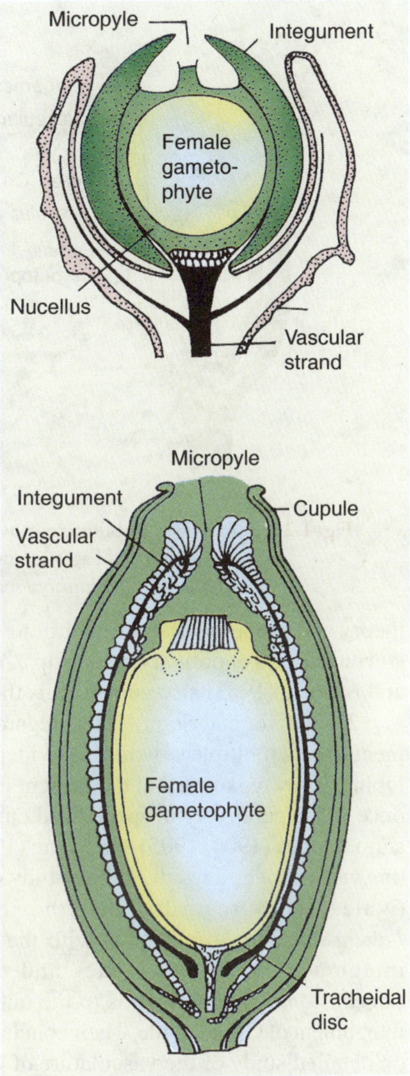

Fig. 1.23. Lagenostomalean ovules. A. *Tyliosperma Orbiculatum*. B. *Sphaerostoma ovule*. (After Andrews).

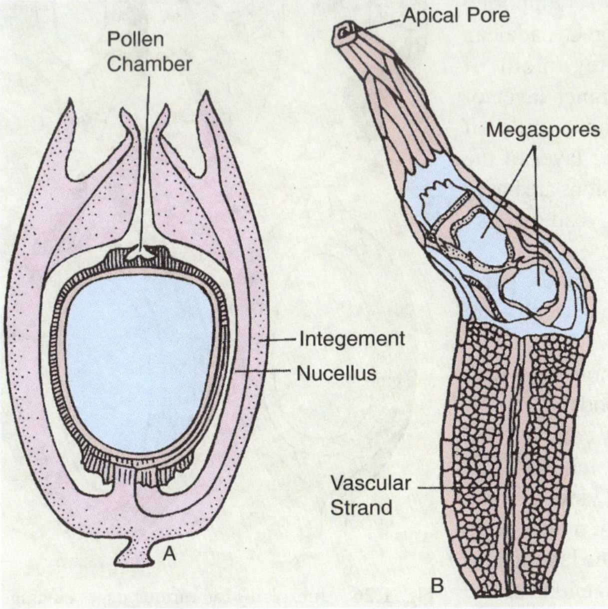

Fig. 1.24. A. L.S. ovule of *Panchytesta illionense*. B. Megasporangium of *Stauropteris burntislandica* (After Andrews).

Theories regarding Origin of Palaeozoic Ovules. A brief discussion of these theories is given in the text:

1. Benson (1940) proposed that a pteridosperm (*Cycadofilicales*) ovule is a **synangium** in which the central sporangium has retained a single megaspore, whereas the surrounding ring of sporangia became sterile and formed the integument. Walton (1953) supported this

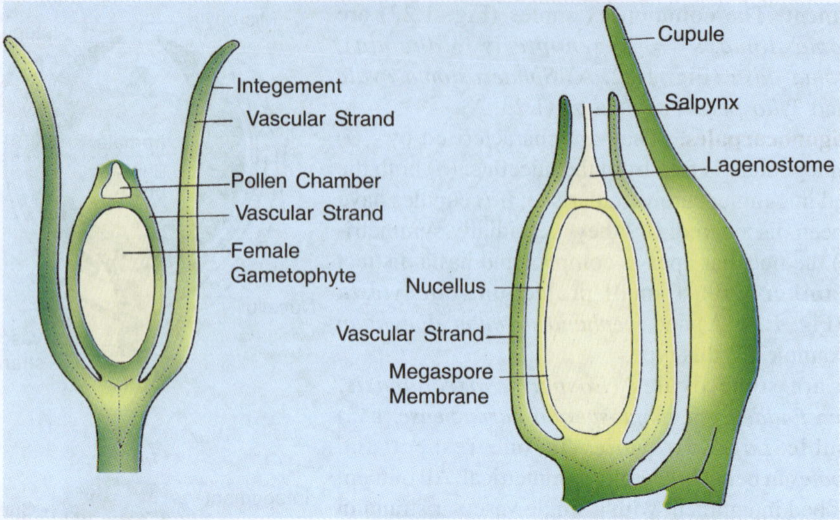

Fig. 1.25. Some lower carboniferous ovules that resemble Cardicarplean and Trigonocarpalean ovules in one respect and Lagenostomalean ovules in other respect. A. Ovules of *Genosperma kidstonii*. B. Ovules of *Geminitheca scotica*. (After Smith, 1959).

theory and proposed that the encircling elements were **sterile telomes** or ordinary branch tips. Andrews (1961) and Meeuse (1963) also support this theory.

2. Oliver and Scott (1904) regarded the integument to be a multiple structure as evidenced by its free lobing. They regarded that the present integument was once a comparatively unspecialised cupule-like indusium. Stopes (1904, 1905) and Matte (1904) supported this viewpoint as a result of their study on Cycadalean ovules. They regarded that the integument of *Lagenostoma* is homologous with the inner layer of integument of Cycad ovules and the cupule of *Lagenostoma* is homologous to the outer layer of the integument of Cycad ovule. Their conclusions are based on detailed study of the vasculature of Cycad ovules.

3. De Haan (1920) concluded:—

- The integuments of pteridospermous, gymnospermous and angiospermous ovules are homologous.
- The integuments are not homologous with the indusia of ferns and lycopods.
- The integuments are originally composed of units that are now regarded as **telomes**.

4. Halle (1933, 1937) was of the view that the cupules and the synangia of pteridosperm ovules were borne on the leaves and were formed by the lateral concrescence of groups of cyclically arranged telomes. The fused telomes in the case of cupule were sterile whereas those of synangia were fertile.

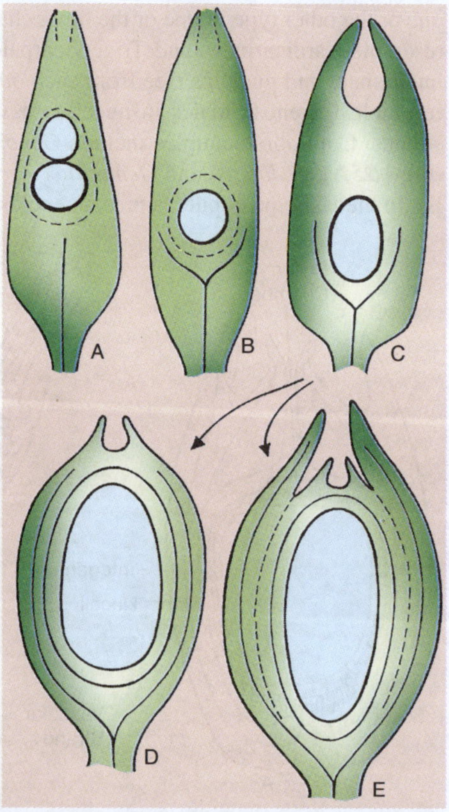

Fig. 1.26. It explains the various stages envisaged by the Nucellar modification concept leading to the origin of Pteridospermic ovule.

5. Walton (1940) opined that cupule and integument of pteridosperm ovules were formed from a frond or its part by inrolling and further fusion to form a cup-like structure. In 1953 Walton again suggested that the cupule of the Lagenostomalean ovules is morphologically equivalent to the integument of the Trigonocarpalean ovules. According to Walton the vascularised outer part of the nucellus of *Trigonocarpales* is equivalent to the integuments of the *Lagenostomales*. Smith (1959), Long (1960) and Lam (1959) have supported Walton.

6. Nucellar Modification Concept (Fig. 1.26). This theory was put forth and later elaborated by Andrews in 1961. According to this theory the Lagenostomalean integument is derivable from the nucellus or sporangium wall by its vascularisation and growth and lobing of the apex. It does not account for the origin of cupule. It fails to explain the origin of Trigonocarpalean integument also. Andrews took *Stauropteris burntislandica* (heterosporous *Coenopterid*) as a starting point to explain his concept (Fig. 1.24, B). The following characteristics of the fern megasporangium attracted Andrews attention :—

- The sterile part of the megasporangium is vascularised.
- Presence of a small opening at the tapering apex of the megasporangium.
- Most of the basal half of the sporangium is sterile and acted as a nutritional reservoir.

Andrews postulated that such a megasporangium must represent an important stage in the evolution of seed. The possible evolutionary steps as underlined by him are:—

- The number of megaspores is reduced to one and its increase in size.
- The functional large megaspore sinks to the bottom of the megasporangium.
- The vascular strand is displaced and divides into two or more branches which are forced to grow up around the megasporangium.
- Slight modification of the sporangial wall and apex. This would result in an organ similar to the vascularised nucellus of the Trigonocarpalean ovule.
- Further modification, presumably growth and lobing at the apex of the sporangium resulting in the ovule of the Lagenostomalean type.

This concept does not explain the origin of Lagenostomalean cupule and the Trigonocarpalean integument. In Andrews own words, "it is not intended as a cure-all to explain the evolution of the pteridospermean seed, in fact it is obvious that at best it only answers some of the problems that confront us". Later in 1963 Andrews reconsidered his concept and supported the view that the integument is derived from a system of telomes.

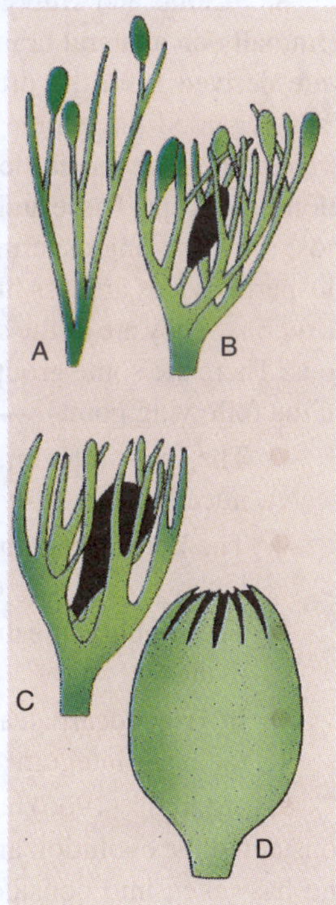

Fig. 1.27. It explains the various stages envisaged by the Telome concept, leading to the origin of Pteridospermic ovule.

7. The Telome Concept (Fig. 1.27). It is the synthesis of the views of Smith (1959), Long (1960) and Andrews (1961). They all consider the derivation of pteridosperm ovule from a terminal fern megasporangium surrounded by sterile and fertile telomes as those of *Hedesia corymbosa* or *Stauropteris burnstislandica*. The following stages explain the concept:

- Reduction in the number of megaspores to one.
- Elaboration of megasporangial apex to form pollen catching device.
- Surrounding sterile telomes underwent modification to form integument. Long (1960) reported a similar organisation in *Genosperma kidstonii*.
- Partial fusion of the encircling telomes to form a cupule like structure that was initially free from the nucellus but later became

fused with its base. This step is illustrated by the ovule *Genosperma latens*, where the lobed integuments are fused basally and are slightly adnate to the nucellar base.

- One or more terminal ovules become enclosed by further sterile telomes to form a cupule or second integument. This stage is illustrated by *Geminitheca scotica* a species which consists of two ovules enclosed by one cupule and each ovule has a cupule like integument that is free from the nucellus.
- The integuments become fused with the nucellus—a stage illustrated by the ovules *Calathospermum scoticum* and *Salpingostoma dasu*.
- The integument becomes simplified by disappearance of lobing and reduction of apical lobes.
- The function of catching pollen grains is transferred from the nucellus apex to the integuments through the micropyle formed by the fusion of integuments. The upper Carboniferous Lagenostomalean ovules have reached this stage.
- Further reduction in the lobing of the cupule and the integument as illustrated by Trigonocarpalean ovules.
- Fusion of the two integuments. The ovules of living cycads represent this stage.

The available fossil record is overwhelmingly in favour of this concept. It can well be relied upon to explain the course of evolution of the ovule in the petridosperms as it can well explain all the structural variations so far recorded in these extinct forms.

8. Camp and Hubbard (1963) postulate that the ovule is derived from a sporangium borne terminally on a lateral branch of a modified dichotomously branching system. The two integuments were derived from the dichotomous branches immediately surrounding the terminal sporangium. They regarded the cupule to have originated by the condensation of the dichotomies bearing the ovules. Their views on the origin of the cupule are based on *Eurystoma angulare* and resemble the telomic concept. These authors have gone up to the extent of comparing the pteridosperm ovules with those of some angiosperms especially *Paeonia* and *Myristica*. They have concluded that the angiosperm ovules are less highly specialised and modified as compared to those of *Cycads* and *Pteridosperms*. They are of the opinion that ovules of all groups possess two integuments and are homologous. There are some groups in which third integument is also present. They have based their theory on the following points :—

- The inner integument of the Lagenostomalean ovules has been mistaken for a part of the nucellus. It is fused with the nucellus and its distal part is recognisable as **Salpynx**.
- The outer integument is fused with the inner integument and is usually interpreted as the inner.
- The cupule is the third investing structure comparable with the aril found in many angiosperm ovules.
- In Trigonocarpalean ovules the outer integument and the cupule are fused together, whereas the inner integument is adnate to the nucellus.

9. Martens (1966) has proposed a theory quite different from all those discussed above. He has considered the evolution and origin of ovule in the gymnosperms from a heterosporous pteridophyte and has taken into consideration the living groups. His ideas are based upon the fact that in the heterosporous pteridophytes the megaspore is free in the sporangial cavity, whereas in the seed-plants the functional megaspore is fused by its own walls with the tissue of the nucellus. As a result of this the spore has lost all the possibilities of an isolated dissemination. It will consequently be disseminated together with the tegumented megasporangium (ovule). According to Martin the megaspore in heterosporous pteridophytes is free from the megasporangial wall due to an isolatory mechanism, which is not found in the seed plants. This isolation is due to the deposition of callose around the megaspore wall. It has been discovered in the pteridophytes, but in the gymnosperms it has been found around the megasporocyte of a cycad *Encephalartos poggei.* It later disappears during sporogenesis. In case it has persisted the megaspore of this cycad would have been free from the nucellus. Martens postulates that there must have been some primitive gymnosperms in which callose deposits persisted and megaspore was free from the nucellus. During the course of evolution this callose secretion became less and ultimately disappeared thus affecting complete fusion of megaspore with the nucellus.

Discussions in the previous pages have made it clear that the pteridospermous ovule has arisen independently a number of times, but the process of evolution has followed a similar pattern, which is explained by the Telome concept. The multiple origin of the pteridospermous ovule suggests that the pteridosperms must have polyphyletic origin from different types of pteridophytic plants. Such a conclusion does not seem to be sound because most of the cardiocarpalean ovules show close resemblance and suggest their origin from a common ancestor, *i.e.*, they have a monophyletic origin. Since now it is believed that they belong to the *Cordaitales* and it will not be wrong to infer that the order as a whole has a monophyletic origin. Florin (1951) is also of the same view. It has been suggested that ovules with radial symmetry were ancestral, *e.g.*, Cordaite ovule. It evolved into a bilaterally symmetrical ovule simultaneous with the adoption of a strobilar habit. In case this presumption is true, the primitive Cordaite ovule must bear a close resemblance with the primitive pteridosperm ovule. Such a proposition is further supported by the existence of some ovule genera that possess characters of both pteridospermous ovules and cardocarpalean ovules. Such a conclusion is further strengthened by evidence from vegetative structure and anatomy (it has been discussed before). The progymnospermous plants like *Archaeopteris* and *Protopitys scotica* possess pteridospermous as well as Cordaitalean vegetative characters, but their method of reproduction is pteridophytic. This group has been regarded by Beck (1960, 1962, 1970) as ancestral to both pteridosperms and Cordaitales. This is suggestive of a monophyletic origin for the gymnosperms.

CLASSIFICATION OF GYMNOSPERMS

Many systems of classification of the gymnosperms have been proposed up to the present day. These will be described briefly in this text.

It was **Robert Brown** (1827) who first recognised Gymnosperms as a separate group of plants. Bentham and Hooker (1883) while classifying Angiosperms placed them in between dicotyledons and monocotyledons in their book *Genera Plantarum*. Gymnosperms were first recognised as a major division of Spermophyta by **Van Tiegham** (1898) when he divided Spermophyta into two divisions: The Gymnosperms (Astigmatae) and Angiosperms (Stigmatae).

Coulter and Chamberlain (1917) divided the gymnosperms directly into seven orders : **Cycadofilicales; Bennettitales; Cycadales, Cordaitales; Ginkgoales; Coniferales** and **Gnetales**. **Chamberlain** (1934) divided the gymnosperms into two classes which were further subdivided into orders:—

A. Class Cycadophyta. It is characterised by unbranched and stumpy stems; leaves large and pinnately divided; male cones large and compact with simple sporophylls; female cones loose and consist of pinnate or simple sporophylls that bear large ovules; stems have wide cortex and manoxylic wood. It includes three orders:—

- **Cycadofilicales :** It includes all extinct forms.
- **Cycadeoidales :** It includes both living and extinct forms.
- **Cycadales :** It includes both living and extinct forms.

B. Class Coniferophyta. The stems are profusely branched and the foliage gives cone-like appearance; leaves simple; male and female strobili are compact and bear complex sporophylls;

wood is pycnoxylic. It includes 4 orders:—

- **Cordaitales :** Extinct order.
- **Ginkgoales :** Includes extinct and one living representative.
- **Coniferales:** Includes both extinct and living genera.
- **Gnetales :** Includes both extinct and living genera.

Another classification which recognises two classes among Gymnosperms is by Jeffrey (1917). He named these classes as :—

A. Class Archigymnospermae. It includes all those gymnosperms that resemble ferns in their general appearance, anatomy and in possessing motile spermatozoids. He included five orders in this class :—

- Order **Pteridospermae:** Includes all extinct forms. They are the primitive most gymnosperms that lived in the Palaeozoic.
- Order **Cycadeoideae :** Includes all extinct forms.
- Order **Cycadales :** Includes extinct as well as extant (living) forms.
- Order **Cordaitales :** Includes all extinct forms.
- Order **Ginkgoales :** Includes extinct and one living genus.

B. Class Metagymnospermae. The members of this class bear no resemblance to the ferns, possess simple leaves, male gametes are non-motile, the pollen grains grow into a pollen tube. It includes two orders :—

- Order **Coniferales.** It includes living genera, *e.g. Pinus, Cedrus, Taxus, Podocarpus, Taxodium*, etc. It also includes fossil genera.
- Order **Gnetales.** It includes *Ephedra, Gnetum* and *Welwitschia*, and many fossil genera.

— Depending upon the **composition of wood,** the Gymnosperms were divided into two classes:— (*i*) **Manoxylic** with loose textured and porous wood; and (*ii*) **Pycnoxylic** with compact wood. The former included orders: **Cycadales, Cycadeoidales**, and **Cycadofilicales**. The latter included the orders : **Cordaitales, Ginkgoales, Coniferales** and **Gnetales**. This classification was proposed by **Seward in** 1919.

Gymnosperms were divided into three classes by **Wieland** (1919). These are : (*i*) **Neostrobileae** including orders **Cycadofilicales** and **Cycadales**; (*ii*) **Anthostrobileae** including orders **Cycadeoidales** and **Gnetales**; and (*iii*) **Metastrobileae** including orders **Cordaitales, Ginkgoales** and **Coniferales**.

Professor Birbal Sahni (1920) divided the gymnosperms into two major groups depending upon the axial or foliar origin of the ovules. These are : (*i*) the **Stachyspermae** in which the ovules arise on the axial organs, *e.g.*, **Cordaitales, Ginkgoales, Coniferales** and **Taxales**; and (*ii*) **Phyllospermae** in which the ovules are borne on leaves, *e.g.*, **Pteridosperms** and **Cycadales**.

The Gymnosperms were divided directly into seven orders that were arranged in the order of increasing complexity as evidenced by reproductive features, by **Engler and Gilg** (1926). These are **Cycadofilicales, Cordaitales, Cycadales, Bennettitales, Ginkgoales, Coniferales** and **Gnetales**. **Arnold (1948)** divided Gymnosperms into three phyla:

- **Cycadophyta :** including **Pteridospermales, Cycadeoidales** and **Cycadales**.
- **Coniferophyta :** including **Cordaitales, Ginkgoales, Taxales** and **Coniferales**.
- **Chlamydospermophyta :** including **Ephedrales** and **Gnetales**.

Lam (1952) included pteridosperms along with the ferns under the **Pteropsida**. He included the orders **Cycadales, Nilssoniales, Pentoxylales** and **Cycadeoideales** under the **Cycadophyta**; whereas in his **Conferopsida** he included **Cordaitales, Pityales, Ginkgoales, Taxales** and **Coniferales**. His group **Chlamydospermae** included the **Gnetales**.

Professor D.D. Pant (1957) divided the Gymnosperms into three divisions each of which is further divided into classes and orders. The classification is represented below in a tabulated form:—

GYMNOSPERMS

Div. **Cycadophya** | Div. **Chlamydospermophyta** | Div. **Coniferophyta**

Class : Gnetopsida

Order *Gnetales* | Order *Welwitschiales*

Class : **Pteridospermopsida** | **Cycadopsida** | **Pentoxylopsida** | **Cycadeoideopsida**

Order *Cycadales* | *Pentoxylales* | *Cyacadeoideales*

Order

Lyginopteridales | *Medullosales* | *Glossopteridoles* | *Peltaspermales* | *Corystospermales* | *Caytoniales*

Classes :

Coniferopsida | **Ephedropsida** | **Czekanowskiopsida** | **Taxopsida**

Order : *Ephedrales* | Order : *Czekanowskiales* | Order : *Taxales*

Orders :

Cordaitales | *Coniferales* | *Ginkogoales*

Raizada and Sahni (1960) divided the Gymnosperms into three divisions : (*i*) **Cycadophytes (Pteridospermales, Cycadeoideales, Cycadales)**; (*ii*) **Pentoxylates (Pentoxylales)** ; and (*iii*) **Coniferophytes (Cordaitales, Ginkgoales, Coniferales** and **Gnetales)**.

Lyman Benson (1957) divided the living Gymnosperms into four classes each of which were further divided into orders and families. This classification is as follows :

Classes	Orders		Families		Examples
1. Conopsida	1.	Ginkgoales	1.	*Ginkgoaceae*	*Ginkgo*
	2.	Pinales	1.	*Pinaceae Pinus,*	*Abies, Cedrus*
			2.	*Taxodiaceae*	*Taxodium*
			3.	*Cupressaceae*	*Cupressus*
			4.	*Araucariaceae*	*Araucasia*
			5.	*Podocarpaceae*	*Podocarpus*
			6.	*Cephalotaxaceae*	*Cephalotaxus*
	3.	Taxales	1.	*Taxaceae*	*Taxus*
2. Ephedropsida	1.	Ephedrales	1.	*Ephedraceae*	*Ephedra*
3. Gnetopsida	1.	Gnetales	1.	*Gnetaceae*	*Gnetum*
	2.	Welwitschiales	1.	*Welwitschiaceae*	*Welwitschia*
4. Cycadopsia	1.	Cycadales	1.	*Cycadaceae*	*Cycas, Zamia* etc.

Andrews (1961) divided the Gymnosperms into six divisions as given below :

Classes	Orders	Families	Examples
1. Pteridospermophyta	1. Pteridospermales	1. *Lygnopteridaceae,*	*Lyginopteris*, etc.
		2. *Medullosaceae*	*Medullosa*
		3. *Corystospermaceae*	*Umkomasia*, etc.
		4. *Peltaspermaceae*	*Lepidopteris*
2. Cycadophyta	1. Cycadales	1. *Cycadaceae*	*Cycas, Zamia*, etc.
	2. Bennettitales	—	*Cycadeoidea*

3. Ginkgophyta	1. Ginkgoales	1. *Ginkgoaceae*	*Ginkgo biloba*
4. Coniferophyta	1. Coniferales	1. *Pinaceae*	*Pinus, Cedrus, Abies*
		2. *Taxodiaceae*	*Taxodium*
		3. *Taxaceae*	*Taxus*
		4. *Cupressaceae*	*Cupressus*
		5. *Podocarpaceae*	*Podocarpus*
		6. *Araucariaceae*	*Araucaria*
		7. *Cephalotaxaceae*	*Cephalotaxus*
	2. Cordaitales	—	*Cordaites*
5. Gnetophyta	1. Gnetales	1. *Gnetaceae*	*Gnetum, Welwitschia*
		2. *Ephedraceae*	*Ephedra*
6. Gymnosperms of uncertain position	1. Pentoxyleae	—	*Czekanowskia*
			Pentoxylon
	2. Vojnovskyales	—	*Vojnovskya*

Delevoryas (1963) agreed with Chamberlain (1934) in recognising two classes among the gymnosperms. His classification resembles that of Chamberlain except that he recognised **Taxales** as a separate order of the **Coniferophyta**.

Sporne (1965) also recognised three classes among the Gymnosperms. His system of classification is as follows :

Classes	Orders	Families	Examples
1. Cycadopsida	1. Pteridospermales	1. *Lyginopteridaceae*	*Lyginopteris*
		2. *Medulosaceae*	*Medulosa*
		3. *Calamopityaceae*	*Calamopitys*
		4. *Glossopteridaceae*	*Glossopteris*
		5. *Peltaspermaceae*	*Lepidopteris*
		6. *Corystospermaceae*	*Xylopteris*
		7. *Caytoniaceae*	*Caytonia*
	2. Bennettitales	1. *Williamsoniaceae*	*Williamsonia*
		2. *Wielandiellaceae*	*Wielandiella*
		3. *Cycadeoideaceae*	*Cycadeoidea*
	3. Pentoxylales	1. *Pentoxylaceae*	*Pentoxylon, Sahnia*
	4. Cycadales	1. *Cycadaceae*	*Cycas, Zamia*, etc.
		2. *Nilssoniaceae*	*Nilssonia*
2. Coniferopsida	1. Cordaitales	1. *Erytophytaceae*	*Eristophyton*
		2. *Cordaitaceae*	*Cordaites*
		3. *Poroxylaceae*	*Poroxylon*
	2. Coniferales	1. *Lebachiaceae*	*Lebachia*
		2. *Voltziaceae*	*Voltziope*
		3. *Palissyaceae*	*Palissya*
		4. *Pinanceae*	*Pinus, Abies, Picea*
		5. *Taxodiaceae*	*Taxodium*
		6. *Cupressaceae*	*Cupressus*
		7. *Podocarpaceae*	*Podocarpus*
		8. *Cephalotaxaceae*	*Cephalotaxus*
		9. *Araucariaceae*	*Araucaria, Agathis*

	3. Taxales	1. *Taxaceae*	*Taxus, Torreya*
	4. Ginkgoales	1. *Trichopityaceae*	*Trichopitys*
		2. *Ginkgoaceae*	*Ginkgo*
3. Gnetopsida	1. Gnetales	1. *Gnetaceae*	*Gnetum*
		2. *Welwitschiaceae*	*Welwitschia*
		3. *Ephedraceae*	*Ephedra*

Takhatajan, Cronquist and Zimmermann (1966) proposed a different classification in which they included the Gymnosperms in a separate division which they called **Pinophyta**. It was further divided into sub-divisions, classes and sub-classes as follows :—

DIVISION: PINOPHYTA

Sub-divisions : **Cycadiceae**

Classes : 1. **Lyginopteridatae**
2. **Cycadatae**
3. **Benettitatae**

Pinicae

Classes :
1. **Ginkgoatae**
2. **Pinatae** — Sub-classes : 1. **Cordaitidae** 2. **Pinidae**

Gneticae

Classes : **Gnetaitae**
Sub-classes :
1. **Eptedridae**
2. **Welwitschiidae**
3. **Gnetidae**

Gangulae (1968) divided the gymnosperms into four classes and fourteen orders as follows :—

GYMNOSPERMS

Classes :

1. **Pteridospermopsida**
Orders :
1. **Cycadofilicales**
2. **Glossopteridales**
3. **Caytoniales**

2. **Cycadopsida**
Orders :
1. **Cycadales**
2. **Nilsoniales**
3. **Cycadeoidales**
4. **Pentoxylales**

3. **Coniferopsida**
Orders :
1. **Cordaitales**
2. **Taxales**
3. **Ginkgoales**
4. **Czekanowskiales**

4. **Chlamydospermopsida**
Orders :
1. **Ephedrales**
2. **Welwitschiales**
3. **Gnetales**

David Bierhorst (1971) proposed the following classification :—

GYMNOSPERMS

Classes :

1. **Cycadopsida**
Orders :
1. **Pteridospermales**
2. **Caytoniales**
3. **Cycadeoidales**
4. **Pnetoxylales**
5. **Glossoptridales**

2. **Coniferopsida**
Orders :
1. **Cordaitales**
2. **Protopityales**
3. **Ginkgoales**
4. **Pentoxylales**
5. **Taxales**

3. **Coniferopsida**
Orders :
1. **Ephedrales**
2. **Gnetales**
3. **Welwitschiales**

DIVISION TRACHAEOPHYTA GYMNOSPERMS

Classes : (3)

- Progymnospermopsida
  - Orders (1)
    - Aneurophytales
    - Archaeopteridales
    - Protopityales
- Gymnospermopsida
  - Orders (2)
    - Pteridospermales
    - Cycadales
    - Cycadeoidales
    - Caytoniales
    - Glossopteridales
    - Pentoxylales
    - Czekanowskiales
    - Ginkgoales
    - Cordiatales
    - Voltziales
    - Coniferales
    - Taxales
- Gnetopsida
  - Orders (3)
    - Gnetales
    - Ephedrales
    - Welwitschiales

Stewart (1983) proposed a system of classification of Gymnosperms (vascular plants). according to which division *Tracheophyta* (vascular plants) have been divided into three classes : *Progymnospermopsida*, *Gymnospermopsida* and *Gnetopsida*. This classification is as follows:

Modifying the classification of **Takhtajan, Cronquist and Zimmerman** (1966), **Meyer** (1984) proposed system of classification for *Gymnosperms* with a single Division *Pinophyta* (*Gymnosperms*). The *Pinophyta* is further divided into three classes *viz.*, *Gingoopsida*, *Cycadopsida* and *Pinopsida*. The class *Gingopsida* has nine orders including *Ginkgoales; class cycadopsida* has six orders like *Cyatoniales*, *Glosspteridales*, *Pentoxylales*, *Ephedrales* etc; class *Pinopsida* has two order namely Cordaitales and *Pinales*. Though Meyens system provides good survey of *Gymnosperms* but the placement of *Ephedrales*, *Gnetales* and *welwetschiale*, stems to be misplaced.

Gifford and Foster (1989) raised important groups to the rank of Divisions. These are seven in total *viz. Progymnospermophyta*, *Pteridospermophyta*, *Cycadophyta*, *Cycadeoidophyta*, *Ginkgophyta*, *Coniferophyta* and *Gnetophyta*.

Kramer and Green (see Kubitzki, 1990) in their treatise proposed a system of classification which is given on pg 61:

S.P. Bhatnagar and Alok Moitra (1996) in their book on Gymnosperms offered a plan of classification which is as follows :

GYMNOSPERMS OF INDIA AND THEIR DISTRIBUTION AND ECONOMIC IMPORTANCE

The total number of living gymnosperms in the world approximately number seventy genera and 725 species. M.B. Raizada and K.C. Sahni (1960) reported 16 genera and 53 species from India.

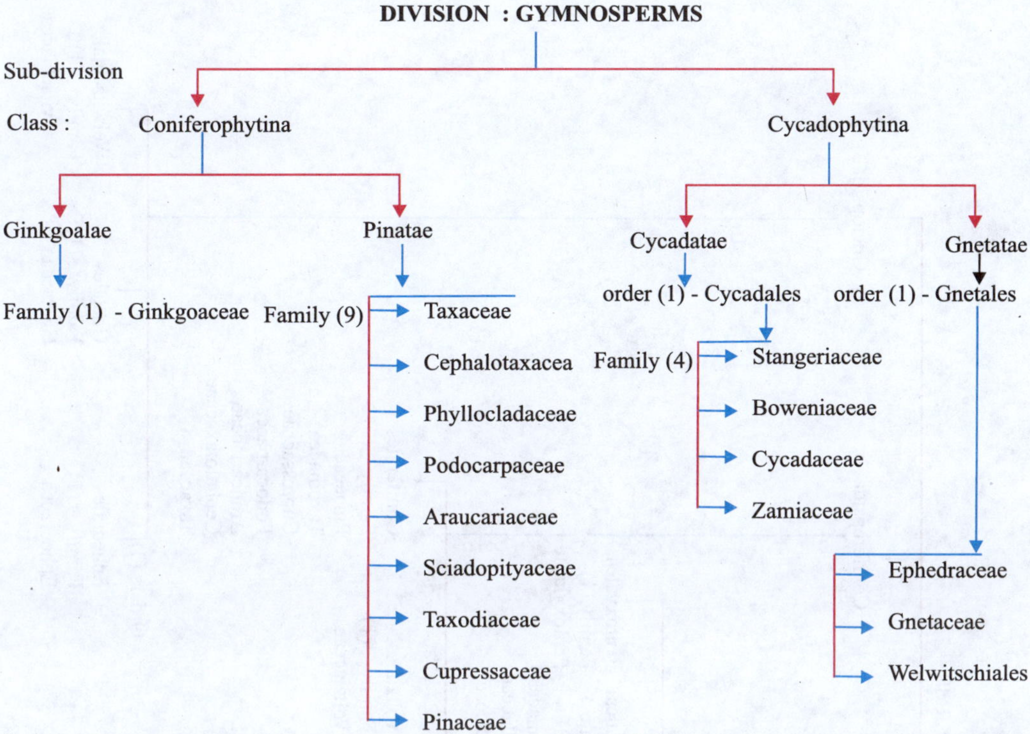

Maheshwari listed only 14 genera. This lesser representation is due to the fact that gymnosperms are mainly dwellers of temperate regions and in India such a climate is afforded only by the Himalayas. They form extensive forests and grow luxuriantly in the various Himalayan ranges. Due to their economic importance silviculturists have paid considrable attention towards this group.

There are six living orders of gymnosperms and out of them four are represented in India. These include : *Cycadales*, *Coniferales*, *Ephedrales* and *Gnetales*. *Ginkgo biloba* (*Ginkgoales*) has been cultivated at many places in India but it originally belongs to China.

The *Cycadales* are represented by 4 species of *Cycas* in India. These are : *C. beddomei*, that is, restricted to dry hills of Cuddapah in the Andhra Pradesh; *C. circinalis* found wild in the dry deciduous forests of South India; *C. pectinata* grows wild in several parts of Eastern India; and *C. rumphii* grows in Andaman and Nicobar islands. *C. revoluta* is a Japanese species, that is, cultivated as an ornamental plant in Indian gardens and is propagated vegetatively by means of bulbils. It is represented only by female plants. The stem and seeds of *C. circinalis* yield a starch called Sago. The resin of *C. rumphii* is applied to ulcers.

The *Coniferales* are represented by eleven genera in India. These include : *Abies*, *Cedrus*, *Cephalotaxus*, *Cupressus*, *Juniperus*, *Larix*, *Picea*, *Pinus*, *Podocarpus*, *Taxus* and *Tsuga*. All of them except *Podocarpus* are restricted to the Himalayas. Some of them occur in both the Eastern and Western Himalayas whereas others are restricted in distribution. They form characteristic formations and their distribution is governed mainly by altitude. *Pinus roxburghii* Sarg., grows luxuriantly at 1,000 metres above sea level, whereas *P. wallichiana* A.B. Jack, *Cedrus deodara* Loud, and *Abies pindrow* Royle grow luxuriantly at 2,500 metres above sea level in the Western Himalayas. The genus *Abies* has four species out of which *A. pindrow* Royle and *A. spectabitis* Spach grow in the Western Himalayas and *A. densa* Griff., and *A. delavayi* Franchet grow in the Eastern Himalayas. They grow luxuriantly at higher altitudes and yield a light useful wood used for making slippers. *A. pindrow* and *A. spectabilis* also grow along the higher ranges of Western and Central Himalayas eastwards to

GYMNOSPERMS

Class

- **Progymnospermopsida**
  - Orders (3)
    - Aneurophytales — Family : Aneurophytaceae
    - Archaeopteridales — Family : Archaeopteridaceae
    - Protopityales — Family : Protopityaceae
- **Cycadopsida**
  - Orders (5)
    - Pteridospermales — Family (4)
      1. Calamopityaceae
      2. Lyginopteridaceae
      3. Medullosaceae
      4. Callistophytaceae
    - Glossopteridales — Family (1) — Glossopteridaceae
    - Caytoniales — Family (3)
      1. Caytoniaceae
      2. Corystospermaceae
      3. Peltaspermaceae
    - Cycadales — Family (1) — Cycadaceae
    - Cycadeoidales (Bennettitales) — Families (3) — Williamsmnaceae, Wielandiellaceae, Cycadeoidaceae
    - Pentoxyales — Family (1) — Pentoxylacae
- **Coniferopsida**
  - Orders (5)
    - Ginkgoales — Family : (1) — Ginkgoaceae
    - Czekanowskiales — Family : (1) — Czekanowskiaceae
    - Cordaitales — Family : (1) — Cordaitaceae
    - Voltziales — Family : (1) — Voltziaceae
    - Coniferales — Family : (7) — Pinaceae, Taxodiaceae, Cupressaceae, Podocarpaceae, Araucariaceae, Cephalotaxaceae, Taxaceae
- **Gnetopsida**
  - Orders (3)
    - Ephedrales — Family : (1) — Ephedraceae
    - Gnetales — Family : (1) — Gnetaceae
    - Welwitschiales — Family : (1) — Welwitschiaceae

Nepal between 2,600 and 3,400 metres. *A. delavayi*, a Chinese species common in Szechuan, occurs in remarkable pure stands, confined to the northern slopes of Piri between 2,750 to 3,350 metres. It is up to 20 metres tall and attains a girth of two metres. *A. densa* and *Tsuga dumosa* Eichler grow abundantly at 2,750 metres in Darjeeling. Some of the *Tsuga* trees attain a diameter of over 6 metres. *Cedrus deodara* Loud., is the only Indian species that grows in Western Himalayas between 1,200-3,300 metres. It grows to a huge size and may attain an age of 704 years. It gives a gigantic coniferous appearance and looks extremely beautiful and yields strongest Indian timber which is resistant to white ants and fungal attacks. The wood is also diuretic, carminative, given in pulmonary and urinary disorders, piles and rheumatism. *Cupressus torulosa* D. Don grows along with *C. deodara*, but also grows elsewhere throughout the outer and middle ranges of Himalayas from Chamba (H.P.) to the Aka Hills (NEFA) at 1,800 to 2,800 metres, rarely in extensive blocks. *C. torulosa* forests are met with most frequently on limestone rocks which provide relatively dry soil conditions, particularly on screes of lime-stone shales. It is very sensitive to fire. *Cupressus sempervirens* and *C. funiberus* are cultivated as ornamentals in Indian gardens. The leaves of former yield an essential oil which has vermifuge properties. The genus *Cephalotaxus* is represented by *C. mannii* Hk. f., and *C. griffihii* Hk. f., in India. Both grow in Eastern Himalayas and are small trees. The genus *Juniperus* is represented by six species, that grow at higher altitudes in the inner valleys of Eastern and Western Himalayas. They grow above tree limit in the sub-alpine and alpine regions. They are all small trees or shrubs whose twigs are usually burnt as incense in the temples. *Juniperus communis* L., is about 0.6 metre high and forms more or less compact patches of a few square metres at 2,900 metres to 4,250 metres above sea level in the Garhwal Himalayas. *J. wallichiana* Hk. f. ex Parl., with low spreading branches is 30 cm. to 1.5 metres high and forms dense patches up to 0.2 hectare in extent at 2,900 to 4,200 metres. *Juniperus recurva* Buch-Ham., and *J. squamata* Buch-Ham., grow from 3,000 to 4,900 metres in the Eastern Himalayas. *J. macropoda* Boiss., grows from 3,000 to 3,600 metres in Ladakh and Kanawar, and from 2,500 to 4,300 metres in Alaknanda valley (U.P.). *J. coxii* A.B. Jack also grows at higher altitudes in both the Western and Eastern Himalayas. *Larix griffithiana* Host. ex Carr., is the only deciduous conifer of India that grows in Sikkim and through Chumbi Valley of Tibet to Bhutan and the Mishmi Hills of NEFA usually above 3,000 metres. It is usually mixed with *Pinus wallichiana* or *Abies* and *Tsuga*. *Picea smithiana* Boiss., grows in West Himalayas along with *Abies pindrow* and grows up 60 metres high. Its wood is used for cheap joinery and for treated railway sleepers. Pinus is represented by five species in India. Of these *P. roxburghii* Sarg., *P. gerardiana* Wall. ex Lamb., and *P.wallichiana* A.B. Jack occur commonly in the Western Himalayas at various altitudes. *P. insularis* Endl., and *P. armandi* Franchet grow luxuriantly in the Eastern Himalayas. *P. roxburghii* grows from 700 to 1,000 metres above sea level and is commonly called Chir. Its wood is used for timber, and is also extensively tapped for turpentine which is distilled to obtain resin and turpentine oil. The resin is also used internally as stomachic and as a remedy for gonorrhoea; it is applied externally as a plaster to buboes and abscess for suppuriation. *P. wallichiana* grows upto 3,000 metres and is widely distributed. Its wood is superior to that of *P. roxburghii* and yields resin. *P.gerardiana* (*Chilgoza*) grows in N.W. Himalayas from 300 to 500 metres above the sea level. Its seeds are edible and very nutritious. In the Western Himalayas it forms forests in Kilba and Ram Pur Bushair (Himachal Pradesh) and extends up to Kishtwar valley but not to Kashmir. Seeds also yield an oil applied as a dressing to wounds and ulcers. *P.insularis* grows in Khasia and Chittagong hills from 800 to 1,600 metres above sea level. Its leaves and wood yield an essential oil and as fuel by the natives. *Podocarpus* has two Indian species— *P. neerifolia* D. Don grows wild in the Andaman islands and in Eastern Himalayas. It is a graceful tree growing up to 900 metres and is commonly found in the ever green climax forests. *P.wallichianus* Presl., occurs from Nilgiris southwards and in Assam. *Taxus baccata* L., grows in west as well as East Himalayas. It loves moist and shady places above 1,800 metres. It grows all along the Khasi-Jaintia hills and Naga hills. The wood is durable. *Cryptomeria japonica* D. Don, is a native of Japan and was introduced into India in the nineteenth century and was grown by its seeds in Darjeeling. It grows in the form of dense forest and yields a valuable light wood. It has also become naturalised in Western Himalayas. *Cupressus cashmeriana* Royle ex Carr., is a native of Tibet, but is now extensively cultivated in India. *Callitris cupressi* Vent was introducd into Nilgiris from Australia in 1885 and is now the finest hedge plant of the area. *Thuja occidentalis* L., has been cultivated as an ornamental plant in North India. *Araucaria cookii* and *A. cunninghamii* are also cultivated as ornamental plants in N. India.

The genus *Ephedra* is represented by six species in India, which are erect or climbing shrubs or perennial herbs. Some of these species, *e.g. E. nebrodensis* Tineo and *E. gerardiana* Wall., contain good quantities of *Ephedrine* and are being commercially exploited for this purpose. This drug is used against asthma and bronchial troubles. Both of them grow at high altitudes in the Himalayas along with *E. intermedia* Schrenk et May, *E. regeliana* Florin and *E. saxatilis* Royle ex Florin., *E. Foliata* Boiss. et Kotsehy is a scrambling shrub growing wild in drier parts of Rajasthan.

Five species of *Gnetum* grow in various parts of India. *G. ula* Brong., is very common in the evergreen forests of western Ghats, Andhra Pradesh and Orissa. It is a woody climber. *G. contractum* Mgf., grows in Nilgiri hills and Kerala and is a scandent shrub. *G. montanum* Mgf., grows wild in Assam, Sikkim and some parts of Orissa. *G. latifolium* Bl., grows in the Andaman islands. *G. gnemon* L., is an erect shrub that grows in the rain forest of Eastern parts of India. The plants of *G. contractum* and *G. latifolium* are used as a fish poison. Their stems and roots are antiperiodic. The bark of *G. gnemon* is utilised for making ropes. Its seeds are edible. The seed oil of *G. ula* is used in rheumatism and for illumination. Its seeds are also edible.

RESEMBLANCES WITH PTERIDOPHYTES

The gymnosperms resemble the pteridophytes in several features. These can be listed below :—

1. Both possess independent sporophytes that are distinguished into root, stem and leaves.
2. The sporophytes possess a well developed vascular system.
3. The xylem in most of the members of both groups lacks vessels and the phloem lacks companion cells.
4. Many cycads resemble the ferns in possessing large compound leaves and in the circinate ptyxis of the leaves.
5. The alternation of generations is **heterologous**.
6. The female sex organs are archegonia in both the groups.
7. The male gametes of Cycads and *Ginkgo* are motile, like the pteridophytes.
8. Many pteridophytes like *Marselia*, *Salvinia*, *Azolla*, *Isoetes* and *Selaginella* are **heterosporous** like the gymnosperms.
9. Some of the heterosporous pteridophytes possess one megaspore in the megasporangium, like the gymnosperms.
10. The gametophytes of gymnosperms and heterosporous pteridophytes are **endosporic**.
11. The gametophytes are highly reduced. The male gametophytes of heterosporous pteridophytes and majority of gymnosperms possess only two male **prothallial cells**.
12. The female prothallus develops before fertilization and there is free nuclear division.
13. Development of a distinct embryo after fertilization. The embryogeny is **endoscopic** in many pteridophytes and all gymnosperms.
14. Germination of spores is **precocious** in gymnosperms and heterosporous pteridophytes.
15. Dichotomous venation in *Ginkgo* and ferns.
16. Presence of **Megaphyllous** leaves.

DIFFERENCES BETWEEN GYMNOSPERMS AND PTERIDOPHYTES

1. Majority of pteridophytes are **homosporous**, whereas all the gymnosperms are **heterosporous**.
2. The roots are adventitious in pteridophytes whereas in most of the gymnosperms the roots are tap roots.
3. The stems are mostly underground rhizomes in most of the ferns, whereas in the gymnosperms the stems are aerial.
4. The gymnospermous stems have a **eustelic** organisation, whereas in the pteridophytes eustele is not found.
5. The gymnosperms produce seeds, whereas pteridophytes do not produce seeds at all.
6. Presence of secondary growth in gymnosperms whereas it is absent in majority of pteridophytes.
7. Development of a pollen tube, in all the living gymnosperms, from the pollen grains after their germination. Such a structure is absent in the heterosporous pteridophytes.
8. The female gametophyte is permanently retained within the megaspore, which is permanently retained within the megasporangium (nucellus).
9. The megasporangium in gymnosperms is protected by an integument a structure absent among the pteridophytes. This protected megasporangium is called the **ovule** and has a small opening or the **micropyle** at one end.
10. The archegonia among gymnosperms lack neck canal and neck **canal cells**, whereas pteridophytes possess them. The archegonia are absent in *Gnetum*.
11. Presence of a free nuclear division in the earlier stage of embryo development.

RESEMBLANCES WITH ANGIOSPERMS

1. Both include trees and shrubs and may be erect, scandent, or climbing.
2. The root system is well developed and the roots may be **diarch, triarch, tetrarch,** or **polyarch**. The roots possess secondary growth. The xylem is **exarch** in the roots.
3. The stem stele is eustelic. Secondary growth is present in the stem. The secondary wood may be **monoxylic** or **polyxylic**.
4. Both produce seeds.
5. Both the groups are heterosporous and possess **endosporic** and reduced gametophytes.
6. The megaspore is in organic connection with the megasporangium and is never shed.
7. The nucellus is surrounded by an integument to form a structure called the ovule. The ovule has a micropyle and may be stalked or sessile.
8. The pollen grains grow into pollen tubes.
9. The male gametes are non-motile in majority of gymnosperms and all angiosperms.
10. The male gametophytes are highly reduced. *Gnetum* lacks prothallus cells.
11. The germination of the spores is **precocious**.
12. The embryogeny is **endoscopic**.

DIFFERENCES BETWEEN GYMNOSPERMS AND ANGIOSPERMS

1. The gymnosperms are not herbaceous, whereas majority of angiosperms are herbaceous.
2. The gymnosperms are unisexual and may be **monoecious** or **dioecious**, whereas angiosperms may be bisexual as well as unisexual, monoecious or dioecious.
3. The gymnosperms are perennial with few exceptions, whereas angiosperms may be **annual**, **biennial** or **perennial**.

4. The gymnosperms rarely reproduce by vegetative means whereas this method is very common among the angiosperms.
5. Pollination and dispersal is anemophilous among gymnosperms, whereas among angiosperms, anemophily, entomophylly, hydrophylly and zoophylly are all prevalent.
6. The gymnosperms lack beautifying devices like petals and sepals from their reproductive organs.
7. The ovules are naked among the gymnosperms, whereas in angiosperms they are protected by the carpels.
8. During pollination the pollen grains directly settle on the micropyle among the gymnosperms, whereas in angiosperms they fall on the stigma and the pollen tubes have to travel through the stylar canal to reach the ovary and the ovules.
9. The angiosperms lack archegonia.
10. The microgametophytes, in angiosperms, lack male prothallial cells.
11. The endosperm is formed after fertilisation among the angiosperms, whereas in gymnosperms it is formed before fertilisation.
12. The endosperm is **triploid** among angiosperms but it is **haploid** among the gymnosperms.
13. There is no double fertilisation among gymnosperms.
14. Cleavage polyembryony is absent in the angiosperms and is prevalent among the gymnosperms.
15. The zygote does not undergo free nuclear division among the angiosperms, whereas this condition is prevalent among the gymnosperms.
16. The phloem in gymnosperms lacks **companion cells**, whereas these are present among the angiosperms.
17. Vessels are usually absent from the xylem of gymnosperms but they are present among the angiosperms. Exceptions are however, there in both the groups.

The above resemblances and differences between gymnosperms and other vascular plants suggest a close affinity between them but also support that gymnosperms are an independent assemblage of plants that have certain characters peculiar to them alone.

In this textbook the authors have followed the classification proposed by Sporne (1965).

Indian Contribution on Gymnosperms.

Studies on living Gymnosperms of India began after the return of M.A. Sampathkumaran from the laboratory of C.J. Chamberlain at the University of Chicago. As Sampathkumaran was based in south India, he was mainly attracted to *Cycas* and *Gnetum.* Unfortunately he did not care to publish his observations except some brief notes in the proceedings of Indian Science Congress in whose sessions he beautifully demonstrated the microphotographs of *Cycas circinales* and *Gnetum ula*. Since then gymnosperms have been the subject of reserach throughout India. Several reviews on Gymnosperms have appeared from time to time. These include those of Agharkar (1938), Maheswari and Kapil (1963), and Datta (1973). In addtion several reviews have also been published dealing with reproductive structures (Konan and Oberoi 1969), pollination mechanism (Dogra, 1964), archegonium (Maheshwari and Sanwal 1963), male gametophytes (Singh, 1964), female gametophytes (Maheshwari and Singh, 1967), embryogeny (Roy Chowdhury, 1962, Dogra, 1970), morphology, systematies and reproductive structures (Sharma, 1994) and morphogenetic studies (Biswas, 1994). Earlier, Maheswari and Singh (1960) have published a detailed account of the economic importance of Gymnosperms.

It was Birbal Sahni (1920) who founded the base for the study of Gymnosperms in India Ref. Box. His observations and interpretations on the nature of epimatium in Podocoapaceae are accepted even today. He also classified seed bearing plants into Phyllospermy and Stachyospermy on the basis of studies on both extinct and extant plants. Soon after his return from Cambridge, Sahni established an active Centre of research at Lucknow University where he and his students started research on both living and fossil Gymnosperms. Sahni earned international recogniton for his brilliant investigations on Gondwana Flora, Rajmahal flora, Saline series, Deccan interappean flora, and continental drift. He established a new order of fossil gymnosperms called Pentoxylales and reconstructed several important fossil plants. His students include K.R. Surange, A.R. Rao, B.S Trivedi etc. Sahni always dreamt of establishing an institute of research for fossil plants and his dream could come true only after his sudden death. This centre is now known as Bribal Sahni Institute of Palaeobotany and is located at Lucknow and is engaged in active research on Gymnosperms. At present at the university of Lucknow Dinesh Kumar and his students are engaged on carrying out researches on palynology and cuticular structures Cupressaceae and *Cycas* respectively.

Studies on development and embryology of Indian Gymnosperms were initiated by Sethi (1929) at Lahore and since then process of reproduction and embryology of several Indian taxa have been studied mainly at Agra college, Punjab University and Delhi University.

Raizada and Sahni (1958) at F.R.I. Dehradun initiated studies on taxonomy, systematics of Indian Gymnosperms. They also reported that species of conifers change with altitude in NorthWest Himalayas. Simultaneously Chowdhury and Ghose (1958) identified several species of conifers on the basis of wood structure.

A strong and active centre of research was established at Botany Dept., University of Delhi by P. Maheshwari. A number of genera were studied by Maheshwari and his coworkers. Three excellent monographs were also published from the Department by Maheshwari and his students. These are : *Gnetum* by P. Maheshwari and V. Basil (1961), *Cedrus* by P. Maheshwari and C. Biswas (1970) and *Pinus* by P. Maheshwari and R.N. Konar (1971). As stated above a number of reviews have also been brought out by Maheshwari and coworkers. S.S. Bhojwani and M. Banerjee are presently engaged on studies on Gymnosperms. SP Bhatnagar has recently published an excellent book on Gymnosperms.

At Agra College, Agra, Johri (1935, 1936) initiated studies on microsporogenesis and male gametophyte of gymnosperms which he continued when he joined Maheshwari at Delhi (1961). Swamy (1946) at Bangalore and N. Rao (1961) at Mysore studied the embryology of spp. of *Cycas*.

A monographic account of the genus Cycas was published by D.D. Pant and Mehra (1962) at Allahabad. Pant (1957) presented a system of classification of gymnosperms after the discovery of new group of fossil plant. He classified gymnosperms into three subdivisions namely Cycadophyta,, Coniferophyta and Gnetophyta.

Mehra (1939), With Prof. Priestley studied anatomy of *Pinus*. After partition, he moved to Punjab University, Chandigarh and began cytological studies on Gymnosperms. He along with TN Khoshoo and P.D. Dogra published a large number of papers on the subject. In 1988 Mehra published a monographic account on Indian Conifers and Gnetales.

Experimental and cultural studies on gymnosperms are rather limited and were initiated by Konar. Since 1960 the chief centre of research on experimental and cultural studies of Gymnosperms has been the Department of Botany. University of Delhi. After the demise of P. Maheswari, Biswas, Konar, Vasil etc. are engaged in such studies as cultures of male cones, pollen graines, male gametophyte, female gametophyte and embryo of several plants have been prepared and maintained. Tissue culture studies have also helped the mankind to grow economically important taxa at a faster rate.

At present active centres of research on gymnosperms include Universities of Delhi, Mysore, Lucknow and Allahabad and F.R.I. Allahabad.

1A. CLASS : PROGYMNOSPERMOPSIA

Progymno spermopsida was created after the discovery by Becks (1960) that there existed organic connection between the leaf of a free sporing fern *Archaeopteris* and the stem of *Callixylon* which shows gymnospermous characters. This discovery of plants having pteridophytic reproduction and gymnospermic secondary wood initiated a process of discussion on the interpretation of the origin and systematics of Gymnosperms. Subsequently Namboodiri and Beck (1968) opined that the *Progymnospermopsida* may be the progenitors of gymnosperms.

The class Progymnospermopsida consists of three orders: Protopityales (genus *Protopitys*), Archaeopteridales (genus *Archaeopteris*) and Aneurophytales (genra *Aneurophyton, Protopteridium, Tetraxylopteris, Triloboxylon* and *Proteokalon*). As a representative of the class Progymnospermopsida, genus *Archaeopteris* is discussed in brief:

Genus : ARCHAEOPTERIS.

Archaeopteris was the late devonian fern-like tree having large compound leaves. The reconstructed plant was of 18 m height with a canopy of spirally arranged bipinnate fronds (Fig. 1.28). About 15 species of the plant have come to the knowledge today of which *A. macilenta* was most studied species.

Archaeopteris leaves were studied by Carluccio *et. al.* (1966) and Beck (1971). Accroding to Carluccio *et. al.* (1966) the "pinnules" or ultimate appendages are basically spirally arranged simple leaves. The pinuules may be fertile or sterile, the fertile pinnule bears one or two rows of adaxial fusiform sporangia which dehisced longitudinally. Sporangia may be both micro-and megasporangia.

Primary xylem is mesarch in stelar anatomy and it is nine ribbed (Fig. 1.28 B). From there ribs branch and leaf traces diverge, the two give rise to branch and the other seven to leaf traces. Before entering the leaf the trace divide into two and proceeds similarly giving a typical dichotomous venation in *Archaeopteris* leaf.

The secondary xylem is abundant and compact like in other coniferes. It consists of only tracheids and vascular rays. Pits are present on the radial walls of tracheids. Ray tracheids are also present. The growth rings may be present or absent. Vascular rays may be uniseriate to multi seriate depending upon the species. The wood is pycnoxylic with groups of circular bordered pits present on radial walls of tracheids.

Arnold (1939) reported heterospory in *A. halliana* for the first time (Fig. 1.29).

Archaeospermia is believed to be the seed of *Archaeopteris.* It was observed that cupule of *Archaeospermia arnoldii* was derived from basal sterile leaves on the fertile branch of *Archaeopteris*. (Phillips et.al., 1972).

Phylogenetic considerations of Archaeopteris.

There are many characters of *Archaeopteris* like tree like habit, megaphyllous leaves (fronds), secondary growth, stelar anatomy and heterospory which suggest the plant to be an intermediate between Psilophytes and seed plants (Beck, 1962). The presence of pycnoxylic wood makes it closer to coniferophytes. Foliage studies, their fructifications and morphology of plant resembles somewhat pteridosperms and to some extent coniferophytes. On the basis of above characters, Beck (1971) proposed a theory that *Archaeopteris* may be considered the ancesters of conifers becauseof presence of pycnoxylic wood and simple decurrent leaves on its branches.

It is considered that Progymnosperms have given rise to Cycadophytes and coniferophytes both. The Aneurophytales have given rise to Archaeopteridales and Cycadophytes and in turn Archaeopteridales gave rise to coniferophytes. With protostelic characters and terminal sporangia Aneurophytes are considered older which gave rise to eustelic condition as in cycadophytes and coniferophytes. The evolution could have been on the lines of Archaeopteridales to Cycadophytes and Coniferophytes. *i.e.* from Protostele to Eustele.

According to Scheckler and Banks, (1971) Gymnosperms could be of monophyletic origin having a common ancestry derived from group of ancient plants showing characters of both the gymnospermous wood and pteridophytic reproduction (Stewart, 1981).

Fig. 1.28. *Archaeopteris* tree, a reconstruction (after Beck, 1962).

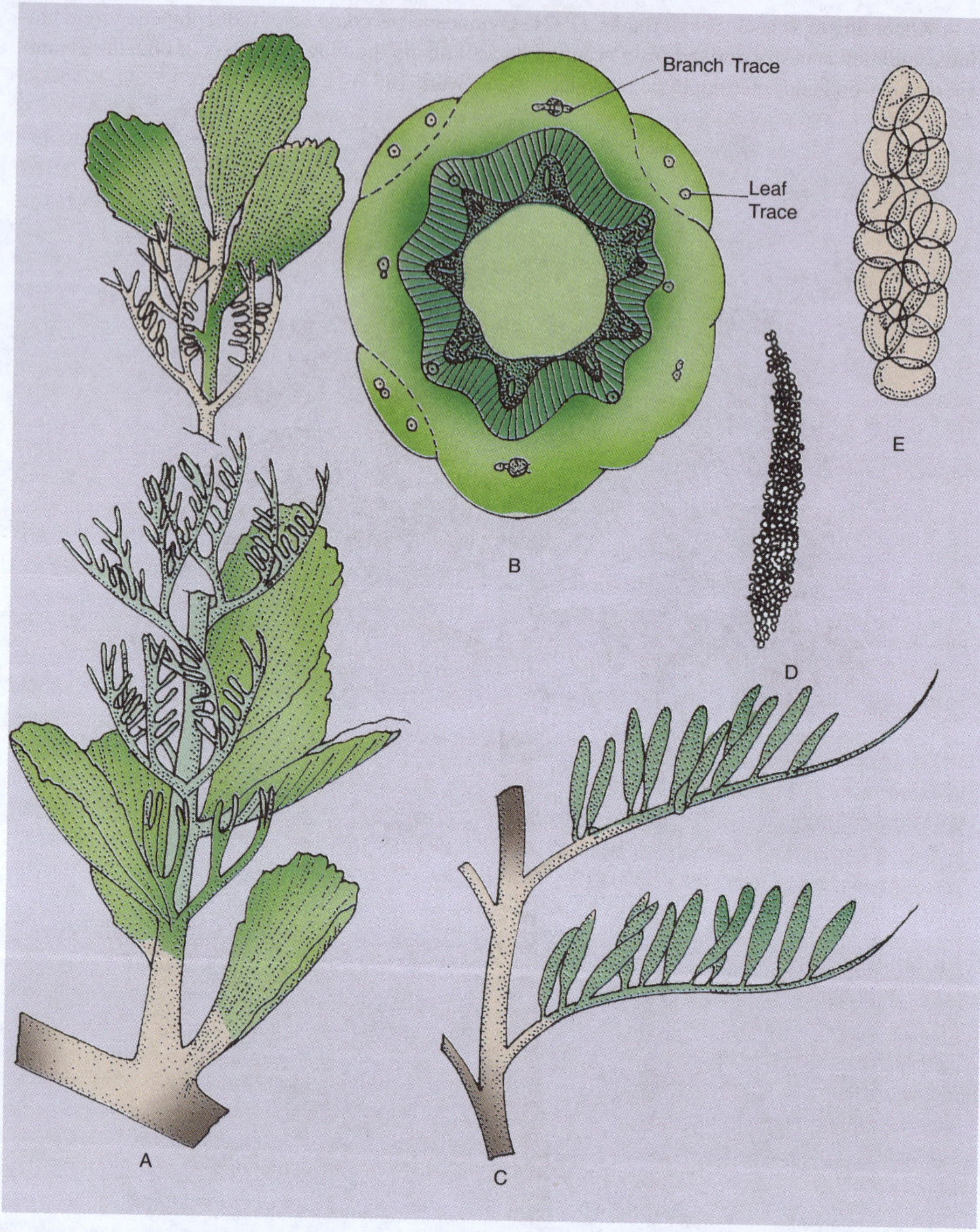

Fig. 1.29. A, C–E, *Archaeopteris halliana*; A. Reconstruction of a branch bearing sterile and fertile leaves. Note the leaves are spirally arranged and four ranked. B. *A. macilenta.* TS stem showing prominent leaf bases, leaf traces and branch traces. C. Fertile leaves bearing micro- and megasporangia. D, E. Cluster of microspores and megaspores, respectively. (A, after Phillips et al., 1972; B, redrawn from Stewart, 1983; C–E, after Arnold, 1939).

QUESTIONS

ESSAY TYPE

1. Give an account of evolutionary tendencies in Gymnosperms. **(Kanpur M.Sc. 1985; 1995)**
2. Write an illustrated account of anyone of the following :
   (*a*) Evolutionary tendencies in Gymnosperms
   (*b*) Distribution of gymnosperms in India. **(Kanpur M.Sc. 1999)**
3. Give a brief account of the male gametophytes of Gymnosperms and discuss their evolutionary tendenies. **(Kanpur, M.Sc. 1992; Rohilkhand, M.Sc. 1989)**
4. Give an account of distribution of gymnosperms in India. **(Kanpur, M.Sc. 1994; Rohilkhand, MSc. 1993; Rohilkhand, 1993)**
5. Give an account of the distribution of living gymnosperms with special reference to India. **(Kanpur, M.Sc. 1996)**
6. Discuss the origin of Gymnosperms. **(Kanpur, M.Sc. 1997)**
7. "There is gradual elimination of female sex organ archegonium in Gymnosperms." Discuss the statement. **(Kanpur, M.Sc. 1999)**
8. What are Gymnosperms? Give an outline classification of this group. Give important features of various groups you have studied. **(Kanpur, M.Sc. 2000)**
9. Give an account of general characters and classification of Gymnosperms. **(Awadh, M.Sc. 1997)**
10. Describe the economic importance of gymnosperms. **(Kumaon, 1995; Rohilkhand, 1993, 1994, 2003; Kumaon, 1995, 1997, 1999; Lucknow, 1997, 1998, 2000; Awadh, 1991)**
11. Describe in brief the salient feature of bryophytes, pteridophytes and Gymnosperms. How will you differentiate them from each other? **(Kumaon, 1996, 1999)**
12. Classify Gymnosperms and add a brief note on the economic importances of this group. **(Rohilkhand, 1997)**
13. Give the outline classification of Gymnosperms upto order level given by David Bierhost (1971) and point out the features of special significance of each class studied by you. **(Kumaon, 2000)**
14. Discuss the general characteristics and affinities of gymnosperms. **(Poorvanchal, 2002)**
15. Describe the broad outline classification of Gymnosperms and distinguishing features of each group. **(Rohilkhand, 1992)**
16. Write an essay on classification of gymnosperms. **(Rohilkhand, 2000; Kanpur, 1999)**
17. What is the Division Chlamydophyta of DD plant? Discuss giving features. **(U.P. College, 1996)**
18. Describe the general characteristic of gymnosperms and write outline classification. **(Meerut, 1999; Awadh, 1993)**
19. Give an account of Prof. D.D. Pant's system of classification of Gymnosperms. **(Gorakhpur, 1999)**
20. Classify Gymnosperms upto orders as proposed by Pant. **(Gorakhpur, 2003)**
21. Give a brief outline and classification of gymnosperms Classify *Cycas, Pinus* and *Ephedra* with distinguishing characters. **(Awadh, 1992)**
22. Write an essay on Economic importance of Gymnosperms. **(Agra, 1994)**
23. Give an account of Prof. DD. Pant's system of classification of Gymnosperms. How far this classification is regarded as natural & Phylogenetic ? Discuss. **(Allahabad, 1995, 2000)**
24. Give a classification of Gymnosperms. Mention important characters and examples of each class. **(Allahabad, 1997; Lucknow 2001)**

25. Discuss briefly the evolutionary significance of Gymnosperms. **(Lucknow, 1992)**

26. According to some botanists Gymnosperms are not heterosporous but heterothallic plants. Discuss. **(Lucknow, 2001)**

27. Give a broad classification of Gymnosperms and describe the salient features of various groups. **(Guru Nanak Deo, 1991)**

28. Give an account of general characters of gymnosperms. Describe its similarities and dissimilarities with that of Pteridophytes. **(Kerala, 2000)**

29. "Gymnosperms constitute an independent intermediate group between Angiosperms and Pteridophytes". Justify it. **(Garhwal, 2000)**

30. Give an account of different systems of classification of Gymnosperms.

31. Describe a general account of the morphology of vegetative organs of the living gymnosperms.

32. Give a general account of the reproductive organs of living gymnosperms.

33. Give a general account of embryology in living gymnosperms.

34. Give a brief account of the Indian Contributions on Gymnosperms.

SHORT ANSWER TYPE

35. Write short notes on :

(*i*) Economic importance of Gymno sperms **(Kanpur, M.Sc. 1999; Rohilkhand, M.Sc. 1993; Awadh, M.Sc. 1998; Rohilkhand, 1998, 1999, 2000; Kanpur 1998, 2000; U.P. College, 1997, 2000; Gorakhpur, 1993, 1996, 1998; Agra, 1993; Allahabad, 1994; Punjab, 1998; Gauhati, 2000)**

(*ii*) Evolutionary tendencies in Gymnosperms **(Rohilkhand, M.Sc. 1992**

(*iii*) Stomatal apparatus in Classification of gymnosperms **(Rohilkhand, M.Sc. 1992, 1994)**

(*iv*) Polyembryony in Gymnosperms **(Rohilkhand, M.Sc. 1994, 2003)**

(*v*) Classification of gymnosperms **(Rohilkhand, M.Sc. 2000; Rohilkhand, 1998; Kanpur 2000; Arga 1993)**

(*vi*) Distribution of Gymnosperms in India **(Rohilkhand, M.Sc 2003; Kumaon, 1996)**

(*vii*) Indian contribution to Gynmosperm **(Awadh, M.Sc. 1999)**

(*viii*) Female gametophyte in Gymnosperms **(Awadh, M.Sc. 2001)**

(*ix*) Similarities and difference between Gymnosperms and Angiosperms **(U.P. College, 1999; Poorvanchal 1999)**

(*x*) Differences of Gymnosperms with pteridophytes **(U.P. College, 2000)**

(*xi*) Classification and Economic importance of Gymnosperms **(Agra, 1991)**

(*xii*) Xeromorphic Characters of Gymnosperms **(Kerela, 2000)**

(*xiii*) Pollination in Gymnosperms

(*xiv*) Gymnosperms in India

(xv) Progymnosperms.

36. Differentiate between Gymnosperms and pteridophyte on the basis of a single character.

37. Who classified gymnosperms into phyllosperms and stachyosperms? **(Kanpur, 1996)**

38. Name an Indian Scientist who has been associated with the study of Gymnosperms. **(Kanpur, 1997, 2000)**

39. Comment upon the economic importance of Gymnosperms. **(Kanpur, 2001)**

40. Gymnosperms form a bridge between Pteridophyta and Angiosperms. Comment. **(U.P. College, 1996)**

41. If the microspore of a gymnosperm has 14 chromosomes, how many chromosomes would by found in (*a*) embryo (*b*) endosperm ? **(Lucknow, 2001)**

42. If the chromosomes number of gymnospermous plant is 20, what would be the chromosome number in its endosperm tissue? **(Poorvanchal, 1999)**

43. What distinguishes Gymnosperms from Pteridophytes? **(Gorakhpur, 1991)**

44. Differentiate between zoidiogamy and siphonogamy. **(Gorakhpur, 1994)**

45. Mention a few economic importance of Gymnosperms. **(Gorakhpur, 1997)**

46. If the microspores of a certain gymnosperm have 12 chromosomes, how many chromosomes will there be in the cells of (*a*) the endosperm (*b*) the nucellus (*c*) the embryo. **(Allahabad, 1999)**

47. Name any gymnospermalogists of the country. **(Awadh, 1993)**

48. Write any two distinguishing characters of Gymnosperms. **(Awadh, 1993)**

49. Mention the name of divisions in Pant's system of classification of Gymnosperms. **(Allahabad, 1998)**

50. Mention two features of phloem in Gymnosperms. **(Mahatma gandhi, 1997)**

51. Highlight the economic importance of Gymnosperms. **(Calicut, M.Sc. 20001)**

52. Differentiate between gymnosperms and angiesperms.

53. In what respects do pteridophytes resemble gymnosperms ?

54. What are Gymnosperms ?

OBJECTIVE TYPE

55. Select the correct answer:

(*i*) The term Gymnosperm was first used by
(*a*) Aristole
(*b*) Theofrastus
(*c*) Tansley
(*d*) Robert Koch.

(*ii*) In which of the following groups of plants, the adult plant is sporophyte?
(*a*) Algae
(*b*) Fungi
(*c*) Bryophyta
(*d*) Gymnosperms

(*iii*) Which of the following does not require external water for fertilization?
(*a*) Mosses (*b*) Cycads
(*c*) Ferus (*d*) Liverworts

(*iv*) Which of the following habits are commonly not found in Gymnosperms?
(*a*) Herbs (*b*) Shrubs
(*c*) Trees (*d*) Lianas

(*v*) Which of the following are not found in Gymnosperms?
(*a*) Antheridia
(*b*) Archegonia
(*c*) Ovules
(*d*) Eggcells

(*vi*) In Gymnosperms, archegonia lack
(*a*) Eggcell
(*b*) Neck canal cells
(*c*) Neck cells
(*d*) Venter canal cells.

(*vii*) A Gymnosperms is a plant which
(*a*) bears flowers
(*b*) produces seeds in ovary
(*c*) does not produce seeds enclosed in ovary,
(*d*) has no vascular tissues.

(*viii*) Gymnosperms differ from the angiosperms
(*a*) in showing heterospory
(*b*) in having naked ovules
(*c*) in having seeds
(*d*) in being smaller in size.

(*ix*) Fruits are not formed in Gymnosperms because
(*a*) the process of fertilization does not take place in them
(*b*) they are not pollinated
(*c*) they are seedless plants
(*d*) they have no ovary.

(*x*) In which of the following groups of plants would you place a plant which produces seeds but lacks flowers?
(*a*) Fungi
(*b*) Bryophytes
(*c*) Pteridophytes
(*d*) Gymnosperms

(*xi*) Gymnosperms resemble angiosperms
(*a*) in having seeds covered with pericarps

(*b*) in having fruits
(*c*) in the process of secondary growth
(*d*) in Phloem with sieve tubes.

(*xii*) Most gymnosperms have
(*a*) both archegonia and antheridia
(*b*) archegonia but no antheridia
(*c*) Neither archegonia or antheridia
(*d*) antheridia but no archegonia.

(*xiii*) A plant produces spores has vascular tissues, and bears seeds without fruits. The plant has rightly been placed in the group
(*a*) Bryophyta
(*b*) Pteridophyta
(*c*) Angiosperm
(*d*) Gymnosperm.

(*xiv*) In gymnosperms pollination is exclusively by
(*a*) Water
(*b*) Insects
(*c*) Animals
(*d*) Wind

(*xv*) Who amongst the following has classified Gymnosperms into phyllosperms and stachyosperms?
(*a*) Coulter and Chamberlain
(*b*) Arnold
(*c*) Birbal Sahni
(*d*) DD. Pant

(*xvi*) Endosperm in Gymnosperms is
(*a*) Haploid
(*b*) Diploid
(*c*) Triploid
(*d*) Tetraploid

(*xvii*) In Gymnosperms, archegonia lack
(*a*) Neck canal cells
(*b*) Ventral canal cells
(*c*) Both of the above
(*d*) None of the above

(*xviii*) The pollen grains in gymnosperms are
(*a*) Haploid
(*b*) Diploid
(*c*) Triploid
(*d*) None of the above

(*xix*) The phloem in gymnosperms lack
(*a*) phloem parenchyma
(*b*) sieve tubes
(*c*) Companion cells
(*d*) all of the above.

(*xx*) Studies on Gymnosperms in India were initiated by
(*a*) M.A. Sampathkumaran
(*b*) M.B. Raizada
(*c*) Birbal Sahni
(*d*) S.C. Bhatnagar.

(*xxi*) In Gymnosperms complete ring of cambium is formed by
(*a*) interfascicular cambium
(*b*) intrafascicular cambium
(*c*) both
(*d*) none

56. Fill in the blanks

(*i*) The term Gymnosperm was first used by ....

(*ii*) .... is the largest order of gymnosperms and has .... families.

(*iii*) Phloem in Gymnosperm is devoid of ....

(*iv*) Nacked seeded plants are called ....

(*v*) In Gymnosperms the seeds are nacked because they lack ....

(*vi*) The most ancient seed plants belong to the ....

(*vii*) Roots in Gymnosperms are .... roots while in Pteridophytes the roots are ....

(*viii*) The Gymnosperms are ....

(*ix*) Most cycads resemble ferns in having .... vernation.

(*x*) Gymnosperms rarely reproduce by .... means.

(*xi*) While the angiosperms lack .... the gymnosperms possess ....

(*xii*) In gymnosperms .... fertilisation is absent.

(*xiii*) Studies on Gymnosperms in India were initiated by ....

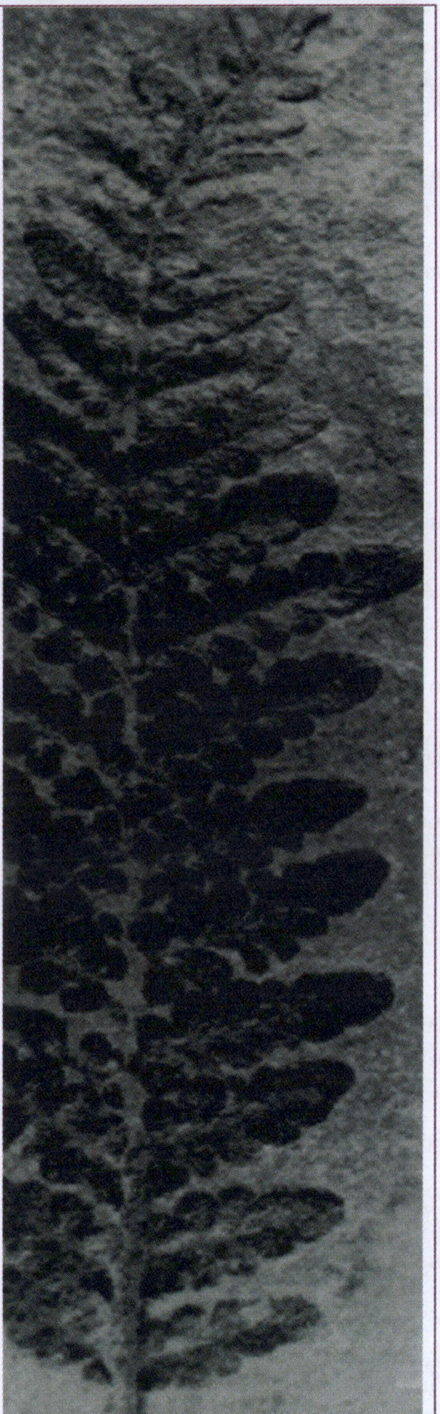
GLOSSOPTERIS LEAF FOSSIL

CHAPTER 2

Cycadopsida – Pteridospermales

- Characteristic Features
- Family : Lyginopteridaceae
- Family : Medullosaceae
- Microsporangiate Organs Attributed to Medullosaceae
- Phylogeny of Medullosaceae
- Family : Calamopityaceae

Class : **CYCADOPSIDA**

Order : **Pteridospermales** (Cycadofilicales)

Families : *Lygniopteridaceae, Medullosaceae, Calamopityaceae*

Introduction

The pteridospermales or *the seed ferns* are characteristic gymnospermous plants that bore fern-like foliage, which in turn bore seeds that were unprotected. It is a very large order that includes a heterogenous assemblage of tree-like or reclining or sprawling plants that came into existence during the Upper Devonian and lived through Carboniferous and Permian periods and reached their climax in the Mesozoic era. This group was discovered in 1877 by Grand Eury. It was later studied by Stur (1883), Williamson (1887), Schenk (1889), Weber and Sterzel (1896), Potonie (1899), Oliver and Scott (1903, 1904), Kidston (1904) and later by several workers (Seward, Walton, Sahni, Florin, Arnold, Andrews, Beck). In India, the Pteridospermales occur in lower, middle and upper Gondwana systems that were formed in the Palaeozoic and the Mesozoic eras. The common genera and species include : *Gangamopteris buriadica*; *G. cyclopteroides*; *G. whittiana; G. srivastavae*; *G. kashmirense*; *Glossopteris* (74 species); *Gondwanidium validium*; *Indotheca sakasarensis*; *Ottakaria kath- wainensis*, *O. benghalensis*; *Pachypteris* (3 species); *Pecopteris concinna*; *Sagenopteris bambanni*; *Palaeovittaria kurzii*; *Sphenopteris* (4 sp.); *Sterocarpus emarginatus*; *Thinnfeldia* (6 species); *Vetebraria* (3 sp.); *Callipteridium*; *Cyclopteris pachyrhacis*.

Characteristic Features

The Palaeozoic pteridosperms have been thoroughly studied as compared to the Mesozoic forms, and taking into consideration the thoroughly studied forms, the group presents the following characteristic features :—

1. The plants have erect slender, or weak stems.
2. The leaves are large, pinnately compound and frond-like.
3. The leaves have resistant cuticle.
4. Primary xylem is in the form of a solid or a medullated protostele and is usually mesarch, rarely exarch. In some forms it is reduced to circummedullary strands. Polystelic condition is also prevalent.
5. Secondary wood and phloem were formed in smaller amounts.
6. Secondary wood was manoxylic.
7. The tracheids of secondary wood bear multiseriate bordered pits on their radial walls.
8. The seeds are not produced in cones or inflorescences, but are borne directly on modified or unmodified foliage. In case of modification the differences were slight.
9. The seeds resemble those of recent cycads.
10. The ovules lack annulus and are surrounded by cupule-like structures.
11. The ovules have integuments either free or fused with the nucellus and have a distinct pollen-chamber and a micropyle.
12. The integuments are vascularised.
13. The nucellus is also vascularised in some ovules.
14. The microsporangia have no annulus and are sometimes grouped into synangia.
15. The pollen grains lacked pollen tubes, and the male gametes were motile and discharged directly into the pollen chamber.
16. The pollen grains on germination produced a number of peripheral cells, enclosing a central cavity (*Stephanospermum akenoides; Physostoma elegans*).
17. The pollen grains may be monolete (*Medullosaceae*) or trilete (*Lyginopteridaceae*).
18. The megaspore is surrounded by a thick wall.
19. The leaf traces are mesarch and may be double or multiple (*Medullosaceae*).

Inter-relationships

The pteridosperms occupy an intermediate position between the ferns and the cycadophytes. They are decidedly gymnosperms because they bear seeds that are exposed. Their resemblances with pteridophytes and cycadophytes are discussed below :—

Resemblances with Pteridophytes

Among pteridophytes, they resemble the ferns more closely in the following points :—

1. The leaves are large and pinnately compound.
2. The lateral veins branch dichotomously.
3. Protostelic stems of some genera. Polystelic condition met within the *Medullosaceae* is also characteristic of some ferns.
4. The xylem is mesarch or rarely exarch.
5. The leaf traces are also mesarch and may be single or multiple.
6. The leaf traces arise by the tangential division of the cauline strand.
7. The microspores are trilete like the pteridophytic spores.
8. The leaves have a circinate ptyxis, which is a fern characteristic.
9. The spermatozoids are motile and multiflagellate.
10. The gametophytes are endosporic like many heterosporous ferns.
11. Presence of archegonia.
12. The xylem lacks vessels and phloem lacks companion cells.
13. The microsporangiate synangia borne on pinnules are quite reminiscent of the recent *Marattiaceae*.
14. The megaspore is surrounded by a thick wall like those of some heterosporous ferns.

The anatomical organisation and the foliage of the pteridosperms suggest their filicalean affinities. So far as fructifications are concerned the pteridosperms do not bear any fructifications of filicalean affinity and it is not possible to establish any connection between the two groups on this basis.

Resemblances with Cycadophytes

The pteridosperms and Cycadales resemble each other in the following respects :—

1. The stems are aerial and erect and bear pinnate compound leaves. Cycadales show medullosan condition of pinnae in some species.
2. The leaf traces are mesarch.
3. The leaf traces arise mostly by tangential divisions.
4. The leaf traces in *Medullosaceae* and Cycadales are multiple.
5. Microspores of some *Medullosaceae* and all Cycadales are monolete.
6. The Cycadales possess free microsporangia with indications of ancestral bisporangiate synangial structure. The sporangia are arranged in indistinct sori and possess indusial hair.
7. Both are heterosporous.
8. The gametophytes are endosporic and quite reduced.
9. The male gametes are multiciliate.
10. The ovules possess a single integument that is usually distinguished into three layers.
11. The pollen chamber is present. The female sex organs are archegonia.
12. There is double vascular system in the seeds of Cycadales and the '*Medullosaceae*.'
13. The vernation of leaves is circinate.
14. The sporangia lack annulus.
15. The xylem and phloem are of similar nature.
16. The megaspore has a thick wall.

The pteridosperms and the Cycadophytes have close affinities and it is believed that the Cycadophyta of the Mesozoic and later periods were probably derived from the great plexus of pteridosperms, which overspread the world in Carboniferous times, while the conifers and ginkgoales may trace their origin to the same stock to which the Palaeozoic cordaitean trees belonged. This stock evidently had relations with the pteridosperms themselves.

Classification

Sporne (1965) sub-divided the **Pteridospermales** into seven families : (*i*) *Lyginopteridaceae*, (*ii*) *Medullosaceae*, (*iii*) *Calamopityaceae*; (*iv*) *Glossopteridaceae*; (*v*) *Peltaspermaceae*; (*vi*) *Corystospermaceae*; and (*vii*) *Caytoniaceae*. The first three are known only from Palaeozoic era whereas the rest have been discovered from the Mesozoic. Of these families, *Lyginopteridaceae Medullosaceae* and *Calamopityaceae* have been discussed in this text.

FAMILY : LYGINOPTERIDACEAE

This group of the **Pteridospermales** was represented both in the Upper and the Lower Carboniferous periods of the Palaeozoic era. They show the following characteristic features :—

1. The stems were not strong enough to bear the weight of the foliage and were either climbers or had a straggling growth habit. They needed support to grow erect.
2. The leaves were compound and the ultimate leaf units were dissected.
3. The stem stele was either of discrete bundles or protostelic with mesarch xylem.
4. The leaf traces were also mesarch.
5. The leaf traces arise by tangential division of cauline strand.
6. The leaf traces were single but become double later.
7. The leaf had circinate ptyxis.
8. The microspores were trilete.
9. The microsporangia occur in paired synangia.
10. The seeds were cupulate.
11. The pollen chamber in the ovule was limited externally by an inverted bell-shaped sheath with a solid nucellar beak internally.
12. The ovule was surrounded by a single massive integument that was fused with the nucellus. The outer fleshy layer of the integument was not very distinct.
13. The ovule had a single vascular system.
14. The sporangia were devoid of annulus.
15. The megaspore had a very thick wall.
16. The secondary wood was soft textured with broad medullary rays.
17. The secondary xylem tracheids were long and tapering. The circular bordered pits are multiseriate and were present only on radial walls.
18. There were no mucilage canals in the stem.

The extensively investigated species *Lyginopteris oldhamia* (*Binney*) H. Potonie will be taken as a classic example to illustrate the family.

LYGINOPTERIS OLDHAMIA (BINNEY) H. POTONIE

Lyginopteris oldhamia (*Calymatotheca hoeninghausi*) was described in detail by Williamson, Scott, Brongniart, Binney, Potonie and Oliver and Scott. It was found in abundance in the coal ball horizon of Lancashire and Yorkshire. Binney (1866) described the stem of this species under the name *Dadoxylon oldhamium* which was later assigned to *Lyginodendron*. Brongniart in 1828 described its frond under the name *Sphenopteris hoeninghausi*. Potonie renamed it as *Lyginopteris oldhamia* in 1899. In 1903, Oliver and Scott showed that the seed *Lagenostoma lomaxi* belonged to *Lyginopteris oldhamia*. Later in 1929, Jongmans showed the presence of capitate glands on the cupulate envelope enclosing the seeds. Such glands were later found on the stem, fronds and the rachis. This showed that all these structures described under separate names belonged to only plant—*Lyginopteris oldhamia*. Its rachis was described under the name '*Rachiopteris aspera*' and the root as *Kaloxylon hookeri*. All these parts are now definitely known to be the various organs of *Lyginopteris oldhamia*, which can, therefore, be regarded as a classical member of the family. The only imperfectly known part is the pollen bearing organ called *Crossotheca hoeninghausi*.

External Features (Fig. 2.1)

The stem of *Lyginopteris oldhamia* was aerial varying in diameter—from 2 mm. to 4 cm. diameter. It was frequently branched and bore adventitious roots that were probably prop roots that supported the weak stem. The leaves were spirally arranged up to 0.5 metre long. Each leaf was bipinnate to tripinnate and was once forked near its base. The pinnae were borne at right angles to the rachis and were arranged opposite to each other. The pinnae bore pinnules or leaflets that were cuneate and were arranged alternately on the ultimate branches (pinnae) of the rachis and have free veins. The petiole and the rachis were studded with capitate glands. The stem was radially symmetrical and it showed that the plants grew erect, although due to its large foliage and comparatively narrow diameter it had to recline or depend upon support on the neighbouring plants or other objects. The prop roots which grew from among the leaves also afforded sufficient support to the stem.

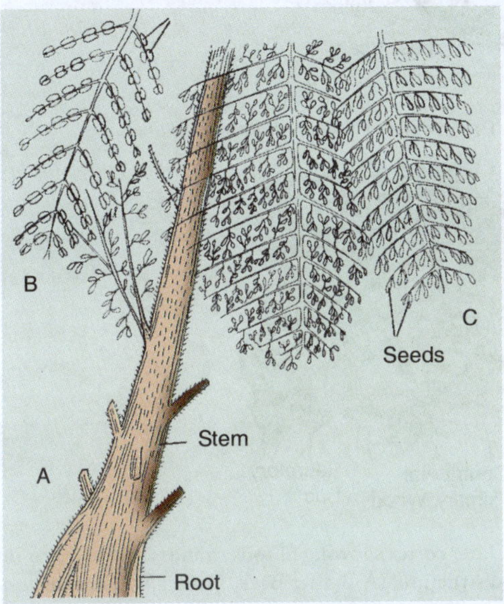

Fig. 2.1. A. Restoration of *Lyginopteris oldhamia* showing external characters. The frond on the left bears the pollen sacs on peltate leaflets (*Crossotheca*); that on the right bears seeds. Also note the stem and roots. (After Mrs. D.H. Scott)

Fig. 2.1(a). B. The frond of *Lyginopteris oldhamia.*

Primary Structure of the Stem

The transverse sections of the stem are nearly circular in outline (Fig. 2.2). Next to the epidermis is the outer cortex which consists of radially broadened fibrous strands that form a vertical network (Fig. 2.2). In the meshes of the network are the parenchymatous cells. This system of fibrous strands affords mechanical strength to the weak stem. The inner cortex consists of ordinary parenchymatous cells. The cortex is many layered. Next to the cortex is the pericycle which consists of short cells and a number of thick walled or sclerotic cells with dark contents.

Next to the pericycle are five strands of primary vascular bundles. These are separated by parenchymatous areas. Each vascular bundle is mesarch and consists of primary phloem towards the outer side. The cells in the phloem are small and irregularly arranged. Next to it is the cambium that is well preserved in some specimens. The xylem consists of tracheids. The tracheids of the protoxylem are spirally thickened, whereas those of the centripetal metaxylem have multiseriate bordered pits arranged on radial walls. The centrifugally developed metaxylem has scalariform tracheids. The protoxylem is surrounded on all sides by metaxylem (Fig. 2.2).

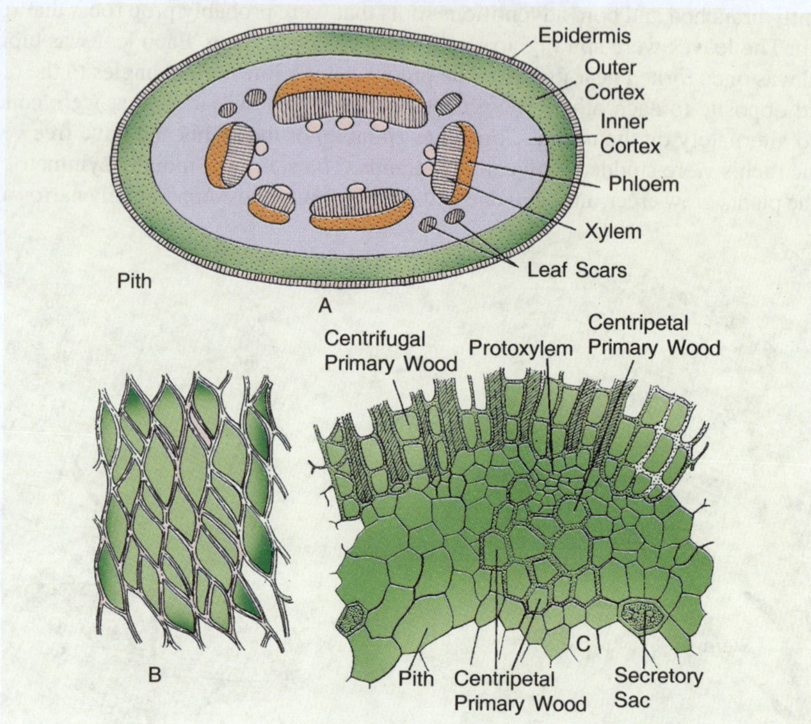

Fig. 2.2. *Lyginopteris oldhamia*, A. T.S. primary stem. B. Outer cortex showing fibrous strands as they appear in L.S. C. Primary vascular bundle showing detailed structure. (A-B after Beck; C after Williamson and Scott).

In the centre there is a large pith made up of parenchymatous cells. Scattered throughout the pith are somewhat thick-walled cells with dark contents. The number of parenchymatous and thick-walled cells is variable. The size of the pith also varies with the diameter of the stem.

Leaf Traces

A leaf trace arises by the tangential division of the cauline strand (vascular bundle of the stem). The strand traverses through the secondary wood and the cortex and divides into two in the inner cortex. These two strands again unite at the base of the petiole to form a V-shaped leaf trace. The trace traverses through the secondary xylem by way of a large ray. On the outer side of the leaf trace is a strip of secondary xylem which is considered to be its own. This wedge-shaped arc of secondary wood of the trace persists beyond the pericycle and then it starts disappearing gradually. The single trace (formed by the union of two branches in the inner cortex) traverses through the petiole and then branches into two and each branch enters the forking rachis. Later it divides to form the veins in each of the leaflets. The xylem strand of the petiole is trough-shaped with the concavity facing upwards. Sometime there are two grooves when the strand is double. The xylem is completely surrounded by phloem and on the lower side of the xylem arc there are many protoxylem groups. The outer cortex in the petiole consists of thick-walled hypodermal strands that are in continuation of those of the stem cortex. The inner cortex in the petiole consists of groups of sclereids.

Secondary Structure of the Stem (Fig. 2.3)

Next to the two cortical layers is the periderm, then pericycle, crushed primary phloem and then the secondary vascular tissue. The secondary cylinder surrounds the primary xylem and it is traversed

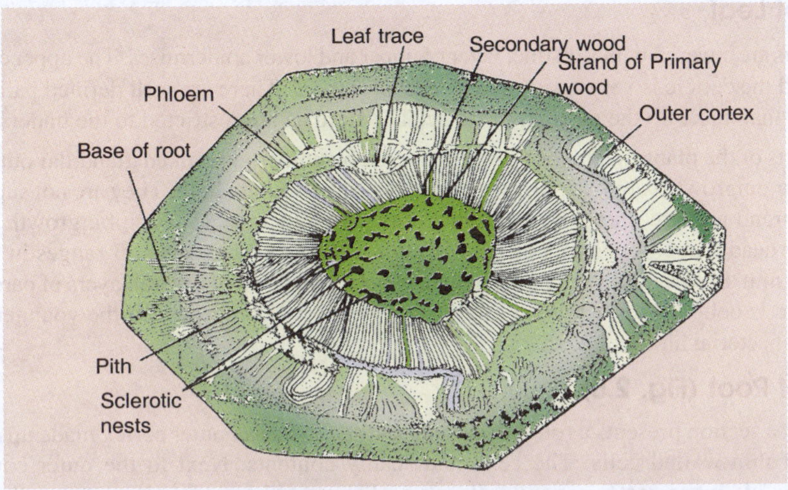

Fig. 2.3. *Lyginopteris oldhamia.* T.S. stem showing secondary structure (After Scott).

by secondary medullary rays that may be narrow or wide. That secondary ring is also interrupted by the passage of leaf trace that originates from the primary xylem strand. The secondary phloem consists of regularly arranged cells and in a transection a line of phloem cells is continuous with the line of secondary xylem tracheids below, formed by the same cambial cell. The cells in the phloem are alternately large and small. The smaller ones divide by a wall at right angles to the vertical row of cells. This results in the formation of a short transverse row of two or more small cells. These

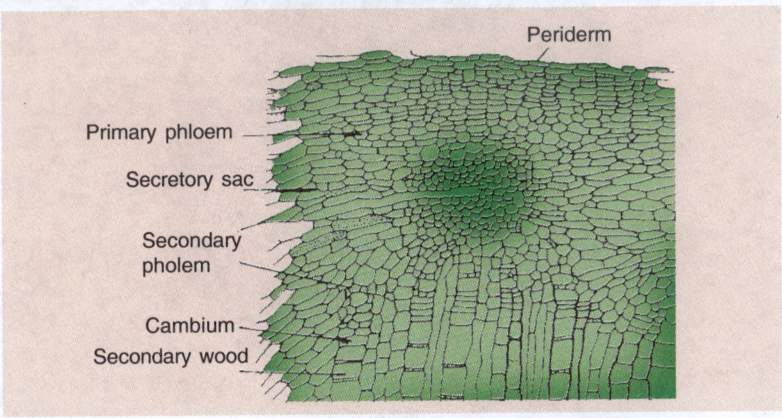

Fig. 2.4. *Lyginopteris oldhamia.* T.S. stem showing details of secondary vascular bundle. (After Williamson and Scott.

cells later become suberised. Next to the secondary phloem is the cambium. The secondary xylem is next to the cambium and forms a narrow zone in stems with smaller diameter and a comparatively wide zone in thicker stems. It consists of large tracheids with multiseriate bordered pits arranged on the radial walls. The bordered pits have angular borders. The secondary xylem rays are 1-12 cells wide and a few to many cells in height. The pits are irregularly arranged.

In some specimens of *Lyginopteris oldhamia,* internal secondary wood has also been reported. It lies internal to the primary xylem strands and is considered to have originated from a cambial strip lying just within the primary strands. It is considered to be an anomalous development.

Anatomy of Leaf

The pinules are covered with a distinct layer of upper and lower epidermises. The upper epidermis is cutinised and the cuticle is very resistant and well preserved. There are well defined palisade and spongy tissues that make up the mesophyll tissue. The stomata are restricted to the underside.

All the parts of the plant except the root are covered by numerous stalked glandular outgrowths. These glandular outgrowths are flask-shaped and may be up to 3 mm. high. They are not supplied by any vascular strand and are, therefore, considered as mere emergences. Each outgrowth bears an apical globular head that consists of a number of thin-walled secretory cells. It ranges in diameter from 0.12 – 0.4 mm. Below the secretory head there is a stalk made up of several layers of parenchyma cells. The gland is believed to secrete an oily or waxy material that protected the younger regions from fungal or bacterial attacks.

Anatomy of Root (Fig. 2.5)

A transverse section presents a roughly circular outline. It has an outer cortex made up of two to three layers of thin-walled cells. The cells had scanty contents. Next to the outer cortex is a comparatively wide inner cortex whose cells were also parenchymatous, but were full of dense contents. These are considered to be mucilaginous in nature. The endodermis and pericycle are also clearly seen in the prepared sections.

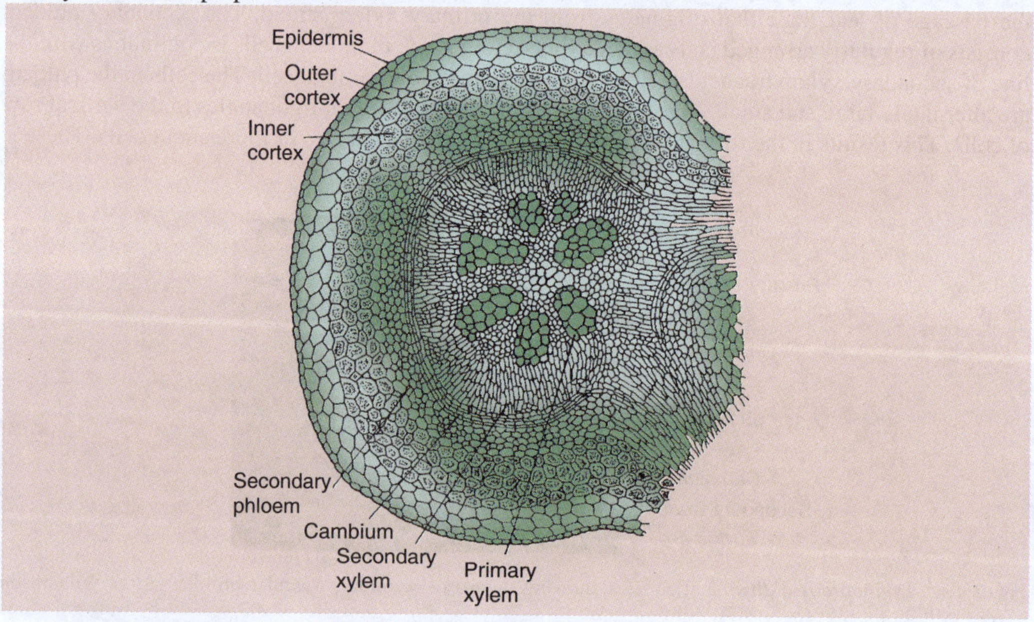

Fig. 2.5. *Lyginopteris oldhamia.*T.S. root.

Regarding its primary vasculature, the root possesses many xylem strands alternating with an equal number of phloem strands. The xylem is exarch *i.e.*, the protoxylem faces outwards and has spirally marked tracheids. The pith is scanty. The branches of root develop opposite to the protoxylem points.

The large roots show a secondary wood. It is traversed by large rays that always lie opposite to the protoxylem groups.

REPRODUCTIVE STRUCTURES

Regarding the reporductive features of *Lyginopteris oldhamia* there is no doubt that it is heterosporous and heterothallic. It bears seeds. The seeds and ovules are enclosed within protective structures called the **cupules.** These cupules have been found in connection with the fronds and also bore capitate glands, which confirms their connecion with the type species. The seeds and ovules were described under a separate name *Lygenostoma lomaxi*. The dispute is regarding the pollen organs or the microsporangia that have been described under the name *Crossotheca*. The *Crossotheca* was discovered by Kidston (1905) in the form of impressions in connection with the fronds genus *Sphenopteris hoeninghausi* which is the frond of *Lyginopteris oldhamia*. Hence *Crossotheca* is now regarded as the microsporangium of this species. Some recent investigations (Andrews, 1961, pp. 135-137) have cast some doubts regarding their validity. Arnold (1948, p. 214) has suggested *Telangium* as the microsporangiate structure of *Lyginopteris*. Irrespective of a doubtful position, the *Crossotheca* is considered as the microsporangiate structure in this text.

The Microsporangium (Fig. 2.6)

The microsporangia are borne on the ultimate branches of the frond. The pinnules in these ultimate branches became flattened with slight development of lamina and bore peripheral, pendent and elongate sporangia. The microsporangium is about 3 mm. long and bilocular. Each loculus is full of several microspores that ranged 50-70 μ in diameter. The microspores show a number of cells whose identity cannot be made out. They have tri-radiate markings (trilete). The sporangia lack annulus and resemble those of *Cycas*. On dehiscence, the microspores were probably carried by wind and some of them were lodged directly on the pollen chamber. They did not develop into pollen tubes as no evidence of their formation is available uptodate. In the absence of pollen tubes the sperms must have been motile and fertilisation followed pollination. There could be no prolonged interval between the two.

The Ovule and the Seed (Fig. 2.7)

Lagenostoma lomaxi (Oliver and Scott) is a barrel-shaped and radially symmetrical ovule with a single stout integument vascularised by nine vascular strands. The integument is not lobed and forms a single complete sheath penetrated at the apex by a narrow micropyle. The integument is fused with the nucellus except near the micropylar end where it is free from the nucellus and forms a canopy that surrounds and overhangs the pollen chamber. This canopy part of the integuments is divided into nine loculi with a vascular strand extending into each. This peculiar locular structure of the canopy is believed to contain water and supply it to the seed. The integument shows two distinct regions, the external **sclerotesta** which is hard and stony and an inner fleshy **endotesta**. The **sarcotesta** or the outer fleshy layer is not distinct. External to the integument, the ovule is partially enclosed by a lobed cupule (Figs. 2.7 and 2.8), which is studded with numerous glandular outgrowths. The seeds are released from the cupule by a basal abscission mechanism. The cupule is also vascularised. The nucellus is elongate and is not vascularised. The nucellus apex has a hollow pollen chamber (Fig. 2.7) or **Lagenostome**. The pollen chamber in this ovule is conical in shape and has a central core of tissue, shaped like an inverted bell (Figs. 2.7 and 2.9). This is known as the **central column** of the pollen chamber. The pollen chamber is formed as a result of formation of flask-shaped prolongation from the nucellar tip. This prolongation extends through the micropyle in the cavity between the free part of the integument and the nucellus. The central column arises from the base of the flask-shaped pollen chamber. The column leaves only a narrow space for the reception of the pollen grains. There is also a possibility that the narrow pollen chamber is formed by the disintegration of the nucellar tissue around the central column. The pollen grains are received in the pollen chamber and perhaps germinated there and liberated the sperms which fertilized the eggs in much the same way as in the living cycads. Long (1944) reported the female gametophytes containing several archegonia in the ovule *Lagenostoma ovoides*. *L. lomaxi* dates back to the Upper Carboniferous.

Pallet, J.M. (1966) observed the megaspore membrane resembling that of the exine of the pteridophytic spore. The outer spore wall was thick.

Fig. 2.6. *Lyginopteris oldhamia*. Part of a frond bearing microsporangia on the flattened pinnules. The microsporangia are pendent. The name *Corssotheca* is also given to the male organs (After Kidston).

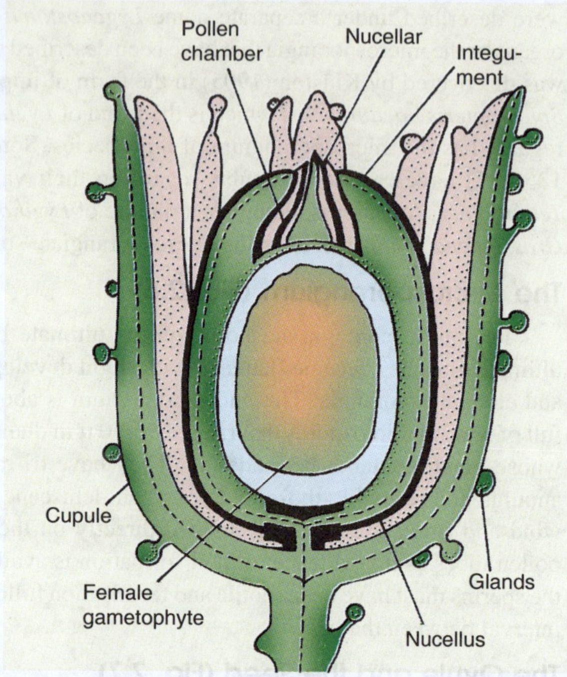

Fig. 2.7. V.S. ovule of *Lyginopteris oldhamia* (=*Langenostoma lomaxi)*. Note the glands on the cupule. (After Oliver and Scott).

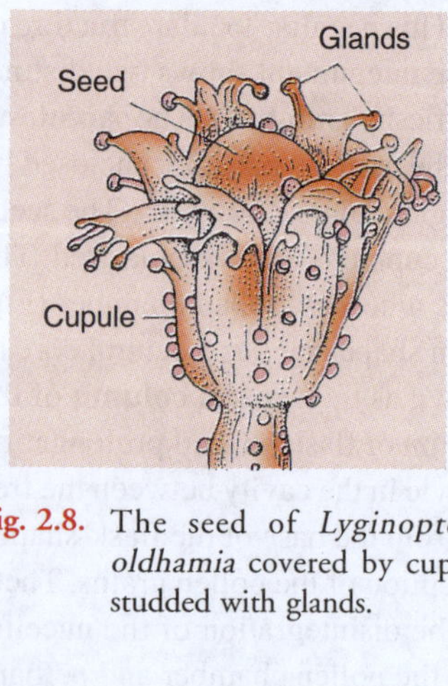

Fig. 2.8. The seed of *Lyginopteris oldhamia* covered by cupule studded with glands.

SOME OTHER FOSSILS ATTRIBUTED TO LYGINOPTERIDACEAE

A number of stems, fronds, seeds and pollen bearing organs have been described from Palaeozoic era that are attributed to the *Lyginopteridaceae*. A few important examples will be discussed here so as to enable the student to form a better idea about the group.

STEMS

The stem genera include : *Tetrastichia*, *Tristichia*, *Rhetinangium*, *Heterangium*, *Lyginopteris*, *Callistophyton* and *Schopfiastrum*. Some of these stem genera are described below : —

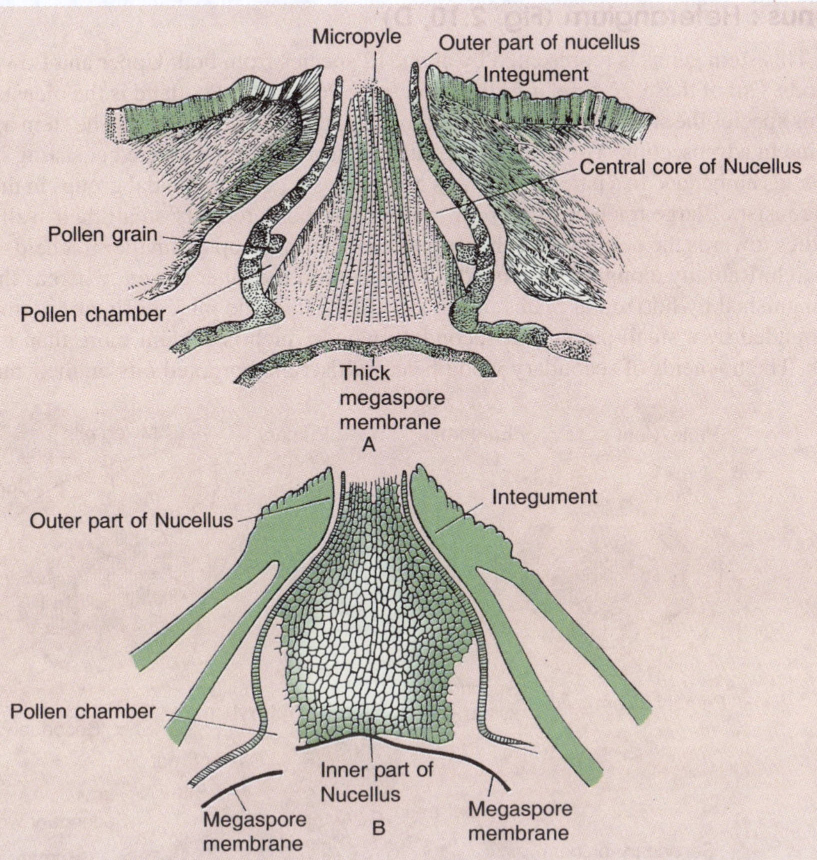

Fig. 2.9. *Lyginopteris oldhamia.* (A—B). Upper portions of the ovule showing central column and pollen chamber (After Williams).

Genus : Tetrastichia (Fig. 2.10, A)

Tetrastichia bupatides was discovered from the oil shale group of Scotland in the Lower Carboniferous. It is petrified petiolate axis. The stem is about 1 cm. in diameter and appears almost elliptical in a cross-section. The primary xylem is cruciform or 4 armed structure, which is made up of scalariform or reticulately pitted tracheids without any xylem parenchyma. There is no pith in the centre, hence the stelar organisation is protostelic. The protoxylem is mesarch and from it the branch trace protoxylems have their origin. The cruciform primary xylem is surrounded by a narrow zone of secondary xylem that has the same cruciate outline. The tracheids of secondary xylem were reticulate. The secondary medullary rays are both uniseriate and multiseriate. The stele is surrounded by a wide cortex that can be distinguished into outer cortex, middle cortex and inner cortex. The middle cortex contains sclereids or thick-walled groups of cells interspersed among thin-walled cells. The outer cortex has longitudinal network of sclerenchymatous fibres. The vascular bundles entering the petioles are butterfly-shaped.

Genus : Tristichia ovensii (Fig. 2.10, B)

The primary wood in this case has three radiating arms. The rest of the structure is similar to *Tetrastichia*. Seed bearing cupules belonging to the genus *Stamnostoma* were associated with this stem genus. These seeds are barrel-shaped and possess unlobed integument fused with the nucellus except at its apex, where it leaves an extremely wide **micropyle**. The seeds are enclosed within a cupule.

Genus : Heterangium (Fig. 2.10, D)

This stem genus is represented by about 12 species from both Upper and Lower Carboniferous periods. Out of these, *H. grievii* Williamson from Pettycur in Scotland is the oldest and best studied. In this species the stem is upto 4 cm. in diameter and is rarely branched. The stem appears angular in outline in a transection. The primary vasculature is a mixed protostele consisting of 6-12 groups of tracheids embedded in a parenchymatous ground tissue. The tracheidal groups in the centre are large and consist of large tracheids that bear multiseriate bordered pits on all their walls. The tracheidal bundles towards the periphery of the central core are made up of smaller tracheids and appear to be mesarch. Radially elongate fibre bands are present in the outer cortex, whereas the inner cortex is distinguished by horizontal plates of thick-walled cells. The mixed primary protostelic cylinder is surrounded by a small amount of secondary wood which is seldom more than a few millimeters thick. The tracheids of secondary wood bear multiseriate bordered pits on their radial walls. The

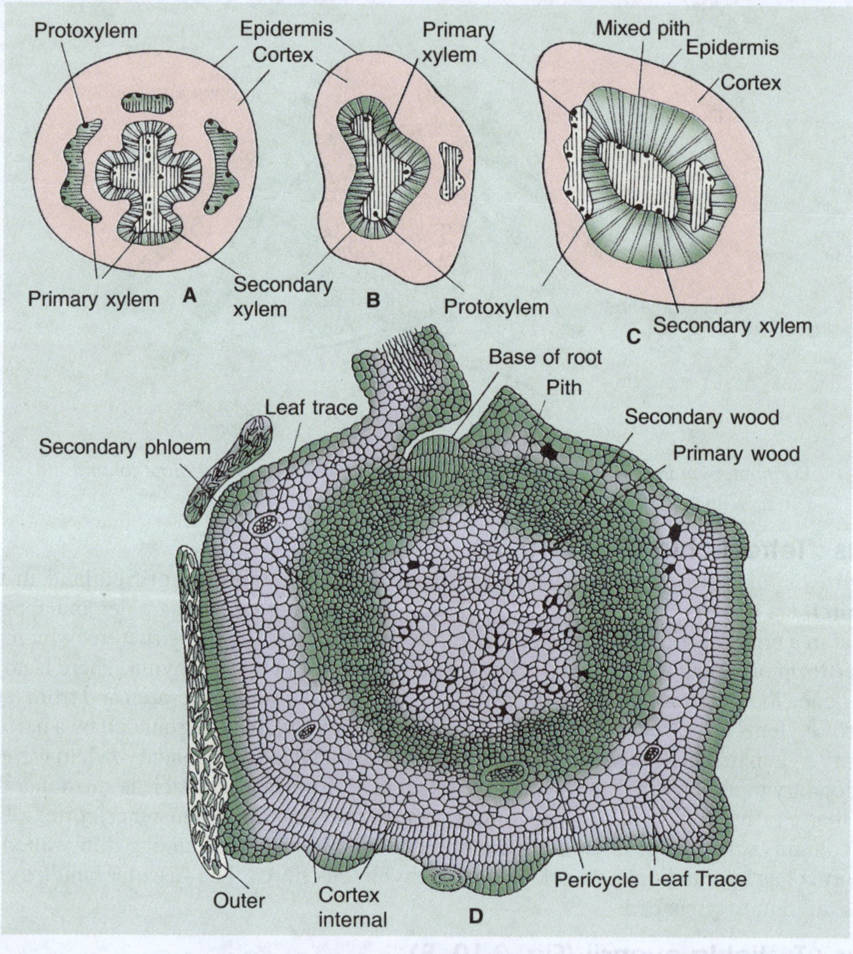

Fig. 2.10 (A—D). Some fossil stems attributed to Lyginopteridaceae A. T.S. stem of *Tetrastichia bupatides*. B. T.S. stem of *Tristichia ovensii*. C. T.S. stem of *Schopfiastrum decussatum*. (After Andrewg). D.T.S. stem of *Heterangium grievii*. (After Majumdar). (A, after Gordon; B, after Long).

secondary medullary rays are quite broad. These rays extend up to the primary xylem and may even merge with the parenchyma. The vascular supply to the leaf originates as a single strand at the periphery of the primary wood. It has a single protoxylem group but soon this protoxylem group divides into groups. The strand remains collateral during its course through the inner cortex but becomes concentric as it enters the base of the petiole. The leaves associated with it belongs to the genus *Diplonema adiantoides* (Andrews, 1961). In *H. schusteri* the leaf trace divides into two immediately on leaving the stele. In *H. shorense* and *H. tilaeoides* the leaf trace is a double structure, from the start and divides further during its outward course towards the leaf base.

Schopfiastrum decussatum. Andrews (Fig. 2.10, C) was discovered from the Des Moines series of Iowa. The stem is about 2 cm. in diameter and the primary vasculature is a mixed protostele surrounded by a wide zone of secondary wood with narrow rays. The tracheids have bordered pits. The fibrous strands in the outer cortex run parallel during their longitudinal course and do not form a network as in the other *Lyginopteris* stems. The two large traces arise on opposite sides of the primary xylem with one arising slightly above the other. The successive pairs of leaf traces alternate. This means that leaves arose in four rows.

Callistophyton poroxyloides. Delevoryas was discovered from the Upper Pennsylvanian coal balls of Illinois has exactly the same internal structure as the stem of *Lyginopteris oldhamia* except the rays in the secondary wood are comparatively narrower. The fibrous bands in the outer cortex do not form a network but run parallel. In this stem, the secondary phloem is well preserved and consists of phloem rays that are continuous with those of the secondary wood. The sieve cells bear simple sieve plates on their radial walls. Phloem parenchyma cells were also well preserved. The primary vasculature has several mesarch bundles surrounding a wide parenchymatous pitch which has scattered secretory cavities. The fronds arose in a spiral manner and bore axillary branches. The venation was circinate.

Seeds

The Lyginopteridaceae include the seed genera : *Sphaerostoma* (Fig. 1.23, A), *Salpingostoma*, *Calathospermum*, *Geminitheca* (Fig. 1.25, B), *Genomosperma* (Fig. 1.25, A), *Eccroustosperma*, *Hydrasperma*, *Stamnostoma*, *Eosperma*, *Dolychosperma*, *Lyrasperma* (Fig. 1.21, A), *Eurystoma*, *Deltasperma*, and *Camptosperma* belong to the Lower Carboniferous; whereas *Lagenostoma* (Fig. 1.22, B), *Physostoma*, *Genostoma* and *Tyliosperma* (Fig. 1.23, A), belong to the Upper Carboniferous. A few representative types will be discussed below.

Salpingostoma dasu Gordon (1941) was found in Lower Carboniferous and has a deeply lobed integument (Fig. 1.22, A), the lobes forming free apical processes above the limit of their fusion to the nucellus. Each lobe has a single vascular strand. The cupule is also lobed and comprises two main tissue layers, an inner thickened layer and an outer parenchymatous layer. Each lobe of the cupule is inervated by one or more vascular strands. The nucellar apex has a hollow, dome-shaped lagenostome or the pollen chamber, whose apex is prolonged into a long tubular structure, named by Gordon as *Salpynx*. The hollow lagenostome and salpynx are suggested to have resulted from the degeneration of the inner tissues of the nucellar apex. The degenerated nucellar tissue forms the pollination drop, in which the pollen grains are caught and ultimately germinate.

Sphaerostoma ovale (Williamson) Benson (Fig. 1.23, B) was discovered from Lower Carboniferous of Petty-cur, Scotland. The cupule closely invests the integument which is lobed at its distal end. Both the cupule and the integument are vascularised. The pollen chamber is almost conical. There are eight vascular bundles in the integument. The integument forms a canopy above the nucellar apex and has eight crested lobes each with a vascular strand. The nucellar apex was modified into a lagenostome in much the same way as described for *Lagenostoma lomaxi*, which is ovule genus of *Lyginopteris oldhamia*. The cupule contained only one seed that measured 3.5 *mm* × 2 *mm*.

Genomosperma kidstoni (Fig. 1.25, A) is a remarkable seed in that the integument is 6-11 lobed and is absolutely free from the nucellus. The nucellus is not vascularised. The integuments are traversed by 6-11 vascular strands. There is distinct lagenostome (pollen chamber) and minute salpynx.

Tyliosperma orbiculatum Mamay (Fig. 1.23A) was discovered from the Upper Carboniferous of the Des Monies series of Kansas. The seed is enveloped within a lobed cupule studded with blunt protuberances that were, of course, not glandular. The cupule is vascularised. The single integument is fused with the nucellus except at the apex where it is divided into seven lobes and is free from the nucellus. A few interesting features of this ovule are :—

(*i*) The integument is not vascularised, whereas the peripheral part of the nucellus is vascularised.

(*ii*) The nucellus and integument possess closely knit unit.

(*iii*) The epidermis of the integument is continuous over the distal end of the nucellus.

There is a distinct lagenostome. The ovule is radially symmetrical.

Physostoma elegans Williamson (Fig. 2.11) was discovered from the Upper Carboniferous of England. It has a single integument that is inseparably fused with the nucellus except at its apex, where it is free from the nucellus and is lobed. The number of lobes is commonly 10 but specimens with 9 and 12 lobes have also been encountered. The lobes are diverging. The outer layer of the integument is covered with numerous, closely set, large and club-shaped hair. The integuments lack sclerenchyma and are vascularised. The micropyle is wide, the pollen chamber is simple and has a characteristic large tent-pole like protuberance of nucellus.

Regarding the theories of origin and evolution of ovule and its phylogenetic importance, see chapter one.

Pollen-Bearing Organs (Fig. 2.12, A—D)

Crossotheca, *Telangium*, *Schuetzia*, *Alcicornopteris*, *Diplotheca*, and *Diplopteridium* are the pollen bearing organs of the *Lyginopteridaceae*. *Crossotheca* has been discussed in detail under *Lyginopteris oldhamia*.

Telangium Benson (Fig. 2.12, A—B) was found on petrified coal ball material of the Lower Carboniferous age along with the frond *Sphenopteris affinis* and *S. bifida*. Telangium *differs* from *Crossotheca* in that the sporangia stand upright on the terminal disc and continue the line of the rachis, whereas in *Crossotheca* they are pendent. In *Telangium,* there is a group of six to eight or more elongated sporangia that are united near their bases to form a synangial disc. The sporangia are bilocular and dehisced by means of a longitudinal split that appeared on the inner side of each sporangium. The sporangia contained many microspores.

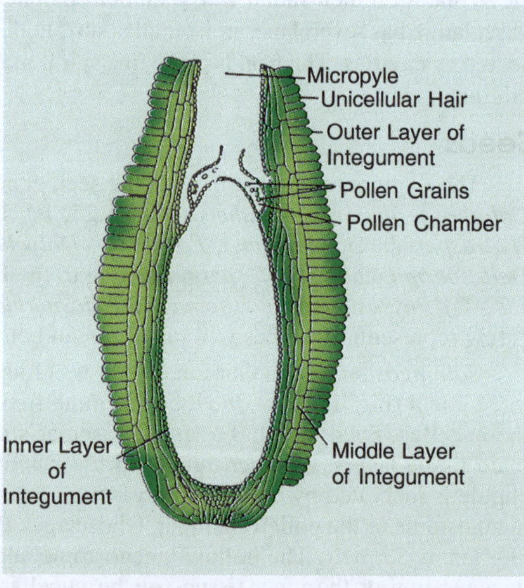

Fig. 2.11. V.S . ovule of *Physostoma elegans.* (After Williamson)

Diplopteridium telianum (Fig. 2.12, C). In this case, the synangial discs are borne terminally on widely forking rachules of a frond. The sporangia arises, from the margin of the discs and hang downwards, that is, they are pendent like those of *Crossotheca.*

Schuetzia bennieana (Fig. 2.12, D). In this case, about 15-20 sporangia were lightly pressed together and fused near their bases to form campanulate synangia. The synangia had stalks of their own and were arranged spirally on the fertile axis. Each sporangium was bilocular and produced numerous microspores with distinct tri-radiate marks. The sporangia stand erect.

In *Alcicornopteris,* there were about 15 sporangia fused at their bases and were attached to a peltate disc borne at the ends of a branching rachis.

In *Diplotheca,* the peltate synangial disc bore 10 sporangia arranged in five pairs.

Family : Medullosaceae

It is one of the families of the order Pteridospermales that was found in great abundance in the Upper Carboniferous and the Permian rocks. Some ancient members of the family have been reported from the Mississippian or Lower Carboniferous but became widespread and abundant during the Permian period. The family is named after the type genus *Medullosa* which is the stem. The family comprises the following organ genus: **Stems**, : *Medullosa colpoeylon* and *Sutchiffia*; **Leaves:** *Alethostesis* and *Neuropteris*; **Petiole** : *Myeloxylon*; **Microsporangiate fructifications** : *Paracalathios, Potoniea, Goldenbergia, Whittleseya, Anlatheca, Dolesotheca, Lacoea* etc.; **Seeds** : *Calathospermum, Pachytesta, Trigonocarpus* and *Stephanospermum.* The generic designation *Medullosa* (after which the family is named) is applied to the stems with or without petioles and other organs. The various organ genera named above occur separately and not in organic connection with the stems. They were discovered in the same or nearby localities and can be easily regarded as belonging to the genus *Medullosa*. Some seeds have been found in organic connection with the leaves.

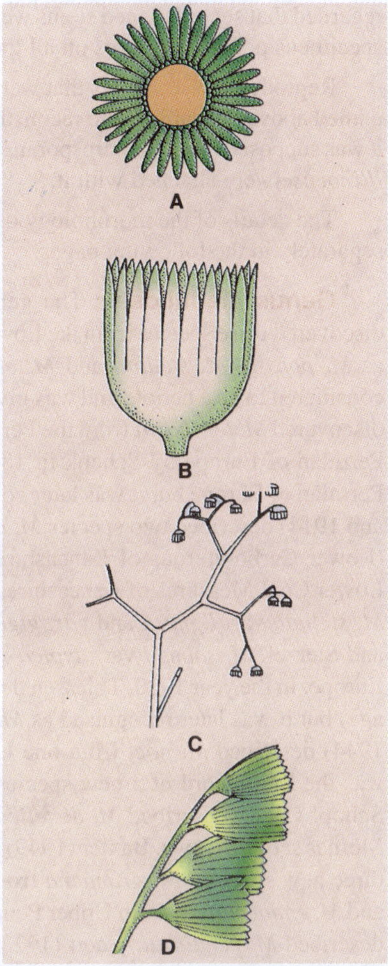

Fig. 2.12. (A-D) Pollen-bearing organs of Lyginopteridaceae. A-B *Telangium* C. *Diplopteridium telianum.* D. *Schuetzia benniena.* (C. after Walton; A, B, D, after Kidston).

Habit

The genus *Medullosa* has been reconstructed as a tree like form (Fig. 1.1. A) bearing a crown of large compound and pinnately divided leaves with long petioles. The older portions of the stem, below the crown, were clothed with large, decurrent and spirally arranged leaf bases of the fallen leaves. These leaf bases got sloughed off with age. The roots were adventitious and arose from the base of the stem. They were prop roots and were usually attached with the stem. The stems were unbranched and varied in length and diameter in various species. The portion of the stem below the crown was covered with periderm that got sloughed off with age or in the older regions of the stem. The pinnate compound leaves were spirally arranged. The diameter of the stem varies from two centimeters (*M. pusilla* from Lower Carboniferous) to 20 centimeters (*M. noei*). In *M. thompsonii,* the stems was 3.6 cm. in diameter whereas in *M. anglica,* it was 7 to 8 cm. In *M. distelica,* the stem measured 4 by 8 cm. Stems with a diameter of about 50 cm. have also been reported (*M. stellata vari. giganta*). Renault has reported stem pieces of *M. gigas* 50 cm. thick. In length the stem may be short and stumpy (*M. anglica*, *M. susilla*, *M. centrofilis*), 12 to 15 feet (*M. noei*) or as long as $3\frac{1}{2}$ meters to 8 meters (*M. stellata*). The long stems in *M. stellata* suggest it to be a liare (woody climber). The stem was distinctly angular in *M. anglica*, less so in *M. noei*, cylindrical in *M. stellata* and flattened in *M. centrofilis*. Delevoryas regarded that such flattened stems were merely crushed mechanically because certain well preserved specimens possessed leaves on all the sides of the stem.

Reproductive structures that were supposed to be attached to the plant include the organ genera named above. Regarding the reconstruction of the whole plant of *M. noei* by Andrews (Fig. 1.1. A) it was supposed that the microsporangiate fructification *Dolerotheca formosa* and the seed *Pachytesta Illinoense* were attached with it.

The details of the morphology of the various parts of the plant (organ genera) will be described separately in the following pages.

Genus Medullosa : The genus *Medullosa* was established by Cotta in 1832, when he discovered three species from the Lower Permian (Rotliegende) of Chemnitz. He named these species as *M. porosa*, *M. stellata* and *M. elegans*. The last named species was later (Brongniart, 1849) considered to be a petiole and was given the generic name *Myeloxylon*. Goeppert and Stenzel (1882) discovered *M. lenckartii* from the Permian of Saxony. *M. solmsii* was later described from the Lower Permian of Europe by Schenk in 1889. Renault (1893, 1896) described *M. gigas* from the lower Permian of Europe but it was later considered to be an "old growth stage" of *M. stellata*. Scot (1899 and 1914) described two species *M. anglica* (Lower Carboniferous of Great Britain) and *M. pusilla* (Lower Carboniferous of Lancashire). Miss de Fraine (1914) described *M. centrofilis* from the Lower Coal Measures of Lancashire. *Medullosa stellata* var. *typica*, *M. stellata* var. *corticata* and *M. stellata* vari *lignosa* and *var. gigantea* were later described from European localities by Weber and Sterzel. *M. solmsii* var. *typica* was also described by Weber and Sterzel from the Permian of Europe. In the year 1920, Thiessen described *M. anglica* from North America (Upper Pennsylvanian age) but it was later recognised as *M. anglica* var. *thiessenii* by Schopf in 1939. Steidtmann (1937, 1944) described *M. noei* from one Upper Pennsylvanian age at Calhoun, Illinois (America). This was the first record of a new species from America. Later from the same age and the same place Schopf (1939) described *M. distelica*. *M. thompsonii* was described by Andrews (1945) from Des Moines series in Iowa. Baxter (1949) gave an illustrated account of the Medullosaceae and described three new species : *M. primaena* from the Des Moines series of Iowa, *M. elongata* also from Iowa and *M. endocentrica* from Upper Pennsylvanian age in Illinois. From the same place, Stewart (1951) described *M. pendurata*, Scott (1923) suggested the name *M. geriensis* for *Heterangium geriense*. This species was later described properly by Boureau (1951), but Delevoryas (1955) regarded it to be synonymous with *M. anglica*. Roberts and Barghoorn (1952) described *M. olseniae* from Permian deposits of Texas. In the same year Stewart and Delevoryas described *M. heterstelica* from the Middle Pennsylvanian of the Kansas. *M. grandis* was described by Andrews and Mamay (1953) from a locality in Kansas.

Andrews and Mamay (1953) estimated that as many as 44 (forty-four) species and varieties of *Medullosa* have been described by various workers from Europe, Britain and America. These species have mainly been distinguished on the basis of number of steles and the presence or absence of secondary xylem in the leaf traces. Regarding the number of steles, it may be pointed out that their number changes at various levels in the stem. It may be distelic at the base and the number may rise to polytelic at higher levels. Such an observation was made by Delevoryas (1955) and reduced the number of American species. As a result of his detailed study of the stem structure of *M. primaeva,* he concluded that the following taxa are merely growth forms of this species :— *M. heterostelica, M. thompsonii, M. distelica, M. grandis, M. elongata* and *M. anglica* var. *ioensis*. These were all regarded synonymous with *M. primaeva*.

M. noei, *M. primaeva*, *M. endocentrica*, *M. olseniae*, *M. anglica* var. *thiessenii* are now recognised as valid North American species.

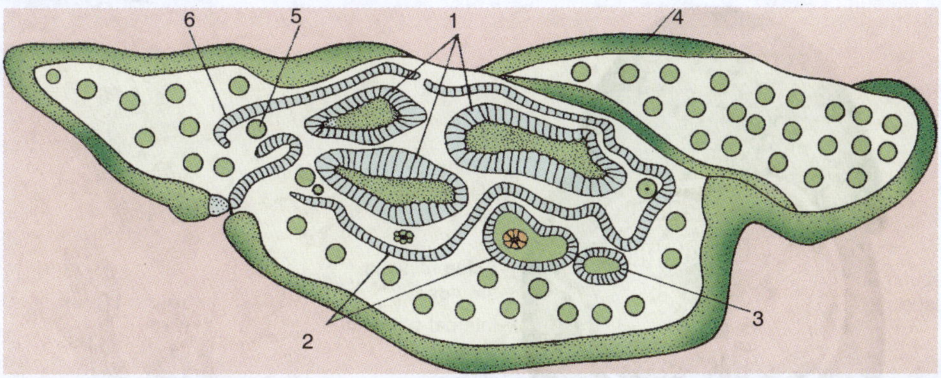

Fig. 2.13. T.S. Stem of *Medullosa anglica* 1. The three steles; 2. Secondary vascular strands; 3. Accessory vascular strands; 4. Outer Cortex containing sclerenchymatous strands; 5. Leaf traces; 6. Periderm; (After Scott)

Out of the large number of species that have been investigated upto now, *M. anglica* Scott can be selected as an example to describe the details of structure. This species was discovered from the Lower and Middle Carboniferous of the British Isles (Lancashire) and possessed an unequally triangular stem 7 to 8 cm. in diameter and almost completely clothed with stout decurrent bases of large, spirally arranged, pinnately compound leaves.

A transverse section of the stem (Fig. 2.13) shows three separate steles (tristelic) that are irregular in shape. Each stele has a central core of **primary xylem** surrounded by secondary xylem and phloem (Fig. 2.13). There is a distinct **cambium** external to secondary xylem. The three steles, as they extend along the length of the stem, fuse at certain places and again redivide (into three separate steles) as individual steles. The tissue between the steles is crushed and disorganised and it is supposed that in the living plants, it was in very small amounts. This tissue cannot be compared to the leaf gaps. Structurally each stele has a structure similar to the stele of *Heterangium*. A.C. Seward has regarded *M. anglica* to be a polystelic *Heterangium*. The central core of each stele consists of primary xylem tracheids and parenchyma. There is no pith. The xylem is **mesarch** *i.e.* the **protoxylem** is surrounded by **metaxylem**. The protoxylem consists of reticulate and spiral tracheids whereas the metaxylem tracheids show pitted and scalariform thickenings. The metaxylem tracheids measure 150 μ in diameter.

The secondary xylem consists of tracheids with multiseriate pittings and are arranged in 2-4 radial series alternating with medullary rays, 1-3 cells broad and several cells in depth. The principal rays are continuous with the parenchyma present in the central core. The secondary phloem is made up of tubular elements that are regarded as sieve-tubes. Each of the three steles increased in diameter due to the activity of a cambium. Each stele had its own **cambium**. The increase in the diameter of the stele leads to the increase in the diameter of the stem. The amount of secondary tissue formed by the activity of the cambium was more on the side of the stele facing the centre of the stem.

The leaf traces arose from the outer face of each stele. Each leaf trace is at first a concentric bundle with primary xylem in the centre surrounded completely or partially by the secondary xylem and secondary phloem. The primary xylem had one or more protoxylem strands. During its course through the cortex towards the leaf, the leaf trace loses its secondary tissue and divides into small collateral bundles with exarch protoxylem. Such bundles were composed mainly of spiral and scalariform tracheids. Each leaf received several leaf trace bundles. It is estimated that a large leaf base received as many as 70 to 80 leaf trace bundles.

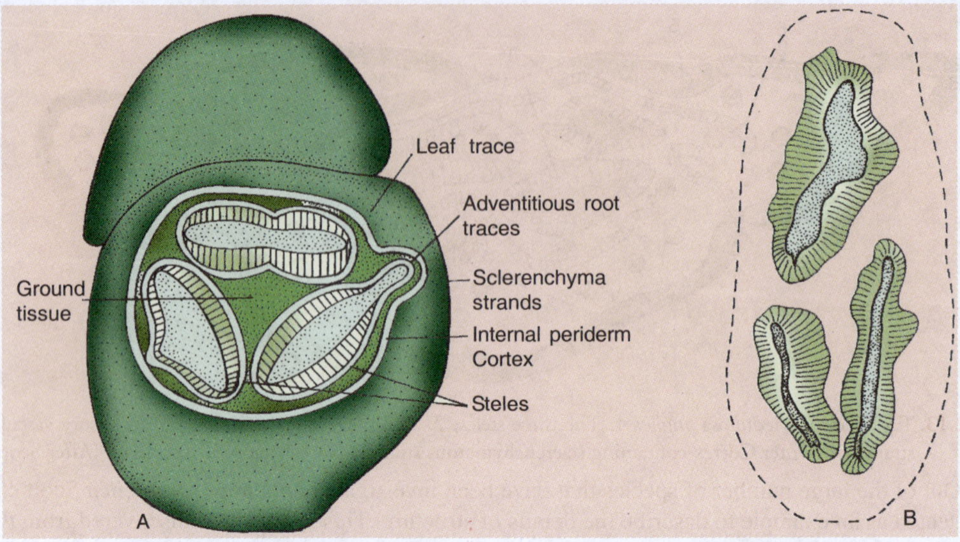

Fig. 2.14. (A—B). A. T.S. of three steled *Medullosa noei*. B. T.S. stem of *Medullosa noei* from Pennsylvanian of Illinois (A. after Stewart and Delevoryas; B. after Andrews)

The **Cortex** is divided into an outer and inner cortex by the presence within it of a layer of internal **periderm** that surrounds the stelar region. The cortex is made up of parenchymatous cells that are seldom crushed. Large number of gum canals are present in the cortex. The outer cortex contains large number of simple leaf trace bundles. Accessory vascular strands are also present in the outer cortex. These strands are supposed to be of secondary origin and are comparable to cortical bundles found in some living cycads. These accessory bundles contain a parenchymatous core with scattered tracheids and surrounded by secondary xylem and phloem. They attain a diameter of 7 mm. Sclerenchymatous strands are also found along the outer boundary of the cortex.

The **periderm** is considered to be a tissue composed of **phellogen** which gave rise to the **phellem** towards its outer side. The periderm is interrupted here and there by the leaf trace strands and the roots. Its suberised nature has not been established. Isolated rings of periderm are also present in the cortex (Fig. 2.13).

Medullosa noei(Fig. 2.14) is another species with tristelic stems. In a transverse section, the three steles appear elongated. Each stele has a central part consisting of primary xylem. The primary xylem is composed of large tracheids and abundant parenchyma. The primary xylem is distinguishable into metaxylem and protoxylem. The protoxylem occupies the edge of each stele *i.e.,* it is exarch in position. The primary xylem is surrounded by secondary xylem tracheids. The secondary xylem usually develops more towards the centre and has many vascular rays. External to the secondary xylem is the cambium and then the secondary phloem. The steles are embedded in the extensive ground tissue which is parenchymatous and is traversed by numerous secretory canals. The cortex is also well developed and surrounds the three steles. It is made up of parenchymatous cells except for its peripheral region which is composed of sclerenchymatous strands. The cortex is traversed by numerous leaf traces which arise from the steles at various levels. At a given level many of them can be observed. Secretory canals are also present in the cortex towards its outer boundary. A distinct periderm develops in the cortex just external to the steles and surrounds them (steles) completely. The periderm develops as a thin band in young stems but in older stems, it is well developed and thick. A number of thick-walled cells form sclerotic-strands that cut across the cortex to mark the position of the separation of the leaf base from the stem.

As the stems of this species grew old the steles increased in size, the cortex sloughed off and the periderm became thick and formed the outer limiting layer of the stem. An interesting observation is that in older stems the steles, besides showing greater development of the secondary xylem also show increased development of the primary xylem in the centre. This greater development of the primary xylem has been regarded to be due to the proliferation of parenchyma. These parenchymatous cells probably retained their meristematic capacity and formed more primary xylem as the stems grew older. The secondary xylem, as in all other species, is formed by the activity of a cambium, which is present in each stele. Another interesting feature in this species is that as the steles undergo secondary growth, the distance between them becomes greater *i.e.* they grow farther apart. Normally the steles should abut on each other in the centre of the stem. This is also due to the proliferation of parenchymatous tissue of the ground tissue in which the steles are embedded. Accessory rings of periderm have also been noticed in the ground tissue. Delevoryas (1962) is of the opinion that every living cell in the stem of *M. noei* retained the capacity to divide and there was ample proliferation in the various regions of the stem. He has also reported the proliferation of ray parenchyma cells which led to the formation of secondary parenchyma in the secondary xylem. He has also opined that the periderm in older stems increased in thickness due to the proliferation of some living parenchyma cells within it. He rules out the presence of a distinct phellogen.

The leaves in this species were spirally arranged as the stem.

Andrews described another tristelic species from the upper carboniferous of Iowa. He named it as *M. thompsonii*. The three steles are surrounded by a continuous ring of periderm which is made up of thin-walled radially elongated cells. Two of the three steles are larger and measure 7 × 3.5 mm. in diameter whereas the third smaller one is 2 mm. in diameter. Each stele has a central core of primary xylem tracheids and some parenchyma surrounded by the large pitted tracheids of secondary xylem and vascular ray cells that are parenchymatous. The secondary xylem is surrounded by secondary phloem. The three steles are embedded in a parenchymatous ground tissue. External to the periderm is the parenchymatous cortex traversed by many leaf trace bundles that arise from the steles at various levels. The cells at the periphery of the cortex are sclerenchymatous. They form circular to radially elongated fibrous strands which run longitudinally and may anastomose at certain levels. This region of the cortex also contains a few secretory canals.

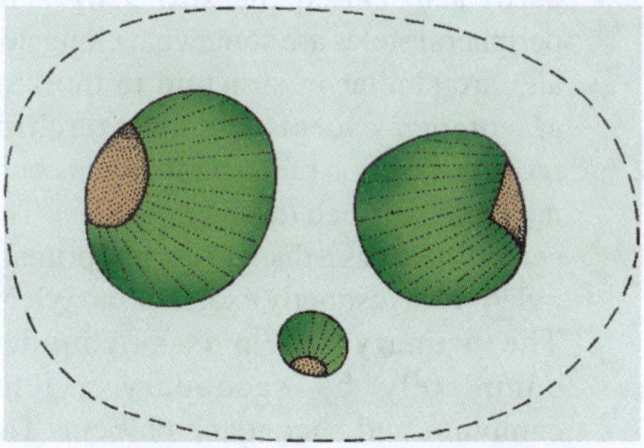

Fig. 2.15. T.S. of stem of *Medullosa endocentrica* showing three steles that are endocentric. Two steles are larger whereas one is smaller.

The petiole of this species consists of large number of small, scattered vascular bundles embedded in a common ground tissue. The outer cortex in the petiole is sclerenchymatous but is traversed by comparatively large number of secretory canals.

The other tristelic species include *M. endocentrica* (Fig. 2.15) discovered by Baxter from Illinois with two larger and one smaller steles that are extremely endocentric with primary xylem not completely surrounded by the secondary xylem (Fig. 2.16). In this species, there is no centrifugal development of the secondary vascular elements. They only develop centripetally. It has been suggested that *M. endocentrica* is merely an apical portion of *M. noei*. Another species with tangentially expanded endocentric steles is *M. elongata* described by Baxter from Des Moines series in Iowa. In this species, the stems are dorsiventrally flattened. *M. pusilla* with stem 2 cm in diameter is also tristelic. It was discovered by D.H. Scott in 1909 and described fully in 1914. The material was found in Lower Coal Measures of Colne in Lancashire. Each stele is 3 mm in diameter and has central core of triangular primary xylem with reticulate tracheids and a small amount of parenchyma, surrounded completely by secondary xylem, cambium and secondary phloem. The protoxylem is either exarch or mesarch. The three steles are surrounded by a ring of periderm. The cortex is traversed by many leaf traces that may possess a little or no secondary xylem. The outer cortex is sclerenchymatous. This species resembles *M. anglica* in many respects. Scott suggested *Allethopteris decurrens* to be the probable foliage of *M. pusilla*.

M. Distelica discovered from the McLeansboro of Illinois is a species with two steles. The stem measures from 4 to 8 cms in diameter and each of the two steles measuring 7 × 18 mm. The steles are asymmetrical and have a distinct primary xylem in the centre. The primary xylem tracheids are mixed with parenchyma. The secondary xylem completely surrounds the primary xylem and develops more towards the centre of the stem. Both the steles are surrounded by a continuous common ring of internal periderm. The secondary xylem towards the cortex or towards the outer side is very thin. The stem is surrounded by three decurrent leaf bases. The internodes are long. The leaf traces depart from the steles and may possess little or no secondary tissue. The cortex is well developed and parenchymatous except for the peripheral portion which is fibrous. The steles are embedded in the parenchymatous ground tissue internal to the periderm ring.

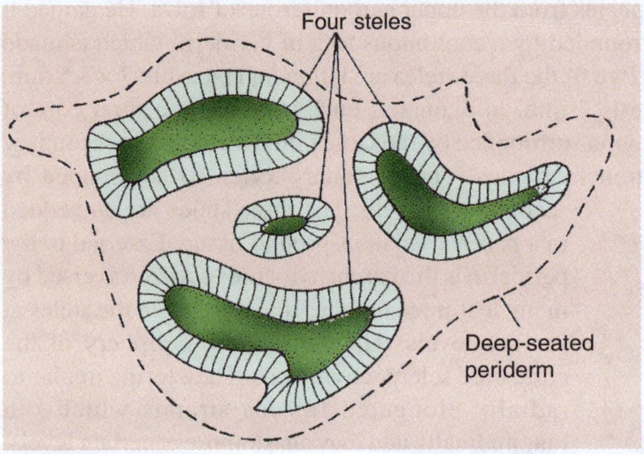

Fig. 2.16. T.S. stem of *Medullosa centrofilis* showing four steles (After de Fraine)

M. centrofilis was discovered by Miss de Fraine from the Lower Coal Measures of Lancashire. The stem was petrified and alongwith the four decurrent leaf bases measured 5 cm in diameter. In transverse section, the stem shows three steles enclosing a small central stele (Fig. 2.16) also called the **star ring.** The peripheral steles are somewhat elongated and are similar in structure to those of *M. anglica* except that the exarch or mesarch position of the protoxylem could not be established (as in *M. pusilla*). The central stele has a distinct core of primary xylem with distinctly exarch protoxylem. The primary xylem is surrounded completely by secondary xylem, cambium and secondary phloem. The peripheral steles may anastomose at certain levels of the stem but the central stele follows an independent course. The leaf traces arise from primary xylem of the peripheral steles and lack secondary tracheids (as in *M. pusilla*). The peripheral or the outer steles are surrounded by a continuous layer of periderm. Miss de Fraine regarded this layer as secondary cortex as its cells are thin-walled and do not contain cork. The outer cortex, besides leaf traces, is traversed by a number of secretory cells. The peripheral region of the cortex is fibrous.

Roberts and Barghoorn (1952) described *M. olseniae* from the Permian of Texas. This species has 3 to 5 steles that are tangentially expanded. The details of structure are not known because the available material is fragmentary.

In 1952, Stewart and Delevoryas described a specimen of *M. primaeva* (which they named as *M. heterostelica* but later referred it back to *M. primaeva*) from a coal ball of south eastern Kansas (Middle Pennsylvanian). They studied the details of internal structure for the whole length of the specimen and found variations in the stelar number. At the lower level, there are two steles but at higher levels (nodal region), the steles divided into as many as 20 to 23 steles (Fig 2.17) and still higher, the steles again fused to form two steles. Some of the branch steles bend outwards into the cortex and function as leaf traces. Secondary vascular elements may or may not be present in the leaf traces. Each stele may be roughly circular or oval in outline and possesses a central core of primary xylem with exarch protoxylem and some parenchyma. It is surrounded by secondary xylem, cambium and secondary phloem. All the steles whether two or many are surrounded by a common periderm ring. External to the periderm is the cortex made up of parenchymatous cells and traversed by numerous leaf trace bundles that are entirely primary in some regions especially where the steles were not undergoing branching, elsewhere secondary tissue may be present. Secretory cells are also present in the outer cortex. The peripheral portions of the cortex are made up of sclerenchymatous cells. Stewart and Delevoryas have reported that enlargement of the primary xylem might have occurred during the development of the stele. This according to them is due to the proliferation of the parenchyma in the primary xylem. In some large specimens of *M. primaeva,* some of the steles become considerably elongated in a transverse section, whereas others remain small and cylindrical.

Stewart and Delevoryas, after studying the sequence of development in *M. primaeva*, suggested the merger of *M. heterostelica*, *M. thompsonii*, *M. distelica*, *M. grandis* (Fig. 2.18), *M. elongata* and *M. anglica var ioensis* with *M. primaeva*. These species are regarded as various growth stages of *M. primaeva*.

Another interesting species is *M. stellata* (Fig. 2.19) described by Cotta from the Lower Permian of Chemnitz (Germany). The specimens of this species collected from this locality, by various authors (Goeppert, Goeppert and Stenzel, Weber and Sterzel), showed a great variation in their anatomical details. They have been described as varieties of *M. stellata*. Four well defined varieties have been described. These are *M. stellata* var. *typica* Cotta, *M. stellata* var. *corticata* Weber & Sterzel, *M. stellata* var. *lignosa* Weber & Sterzel and *M. stellata* var. *gigantea* Weber & Sterzel.

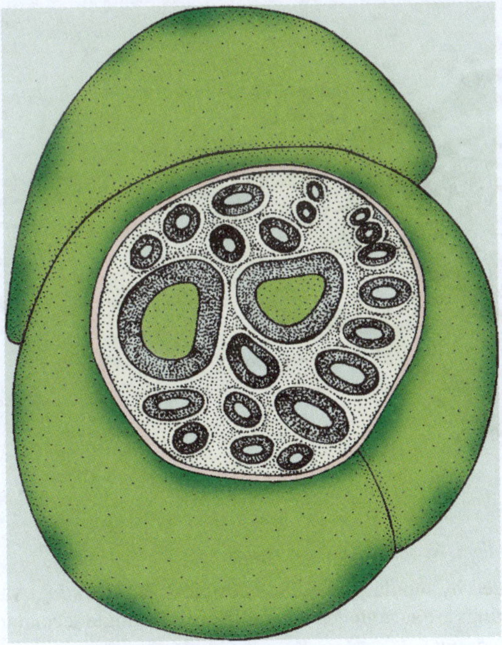

Fig. 2.17. *Medullosa primaeva.* T.S. stem showing many steles. (After Stewart and Delevoryas).

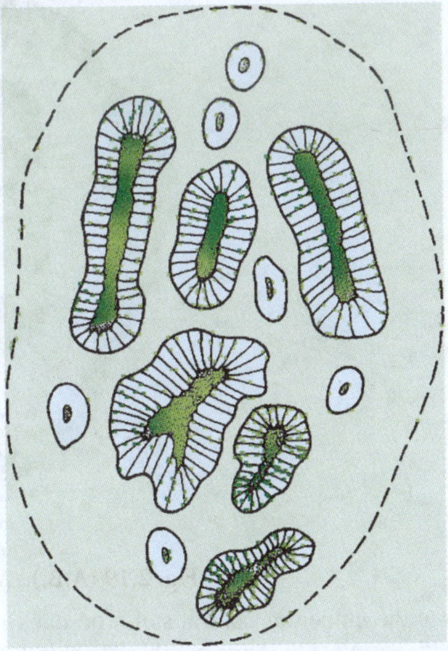

Fig. 2.18. T.S. stem of *Medullosa grandis* showing many steles from Pennsylvanian of Kansas (After Andrews).

Medullosa stellata var. *typica* Cotta represents the type species (Fig. 2.19 A). The stem in this variety is almost cylindrical measuring 24 × 34 cm. A transverse section of the stem shows many small cylindrical steles embedded in the centre. These are surrounded completely or sometimes incompletely (Fig. 2.19 A) by a tubular stele that forms a complete or incomplete cylinder around them. The central steles are also known as **star rings** and each one of them consists of a central core of primary xylem mixed with parenchyma. Primary xylem is surrounded by secondary xylem which in turn is surrounded by cambium and an external ring of secondary phloem. In size these smaller central steles vary in diameter from 3 mm to 2.5 mm. The secondary xylem is well developed and is

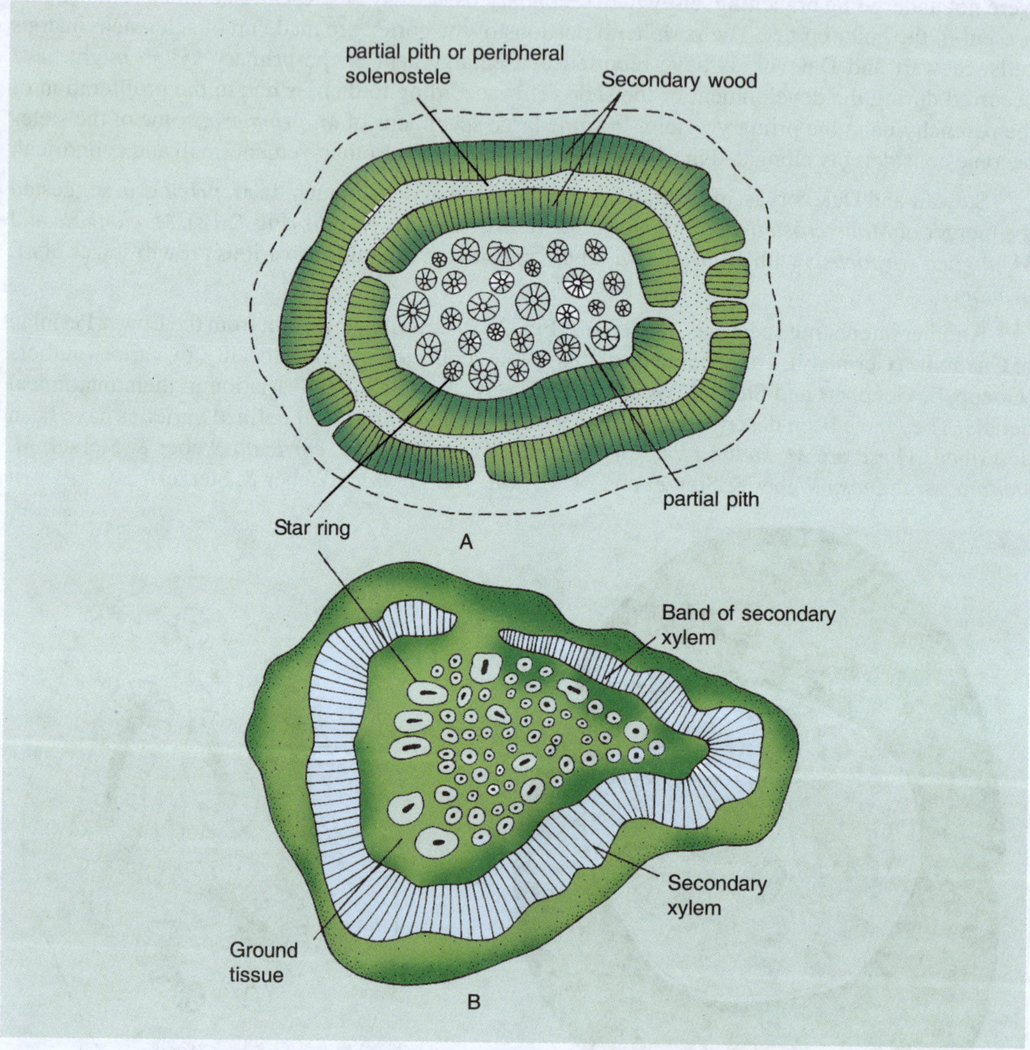

Fig. 2.19 (A,B.). *Medullosa stellata*. A. *M. stellata* var. *typica*.
Note the numerous central steles or star rings surrounded by tubular steles (Solenosteles). B. *M.stellata* var. *corticata*. Note the large number of central steles or star rings are surrounded incompletely by a single secondary xylem ring. (A. after Stewart & Delevoryas; B. after Andrews)

traversed by secondary medullary rays. They are embedded in the central ground tissue which is always parenchymatous. These central steles are surrounded by a narrow cylinder of inversely oriented secondary xylem and phloem (Fig. 2.19 A), the phloem being on the inner side of the xylem. External to the secondary xylem is a ring of parenchyma mixed with tracheids of the primary xylem that have spiral, scalariform or reticulate thickenings. This ring has been designated as '**partial pith**'. Actually it represents the primary xylem. External to this partial pith is the normally oriented ring of secondary xylem and secondary phloem, with phloem external to the xylem. Cambium is present in between the phloem and xylem rings. These two concentric rings separated by the 'partial pith' or the primary xylem are similar in organisation to a **solenostele** of the modern ferns, except that they are secondary in nature. The frequent breaks in the vascular cylinder are not connected with the departure of leaf traces and should not be regarded as foliar gaps. Delevoryas is of the view that "A tangential fusion (phyllogenetically) of the peripheral steles of some polystelic forms like *M. leuckarti* (Fig. 2.20) would result in a continuous cylinder with a ring of primary xylem and both centripetal and centrifugal secondary xylem."

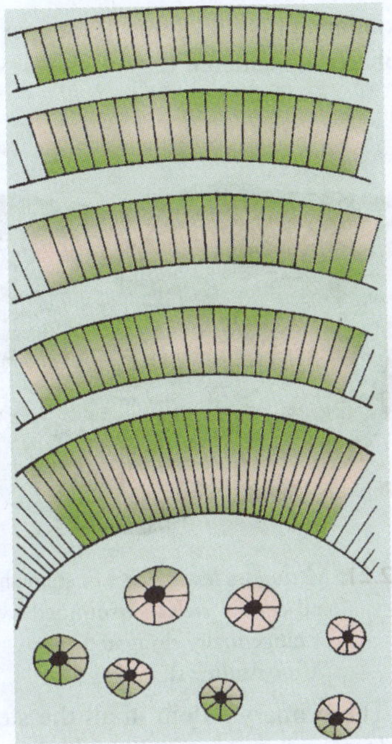

Fig. 2.20. *Medullosa stellata var. gigantea* showing concentric stele with outer centrifugal secondary steles. In the centre are star rings.

The leaf traces arise from the primary xylem at the edge of the so called "partial pith" of the outer cylinder. During its outward passage through the secondary xylem of the outer stellar ring, the leaf trace is surrounded by cambium, secondary xylem and secondary phloem. As it traverses through the cortex , the leaf trace looses its secondary xylem and phloem elements and is only primary in nature. Before entering the petiole, it branches successively into many small collateral bundles which enters the petiole.

The cortex is parenchymatous and contains secretory canals and large number of leaf traces. Strands of sclerenchyma are also present in the cortex.

In *M. stellata* var. *corticata* (Fig. 2.19B) there are numerous central steles as compared to var. *typica*. Some of these central steles are cylindrical (star rings) whereas some of them may be large and tangentially flattened (plate rings). Their structure is similar to that of var. *typica* and are surrounded by an outer main stele which forms an incomplete ring (Fig. 2.19 B). The rest of the structure is similar to the other variety discussed above.

In *M. stellata* var. *lignosa* the number of central steles is reduced to one whereas the rest of the structure is similar to var. *typica*.

In *M. stellata* var. *gigantea* (Fig. 2.20) stem measures upto 50 cm in diameter with as many as 43 large and small central steles (Fig. 2.20) that are surrounded by a complete cylindrical stele as in var. *typica*. This stele is further surrounded by successive concentric bands of normally oriented secondary xylem and secondary phloem (secondary phloem is present only towards the outer side of xylem) that were produced by successive cambial rings either cortical or pericyclic in origin. These bear resemblance to a similar condition in the modern cycads *e.g*. species of *Cycas* and *Macrozamia*.

In *M. leuckarti* (Fig. 2.21) described by Goeppert and Stenzel there are several cylindrical steles in the centre. These are surrounded by a varying number of large tangentially elongated steles. They form a wavy ring (Fig. 2.20) also called a snake ring. In some forms there is a single set of snake rings whereas in others there may be two series of snake rings. Leaf traces are given off from the outer edge of the primary xylem. Radial rows of thick-walled cells have been reported in the secondary phloem. Some (Goeppert and Stenzel) regard them as fibres whereas others (Solms-Laubach) considered them to be sieve tubes.

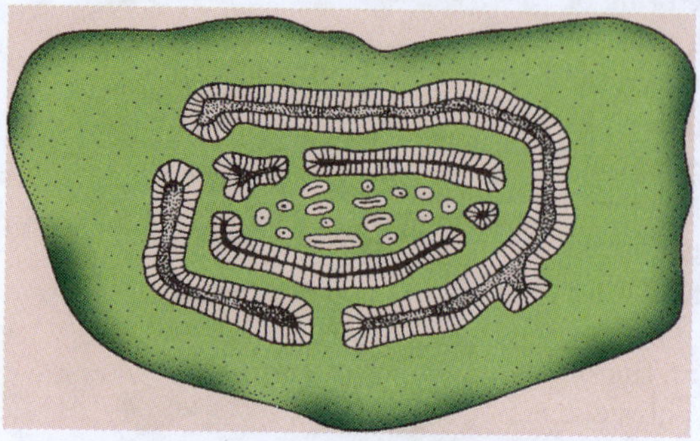

Fig. 2.21. *Medullosa leuckarti.* T.S. stem showing several central steles surrounded by two rings of tangentially elongated called snake rings. (After Andrews).

Fig. 2.22. T.S. stem of Medullosa leuckarti showing central steles surrounded by a single ring of large and tangential steles. (After Stewart).

The primary xylem in all the steles is very clearly preserved and may be exarch or mesarch. Periderm ring encircles the stelar region and several leaf traces traverse the cortex. *Myeloxylon landrioti* is the petiole.

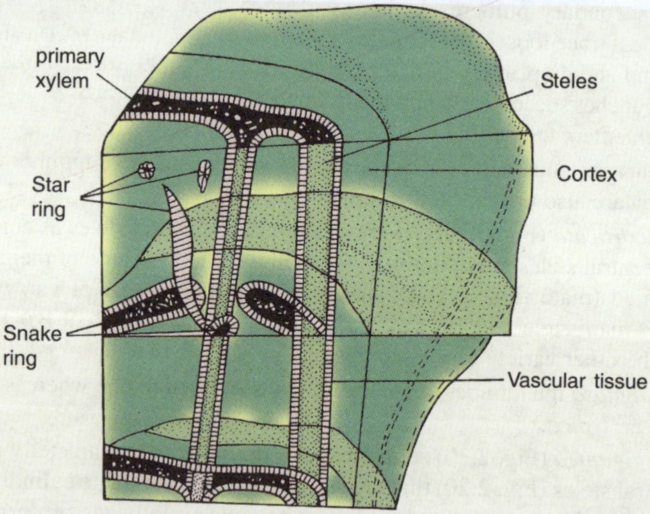

Fig. 2.23. *Medullosa solmsi.*

In *M. solmsi* (Fig. 2.24) the stem has a large pith or axial (central) region containing a number of inconspicuous central cylindrical steles, which are surrounded by two concentric zones of separate tangentially flattened and elongate steles (also called plate rings). Each stele has its own primary xylem in the centre surrounded by secondary xylem, cambium and phloem. It can be compared to the dictyostele of many ferns (in ferns, the stele is primary in nature but in *Medullosa,* it is always secondary). The leaf trace bundles arise as strands from the inner circle of plate rings and pass through the gaps in the outer rings, eventually breaking up in the cortex into numerous collateral bundles. In *M. solmsi* var. *lignosa,* the central steles are surrounded by an inner circle of larger steles which are plate like flattened steles like the type species, but the outermost ring of steles is markedly asymmetrical and larger in size. In them the secondary xylem is excentrically developed *i.e.*, the outer xylem and phloem are much broader than the inner. In addition a number of concentric bands of normally oriented vascular tissue are developed as is found in *M. stellata* var. *gigantea*. The secondary wood in this species is much more compact than in other species.

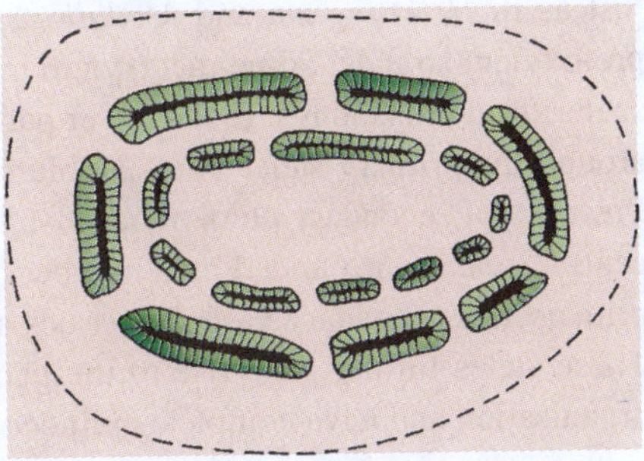

Fig. 2.24. T.S. stem of *Medullosa solmsi.*

In *M. gigas* which was discovered by Renault from the Permian of Artun, the stem reaches the diameter of more than one metre (See Andrews 1967, page 146). A transverse section of the stem reveals a few steles in the centre that are surrounded by a cylinder of secondary vascular tissue in which the secondary xylem of the outer ring is highly developed and it appears as if the entire stem is occupied by it. The internal xylem cylinder is feebly developed and is hardly recognisable.

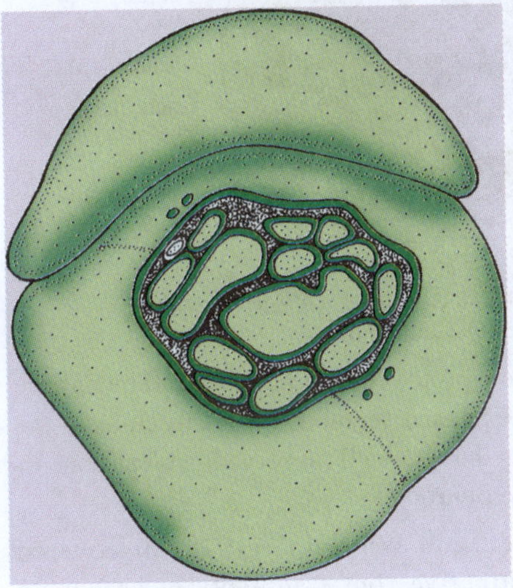

Fig. 2.25. T.S. Stem of *Sutcliffia insignis* (After Scott).

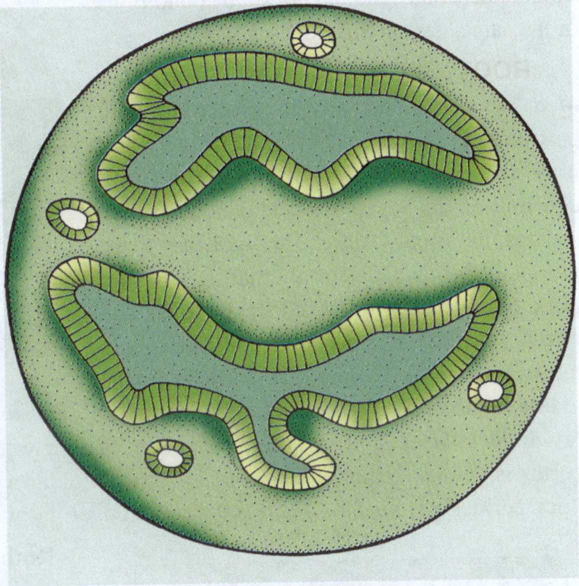

Fig. 2.26. T.S. Stem of *Colpoxylon acudense* (After Renault).

In *M. porosa,* the anatomy of the stem resembles to some extent that of *M. stellata*. In this species the numerous cylindrical central steles are surrounded by an interrupted ring of larger steles, which have feebly developed secondary tissues. This ring is further completely enveloped by continuous cylinder of secondary vascular tissue as in *M. stellata*.

Sutcliffia insignis Scott. (Fig. 2.25). It was first discovered by Dr. D.H. Scott from the Lower Coal Measures of Shore, Lancashire in 1906. The specimen was named after W.H. Sutcliffe who was the owner of the colliery. Later it was described by Miss de Fraine from the same locality. Phillips and Andrews (1963) described it from U.S.A. The stem as described by Scott, attained a diameter of 12 cms. A transverse section reveals a broad cortex composed of parenchymatous cell and contains several secretory sacs and ducts and strands of mechanical tissue. A deep seated periderm has been recorded by Miss de Fraine. In the centre of the stem, there is single large stele consisting of primary exarch protostele surrounded by a cylinder of secondary wood. The primary protostele is made up of large tracheids with bordered pits embedded in an "anastomozing system of parenchyma" just as in *Heterangium* and *Medullosa anglica*. It differs from these two in possessing exarch protoxylem strands. Some metaxylem tracheids bordering the protoxylem tracheids have spiral or scalariform thickenings. In the lower parts of the stem, the presence of cambium has been recorded around the primary stele. This cambium added lot of secondary xylem around the primary core. Presence of secondary phloem has also been recorded at places along the length of the stem. The central stele is surrounded by a number of meristeles of varying sizes. Each meristele, has its own primary and secondary vascular tissues and arose from the main stele. All of them are concentric. These steles further gave rise to the petiole traces and leaf traces. All of them are concentric in organisation and have complete cylinder of secondary wood.

Colpoxylon (Fig. 2.26) is another stem genus described from the Permian of France. It differs from *Medullosa* in having single, large and irregular central stele throughout the length of stem. At certain places, this central stele divides to give rise to smaller meristeles. The genus presents a mixture of monostelic and polystelic conditions in the same stem at different levels.

ROOT (Figs. 2.27 and 2.28). The roots in *Medullosa* developed adventitiously and may be triarch or tetrarch. They branch freely and may attain a diameter of more than one centimeter. In *M. anglica* and some other species, the root is triarch that is, it has three groups of primary xylem alternating with the same number of primary phloem groups. The primary xylem is exarch. In *M. noei,* the root is tetrarch (Figs. 2.27 and 2.28). The primary xylem is surrounded by secondary xylem except opposite the protoxylem, where a broad medullary ray divides the secondary xylem into three or four wedge-shaped segments. The structure of the cortex is similar to that of *Lyginopteris* root. In older roots that are well preserved, a clear zone of thick periderm is visible. It forms the outer covering of the old roots.

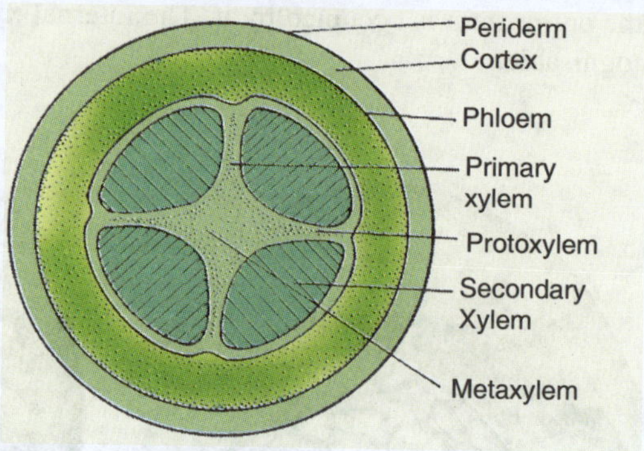

Fig. 2.27. T.S. tetrarch root of *Medullosa noei* (Diagrammatic)

Rothwell and Whiteside (1972) discovered several distinctive root fragments in carbonate petrified material from Middle Pennsylvanian of Illinois. The fragments range from minute rootlets with incomplete primary development to mature axes over 2 cms in diameter with well developed secondary tissue. The axes are characterised by an exarch actinostele with a variable number of protoxylem groups. The **Pericycle** in small rootlets is surrounded by **endodermis** and a well developed parenchymatous **cortex**. The mature fragments exhibit a **periderm** at the outer margin of pericycle. In largest specimens the periderm forms the outermost boundary. **Secondary xylem** is extremely well developed in large specimens and completely surrounds the primary

Fig. 2.28. T.S. small root of *Medullosa noei* (After Arnold).

xylem. Branching is initiated opposite the protoxylem groups.

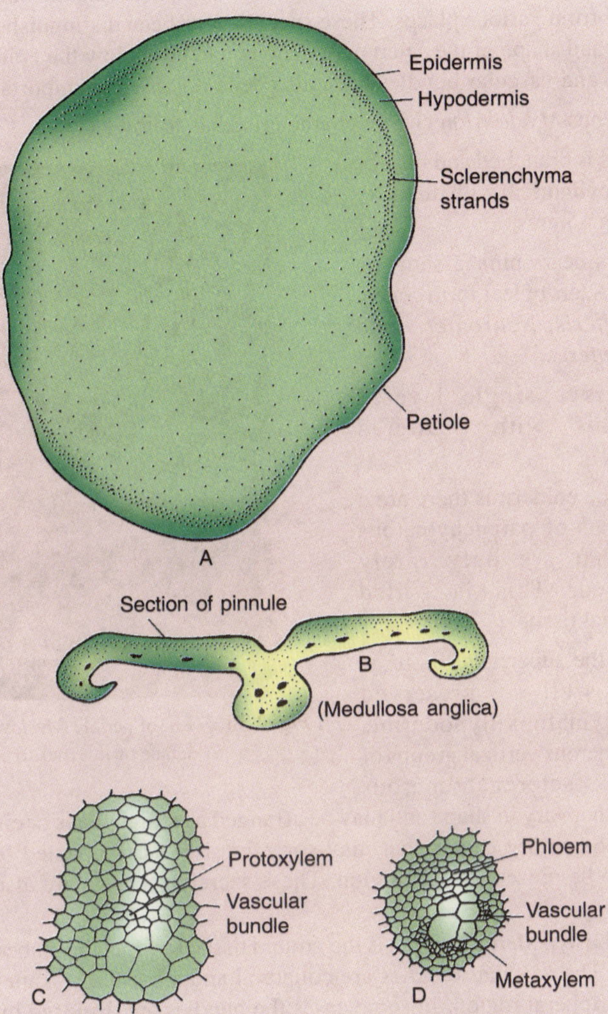

Fig. 2.29. (A-D). A.T.S. *Myeloxylon radiatum* (Petiole of *Medullosa leuckarti*) ; C-D. vascular bundles of the petiole showing detailed structure (After Zeller); B. T.S. pinnule of *Medullosa anglica.* (After Scott).

PETIOLE : The form genus *Myeloxylon* represents the petiole of *Medullosa*. This genus was established by Brongniart in 1849 as a result of re-examination of Cotta's species *Medullosa elegans*. Brongniart found that the internal structure of Cotta's (1832) *Medullosa elegans* resembled that of a monocotyledonous stem and hence changed the generic name to *Myeloxylon*. At that time he did not realise that it was actually the petiole of *Medullosa*. Later Goeppert (1864, 65) renamed it as *Stenzelia* and Renault in 1876 suggested the name *Myelopteris radiatum*. Schenk (1882) suggested that the name *Myeloxylon* should be retained as there was no valid reason for changing it. The fact that *Myeloxylon* was actually the petiole of *Medullosa* was suggested by Solms-Laubach in 1891. The species *Myeloxylon radiatum* was established to include *Medullosa elegans*. Delevoryas and Stewart regarded Schenk as the author of *Myeloxylon radiatum* (Renault) Schenk. Petioles of *Myeloxylon radiatum* have often been found attached to the stems of *Medullosa leuckartii*. Similarly *Myeloxylon* *landriotii* is regarded as petiole of *Medullosa anglica* and *Myeloxylon zonatum* as that of *Medullosa noei*. *Myeloxylon topekense*, *M. missouriensis* and *M. bendixenii* are the other species of medullosan petioles reported from various places. These species have been distinguished mainly on the basis of the arrangement and shape of the sclerenchymatous patches below the epidermis. The arrangement of secretory sacs and vascular bundles have also been regarded as important features.

The form genus *Myeloxylon* can be characterised as follows :

1. The axis is branched and may be oval or cylindrical with diameter reaching 15 cms.
2. The axis bears pinnae showing the characters of leaf form genera *Alethopteris*, *Neuropteris* and *Odontopteris*.
3. There is a single layered epidermis with scattered stomata.
4. Below the epidermis there are a few layers of parenchymatous cells that are only rarely preserved. This is called superficial tissue.
5. Next to the superficial tissue is the well preserved parenchymatous hypodermis with numerous vertical groups of narrow sclerenchymatous patches that vary in shape and may be arranged as radial plates or circular, oval, or reniform strands. Secretory canals that may sometimes be accompanied by thick-walled patches may also be present in this region. These secretory canals are also present in the ground tissue.
6. Next to the hypodermal region is the ground tissue in which lie embedded numerous vascular bundles. The vascular bundles are collateral and are usually scattered. They are abundant in the peripheral region. In some cases the bundles are disposed in concentric circles.
7. The vascular bundles are partially enclosed by a sheath of thick-walled cells. Each bundle consists of both xylem and phloem. The xylem is composed of a small group of tracheids that may have spiral or scalariform or reticulate thickening bands. There is a single protoxylem group next to the phloem. The xylem is as a rule entirely centripetal. In some cases, a few centrifugal tracheids develop and the normally exarch condition appears mesarch.
8. The phloem is made up of narrow sieve tubes with some parenchyma. It is rarely preserved.

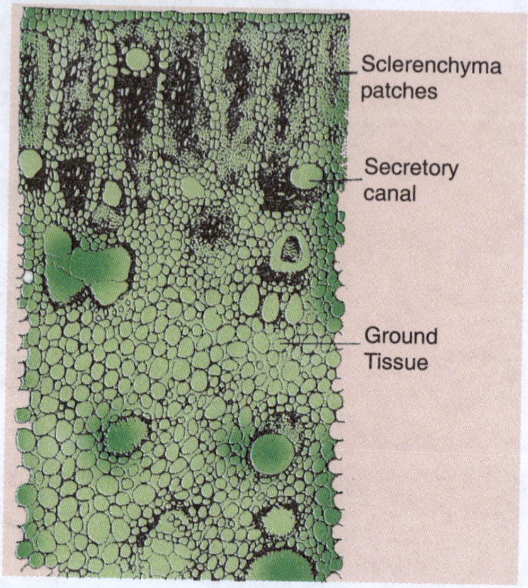

Fig. 2.30. T.S. of petiole *Myeloxylon radiatum* showing details of internal structure (After Kidston).

The collateral form of vascular bundles, the exarch position of protoxylem and the presence of hypodermal patches of sclerenchyma are cycadean features.

Some species of *Myeloxylon* are described below to illustrate their internal structure and anatomical differences.

M. radiatum (Figs. 2.29A and 2.30). This species is regarded as the petiole of *Medullosa leuckartii*. A transverse section of the petiole (Fig 2.29A) shows an outer epidermal layer followed by a few layers of parenchyma below which is the hypodermis with characteristic patches of sclerenchymatous fibres that are radially elongated. Next to the hypodermis is the ground tissue in which are embedded numerous vascular bundles. The vascular bundles are scattered throughout the ground tissue and are collateral and exarch. Each vascular bundle (Fig. 2.29, C-D) is partially enclosed by a sclerenchymatous sheath and consists of a mass of tracheids that may be small or large and have closely or spaced spiral or scalariform thickenings. The protoxylem is composed of smallest tracheids and lies towards the phloem. The phloem is separated from the xylem by a thin layer of parenchyma cells and consists of narrow sieve tubes with scanty phloem parenchyma. The secretory canals or the gum canals are also scattered in the ground tissue. Each canal is surrounded by a layer of secretory cells and may sometimes be filled with a dark material.

The pinnules of this species have been shown by Renault to be of *Neuropteris* type.

M. landrioti differs from *M. radiatum* in possessing peripheral sclerenchymatous patches that were almost circular in outline. This petiole has been found in organic connection with the stems of *Medullosa anglica*.

In *M. topekense,* the sclerenchyma patches are somewhat tangentially expanded. This species was discovered by Penhallow (1897) from the Upper Carboniferous strata at Topeka, Kansas.

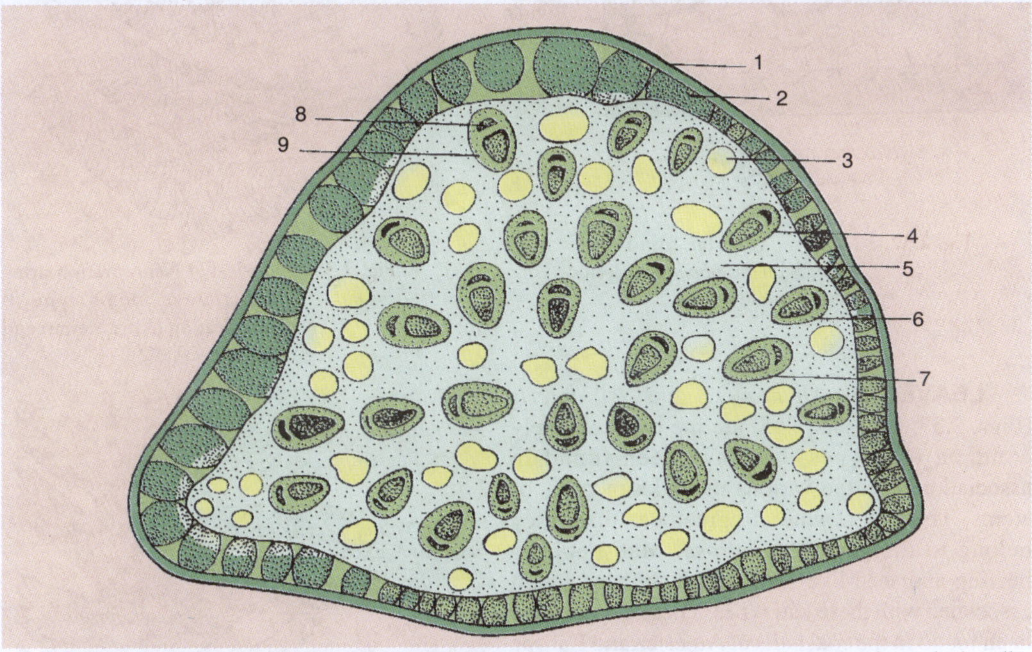

Fig. 2.31. T.S. *Myeloxylon missouriensis.* Note the peripheral sclerenchymatous strands, scattered vascular bundles and gum canals. 1. Epidermis 2. Sclerenchyma strands; 3. Gum canal; 4. Vascular bundle; 5. Ground tissue; 6. Bundle sheath; 7. Xylem Parenchyma; 8. Phloem; 9. Xylem. (After Arnold)

M. missouriensis (Arnold and Steidtmann, 1937) has small sclerenchyma strands (Figs. 2.31, 2.32) that are usually arranged in double rows, are partly coalescent, rounded or slightly elongated in a radial direction. The central part of the petiole is filled with numerous scattered vascular bundles and secretory or gum canals. The epidermis is well preserved in this species.

In *M. zonatum* (Steidtmann, 1944) which is regarded as the petiole of *Medullosa noei*, the sclerenchyma strands are irregular with a distinct ring of vascular bundles bounding the inner surface, and a distinct row of secretory canals immediately below the ring of bundles.

Andrews (1945) recognised *M. bendixenii* on the basis of a thick "rind" of vascular bundles and the presence of abundant secretory canals in association with the sclerenchyma patches. The inner bundles of the "rind" are not sharply delimited.

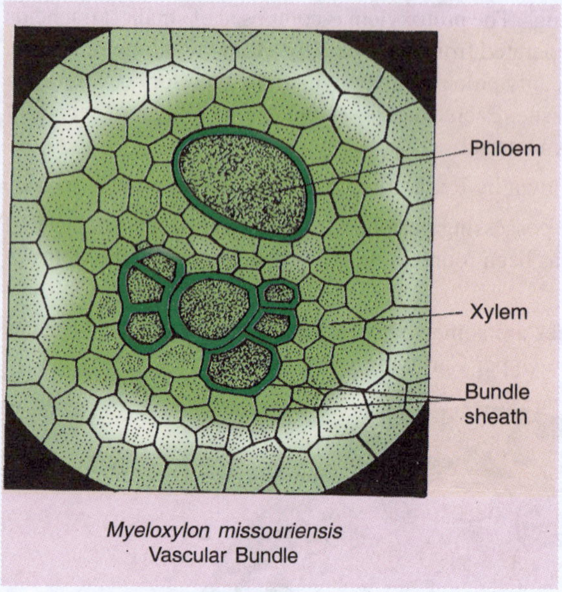

Fig. 2.32. It shows vascular bundle of *Myeloxylon missouriensis* in detail (After Arnold)

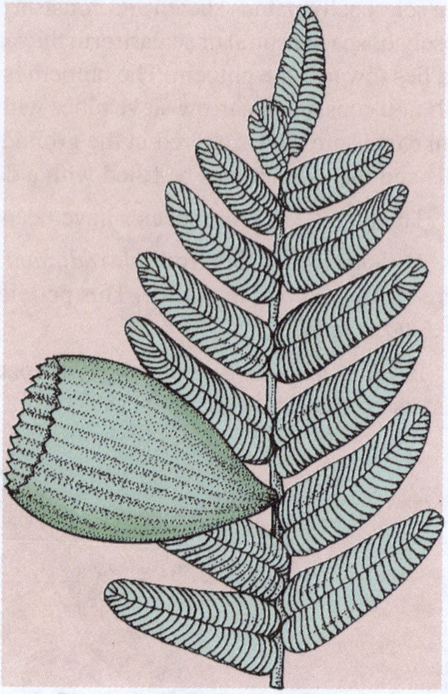

Fig. 2.33. The leaf of *Neuropteris* bearing *Whittleseya media* type of fructification (After Stewart and Delevoryas).

LEAVES :— *Alethopteris* and *Neuropteris* (Figs. 2.33, 2.34 and 2.35) are the two extremely common leaf types that have been found in association with the various medullosan types of stems. There is a general agreement that they belong to the genus *Medullosa*. Many pollen-bearing and seed-like structures have been found associated with these leaf types. These genera have been found in the coal balls of America and Europe in the form of compressions. The other leaf genera that are regarded as medullosan include *Odontopteris* (Seward, 1917) whose rachises resemble *Myeloxylon*, *Linopteris*, *Lonchopteris* (Arnold, 1947), *Callipteris* and *Taeniopteris* (Seward, 1917). In order to have a clear idea of the reconstruction of the entire frond, it is necessary that a compression revealing it in its entirety is discovered. No such specimen is available at present. Andrews (1945), figured a drawing of *Alethopteris* showing unequal branching of the rachis. Delevoryas (1955) observed that axes of *Myeloxylon* obtained from the coal balls showed unequal branching in the lower parts and stated that "regular pinnate arrangement is attained" in the distal regions of the frond. He further stated, "Until there is a correlation of internal structure of

Fig. 2.34. The foliage of *Alethopteris nornii* bearing a fructification (After Andrews).

leaves in petrified material with many types of compressed foliage, any attempts to list the types of leaves borne by the medullosa must be only tentative". A brief description of each frond genus is given below :

Genus 1. Alethopteris (Figs. 2.34 and 2.35 C): This genus was named by Sternberg. It is a compound frond constructed on the pinnate plan and borne on a main rachis which is occasionally dichotomously branched. The fronds bear pinnae that are divided into linear pinnules with a decurrent base. The pinnules have an acute or a rounded apex with entire margin or slightly serrate in the lower parts. Each pinnule is traversed by a prominent mid-rib that reaches its apex and gives off secondary veins that are forked at ends. The decurrent portion is supplied directly by veins arising from the axis of the pinna. In the upper part of the frond the pinnules are replaced by a simple, entire, or lobed lamina. Zeiller examined a piece of cuticle from the upper surface of pinnule of *A. grandini* and observed polygonal epidermal cells with straight walls. The cells above the veins were rectangular. At places, groups of cells encircling a small pore were observed. These were considered by Zeiller to indicate the position of sunken stomata. Renault described petrified specimens of *Alethopteris* fronds which showed anatomical details resembling those of *Myeloxylon*. Numerous fragments of *Myeloxylon* type of rachis bearing *Alethopteris* pinnules have been discovered from the calcareous nodules from English Coal-Seams.

Hemingway discovered from Coal-Measures of Yorkshire several specimens of *Alethopteris lonchitica* (Fig. 2.35, C) in organic connection with *Trigonocarpon* seeds. This foliage has, since been regarded as the bay of *Medullosa anglica*. Grand Eury has discovered some species of *Alethopteris* with *Trigonocarpon* and *Pachytesta* types of seeds in the French Coal fields.

Alethopteris lonchitica and *A. serlii* have been recorded from the Middle and Lower Coal-Measures of Britain, and Coal-Measures of St. Etienne, respectively. The former was associated with the seed *Trigonocarpon* whereas the latter was recorded with the seed *Pachytesta*. *A. serlii* is very common in Upper Coal-Measures of Britain but rare in Middle Coal-Measures. *A. norinii* Halle was described from the Lower Permian of Shansi with a seed attached to its rachis or mid-rib of pinna. The seed was 40 mm long and 12 mm broad. Arnold has described abundant seeds associated with *A. grandifolia* from the Upper Carboniferous of Summit County, Ohio and *A. norinii*.

Genus 2. Neuropteris (Figs. 2.33, 2.36) : This leaf genus was first created by Brongniart and was placed under the ferns but later it was transferred to the Pteriodosperms. Now it is regarded as a leaf genus belonging to the Medullosaceae. A number of species have been reported with seeds attached to them. W.J. Longmans discovered a microsporangiate fructification called *Whittle seya* in connection with *N. schlehani* (Fig. 2.36,A). The specimen was discovered from the Limburg Coal fields of the Netherlands. Seed-like structures resembling *Pachytesta vera* have also been found attached to *N. schlehani* in the same locality. *Neuropteris obliqua* was discovered by Kidston and Jongmans in 1911 along with a seed about 4.5 cm long and two centimeters in diameter. These seeds were borne on special peduncles that were 1 cm long. In *N. tennifolia* the seeds were terminal on, pinnae. It was discovered from the Upper Carboniferous of Yorkshire. In 1905, Kidston discovered seeds attached to the fronds of *N. heterophylla* that were discovered from the iron stone nodules from the Middle Coal-Measures near Dudley, England. Kidston named these seeds as *Rhabdocarpus* because they were covered with fibres. Some authors referred them to a new genus *Neurospermum* because of their radiospermic character. Restoration of *N. decipiens* with attached seeds was figured by Chamberlain in his book "Gymnoperms : Structure and Evolution". Such an attachment has not been observed by any author and Arnold considers that this restoration is based on various published figures of *N. heterophylla*. Grand Eury also recorded the association of this species with seeds from French Coal-fields.

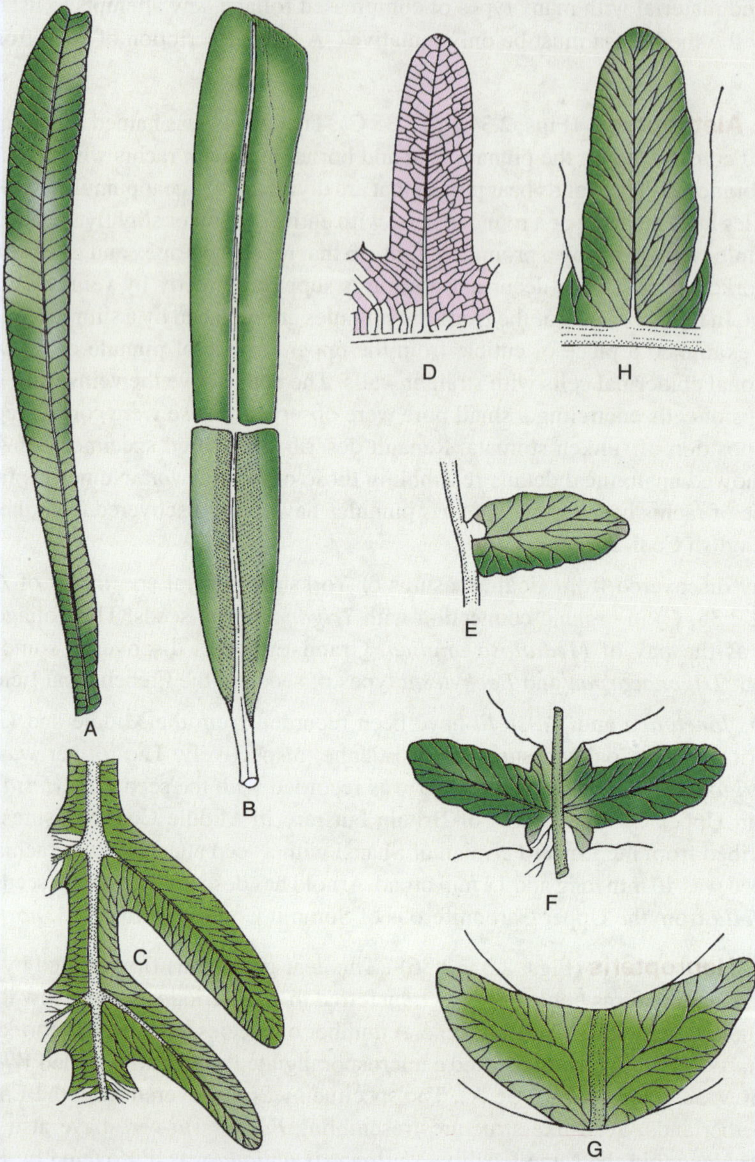

Fig. 2.35. A. *Taenioptesis spatulata* (After Zeller) B. *Taeniopteris vittata* (After Seward) C. *Alethopteris lonchitica* (After Zeller) D. *Lonchopteris surgosa* (After Zeller) E. *Pecopteris plumosa* (After Zeller) F. *Callipteris* sp (After Zeller) G. *Odontopteris genuina* (After Zeller) H. *Callipteris pelatti* (After Zeller)

Renault pointed out that the petioles of *Neuropteris* resemble anatomically to those of *Myeloxylon*. He also described some petrified pinnules of *Neuropteris* and reported the presence of upper palisade tissue with a distinct spongy tissue below. He also noted the presence of a distinct lower epidermis with sunken stomata like those of *Nerium oleander* (an angiosperm). Potonie described the presence of reticulate venation in the pinnules of *N. pseudogigantea*.

The genus *Neuropteris* is characterised as below :

The fronds are of considerable size, sometimes measuring upto ten metres; bi- or tri-pinnate; the rachis may be dichotomously branched; both rachis and petiole bear single pinnules (Figs. 2.34 and 2.36); the single pinnules on the petiole are larger in size and have cyclopteroid laminae (Figs. 2.34, and 2.36). Pinnules are mostly entire, rarely slightly lobed, broadly linear

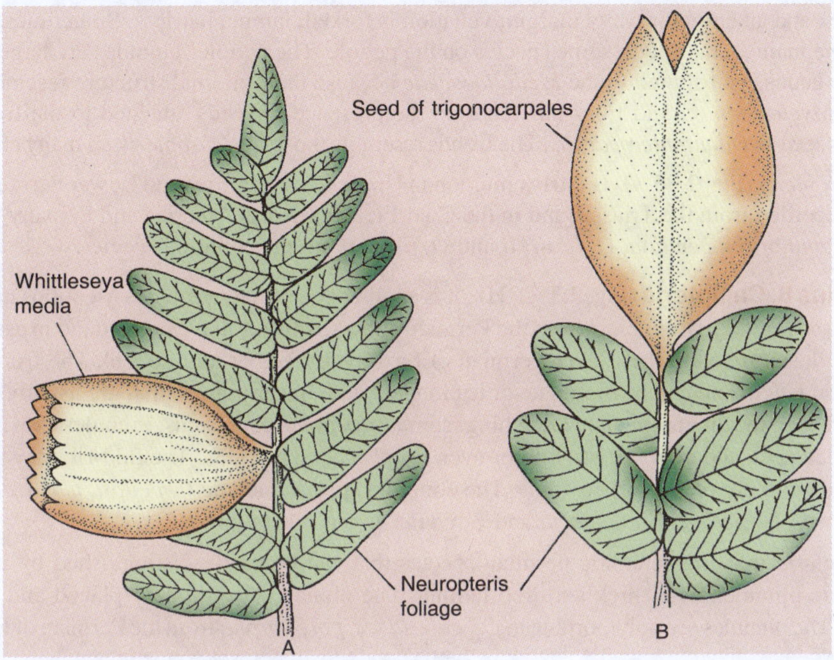

Fig. 2.36. The foliage of *Neuropteris* bearing two types (A. *Whittleseya media,* B. Tsigonocarpalean seed) of fructifications [After Jongman (A) and Arnold (B)].

with cordate base. Brongniart reported the presence of a small pedicel in the pinnules of *N. grangeri* Brongn.; mid-rib does not reach the tip of pinnules; apex blunt or pointed; secondary veins arched and forking repeatedly. In *N. scheuchzeri* the pinnules are hairy, large in size and have a pointed apex. The oval, linear and bluntly rounded pinnules of *N. horrida* also possess a median band of hair on either surface. The margins of the lamina are also hairy in this species.

Genus 3. Linopteris : This name was first proposed by Presl in 1838. The genus resembles *Neuropteris* in many respects except that the veins in the pinnules anastomose with each other and form a network. As in some species of *Neuropteris*, cyclopteroid pinnules occur on the petioles of *Linopteris*. Grand Eury recorded the attachment of seeds called *Hexagonocarpon* with the specimens of *L. brongniarti*. Zeiller recorded the presence of two rows of long and pointed sporangia on the ventral side of the pinnules of *L. schutzei*. They are supposed to be microsporangia. The other known species is *L. obliqua* (from Coal-Measures of Nova Scotia; Upper and Middle Coal-Measures of Britain, Asia Minor, Pennsylvania and many other European localities). The genus is rare in Britain. Leisman (1972) discovered several isolated pinnules of *Linopteris* in the coal balls from the Sahara mine in Southern Illinois. The pinnules possess upper and lower epidermises with stomata restricted to lower epidermis. Hair bases were present on both the layers but were numerous on the lower epidermis. Distinct palisade and spongy layers were present.

Genus 4. Lonchopteris : The genus was named by Brongniart and resembles in many respects with *Alethopteris* except in its reticulate venation. It was discovered from the Upper Carboniferous rocks. *L.rugosa* (Fig. 2.35, D) and *L.bricei* are known from Britain.

Genus 5. Odontopteris : The genus was established by Brongniart for compound fronds (Fig. 2.35 G) obtained from the Coal-Measures. The pinnules are attached to the bipinnate or tripinnate fronds with dichotomously branched rachis (Fig. 2.35, H). The pinnules are sessile with acute or obtuse apex and an entire or wavy margin; venation is forked, lamina hairless. Sometimes pinnules occur on the main rachis and in some species on the petiole. The petioler pinnules are larger in size. This foliar genus is attributed to the *Medullosaceae* because their internal structure resembles very much the *Myeloxylon*. Grand Eury has reported the presence of seeds attached to the fronds. He named this seed as *Odontopterocarpus*. The fronds resemble those of *Neuropteris*, in many characters.

Odontopteris (Fig. 2.35, G) is fairly common in Upper Carboniferous and Lower Permian rocks. It is also recorded from the Triassic and in the Coal-Fields of North America and in many places in Europe. *O. minor*, *O. genuina*, *O. brownii* and *O. fischeri* are the common species.

Gennus 6. Callipteris (Fig. 2.35, F,H) : It is another genus that resembles Medullosan foliages and was discovered by Brongniart from the Permian strata. Its fronds have been found in association with Medullosan stems in the Lower Permian of Saxony, by Weber and Sterzel. The fronds reach one meter in length and can be bipinnate or tripinnate, with linear pinnate that are usually crowded and decurrent on the dichotomously branching (sometimes pinnately) rachis. Pinnules oval or oblong with rounded apex and entire or slightly or even deeply lobed margins. The mid-rib may reach the apex and give off simple or forked veins. The common species include *C. pellati*, *C. lyratifolia*, *C. flabellifera*, *C. bergeronii*, *C. conferta* and *C. conferta* var. *polymorpha*.

C. conferta is a characteristic permian species that can be easily distinguished by its linear, oval, sessile pinnules with thick textured lamina. The pinnae are obliquely placed and are long and linear. The pinnules are polymorphic in *C. conferta* var. *polymorpha* from the Permian of Germany.

Genus 7. Taeniopteris (Fig. 2.35, A,B) : It has been regarded as a leaf genus of the *Medullosaceae* but this identity is extremely doubtful because it has been found without any reproductive organs attached to it. It neither bears any resemblance to the vegetative parts of other leaf types. Even the various species of this genus show lot of variations such as in some cases they are simple (*T. vittata*) and resemble marrattiaceous ferns. In other species, the fronds are pinnate or bi-pinnae as in *T. carnoti* Ren. & Zeill., *T. jejunata* Grand Eury, and *T. missouriensis*. The venation in almost all the species is forked. The secondary veins fork dichotomously. Several species of *Taeniopteris* have been described from the Triasso-Rhaetic rocks in Europe, India, Tonkin etc. From Damuda and Panchet series of India (Triasso-Rhaetic), *T. lata* has been described in detail. The simple leaves in *T. multinervis* Weiss., described from various places such as Lower Permian of Lebach, Permian of Trienbach, and Upper Carboniferous of Autun, are 40 cm long and 6 cm broad. The lamina is traversed by numerous (25-36 per cm of lamina) forked secondary veins. Its mesophyll has been described to be made up of polygonal cells, some of which may become columnar and resemble palisade cells. Other species with simple lamina include *T. jourdyi*, *T. caruthersii*, *T. tenuinervis*, *T. vittata*, *T. spatulata*, *T. major* and *T. beyrichii*. Fontaine and White and Sellards reported the presence of sori in *T. newberriana* and *T. coriacea*, but they are not regarded as sori by majority of the authors because of the absence of cellular structure and lack of direct connection with the veins. *T. jejunata*, *T. carnoti*, *T missouriensis* and *T.coriacea* are species with pinnate fronds. These species have oval or oval-linear pinnae with short or long stalks and reach a length of 25-30 cm and a breadth of about 25-30 millimetres.

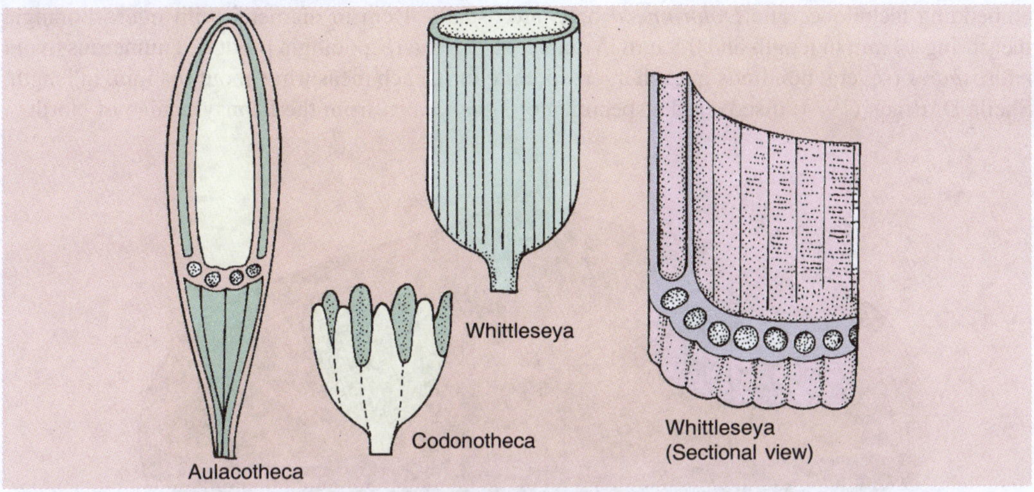

Fig. 2.37. Some Microsporangia organs of *Medullosaceae* (After Halle).

FRUCTIFICATIONS

MICROSPORANGIATE ORGANS ATTRIBUTED TO MEDULLOSACEAE

(Figs. 2.37 to 2.41)

The microsporangiate organs attributed to the Medullosaceae include (*i*) *Whittleseya*, (*ii*) *Aulacotheca*, (*iii*) *Codonotheca*, (*iv*) *Dolerotheca*, (*v*) *Potoniea* and (*vi*) *Goldenbergia*. A brief description of each type will be given below :

Genus 1. WHITTLESEYA (Fig. 2.37) : The genus *Whittleseya* was earlier referred to the Ginkgoales but later studies revealed that it is a microsporangiate organ of the Medullosacea, because it was reported in organic connection with *Neuropteris schlehani* which is the foliage of the Medullosaceae. This discovery was made by W.J. Jongmans from the Limburg Coal fields in the Netherlands. *Whittleseya elegans* Newberry was discovered from the Coal-Measures of North America and Europe (Staffordshire). Halle's described it as a hollow and bell-shaped (compannulate) structure with wall composed of a single ring of tubular (elongated) **microsporangia** that are concrescent *i.e.*, are fused laterally throughout their length. The distal margin of the bell-shaped **microsporophyll** appears twined. The microsporophyll is 3 to 6 cms. long with ribbed surface. The term **microsporophyll** is rightly used here to represent the leaf-like organ that bears the **microsporangia**. The microspores are arranged in more or less longitudinal rows along the inner sides of the bell-shaped microsporophylls. The microspores measure 0.25 mm in length. Majority of the **microspores** possess an oval slit (Fig. 2.40 E) which points to their longitudinal dehiscence and bilateral nature.

Kidston described another species *W. fertilis* from the same locality as *W. elegans*. In this case, the actual sporangia have not been preserved. The spores are elliptical and measure 210-222 μ in length. They also possess an oval slit. *W. media*, *W. brevifolia*, *W. microphylla* and *W. concinna* are some other little known species.

Genus 2. DOLEROTHECA (Fig. 2.38) : These fructifications have been found in connection with medullosan stems and fronds from the American Coal balls. These are large and nicely preserved **microsporangiate organs** that are bell-shaped (like *Whittleseya*) with numerous tubular microsporangia embedded in the inner side of the fructification. *Dolerotheca formosa* Schopf has been found in abundance in the Carboniferous forests of America in the form of compressions. Halle (1933) described in detail the morphology of this fructification from the sections prepared by special embedding technique. The *Dolorotheca* bells are usually 4 cm in diameter with microsporangia measuring 14 mm in length and 0.8 mm in diameter. The microsporangia produced numerous ovoid *microspores* (several hundreds in each microsporangium), each measuring about 0.4 mm. in length. Sheila D. Brack (1973) discovered a specimen of *Dolerotheca* from the Pennsylvanian of North

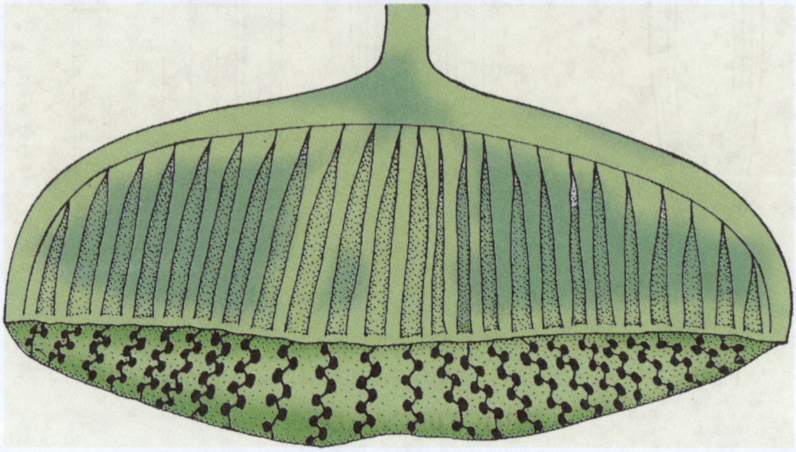

Fig. 2.38. *Dolerotheca formosa.*

America. It is a small villose synangiate pollen organ that was identified as a species of *Dolerotheca*. The synangium has distinct epidermis with conspicuous sclerenchymatous patches below it. The cortex is parenchymatous and contains numerous resinous rodlets and cells with dark contents. The sporangia are arranged in pairs and are either separated lysigenous tubes or fibres. The sporangia are associated with distinct vascular bundles. The spores are **monolete**.

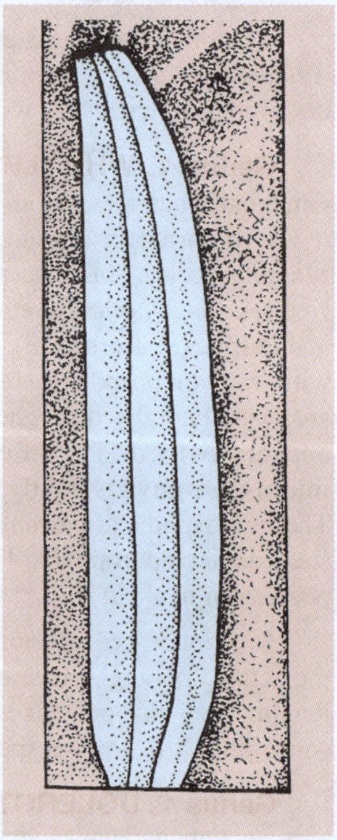

Fig. 2.39. It shows *Aulacotheca campbelli* (After Arnold).

Genus 3. AULACOTHECA (Figs 2.37 A and 2.39) : It occurs in the Lower Coal-Measures of Great Britain and in the Pottsville of Eastern North America. They also occur in Upper Carboniferous of Scotland. The microsproangiate organs are elongated, oval and slender structures with acute apex. They also have a hollow central cavity. The fructification is composed of nine tubular microsporangia that are laterally fused and run along the entire inner surface of the organ. They are embedded in the cellular tissue of the wall. The whole structure is about two cm long. *A. elongata* (Kidston) Halle was found in association with a species of *Alethopteris* and in one instance it was found attached to the fronds of *Neuropteris schlehani*. The spores are **monolete** and **bilateral**.

Genus 4. POTONIEA (Fig. 2.42 F) : In this microsporangiate organ an open cup-like structure enclosed numerous microsporangia that were free from each other and were elongate and tubular. There were several dozen microsporangia within the cup and were not embedded and laterally fused but were only attached to the inner surface. The microspores were also found intact.

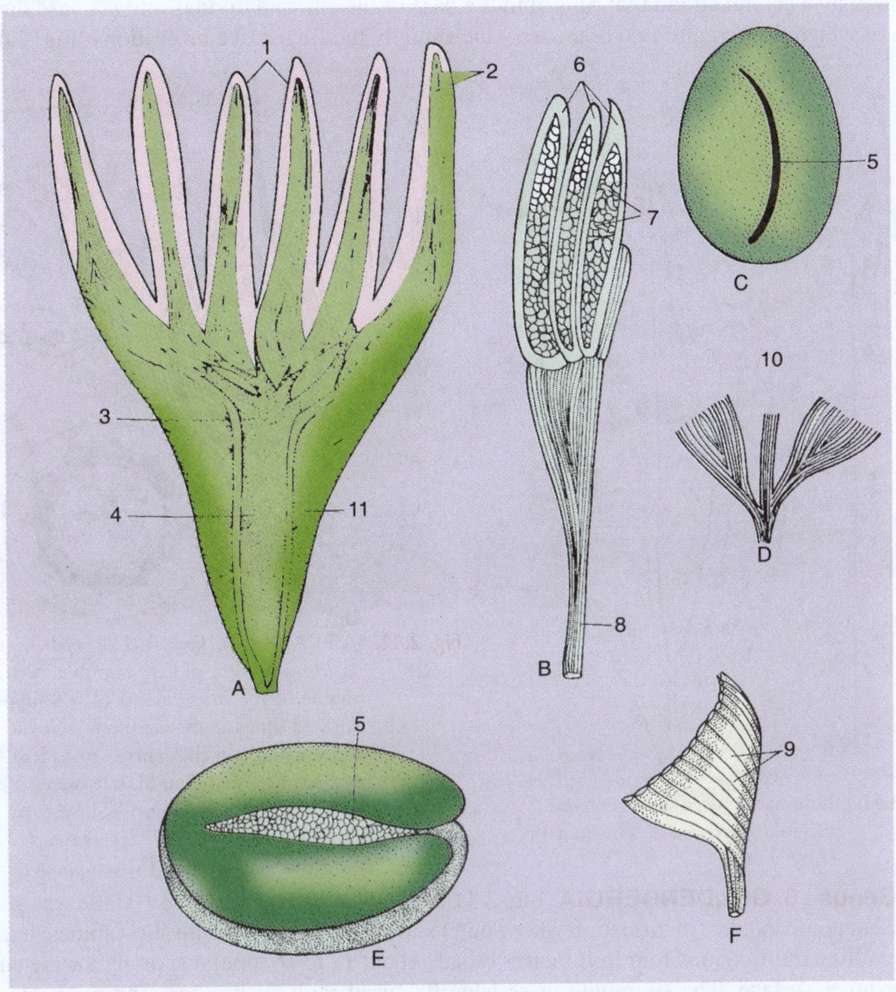

Fig. 2.40. (A-F). Some microsporangiate organs of Medullosaceae. A-C. *Codonotheca caduca*. A. Plan of whole organ opened out flat. (1. Lobes of the cup; 2. Vascular bundles 3. Vascular tissue as a cone in the middle of a fleshy base; 4. Central strand; 11. Stalk of the cupule) B. It shows segments with spores. (6. Segments; 7. Spores) C. A pollen grain with a median slit (5). D. *Potoniea adiantiformis* showing microsporophylls (10) attached to the rachis. E. A single microspore of *Whittleseya elegans* with a median slit (5). F. Side of view of microsporophyll of *Linopteris obliqua* (9. sporangia)

Paraclathiops and *Lacoea* are other microsporangiate organs attributed to Medullosaceae, that resemble *Potoniea* in being open cup-shaped structure with several laterally free microsporangia.

Genus 5. CODONOTHECA (Fig 2.37 and 2.40 A-C) : Like *Whittleseya* it is also an open, distinctly stalked, cup-shaped fructification but is more slender and has a deeply cut rim. The finger like projections (six in number) represent the microsporangia that appear as projecting from a companulate lower portion. It has been discovered along with some species of *Neuropteris* in the Mazon Creek locality in Illinois. *C. caduca* Sellards is the species found along with *Neuropteris decipiens* Lesoq. The microspores were elliptical, 0.29 to 0.31 mm long and 0.18-0.19 mm broad

(Fig. 2.40 C). The presence of a median ridge suggests its bilateral origin. The stalk of the cupule is supplied by a central strand that runs upto the base of the cup and divides into six vascular strands, each of which divides into two branches which supply the finger-like projections (Fig. 2.40, A).

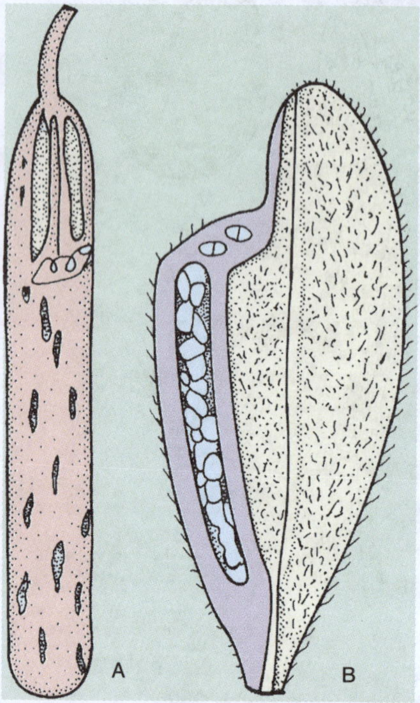

Fig. 2.41. It shows microsporangiate organs of *Goldenbergia* and *Thuringia* (After Remy)

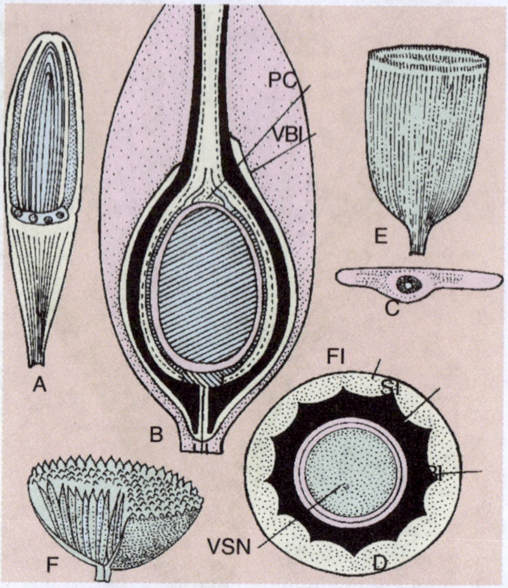

Fig. 2.42. (A-F) A. *Aulacotheca* B. L.S. seed *Trigonocarpus parkinsonii* (Pc, pollen chamber, VBI, vascular bundle of the integument) C. T.S. upper part of the seed showing the sacrotesta, sclerotesta (black) and micropile in the centre (unshaded) D. T.S. of the seed FI, sarcotesta, SI, sclerotesta; VBI, Vascular bundle of integument; VSi, vascular supply of the nucellus) E. Whittleseya (Microsporangium) F. Potoniea (Microsporangium)

Genus : 6. GOLDENBERGIA (Fig. 2.41, B) : *Goldenbergia glomerata* Halle was also a cup-shaped or companulate fructification whose outer surface was covered with short spines. It measured eight millimetres long and four millimetres broad. About 12 to 16 tubular sporangia were embedded in the inner surface. The sporangia were laterally fused with each other. The microspores were almost oval, bilateral and monolete. They were 0.4 mm long. The cupules or the fructifications were stalked and "occurred in a single row along an ultimate pinna axis." It differs from *Aulacotheca* in being shorter in size and in possessing spines along the outer surface.

In *Thuringia*, which resembles *Goldenbergia*, the central space was filled with a soft tissue and has number of long, slender sporangia (Fig 2.41 A). *Boulaya* differs from *Aulacotheca* in having a continuous sporangial region on its inner side. Dennis and Eggert (1972) discovered a new petrified pollen organ in attachment to frond axis from the Upper Pennsylvanian Coal ball in Illinois. In this specimen, several orders of frond axes with numerous branchings terminate in the reproductive structures. The pollen organs that were imperfectly preserved were massive sporangia that contained concentric zones of radially elongated microsporangia and interconnecting sterile tissues. The microsporangia were vascularised and produced prepollen comparable to *Parasporiles*.

MEDULLOSAN SEEDS

The seeds assigned to Medullosaceae belong to the genera *Pachytesta*, *Stephanospermum*, *Trigonocarpus* (Casts only), *Ptychotesta*, *Hexapterospermum*, *Polylophospermum*, and *Codonospermum*.

These seeds possess the following common features :

1. They vary in size from 2 to 5 cms long and upto two centimetres in diameter.
2. They are derived from an ovule with a single massive integument that distinguished into two seed coats, namely outer fleshy *sarcotesta* and an inner hard *sclerotesta*.
3. The sclerotesta is usually distinctly ribbed and sometimes extended into the sarcotesta.
4. The **nucellus** is fused with the *integument* only in the basal portion but is totally free in the upper part.
5. The vascular supply is partly integumentary and partly nucellar.
6. The pollen-chamber is present and is usually dome-shaped.
7. The pollen-chamber is covered by a hollow bell-shaped "upgrowth" of the nucellus.
8. It is devoid of internal nucellar beak as is found in the Lyginopteridaceae and the modern cycads.

In order to illustrate the structure of medullosan seeds, *Pachytesta* is taken as a typical example. These seeds are found abundantly in petrified form. Some of them occur as compressions and are attached to the fronds of *Neuropteris* (Fig. 2.36).

Pachytesta illinoense (Stewart, 1954) (formerly known as *Rotodontiospermum illinoense* Arnold and Steidtmann) is mostly found in association with alethopterid foliage and with the remains of *Medullosa noei*. These seeds are up to 4.5 cm long and appear to be circular in a median transverse section. Near the micropylar end, the transverse section appears triangular in outline. In longitudinal section, it appears almost oval in outline (Fig. 2.44). The seed is covered by a well defined testa that can be distinguished into two layers. The outer layer called the **sarcotesta** is made up of thin-walled and elongated cells with dark contents. The inner layer is composed of thick-walled fibrous cells. This layer is called **sclerotesta**. This layer develops conspicuous longitudinal ribs that number twenty (Arnold) and are of similar height. The ribs alternate with furrows that are made up of thin-walled cells supposed to be spongy in nature. Each furrow is supplied by a vascular strand. Next to the **sclerotesta,** there is an indication of the presence of inner fleshy layer which is not well preserved. The testa encloses the **nucellus** which is not fused with it except at the base. The nucellus is produced into a "Pendent skirt of tissue" at its base. The micropyle is very distinct and is well preserved. It is surrounded by the testa and appears to be "dough-nut shaped". The pollen chamber is clearly preserved and is produced into a short beak at its base. From the basal skirt of tissue of the nucellus there arise about 24 vascular bundles that travel up the nucellus. The sarcotesta is supplied by forty vascular strands that alternate with the ribs of the sclerotesta.

Pachytesta olivaeformis (= *Trigonocarpus parkinsoni* (Figs. 2.42, B,C,D; 2.43)) : The seeds are radiospermic and measure five centimetres in length and two centimetres in width. The **micropyle** is long and tube-like and is about half the length of the total seed (Fig. 2.43). In the middle portion, the seed appears almost circular in a transverse section whereas in the micropylar region, the outline becomes almost flattened (Fig. 2.42 C). The seed is protected by a distinct testa that can be clearly distinguished into an outer fleshy **sarcotesta** and an inner hard **sclerotesta.** Remains of the third fleshy layer next to the sclerotesta are also observed (Fig. 2.43). It extends into the micropyle. There are three prominent and nine less prominent ridges on the sclerotesta. The seed coat or the testa is free from the nucellus except at its base where it is fused with it. The nucellus has a distinct epidermal layer. The sarcotesta is made of thin-walled cells passing outwards into a spongy tissue with hypodermis of columnar or palisade-like cells. The cells in the sclerotesta are thick-walled and form an effective protective layers. Both these layers are continued upwards into the apical region as the wall of the long micropyle. The inner layer of the nucellus consists mainly of tracheal tissue surrounding a large megaspore or female gametophyte which is preserved as a "contracted membrane" detached from the nucellus. Towards the micropylar end of the nucellus, there is a relatively small **pollen chamber** which is dome-shaped and is covered by a hollow, inverted, bell-shaped outgrowth of the nucellus. The characteristic inner nucellar beak found in the lyginopterids and cycads, is absent in the medullosan seeds. The pedicel of the seed is traversed by a central strand of thick-walled cell penetrated by a concentric vascular bundle (Fig. 2.43). The vascular bundle gives off six vascular strands to supply the sarcotesta and then passes into the nucellus to form a sheath of tracheal tissue enclosing the chalazal end of the female gametophyte. Later it passes upwards and divides into anastomosing strands of tracheids extending to the level of the pollen chamber. The apex of the seed is flattened.

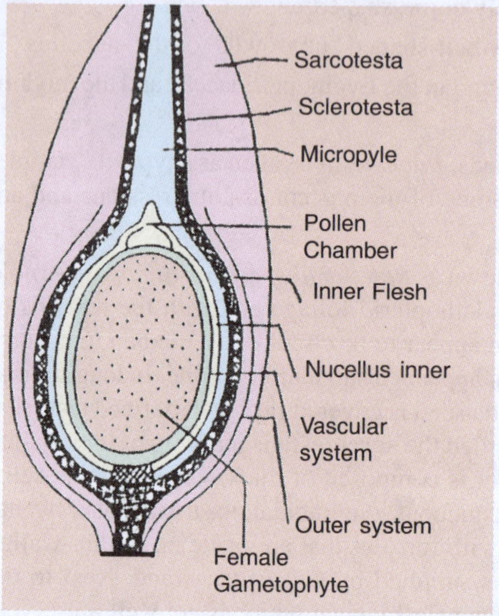

Fig. 2.43. L.S. seed of *Pachytesta olivaeformis* (= *Trigonocarpus parkinsoni*, Fig. 2.42 B-D). (After Scott).

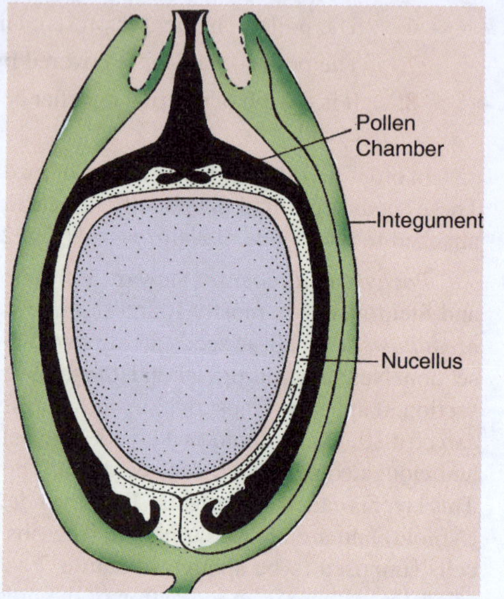

Fig. 2.44. L.S. seed of *Pachytesta illinoensis* (After Stewart).

This seed belongs to *Medullosa anglica*. Majority of the seeds belonging to the genus *Trigonocarpus* have been placed in the genus *Pachytesta* by Hoskins and Cross (1946). They are abundant in Europe and North America. In some seeds belonging to the genus *Pachytesta*, the sclerotesta has only three main ribs and no secondary ones, whereas in others there may be upto thirty smaller ribs. Embryos have not been reported in any of the seeds but gametophytes with archegonia and egg cells have been reported. *Pachytesta composita* is another species which is well preserved (Fig. 2.45 B.C).

Pachytesta incrassata is the largest species measuring 11 cm in length and 6 cm in diameter.

Stewart (1951) described *P. hexangulata* from Lower Coal-Measures. This species was three centimetres long and the fructifications revealed the presence of a distinct female gametophyte with three archegonia. The pollen chamber contained pollen grains liberated from the *Codonotheca* type of male fructification. These pollen grains were 365 μ in diameter and had already germinated with a multicellular male gametophyte within them.

Hoskins and Cross (1946) reported the presence of pollen grains (63 × 54 μ) in the pollen chamber of *P. vera*. In this case, the male gametophytes were not clearly preserved.

Stephanospermum (Figs. 2.45 A and 2.46) is another properly preserved and well studied genus of seeds that can be attributed to the Medullosaceae. Three species, *S. elongatum*, *S. ovoides* Hall and *S.stewartii* Hall have been reported from the Upper Pennsylvanian of Illinois.

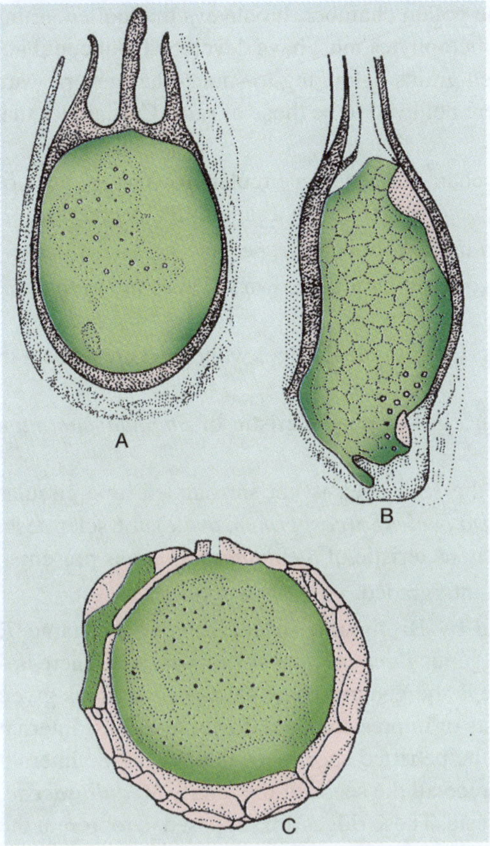

Fig. 2.45. (A—C). Some Medullosan seeds A. *Stephanospermum elongatum*. L.S of seed B. *Pachytesta composita* L.S. Seed. C.T.S. seed of *P. composita*

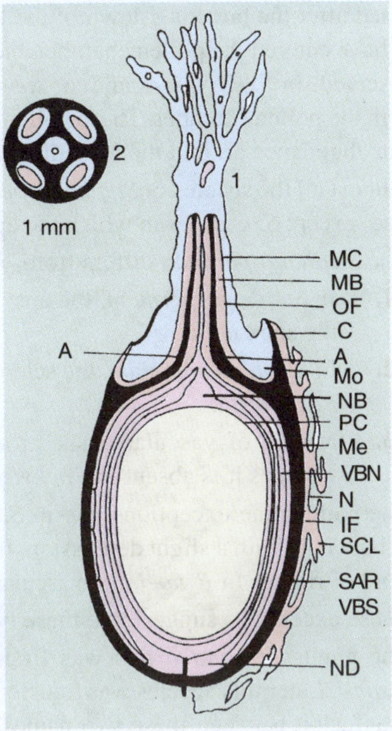

Fig. 2.46. *Stephanospermum elongatum* I. L.S seed, M.C. = Micropylar canal, M.B. = Micropylar bundle, O.F. = Outer fleshy layer I.F = Inner fleshy layer, C = Crown like flange, N.B = Nuclellar beak, P.C = Pollen Chamber, Me. = Megaspore, VBN = Vascular bundle of the nucellus, N = Nucellus, SCL = Sclerotesta, Sar = Sarcotesta, VBS = Vascular bundle of sarcotesta, ND = Tracheal mantle. 2. T.S. of seed at level A—A.

S.elongatum (Figs. 2.45A and 2.46) measures 2 to 3 cms long and about three-fourth of a centimeter in diameter. The seed has a distinct beak at its distal lend, which occupies almost half of the length of the seed. It is surrounded by both the layers of the testa and encloses the micropyle. As in other seeds, the testa is distinguished into an outer *sarcotesta*.

Sclerotesta. A thin-walled inner layer is also present next to the sclerotesta. The sclerotesta completely lacks ridges. The sclerotesta is distally elongated into a beak-like structure. A little distance behind *i.e.* just near the base of the micropylar elongation, the sclerotesta is produced into six toothed buttress or a crown-like flange. (Fig. 2.46). The testa is free from the nucellus except at the base. The sarcotesta is traversed by several vascular strands. The nucellus has an inverted bell-shaped pollen chamber at its micropylar end. The nucellus is characteristic that is has **tracheal mantle** that invests it on the inner side. The tracheal mantle is two to three cells wide. The cells are elongated, slender tracheids with reticulate to scalariform thickenings. In *S. stewartii* Hall, this mantle is made up of three to four layers of elongated tracheids. These tracheids are two to three times as long as wide. This mantle of vascular tracheids abruptly stops short before reaching the pollen chamber. Oliver reported that in *S. akenoides,* the vascular mantle reached the base of pollen chamber and extended beyond it. In *S. elongatum,* there is one floor to the pollen chamber, which means that the seed got fossilised after the breaking down of the floor of the pollen chamber. It conveys that pollen grains might have entered the pollen chamber and male gametophytes must have developed, but could not be preserved. In *S. caryoides* and *S. stewartii* pollen grains belonging to some other genera were found in the pollen chamber. These pollen grains were believed to be those of same Cordaitales that grew in abundance around the Medullosaceae.

Almost all the species of *Stephanospermum* are characterised by a circular crown at the apex of the seed except *S. ovoides* in which the apex of the seed was flat with a saucer-like depression.

The *Stephanospermum* differs from *Pachytesta* in the following respects :

1. In *Stephanospermum,* the apex of the seed has a circular crown whereas in *Pachytesta,* the apex is flat.
2. In *Stephanospermum,* the sclerotesta lacks longitundinal ridges white, that are present in *Pachytesta.*
3. Mantle of vascular tissue around the nucellus is characteristic of *Stephanospermum* whereas it is absent in *Pachytesta.*

There are some exceptions *e.g.* in *S. ovoides,* the seed apex is not surrounded by a circular crown but is flat with a slight depression. In *Pachytesta ovale* (*Sarcospermum ovale*), the sclerotesta is devoid of ridges. In *P. noei,* the vascular mantle characteristic of *Stephanospermum* is present.

These exceptions suggest that these two genera intergraded.

The genus *Trigonocarpus* was first described by Brongniart in 1828 under the name *T. parkinsonii*. Later this species was transferred to the genus *Pachytesta* as *P. olivaeformis*. There is a little confusion between these two names. These days the generic name *Trignonocarpus* is given only to those specimens that occur in the form of casts or impressions and do not reveal the internal structure. The generic name *Pachytesta* is used for the petrified specimens that reveal the internal structure. The name *Trigonocarpales* is used to embrace all the seeds of the family *Medullosaceae*, because of the presence of three ridges on the sclerotesta. These ridges are supposed to represent the lines of fusion of three sectors.

Ptychotesta, *Hexapterospermum*, *Polylophospermum* and *Codonospermum* are some of the other seeds that have been ascribed to the Medullosaceae. In *Ptychotesta,* the inner surface of the sclerotesta is studied with peltate scales, whereas in *Hexapterospermum,* the sclerotesta is lined with, six equal longitudinal ridges. In *Polylophospermum* the sarcotesta is produced into a distal cup-like extension around the micropyle. *Codonospermum* seed has sclerotesta with eight longitudinal ridges and a distinct air chamber at its base. There is a doubt about this seed belonging to the Medullosaceae because it is not trimerous.

The organic attachment of some seeds with the medullosan foliage has been reported in many cases. In all these cases, the preservation was in the form of compressions. Four kinds of attachment have been observed. These are :

1. The seed replaces the terminal pinnule *e.g.* in *Neuropteris tenuifolia* and *N. heterophylla* (Fig. 2.36)
2. The seed attached to the margin of the pinnule at the vein ending *e.g. Pecopteris pluckenetii*.
3. The seed attached to the rachis (Fig. 2.34) of the pinna *e.g. Alethopteris norinii*.
4. The seed attached to the lamina of the frond *e.g. Emplectopteris triangularis* (Fig. 2.46 *a*).

The first two species have been found from the Carboniferous period, whereas the last two belong to the Permian. This suggests that the terminal and marginal position of the seeds is a primitive condition whereas the other two conditions are derived ones. This cannot be said with certainty because there is every possibility that in future, the carboniferous genera may be discovered from the Permian deposits or even *vice-versa*.

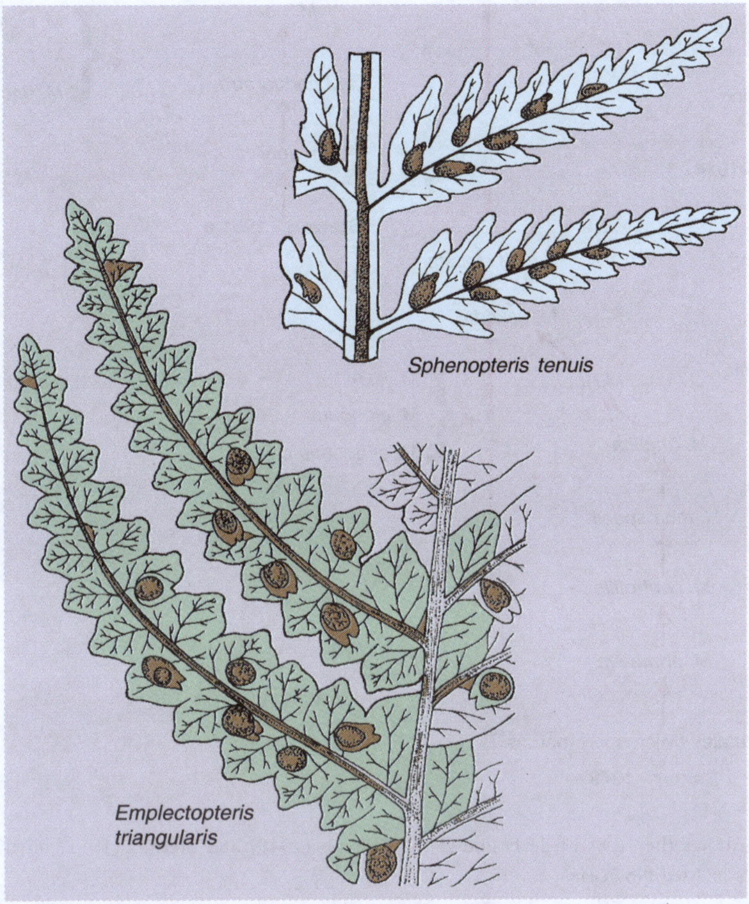

Fig. 2.46a. It shows the attachment of seeds in *Emplectopteris* and *Sphenopteris* (From Andrews).

PHYLOGENY OF MEDULLOSACEAE

A number of views have been put forward to explain the phylogeny of the medullosaceae. Steward and Delevoryas (1956) have summarised all these views and have presented their own scheme of evolutionary developments within the family. A brief description of the various views is attempted in this text.

Polystelic condition of the medullosan stems has evinced great interest among the palaeobotanists engaged in interpreting the phylogeny of the group. It began as far back as 1882 when Goeppert and Stenzel expressed their view that the peripheral steles represent the "flattened plate rings" that encircle the central "star rings," that differed from each other only in shape. The presence of primary wood in the centre of each stele was demonstrated by Schenk in 1889. Later workers like Zeiller (1890), Weber and Sterzel (1896), Solms-Laubach (1896) and Scott (1899) compared the polystelic condition of the medullosan stems with the ferns (*Psaronius* and *Cladoxylon*). Scott suggested a fern origin for the medullosaceae and further compared the medullosan steles with the single stele in *Heterangium* that was regarded as more like the ferns because of superficial resemblance with the stele of some species of *Gleichersia* and their frond-like leaves.

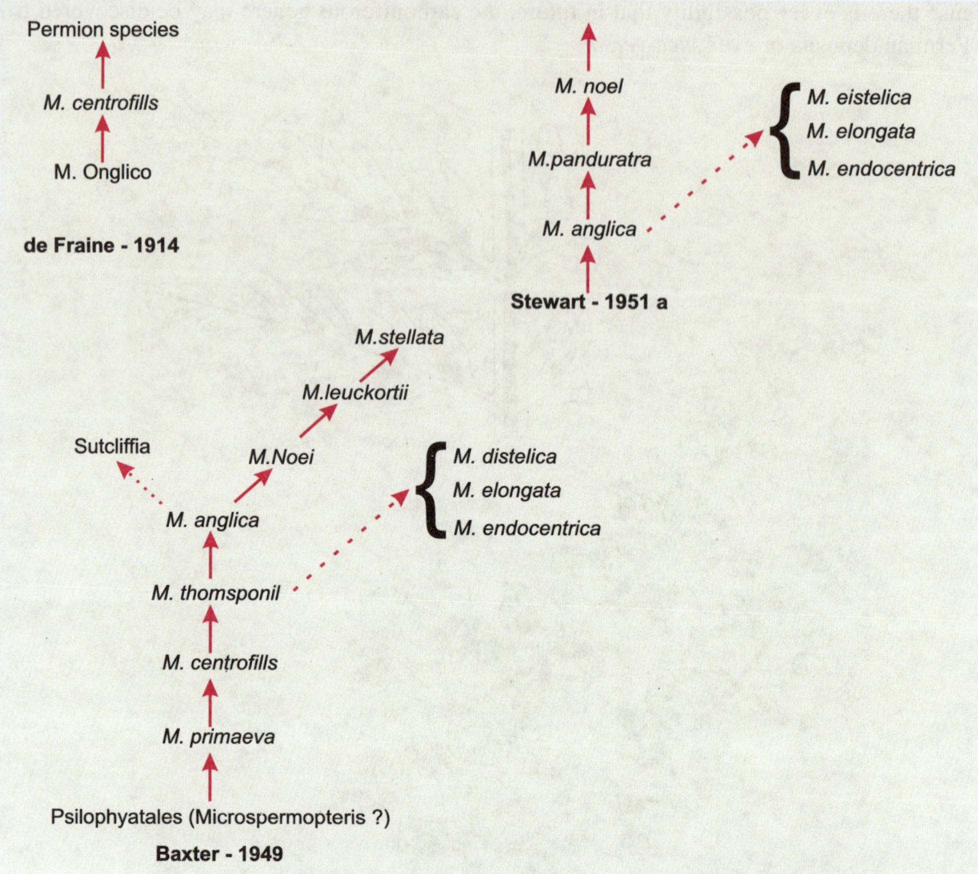

Fig. 2.47. It illustrates the views of de Fraine (1914), Baxter (1949) and Stewart (1951) on the evolutionary trends in Medullosaceae.

The discovery of the genus *Sutcliffia* (Scott, 1906 and Miss de Fraine, 1912) changed the earlier observations and a new interpretation came forth. This genus was regarded as primitive member of the Medullosaceae which gave rise to stelar structure like that of *M. anglica*. From *M. anglica* were supposed to arise all the stelar modifications of the various medullosan stems discovered from the Permian era. In the year 1914, Miss de Fraine discovered *M. centrofilis* in which a small central stele was surrounded by three larger ones. This small central stele was considered to be homologous with the central starrings of some medullosan stems. She proposed that stems like *Sutcliffia* were primitive and *M. anglica* and *M. centrofilis* were derived from it. The series is represented as :

Sutcliffia → *M. anglica* → *M. centrofilis* → *Permian species*

In the year 1939, Schopf proposed that endocentric arrangement of vascular tissue and reduction in the number of steles should be regarded as a derived condition. He suggested that *M. distelica* which is distelic medullosan with endocentric arrangement of vascular tissue should be treated as more advanced to the tristelic *M. anglica*.

Baxter (1949) proposed that the polystelic condition originated from monostelic condition by the proliferation of the single stele. In his evolutionary scheme (Fig. 2.47), he regarded the Psilophytales as the possible ancestors of Medullosaceae and considered polystelic condition to be more primitive than the distelic and tristelic condition. He placed *M. primaeva* as the most primitive member that evolved from the psilophytalean ancestors. It was followed by *M. centrofilis*, *M.thompsonii* and *M. anglica*. *M. distelica*, *M. elongata* and *M. endocentrica* stemmed from somewhere between *M. thompsonii* and *M. anglica* (Fig. 2.47) as a dead end side chain. He supposed the larger Permian forms like *M. noei*, *M. leuckartii* and *M. stellata* to have evolved from *M. anglica*. *Sutcliffia* was regarded as an offshoot from *M. anglica*. Baxter believed that *Sutcliffia* was in reality a protostelic form that resulted from the fusion of a large number of vascular axes. This view regarded *Sutcliffia* to be a derived form.

Stewart (1951) regarded the side chain *M. distelica*, *M. elongata* and *M. endocentrica* to have evolved from tristelic forms like *M. anglica* (Fig. 2.47). In the main chain, he considered his *M. pandurata* to be intermediate between *M. anglica* and *M. noei*. Stewart *suggested* that decrease in the amount of secondary vascular tissue in the leaf traces is an advanced condition and its absence, as in *M. noei*, and other species, is indicative of the most advanced stage. *M. anglica* has secondary tissue in the leaf traces whereas *M. pandurata* has only a slight amount whereas *M. noei* has no secondary tissue in the leaf traces. The "dead end side chain" (*M. distelica* etc.) represents a line of evolution which has undergone reduction in the number of steles, whereas the main line represents reduction in the amount of secondary vascular tissue in the leaf traces.

In the year 1952, Stewart and Delevoryas discovered *M. heterostelica* and confirmed the view that species with many steles are primitive and those with two or three steles are advanced. They suggested that the evolution among Medullosaceae has proceeded along two main lines of evolution. These are :

1. Reduction in the number of steles either by fusion or by the failure of the steles to divide.
2. Reduction in the amount of secondary vascular tissue in the leaf traces.

There is another line of evolution represented by the species in which the secondary vascular tissue in the leaf traces is retained. This line is believed to have led to the evolution of the cycads. These species also retained polystelic condition in the form of accessory steles (*M. anglica*), periderm rings (*M. noei*), and the star rings found in some permian species. In their scheme representing the phylogeny of the Medullosaceae (Fig. 2.48), Stewart and Delevoryas regarded *Sutcliffia* as the most primitive member from which arose species like *M. heterostelica*, *M. anglica*, *M. noei*, *M. leuckartii* and *M. stellata*. They are all polystelic forms that show reduction of the secondary vascular tissue in the leaf traces. The other line which stems from *M. heterostelica* shows reduction in the number of steles and ends with *M. distelica* at the top. *M. heterostelica* is quite interesting that it shows progressive reduction of secondary vascular tissue in the leaf traces as we proceed from the base to the upper portions of the stem.

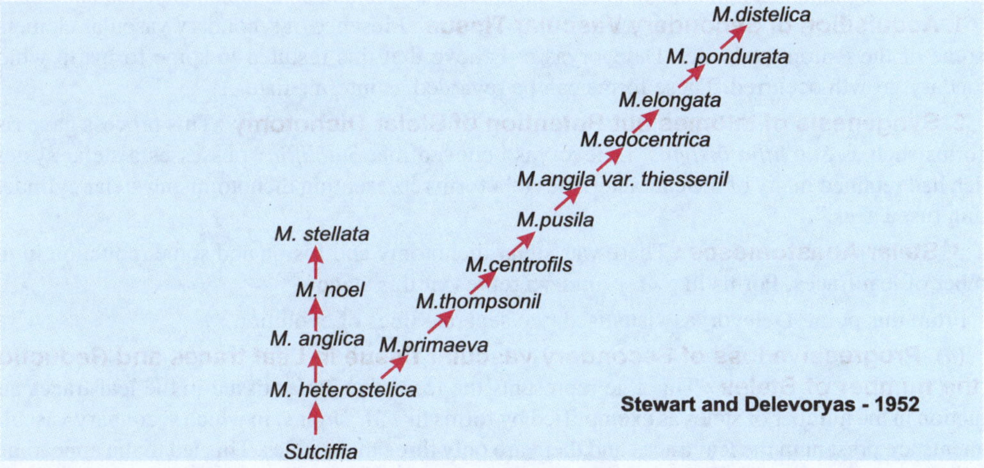

Fig. 2.48. It illustrates the views of Stewart and Delevoryas regarding the Phylogeny of Medullosans.

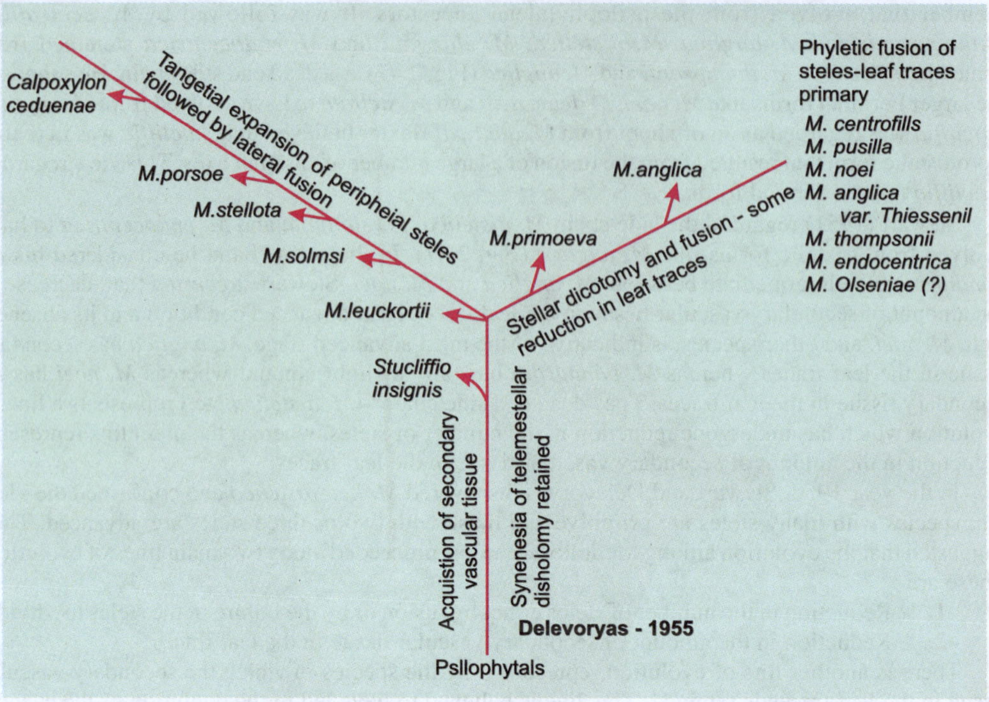

Fig. 2.49. Evolutionary developments in the Medullosaceae presented by Delevoryas (1955).

Andrews and Mamay (1953) discovered some species of *Medullosa* from American horizons and suggested that polystelic condition can arise in two ways :

1. From the distelic forms.
2. From polystelic ancestors such as *Cladoxylon*, *Stetoxylon*, *Pietzschia* and *Xenocladia*.

In the year 1955, Delevoryas reduced (Fig. 2.49) the number of species and varieties of *Medullosa* and presented his own scheme of phylogeny of Medullosaceae. He considered that the Medullosaceae originated from the Psilophytalean ancestors (as was supposed by Arnold and Baxter). Delevoryas view can be conveniently considered under the following headings :

1. Acquisition of Secondary Vascular Tissue : Presence of secondary vascular elements in some of the Psilophytales led Delevoryas to believe that this resulted in some forms in which secondary growth occurred. These forms can be regarded as intermediate.

2. Syngenesis of telomes but Retention of Stelar Dichotomy : This process gave rise to forms such as *Sutcliffia insignis*. Delevoryas believed that *Sutcliffia* "possesses a stelar system which had retained many of the characteristics of a form representing dichotomising stelar cylinders within fused axes.".

3. Stelar Anastomoses : There was stelar dichotomy and fusion and some reduction in the number of leaf traces. Forms like *M. primaeva* represent this stage.

From this point, Delevoryas visualised two separate lines of evolution :

(*a*) Progressive loss of Secondary vascular tissue in Leaf traces and Reduction in the number of Steles : This line represents the loss of vascular tissue in the leaf traces and reduction in the number of steles as exemplified by forms like *M. anglica* in which secondary vascular elements are present in the leaf traces and there are only three main steles. This led to the appearance of forms like *M. centrofilis*, *M. pusilla*, *M. noei*, *M. anglica* var. *thiesseni*, *M. thompsonii*, *M. endocentrica* and *M. olseniae*.

(*b*) Tangential expansion of Peripheral Steles followed by Lateral Fusion : This line of evolution visualises the retention of secondary vascular elements by the leaf traces and reduction in the number of steles by tangential expansion of peripheral steles. In this manner, the peripheral steles fused with each other and there was consequent decrease in their number. This trend is exemplified by forms like *M. leuckartii*, *M. solmsii*, *M. stellata*, *M. porosa* and *Colpoxylon aeduense*. Delevoryas (1955) regarded *Colpoxylon aeduense* Brongniart "as having a stelar system formed by a phylelic fusion (tangentially) of the several peripheral steles of some of the Permian forms originating from the European continent.". He further stated, "the internal star rings" which in some of the Permian medullosae showed only little primary xylem and were often small in size and number, appear to have disappeared entirely in *Colpoxylon*".

To sum up the medullosans may have evolved from the dichotomously branched psilophytic plant which at a certain stage developed secondary vascular tissues. This achievement of secondary vascular tissue was probably followed by dichotomy of the stelar axes and fusion of the cortices of individual axes. This resulted in the polystelic condition, each stele having primary as well as secondary vascular tissues. This polystelic condition later underwent reduction in the number of steles to three or two as a result of anastomosis. The differentiation between stem steles and leaf traces also became well marked.

Presence of many steles with secondary vascular tissue in some Cladoxylales like *Cladoxylon taeniatum* and *Voelkelia refracta* lead us to believe that *Medullosa* might have originated from such forms. Delevoryas has suggested such a possibility. In his own words, "*Xenocladia medullosina*, a related Upper Devonian Species is polystelic, with peripheral radiating steles surrounding a central, more nearly cylindrical ones. It is probably closer to being a fern than a pteridosperm and has a vegetative structure like that of the *Cladoxylales*. In fact, the most promising ancestral types to the pteridosperms appear to be in the group of late Devonian—early Mississippian forms about which more needs to be known concerning foliage and reproductive structures."

Andrews also agreed with Delevoryas in his proposition of the trend involving tangential expansion and lateral fusion of the peripheral steles which led to forms like *M. leukartii*, *M. stellata* and *M. gigas*. In his own words, "I am personally inclined towards this latter theory, in which case it would seem as though one may look to late Devonian and early carboniferous fossils such as *xenocladia—Steloxylon—Cladoxylon* as possible ancestral types." He further states, "At the other end of the scale, some of the Permian medullosans begin to approach a type of stem anatomy that is not distantly removed from that of the cycads". In his concluding remark Andrews states, "....... that *Medullosa* is a genus of convenience. It is difficult to draw sharp lines between the different species and we are in fact, just beginning to learn something of the degree of stellar variation that may be present in the ontogeny of a single plant, but it seems virtually certain that *Medullosa* will eventually be divided into several distinct genera."

RELATIONSHIPS

The Medullosaceae have been related to the ferns and the cycads. Their polystelic condition and the foliage has been mainly regarded as characters that relate them to the ferns. Zeiller (1890), Weber and Sterzel (1896), Scott (1899, 1923) and Seward (1917) have greatly emphasized such a relationship. Seward (1917) pointed out a significant difference between the stelar organization of ferns and the medullosans. He pointed out that in ferns, the leaf is the determining factor in the evolution of the stellar system whereas this is not the case in the genus *Medullosa*. A single stele of *Medullosa* has also been compared to the vascular cylinder of same species of *Gleichenia* and *Lygodium*. The protostelic nature of each stele on *Medullosa* with primary xylem tracheids embedded in parenchyma is likened to a similar situation in *Gleichenia dichotoma*. Such a comparison has also been extended to *Heterangium* but it must be noted here that in *Heterangium*, the primary xylem is radially symmetrical and this is not so in *Medullosa*. In *Medullosa*, the protoxylem is exarch.

According to Delevoryas (1955), such similarities between pteridospermic steles and fern steles are superficial and do not form any concrete basis for a relationship.

The frond-like nature of the leaves in Medullosae is another close resemblance between ferns and the medullosaceae. The pinnae and the pinnules, their arrangement and even their variation patterns are quite similar. The presence of hair, their position and structure are also very similar. It is often very difficult to distinguish between the leaves of ferns and pteridosperms. It is only when seeds are attached to the fronds that a clear distinction can be made between the two.

The chief characters that point towards cycadean relationship are the structure of the leaf, root, the presence of accessory bundles in the stem and structure of the seed. The nature of the secondary xylem and phloem of the stem also point towards cycadean affinity (Seward, 1917). Worsdell (1906), Miss De Fraine (1912), Miss Bancroft (1914) and Delevoryas (1955, 56) are the chief proponents of cycadean affinity of the medullosae. Worsdell (1906) put forth an explanation that the presence of small patches of vascular tissue (with phloem pointing towards the centre) internal to the main stele in *Cycas* are the remains of the centripetal portion of a medullosan stele. He cites *Medullosa stellata* as an example in which hollow stellar cylinder is found. He derived cycads from such a form by a progressive disappearance of centripetal secondary vascular elements and the so called star rings. The presence of concentric rings of secondary vascular tissue in some cycads has also been regarded by Worsdell to have originated from some medullosan steles (*M. stellata* var *gigantea*). The anatomy of some seedlings in cycads and the presence of scattered vascular bundles in the peduncle of *Stangeria* (Cycadales) are also regarded as medullosan characters derived by the progressive disappearance of centripetal xylem and phloem and breaking up of sinuous steles of forms like *Medullosa leuckartii*.

Scott (1909, 1923) and Arnold (1953) differed from Worsdell's views and regarded Cycads to be nearer to lyginopterid stems rather than the Medullosae. The discovery of the genus *Sutcliffia* led Miss de Fraine (1912) to postulate that cycads bear affinities to both lyginopterids and medullosae. *Sutcliffia* exhibited anatomical features that were cycadean (monostelic vasculature) as well as medullosan (polystelic). It also resembled *Heterangium* in some respects. According to her, it combined the characters of cycads, lyginopterids and medullosae. Miss Bancroft (1914) also agreed with Miss de Fraine, but Delevoryas (1955) differed from them and regarded *Sutcliffia* to be a primitive stem that showed primitive medullosan features. However, Delevoryas believed that some advanced medullosae from the Permian showed certain anatomical features that show cycadean affinities (*M. stellata*). He agreed with Worsdell that loss of star rings in such forms led to the "cycadean anatomy" and further added, "indeed, *Colpoxylon acduense* represented just such a type in which the star rings had disappeared."

Pointing towards great similarity between the seeds of medullosae and cycads, Delevoryas states, "a fact of some strength in supporting Worsdell's ideas is the strong similarity in seed types between the medullosae and the cycads. Although the actual means of fertilization were perhaps different (sperm cells may have been formed directly within the pollen grain from which they were released) (Renault, 1887; Stewart; 1951), seeds structure was quite similar. The method of seed attachment to obviously reduced leaves in some species of *Cycas*, furthermore, seems to parallel the attachment of trigonocarp seeds on medullosan fronds."

Regarding pollen bearing organs he states, "at present there is difficulty in deriving cycadean male fructifications from medullosan pollen bearing organs, and further investigation in that line may be promising."

Delevoryas (1955) and Delevoryas and Stewart (1956) have hinted at a close relationship between the stem structure of Pentoxyleae and the medullosan stems. Sahni (1948), Mittre (1957) and Sharma (1969 and 1971) also agree with Delevoryas and Stewart. Sharma (1971) states, "So a line of evolution can be drawn from the Palaeozoic stem genus *Medullosa* through *Guptioxylon* to *Pentoxylon*. During this course, loss of medullary bundles took place besides an overgrowth on the centripetal side of the secondary xylem. This line of evolution must have diverged from the primitive type of medullosan stems as the leaf traces are still possessing well developed secondary xylem in both *Guptioxylon* and *Pentaxylon*. Presence of secondary xylem in a leaf trace is regarded as a primitive feature (Stewart and Delevoryas, 1956)".

Bierhorst (1971) has also supported the derivation of cycads from the Pteridopermales. In his own words (1971, page 400), "It is very difficult not to entertain the hypothesis of direct descent of the Cycadales from the Pteridospermales. One preceedes the other in the fossil record with no intervening time gap. Structural parallels between the two orders are multiple and many are shared by no other taxa." He further states, "A presumed derivation of cycads from medullosans is particularly well supported and seems contradicted only by differences in stelar morphology. Concentric steles in some species of *Medullosa* have been likened to similar—appearing ones in living cycads, but in the latter the outer vascular zones are derived from accessory cambia and are in no way comparable to those of the former. A hollow cylindrical medullosan stele must loose its internal phloem and centripetally developed xylem to approach the cycadean type. Changes comparable to these are assumed to have occurred within the articulates."

A careful perusal of the various views put forward by stalwarts like Scott, Seward, Worsdell, Miss de Fraine, Miss Bancroft, Delevoryas, Delevoryas and Scott, Andrews, Arnold, Bierhost, Sahni, Vishnu Mitre and Sharma, one comes to the conclusion that cycads had closer affinities with the medullosans. A comparison of the Lyginopteridaceae, Medullosaceae and Cycadales will be worthwhile to realise the similarities and differences between them :

		Lyginopteridaceae	Medullosaceae	Cycadales
Habit	1.	Tree like	Tree like	Medium sized trees.
	2.	Stems weak and aerial, frequency branched.	Stems aerial and erect, not branched	Stems aerial, columnar, erect and slow growing, unbranched or rarely branched.
	3.	Leaves large bipinnate or tripinnate. Vernation circinate.	Leaves large, bippinate or tripinnate. Vernation not known.	Leaves reduced and usually once pinnate, or bipinnate circinate.
	4.	The pinnules show free venation *i.e.* the veins were free.	Venation may be pinnate reticulate, dichotomous or even free.	Venation is pinnate or open dichotomous.
	5.	The leaves spirally arranged.	Leaves spirally arranged.	Leaves spirally arranged.
	6.	Roots adventitious	Roots adventitious	Roots adventitious.
Anatomy	7.	The stem stele consists of discrete vascular bundles. In some cases it is protostelic.	The stems are polystelic. In some advanced members a complete vascular cylinder surrounding smaller ones in the centre (Star rings).	The stem stele consists of a rings of discrete vascular bundles.
	8.	The xylem is mesarch.	The xylem is mesarch.	The xylem is endarch.
	9.	A distinct pith is present.	The steles are embedded in well developed ground tissue.	Pith is present.

	Lyginopteridaceae	Medullosaceae	Cycadales
10.	The cortex is well developed and consists of thick-walled or fibrous cells in the outer cortex. The inner cortex is parenchymatous.	The cortex is distinct and consists of hypodermal patches of thick-walled cells and parenchymatous inner cortex.	The cortex is well developed and may consist of outer sclerenchymatous and inner parenchymatous cortex.
11.	Mucilage canals absent.	Mucilage canals present.	Resin passage are present.
12.	Pericycle is distinct and surrounds the entire vascular region.	It is recognizable in each stele separately.	Pericycle is not distinct.
13.	Endodermis indistinct.	Endodermis indistinct.	Endodermis indistinct.
14.	Cambium is distinct.	Cambium present in each stele. Due to polystelic condition individual steles have, their own cambium ring.	Cambium is distinct.
15.	The leaf traces are mesarch.	Leaf traces are mesarch.	Leaf traces are mesarch.
16.	The leaf traces arise by the tangential division of the vascular bundle of the stem.	Their mode of development not very clear.	They arise by tangential division of stem bundle.
17.	The leaf traces are single but become double later.	The leaf traces are multiple.	The leaf traces are multiple.
18.	The leaf trace has a strip of secondary xylem on its outer face. This strip is gradually lost during its course through the cortex.	The leaf traces in some species possess secondary xylem which may or may not persist during their course towards the petiol. In some advanced (Permian) species it may be entirely primary in nature.	The leaf traces are diploxylic *i.e.* they possess both centripetal and centrifugal xylem. Whether the centrifugal xylem is secondary in origin or primary is a debatable question.
19.	Seeding anatomy is not known.	Seeding anatomy is not known.	In *Ceratozamia*, *Dioon edule*, *D. spinulosum* and *Zamia*, the vascular cylinder in the cotyledonary node is a protostele. Later it becomes a siphonostele. The cotyledonary traces are endarch first and later they become mesarch. It is reverse in *Zamia*. In Microcycas, the seedling has an endarch siphonostele which later becomes eustelic. The cotyledonary trace is endarch first but later becomes mesarch or exarch.

	Lyginopteridaceae	Medullosaceae	Cycadales
20.	The secondary wood is manoxylic *i.e.* it is soft textured.	The secondary wood is manoxylic.	The secondary wood is manoxylic.
21.	The secondary growth is monoxylic.	It is monoxylic in each stele.	It is polyxylic in Cycas and some species of *Macrozamia* and *Encephalartos*. In other genera, it is monoxylic.
22.	The development of the secondary xylem is usually exocentric.	The secondary xylem is endocentric in most of the species and exocentric in some.	The secondary xylem is usually exocentric.
23.	Internal secondary xylem patches are reported in some specimens of *Lyginopteris oldhamia*. In *Heterangium* and *Schopfiastrum,* the stele is a mixed protostele.	The star rings and central smaller steles may be regarded as comparable to patches of internal xylem present in some lyginopterids.	Internal xylem is absent.
Reproductive Structure			
24.	The microsporangia are borne on the ultimate branches of the frond.	They are borne on the fronds.	The microsporangia are borne on microsporophylls that are spirally arranged along a central axis to form a definite cone.
25.	The microsporangia are bisporangiate.	The microsporangia are free and not bisporangiate They may be laterally fused.	The microsporangia are free, but there are indications of slight fusions.
26.	The microspores are trilete.	The microspores are monolete or trilete.	The microspores are monolete.
27.	The ovules and seeds are cupulate.	The ovules and seeds are not cupulate.	The ovules and seeds are not cupulate.
28.	The cupules in *Lyginopteris* are covered by distinct glandular outgrowths that are stalked. These glands are present on all other parts also.	Glands are absent.	Glands are absent.

	Lyginopteridaceae	Medullosaceae	Cycadales
29.	Inverted bell-shaped sheath encloses a solid nucellar beak.	Bell-shaped sheath present, but nucellar beak is absent.	The external bell-shaped sheath is present but in internal nucellar beak is usually absent except in some cases.
30.	Sclerotesta is not distinct.	Sclerotesta is distinct and is covered by outer sarcotesta. In some cases an inner fleshy layer is distinct.	The same as in Medullosaceae.
31.	There is a single vascular system in the seed.	There is double vascular system in the seed.	The vascular system is double.
32.	The seed coat develops decomposition Iacunae.	Decomposition lacunae are present.	No definite Iacuane but gelatinisation of the tissue in fleshy layer is reported.
33.	The sporangia do not possess an annulus.	Annulus is absent.	Annulus is absent.

The above comparison clearly brings out the similarities and differences between the three groups. This shows that they are closely related to each other but are in no way derived from each other. The cycads might have originated from some palaeozoic pteridospermous group of plants, but not the Medullosaceae. The Medullosaceae appear to be a group of plants with abnormal features which ended blindly in the course of further evolution. The manocyclic nature of the wood, presence of mucilage canals, girdling leaf traces that are mesarch, persistent cortex, similarity in the structure of the ovules, presence of centripetal xylem in both the cycads and the pteridosperms, point towards the origin of Cycadales from some Palaeozoic pteridospermous stock which also gave rise to lyginopterids and medullosoids, along separate lines of evolution (Taylor, 1969).

Mamay (1969) described some megasporophylls from Permian that were convincingly cycadalean. Each such organ bore two rows of ovules on its proximal end and a laminate structure distally. They were more leaf like and led Mamay to propose a Palaeozic origin for the cycads. But whether the ancestors were really pteridospermous or some more primitive group of plants *e.g.*, the progymnosperms cannot be definitely stated.

Prof. Mehra (1974, 1988) envisaged the origin of gymnosperms from a group named by him as *Protoseminales*, which probably existed in Late Devonian and Early Carboniferous. He characterised the group as follows :

1. The habit of these plants was presumably shrubby or small trees.
2. The secondary wood was *manoxylic*.
3. They had three dimensional branched structure without any distinction into leaves and stems, and made up of *phyllophores* the like of which we find in some *Aneurophytales* or *Coenopteridales*.
4. These phyllophores bore seeds at their ultimate branches.
5. The seeds were of the most primitive type. They were *radiospermic* or in some cases *platyspermic* below and radiospermic above (*e.g., Hydrasperma*).
6. The seeds were surrounded by free *telomes* (*Genomosperma*) or the telomes were fused partly in some relatively advanced members to form the integument but with apical lobes still free from each other (*Hydrasperma*, *Eurystoma*).
7. The seeds lacked a well organised *Micropyle* but possessed a **salpinx** for catching pollen and a '*crevice-like*' pollen chamber.

This group was proposed by Mehra (1974, 1988) on the basis of discovery of some primitive seeds by Long (1958-1960). From this group (Protoseminales), Mehra proposed that there emerged four main lines of evolution in the Carboniferous period.

1. The first line resulted in the origin of Pteridorspermales that gave rise to the sub-lines *Caytoniales* and *Cycadales*. The *Lyginopteridaceae* and *Medullosaceae* were also believed to have arisen from the primitive pteridospermales as two separate offshoots that died down in the Carboniferous and Permian respectively.
2. The second independent line was to the *Bennettitales*.
3. The third line was to the *Ginkgoales*.
4. The fourth line was to the *Cordaitales* which had a span of 115 million years and was responsible for the origin of *Coniferales* along with *Taxales* and the *Gnetophytes* (*Gnetum*, *Ephedra* and *Welwitschia*).

Mehra regarded *Glossopteridales* as an isolated group with no "indication of its affinities or origin."

FAMILY : CALAMOPITYACEAE

This family is considered to be an unnatural assemblage of some genera that could not be placed elsewhere. All of them resemble *Lyginopteridaceae* in being *monostelic* and in possessing *manoxylic* secondary wood (except four genera that have been transferred to Cordaitales by Lacey, 1953). The foliage and the reproductive structures are not known. The family is named after the stem genus *Calamopitys* which was discovered by Unger in 1856 from the Thuringian strata of Upper Devonian age in the form of petrifactions. He named the type species as *Calamopitys saturnii*. The family name *Calamopityeae* was given by Solms—Laubach, (1896). The family is entirely Palaeozoic, restricted to the Upper Devonian and Lower Carboniferous periods. It includes the stem genera *Calamopitys*, *Stenomyelon* and *Diichinia* and the petioles belonging to the genus *Kalymma*. Later (Read, 1936) four more stem genera were discovered and the number rose to seven genera and sixteen species (*Bilignea*, *Eristophyton*, *Endoxylon* and *Sphenoxylon*). These four genera had pycenoxylic secondary wood and were transferred to the *Cordaitales* (Lacey, 1953). Only genera with *monoxylic* secondary wood are retained in the family (*Calamopitys*, *Stenomyelon*, *Diichnia*).

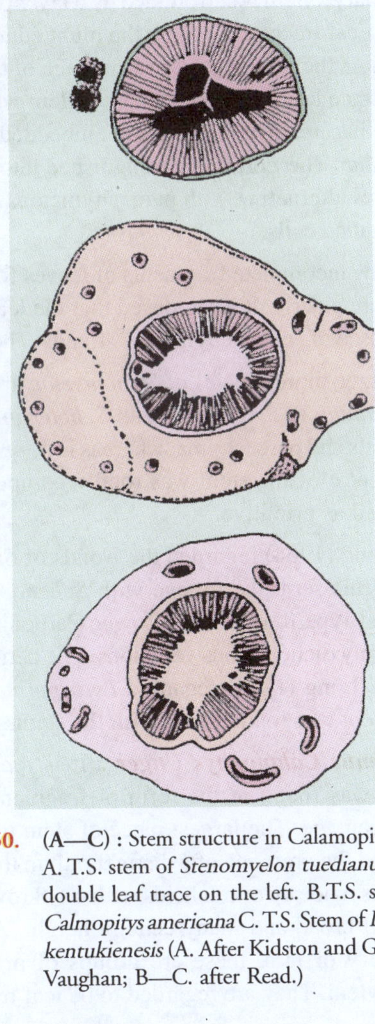

Fig. 2.50. (A—C) : Stem structure in Calamopityaceae. A. T.S. stem of *Stenomyelon tuedianum* with double leaf trace on the left. B.T.S. stem of *Calmopitys americana* C. T.S. Stem of *Dilchnia kentukienesis* (A. After Kidston and Gwynne-Vaughan; B—C. after Read.)

Stenomyelon (Fig. 2.50A) was originally discovered by a geologist Matheson in the fifties of the nineteenth century from the late Devonian—Lower Carboniferous and named it as a "Tweed Mill fossil", after the name of the place where it was discovered . Dr. Kidston later rediscovered it from the same bed and gave a full description of the specimen with the help of Prof. Gwynne—Vaughan. The type species was named as *Stenomyelon tuedianum* Kidston. It was found in the form of petrified pieces of a flattened stem. Fragment of rachis and portions of laminae were also found. A section of the stem (Fig. 2.50A) reveals a *protostele* that is bluntly triangular in outline and is composed of exarch primary xylem. The primary xylem consists of reticulate pitted tracheids. Some parenchymatous cells have been reported to occur along the periphery of the protostele. The primary phloem has not been reported. A band of parenchyma extends from the middle of each of the three sides of the primary xylem towards the centre of the stele. It divides the primary xylem into three groups. These parenchymatous bands are supposed to be **pith.** The tracheids at the outer face of primary xylem are scalariform and narrow. They represent the **protoxylem.** During secondary growth, the secondary xylem first appears along the slightly concave sides of the primary stele. Later it extends all round and completely surrounds the triangular primary stele. Later the secondary xylem assumes a cylindrical form. It is made up of tracheids with multiseriate pits on the radial walls. The secondary xylem is traversed by many medullary rays that are 1—6 cells broad. The secondary xylem is surrounded by a layer or more of thick-walled cells that are regarded as **periderm** layer. Leaf traces arise from the blunt edges of primary xylem. They are more or less cylindrical in shape. As they traverse outwards, each of them bifurcates into two and later into many in the cortex. Each trace has a little secondary xylem which persists for some time. In the petiole the leaf traces divide into many traces that are embedded in the ground tissue. The generic name for the petiole is *Kalymma.* The cortex is distinguished into an outer cortex which is made up of long vertical strands of fibres alternating with parenchymatous cells. The rest of the cortex is supposed to be made up of thin-walled cells.

The incomplete fragments of leaves found in association with the stem pieces do not reveal the structure clearly. It is supposed that the leaves were simple with thick leaf blade and a hypodermis composed of sclerenchymatous strands. There is an evidence of presence of several vascular bundles.

Three more species of *Stenomyelon* are known. These are *S. primaevum*, S. *heterangioides* and *S. muratum*. In *S. muratum* and *S. heterangioides,* the primary xylem was made up of a mixture of tracheids and parenchyma, whereas in *S. primeavum,* there was no parenchyma in the primary xylem. The centre of the stem was totally composed of tracheids of the primary xylem. This species is regarded as primitive.

Long (1964) regarded the fronds of *Stenomyelon* to be like those of *Diplothmenea* which is a Palaeozoic fern-like foliage with *Sphenopteridium* like pinnules. *Sphenopteridium* is also a fern-like leaf type found in the Lower Carboniferous having wedge-shaped or dissected pinnules with uniformly dichotomous venations. The petioles in *Diplothmema* are equally forked below the lowest pinnae. Long (1964) regarded *Lyrasperma scotica* (Fig. 1.21. A) to be the seed of *S. tuedianum* because it was found along with the stems and petioles of this species.

Genus Calanopitys Unger : It is represented by four species. The type species *C. saturnii* Unger was found in the form of fragmentary petrifactions from the Thuringian strata of Upper Devonian age. The largest piece of stem measured about 1.5 cms in diameter and in cross-section revealed the presence of a distinct pith in the centre. The pith is surrounded by the secondary xylem which is made up of tracheids with 4—8 rows of bordered pits and medullary rays that are more than one cell broad and many cells in height. Along the inner border of the secondary xylem and the periphery of pith, these are groups of primary xylem strands (6) that have a distinct internal protoxylem. They are regarded to be leaf traces as found in *Lyginopteris*. These leaf traces traverse through the secondary xylem as single traces and on entering the cortex divide into two (as in *Lyginopteris*). On entering the leaf base, each trace divides into six. The decurrent leaf base is separated from the stem proper by a distinct band of thick-walled cells. The cortex in the stem is wide and consists of an outer cortex made up of sclerenchymatous strands alternating with parenchymatous cells (Sparganum type). The inner cortex is parenchymatous.

Calamopitys annularis (Unger), was discovered from the Upper Devonian deposits of Kentucky. If differs from *C. saturnii* in possessing tracheids mixed with the parenchyma of the pith. The periphery of the pith is occupied by a comparatively well developed primary xylem which exhibits a strong tendency towards the formation of a continuous zone except at places where the protoxylem elements occur. At such places the metaxylem forms a definite strand. The secondary xylem is **manoxylic** and completely surrounds the primary xylem and the mixed pith. The formation of leaf traces is not clear but the strands seem to divide before entering the cortex. The leaf traces in the cortex are concentric.

Calamopitys americana Scott and Jeffrey (Fig. 2.50, B) was also discovered from Upper Devonian deposits from Waverley Shales of Kentucky. It is represented by petrified portions of stems with leaf bases and detached petioles. The centre of the stem is occupied by parenchymatous pith with mixed tracheids. At the periphery of the pith are many primary xylem strands that appear to be contiguous and look like a continuous band of primary xylem. These strands are mesarch. The tracheids are large and are separated from one another by narrow strips of parenchyma. The monoxylic secondary xylem completely surrounds the primary xylem and pith. Its tracheids are 30—60 μ in diameter and have 5—6 alternating rows of small pits. The cambium and phloem are indistinct. The outer cortex is of Sparganum type. The leaf traces arise from the peripheral primary xylem strands. The primary xylem strand divides into two as it departs from the peripheral zone and traverses through the secondary xylem as two separate bundles accompanied by an arc of secondary xylem tracheids. On entering the cortex the, leaf trace bundles are completely surrounded by the secondary xylem. The strands are mesarch and divide further on entering the leaf base.

C. foerstes Read differs from other species in the manner of origin of the leaf traces. In this case the leaf trace originated from two adjacent vascular bundles of the perimedullary primary xylem, as compared to one in all other species. These two become fused with one another during their course towards the cortex through secondary xylem. Each of the fused traces gives out a branch which again fuses to form a single leaf trace in the cortex. This leaf trace has two protoxylem groups.

Genus Diichnia kentuckiensis Read (Fig. 2.50 C and 2.51A) was discovered from the Upper Devonian of the New Albany Shale. The section of the petrified stem reveals a wide cortex with outer cortex traversed by a network of hypodermal strands. The rest of the cortex is parenchymatous and is delimited from the vascular region by a periderm-like layer of thick-walled cells. The centre of the stem is occupied by a five-angled pith of mixed parenchyma and tracheids (Fig. 2.50 C). The angles are occupied by five strands of mesarch primary xylem that are separated from the secondary xylem by a narrow band of xylem parenchyma. The secondary xylem is manoxylic and completely surrounds the pith and the primary xylem. The leaf trace is double and originates as a pair of strands from the adjacent primary vascular bundles at the pith angles. These traces do not fuse and traverse separately through the secondary xylem and enter the cortex where they slightly increase in size. They divide once or more times and finally enter the leaf base and the petiole. The unilacunar node with a double trace is considered to be a primitive feature (Bailey, 1956) and has been described from a number of living and extinct genera that are not related otherwise. They are known in the living members like *Ginkgo* (Gunkle and Wetmore, 1946) and *Ephedra* (Marsden and Steaves, 1955). It has also been described in some angiosperms (Bailey, 1956).

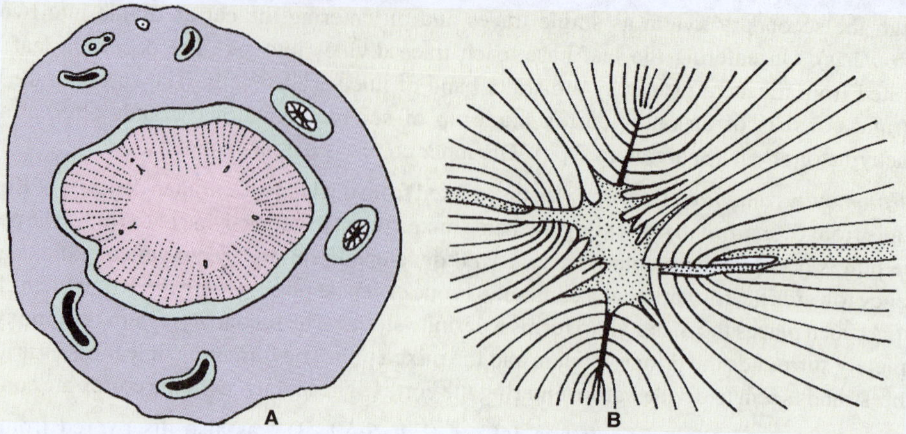

Fig. 2.51. Calamopityaceae. A T.S. stem of *Dilchnia kentuckinensis.* Note the five angled pith and small primary xylem strands in the angles. Two strands on the left have divided. (After Read) B. *Sphenoxylon enpuncate* T.S. stem showing irregular pith, four inwardly projecting secondary wood wedges each having a departing trace. Primary xylem is solid black (After Thomas)

Genus Kalymma (Fig. 2.52). It was originally discovered by Unger from Thuringia of Upper Devonian age. He regarded it to be a stem and described two species which he named as *K. grandis* Unger and K.*striatum* Unger. Later Solms—Laubach regarded it as a petiole of *Calamopitys* and his findings were confirmed by Scott and Jeffrey who found a leaf base with *Kalymma* structure in connection with a piece of *Calamopitys* (believed to be *C. americana*). Seward also confirmed the earlier findings and regarded it as a petiole. It was also collected from Kentucky and Genessee shales of Upper Devonian age. It was also collected from Waverley shales believed to be Lower Carboniferous. Now *Kalymma lirata* is regarded as the petiole of *Calamopitys americana* and *K. resinosa* is believed to belong to *Diichnia kentuckiensis*. The section of the petiole shows a layer of epidermis followed by hypodermal layer of thick walled sclerenchymatous cells. The rest of the tissue is called ground tissue in which is embedded a ring of vascular bundles. In some cases, the bundles may be arranged in a horse- shoe manner. The vascular bundles may be radially elongated or circular in outline. The ground tissue consists of homogenous parenchyma. The xylem in the bundle is entirely primary in nature with a clear distinction into metaxylem and protoxylem. The protoxylem may occupy the centre (in circular bundles) or it may be found at the ends. Phloem elements are not well preserved.

The affinities and the phylogenetic portion of the Calamopityaceae are not yet clear because we are not certain about their foliage and reproductive organs. Long (1964) has no doubt speculated that the fronds of some members are like those of *Diplothemma* and the pinnules like *Spehnopteridium* but it is not certain. He also believed that the seeds called *Lyrasperma scotica* also belong to *Stenomyelon tuedianum* because they were usually associated with stems and petioles of this species. In case these findings are further substantiated by their occurrence in organic connections, then certainly the Calamopityaceae have close relationship with the Lyginopteridaceae. The Calamopityaceae resemble the other Pteridospermales in possessing "Sparganum" type of hypodermis, monostelic primary vasculature and manoxylic secondary wood. The structure of the petiole and the mesarch leaf traces in some species and the association of secondary xylem with the leaf trace bundles are also characters that bring them closer to the Lyginopteridaceae and Pteridosperms in general.

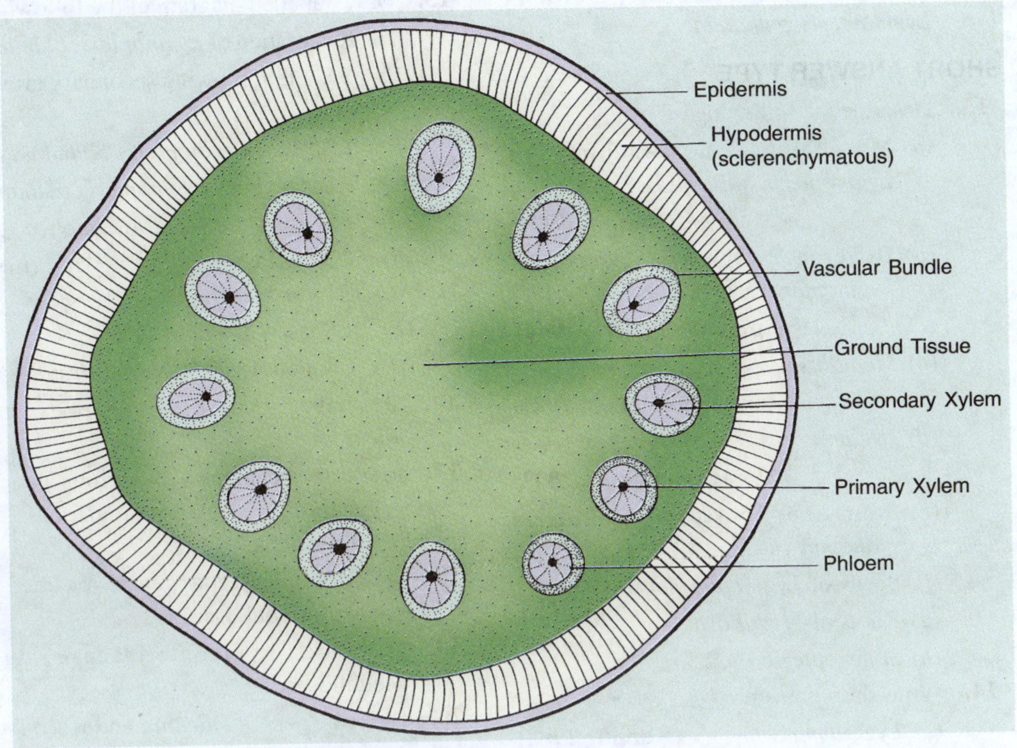

Fig. 2.52. T.S. petiole of *Kalymma lirata* showing internal structure. (1. Epidermis; 2. Hypodermis of sclerenchyma cells; 3. Vascular bundles; 4. Ground tissue; 5. Secondary xylem; 6. Primary xylem; 7. Phloem

QUESTIONS

Essay Type

1. Discuss general characters of Pteridospermales. **(Kanpur, 1986)**
2. Describe the characteristic features of Pteridospermales and discuss their affinites. **(Kanpur, 1997)**
3. Give a brief account of important features of Pteridospermales. Discuss its affinities with other groups. **(Kanpur 2000)**
4. Write brief account of the morphology, anatomy and fructifications of Pteridospermopsida with particular reference to two genera studied by you. **(Awadh, 1990)**
5. Write in detail the reproductive biology of the Pteridospermales. **(Rohilkhand, 1996)**
6. Discuss the anatomy of early Pteridosperms with suitable examples. **(Rohilkhand, 1998)**
7. What are Pteridosperms ? Mention their distribution, vegetative and reproductive characteristics and the phylogeny of the group. **(Poorvanchal, 1998)**
8. What are Pteridospermales? Describe briefly the fossil flora of Lyginopteridaceae. **(Rohilkhand, 2002)**
9. With labelled diagrams, describe the plants attributed to the family Medullosaceae.
10. Give an illustrated account of Calamopityaceae.
11. With diagrams describe *Lyginopteris oldhamia*.

12. Give an account of secondary growth in *Lyginopteris oldhamia.*

SHORT ANSWER TYPE

13. Describe the following :
   (*a*) Morphology and Anatomy of *Lyginopteris oldhamia.* **(Kanpur 1994)**
   (*b*) Differences between seed of *Lyginopteris* and *Trigonocarpus.* **(Kanpur, 1990, 1992, 1994)**
   (*c*) *Medullosa* **(Rohilkhand, 2003, 1989)**
   (*d*) *Lyginopteris* **(Rohilkhand, 2003)**
   (*e*) Stem and Root of *Lyginopteris oldhamia*
   (*f*) Seeds of *Lyginopteridaceae*
   (*g*) Roots of *Medullosa.*
   (*h*) *Sphenopteris*
14. Write short notes on :
   (*a*) *Lyginopteris* **(Kanpur 1988)**
   (*b*) Characteristic features of *Medullosa.* **(Kanpur, 1998)**
   (*c*) *Medullosa* **(Awadh, 2001; Rohilkhand, 2003)**
   (*d*) Pteridospermales **(Rohilkhand, 2003)**
   (*e*) Fructifications of Pteridosperms **(Rohilkhand, 1992)**
   (*f*) Vascular tissues in *Medullosa* **(Poorvanchal, 2001)**
   (*g*) Fructification of *Lyginopteris* **(Gauhati, 2000)**
   (*h*) Ovule of *Lyginopteris* **(Himachal Pradesh, 2000)**
   (*i*) Affinities of Pteridosperms **(Poorvanchal, 2002)**
   (*j*) *Heterangium*
   (*k*) Fructifications of Medullosaceae
   (*l*) Phylogeny of *Medullosaceae*
   (*m*) Calamopityaceae
15. Draw labelled diagram of the following :
   (*i*) Restoration of *Lyginopteris oldhamia*
   (*ii*) T.S. Stem showing secondary growth of
   (*iii*) T.S. Root *Lyginopteris oldhamia*
   (*iv*) V.S. Ovule of *Lyginopteris oldhamia*
   (*v*) T.S. Stem of *Medullosa anglica*
   (*vi*) T.S. Petiole of *Medullosa* (*Mycloxylos*)
   (*vii*) L.S. seed of *Pachytesta*
   (*viii*) T.S. Petiole of *Kalymma lirata*
16. Describe genera *Neuropteris* and *Taeniopteris.*
17. Mention the names of various genera regarded as fructifications of Medullosaceae.
18. Describe at least four genera regarded as seeds of Medullosaceae.
19. Compare Lyginopteridaceae with Medullosaceae.
20. Describe genera *Dilchnia* and *Kalymma.*

OBJECTIVE TYPE

21. **Select the correct answer:**
   (*i*) The Pteridosperms were discovered by
   (*a*) A.C. Seward (*b*) Grand Eury
   (*c*) Birbal Sahni (*d*) K.R. Surange.
   (*ii*) Pteridosperms reached their climax in
   (*a*) Palaeozoic (*b*) Mesozoic
   (*c*) Coenozoic (*d*) Permian
   (*iii*) Pteridospermales have been divided into the families which number
   (*a*) 2 (*b*) 3 (*c*) 5 (*d*) 7
   (*iv*) The plants of Lyginopteridaceae were represented in
   (*a*) Upper and lower carboniferous
   (*b*) Middle Carboniferous
   (*c*) Mesozoic (*d*) Devonian
   (*v*) *Crossotheca* is the pollen bearing organ of the family
   (*a*) Lyginopteridaceae
   (*b*) Medullosaceae

(*c*) Calamopityaceae

(*d*) Glossopteridaceae

(*vi*) Which of the following was established by Cotta?

(*a*) *Lyginopteris* (*b*) *Crossotheca*

(*c*) *Medullosa* (*d*) *Cycas*

(*vii*) Pteridosperms occupy an intermediate position between

(*a*) Ferns an Cycadophytes

(*b*) Bryophytes and Pteridophytes

(*c*) Pteridophytes and Conifers

(*d*) Pteridophytes and angiosperms

(*viii*) The pteridosperms resemble the ferns in being

(*a*) Homosporous

(*b*) Heterosporous

(*c*) Exarch in position

(*d*) Large arboreal habit

(*ix*) *Dolerotheca formosa* was described in detail by

(*a*) Cotta (*b*) Halle

(*c*) Presl (*d*) Seward

22. Fill in the blanks :

(*a*) Pteridosperms belong to the class ....

(*b*) Mesarch leaf traces are found in ....

(*c*) Monolete microspores are present in .... and ....

(*d*) The group Cycadopsida was discovered in ....

(*e*) *Lyginopteris oldhamia* was found in abundance in coal ball horizon of .... and ....

CHAPTER 3

Pteridospermales Family–Glossopteridaceae

- Glossopteris
- Gangamopteris
- Vertebraria–the Stem or Rhizomes
- Male Reproductive Organs
- Female Reproductiveorgans – Fructifications
- Habit or Reconstruction of Glossopteris
- Affinities of Glossopteris

PECOPTERIS FOSSIL

Introduction

The Glossopteridaceae flourished well on the **Gondwanaland.** In this introduction, attempt has been made to describe as to how Gondwanaland came into existence.

The transmigration of flora from sea to land took place sometimes in the **Upper Silurian** to the **Lower Carboniferous.** During this time the entire world consisted of a single land mass called the **Pangaea** (name given by Wagner and means all lands). It was surrounded by a universal sea called the **Panthalsa** (it means all seas). Pangaea was inhabited by a uniform flora called the *Rhacopteris–Lepidodendropsis* flora. It is believed that later as we proceed towards the **Upper Carboniferous**, this uniform land mass was split into a **northern continental mass** and a **southern land mass** by a median sea called the **Tethys Ocean.** The former was given the name **Lauracea** (North America and Asia) and the latter was called the *Gondwanaland* (name given by Suess after the Gond tribe). At that time the Laurasia land mass had a uniform climate that was warm and humid. This region had lots of swampy area. It afforded a rich vegetation consisting of evergreen forests of *Pecopteris* type. The fossil remains of this flora are found uptodate in the coal seams of North America and Europe. During this period *Gondwanaland* (Southern land mass) was in the grip of the greatest **ice age** that started in the Upper Carboniferous and perhaps persisted upto the Lower Permian. During this icy age, the *Gondwanaland* had very little of vegetation. It was mostly barren, the remains of this ice age are now evident throughout the *Gondwanaland* by the occurrence of boulder beds covered by **Glacial Conglomerates.** Such boulder beds are now called **Talchir Boulder Beds** and **Tillites** because they were discovered at Talchir in India. Elsewhere in India such beds are also found in Umaria in Madhya Pradesh, Kumaon and Kashmere in Himalayas, and Salt range in Pakistan. Such beds have also been found in Africa (Dwyka tillite), Australia (Bachhus Marsh tillite) and South America. There is some evidence that during this ice age the winters alternated with mild summer (Sussmilch, 1923). There are reports about the occurrence of some, mega- and microfossils in the beds of this ice age. Lele (1966) reported the presence of *Gangamopteris* (Leaf of Glossopteridales) in most of the Indian localities, belonging to Talchir series (Ten) whereas *Glossopteris* (Leaf genus) was reported from only two localities of the same series. A large number of *miospores* have also been described from the ice age. It is now believed (Pant, 1987) that the earliest members of *Glossopteris* flora and the Gondwana ice age appeared simultaneously. Prof. Pant remarks, "We can safely conclude that the pioneers of the *Glossopteris* Flora had come into existence during the Gondwana Ice Age". This conclusion is confirmed by the fact that many vascular plants grow under similar situations in the present times in India and other countries (Green land, New Zealand).

The Gondwana system has been classified into three principal divisions:

1. Lower Gondwana System 2. Middle Gondwana System 3. Upper Gondwana System.

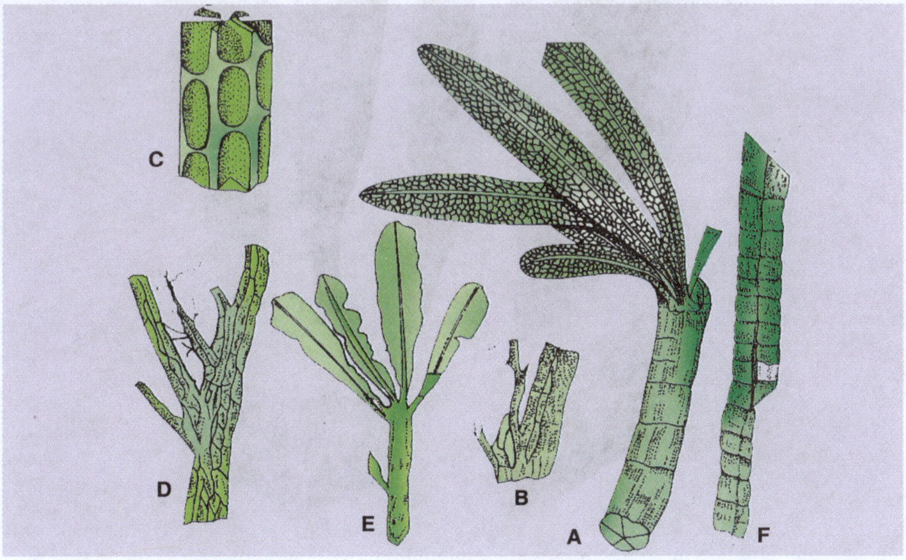

Fig. 3.1. (A—F). Glossopteridaceae. A. Reconstruction of a branch of *Glossopteris* bearing axis (*Vertebraria*) and leaves (After Edwards). B and D. Branched *Vertebraria* axis bearing slender sorts (After D.D. Pant) C. Reconstruction of xylem of the axis called *Vertebraria* (After Walton and Wilson). E. *Glossopteris* branch (*Vertebraria*) bearing leaves (After Walton and Wilson). F. *Vertebraria* as drawn by D.D. Pant (1956).

The lower Gondwana system is further divided into two series:

(*a*) *Talchir Series*. This series is named after Talchir district of Orissa, where it was first discovered. These series also exist in the Aravallis, Rajputana, Salt ranges, Hazara and Simla. They constitute the lowest beds of Lower Gondwana and extend from the Rajmahal hills to the Godavari and from Raniganj to Nagpur. They existed during the Upper Carboniferous and afforded luxuriant growth of gymnospermous plants belonging to the Pteridosperms, Cordaitales and Equisetales. The Glossopteris genera include several species of *Gangamopteris* (in the Karharbari stage), *Glossopteris* and its stems *Vertebraria* and *Gondwanidium*.

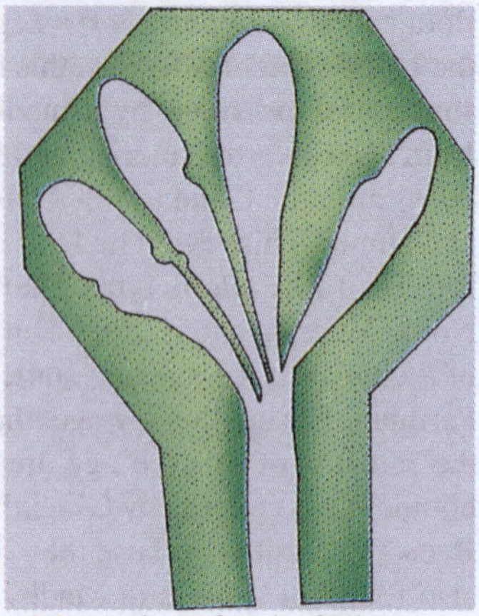

Fig. 3.2. It shows fronds of *Glossopteris* sp. attached to the rhizone (From A.C. Seward).

(*b*) *Damuda Series*. It includes Raniganj (5000 ft.), Iron stone shales or Barren measures (1400 feet) and Barakar (2000 feet) stages that existed during the Middle and Upper Permian ages. The Pteridosperm flora that flourished in this series include *Glossopteris* with *Vertebraria* (mostly found in Raniganj stage), *Gangamopteris, Sphenopteris, Pecopteris, Palaeovittaria, Blemnopteris* and *Merianopteris*. Besides these, many other gymnospermous genera also flourished. Many pteridophytic plants also existed.

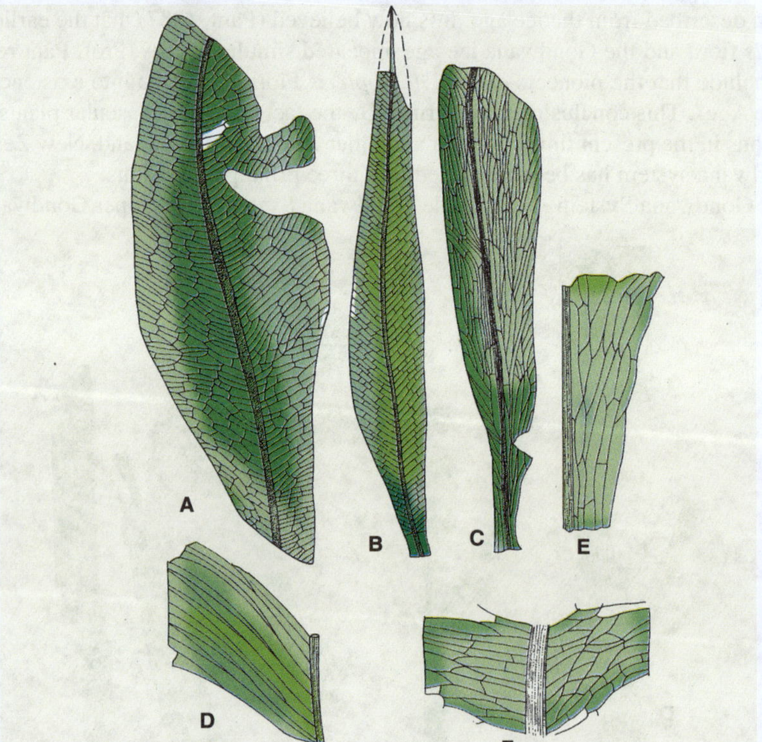

Fig. 3.3. (A—F). *Glossopteris* leaves and their venation patterns. A.Leaf of *Glossopteris retifera* (After Fiestmentle). B.Leaf of *Glossopteris angustifolia* (After Fiestmentle). C. *Glossopteris browniana*. A leaf with mid rib and venation pattern (After Seward). D.Venation pattern in *G. angustifolia* var. *taeniopteroides* (After Seward). E. Details of venation pattern in *Glossopteris browniana* (After Steward). F. Details of venation in *G. indica* (After Zeller).

Strips of Lower Gondwana rocks are also found along the foot of the Himalayas. In Kashmere and Darjeeling area, these rocks are well represented. *Glossopteris* species and *Vartebraria* occur in these areas.

2. Middle Gondwana. The rocks of this stage belong to the Triassic age. It includes the Maleri, Parsora, Mahadev or Pachmarhi and Panchet series. The glossopterid genera *Glossopteris, Vertebraria* (*V. indica*), *Gangamopteris* flourished in these series.

3. Upper Gondwana. These rocks are found in a number of distant and widely separated areas in the Indian Peninsula *e.g.*, from Rajmahal hills in Bengal to Madras. The glossopterids are rare in these rocks but large number of pteriodophytes and other gymnospermous plants have been reported. The middle Jurassic age was ascribed to these rocks.

Floristics : The *Glossopteridaceae* include the following organs genera representing its leaves, stems, roots, male fructifications and female fructifications:

Leaves : *Glossopteris, Gangamopteris, Palaeovittaria* and *Rhabdotaenia* and *Rubidgea*.

Stems and Roots : *Vertebraria.*

Male Fructifications : *Eretmonia, Glossotheca, Kendostrobus, Arberiella, Pteruchus.*

Female Fructification : They are usually described under two separate headings (Surange and Chandra, 1978):

1. *Cupulate Fructifications*, which include form genera like *Lidgettonia, Partha, Russangea, Mooia, Rigbya, Denkania, Stephanostoma, Pterigospermum, Platycardia.*

2. *Multiovulate Fructifications*, which include form genera like *Scutum, Lanceolatus, Hirsutum, Cistella, Pluma* and *Ottokaria.*

Fig. 3.4. *Glossopteris shailae.* [The upper part of leaf showing mid rib and venation (After Usha Bajpai)].

The literature on *Glossopteris* flora in India and abroad has gathered so much and huge amount of work has been done that it is almost impossible to discuss it entirely in text like this. A separate volume is required to describe and discuss this group of plants. It is now not proper to regard it as family of the pteridospermales. It should be regarded as a distinct order if not a class. A brief account of this flora is attempted here so as to acquaint the students about the morphology of this interesting group of fossils.

GLOSSOPTERIS Brongniart (1828)

The name *Glossopteris* (Figs. 3.1—3.9) was first used by Brongniart (1828-1832) to describe some Indian and Australian leaves that were spathulate and tongue-shaped (*Glossa*-tongue). These were discovered from the Permo-Carboniferous strata or rocks. These leaves have a distinct midrib and a reticulate venation. *Glossopteris browniana* Brong and *G. angustifolia* Brong are two species that were first described by Brongniart from India and Australia. Later he described two varieties of *G. browniana*. These were named as *G. browniana* var. *australasica* from Australia and *G. browniana* var. *indica* from India. Some authors (Schimper) later raised them to the rank of distinct species. At present, large number of species of this genus have been described from the Indian Gondwanaland, from the Antarctica, Australia, Africa and South America (the non-separated parts of Gondwana Land). Such leaves have also been described from Russia in the Upper Permian (Zimina, 1967 ; Meyen 1969). Wieland (1913, 1914-16) described some leaves from the Jurassic of South-Eastern Mexico. Delevoryas and Person (1975) referred them to a new genus called *Mexiglossa*. The leaves from the Late Triassic or Early Jurassic of Honduras which resemble *Glossopteris* have also been put under the genus *Mexiglossa* (*M. varia*). Sidney Ash (1981) discussed such *Glossopteris* like leaves from Northern Mexico and Honduras in detail and agreed with Delevoryas and Person (1975) to include specimens from Honduras under their genus *Mexiglossa*, but opined that North Mexican specimens should be treated on species of *Glossopteris*. These reports are from outside the boundary of Gondwanaland.

Fig. 3.5. Lower part of the leaf of *Glossopteris shailae* showing evanescent mid-rib and venation (Afer Usha Bajpai).

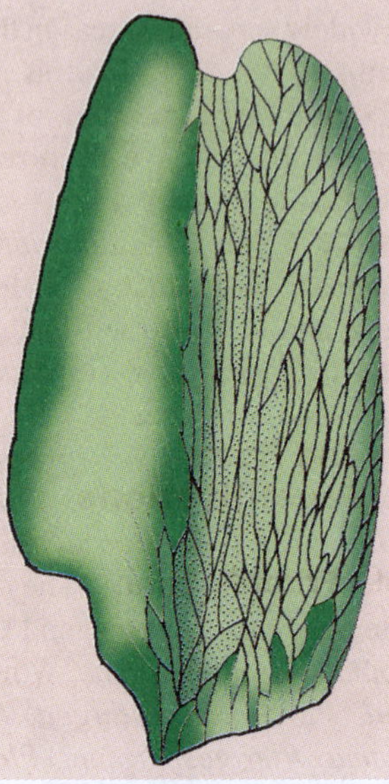

Fig. 3.6. Scale leaf of *Glossopteris*. Enlarged to show details of venation (After Srivastava).

Maheshwari (1965) has given a vivid account of the genus *Glossopteris* and has given a list of the various species found in different parts of the world including Africa (21 species), Antarctica (17 species), Argentina (9 species), Australia (21 species), Brazil (9 species), Falkland Islands (8 species), Madagascar (1 species, *G. indica*), Turkey (1 species, *G. cf. stricta*) and India (32 species). After the publication of his work, many more species have been reported from India and other countries. More than forty species are known from India alone. He studied large number of species of *Glossopteris* from the Barker and Raniganj stages of the Damuda series and came to the conclusion that an aggregate of characters should be taken into consideration before naming a new species. He has given beautiful and most acceptable list of morphological characters that must be studied and taken into considerations while determining a species. These characters (after Maheshwari, 1965) are listed below:

1. Shape of the Lamina. The leaves are simple and there is great variation in the shape of lamina of the various species. It may be linear (*G. formosa, G. leptoneura*) ; linear-lanceolate (*G. angustifolia, G. fuchsii*), lanceolate (*G. indica, G. shailae, G. duocaudata*) ; lanceolate spathulate (*G. indica, G. barakarensis*) ; Spathulate (*G. brownii., G. churiensis, G. intermittens*) ; pandurate (*G. pandurata*) ; Oblong (*G. emarginata*) ; Obovate (*G. retusa*) ; Cordate (*G. cordata*) ; Obcordate (*G. spathulatocordata*) ; Elliptical (*G. conspicua*) ; Oblong-lanceolate (*G. communis*) ; Oblong-spathulate (*G. decipiens*) etc.

2. The Apex of Lamina. The apex is usually spathulate (*G. brownii, G. indica, G. churiensis, G. intermittens, G. barakarrensis, G. intermittens*) ; acute (*G. communis, G. leptoneura, G. conspicua, G. fuchsii*) ; obtuse (*G. shailae, G. duocaudata*) ; retuse (*G. retusa*) ; emarginate (*G. emarginata*).

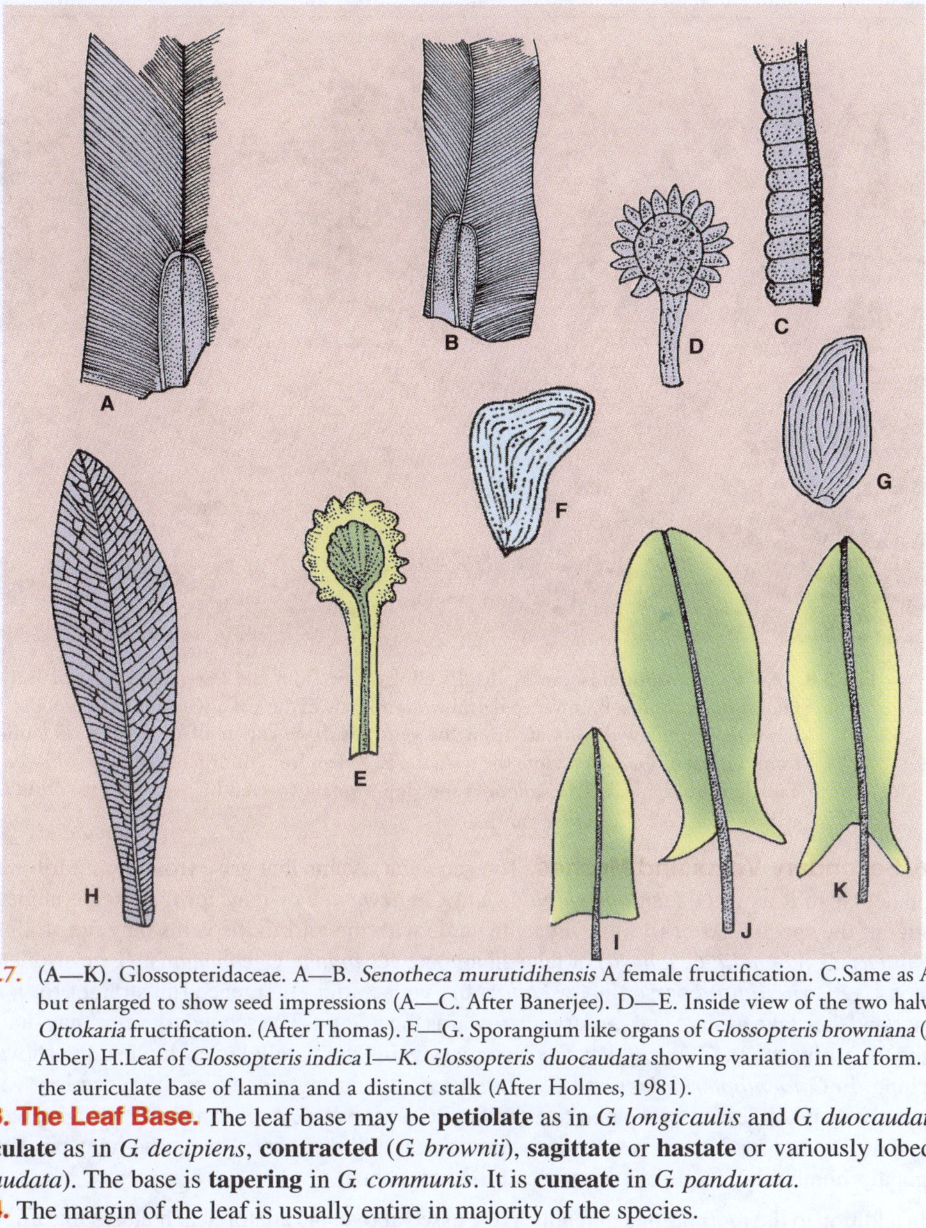

Fig. 3.7. (A—K). Glossopteridaceae. A—B. *Senotheca murutidihensis* A female fructification. C.Same as A—B but enlarged to show seed impressions (A—C. After Banerjee). D—E. Inside view of the two halves of *Ottokaria* fructification. (After Thomas). F—G. Sporangium like organs of *Glossopteris browniana* (After Arber) H.Leaf of *Glossopteris indica* I—K. *Glossopteris duocaudata* showing variation in leaf form. Note the auriculate base of lamina and a distinct stalk (After Holmes, 1981).

3. The Leaf Base. The leaf base may be **petiolate** as in *G. longicaulis* and *G. duocaudata* or **auriculate** as in *G. decipiens*, **contracted** (*G. brownii*), **sagittate** or **hastate** or variously lobed (*G. dyscaudata*). The base is **tapering** in *G. communis*. It is **cuneate** in *G. pandurata*.

4. The margin of the leaf is usually entire in majority of the species.

5. The Midrib. The midrib is present in all the species and is a diagnostic character of the genus *Glossopteris*. The midrib in many species is distinct and well marked but in some cases, it may be evanescent or indistinct. Sometimes as in *G. shailae* and other species, it may be evanescent in the upper part and distinct in the lower part. It is well marked and distinct in *G. divergens, G. communis, G. nidupurensis, G. retifera, G. browniana, G. angustifolia, G. indica* etc. In some species, it is distinctly grooved *e.g., G. stricta, G. communis*, etc. In *G. decipiens* the midrib is clear upto a length of 4.3 cm and later becomes diffused into fine secondary veins near the apex. In *G. shailae* distinct circular depressions are found on the midrib. In many species, the mid rib is traversed by one or more lines or striations. These striations do not anastomose with one another and are therefore different from the median veins of *Gangamopteris* which show distinct anastomoses.

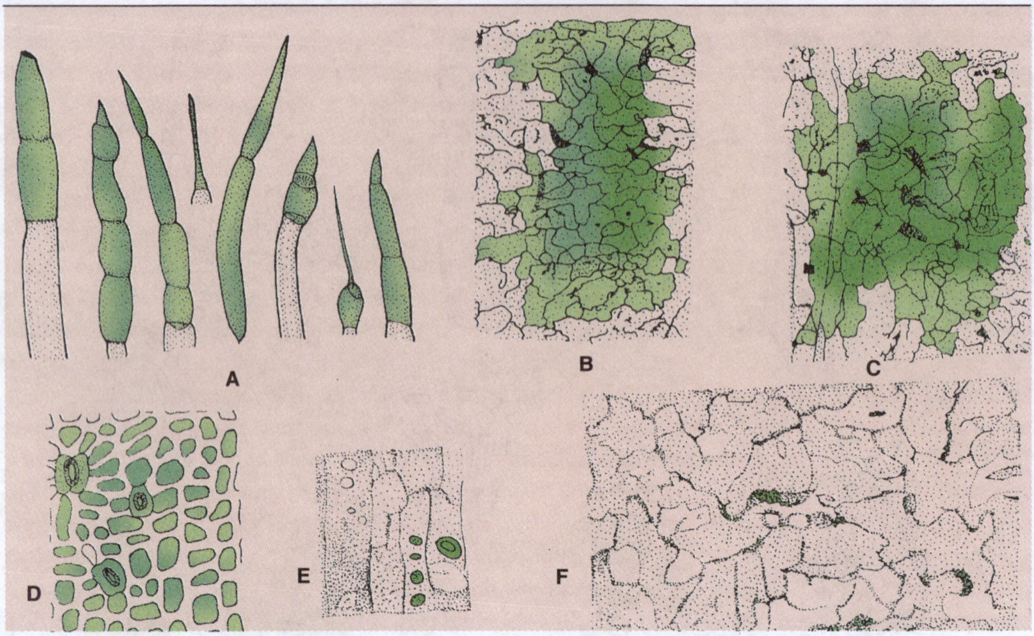

Fig. 3.8. (A—F). Glossopteridaceae. A. Multicellular hair from the lower epidermis of leaf of *Glossopteris hispida*. B.Lower epidermis with stomata in the leaf of *Glossopteris colpodes*. C. Lower epidermis with stomata from the ventral surface cuticle of *G. hispida*. D.Cuticle from the leaf of *G. indica*. Note the stomata. E. Xylem from midrib of leaf of *G. fibroser*. F. Cuticle from the leaf of *G. colpodes* showing stomata covered by over-arching subsidiary cell papiller. (All after D.D. Pant).

6. Secondary Veins and Meshes. The secondary veins that arise from the midrib may be perpendicular to it as in *G. taeniopteridoides* and *G. euryneura* or may form an acute angle as in majority of the species. After forming an acute angle with the midrib the veins may run straight (*G. intermittens, G. mitchelii, G. jonesii*) or take an arcuate (*G. indica, G. communis, G. brownii* etc.) or a sinuous (*G. verticillata*) course. In *G. shailae,* the veins, after emergence run straight for a short, distance and then take a curve and run straight towards the margin. The secondary veins may branched sparsely or profusely. In *G. parallelae,* the secondary veins run parallel to each other and show little branching. In *G. taeniopteridoides* and *G. intermittens,* the anastomosing of the secondary veins is comparatively much less as compared to *G. indica, G. brownii, G. angustifolia, G. retifera*, etc. The shape of the meshes formed as a result of anastomosing is also different. The meshes may be elongate, polygonal, rhomboidal, trapezoid or may be broad and open.

In addition to the morphological features discussed above, the anatomical features like thickness of the upper and lower cuticles, the nature of the stomata or whether they are sunken or not, the shape of the cells of the upper and lower epidermises and the nature of their cell walls (straight or undulating) are also of great importance. In some cases other anatomical features (if available or clear) like hypodermis, mesophyll, and structure vascular bundles are also taken into considerations. Cuticular studies on a large number of leaves have been conducted by botanists in India and abroad. The details are not within the scope of this text but reference can be made to the work done by Sahni, Zeiller, Pant, Pant and his students, Miss Chinna Virkki (Ph. D. Thesis), Srivastava, Surange, Surange and his students, and many other workers. Pant (1958) described three new species of *Glossopteris* based both on morphological and anatomical features, like the presence or absence of interstitial fibres, presence or absence of multicellular-unbranched hair, nature of epidermal cell walls, type of vascular strands in the midrib, kinds of tracheids and their thickening pattern and stomatal type. He also described the hypodermics and mesophyll cells in the leaves of these species (*G. fibrosa, G. hispida* and *G. colpodes*). All the three species resemble each other in :

1. In having a midrib composed of a number of anastomosing and parallel vascular strands, which give off lateral veins that traverse the leaf blade.
2. The vascular bundle in the midrib possesses tracheids with scalariform as well as pitted tickenings.
3. The leaves are dorsiventral.
4. The lamina is flat on the dorsal side whereas on the ventral side the midrib and the veins are raised.
5. The mesophyll is distinguished into spongy parenchyma and palisade layers.
6. The stomata on the ventral side are hypochelic and are present between the veins.
7. The stomata are sunken and possess subsidiary cells with papillae.

These species differ in the following respects:

1. *G. fibrosa* has interstitial fibres (Fig. 3.9 A&E) but no hair on the ventral side.
2. *G. hispida* has multicellular and unbranched hair (Fig. 3.8 A) on the ventral side and also interstitial fibres.
3. *G. colpodes* lacks both hair and the interstitial fibres. The walls of upper and lower epidermal cells are sinuous (Fig. 3.8 F).

The examples given above are sufficient to prove the importance of anatomical characters and cuticular

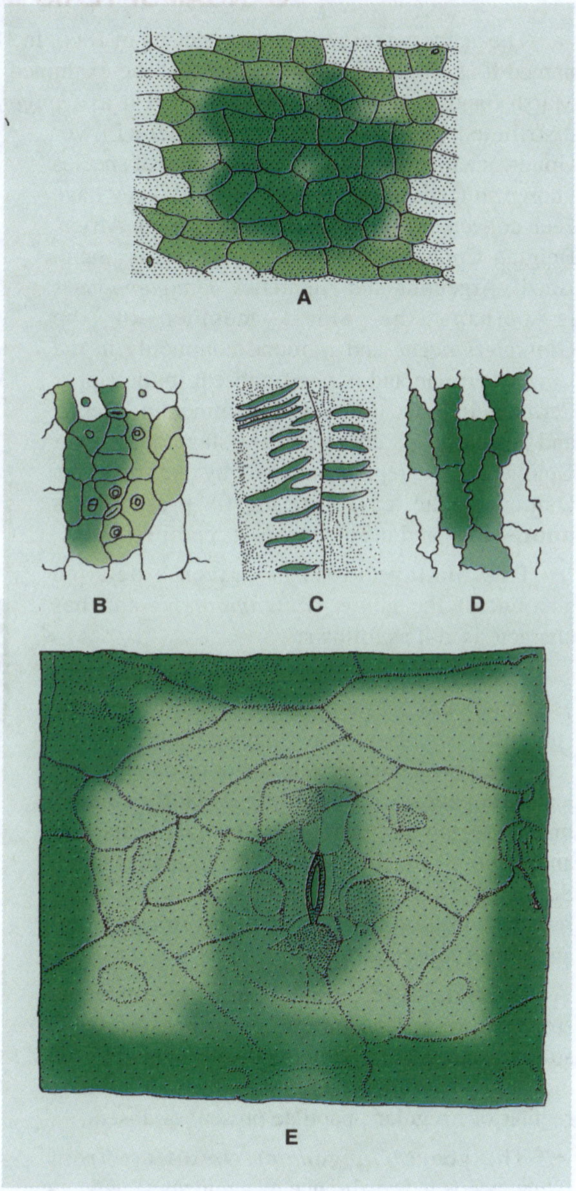

Fig. 3.9. (A— E). Glossopteridaceae. A. Upper cuticle of *Glossopteris fibrosa*. B. Lower cuticle of *Glossopteris augustifolia*. Note two stomata. C. Xylem from midrib of leaf of *Glossopteris fibrosa*. D. Upper cuticle of *G. augustifolia*. E. Lower cuticle of *G. fibrosa* with stomata. (All after Pant).

studies in distinguishing between the various species of *Glossopteris*. For details the students are advised to consult a monograph titled "Revision of the Indian species of *Glossopteris*, Monograph No. 2" (Chandra & Surange, 1979).

GANGAMOPTERIS McCoy (Fig. 3.10)

The genus was established by McCoy in 1860. In 1847, McCoy described it from Australia and named it as *Augustifolia*. Later in 186,0 he examined large number of specimens from Bachhus-Marsh Sandstone, Victoria and renamed it as *Gangamopteris augustifolia*. The genus is widely distributed in the southern hemisphere Lower Gondwana formations. The total number of species known so far is approximately twenty. They have been collected from India, Australia, South Africa, Belgian Congo, Angola, Rhodesia, Nyassaland, Brazil, Argentina and Antarctica. *Gangamopteris* is perhaps the oldest member of the *Glosspteridaceae* and is found commonly in the Lower Permian and is rare or absent in the Upper Permian. In India, the genus is common in Talchirs and Karharbasris and Raniganj stages of Lower Gondwana. It is represented only by one species in Damudas. The leaves are mostly presented as **impressions** and in some cases as **compressions**.

Fig. 3.10. The leaf of *Gangamopteris cyclopteroides*, the midrib is absent.

P.K. Maithy (1981) has given a detailed account of the genus *Gangamopteris* and has characterised it as follows:

"Leaves simple ; entire ; symmetrical or asymmetrical ; linear, lanceolate, elliptical, spathulate or obovate in shape ; apex broadly rounded, obtuse, acute, acuminate or mucronate ; leaf base petiolate, or contracted. Midrib absent ; median region occupied by sub-parallel veins with anastomoses of elongate or hexagonal outline. Secondary veins arise from median veins by repeated dichotomy ; arched, bifurcating and anastomosing network."

"Cuticle differentiated into two surfaces, stomata present on one or both the surfaces ; stomatal apparatus haplocheilic, monocyclic or dicyclic ; distribution and orientation of stomata regular or irregular ; papillae present or absent."

The genus *Gangamopteris* differs from *Glossopteris* in the absence of a midrib.

Maithy (1981) sub-divided the genus *Gangamopteris* into two groups, depending mainly on morphological characters. These are:

Group I. The median region is composed of a number of sub-parallel veins that anastomose and form elongate-rectangular meshes. The shape of leaves may be linear, lanceolate, elliptical or spathulate. It includes 16 species that are categorised as:

A. Leaf Asymmetrical with linear shape and acute apex

1. *G. augustifolia* McCoy.

Leaf Asymmetrial with lanceolate-spathulate shape, apex acuminate and with deltoid upper portion.

2. *G. buriadica* var *arodeltoides* Dolianiti.

B. Leaf symmetrical:

(*a*) Leaf with rounded-oval shape, obtuse apex and a distinct groove in the median portion.

3. *G. stephensoni* Plumstead.

(*b*) The median portion is traversed by sub-parallel veins. Veins distinct from base to apex, leaf shape lanceolate with acute apex.

4. *G. kashmiriensis* Seward.

The median vein distinct only in the basal part and gets gradually diffused.

(*i*) Fibres present in between the meshes. Leaf shape lanceolate, obtuse apex, secondary veins fleuxuous.

5. *G. fibrosa* Maithy.

(*ii*) Fibres absent, apex pointed mucronate, shape elliptical, veins run obliquely forming rectangular meshes.

6. *G. mucronata* Maithy.

(*iii*) Fibres absent, shape lanceolate, veins erect-oblique, meshes rectangular.

7. *G. mosesi* Dolianiti.

(*iv*) Fibres absent, apex acuminate, lanceolate shape, secondary veins broad and elongate near median portion.

8. *G. cyclopteroides* var *acuminate* Feistm.

Secondary veins form polygonal meshes of more or less uniform size.

9. *G. conspicua* Feistm. (Sym. *G. mersyensis* Fiestm)

(*v*) Fibres absent, shape spathulate, secondary veins arcuate, forming elongate-rectangular meshes.

10. *G. spathulata* McCoy.

Secondary veins erect forming elongate rectangular meshes.

11. *G. major* Feistm.

(*vi*) Fibres absent, apex obtuse, shape lanceolate.

12. *G. cyclopteroides* Feistm.

(*vii*) Fibres absent, apex obtuse, shape elongate obovate.

13. *G. obovata* (Carr.) White.

(*viii*) Apex broadly rounded, spathulate shape.

14. *G. major* (Fiestm) Maithy.

(*ix*) Apex broadly rounded, shape obovate.

15. *G. clarkeana* Feistm.

(*x*) Apex truncated with lanceolate leaf shape.

16. *G. buriadica* Feistm.

Group II. The median part of the leaf traversed by veins that are hexagonal in shape. Leaves generally obovate.

(*a*) Apex broadly rounded

(*i*) Shape oblong.

17. *G. castellanosi* Maithy.

(*ii*) Shape obovate

18. *G. intermedia* Maithy.

(*a*) Apex acute

(*i*) Shape obovate.

19. *G. obliqua* McCoy.

The other three leaf genera of the *Glossopteridaceae* are:

1. *Palaeovittaria.* It possesses a distinct midrib in the proximal part of the leaf only. It is absent in the distal portion. The secondary veins do not anastomose to form meshes. The secondary veins follow an oblique course.

2. *Rhabdotaenia.* It possesses a distinct midrib like *Glossopteris*. The secondary veins arise at right angles and follow a straight course. The lateral veins fork but anastomoses are rare. Srivastava (1971) reported a species from the Triassic of Nidpur, Sidhi district of Madhya Pradesh.

3. *Rubidgea* resembles *Gangamopteris* in the total absence of a midrib but differs in the absence of anastomosis of the secondary veins.

These genera can no doubt be separated from each other if the specimens are well preserved and all the details are clear. The difficulty arises when specimens are ill-preserved and in some cases where transitional forms are encountered. In *Glossopteris parallela* Plumstead, the secondary veins run parallel to each other and show very little anastomosis. It almost resembles *Rhabdotaenia*. Similarly in some species of *Glossopteris,* the midrib is clear only in the basal portion and is evanescent or even absent in the upper part, as in *Palaeovittaria*. Surange and Srivastava (1956) have expressed their opinion that these genera grade into one another and sharp delimitation becomes difficult. Maheshwari (1965) is of the opinion that careful and detailed observations of morphological characters can overcome the difficulties encountered during identification. He further states, "on the whole the generic separation based on morphographic studies is quite satisfactory and easily accomplished than that based on the characters of the cuticle or the fructifications, which are rarely found." While examining fossilised material, especially in the form of impressions, one has to depend mainly on external features. In the case of availability of compressed forms, cuticular and epidermal studies can be conducted. Such studies are extremely useful and reliable for the proper identification and circumscription of a speices. In recent years, such studies have been vastly carried out on the various form genera of the Glossopterids (including vegetative as well as reproductive structures). The leaves and other organs that show similar morphological characters generally possess a particular type of cuticular and other epidermal features. There are, no doubt, evidences that in some cases, the leaves exhibiting similar morphological features show marked differences in their cuticular structures and *vice versa*. Maithy (1981) has pointed out such cases *e.g.*, *Glossopteris indica* Schimper type of leaves have yielded three different types of cuticles, *viz.*, *Glossopteris indica* Zeiller (1896), *G. arberii* Srivastava (1956) and *G. jamotii* Heg & Bose (1960). On the other hand, leaves with different morphological features *viz.*, *Palaeovittaria kurzi* and *Glossopteris intermittens* have similar cuticles.

THE SCALE LEAVES (Fig. 3.6)

The scale leaves differ from the foliage leaves in their smaller size and in the absence of a midrib. They vary in shape and may be ovate, triangular, diamond-shaped, oval, cordate, deltoid and spathulate in shape. The apex may be acute, obtuse and even notched. The margin may be entire or deep lobed. The surface may be smooth or rugose. The leaf base broad or narrow and contracted. They vary in length from 1-6 cms and 1-3 cms broad. The leaves may be sessile or petiolate. The veins usually arise from the base and anastomose to form meshes of various shapes. There may be present a few thick parallel inter-connecting veins in the middle. They bear fructifications and act as sporophylls. Bose and Srivastava (1970) described a scale leaf from Triassic beds of Nagpur. They named it as *Glottolepis rugosa*. Maithy (1965) described *Palmatophyllites lacerata* from the Karharbari beds, Giridih Coal field. Srivastava, A.K. (1979) described two completely preserved scale leaves from the Auranga Coal fields, Bihar. The scale leaves will be further discussed along with the reproductive organs.

THE STEMS OR RHIZOMES

VERTEBRARIA (Fig. 3.1). It is regarded as the stem or the rhizome of *Glossopteris* and was first described by Royle. It is found in abundance, in the form of casts, in Permo-Carboniferous strata in association with *Glossopteris*. The stems are usually fossilised in the form of flattened, branched or unbranched axes, which in some specimens, bear lateral, slender and branched roots. The stem is usually traversed by two or three longitudinal grooves and ridges. The longitudinal grooves give off lateral grooves and ridges that arise at right angles and divide the surface of the stem into more or less rectangular areas (Fig. 3.1 F) 1 centimeter or a little more in length. The surface of these rectangular areas appears to be convex and in some cases, faint outlines of cells are also visible. Oldham (1897) described some specimens of *Vertebraria* from India in which he reported the occurrence of dark brown ferruginous substance or carbonised remains of plant tissues in the grooves (Fig. 3.1). A cross section of the *Vertebraria* cast as figured by Zeiller (1896) shows a solid central axis from which radiate a number of wedge-shaped segments called the **stem ribs** or **arms.** In between these ribs are intrusions of sedimentary material. The rectangular areas on the surface of the stem (Fig. 3.1 C) represent the intruded segments of rocks separated at intervals by longitudinal and transverse grooves. The transverse grooves represent the vascular strands given off at each anastomose of the longitudinal wings. These strands supply the leaves. Oldham (1897) expresses a view quite different from that of Zeiller (1896). He regards the stem to be a cylindrical organ and not an irregullary winged axis. He believes that the stem consists of a central axis "joined to an outer rind by a series of radial septa," the spaces between the septa are divided into chambers by transverse partitions. Seward (Fossil Plants Vol. II, p.505) expressed his opinion, "It may, however, be possible that the spaces between the longitudinal and transverse grooves on a *Vertebraria* axis, which have been filled with the surrounding rock, were originally occupied in part at least by secondary wood, and the transverse strips of carbonaceous material lying in the grooves may represent medullary ray tissue and accompanying leaf traces. The longitudinal striations seen in some specimens of *Vertebraria* on the areas between the grooves may be the impressions of woody tissue." A *Vertebraria* axis along with the leaves of *Glossopteris browoniana*, collected from the Permo-Carboniferous of Transwal by Dr. Mohlengraff did not show any surface features (Fig. 3.2). Etheridge (1894) in a material collected from Australia showed the presence of transversely elongated or rhomboidal leaf scars (Fig. 3.2) on the stem surface, in place of rectangular areas. He also noted imbricate rod-like projections, which according to him represent vascular bundles. In all these cases, the stem was attached with the leaves of *Glossopteris* that formed a crown at the apex. Walton and Wilson (1932) described *Vertebraria* axis with a whorl of *Glossopteris* leaves at the apex and two whorls of leaf bases below. These whorls were separated by distinct internodes. The axis lacked rectangular areas. In 1954, Dolianiti reported a *Vertebraria* from South America with four pairs of attached leaves. The axis showed distinct longitudinal and transverse grooves. Pant (1956) reported *Vertebraria* axis with roots (Fig. 3.1). These differences in the external features of the stem may be due to preservation. Sometimes some features are lost during the process of preservation.

Walton and Wilson (1932), Pant (1956, 1958, 1960), Sen (1958) and Plumstead (1958) have described the internal structure of the *Vertebraria* axis and roots (1958). It is a decorticated axis showing only secondary wood with multiseriately pitted tracheids and parenchymatous rays. There is hardly any pith in the centre, which is occupied by xylem that sends radiating xylem plates towards the periphery. These plates anastomose during their course towards the periphery. In between the xylem rays that is parenchyma it has been demonstrated that the rectangular areas have a thin coaly layer on their surface. This layer consists of elements of secondary xylem. The broad medullary rays of secondary xylem were probably crushed during preservation and got filled with mud.

A number of fossil woods have been described from Lower Gondwana formations. They probably belong to the genus *Glossopteris*. Most of them are only pieces of secondary wood but in some

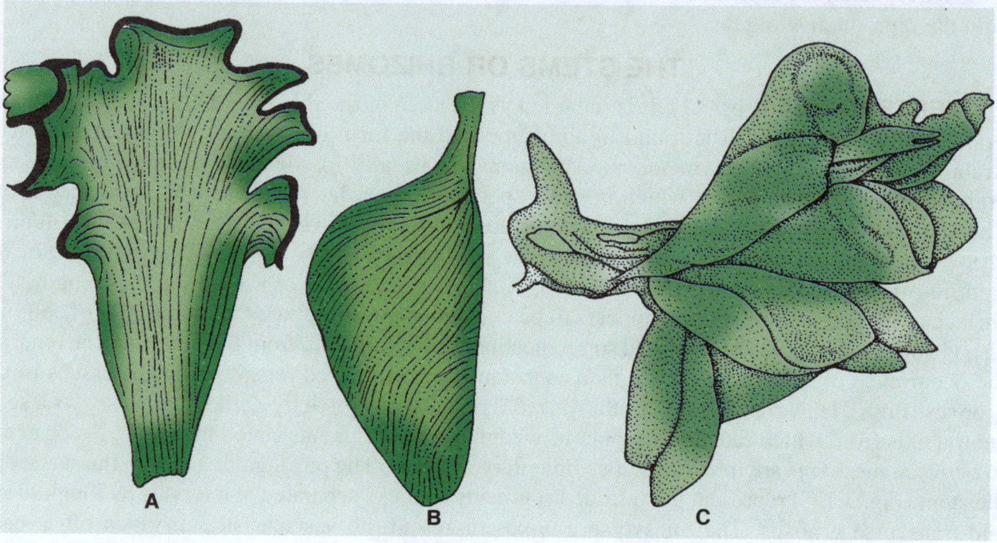

Fig. 3.11. A. *Arberia minasica*;. B. Sporangium of *Arberiella vulgaris* ; C. sporangia of *Arberiella africana* attached to a branching axis.

specimens, pith and primary xylem are also preserved. Maheshwari (1966) has given a vivid account of some such woods collected from the Raniganj stage of Raniganj Coalfields, Bengal. He described *Trigonomyelon raniganjensis* Maheshwari; *Damudoxylem waltoni* Maheshwari; *Megaporoxylon kraeuselii* Maheshwari and *Kookoxylon zalesskyi* (Sahni) Maheshwari, Surange and Maithy (1962) described *Barakaroxylon* (= *Dudoxylon jhariense*) from Barakar stage of Jharia Coal Field. Many species of *Dadoxylon* have been reported from Lower Gondwana by Holden (1917) ; Sahni (1932) ; Rao (1935) ; Surange and Sah (1957) ; Maheshwari (1964 and 1965) and Maithy (1965). Mehta (1950) described *Spiroxylon indicum* from Singrauli Coal Field, Madhya Pradesh and *Dadoxylon* Sahni (Singh, 1958) from Olive Shales of Punjab Salt Range. Their detailed descriptions are not within the scope of this text. Growth rings are common in these fossils woods. It points to the prevailing seasonal variations and growth periodicity.

REPRODUCTIVE ORGANS

A large number of both male and female fructifications have been described from the Lower Gondwana of India and other widely separated parts of Gondwanaland *e.g.*, Australia, Africa and Brazil. These reproductive organs are found either associated or attached to the leaves of *Glossopteris, Gangamopteris* and *Palaeovittaria*. Arber (1905) discovered some sporongia like organs (Fig. 3.11 A) which he attributed to *Glossopteris*. Later David White (1908) described two Aphlebia-like leaves from the Palaeozoic flora of South Brazil. These were named as *Arberia* and *Derbyella*. These were scale leaves found associated with *Gangamopteris*. These scale-like leaves had abruptly truncated lobes which point towards an evidence of separation from reproductive bodies. Pant and Nautyal (1960) described the genus *Arberiella* from India. *A. vulgaris* (Fig. 3.11B) and *A. africana* (Fig. 3.11 C) were described from India and Tanganyika (Pant, 1958), respectively. Walkom (1921, 1925) ; Thomas (1921, 1925, 1958) ; du Toit (1932) ; Virkki (1937, 1946) ; Zeiller (1902) ; Plumstead (1952, 1956, 1958) ; K.R. Surange and Shaila Chandra (1971, 1973, 1974, 1978) ; Sen (1955) ; Chatterjee and Sen (1963) ; Varma (1963) ; Mukherjee (1966) ; Surange and Maheshwari (1970) ; Maheshwari (1965) ; Chandra and Surange (1976, 1977); Pant and Gupta (1968, 1971) ; Lacey *et al* (1975) ; Fiestmantle (1881) ; Banerjee (1969) ; Saksena (1962) ; and Gould and Delevoryas (1977) have contributed a lot to our knowledge about the reproductive organs of *Glossopteridaceae*.

A brief account of the male and female reproduction organs of the Glossopteridaceae is given below:

Male Reproductive Organs. There are four common sporangia bearing organs of *Glossopteridaceae viz. Arberiella* Pant (1958) ; *Eretmonia* du Toit (1932) ; *Glossotheca* Surange and Maheshwari (1970) ; and *Kendostrobus* Surange and Chandra (1974).

Arberiella vulgaris (Fig. 3.11 B) was described from India by Pant and Nautiyal (1960) from the Lower Gondwana of Raniganj Coal Fields. They are stalked sporangia, more or less oval in shape with abundant striations on the wall. They have been found attached to fertile scales that were concave on one side. The sporangia were attached to the concave side and contained two winged spores. Pant (1958) described *A. africana* from Tanganyika. In this case groups of sporangia were found attached to a branching axis (Fig. 3.11 C). Detached *Arberiella* type of sporangia are quite common in the Permian deposits of Gondwanaland. They look alike externally but contain different types of two-winged spores. In two such specimens, one contained **striatites** type (Fig. 3.12 B) of spores whereas the other contained **Faunipollenites** type (*A. africana*). In *A. vulgaris,* the spores were of **Lunatisporites** type.

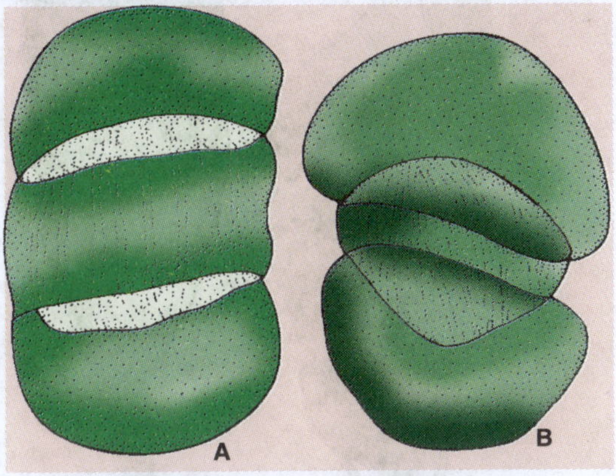

Fig. 3.12. (A—B). Microspores of *Glossopteris* A. *Striatodocarptes haploxylanoides*. B. *Straitites obliqunis* (After Srivastva).

Eretmonia Du Toit (Figs. 3.13-3.14) is also attached to variously shaped fertile scale leaves that are always net-veined and lacked a mid-rib. According to the size and shape of the scale leaves, five species have been recognised. These are *E. ovoides* (Fig. 3.13 B, C) Surange and Chandra (1974) with ovate fertile bracts ; *E. hingridaensis* Surange and Maheshwari (1970) (Fig. 3.14 D) with spathulate bracts having acute apex ; *E. emarginata* Chandra and Surange (1977) with triangular (Fig. 3.14 E) bract ; *E. utkalenisis* Surange and Maheshwari (1970) with diamond-shaped bracts ; and *E. karanpurensis* Surange and Maheshwari (1970) with orbicular bracts. In all these species the sporangia are of *Arberiella* type, and oval or elliptical in shape with fine striations on the external wall. The fertile scales are stalked (Fig. 3.13 C). From the stalk of the fertile scale arises a pedicel which divides into two equal *branches* (Fig. 3.13 C). Each such branch undergoes repeated dichotomy and the ultimate branches bear one sporangium each at their tips and form a sporangial cluster (Figs. 3.14, 3.13). So one pedicel of a fertile bract bears two sporangial clusters (Fig. 3.13 B, C). The fertile scales are mostly convex on one side and concave on the other and are thick in the middle and comparatively thinner at the margins. It is presumed that they might have acted as protective covers for the sporangial clusters. Each sporangium contains two winged spores.

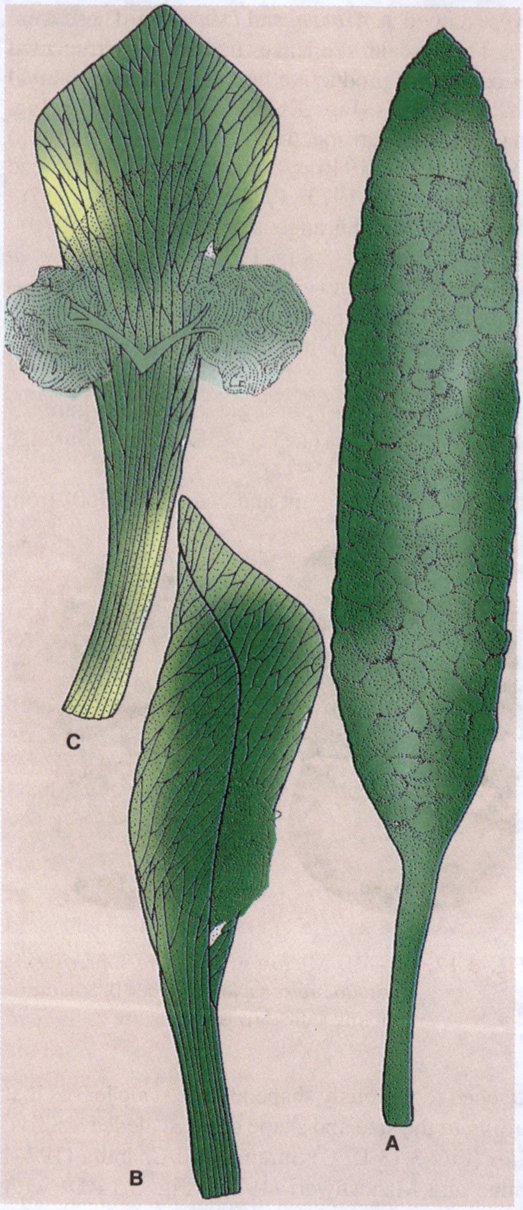

Fig. 3.13. (A—C) Male fructifications of *Glossopteris*. A *Kendostrobus cylindricus* B. *Eretmonia ovoides*. C. *Eretmonia ovoides*. (After Surange & Chandra 1978).

Fig. 3.14. (A—E). Male fructifications of *Glossopteris*. A—C *Glossotheca* (A. *G. immanis* Chandra & Surange), B. *G. utkalensis* (Surange & Maheshwari), C. *G. arrisiana* (Surange & Chandra), D. *Eretmonia hingridaensis* (Surange & Maheshwari), E. *Eretmonia emarginata* (Chandra & Surange).

Glossotheca Surange & Maheshwari (Fig. 3.14 A—C) differs from *Eretmonia* in that the fertile scale bears more than one pedicels that bear sporangia on their ultimate branches. *G. utkalensis* has spathulate type of fertile scales or bracts that have long and stout stalks. In *G. orissiana,* the scales or bracts are rhomboidal with short stalks. In *G. immanis,* the bracts are large and oblanceolate with broad and short stalks. In the last named two species the fertile scale or the bract bears four slender pedicels on its abaxial side. They arise one below the other in a single row. In *G. utkalensis,* three pedicels arise on the abaxial side of the bract (Fig. 3.14 B). Each pedicel divides into two equal branches which divide further by repeated dichotomies whose ultimate branches bear *Arberiella* type of sporangia. Each fertile bract may bear light (*G. immanis* and *G. orissiana*) or six (*G. utkalensis*) clusters of sporangia. The sporangia are elliptical and striated. They contain two winged spores. The outer cells of the sporangial wall are longer than broad and are thick-walled. They are twisted in various ways. A layer of thin-walled elongated cells is below the thick-walled cells. The spore mass in some specimens is covered by a thin and delicate layer of elongated cells. It appears the sporangial wall is made up of three layers. The dehiscence of the sporangium is believed to be longitudinal.

Kendostrobus Surange & Chandra (Fig. 3.13 A) differs from other genera in its organisation and has been found scattered from a number of places in Raniganj. It is neither attached to a bract nor is found in connection with *Glossopteris* leaves. It is a long and cylindrical cone-like structure with distinct central axis bearing whorls of sporangia in close spirals. The sporangia are large, oval to elongate and are without an annulus. Their surface bears oval to round depressions and the contained spores are stripped and monolete. Pant and Nautiyal (1960) have described such detached sporangia under the form genus *Lithangium*.

Female Reproductive Organs. They can be discussed under two headings:

1. **Cupulate Fructifications**
2. **Multiovulate Fructifications.**

1. Cupulate Fructifications. In such fructifications, the seeds are protected by a cup-like or circular and disc like structure called the *cupule*. Each cupule contains only one seed. These cupules are borne on pedicels which arise from the region of mid veins on the lamina of the fertile leaves or the bracts which may be of various shapes and sizes and are usually stalked. The cupules are borne terminally on each pedicel. Each pedicel may bear 4-8 or even 14 (Lacey *et al*, 1975) cupules. The number and arrangement of pedicels on a fertile bract also varies with genera and species. Such fructifications have been reported from the Permian deposits in India, Australia and South Africa. The common genera include *Lidgettonia, Partha, Denkania, Mooia, Rigbya* and *Russangea*. A brief description of each genus is given below :

Lidgethonia Thomas (Fig. 3.15 A) :— Thomas (1958) described *L. africana* from Natal and later Lacey (1975) described it from Mooi River, both the localities from South Africa. Surange and Chandra (1974) described *L. mucronata* from Orissa in India. It is a female fructification that arises from the abaxial side of a stalked fertile bract that is oblanceolate or spathulate-lanceolate or ovate-lanceolate in shape with reticulate venation. Two longitudinal rows of pedicels arise from the abaxial side of the fertile bract. The pedicels arise from the median portion of the bract which lacks midrib. Each pedicel bears at its tip 4 or more cupules that may be semi-circular, circular or discoid or companulate or cup-like in shape. Each cupule contains one seed that may be winged (*L. africana*) or un-winged (*L. mucronata*).

Fig. 3.15. (A—D). Female fructifications of *Glossopteris.*

A.*Lidgettonia mucronata* ; B. *Jambadostrobus pretioous* ; C. *Venustostrobus diademus* D. *Plumsteadiortrobus ellipticus*. (All after Chandra and Surange).

Partha Surange & Chandra (Fig. 3.16 A—D):— It differs from *Lidgettonia* in possessing only one row of pedicels as compared to two in *Lidgettonia*. The pedicels are short and each bears four terminal cupules that contain one winged seed.

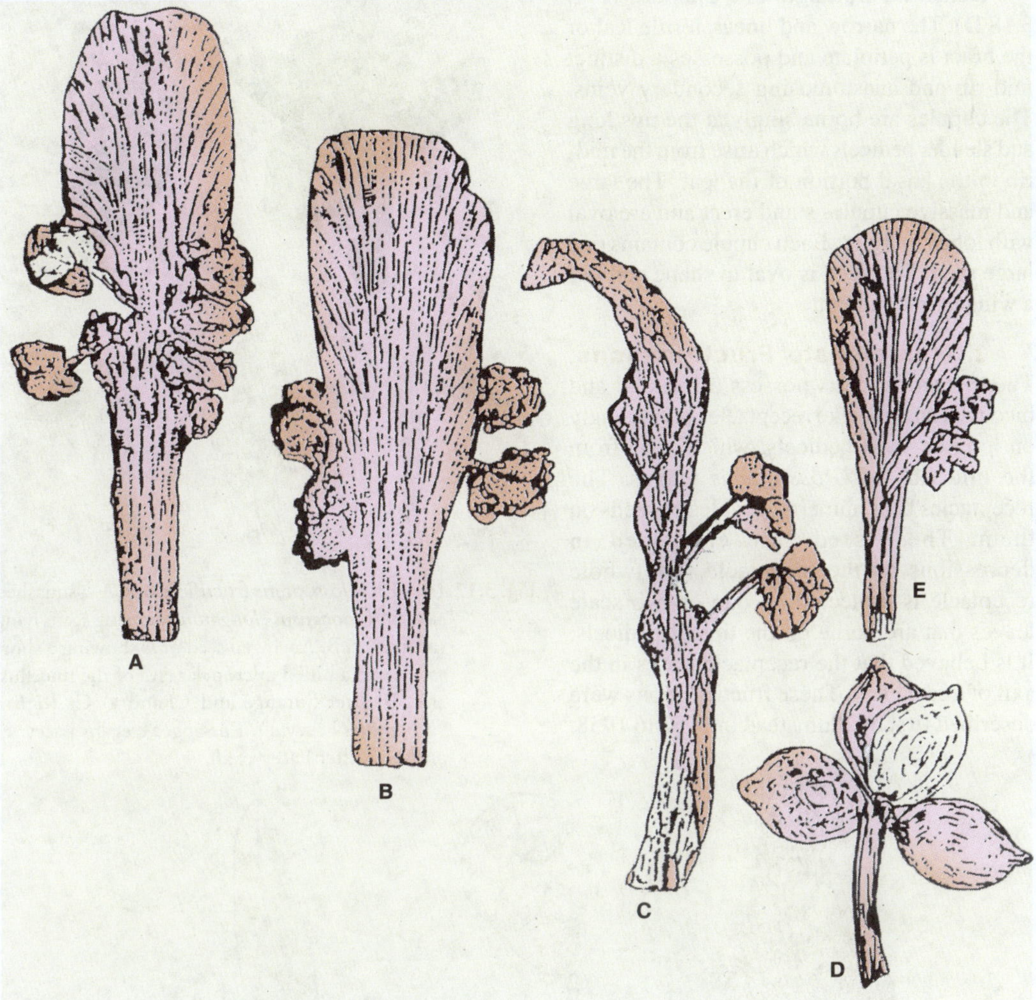

Fig. 3.16. (A—E). *Partha spathulata.* A female cupulate fructification of *Glossopteris.* In D the attachment of cupules to the pedicel is shown.

Russangea Lacey, Dijk & Gordon-Gray (Fig. 3.17 D). The fertile bract is stalked and narrow-ovate. It bears on its adaxial side, near its base, two pedicels one after the other. Each pedicel bears a expanded seed bearing scale that extends beyond the point of attachment of the seed and covers it. The seeds are oval and un-winged. In this case, one fertile leaf bears two seeds.

Mooia Lacy, Dijk and Gordon-Gray (Fig. 3.17 E). It has been described from South Africa by Lacey *et al* and the seeds in this case are surrounded, each by a companulate cupule, which is borne terminally and singly at the tip of a pedicel. Two to four pedicels arise from the adaxial side of a spathulate or rhomboidal fertile bract. The bract is petiolate. The cupules are often reflexed and drooping. Each cupule encloses a single flattened and un-winged or scarcely winged seed.

Rigbya arberoides Lacey, Dijk & Gordon-Gray (Fig. 3.17 C). In this case, the fructification consists of a stalk that bears at its apex an aggregation of scales. Each scale bears a single flattened, oval and un-winged seed.

Denkania Surange and Chandra (Fig. 3.18 D). The narrow and linear, fertile leaf or the bract is petiolate and possesses a distinct mid-rib and anastomosing secondary veins. The cupules are borne singly at the tips long and slender pedicels which arise from the mid-rib in the basal portion of the leaf. The large and massive cupules stand erect and are oval with lobed margins. Each cupule contains one large seed. The seed is oval in shape and has a winged integument.

2. Multiovulate Fructifications.

These fructifications possess a bilateral and biconvex cushion like receptacles borne singly on long or short pedicels, which arise from the mid ribs of *Glossopteris* leaves. The receptacles bear numerous ovules or seeds on them. These seeds are embedded in depressions on the receptacle. The whole receptacle is protected by one or two scale leaves that are borne on the tips of pedicels. It is believed that the receptacle arises in the axil of a scale leaf. These fructifications were described first by Plumstead in 1952 to 1958.

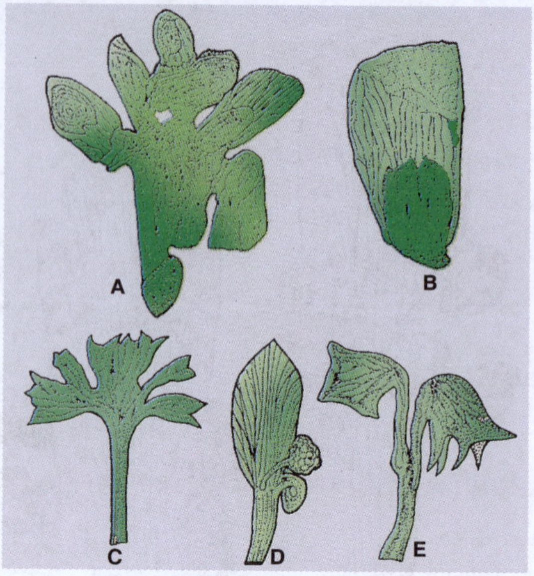

Fig. 3.17. (A—E). *Glossopteris fructifications.* A. a squashed axis of *Indocarpus elongatous* showing seeds lying around it. B. A winged seed of A showing a short stalk and a bified micropylar end of the nucellus. (A, B After Surange and Chandra) C. *Rigbya arberoides* Lacey. D. *Russangea elegans* Lacey. E. *Mooia* (After Lacey *et al*).

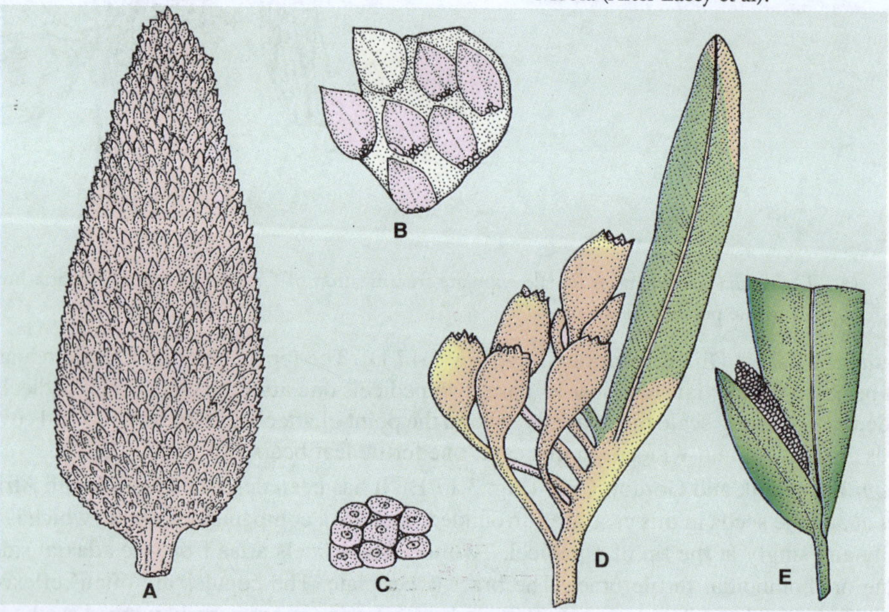

Fig. 3.18. (A—E). Fructifications of *Glossopteris.* A—C. *Dictyopterictium aposiferum* Fiestmantle. A. Reconstruction of the cone. B. Showing attachment of seeds to the cushion. C. Showing enlarged cushions to which the seeds are attached. E. Female reproductive organ of *Glossopteris.* It is called *Denkania indica* Surange and Chandra. E. *Dictyopteridium fiestmanteli* After Chandra and Surange (1978) showing bract and attachment with the *Glossopteris* leaf.

She considered them to be bisexual *i.e.*, bearing ovules on one side and pollen grains on the other side. She regarded the fertile receptacle as a bisexual flower. Later Surange and Chandra (1974) studied such fructifications from India and cleared all doubts regarding the morphology of these multiovulate fructifications. They are now regarded as unisexual. Plumstead (1958) named six genera as belonging to this category. These are *Scutum* (9 species), *Hirsutum* (2 species), *Lanceolatus* (3 species), *Ottokaria* (3 species), *Cistella* (2 species) and *Pluma* (2 species). Except *Pluma,* all other types of fructifications have been described from India. Some new types have been added to this list by Surange and Chandra (1971, 1974, 1978). They have described some genera *e.g.*, *Indocarpus, Venustostrobus, Plumsteadiostrobus, Jambadotrobus* and *Dictyopteridium feistmantelli*. Sen (1955) recorded an ovulate fructification from Mohuda Coal seam, which was later included under *Lanceolatus communis* by Plumstead. Chatterjee and Sen (1963) recorded another species of *Lanceolatus* from Barakars of Bihar. Varma (1963), Mukherjee *et al* (1966), Surange and Maheshwari (1970) and Maheshwari (1965) reported some species of *Scutum, Hirsutum* and *Cistella* from various places in India. *Ottokaria bengalensis* was described by Zeiller (1902) from Karharbari beds. Mukherjee *et al* (1966) found it attached to the leaf of *Glossopteis indica*.

A brief description of some of these genera is given below :

1. *Scutum* Plumstead (Figs. 3.19—3.21). "It is a female reproductive organ of *Glossopteris*, which consists of a bilaterally symmetrical lens-shaped receptacle bearing a large number of naked ovules (without ovuliferous scales or megasporophylls) all round it (Fig. 3.20). This ovule bearing receptacle is borne in the axil of a scale leaf, which, being protective in function, is closely appressed to it in immature stage. Both the ovule bearing receptacle and the scale leaf are carried on a common pedicel (Fig. 3.19), which, in its turn, is attached to the petiole of a vegetative leaf (Fig. 3.19). In other words, this *scale leaf-ovule bearing complex* is itself borne in the axil of a vegetative leaf of *Glossopteris*. In mature stage, the scale leaf opened out or fell off, exposing the ovules for fertilization by wind borne winged pollen grains." This is the best and extremely simple explanation of the morphology of *Scutum* in the words of Surange and Chandra (1974, page 15). In the same paper, they have reported three new species of *Scutum* from India. These are *Scutum sahnii* Surange and Chandra from Raniganj stage of Handappa in Orissa ; *S. elongahim* Surange and Chandra from the same place and *S. indicum* Surange and Chandra also from the same place. *S. leslium* and *S. dutoitides* were described by Mukherjee and others (1966) from India. They were found attached to the leaves of *Glossopteris browniana* and *G. indica*, respectively. *S. sahnii* was found attached to the leaf of *G. longicaulis*. In all the species of *Scutum,* there is a wing-like rim all round the ovule bearing receptacle.

2. *Cistella* Plumstead (Fig. 3.23 C). It differs from *Scutum* on the absence of a rim around the receptacle and of the scale leaf. The receptacle is more convex in the middle and would appear oval in a transverse section. The ovules are small and arranged in "close spirals" all round the receptacle.

3. *Ottokaria bengalensis* Zeiller (Fig. 3.7 D, E). It was originally described by Zeiller from the Karharbari beds. Mukherjee and others (1966) reported that it is attached to the leaf of *Glossopteris indica*. It has also been found associated with the leaves of *Gangamopteris*. Plumstead (1956) reported the asssociation of *Ottokaria* type of fructification with *Gangamopteris obovata* and *G. buriadica*. It consists of a slender pedicel arising from the petiole of the vegetative leaf. The pedicel bears a terminal globular receptacle embedded with laterally flattened seeds. The receptacle has a wing-like rim all around it.

4. *Lanceolatus* Plumstead (Fig. 3.24 A—C). It was found attached to the midrib of *Glossopteris communis* and consisted of a cylindrical and swollen axis studded with numerous oval sacs that could be seeds. Plumstead reported its attachment to *Glossopteris retifera* and *Palaeovittaria Kurzi.*

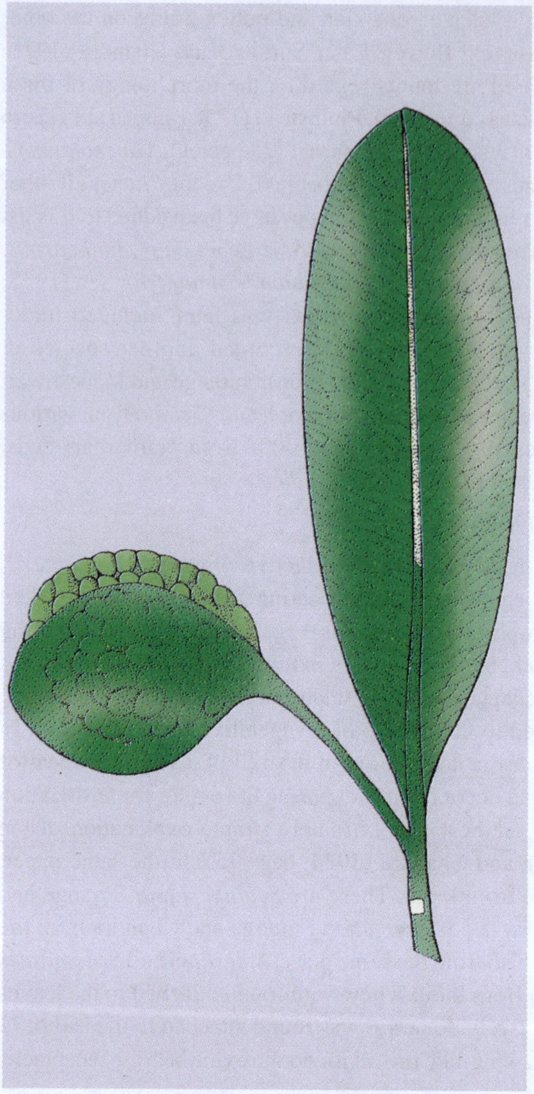

Fig. 3.19. *Scutum sahnii* shown attached to the stalk of *Glossopteris conspicua* by a long pedicel. The scale leaf is shown here to partially cover the receptacle bearing seeds. (After Surange and Chandra, 1974).

Fig. 3.20. *Glossopteris* fructification called *Scutum sahnii* Surange and Chandra. The scale leaf is completely opened up. Note the veins. The receptacle is clearly visible showing the marginal seeds arranged in a distinct row. Also note the wing-like impressions of these seeds on the scale leaf. The scale leaf covers the receptacle in natural position. (After Surange and Chandra, 1974).

5. *Senotheca murulidihensis* Banerjee (1969) :– It was found attached to *Glossopteris taeniopteroides* Fiestm (Fig. 3.7, A, C) and consists of a linear receptacle with faint circular marks of attached seeds. The receptacle is covered by two raised flaps. It is believed that the seeds borne by this fructification were platyspermic.

6. *Indocarpus* Surange and Chandra (Fig. 3.17 A, B). *Indocarpus elongatus* is the type of species found in the Raniganj stage at Handapa in Orissa. The entire structure of the fructification is not clear due to faulty preservation. It seems that the one-winged seeds, that were nicely preserved,

were borne on a stalked receptacle (Fig. 3.17). Each seed (Fig. 3.17) has a short stalk towards the chalazal end. The stalk is obscure and may not be well preserved in some cases. The seeds are almost oval in outline with a well preserved nucellus. The nucellus is round at chalazal end and pointed or bifid at the micropylar end. The seed is winged. There is only one wing which is expanded, flat and projected on one side far beyond the micropylar end. The wing exhibits longitudinal markings. The whole seed gives an appearance of an ovule attached to a megasporophyll, which is not the reality.

7. *Venustostrobus* Surange and Chandra (Figs. 3.25 and 3.15 C). The genus was erected by Surange and Chandra (1977) and includes two species. It is a *Scutum* type of fructification consisting of a seed bearing receptacle borne in the axil of a scale leaf or fertile bract. Both of them are borne on a common pedicel like *Scutum*. The pedicel is attached to the mid rib of a vegetative leaf of *Glossopteris*, near its base and appears to be axillary in position. The receptacle is thus placed between the net veined fertile bract and the vegetative leaf. The receptacle may be orbicular or circular in outline with small oval seeds embedded in it. The seeds are sessile, non-winged and are spirally arranged. The seeds are covered with a thick integument and possess a nucellus. *V. indicus* and *V. diademus* are the two known species from India. The former was attached to a new species of *Glossopteris*, whereas the latter was found attached to *Glossopteris ghusikensis* (Pant and Gupta 1968).

8. *Jambadostrobus pretiosus* Chandra & Surange (1977). It differs from all other types of fructifications in that more than one receptacles arise from a single vegetative leaf and there is no bract *i.e.*, the receptacle is naked. The pedicels bearing naked receptacles arise separately from the midrib (Fig. 3.15 B). The pedicels are short. The fructification was found attached to *Glossopteris contracta* (Pant & Gupta 1971). The seeds are large and non-winged. The receptacle is elliptical or lenticular in shape. The seeds are arranged spirally on the receptacle. It is a *Hirsutum* type of fructification.

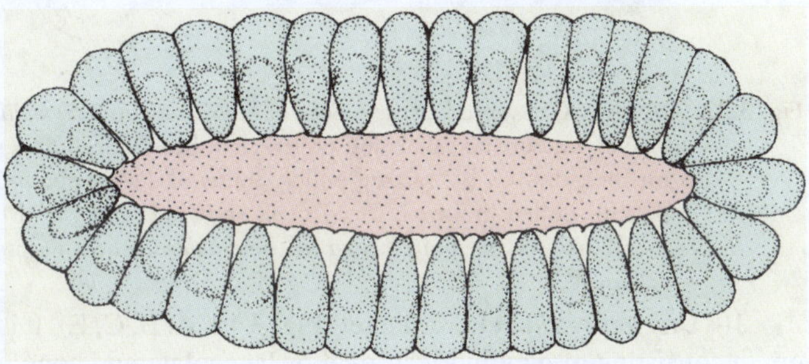

Fig. 3.21. A transverse section of the receptacle of *Scutum sahnii* showing ovules attached to the marginal cushions all around. (After Surange & Chandra).

9. *Plumsteadiostrobus ellipticus* Chandra & Surange (Figs. 3.15 D, 3.26, 3.27) is a *Cistella* type of fructification attached to the leaf of *Glossopteris gondwanensis* (Pant and Gupta, 1971). It consists of an oval seed bearing receptacle borne in the axil of a spathe-like protective bract both of which are borne terminally on a common pedicel that arise from the midrib of a vegetative leaf. The seeds are winged (uni-winged). The wing is an expanded outer overing of the integument and is broad at the micropylar end and narrows down towards the chalazal end. The sessile seeds are arranged in close spirals on the receptacle. The receptacle is beset with lenticular holes to which the seeds are attached. The nucellus and the pollen chamber are preserved. The pollen chamber contains winged pollen grains of *Faunipollenites* type.

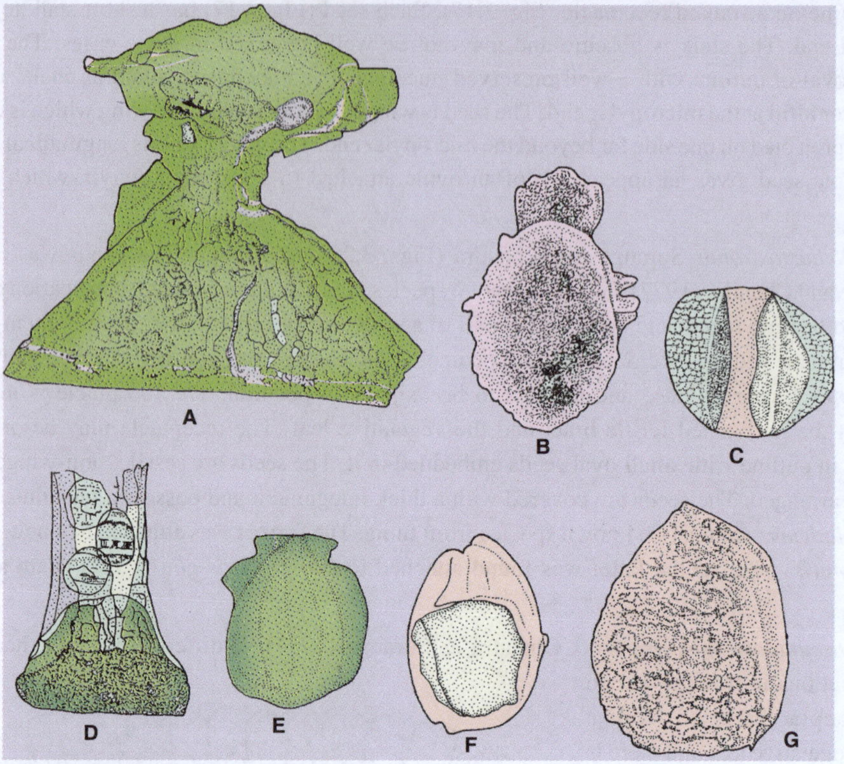

Fig. 3.22. (A—G). *Glossopteridaceae* A.Seed of *Stephanostoma crystallinum* showing two winged pollen grains in funnel, micropylar canal, and pollen chamber. B. Seed of A showing micropylar funnel and processes on the sides. C. Bi- saccate pollen grains of *Pterigospermum ranigunjense*. D. Micropylar tube of C with pollen grains. E. Seed of C showing wings. F. Macerated seed of C showing megaspore membrane. G. Seed of *Platycardia benghalensis* showing processes all over the surface. (After Pant & Nautiyal).

10. *Dictyopteridium* Feistmantel (Fig. 3.18, A, B, C, E). It is a multiovulate female reproductive organ consisting of a cylindrical oval or lanceolate and cone like (Fig. 3.18, A) receptacle or axis borne in the axil of a scale leaf or a bract, which is protective in function and fits closely on one side of the fructification. Both the receptacle and the scale leaf are borne on short pedicel which arises from the midrib of the vegetative leaf. The receptacle is covered with small round seed-cushions which are arranged in close spirals. The seeds are attached on seed cushions and are small, unwinged with pointed micropylar end. The nucellar membrane is thin without any cellular markings. The fertile scale is thick and stalked and is net veined. *Dictyopteridium sporiferum* Fiestmentel (Fig. 3.18, A) and *D. feistmanteli* Chandra & Surange (Fig. 3.18 E) are the two species known from India.

Pant and Nautiyal (1960) described some seeds from Raniganj Coal fields in India. These are *Stephanostoma crystallinum* (Fig. 3.22, A, B) ; *Pterigospermum raniganjense* (Fig. 3.22, C—F) ; and *Platycardia benghalensis* (Fig. 3.22, G).

Plumstead (1963) described fructification attached to the leaves of *Gangamopteris clarkeana*. She named it as *Vanus gondwanensis* Plumstead (Fig. 3.23 A). She was not able to determine whether it was male or a female fructification as neither pollen grains nor the seeds were observed. It consists of two parts, the lower tapering cup-like part and an upper fan-shaped collection of stiff bract-like organs. The outer surface of the cup is rough and "harsh" whereas the inner surface is hairy. It is of axillary growth. The cup is hollow and its

Fig. 3.23. (A—E). Fructifications of *Glossopteridaceae.* A. *Vanus gondwanensis* Plumstead. It is regarded as a fructification of *Gangamopteris.* B. Inside of female half of the fructification called Hirsutum. It belongs to *Glossopteris.* C.Inside of female half of sessile cupule of the fructification called *Cistella.* It belongs to *Glossopteris.* D. Inside of male half of *Hirsutum* after shedding pollen. E. Male half of Hirsutum with pollen organs. (All after Plumstead).

function, according to Plumstead, is not known. She is inclined towards regarding the fructification as male. She regarded the bract as a thick, flat, petaloid stamens on which anthers occupied the small hollows near the top, in much the same way as in *Williamsoniella* (Thomas, 1915). She has gone upto the extent of regarding *Vanus* as representing a stage anticipating a capitulum type of inflorescence. In her own words, "It can, with little difficulty, be fitted into the scheme recently outlined by Melvile (1960), as the Gonophyll theory of the origin of Angiosperms from a possible early stock of Glossopteridae. Although *Vanus* is not yet well enough known to be allocated to any exact stage of development there are possibilities that it may prove to be the most advanced form yet described, for in several ways it anticipates a capitulum type of inflorescence." This view seems to be a hasty conclusion. Maheshwari (1965, page 38) is of the opinion, "until, enclosed nature of the glossopterid fructifications is definitely proved any suggestion of a relationship between *Glossopteris* and the angiosperms will be highly conjectural. In fact most of the evidence available till now seems to favour a true gymnospermous affinity of this genus."

Nature of the Fructification

It will be worthwhile here to discuss the nature of glossopterid fructifications. This subject has been dealt with in detail by Surange and Chandra (1974, 1978). They have given a correct and a plausible explanation, based on facts, regarding the morphological nature of the glossopterid fructifications (especially *Scutum, Cistella* and *Dictyopteridium*). It is now certain that the glossopterids are seed plants and belong to the gymnosperms. The current interpretation of their reproductive structures will help us to determine their affinities and position among the various groups of gymnosperms. They have been variously interpreted by different authors. Plumstead (1952) interpreted these fructifications as stalked, dorsiventral and flattened cupules with a raised fertile head containing small, oval sacs and a fluted or striated wing. In the year 1956, she regarded the cupules as hermaphrodite flowers with the fertile half containing the seeds and the other half bearing the bract-like staminate organs. Her observations were later supported by Schoff (1976) who also regarded

Fig. 3.24. (A—G). Fructifications of *Glossopteris*. A—C. *Lanceolatus*. A.Cupule on the leaf of *Glossopteris* covering swollen surface. B. Outer side of Cupule. C. Inner side of Cupule. D. Mature female fructification called *Pluma*. E. Mature male fructification called *Pluma*. F. Seed of *Plumsteadiostrobus* showing a wing broad at the micropylar end and narrow at the chalazal end. Note the thin-walled cells of the wing. G. Seed of *Plumsteadiostrobus* opened up to show nucellus and the integument. (A—E, After Plumstead ; F—G After Chandra and Surange).

the glossopterid fructifications as bilaterally symmetrical, dorsiventral and flattened structures showing different features on both the sides. He regarded the vegetative leaf as a bract bearing the pedicellate and dorsiventral fertile head or capitulum. The term "fertiliger" was suggested by him for the reproductive structure. Edward (1928) and Walton (1929) regarded the fertile receptacles as cone-like or strobilar in nature. Pant (1962) and some other workers have pointed out that there is yet no concrete evidence to prove that the glossopteridian fructifications were bisexual or unisexual or even what was their actual nature. Regarding Plumstead's interpretation, Pant has remarked (Pant, 1962, page 312), "Her interpretations are no doubt ingenuous but ingenuity alone will not solve problems created by lack of factual details. The wingless pollen grains reported by her were isolated by macerating the rock matrix scrapings from a number of specimens. The structures regarded as stigmatic scars could as well represent the micropyler supra-micropylar structures of seeds. The interpretation of the two counterparts of her supposed bisexual fructifications as male and female halves of bivalved structures is doubtful, for if so there should at least be a thin layer of matrix between the two counterparts."

Surange and Chandra (1974, 1978) also support the interpretation that the seed bearing central axis or the receptacle is strobilar or cone like in nature. They regard it as a cylindrical, oval, lenticular or even orbicular receptacular surface bearing ovules in close spirals all around it. According to them, the fertile ovule bearing strobilar structure arises in the axil of a scale-like bract, which is net-veined and functions as a protective structure or **spathe.** The bract and the receptacle or the cone-like or strobilar fertile head are borne on a common pedicel, which arises from the midrib of an ordinary vegetative leaf and is also partly adnate to it. To sum up, the fertile receptacle or a strobilus is borne in the axil of a bract and occupies a position in between the fertile bract and the vegetative leaf. They advance the following reasons in support of the strobilar nature of the fertile receptacle.

1. The compressed material of the reproductive organs has clearly revealed that the veined scale leaf or the bract and the axis bearing the seeds or ovules are two distinct organs and not a single dorsiventral structure with veined adaxial and fertile abaxial surface. Cuticular studies support this view because the cuticles obtained from the upper and lower surface of the fertile bract are of two distinct types. The cuticle obtained from the receptacle or the strobilus is quite different from the cuticles of the bract.
2. The bracts and receptacles also occur detached and are found preserved separately in the form of compressions and impressions. This also proves that these are two separate organs.
3. The cuticles of detached bracts and receptacles are identical to those of the attached ones.
4. Successive cellulose pulls from a compression enabled them (Surange and Chandra, 1974) to remove the vegetative leaf, seed mass, the receptacle and the bract separately. This also confirms that they are separate organs.
5. Spiral arrangement of the seeds is not possible in a bifacial or a dorsiventral organ.
6. "Striated wing is nothing but the impressions left by the seeds on the margin of the receptacle."

These observations made on the preserved material clearly indicate that the fertile head or receptacle is a strobilus or a cone without ovuliferous scales and can in no case be compared with a pteridospermous cupule. "*Fertiliger*", a new term coined by Schopf (1976) is in reality an ordinary

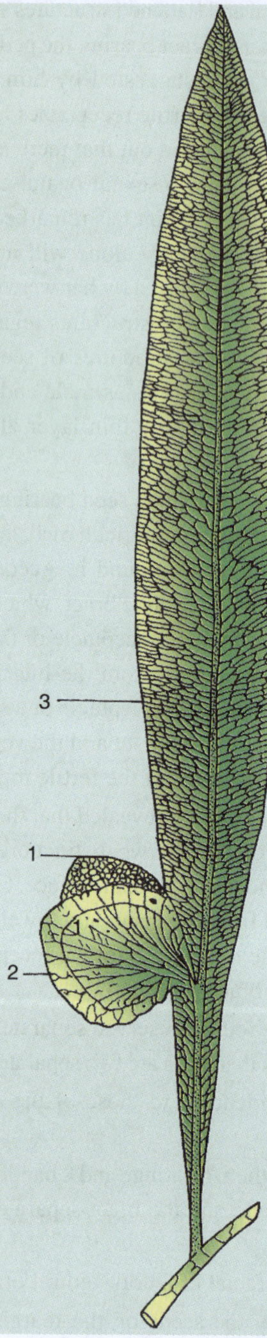

Fig. 3.25. Female reproductive fructification of *Glossopteris attached* to leaf mid-rib near its base. It is named as *Venustostrobus indicus* 1. Seed bearing receptacle 2. Protective bract 3. Leaf showing venation and midrib (After Chandra and Surange, 1975).

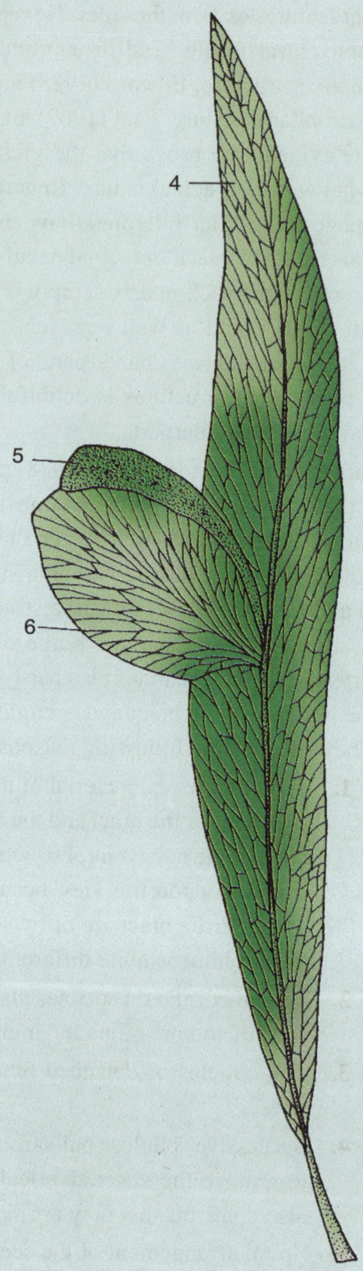

Fig. 3.26. *Glossopteris taenioides.* The figure shows the fructification *Plumsteadiostrobus* attached to the midrib of the leaf. The seed bearing receptacle is situated between the vegetative leaf and the protective bract. (After Chandra and Surange, 1975).

vegetative leaf that resembles other such leaves in its external as well as cuticular structure. The veined portion adpressed to the fertile receptacle is in reality the fertile bract which differs from the vegetative leaves in structure.

Fig. 3.27. It shows a branch of *Glossopteris taenioides* bearing sterile leaves and a fertile branch bearing the fructification called *Plumsteadiostrobus*.

It must be properly understood that the glossopterids possessed two distinct types of fructifications—the **cupulate type** in which the ovules were attached to megasporophylls and may be dorsiventral structures, and the **multiovulate type** that were strobilar in organisation and were radial in organisation. Both these types were found together in various localities of lower Gondwanaland. The cupulate fructifications have a quite different organisation and resemble the pteridosperms in many features. The *Ottokaria* (Pant, 1977) and some petrified cupulate fructifications described by Gould and Delevoryas (1977) have seeds borne on one side of the megasporophyll. They resemble the cupulate fructifications like *Lidgettonia, Partha, Russangea, Mooia, Denkania* and *Rigbya*, in being dorsiventral. The megasporophyll in these fructifications is a specialised structure with *Gangamopteris* like venation. It cannot be compared with the ordinary vegetative leaves, that bear the multiovulate fructifications. The mode of attachment of the two types of fructifications is almost similar. In the cupulate forms the cupules are either attached individually to the midrib of the fertile bract or they are borne on a cupulate disc which is attached, by means of a stalk, to the midrib of the fertile leaf. The naked ovules of the cupulate fructifications are not aggregated on a special central axis as is the case in multiovulate fructifications. The cupulate types are attached to fertile leaves whose venation pattern is exactly similar to the *Gangamopteris* leaves. These have not so far been reported on *Glossopteris* leaves (that have a distinct midrib and are net veined). The multiovulate forms always occur on *Glossopteris* leaves. On the basis of these observations, Surange and Chandra

(1978) recognised two distinct orders of gymonosperms that dominated the southern hemisphere (Gondwanaland) during those times. They included the cupulate forms (in which the naked ovules are borne on a modified frond called the megasprophylls) under the order *Pteridospermales* and the multiovulate ones under the order *Glossopteridales*. Their scheme is outlined below:

Order—**Pteridospermales**

Family-1. *Parthaceae* (*Partha, Russangea*)

Family-2. *Lidgettoniaceae* (Lidgettonia)

Family-3. *Denkaniaceae* (*Denkania, Mooia*)

Order—**Glossopteridales**

Family-1. *Cistellaceae* (*Dictyopteridium, Plumsteadiostrobus*)

Family-2. *Scutaceae* (*Ottokaria, Venustostrobus*)

Family-3. *Hirsataceae* (*Jambodostrobus*)

MICROSPORES (Fig. 3.12). Large number of microspores have been discovered and described from the Lower Gondwanaland. Srivastava (1979) reported the presence of 18 genera and 36 species of microspores from the Raniganj assemblage of the Auranga Coal fields. These have been described by a number of Indian workers like Bharadwaja and Salujaha (1964, 1965) ; Bharadwaja and Srivavastva (1969) ; Bharadwaja (1962) ; Bharadwaja and Sinha (1969) ; Tiwari (1965, 1973) ; Lele and Srivastava (1977) ; Pant (1949, 1955, 1942, 1943) ; Virkki (1939, 1945) ; Pant and Mehra (1963). Pant described as many as 1200 microspores from a single rock sample from Bacchus Marsh in Australia. These samples included about 41% monosaccates, 8% striate or non-striate disaccates, 6.5% asaccate monoletes, 6.5% trilobate triletes, 5% monocolpates and about 24% alleles and other indeterminate types (Pant, 1987). These microspores have been put under different genera and species. All of them do not belong to *Glossopteris*. Pant (1977) recovered large number of microspores belonging to the genus *Striatites* (Fig. 3.12B) from the microsporangia named *Arberiella* and also from the pollen chambers of some glossopteris ovules called *Platycardia, Pterigospermum raniganjensis* and *Stephanostoma crystallinus*. J. Sen (1955) described *Pityosporites* type of pollen grains from the Lower Gondwana Coal and considered it to be the pollen of *Glossopteris*, "on grounds of their close association in the Lower Gondwana Rocks in different parts of the southern hemisphere and particularly India." These spores were obtained by him from the sporangia-like bodies discovered from Raniganj and Karharbari stages.

HABIT OR RECONSTRUCTION OF GLOSSOPTERIS

A number of authors have described the leaves of *Glossopteris* attached to the axis of *Vertebraria*. These are called **foliage shoots** and may bear a number of leaves arranged spirally or in apical clusters. Pant and Singh (1974) suggested, "that they could be of the nature of short shoots and the plants of *Glossopteris* and its allies were possibly large trees." This observation is supported by the fact that Pant has collected a *Vertebraria* axis as thick as thirty centimeters in diameters from the Raniganj area. There are other authors who regard *Glossopteris* as a herb or a shrub or even a small tree. The trees were mostly deciduous as is evident from the occurrence of large number of isolated leaves in Permian deposits of the Gondwanaland. Surange and Chandra (1978) state that these were arborescent plants and many petrified woods described from Gondwanaland represent their stems. The occurrence of distinct growth rings suggests the possibility of a seasonal growth periodicity. Surange and Chandra also point out that the male and the female plants were separated as in most of the gymnosperms. Some authors like Schopf (1965) regard *Vertebraria* as a root rather than a stem, but recent collections and observations on the internal structure of *Vertebraria* confirm its stem-like nature.

Pant (1958) described the roots of *Vertebraria* under the generic name *Lithorhiza tenuirama*. These roots were extensively branched and their finer branches showed root tips. The lateral roots originated endogenously. Some of these roots were found attached to the *Vertebraria* axis, whereas most of them were attached to thicker "root like axes" that were devoid of rectangular areas characteristic of *Vertebraria* but possessed xylem and parenchyma like that of *Vertebraria*.

The vegetative leaves are dorsiventral and their internal structure (Fig. 3.9) has been worked out in detail by Pant (1958) ; Pant and Gupta (1968, 1971) ; Pant and Singh (!974). The leaves show a distinctly raised mid-rib on the ventral side with side veins and slightly recurved margins. The dorsal side is flat with a slight depression in the region of the mid-rib. The stomata are mostly haplocheilic and are found on the ventral side. The epidermis is made up of straight-walled or sinuous-walled parenchymatous cells with a thick cuticle. The mesophyll is clearly distinguished into elongated palisade cells and loosely arranged spongy parenchyma cells. Hypodermis is present in some species below the upper epidermis. In some cases, it is restricted to the mid-rib region only. These are distinct mid-rib bundles and the bundles of the side veins. The tracheids have mostly scalariform thickenings. Bordered pits may be present in the tracheids of the mid-rib bundles.

A number of rounded and lanceolate scale leaves have been reported isolated or in close association with the vegetative leaves. Pant (1958) found some scale leaves in close proximity with *Glossopteris fibrosa*. He has confirmed that such scale leaves from India and Australia did not bear any indication of the presence of reproductive organs. They are entirely vegetative and are regarded by Pant (1977) to be of the nature of **bud scales** that protected younger regions of the plant and were shed during the growing season. He found some scales in a "bud-like" group.

The fructifications of *Glossopteris* have been described in detail in the previous pages.

AFFINITIES OF GLOSSOPTERIS

As is clear from the account given in the previous pages, the *Glossopteris* leaves bear different types of fructifications that have been classified as cupulate and multiovulate types by Surange and Chandra. Pant has considered them under two headings: (*a*) associated detached fructifications which have never been reported in organic connection with *Glossopteris* leaves ; and (*b*)attached fructifications (some of which may also occur detached). The occurrence of such diverse fructifications in association or attached to the leaves of *Glossopteris* clearly suggests that this genus is an unnatural assemblage of a group of natural genera. Plumstead (1958) while commenting upon the fructifications of Glossopteridae has suggested that in some cases the fructifications borne upon *Glossopteris*, *Gangamopteris* and *Palaeovittaria* are so similar that these three form genera need not be treated as different. Their boundaries seem to be unnatural. Even morphologically, there are some intergrading forms which suggest their merger into one genus. However, some further investigations are required before final consensus of their merger is arrived at. At present, it can be finally and certainly said that Glossopterids are gymnosperms because of the following characters:

1. The occurrence of simple leaves that are covered with a resistant cuticle.
2. The structure and arrangement of stomata.
3. The presence of gymnosperm type of lignin lamellae in their stomata.
4. The structure of primary and secondary xylem in *Vertebraria*.
5. The seeds are naked in both and they are pollinated in a similar fashion. The occurrence of winged pollen grains in the micropylar tubes of some ovules reported by Pant, and Pant and Nautiyal clearly suggest a gymnospermous type of pollination. This has also been demonstrated in the case of *Ottokaria*.
6. The structure of the ovule, where known, is also gymnospermous in all characters.
7. The fructifications are unisexual. Plumstead has described some bisexual fructifications from African Gondwanaland but her observations are regarded as faulty.

The relationship between the *Glossopteris* and other gymnospermous groups is a debatable question, yet unsettled. It is considered to be more akin to the Pteridospermales and is regarded as one of its families called the *Glossopteridaceae*.

Surange and Chandra (1975, 1978) have suggested that the leaves of *Glossopteris* belong to two different orders called the Pteridospermales and the Glossopteridales. The former include the cupulate fructification bearing genera like *Mooia, Denkania, Partha, Russangea* and *Lidgettonia* (included under families *Denkaniaceae, Parthaceae* and *Lidgetteniaceae*). The latter included the multiovulate genera like *Dictyopteridium, Plumsteadiostrobus* (*Cistellaceae*) ; *Ottokaria, Venustostrobus* (*Scuttaceae*) ; and *Jambostrobus* (*Hirsutaceae*). Pant (1977) has favoured the inclusion of *Glossopteris* under the pteridosperms, "even though similar attachment of a cupule stalk to the midrib of a simple leaf is unknown in other seed ferns its being borne on a leaf is a pteridospermous feature. Perhaps we could continue to keep these plants among the pteridosperms in the manner of the *Caytoniales*, *Coriptospermaceae* and *Peltaspermaceae*, the attachment of whose "megasporophylls" and "microsporophylls" is unknown." He further states, "however, if the fructification is axillary as seen in *G. taeniodes* (Pant and Singh, 1974) and as sometimes assumed for *Scuttum*, etc. it would undoubtedly suggest that the Glossopteridales are very different." Harris has suggested (See Plumstead, 1952) that strange plants like those of *Glossopteris* should not be included under the pteridosperms. He expressed his opinion that such plants could as well be included under the Angiosperms. These remarks lead Plumstead (1958) to suggest a possible origin of Angiosperms from the Glossopterids.

Schopf (1976) advanced an hypothesis that glossopterids were derived from the Cordaitales. His ideas were mainly based on ; (*i*) apparent similarities between the supposed fructifications of *Noeggerathiopsis* (*Arberia* is attributed by some authors to *Neoggerathiopsis*), *Cordaianthus* and fructifications of *Glossopteris* ; (*ii*) similarities between the foliar bundles of glossopterids and Cordaitales, and (*iii*) similarities between the seeds of glossopterids and Cordaitales. He attributed *Noeggerathiopsis* to Cordaitales on the basis of similarities in vegetative leaves. Schopf (1976) also went upto the extent of deriving Gnetales from the Glossopteridales, without any convincing arguments. Moreover the strongest argument against the derivation of *Glossopteris* from Cordaitales is the absence of the latter in the southern hemisphere, which is regarded as the home of the glossopterids. The geological records clearly reveal that *Arberia, Ottokaria, Noeggerathiopsis, Glossopteris* and *Gangamopteris* all existed in the same age. Their coexistence in the same age does not suggest that any one of them is derived from the other. According to Surange and Chandra (1978), "It clearly shows that the two distinct types of reproductive structures, one cupulate type represented by *Arberia* and the strobilar multiovulate type as represented by *Ottokaria* were already present in the Lower Permian. The same two types, representing two orders, diversified and multiplied in the latter part of the Permian (Raniganj Stage)." These authors included all the cupulate fructifications under the order Pteridospermales and the multiovulate strobilar types under the order Glossopteridales. While commenting upon the origin of Glossopterids they further added, "the origin of two types of reproduction therefore has to be traced back to the Middle Carboniferous period, particularly in the early seed plants, the fossil records of which are unfortunately missing in the Gondwanaland."

Depending upon the information made available by the various types of reproductive structures belonging to fossil as well as living gymnosperms, Surange and Chandra (1978) suggested, "it appears likely that the evolution in seed plants diversified at least in three directions quite early in its history (Lower Carboniferous), almost at its beginning." They suggested the three lines as.

1. Pteridosperm—Cycad complex.
2. Glossopterid complex, possibly giving rise to Pentoxylales and Cycadofilicales.
3. Cordaitean—Conifer complex.

These three complexes originated from the primitive seed plants and further gave rise to the various gymnospermous groups as follows:

1. The pteridosperm complex gave rise to the various groups of the pteridosperms *e.g.*, Lidgettoniales, Corystispermales etc. This complex also gave rise to the Cycadales.
2. The Cordaitean complex gave rise to the Coniferales and the Ginkgoales.
3. The glossopterid complex gave rise to the *Rajmahalia* type of plants, *Pentixyleae* and the *Bennettitales*. They speculated the origin of bennettitalean "flower" from the multiovulate and strobilar fructifications of the glossopterids because there is a possibility that such an aggregation of ovules was first achieved by the glossopterids.

Pant (1987, p.112) while commenting upon the ancestral stock and origin of the *Glossopteris* flora stated, "The sudden appearance of an entirely new flora leaves us guessing about its ancestral stock and about the manner of its origin. No pre-Gondwana plants of the area are closely comparable with the pioneers or even the later elements of the *Glossopteris* Flora. There is also no possibility of the glossopterids having invaded Gondwanaland from any other territory because nothing like its elements was existing in any previous or contemporaneous floras. The origin of the *Glossopteris* Flora with its pre-dominantly simple leaved gymnosperms is therefore entirely unknown. In this connection, Seward (1922, 1924) suggested that the onset of catastrophic climatic changes in the geological history of the earth like the sudden chilling of Gondwanaland during the Carbo-Permian glaciation or the sudden rise of temperatures during widespread volcanic upheavals during some periods, etc., may have been responsible for large-scale extinctions of old forms and the abrupt appearance of new forms. Sahni (1937, 1939) suggested that since it had become well known that radiations, chemicals, sudden heating and chilling, etc., can induce mutations, the Gondwana glaciation may have been itself responsible for the disappearance of the previous vegetation and the appearance of the *Glossopteris* Flora. Such mutations must have been induced on a large scale in pre-existing plants in a widespread area so that the flora spread rapidly all over Gondwanaland. At present this is the only plausible explanation known to us about the origin of the *Glossopteris* Flora."

Plumstead (1962) described a new glossopteris fructification from the Permo-Carboniferous of Transvaal (South Africa). She named it as *Vanus gondwanensis* and found it to be axillary in position on the leaf of *Gangamopteris* c.f. *Clarkeana*. She has suggested some "striking but possibly superficial" resemblances between *Vanus gondwanensis* and *Vojnovskya paradoxa* Neiberg (1955). She states, "it is not yet possible to suggest any relationship between the vojnovskyales and some of the glossopterideae, but the present resemblances are important especially since they are not known to have been shared by any Palaeozoic representatives of either Euramerican or Cathaysian floras. The similarities moreover are not confined to the plants under immediate discussion, but include plants like several species of *Noeggerathiopsis*, *Ginkgophyllum*, *Barkaria*, *Gondwanidium*, *"Glossopteris" uralica* Zallesky (Neiburg, 1948,1954) and others. Not all of these exhibit close resemblances to the Gondwana genera of the same name and further investigations may either increase or refute the relationship suggested by the common use of these generic names but the present evidence is sufficient to indicate that a different direction or possibly a different rate of evolution existed in the floral provinces of Gondwanaland and Angara land from those elsewhere in late Palaeozoic times."

Carrying forward her discussion she has suggested that the *Glossopterideae* provide a direct link towards the Angiosperms and have a close relationship with the flowering plants than with any other known "class of plants". Neiburg (1955) has also pointed out that her order Vognovskyales is placed under gymnosperms because of the seeds being naked, otherwise the fructifications point towards primitive bisexual flowers that might have developed during carboniferous times. Plumstead is of the opinion that Glossopterideae appeared first in South America, Africa and Queensland during the Carboniferous period.

So far as the occurrence of *Glossopteris* only from the Gondwanaland is concerned, it may here be pointed out that there are reports of its occurrence beyond the boundaries of Gondwanaland. These reports may be erroneous but deserve some consideration. Restricted and isolated occurrence of *Glossopteris* only in the southern continents that constitute Gondwanaland, has shrouded the affinities and inter-relationships of these remarkable plants. There are a number of reports of the occurrence of *Glossopteris* or *Glossopteris* like leaves from areas beyond what are thought to be the boundaries of Gondwanaland, and in some cases in younger rocks. Newberry (1876, 1888) described *Glossopteris* like leaves from early Mesozoic of Northern Mexico and Honduras. Wieland (1913, 1914) described such leaves from South-eastern Mexico. Zeiller (1902, 1903) described such leaves from uppermost Triassic of Vietnam. Harris (1932) described *Glossopteris* leaves from uppermost Triassic of Greenland. Zimmina (1967) and Meyen (1969a) described typical *Glossopteris* leaves from Upper Permian of Russia. Such reports have been considered to be erroneous and many authors have disregarded them. Delevoryas and Person (1875) have placed some of these leaves (described from Mexico) under the new genus *Mexiglossa* but at the same time admit that it is very difficult to separate them from some specimens of *Glossopteris*. Sidney Ash (1981) has re-examined the *Glossopteris* like leaves from North-Western Mexico and Honduras both of which are outside the boundaries of Gondwanaland. He tentatively assigned the *Glossopteris* like leaves from North-Western Mexico to the genus *Glossopteris* and those from Honduras to *Mexiglossa* (*M. varia*). The Honduran specimens closely resemble *Glossopteris communis* and *G. browniana* on one hand and *Mexiglossa varia* on the other hand. The *Glossopteris* from Mexico looks like *G. indica* and another species described by Maheshwari (1965). This species also resembles *Mexiglossa* but differs from it in possessing a midrib made up of many veins and not merely a groove as in the case in *Mexiglosa*. The disposition of lateral veins also differs from *Mexiglossa*. Sidney Ash (1981) concludes, "The occurrence of *Mexiglossa varia* in the Late Triassic or Early Jurrassic of Honduras provides a link between the typical *Glossopteris* of the southern continents and the *Mexiglossa* in the Middle Jurassic of Southern Mexico. The significance of the *Glossopteris*-like leaf in the Upper Triassic or Lower Jurassic of North-Western Mexico is difficult to assess." The occurrence of plants bearing *Glossopteris* like leaves, irrespective of their affinities and relationships, was found in abundance and was widespread in the Early Mesozoic of Mexico and Honduras. The reports of their occurrence in the Uppermost Triassic of Vietnam (Zeiller, 1902, 1903) should also not be disregarded. Their reports from the Permian of Russia along with many somewhat similar leaves may be of use at a later stage to decide their affinities and inter-relationships.

The glossopterids, in this text, are considered as more akin to the Pteridosperms and are treated as one of the families of the order Pteridospermales. It is however, realised that the rank of the order Pteridospermales should be raised to that of a division pteridospermophyta and the family *glossopteridaceae* should be regarded as an order. Some other families like *Medullosaceae*, *Lygenopteridaceae* and *Caytoniaceae* should also be treated as orders.

QUESTIONS

ESSAY TYPE

1. Write an illustrated essay on *Glossopteris* flora of India. Give suitable examples. **(Allahabad, 2001)**
2. Give a general account of the order Glossopteridales (family Glossopteridaceae). Discuss their affinities. **(Awadh, 1997)**
3. Give a comprehensive account of *Glossopteris* an d comment on its recent reconstruction. **(Kerrla, 2001)**
4. Describe the affinities of *Glossopteris*.
5. Describe the following with suitable sketches :

(*a*) *Glossopteris*

(*b*) *Vertebraria*
(*c*) *Gangamopteris*
(*d*) *Kendrostrobus*
(*e*) *Glossotheca.*

6. Give a brief account of the family Glossopteridaceae.
7. Give an account of male reproductive organs of Glossopteridaceae.
8. Give an account of female fructifications of Glossopteridaceae.

SHORT ANSWER TYPE

9. Write short notes on
   (*a*) *Glossopteris* flora **(Kanpur, 1992)**
   (*b*) *Glossopteris* **(Kanpur, 1994; Rohilkhand, 1992; Poorvanchal, 1998)**
   (*c*) Fructifications of *Glossopteris* **(Awadh, 1998)**
   (*d*) Affinities of Glossopteridales (Glossopteridaceae) **(Awadh, 2000)**
   (*e*) *Glossopteris* leaf **(Poorvanchal, 2002)**
10. Briefly describe the characters of *Glossopteris*
11. Give the names of contributors to the family Glossopteridaceae.
12. Draw labelled diagrams of the following:
    (*a*) Variation pattern of *Glossopteris*
    (*b*) Shape of the lamina
    (*c*) Multicellular hair of *Glossopteris*
13. Assign the following to their systematic position and point of features of special interest :
    (*i*) *Glossopteris* **(Rohilkhand, 1994)**
    (*ii*) *Gangamopteris*
    (*iii*) *Vertebraria*
    (*iv*) *Kendrostrobus*

OBJECTIVE TYPE

14. Select the correct answer :
    (*i*) The name Gondwanaland was given after the prepondenance of the
    (*a*) Gond tribe (*b*) Bhil tribe
    (*c*) Tribes of the South America
    (*d*) Tribes of India.
    (*ii*) Talchir Boulder Beds were first discovered at
    (*a*) Umaria (*b*) Kashipur
    (*c*) Jammu (*d*) Talchir
    (*iii*) The name *Glossopteris* was first used by
    (*a*) Lele (*b*) Sahni
    (*c*) Brongniart (*d*) Pant
    (*iv*) The leaves of *Glossopteris* are
    (*a*) tongue shaped (*b*) spoon shaped
    (*c*) fan shaped (*d*) Oval in shape.
    (*v*) *Gangamopteris* is usually preserved as
    (*a*) Petrifactions (*b*) Impressions
    (*c*) Both (*d*) None.
    (*vi*) The stem or rhizome of *Glossopteris* is known as
    (*a*) *Gangamopteris* (*b*) *Vertebraria*
    (*c*) *Glossopteris* (*d*) *Rubidgea*
    (*vii*) The male reproductive organ of Glossopteridaceae is known as
    (*a*) *Kendrostrobus* (*b*) *Rubidgea*
    (*c*) *Glossopteris* (*d*) *Gangamopteris*
    (*viii*) The female fructification of Glossopteridaceae is
    (*a*) *Kendrostrobus* (*b*) *Lidgethonia*
    (*c*) *Rubidgea* (*d*) *Gangamopteris*
    (*ix*) *Scutum* is the name given to
    (*a*) Male reproductive organ
    (*b*) Female reproductive organ
    (*c*) Leaf (*d*) Stem.
    (*x*) In which plant of Glossopteridaceae mid rib is absent?
    (*a*) *Glossopteris browniana*
    (*b*) *Glossopteris Indica*
    (*c*) *Gangamopteris cyclopterioides*
    (*d*) *Glossopteris fibrosa*

CHAPTER 4

Pteridospermales (Contd.) Families — Peltaspermaceae And Corystospermaceae

- Lepidopteris–general Structure
- Seed Bearing Organ
- Pollen Bearing Organs
- Family Corystospermaceae
- Female Reproductive Organs
- Male Reproductive Organs

LEPIDOPTERIS

Family PELTASPERMACEAE

This family includes form genera *Lepidopteris* (for the leaves or fronds), *Peltaspermum* (for seed bearing organs) and *Antevsia* (for pollen bearing organs). The stem fragments were also found alongwith them (Andrews, 1961) but have not been fully described. These form genera were collected from the Late Triassic of Natal (Thomas, 1933) and Sweden (Antevs, 1914) and Rhaetic of Greenland (Harris, 1932). The family is also represented in the Upper Permian of China, Madagascar, Australia and Argentina.

Antevs (1914) was the first person who described the structure of the cuticle of the form genus *Lepidopteris ottonis* Schimper and of the microsporophylls found along with it. Their cuticular structure was exactly similar. Both the foliar and the reproductive structure were found in Triassic of Sweden. In both the cuticle was thick and the sunken stomata were surrounded by a ring of subsidiary cells. This similarity was a clear proof of both the foliar and male reproductive organs as belonging to the same plant. Antevs concluded that both *Lepidopteris ottonis* and the pollen sacs that were attached to pinnately divided fertile organs were seed plants. This was later confirmed by Harris (1932) who found both these forms occurring together in the Rhaetic rocks of East Greenland. Harris also found peltate disc like structures to which the seeds might have been attached. The cuticular and epidermal structure of the cupular peltate disc was similar to that of *Lepidopteris ottonis*. This confirmed that all the three structures (foliage, male fructifictions and cupular discs) belong to the same plant. In the year 1933 Thomas collected *Lepidopteris* and a spirally branched fertile axis bearing terminal peltate heads or discs to which seeds were found attached on the lower side. Their cuticular structure also resembled those found in Sweden and Greenland. Thomas collected his specimens from the Triassic beds of Natal (South Africa). The evidence collected from three widely separated regions confirmed that they belong to a same family which was named as **Peltaspermaceae** by Thomas (1933).

Lepidopteris (Fig. 4.1 A). This is the generic name given to the sterile and **bipinnate** leaves that were 8 to 30 cms long and appeared **lanceolate** in shape. Each leaf or frond has a stout **rachis** which is unforked and bears **pinnae** on either side. The pinnae bear **pinnules** that are closely set (Fig. 4.1A) and linear or lanceolate in shape. Each pinnule is sessile and attached to the pinnae with a broad base. The pinnules are traversed by a distinct **midrib** and forking **secondary veins.** The rachis also bears pinnules directly on it between the pinnae (Fig. 4.1 A). A close series of blister-like swellings are also present on the rachis giving a scale like appearance. These swellings are not so conspicuous on the South African forms (*L. natalensis*). Such swellings are also present on the axes of the reproductive structures, thus providing an evidence that they belong to the same plant. The upper and the lower surfaces of the pinnules are covered with thick **cuticle.** The **stomata** occur on both the surfaces but are abundant on the lower epidermis. They are sunken in pits and are surrounded

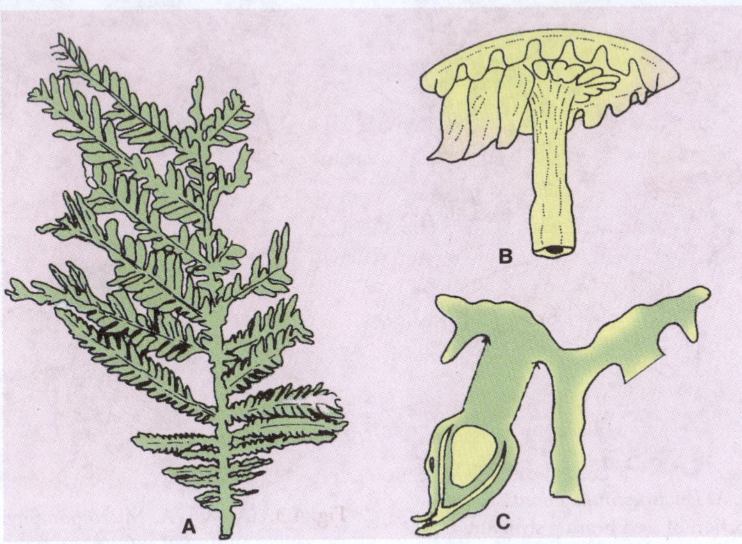

Fig. 4.1. (A—C). Peltaspermaceae. A. *Lepidopteris natalensis, a frond showing rachis and pinnules.* B. *Lepidopteris ottoris,* A cupulate fructification. It is peltate. The disc bears three seeds intact on its ventral side. C. *Lepidopteris ottonis,* A L.S. of B with one seed. (*A after Thomas, B, C, after Harris*).

by radially arranged **subsidiary cells** that number five or six around each stomatum. The common species of *Lepidopteris* include *L. ottonis*, *L. natalensis* and *L. stormbergensis*. In all five species have been recognised. *Lepidopteris* has also been reported from Triassic deposits of Gondwana. It was found along with *Pachypteris* (Srivastava, 1974).

Stem pieces measuring 10 cms long and 5.5 cms in diameter have also been found associated with the leaves. These stem pieces also bear characteristic blister-like swellings found on other organs.

Seed Bearing Organ (Fig. 4.1, B, C). The seed bearing axis as described by Thomas was slender and several centimeters long and bore spirally arranged branches that terminated in peltate heads bearing a number of more or less oval seeds on their lower side near the margin (Fig. 4.1, B). These peltate heads were five millimeters or more in diameter. These seed bearing axes are put under the genus *Peltaspermum. P. Thomasii* is the name given to the species found in Natal. It occurred in association with *L. stormbergensis* and bore only two seeds on each peltate head. Thomas (1933) was of opinion that only two seeds reached maturity, the rest either withered or remained abortive. Townrow (1960) is of the opinion that these peltate heads were bilaterally symmetrical and not radially symmetrical and were arranged in a pinnate manner on the main axis. He regarded the whole structure as a fertile frond or a sporophyll and not a branch bearing the female fructifications or **sporangiophores**. Harris (1932) described *Peltaspermum rotula* found in association with *Lepidopteris ottonis* from eastern Greenland. It was a comparatively well preserved specimen with distinctly **pelatate** and radially symmetrical heads that bore 10–12 elongated seeds (Fig. 4.2 A) on their lower surface. The peltate disc had 10-15 marginal lobes and were about 1.5 cms in diameter. The seeds were 7 mm long and had a curved **micropyle** (Fig. 4.1, C). "The **nucellus**

Fig. 4.2. (A—C). A. *Peltaspermum Thomasii*, showing a portion of seed bearing structure (After Thomas) B. Part of *Lepidopteris natalensis*. C. Cupulate disc of *Peltaspermum sotula*.

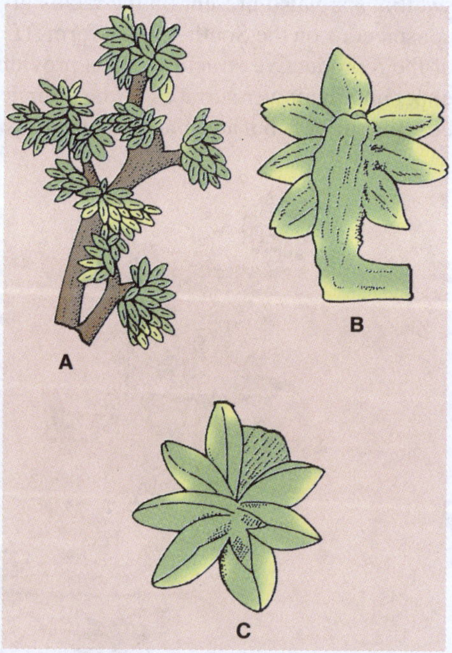

Fig. 4.3. (A—C). A. Microsporophyll of *Antevsia Zeilleri* B—C. Opposite sides of the microsporangiophores. (After Harris and Townrow).

was free from the **integuments** near to the base. The nucellus formed a short beak-like structure extending into the base of the **micropyle** (Fig. 4.1, C), and probably breaking down to form a **pollen chamber** in which many pollen grains were observed (Thomas, 1955)."

Pollen Bearing Organs (Fig. 4.3, A–C). The pollen bearing organs are put under the genus *Antevsia*, which is a pinnately branching structure with three dimensional pattern of branching. It had a stout main axis from which arose short lateral branches in an alternate fashion. These primary branches bore secondary branches that were not regular in arrangement and each bore upto 12 pollen sacs or microsporangia in *A. zeilleri* and only four in *A. extans*. The **microsporangia** were about two millimeters long and were dorsiventral structure with a distinct mid rib and a few stomata on the dorsal side. The longitudinal line of dehiscence marked the ventral side. The presence of stomata on the thick and massive **microsporangial wall** is suggestive of its being photosynthetic and is considered to be an unusual feature. The **microspores** were oval in shape and were without wings. Each had a single longitudinal furrow on the outer wall. The pollen bearing organs of *Antevsia* type are unknown in any other Mesozoic plant. They do resemble *Antholithus wettsteinii* and *sphenobaiera furcata* but differ markedly in their shape and presence of stomata in the wall. Townrow (1960) regarded the pollen bearing organs as **sporophylls** because they were bipinnate in organisation with alternate primary branches that bore secondary branches with microsporangia terminating them.

The Peltaspermaceae are pteridospermous so far as their affinities are concerned except the seed bearing axes which show striking departure. Bierhorst (1971) states, "in any consideration of the origin (S) of the angiosperms, the ovulate disc of the peltasperms bearing a ring of ovules should not be overlooked. If the disc is folded medianlly its resemblance to certain winteraceous carpels is apparent. In these, the ovules are inserted, not in two marginal or sub-marginal series, but in a complete ring, and the application of the descriptive term peltate can be defended."

H.H. Thomas (1955) states, "The fact that several of the mesozoic inflorescences were branched in one plane while others had a spiral arrangement of branches, is a point of some interest since it is seen in both the pollen– and seed–bearing structures. It suggests that the spiral mode of branching in the fertile parts may have arisen from a system in which the branching resembled that of a photosynthetic frond. This removes one of the objections to the view that the angiosperms may have evolved from one or more types of pteridosperms."

Family CORYSTOSPERMACEAE

The families **Peltaspermaceae** and **Corystospermaceae** are regarded as Mesozoic pteridosperms in comparison to the Palaeozoic or Carboniferous pteridosperms from which they differ in some respects. The family *Corystospermaceae* was also constituted by Thomas (1933) to include some fossils collected by him from the Triassic plant beds in Drakensberg in Natal (South Africa). Later in 1943 and 1944 Franguelli collected and described them from Australia and Argentina. They have also been reported from India in a vegetative condition (only leaves). The family includes the following form genera:

Leaves or Fronds: *Dicroidium* and *Xylopteris*

Female reproductive organs: *Umkomasia*, *Spermatocodon* and *Pilophorosperma*.

Male reproductive organs: *Pteruchus* and *Pterorachis*.

The form genera listed above have so far been collected and described from the Southern Hemisphere.

Fronds (Figs. 4.4, F, G, H, I and Fig. 4.5 A). They have been collected from the Triassic plant beds in Natal, Argentina and Australia. In all of them the rachis forks equally into two equal branches as is the case in the Palaeozoic Pteridosperms. Due to variations in their general appearance and form they were assigned to a number of different genera like *Thinnfeldia*, *Stenopteris*, *Odontopteris* and *Pachypteris*. As result of cuticular studies, it was found that these were differences between them and as a consequence the Corystospermous fronds were included under two genera named *Dicroidium* and *Xylopteris* (Townrow, 1957). The rachis, in both these genera, forked dichotomously into two equal branches near the base. In this character they resemble the Palaeozoic pteridosperms. *Dicroidium odontopteroides* and *D. fiestmantelli* (= *Zuberia zuberi* Frenguelli) are the two common species, the former described from Natal in South Africa (Thomas, 1933) and latter from Argentina (under the name *Zuberia* by Frenguelli in 1944). Both the species also occur in Australia. The fronds in *D. odontopteroides* are unipinnate and the pinnules have open venation. The leaves are upto 10 cms in length. The fronds in *D. fiestmantelli* are bipinnate and upto 100 cms long. The pinnules have open venation. In both the cases the pinnules are also borne on the basal undivided portion of the rachis.

In *Xylopteris densifolia* (= *Stenopteris densifolia*) the leaves are once pinnate (Fig.4.4, G) with linear pinnules having a single unbranched midrib.

In India *Dicroidium* (Fig. 4.4, F, H, I) has been reported from the early Triassic age of the Middle Gondwana period. It was collected from the Parsora beds by Lele (1961) along with

Fig. 4.4. (A—I). A. Female inflorescence of *Umkomasia macleani* bearing seeds B. Cupule of *Pilophorosperma granulatum* showing micropyle of the seed. C. Cupule with seed. D. Microsporangiate inflorescence of *Pteruchus africanus.* E. Portion of main axis of *Umkomasia macleani* showing a lateral branch and a bract and bracteole. F, H, I. Fronds of *Dicroidium* sp. G. Frond of *Stenopteris densifolia* (= *Xylopteris densifolia*) (After Thomas, 1933).

Pterophyllum, *Pseudoctenis*, *Bairea*, *Araucarites*, *Noeggerathiopsis*, *Samaropsis*, *Cordaicarpus* and *Glossopteris*. This flora is usually called the *Dicroidium* flora because it occurs in abundance. It was also reported from the Lower Triassic Gopad river valley near Nidpur, Sidhi district of Madhya Pradesh by Srivastava (1969, 71) and Bose and Srivastava (1971). This species was named as *D. nidpurensis*. The frond has a rachis forked dichotomously like the species from Natal, Australia and Argentina. The fronds are unipinnate with sessile pinnae attached by their broad bases. Each pinna possesses a single midrib with numerous forked lateral veins. The pinnae have sinuous and papillate epidermal cells. The stomata are amphistomatic but are found in greater numbers on the lower epidermis. Each stoma is surrounded by four subsidiary cells. In these characters it differs from *Thinnfeldia*. The fertile appendages are unknown, except for the male fructifications called *Nidostrobus harrisiana* and the female ones called *Nidia ovalis* from the Nidpur beds in close association with *D. nidpurensis*. The cuticles of these fructifications resemble closely with *D. nidpurensis*. These were described by Bose and Srivastva in 1973. Srivastava (1974) also recorded *Pteruchus nidpurensis* from Triassic beds of Nidpur.

Female Reproductive Organs (Fig. 4.4, A–C). Thomas described three seed bearing genera. These were named as *Umkomasia* (with two species, *Pilophorosperma* (with eight species) and *Spermatocodon* (with one species). The structure of their cuticles and stomata resembled those of the frond genus thereby indicating that they belonged to the same plant. In all the three genera the seeds were enclosed within cupules that were helmet-shaped in *Umkomasia* and

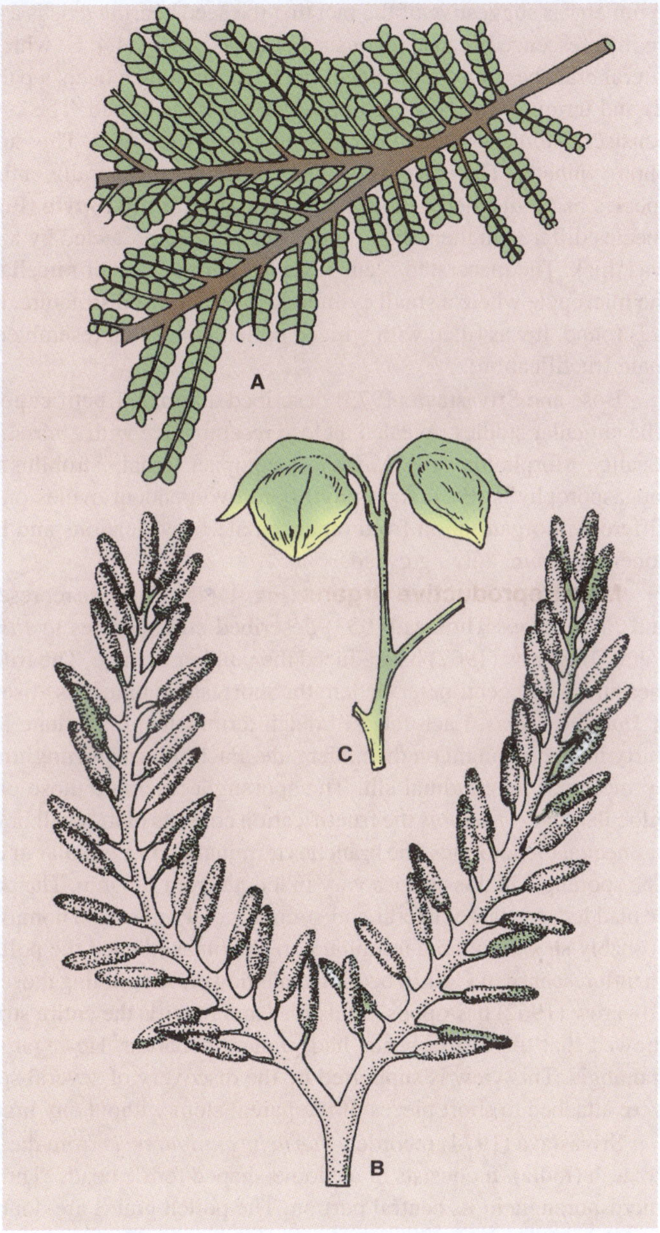

Fig. 4.5. (A—C). A. Portion of a sterile frond of *Zuberia zuberi*. B. Male fructification of *Zuberia zuberi*. C. Cupulate structures of A. (After Franguelli. 1944).

Pilophorosperma and bell-shaped in *Spermatocodon*. The cupules in *Umkomasia* were distinctly bivalved structures without any hair. In *Pilophorosperma* the cupule is lined with hair produced from the epidermal cells of the inner surface. In some species the hair were cuticularised, whereas in others only bases of the hair are seen. In *spermatocodon* the hair are absent but the shape of cupule is like a bell. In all the three genera the cupules were borne on an inflorescence axis which is distinctly dorsiventral. It has different type of cuticle on the dorsal and ventral surfaces. This dorsiventral symmetry is suggestive of the fact that the seed bearing axis is a fertile frond. Thomas regarded it as an inflorescence because it consists of an axis (Fig. 4.4, E) which bears **bracts** in whose axils arise lateral branches in a single plane. Each lateral branch bears a pair of scale-like **bracteoles** (Fig. 4.4, E) and terminal **cupules** each of which encloses a seed. The cupules in all the genera and species consist of stomata on the outer as well as inner surfaces. This suggests that cupules were green and photosynthetic. The margins of the cupules were generally entire. The seeds in all the genera and species had a distinctly short-curved and a bifid **micropyle** (Fig. 4.4, B, C). The seeds of various species differed in their shape and size and were protected by a smooth and hard **testa**. The cuticle was thick. The macerated seeds revealed a little trace of **nucellar membrane** except at the base of the micropyle where a small cylindrical and cutinised structure, believed to be the **pollen chamber**, was found. It was filled with winged pollen grains that resembled those produced by *Pteruchus* (the male fructification).

Bose and Srivastava (1973) described a female fructification called *Nidia ovalis* from Nidpur. The cuticular studies revealed a close resemblance with *Dicroidium nidpurensis* found in the same locality. Morphologically *Nidia* is a compact female strobilus that bore spirally arranged peltate megasporophylls. Each sporophyll bore two pendent ovules on either side of the stalk. It is quite different in organisation from other female fructifications and bears a close resemblance with the cones of *Zamia*, a living cycad.

Male Reproductive Organs (Fig. 4.4 D). They are represented by the organ genera *Pteruchus* and *Pterorachis*. Thomas (1933) described eight species to *Pteruchus* from the Triassic rocks of Natal. Townrow (1962) has reduced the number to three. The fructification consists of a central axis measuring four centimeters in length. Short lateral branches arise from the central axis (Fig. 4.4D) in *P. stormbergensis*. Each lateral branch terminates in a peltate head. Each peltate head bears upto thirty microsporangia on the under side. Each **microsporangium** is 1.5 to 3 mm long and dehisces by means of longitudinal slit. The sporangia resemble those of Palaeozoic *Crossotheca* in being bilocular. In *P. africanus* the fructification consists of a "small branch system which divides by equal or unequal dichotomies, the branches terminating in a circular or elliptical lamina (Andrews, 1961)." The sporangia in this species vary in length from 1-4 mm. The pollen grains are winged. The wings or bladders are large, lateral and symmetrically placed. Thomas (1955) states that the pollen were probably shed through a terminal pore. Thomas likened the pollen bearing organs of *Pteruchus* to an inflorescence of **Catkin** bearing male flowers resembling those of an angiospermic genus *Populus*. Townrow (1962) has opposed this idea and regards the entire structure as a fertile frond because he showed that the fertile heads had pinnate venation. He regarded them as fertile pinnae bearing sporangia. This view is supported by the discovery of several specimens in which the male organs were attached to short pieces of the parent stem without any bract or a subtending structure.

Srivastava (1974) recorded *Pteruchus nidpurensis* from the Triassic beds of Nidpur in Madhya Pradesh (India). It consists of a "dome-shaped fertile head." The head bears a number of ellipsoidal microsporangia in its central portion. The pollen grains are double winged. The wings are distinct, lateral and symmetrically placed.

Frenguelli (1944) while describing Argentinian Corystosperms recorded a new genus called *Zuberia zuberi* which was previously called *Thinnfeldia fiestmantelli* (Ehinghausen, 1852) and was regarded as fern. Later Gothan (1912) transferred this genus to *Dicroidium*. It is now regarded as *Dicroidium fiestmantelli* (= *Zuberia zubei* Frenguelli). Its fronds were large and bipinnate. Frenguelli (1944) found these fronds from Triassic rocks of Argentina. Associated with these fronds were found forking axis (Fig. 4.5) whose ultimate branches terminated in a pair of small recurved cupules (Fig. 4.5 C) showing three equal marginal segments. The ovules or the seeds were not clearly visible. These cupules closely resemble some of the South African Corystosperms (most probably *Umkomasia*). Some pollen bearing structures were also found in the same beds. They were named as *Pterorachis* by Frenguelli. These male fructifications consisted of a forking rachis (Fig. 4.5, B). Each branch bore short lateral branches. These short branches bore a cylindrical mass of small elongated microsporangia. The microsporangia resemble those of *Pteruchus*. The structure of spores and cuticles is not known so far, but Thomas (1955) is of the opinion that this plant is a distinct member of the Corystospermaceae. Andrews (1961) is also of the same opinion.

QUESTIONS

ESSAY TYPE

1. Give an account of family Peltaspermaceae.
2. Describe the genera belonging to the family corystospermaceae.
3. Give a detailed account of *Lepidopteris*.
4. Describe briefly the pollen bearing genera of family Corystospermaceae.

SHORT ANSWERS TYPE

5. **Write short notes on :**
   (*i*) *Lepidopteris*
   (*ii*) *Peltaspermum*
   (*iii*) *Antevsia*
   (*iv*) Affunties of Peltaspermaceae
   (*v*) Fronds of *Xylopteris*
   (*vi*) *Pteruchus*
   (*vii*) Female reproductive organs of *Umkomasia*
   (*viii*) Male reproductive organs of Corystospermaceae
   (*ix*) Characteristics of Corystospsmaceae.
6. Draw labelled diagrams of the following :
   (*a*) *Lepidopteris* (*b*) *Xylopteris*
   (*c*) *Umkomasia* (*d*) *Antevsia*

OBJECTIVE TYPE

7. **Select the correct answer :**

(*i*) The family Peltaspermaceae is represented by
(*a*) *Lepidopteris* (*b*) *Peltaspermum*
(*c*) *Antevsia* (*d*) All of the above.

(*ii*) Seed bearing organ of the family Peltaspermaceae known under the generic name is
(*a*) *Lepidopteris* (*b*) *Peltaspermum*
(*c*) *Antevsia* (*d*) *Xylopteris*

(*iii*) Under genus *Antevsia* is described the
(*a*) Fronds (*b*) Stem
(*c*) Seeds (*d*) Pollen bearing organs.

(*iv*) Male reproductive organ of the family Corystospermaceae is described under the genric name
(*a*) *Xylopteris* (*b*) *Lepidopteris*
(*c*) *Umkomasia* (*d*) *Pteruchus.*

(*v*) Lele collected the genus *Dicroidium* from
(*a*) Parsora beds (*b*) Rajmahal hills
(*c*) Kumaon range (*d*) Shivalik beds.

(*vi*) The genus *Umkomasia* represents the
(*a*) Fronds (*b*) Rhizome
(*c*) Female reproductive organ
(*d*) Pollen bearing organ.

(*vii*) *Pteruchus nidpurensis* is described from triassic beds of Nidpur by
(*a*) Srivastava
(*b*) Lele
(*c*) Surange
(*d*) Chandra and Surange.

CHAPTER 5

Pteridospermales (Contd.) Family — Caytoniaceae

- Caytonia (Megasporophyll and Microsporophyll)
- Structure of the Seed
- Microsporophyll (Caytonanthus)
- Phylogenetic Relationships

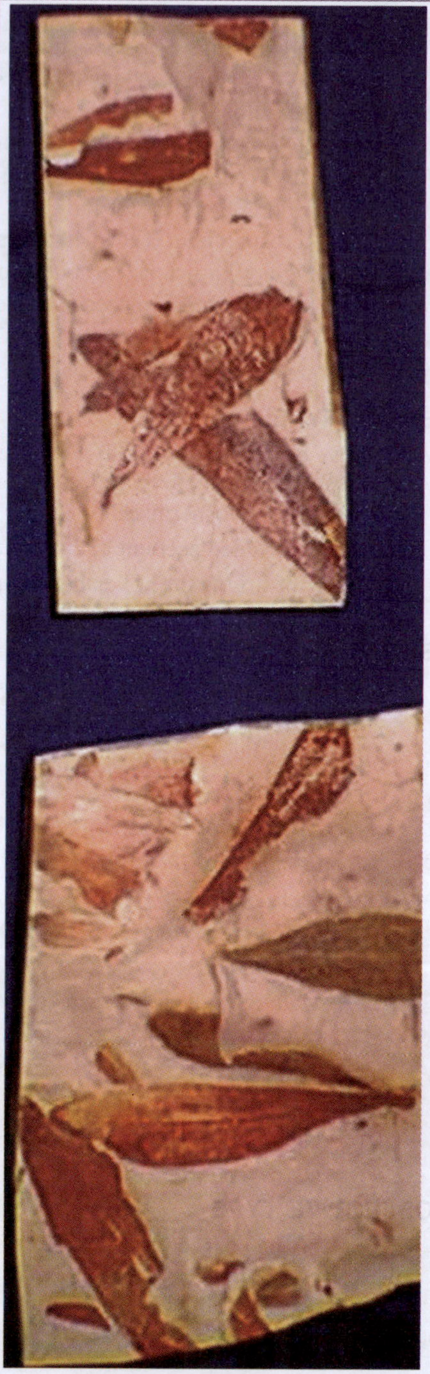

SAGENOPTERIS

It is a small group of very interesting and controversial plants that were first described by Thomas (1925) from the Middle Jurassic plant bed at Cayton Bay in Yorkshire. These extinct plants range from the Upper Triassic to the Upper Cretaceous. The various organs have been collected in the form of compressions. Apart from their external appearance, much is known about their cuticles, epidermis, stomata and the hardest internal parts. The vascular tissues are yet unknown. The different organ genera, that belong to the same plant, can be listed as under:

1. LEAF: *Sagenopteris*

S. colpodes; *S. phillipsi* and *S. nilssoniana*. The first two species are known from the Middle Jurassic of Yorkshire, whereas the third is known from the basal Lias of Greenland.

2. MEGASPOROPHYLL: *Caytonia sewardii, C. nathorsti* (= *Gristhorpia nathorsti*), *C. thomasi.*

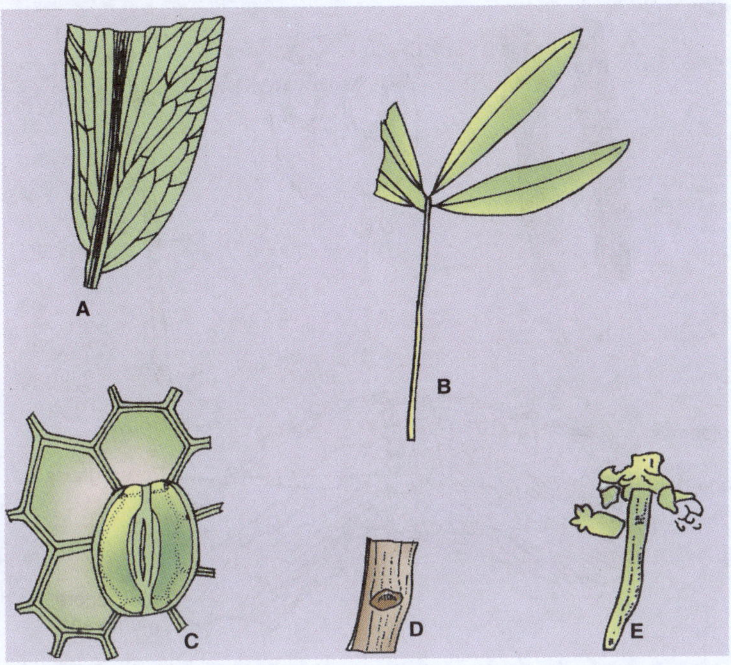

Fig. 5.1. (A-E). *Sagenopteris phillipsi.* Vegetative organs. A.Base of leaflet showing veins, restoration, × 4. B. Leaf showing shedding of leaflets, restoration, × 1. C. Stoma (from outside), restoration, × 800. D.Stem with leaf scar, × 4. E.Shoot with leaf scars and scale leaves, × 2. (D & E, simplified from original drawings in Harris, 1940. A—C after Harris 1951).

3. MICROSPOROPHYLL: *Caytonanthus oncodes, C. arberi* (= *Antholithum arberi*) (= *Ginkgoanthus phillipsi*), *C. kochi.*

It is now established that all the three organs belong to the same plant, because they have been found in close association in the Middle Jurassic of Sardinia (Edwards, 1929) and Greenland (Harris, 1926). Occurrence of same type of *capitate glands* on all three organ genera collected from Greenland confirms the view that they belong to the same plant.

STEM: All that we know about the stem is a small, branched twig bearing "foliar bed scales" (Fig. 5.1, E) and a C-shaped (Fig. 5.1, D) leaf trace of *Sagenopteris phillipsi.* It indicates that the plants were either branched trees or shrubs and not that the stem was thick, stumpy or columnar and unbranched like most of the cycads and ferns.

LEAF (Fig. 5.1, A—C). The leaf which belongs to the organ genus *Sagenopteris* is known since 1828 and was given the generic name in 1838. It consists of a slender petiole bearing four terminal leaflets (Fig. 5.1, B). Each leaflet consists of a distinct midrib from arched side veins that give off (Fig. 5.1, A) finer branches that fork freely and form a network of veins with any free endings except at the margin. The venation pattern resembles *Linopteris* and some Glossopterids. The margin of the leaflets is smooth with acute apex and more or less elliptic or vaguely lanceolate shape. The leaflets are in two pairs and not in a whorl. Leaves with three pairs of leaflets have been recorded by Harris (1926). Thomas (1925) recorded specimens with a single pair or even a single leaflet. Transitional forms between leaves and **bud-scales** have been recorded by Halle (1910). The interesting feature is that the leaflets or even the whole leaf falls by the formation of an **abscission layer** which is an angiospermous character. Besides *Glossopteris* and some *Pteridosperms* (*Linopteris*) the venation of the leaflets, in some cases, resembles a Bennettitalean genus called *Dictyozamites* and a cycad called *Ctenis*.

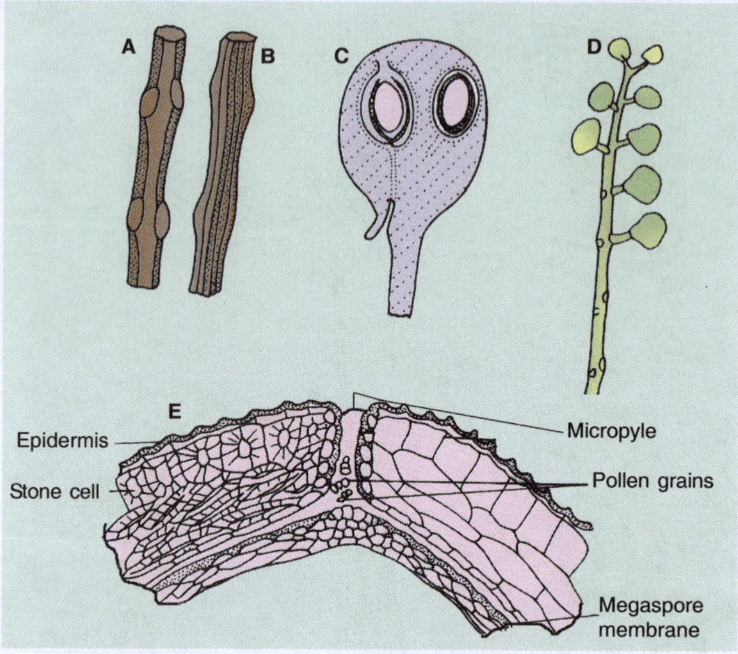

Fig. 5.2. (A-E) *Caytonia nathorsti*, restorations. A, B, part of the fruiting rachis from above and below, showing fruit scars and surface ribs ×8. C.Young fruit in ls showing the canal running from the mouth to a seed (a second seed is not cut through the middle). D.Fruiting rachis, the lower fruits have been shed ×3. E.L.S through seed micropyle, cutinized layers are stippled. The integument has a cutinized epidermis, all around it and contains two layers of stone cells. The nucellus has a thickly cutinized epidermis; the next two cell-layers are imaginary; the innermost layer is the megaspore membrane. There are pollen grains in the micropyle. This figure is based partly on *C. thomasi* and *C. sewardi.* (After Harris, 1951).

The dorsiventral leaflets have an upper and lower epidermis covered with a thick cuticle. The stomata abound on the lower surface (Fig. 5.1, C). Each stoma has a central pore surrounded by two **guard cells**.The **subsidiary cells** have an independent origin (haplocheilic) as in most of the gymnosperms except the Bennettitales and *Gnetum*. The outer surface of the guard cells seems to be flat and overlaps slightly the wall of the subsidiary cells (Fig. 5.1, C). The mesophyll is distinguished into **palisade** and **transfusion mesophyll** whose cells are transversely elongated. The three species of *Sagenopteris* named above resemble each other very closely and differ only in the shape of the epidermal cells and in their general form. Filamentous hair with spherical apical cells are found on the petioles, margins and main veins of the leaf.

MEGASPOROPHYLL (*Caytonia*). *Gristhorpia* and *Caytonia* were the two fruit genera described by Thomas (1925). Later he expressed the view that they can be merged into one. Harris (1940) showed that most of the differences were inconstant and a third species combined the characters of the other two. So it was named as *Caytonia nathorsti* (Thomas) Harris.

Caytonia consists of a dorsiventral rachis measuring about five centimeters in length bearing a number of lateral outgrowths that were distinctly stalked and were swollen at tips like the berries. These outgrowths were called fruits by Harris who did not use the term in its strict morphological sense but "in its normal use for a sweet juicy pulp surrounding the seeds." Harris guessed that these fruits were eaten by a small animal for its juicy pulp. The term **cupule** has also been used for them, but since the seeds are completely enclosed, this term is not justifiable. The fruits often get abscissed from rachis in just the same way as the pinnae from the petiole. *Caytonia* has not so far been found attached to the stem. The megasporophyll (Fig. 5.2, D) is regarded as a **pinnate** structure and not a **fruiting axis** bearing berries or an **elongated floral** axis with ovaries. In its pinnate nature it resembles pteridosperms.

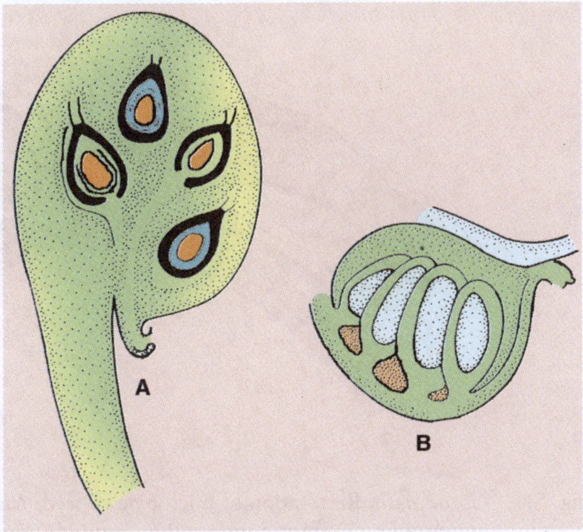

Fig. 5.3. (A—B) A Restoration of fruit of *Caytonia nathorsti* in longitudinal section. Fruit wall dotted, stone of seed solid black, cuticles continuous black lines. The inner epidermis of the fruit forms a pocket round each seed connected by a narrow channel with the stigma ; three of the seeds are in non-median section and the channel is not seen. B. L.S. Fruit showing four seeds.

The fruit, when removed from the rachis consists of stalk which broadens terminally into a pouch like structure (Fig. 5.2, 5.3C). The stalk of the fruit bears a conspicuous flange like structure on one side near its upper part where it expands into a fruit. This structure is called the **lip** by Harris (Fig. 5.3). Thomas termed it as **stigma** because he believed that pollen grains germinated on it. Later it was shown that pollen grains reach the **micropyle** of the ovules and the term stigma cannot be applied to it. The term **mouth** was used by Harris for a small crack or a slit running between the lip and the stalk. The mouth is the only discontinuity in the fruit wall, which is covered all over by a cutinised epidermis except for some stomata. The mouth, according to Harris, is closed in a ripe fruit. He believes that it must have been open during early stages of development and upto the time after **pollination**. A series of **canals** (Fig. 5.2, C and 5.3, A) lead from the mouth to the micropyles of the seeds. The number of canals corresponds to the number of ovules or seeds in the fruit. In a compressed state the position of seeds is indicated by a series of transversely arranged **bars** (Fig. 5.2, C) on the fruit wall. The number of bars also indicates the number of seeds enclosed. In *C. sewardii* there were eight seeds in each fruit, in *C. nathorsti* there were fifteen seeds. In both these species the seeds were arranged in a single arched row. In *C. thomasii* there were two arched rows of thirty seeds. The canals extend in between these bars. Harris thought that after pollination the cells around these canals grew as **papillae** which become cutinised and obliterate the mouths of the canals. The cutinised **stigmatic papillae** of Thomas may be the outward extensions of the papillae that close the mouth. It is now known that no pollen grains ever germinated in these so called stigmatic papillae. Harris is of the opinion that the entire cavity of the fruit, except in the region of canals, was full of large fleshy cells and there is no trace of any **vascular tissue**. He observed such fleshy cells in some fruits of *C. sewardii* (1951). This fleshy tissue consists of a few layers of large rounded or oval cells separated by large spaces. The fruit is covered by a thick cuticle and the epidermal cells are roughly isodiametric. The walls of the cells are straight and thickened, their surface is obscurely mottled. Hair and stomata are absent but at certain places along the surface of the fruit, small groups of cells with very thin cuticle or even devoid of cuticle, are found. These areas have been likened to **lenticels** by Harris (1940). The arrangement of the cells on the wall of the sac is in distinct rows which curve round to the lip thereby suggesting its terminal position.

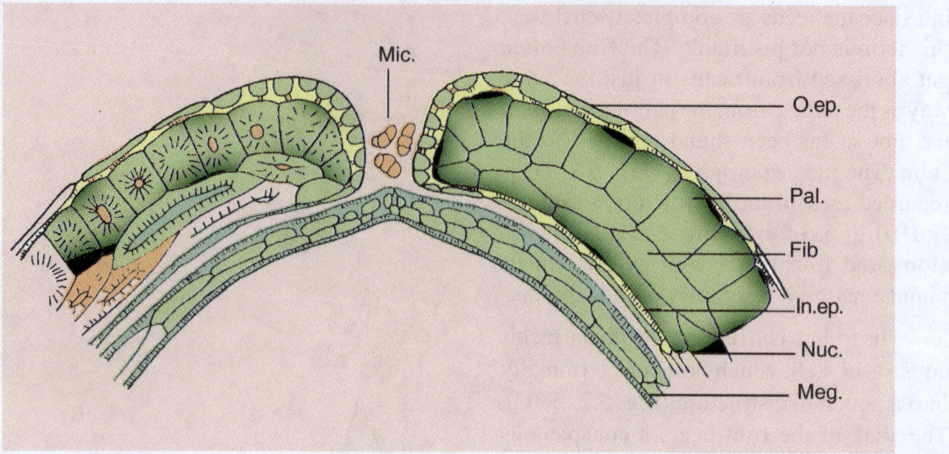

Fig. 5.4. *C. thomasii.* Restoration of micropyle of seed: *Mic.* = micropyle with pollen grains; *O.ep.* = cuticle of outer epidermis of integument; *Pal.* = palisade layer of stone of integument; *Fib.* = fibrous layer of stone; *1 n.ep.* = cuticle of inner epidermis of integument; *Nuc.* = cuticle of nucellus; *Meg.* = cuticle of megaspore.

The stalk of the fruit, which is usually short and slender, is also dorsiventral like the rachis on which it is borne. The cuticles and the epidermal cells show slight differences on the upper and the lower side (*i.e.*, adaxial and abaxial sides). The upper side bears numerous unicellular hair that are rare on the lower side. In *C. nathorsti*, the hair may be even two or three celled. In this species the hair are absent on the lower side.

The rachis, on which the stalked fruits are borne, also shows marked dorsiventrality. In *C. nathorsti* (Fig. 5.2, A, B) it bears a number of longitudinal **ridges** on the ventral or lower side whereas the upper or dorsal side it shows distinct oval **scars** where fruits have fallen off. On the upper side there are two faint ridges which extend only between the fruit scars whereas on the lower side there is a broad ridge along the middle and smaller ones on either side. These ridges on the lower side run uninterruptedly throughout the length of rachis. Their number is usually increased by supplementary ridges. Walton (1936) regarded these ridges as original features due to ribs of mechanical tissue.

The early stages of the development of the fruit are not fully known but Harris (1940) has suggested the following stages that might have led to its development:

1. A flat outgrowth from the rachis developed and assumed a spoon-shaped or strap-shaped form.
2. It developed a hollow on its upper surface just behind the apex.
3. The hollow developed into a sac-like or a pouch-like structure due to growth at its sides and base.
4. This sac-like structure underwent bending at its point of attachment and came into line with the stalk and "approximated" the former apex to the stalk. This apex developed into the **lip**.
5. During these stages of growth orthotropus ovules started developing along separate longitudinal furrows at the base of the sac.
6. Due to curvature of sac the developing ovules assumed marginal portion with their **micropyles** facing the mouth of the sac.
7. The furrows or the **canals** develop leading from the mouth to the micropyles of the ovules.
8. The wall of the fruit becomes fleshy thus obliterating nearly all the space in the fruit.

9. The sac and the ovules grow to maturity and at some stage pollination occurs and the pollen grains from the lip are drawn to the mouths of the ovules through the canals. The mouth closes and the canals are also obliterated by cell growth.
10. The fertilization occurs and the seeds complete their development.
11. The micropyles of the seeds are also closed and the **nucellus** breaks down or collapses.

Structure of the Seed (Figs. 5.2, C, E; 5.3, A, B; 5.4). The seeds are pendulous, **orthotropous** and ovoid or oval or even flattened (due to compression). It varies in size from 1.4 mm to 1.5 mm long and upto one millimeter in width. The surface of the seed is smooth. The **chalaza** is at the sharp end of the fruit and the **micropyle** at the blunt end. Usually a small spike of tissue projects from the chalaza. It is supposed to be a remnant of the **funicle**. The seeds (Fig. 5.2, E and 5.4) are composed of three parts:

1. The outer hard layer called the **testa** or the **integument** which is free from the **nucellus** upto the base. The integument is lined with distinct outer and inner cuticles. These cuticles cover the epidermal cells whose outlines are not very clear. The single **epidermis** covers the whole of integument. Its cells are smaller in size in the micropylar region and are comparatively larger elsewhere. It is also called **blow off layer**. In between the two layers of cuticle (inner and outer), there is a *stone layer*. The **micropylar canal** is also cutinised and lies at the apex of the seed. The **hilum** is at the base just opposite to the chalaza. The stone layer is made up of a layer of **palisade** like cells and a few layers of **fibrous cells** (Fig. 5.2, E). So the integument is many layers in thickness. The surface of the integument is pitted (honey comb like pits) in *C. sewardii* and smooth in *C. thomasi* and *C. nathorsti*.

2. The nucellus is a thickly cutinised body almost oval in shape and is free from the integument right upto its base. The cell outlines are clear. In the micropylar and chalazal regions of *C. thomasi* the cells are isodiametric, elsewhere they are elongated. At the base of the micropyle the nucellar cuticle is absent leaving a very small hole. Same is the case at the chalazal end where the gap is about 120 μ in diameter. In *C. sewardi* and *C. nathorsti* the cells are large and rectangular in the middle becoming smaller and shorter and finally square-shaped at the micropylar end and the chalazal end.

3. The **megaspore membrane** is completely surrounded by the nucellus. It is made up of a cuticle of varied thickness in different seeds. It shows no indication of cells but its surface appears to be pitted or granulated. In some seeds the megaspore membrane becomes transversely folded. In some seeds as in *C. nathorsti* and *C. thomasi*, the megaspore projects upwards a short distance in the micropylar region forming a structure called the **plinth**. It is closely invested with a cellular cuticle. This structure has not so far been reported in *C. sewardi*.

Harris (1933) demonstrated the presence of structures, which he called *spicules* in the seeds of *C. thomasi* and *C. sewardi*. He observed them sticking to the inner side of the integument cuticle and considered them to be derived from the rectangular or palisade like layer of cells present immediately below the outer cuticle. He called them **outer spicules** to distinguish them from the **inner spicules** that were present in all other parts of the seed and were derived from the elongated or fibrous cells of the integument. The inner spicules were abundant in the chalazal region. Spicules are absent in the seeds of *C. nathorsti*. The spicules have been found in thick-walled cells in the testa and are considered to be the result of peculiarity in the "chemical nature of the original content of the cells of the testa rather than from a structural peculiarity."

No trace of any vascular element has been found in the seed. Nothing is known about the embryo or the contents of the megaspore.

The micropyle leads into a micropylar canal that is narrow at the base and widens above and is lined with a single layer of epidermal cells with a comparatively thin cuticle. Pollen grains have usually been found at the base of the micropyle.

MICROSPOROPHYLL (*Caytonanthus*, Fig. 5.5). It was first described by Phillips (1829) under the heading "unknown leaves". Seward (1919) figured them under the name *Antholithus* sp. Later Johnson (1970) described the specimens collected by Phillips (1829) under the name "*Ginkgoanthus phillipsi*." Thomas (1925) described it under the name *Antholithus arberi*. The name *Caytonanthus* was instituted by Harris in 1937 for the microsporophylls, when some more species were described (See Harris, 1937, 1941) by him. Three species of *Caytonanthus* are known uptodate. These are *C. arberi* (Thomas) Harris, *C. oncodes* Harris, and *C. kochi* Harris.

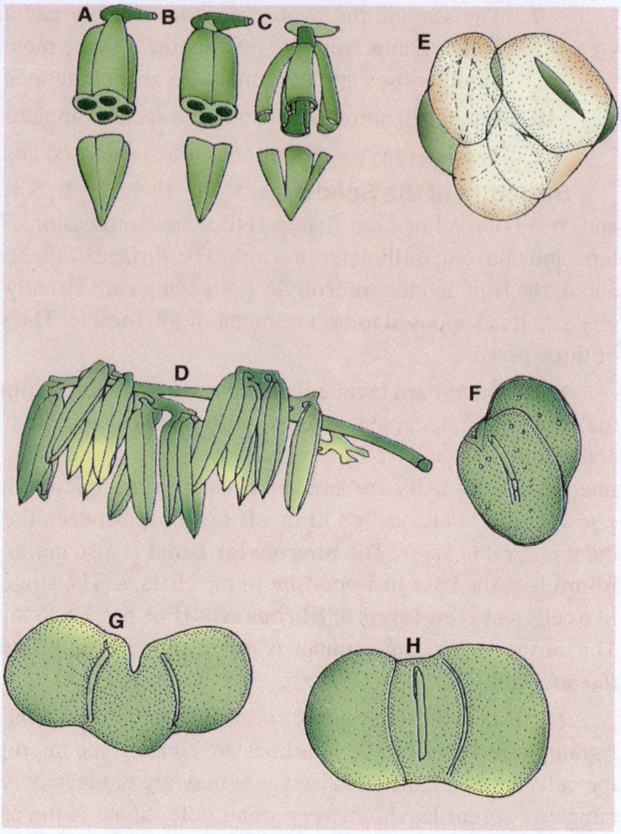

Fig. 5.5. (A-H). *Caytonanthus arberi.* A, B, C, restorations of synangia; A.the normal 4-locular; B.the less common 3-locular; C. dehisced × 20. D. restoration of part of microsporophyll (assuming pendulous synangia) × 15 E-H. pollen from a preparation of a coprolite (original slide deposited in British Museum) all × 2000. *E.immature* tetrad showing the germ groves on the two upper body cells. F.grain seen in end view. G.grain with the germ grove split open. *H.commonest* view of the grain grove in surface view. (After Harris, 1951).

Caytonanthus consists of a **rachis** (Fig. 5.5, D) that is dorsiventral in symmetry and is pinnately branched. The rachis bears opposite **pinnae** each of which branches freely in an irregular manner. The ultimate branches bear **synangia** on their surface. Each branch bears one synangium terminally or a little short of the apex. All the synangia are borne on the same side, but it cannot be definitely said which. Each synangium consists of four **pollen sacs**. There are cases with two, three and five pollen sacs per synangium but four is the usual number. The synangia are radially symmetrical. The dehiscence seems to be longitudinal and separates the four pollen sacs from one another except at the base and the apex (Fig. 5.5, A—C). There is no central core of tissue comparable to the **connective**. The central vascular strand is also absent. The outer wall of the pollen sac is dense and opaque and consists of thick-walled and elongated cells covered with a thin cuticle. In some cases the apex of the synangium bears cutinised and unicellular hair. In some specimens Harris (1941) noted the presence of "obscurely cellular granular membrane to which the pollen grains often adhere." He regarded it as the fatty matter of **tapetum** that became hardened during the process of preservation.

The cuticle of the rachis was of moderate thickness, the thickness being more pronounced on one side than the other. The side with thicker cuticle consists of cells with longitudinal walls thicker than the transverse walls. On the other side the cells were evenly thick and were covered with bases of trichomes that were fewer in number. These observations confirm the dorsiventral nature of the rachis.

Pollen Grains (Fig. 5.5, E—H). The pollen grains are unicellular and **winged**. There are two wings, placed opposite to each other (Fig. 5.5 G). There is a **grove** or a slit (Fig. 5.5 E) found on almost every pollen grain. It runs along the length of the **body cell** and is situated on the distal surface. It appears in the form of two ridges separating a thin strip, in younger pollen grains. In mature grains, the wall is split here and there by 'gaping crack', Tri-radiate scar is absent. The pollen grains are covered by a distinctly cutinized wall. In some cases pollen grains with three wings have been noted (Harris, 1951). They have a large body cell. Pollen grain tetrads (Fig. 5.5 E) with distinct **germ slits** have also been recorded by Harris. In the three species of *Caytonanthus*, the pollen grains are the largest with smooth wings in *C. kochi*, in *C. oncodes* the medium sized grains have pitted wings and in *C. arberi* the grains are smallest in size and show obscure pitting.

Pollination. Thomas (1925) found numerous pollen grains of *Caytonanthus arberi* on the lip of the *Caytonia nathorsti* and concluded that the lip functioned as **stigma**. Thomas, at that time, perhaps did not find pollen grains in the **micropyles** of seeds. Harris (1933) found pollen grains inside the micropyles of seeds of *Caytonia thomasi* and later (1941) inside the seeds of other species (*C. sewardi*, *C. nathorsti*). All these seeds contained pollen grains of *Caytonanthus* type. Presence of seeds in the micropyles suggests that pollination is of gymnospermous types. It was thought that the seeds secreted a liquid called **pollination drop** which passed down the canals and reached the openings of the canals on the lip. At this stage the sporophylls were supposed to stand erect. The pollen grains were supposed to be caught in this pollination drop and travelled through the canals by means of the wings, as in some conifers. It reached the micropyles and got deposited there. After pollination the canals were sealed by ingrowth of the cells lining their walls. The fruit then got securely sealed. Later the seed micropyles also got "closed by swellings in their outer parts."

The process of fertilisation and subsequent changes are not known so far.

Relationship of Various Organs. Harris (1941) has discussed in detail the structural and morphological features of the various species of the three organ genera described above. He related them as follows:

1. *Caytonia thomasi* and *Caytonanthus kochi* were most probably borne on the leaf *Sagenopteris nilssoniana*. All the three belong to the same age *i.e*., Lower Lias.
2. *Caytonia nathorsti* and *Caytonanthus arberi* were borne as the leaf *Sagenopteris phillipsi*. All the three belong to the Lower Oolite age.
3. *Caytonia sewardi* and *Caytonanthus oncodes* belong to the leaf *Sagenopteris colopodes*. They also belong to Lower Oolite.

Some other species of *Sagenopteris e.g*., *S. dentata* and *S. hallei* have been found associated *Caytonanthus* type of pollen (Harris 1932). Many species of *Sagenopteris* type of leaves have been described but their association with proper reproductive organs has not been found.

PHYLOGENETIC RELATIONSHIPS

The Caytoniaceae are not fully discovered so far and we know only a little about the stem fragment and nothing about the root, the gametophytes and the embryos. About the stem too we have no knowledge about its anatomy and other histological details (only a small fragment was found). The fossils are available only in the form of compressions and hence anatomical and histological details of sporophylls are available in scanty. It is not possible to form an idea of the whole plant and since phylogenetic relationships are decided only when we know completely about the entire plants, it is difficult to decide about such relationships of the Caytoniaceae. Attempts have, however, been made by several authors to compare the individual organs with other groups. Such individual comparisons are liable to suggest false relationships. This is the reason why exact systematic position of the Caytoniaceae is a question mark.

Each organ of the Caytoniaceae has been compared to those of several other groups including the angiosperms, Thomas believed that the **synangia** of *Caytonanthus* are in close agreement with the **stamens** of angiosperms. This view has been challenged by Harris (1941) because the stamens in angiosperms are bilateral and the synangia of *Caytonanthus* are radial in symmetry. Moreover the synangia of *Caytonanthus* do not possess a vascular connective and a filament. Harris (1941) has advanced a view that the attachment of the synangia to the surface and their structure and mode of dehiscence brings them into relation with the synangia of fossil Marattiales and Pteridosperms. He has likened *Caytonanthus* with *Antevisia* which is the microsporophyll of *Lepidopteris* (1941) and *Harrisia* (microsporophyll of *Ptilozamites*) both of which are Mesozoic Pteridosperms. Their pollen grains are also very similar. He has strongly supported the connection of *Caytonanthus* with the Pteridosperms. In its dorsiventral symmetry and pinnate nature *Caytonanthus* resembles very closely the pteridosperms especially the *Corystospermaceae* but in the radial symmetry of its synangia it stands apart from all gymnospermous and angiospermous fossils. Krausel (1949) described *Pramelrenthia* a microsporophyll from Triassic of Austria and regarded it as a member of the caytoniaceae. In this specimen the so called synangia arose laterally from an axis and contained oval or elongated pollen sacs. The pollen grains were similar to those of *Caytonanthus*, but the so called synangia were very much different in structure. In *Harrisia* the pollen sacs are elongated and are borne in oval masses that can be called synangia but they are two valved and bilateral in symmetry. The same can be the case with *Pramelreuthia*.

Regarding the megasprophyll, Thomas (1925) considered that they provide clues to the origin of angiosperms. He regarded the ovule bearing organs as representing an early stage in the evolution of the carpel. Eames (1961) states, "The Caytonalian megasporophyll (Thomas, 1925) is a frond-like, pinnate, dorsiventral organ, consisting of an axis, which bears two lateral rows of globose "cupules" or "ovaries." Each cupule has a median longitudinal vein, along which ovules are borne. A follicle like carpel is considered derived by the reduction and condensation of this fertile organ; the axis is greatly shortened and the cupules (as Thomas called them) reduced to the lowest member of each row, and these fused to the axis and to each other. In the complex organ so formed, the axis of the ancestral sporophyll would form the midvein of the follicle, the vascular bundles of the two cupules would form the vascular bundles along which the ovules are borne, the ovules of each cupule form one of the two rows. This derivation, as suggested by Thomas, seems farfetched and complex derivation for so simple a structure as the follicle." He further commented that "the Caytonalian sporophyll are wholly unlike; the carpel is palmately veined, Caytonalian organs are strongly pinnate. The ovule of the angiosperms has two integuments that of the Caytoniales one." Later Thomas (1931) expressed his views that Caytoniales are late descendants of the Pteridospermae and never had any close connection with the angiosperms. He did not suggest the angiospermous carpel has originated from the Caytonalion megasprophyll but thought that it may suggest a possible method of the origin of the carpel from a pteridosperm sporophyll. Andrews (1961) states, "Thomas has interpreted the apparent inverted *Caytonia* as having evolved from earlier pteridospermous plants. Beginning with the Upper Carboniferous *Lagenostoma*, in which the single seed was contained in a lobed cupule, one may regard the Triassic *Pilophorosperma* with its cupule lined with hairs as a later type, and in *Caytonia* the seeds are very nearly enclosed. According to this view the carpel found in primitive modern Angiosperms, such as *Degeneria* and *Drimys*, is the result of a lateral fusion of the two carpels of the *Pilophorosperma* or *Caytonia* types."

Andrews (1961) further states, "the **fused cupule concept** is admittedly speculative since the later stages are not known, whereas the concept of the carpel as an enfolded leaf like structure, although long favoured by many botanists, finds no supporting evidence in the form of gymnospermous plants, living or fossil, which may be regarded as ancestral." Moreover the occurrence of seeds in a closed structure called the fruit or the carpel cannot be regarded as a cogent reason of its being ancestral to the angiosperms, because in many modern angiosperms the carpels are incompletely closed.

Bierhorst (1971), while discussing the relationships of Caytoniaceae states, "the structure of the ovule is consistent with that of pteridosperms. There are suggestions of pinnate branching in the fructification axis. The microsporangiate organs could conceivably have been derived from types known in several pteridosperm families. The **spermocarp** (fruit) is reasonably consistent in its morphology with **ovuliferous** organs of corystosperms and peltasperms. If the **helmet** of the corystosperms contained several **ovules** instead of one, the morphological condition found in *Caytonia* might be closely approached. Similarly, a reduction of ovule number and an overgrowth of the disc in the peltasperms would result in a similarity."

Harris (1951) has stated that the Caytoniaceae resemble the angiosperms in possessing two characters:

1. The structure of the stomata.
2. The distribution of the cuticles in the seed.

He is of the opinion that in possessing these two characters the Caytoniaceae agree more with the angiosperms than with any known gymnosperms.

Since *Sagenopteris* (Leaf) has four leaflets, it bears some resemblance with *Marsilea*, an aquatic pteridophyte.

It becomes clear from the above discussion that Caytoniaceae are decidedly gymnosperms and have closer affinity with the pteridosperms, because of the following characteristics:-

1. They have pinnate leaves.
2. The pollen bearing organs are also dorsiventral and pinnate.
3. The megasporophylls are dorsiventral and pinnate.
4. The reproductive organs are not organised into cones and flowers.

It is almost agreed upon by majority of botanists that Caytoniaceae evolved from the pteridospermous ancestors.

To conclude this discussion it will be worthwhile to quote Thomas (1955), "The occurrence of pteridrosperms from many parts of the world prior to the Cretaceous period would lead one to expect that this group should be represented in the flora of the present day. Among the modern ferns, conifers, and the cycads are types closely similar to certain Mesozoic genera; the Ginkgoales are still represented by one species. The Bennettitales, already highly specialised by Lower Cretaceous times, are extinct. But the relatively unspecialised pteridosperms seem to have died out also, unless *Gnetum* and *Welwitschia* belong to this group. This may be because the angiosperms, a race of plants with better equipments for competition, needed the same habitats and displaced the pteridosperms from them. Future work will probably show that angiosperms themselves have evolved from some members of the pteridosperm alliance, at present unknown; a thorough search should be intensified in India and the Southern Hemisphere for such forms."

The Mesozoic Pteridosperms included in the families Peltaspermaceae, Corystospermaceae and Caytoniaceae flourished well during the Jurassic period, which lasted 46 million years and continued into the Lower Cretaceous, which lasted for 36 million years. They became extinct by the Upper Cretaceous Period which also lasted 36 million years. The entire Mesozoic lasted for about 167 million years. Out of this total span the Mesozoic pteridosperms flourished for about 82 million years. The Mesozoic Era afforded a very genial climate and is considered to be an era of great stability with no cataclysmic crustal changes or volcanic eruptions excepting at its closing phase during the late Cretaceous when the earth was rocked by **Laramide Revolution**. It is considered to be an age of gymnosperms and primitive farms. The vegetation was rich, but of an entirely different type, in comparison to that of the Carboniferons Period. During the Jurassic Period many prominent and well preserved gymnospermous plants belonging to the Cycadales, Bennettitales, Coniferales, Cycadeoidales and Ginkgoales, also flourished along with the Mesozoic Pteridospermales belonging to the families Peltaspermaceae, Corystospermaceae and Caytoniaceae. They were discovered in the sedimentary rocks from Scotland, France, Germany, Italy, Grahamland bordering Antarctica, New Zealand and India (Middle Gondwana and Upper Gondwana). Members of Caytoniaceae have not been discovered from India, whereas the families Peltaspermaceae and Corystospermaceae are well represented in Triassic beds (Middle Gondwana) of India. In India, Jurassic to Lower Cretaceous (Upper Gondwana) is considered to be the best period when gymnosperms flourished. In the Lower Cretaceous abundant fossiliferous beds have been discovered in Greenland, Sakhalin, England, Belgium, Germany, France, India (Upper Gondwana), Portugal and along the Atlantic coast in North America. During this period a rich angiospermic flora also flourished. It included such plants as *Platanus*, *Quercus*, *Artocarpus*, *Magnolia*, *Dalbergia*, *Cinnamomum* and some members of the Menispermaceae. The luxuriance of fern vegetation was a noteworthy feature of Greenland and Sakhalin flora (Arctic regions). In the present age the arctic region is treeless with hardly any fern vegetation, in strong contrast to the Lower Cretaceous flora of those times. This suggests that mild climatic conditions prevailed in the Cretaceous times in these areas. The caytoniaceae alongwith other Mesozoic pteridosperms were living during this period and were on the verge of extinction. They alongwith Bennitittales and some primitive coniferales became completely extinct in the Upper Cretaceous. During this period angiosperms started dominating the landscape. Dakota Flora of North America during this time constituted about 90% angiosperms. The Cycadales, Ginkgoales and Coniferales (including Taxodiaceae) continued beyond the Cretaceous.

Why the genial and mild climate of late Mesozoic could not afford the existence of Caytoniaceae, Peltaspermaceae, Corystospermaceae, Pentoxylales and Bennetittales? The possible causes of their extinction can be attributed to the following reasons:

1. The struggle for existence and competition with the surrounding flora.
2. Harmful exudations from the flora around them.
3. Destruction caused by disease causing fungi, bacteria, viruses and harmful insects.
4. Due to wholesale grazing and devouring by animals.
5. Lowering or complete running out of the gene potential for environmental adaptation. This deficiency is the major factor for the extermination of entire groups of plants during cataclysmic crustal revolutions and other environmental changes like lowering of temperature during periodic glaciations.
6. The gene mutations and certain chromosomal aberrations lead to the operation of auto-annhilation processes due to over specialisation of some morphological organ, or organs or physiological processes, out of pace with the rest of the system. The recently discovered **jumping genes** can be the major causes of appearance of harmful mutations and abnormalities in the choromosomes.

The above listed factors may act individually or mutually to exterminate a particular group or groups of plants from the globe. The last two factors are considered to be very effective or even most vital for the extinction of plants belonging to the higher categories.

It is difficult to explain the absence of Caytoniaceae from the Indian fossiliferous beds especially when the closely related families like Peltaspermaceae and Corystospermaceae are well represented.

QUESTIONS

ESSAY TYPE

1. Explain the fructifications of Caytoniales (Caytoniaceae). **(Kanpur, 1985, 1991)**
2. Describe the characteristic features of Caytoniales (Caytoniaceae) and discuss their affinities. **(Kanpur, 1998)**
3. Write an account of Caytoniales (Caytoniaceae). Discuss its affinities. **(Rohilkhand, 1991)**
4. Explain why Caytonia was considered to be closely related to angiosperms? **(Kerala, 2001)**
5. Give an account of phylogenetic relationships of the family Caytoniaceae.
6. Explain the microsporophylls of the family Caytoniaceae
7. Describe the fronds of Caytoniaceae with diagrams.

SHORT ANSWER TYPE

8. **Write short notes on :**
   (*a*) *Caytonia* **(Kanpur 1988, 1996; Rohilkhand, 2003, 1999)**
   (*b*) Differences between *Caytonia* and *Williamsonia*. **(Kanpur, 1993, 1995)**
   (*c*) Ancesters of Angiosperms-Caytoniales **(RohilKhand, 1995)**
   (*d*) *Caytonia* megasporophyll **(Rohilkhand, 1996)**
   (*e*) Caytoniaceae **(Rohilkhand, 2003)**
9. **Draw labelled diagrams of**
   (*a*) Fructifications of Caytoniales (Caytoniaceae) **(Kanpur, 1990)**
   (*b*) *Sagenopteris*
   (*c*) Caytonia
   (*d*) Pollen grain of Caytoniaceae
10. "Caytoniales are ancestors of Angiosperms". Describe in brief. **(Rohilkhand, 2001)**
11. Caytoniaceae is regarded to be Gymnosperms. Why ?
12. What are the genera bearing microsporophylls placed under Caytoniaceae? Describe at least one genus with diagrams.
13. Give a brief account of pollination in Caytoniaceae.

OBJECTIVE TYPE

14. **Select the correct answer:**

(*i*) The fossils described from Cayton Bay in Yorkshire have been described under the family
   (*a*) Peltaspermaceae
   (*b*) Caytoniaceae
   (*c*) Pteridospermaceae
   (*d*) Corystospermaceae

(*ii*) *Sagenopteris* is the genera that represents
   (*a*) the leaves
   (*b*) the fruits
   (*c*) the pollen grains
   (*d*) None of the above.

(*iii*) Since the genus *Sagenopteris* has four leaves it resembles the pteridophyte
   (*a*) *Marsilea* (*b*) *Azolla*
   (*c*) *Salvinia* (*d*) *Dryopteris*

(*iv*) *Caytonia Sewardi* is the name given for the
   (*a*) Microsporophylls
   (*b*) Megasporophylls
   (*c*) Prophylls
   (*d*) Rhizomes

(*v*) Caytoniaceae were first described by
   (*a*) Thomas (*b*) Harris
   (*c*) Lele (*d*) Sahni.

CHAPTER 6

Bennettitales (Cycadeoideales)

- History and Distribution of Bennettiales
- Characteristic Features of Bennettiales
- Features of Cycadeoideaceae
- Features of Williamsoniaceae
- Features of Wielandiellaceae

Introduction

The Cycadeoideales or the Bennettitales is a fossil group that has greatly attracted the attention of palaeobotanists since 1825 when the genus *Bucklandia* was described in the form of sandstone cast from Great Britain. This magnificent group of fossil plants, which mostly consisted of petrified trunks, flourished well during the Triassic to Cretaceous periods of the Mesozoic era well during the Triassic to Cretaceous periods of the Mesozoic era. They resembled Cycads to such an extent that Mesozoic was described as the 'age of Cycads'. This resemblance was, however, superficial because the *Bennettitales* differed in many respects from Cycads.

OTOZAMITES BENGHALENSIS

They are entirely extinct and resembled cycads in the outward stumpy appearance of the trunk and the apical crown of large pinnate compound leaves. In the year 1913, Thomas and Bancroft pointed out significant anatomical differences between these plants and the Cycads. The main difference lies in the stomata. The *Bennettitales* have syndetocheilic whereas the Cycads have haplocheilic stomata. In the year 1933 Florin also confirmed these findings. The secondary wood in the *Bennettiales* is **monoxylic** as compared to the **polyxylic** wood among the cycads. The leaf trace always arises strands during its course through the cortex; whereas in Cycads the traces arise as two independent units from the very beginning and their behaviour in the cortex (*i.e.,* complete reversion and girdling around the stele) is very different and characteristic. The reproductive organs of the *Bennettitales* also differ appreciably from those of Cycads. In the *Cycadeoidales* the fructifications are flower-like organs consisting of a whorl of stamens surrounding a central ovuliferous receptacle, whereas in the Cycads the fructifications are strobilar in organisation like other gymnosperms.There are only two main characteristics in which the two groups resemble each other : (*i*) both are naked seeded and (*ii*) the leaves in both are pinnately compound. These major differences in fundamental morphological features lead to the conclusion that the *Cycadales* and the *Bennettitales* represent parallel lines independently derived from the palaeozoic pteridosperms.

The reproductive organs of *Bennettitales* were flower-like in organisation with an elongated thalamus or receptacle bearing a central ovuliferous region surrounded by microsporangiate microsporophylls that in turn were surrounded by a whorl of perianth-like bracts. This superficial resemblance led some botanists (Arber and Parkin, 1907) to suggest that angiosperms originated from the *Bennettitales.* The differences were later brought to the surface and such an interpretation became obsolete.

The *Bennettiitales* became fossilised either in the form of compressions or petrifactions. Except for the roots the rest of the plant organs have been discovered in a petrified form.

History and Distribution

Bucklandia the first member of this order that was collected and named in 1825 It was found in the form of sandstone casts in Great Britain. Later as many as 10 petrified trunks were collected from the iron ore beds of the Potomac formation between Baltimore and Washington. These were described about 29 years after when in 1889 the first among them was named as *Cycadeoidea marylandica. Cycadeoidea etrusca* was discovered from an Etruscan's tomb in 1860 where it was placed about 4,000 years ago. This was a silicified trunk and the oldest known fossil collected by man. Wieland (1906) described that it was, "Placed with vases and other objects of superstitious reverence on one of the sepulchral chambers of the ancient Necropolis at Marzabotto by the Etruscans more than four thousand years ago". *Williamsonia gigas (= Zamia gigas)* was discovered in 1868 by Williamson from the Jurassic of the Yorkshire. *Bennettites gibsonianus* of the Lower Carboniferous was found in the form of silicified petrefactions from the Isle of Wight and Portland off the South Coast of England. Later studies from thin sections of these trunks it was found that, *Bennettites* resembled *Cycadeoidea* in all respects and hence the two were declared to belong to the same genus. The name *Cycadeoidea* being older is retained. Other places from where cycadeoid trunks have been discovered include the Jurassic rock of Wyoming, early Cretaceous of Black Hills of South Dacota, Potomac formation in Maryland, Mesozoic Horizons of Kansas, Colorado, Prince Edward Islands, France Belgium, USSR, Poland, Germany and India. In India the *Bennettitales* are found in abundance in Rajmahal Hills in Bihar. The fossils recovered from there include : *Bucklandia indica. B. sahnii; Dictyzamites* (4 species), *Ontheanthus sessilis* and *O. polyandra, Otozamites benghalensis, Pterophyllum incisum; Cycandinocarpus rajmahalensis; Beaniopsis rajmahalensis; Ptilophyllum* (3 species); *Sahnioxylon rajmahalensis, Williamsonia indica, W. microps, W sahnii, W. santalensis,*

W. sewardiana, Otozamites, Zamites. In the Lower Gondwana, other *Bennettitales* are available in states of Talchir and Orissa. These include species of *Noeggerathiopsis, Samaropsis and Dadoxylon.*

Characteristic Features

The members of this order possess the following distinguishing characteristics:

1. The petrified stems reveal them to be upright columnar forms, of conical shape, and under a metre in height. In some specimens of *Cycadeoidea jenneyana* the stems have attained a height of several meters.
2. The stems grow very slowly and branch profusely.
3. The leaves are large and pinnate compound. In *Williamsoniella,* they are, however, simple and linear. The pinnae in *Williamsonia* are attached by a broad base to the upper side of rachis whereas in *Pterophyllum* and others they are attached to the lateral sides of the rachis. In *Zamites,* the pinnae are constricted at the base.
4. The venation is unicostate parallel. In *Dictyozamites,* the venation is reticluate unicostate.
5. The stomata are syndetocheilic.
6. The stems have large pith surrounded by a ring of collateral conjoint, endarch and open primary vascular bundles. The cortex is parenchymatous and possesses mucilage canals.
7. Presence of a ramentum of scale-like hairs in between the leaf bases of some genera (*Cycadeoidea*).
8. The reproductive organs are bisexual (*Cycadeoidea, Williamsoniella*) in majority of the genera but unisexual condition (*Wielandiella angustifolia*) is not rare. The reproductive organs are called the flowers. In *Cycadeoidea* the flowers present some similarity in their general organisation to the flowers of the genus *Magnolia* among the dicotyledonous angiosperms. The flower in the Bennettitales, as in *Magnolias,* consists of a basal aggregation of microsporophylls surmounted by a receptacle that may be flattened, dome-shaped or conical (*Cycadeoidea*).
9. Presence of interseminal scales on the receptacle. Seeds and ovules are distributed among them.
10. The flowers arise in the axils of leaves in all the *Bennettitales.*
11. The ovules and seeds are stalked (*Cycadeoidea. Williamsonia*) or sessile (*Williamsoniella*).
12. The flowers are surrounded by numerous hairy bracts that arise from the base of the receptacle.
13. The ovules are not enclosed in a cupule, *i.e.,* the cupule is absent, but Harris (1932) reported its presence in *Bennetticarpus crossospermus.* It was also reported in *Vardekloeftia sulcata.*
14. The microsporophylls or pollen bearing organs arise in whorls and may be pinnate (*Cycadeoidea*) or entire (*Williamsoniella*), free or fused (*Williamsonia*).
15. The pollen grains are borne in bilocular synangia that may be sessile or stalked (*Cycadeoidea*).
16. The seeds possess two cotyledons.

Classification

Arnold (1948) classified *Bennettitales* into two families : (*i*) *Williamsoniaceae (Willamsonia, Williamsoniella)* and (*ii*) *Cycadeoideace (Cycadeoidea).* Sporne (1965) divided it into three families:

(*i*) *Williamsoniaceae;* (*ii*) *Wielandiellacea;* and (*iii*) *Cycadeoideaceae.* Andrews (1961) does not recognise any distinction of the order into families.

FAMILY : CYCADEOIDEACEAE

The family as constituted by Sporne is represented by about thirty species of a single genus *Cycadeoidea.* These species have been collected from various places in the world from the Upper Jurassic period to the Upper Cretaceous of the Mesozoic era. The species have been reported from U.S.A., USSR, Europe, Belgium, France, Isle of Wight and Portland. As stated above, the petrified trunks of *Cycadeoidea entrusca* is the oldest fossil ever collected by man. It was found in the entruscan tomb. *Cycadeoidea gibsonianiana,* was discovered from the Isle of Wight by Carruthers who named it as *Bennettites gibsonianus,* but later it was placed under the genus *Cycadeoidea. C. marylandica* (from Baltimore and Washington) was described in 1889 by Fontaine. *C. dacotensis* was described by H. Macbride (1893) from the Lower Cretaceous of Black Hills of South Dakota, *C. jenneyana* was also described from the same locality by L.F. Ward *C. blanca, C. ingens, C. dartoni, C. clossalis, C. marshiana, C. wielendii, C.morrieri, C. equalis* and *C. magnifica* are some of the other species that have been properly studied by various authors.

Fig. 6.1. *Cycadeoidea dacotensis* showing external features (After Macbride).

External Features

The genus *Cycadeoidea* has a short, branched or unbranched, spherical to conical to irregular trunk (Fig.6.1) that reached a diameter of about 50 centimetres. The surface of these petrified trunks was marked by several rhomboidal leaf bases with multicellular hair in between. The height of the trunk rarely reaches a meter. In *C. jenneyana* it attained a height of several meters. The unique trunks have been nicely illustrated by Wieland in his book 'American Fossil Cycads,' and by Ward in his book 'Status of the Mesozoic Flora.' The trunks are low growing and those that profusely branched give the effect of a bunch of pineapples. The trunk bears at its apex a crown of pinnately compound leaves that reached the length of 10 feet in some specimens. These leaves were found in a partially developed condition in organic connection with some silicified trunks.

Anatomy

A cross section of the stem shows roughly a circular outline. The epidermis is not well preserved due to the heavy armour of leaf bases with hair-like ramenta wedged in between them. The wide cortex is entirely parenchymatous and is traversed by numerous leaf traces and mucilage canals. The primary vasculature consists (Fig.6.2, A) of a ring of endarch, collateral, conjoint and open vascular bundles encircling the wide parenchymatous pith. Ray-like extensions of the pith pass between the vascular bundles, which therefore appear discrete. The primary vasculature in its longitudinal course forms a network with wide meshes, with a single leaf trace arising from the lower angle.

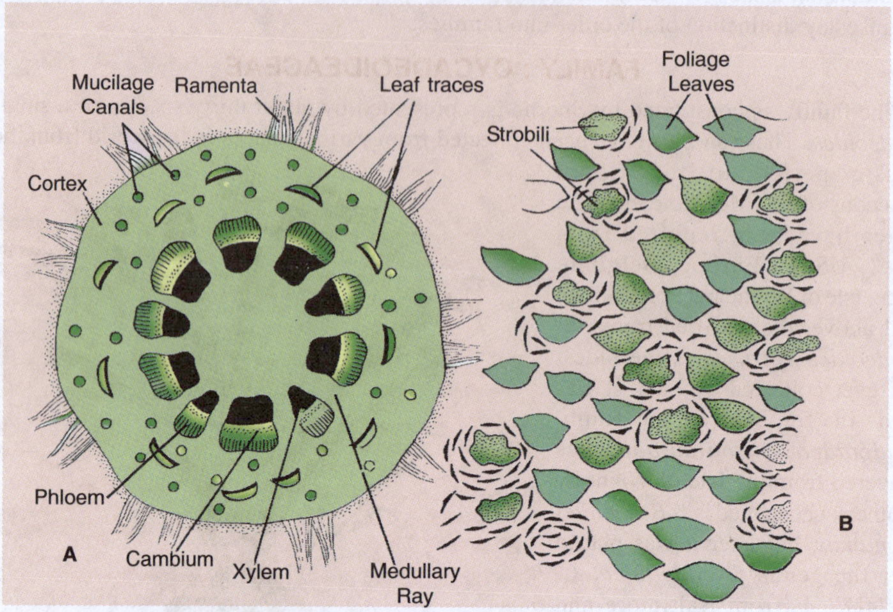

Fig. 6.2. (A—B). *Cycadeoidea* sp. A. T.S. stem showing primary structure; B. Tangential section of a trunk.

The secondary wood encircles the primary xylem and consists of tracheids with scalariform and circular bordered pits. The secondary medullary rays traverse the secondary xylem and the secondary phloem. They are uniseriate as well as biseriate. Large rays that form a network and through which the leaf traces ran are also found in the secondary wood. The tracheids appear rectangular (Fig. 6.3, B) in a cross section. A distinct cambium is visible between the secondary xylem and the phloem.

The leaf trace arises singly from the primary vascular strand and passes through broad ray of the secondary wood and on entering the cortex divides into several mesarch strands that are disposed in a horse-shoe-shaped pattern. The leaf trace is c-shaped at the time of its origin. The leaf traces enter straight into the leaf without circling around the cortex, *i.e.,* there are no girdling traces like those of the cycads.

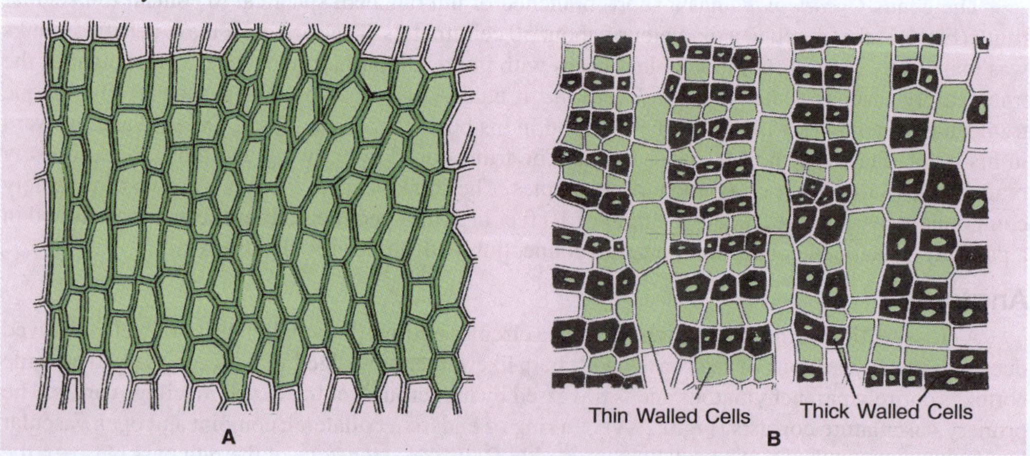

Fig. 6.3. (A—B). *Cycadeoidea sp.* T.S. of secondary wood (B) and secondary phloem (A).(After Chamberlain)

A vertical section through the pinnule (Fig. 6.4) shows a distinct upper and lower epidermal layer. The cells of epidermis are thick-walled. The mesophyll is distinguished into palisade parenchyma and spongy parenchyma. The vascular bundles have phloem facing the lower epidermis and xylem facing the upper epidermis. Each bundle is surrounded by a distinct bundle sheath. Bundle sheath extensions are also clearly preserved.

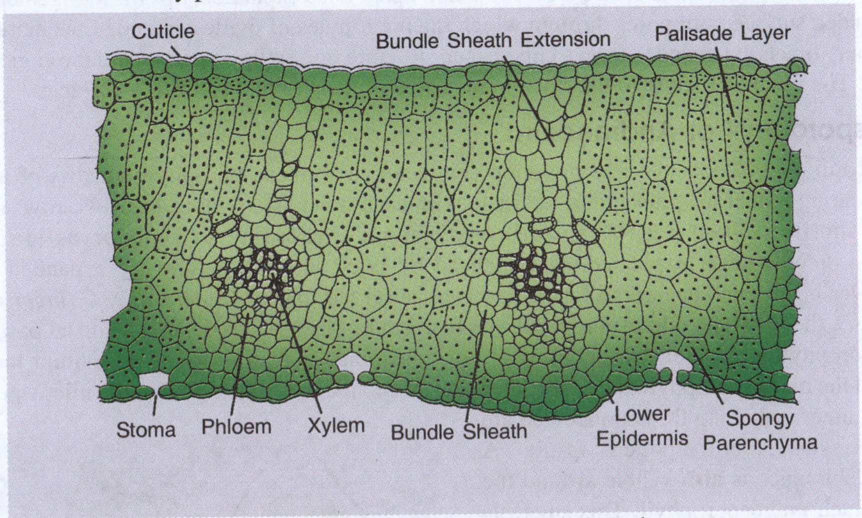

Fig. 6.4. *Cycadeoidea* sp. T.S. through the pinnule.

Reproduction

The reproductive structures of this order are designated as flowers. In *Cycadeoidea* the flowers are bisexual in majority of species and arise on short stalks, singly in the axil of each leaf. In *C. dartonii* as many as 500 floral buds were found on a single plant (each arising singly in a leaf axil) and all of them were at the same stage of development. It is presumed that all of them opened simultaneously and the plant flowered only once in its life time *C. wielendii* is unisexual.

Structure of Flower

The bisexual flower (Fig.6.5) consist of a short stalk that extends laterally from the main stem (leaf axils) into the armour of leaf bases and is 5-10 centimetres in diameter and up to 6 centimetres long at maturity. About 100-150 elongated hairy bracts arise in close spirals little below the apex of this shoot. These bracts protect the microsporangiate and megasporangiate parts of the flower. The apical region of the stalk behaves as a receptacle that may be slightly convex in *C. gibsoniana* and *C. wielendii* or conical as in *C. dacotensis* (Fig. 6.5). The micro sporangiate organs or microsporophylls or androecium consists of 19-20 once pinnate appendages that arise in whorls from the base of the conical receptacle in *C. dacotensis*. The microsporophylls were usually seen in a young state when they were folded in around the central gynnoecium. The microporophylls were united at their base to form a disc. The gynoecium consisted of numerous stalked ovules borne around the entire conical or convex protuberance of the receptacle. In *C. dacotensis,* the stalks of ovules were

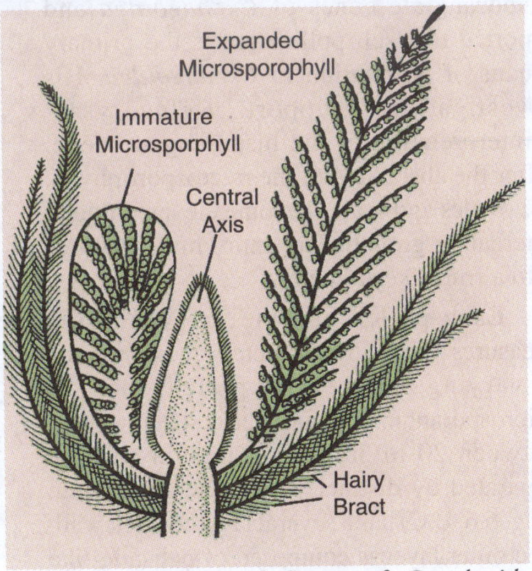

Fig. 6.5. Bisexual flower of *Cycadeoidea dacotensis*.(After Weiland)

short and arose at right angles to the conical receptacle. Whereas in *C. wielendii* and *C. gibsoniana* the stalks were long and arose vertically upwards from the cushioned receptacle. Interspersed among the ovules were interseminal scales with expanded tips. Those expanded tips fused together to form an expanded surface with pores through which the micropyles of ovules extended. At maturity, the sterile hairy bracts expanded to form a cup-shaped structure resembling a perianth of the angiospermic flowers. The vascular supply to the flower consisted of many branches from leaf traces.

Microsporophyll or Androecium

According to Wieland's interpretation (Fig 6.5.) each microsporophyll consists of a central rachis bearing numerous pinnae that alternate with each other. The pinnae bear two rows of bean-shaped shortly stalked pollen capsules or synangia. According to Delevoryas (1963), the microsporophylls never expanded, as visualised by Wieland (1906). Actually expanded flowers have never been encountered so far. Delevoryas observed that in *C. colosalis, C. ingens* and *C. dacotensis,* the tip of the microsporophyll was fused to its base (Fig. 6.6. A). The distal part of each microsporophyll was parenchymatous with no indications of fusion or canals. Within the fertile region of the microsporophyll there are trabeculae that connect the outer wall of the microsporophyll with the inner and along these trabeculae are borne the synangia or the pollen capsules. A line of dehiscence is also visible around the base of each microsporophyll. This suggests that the entire microsporophyll might have been shed as a single unit. Such an interpretaion presents difficulties regarding the liberation of the pollen grains as to whether they were liberated before the detachment of the microsporophyll or after. William. L. Crepet (1972) investigated the bisporangiate cones of *Cycadeoidea* and reported that self-pollination is the primary means of pollination in *Cycadeoidea.* His investigations support Delevorysas' reinterpretation of the bisporangiate cone. After the abscission of the microsporophylls, the ovules and scales become elevated due to intercalary growth. The entire mass matures into a fruit-like structure.

Each pollen capsule or synangium measures 3.5 × 2.5 millimeters and contained a varying number of pollen sacs or microsporangia. The number is stated to vary between 20 to 30 per synangium and are separated by distinct walls. The synangium (Fig. 6.6, B,C) has a several layered thick wall. The outer layer is composed of palisade-like (Fig. 6.6, B,C) cells that have thick walls. The inner layers are made up of thin-walled cells followed by a ring of microsporangia arranged around the periphery. The synangia later opened at maturity to release them. The

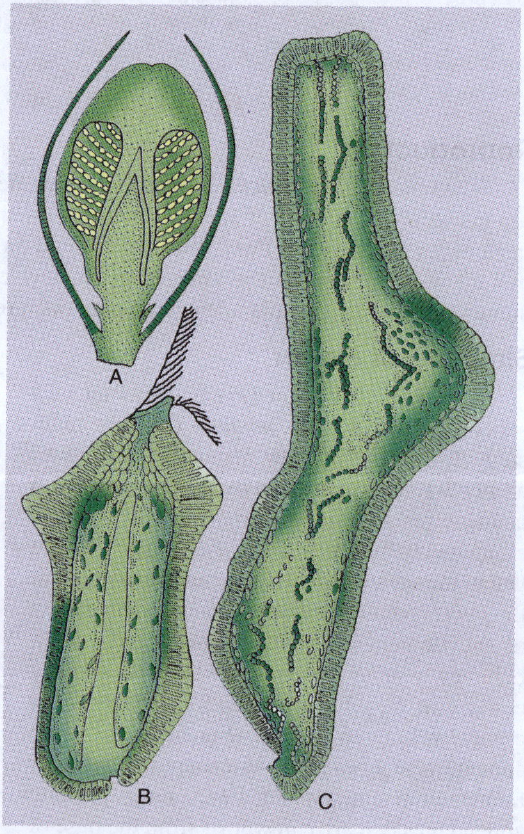

Fig. 6.6. (A—C). *Cycadeoidea*. A. *Cycadeoidea* sp. Flower as visualised by Develoryas; B.L.S. synangium showing two pollen sacs containing pollen grains. Also note the structure of wall. The outer palisade-like wall layer is clearly visible; C.T.S of the synangium. (B—C. After Wieland).

synangia dehisced by means of an apical slit and separated into two valves. The microspores have been seen to contain many cells, which perhaps represent the male gametophyte. Taylor (1993) has described a number of multicellular pollen grains in *Cycadeoidea* and has discussed its morphology and ultrastructure.

The Gynoecium

The gynoecium (Fig.6.7, A) consists of a conical or spherical receptacle that bears numerous stalked ovules and interseminal scales. Each ovule is about 1 mm long and consists of a single integument that according to Harris (1932) is fused with the nucellus except at its apex. Lignier has pictured the ovule of *C. morieri* (Fig. 6.7, B) as having a nucellus free from the integument. Each ovule has a pollen chamber and an external nucellar beak. The ovules are orthotropous and have long micropylar beaks that extend beyond the pores in the surface plate formed by the union of distal flattened ends of interseminal scales. Actually the integument extends through these pores to form the micropylar beaks. These fused tips. of the interseminal scales form a sort of external protective covering or pericarp surrounding the seeds. Harris (1932) described a cupule-like outer covering around the stalk and the body of the ovule in a species of *Cycadeoidea.* It has elongated tubular cells and represents a vestigeal structure that might have developed fully in the ancestors of the order.

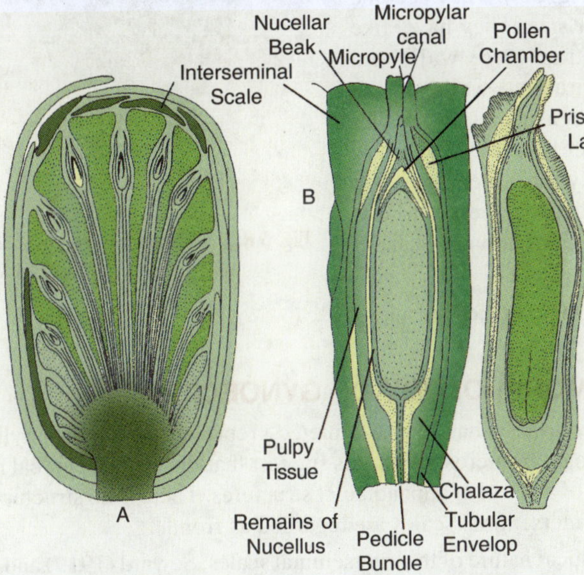

Fig. 6.7. (A—C). *Cycadeoidea*. A. V.S. gynoecium of *C. gibbsoniana* showing convex thalamus, stalked ovules and interseminal scales. B. V.S. ovule of *C. morieri*. C.L.S. seed of *C. dartonii*.

Crepet and Delevoryas (1972) discovered many bisporangiate cones of *Cycadeoidea* from the cretaceous of the Black hills. They yielded very young ovules some of whom were in the pregametophytic stage of nucellar development. They have studied the structure of such ovules in detail. These ovules resembled those of *C. wellsii* and are stalked and almost urn-shaped and were widest in the middle. There is a distinct constriction below the flaring and conspicuous funnel-shaped micropyle (Fig. 6.8). The micropyle is lined with large cells of the innermost layer of the integument. These large cells (Fig. 6.8) are considered to be the epidermal cells. The single integument develops rapidly and has three distinct layers. The nucellus is free from the integument and has a distinct stalk. At a place where the integument merges into the ovular stalk, its cells are small and appear polygonal in a radial section. The cells of the integument in the micropylar region are sufficiently radially elongated. Perhaps they closed the micropyle after pollination. The outer layers of the integument have axially elongated cells. Those of the middle layer have thick walls. They later develop into the outer fleshy layer, the middle stony layer and the inner fleshy layer.

The young nucellus is made up of thin-walled cells, those at the chalazal end being much smaller than those at the micropylar end. The cells in the micropylar region are much elongated and are up to 80 μ long. A single cells with a pointed tip appears to terminate the nucellus in early stages of development (Fig. 6.8.). On either side of the nucellar tip, the cells bend outwards with their outer curved in a manner to confirm to the outline of the nucellus. The cells at the chalazal end are smaller.

Crepet and Delevoryas (1972) also observed a linear tetrad or a row of three cells, a little below the middle of the nucellus (Fig. 6.8). They become conspicuous by their smaller size. They have, like other gymnosperms, thin transverse walls and comparatively thicker longitudinal walls. In the living gymnosperms, the megaspores are smaller than the surrounding cells in *Stangeria (Cycadales).There* are usually three megaspores in majority of the values studied by these authors. Further stages of development have not been described as yet.

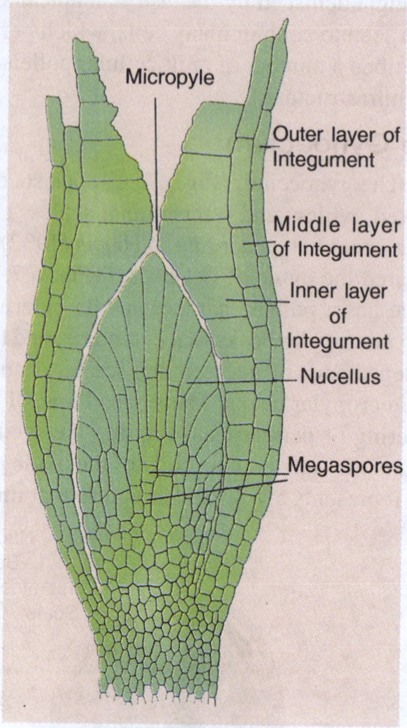

Fig. 6.8. V.S. through young ovule of *Cycadeoidea* sp. (After Crept and Delevoryas)

The seed has two cotyledons (Fig. 6.7C) and is almost oval in shape.

MORPHOLOGY OF GYNOECIUM

The stalked ovule of *Cycadeoidea* has been interpreted s representing a sporophyll. Delevorya's observations on the morphology of microsporophylls suggest that the central conical receptacle is a compound structure made up of a whorl of appendicular structures. There is no structural observation to support such a view. The interseminal scales are regarded as frond.

Regarding the morphological nature of the interseminal scales, Seward (1919) and Harris (1932) support the view that the interseminal scales are in fact formed by the diverted development of ovule intitials.

FAMILY : WILLIAMSONIACEAE

As compared to *Cycadeoideaceae,* the *Williamsoniaceae* have comparatively taller and branched (*W. sewardiana*) or unbranched (*W. gigas*) stems covered with persistent, rhomboidal leaf bases. The stems are two meters or more tall. The family as constituted by Sporne (1965) includes the genus *Williamsonia* which represents the whole plant including the reproductive organs. Several generic names have been given to the isolated fronds. *e.g., Ptilophyllum, Otozamites, Dictyozamites* and *Pterophyllum.* Similarly seed bearing organs are known by generic names such as *Benettiticarpus, Vardekloeftia,* and *Westersheimia.* The pollen bearing organs are *Bennettistemon.* The genus *Sturiella* is the name given to isolated specimens of hermaphrodite flowers. The only completely reconstructed gens is *Williamsonia.* All these genera have been discovered from the Mesozoic era.

Detailed description of *Willamsonia* is given in this text.

GENUS WILLIAMSONIA

Williamson (1870) discovered from the Jurassic of Yorkshire the fossil remains of a plant which he named as *Zamia gigas.* It had a cycad-like habit, but later studies and presence of flower-like reproductive bodies showed that it actually belongs to the *Bennettitales.* The name *Williamsonia gigas* was given to it. It is not found in petrified remains. Many species of *Williamsonia* have been discovered in the Triassic, Jurassic and Cretaceous periods of the Mesozoic. *Williamsonia spectabilis* and *W. scotica* are the fructifications found isolated (*i.e.,* not attached to the trunk) from the Oolite or Yorkshire and from the Jurassic of Scotland, respectively. Five species of *Williamsonia* have been reported from the Rajmahal hills of Bihar. These are : *W. sewardiana; W. Sahni; W. microps;* and *W. santalensis. Bucklandia indica* whose petrified trunks were discovered from Rajmahal hills is actually *W. sewardiana.*

External Features

Sahni (1932) reconstructed *W. sewardiana* (Fig. 6.9, A) from Jurassic period of Upper Gondwana of the Rajmahal hills (Bihar).His reconstruction showed it to be a branched. slender, and columnar

Fig. 6.9. *Williamsoniaceae* (A—E). A. Reconstruction of *Williamsonia sewardiana* (After Sahni).B. male flower of *W. whitlensis* showing the whorl of stamens fused below, and bearing pollen sacs at their free portions (After Nathorst). C. One stamen as seen from above with a double row of pollen sacs (After Nathorst). D. Male flower of *W. spectabilts* (After Wieland). E. *Williamsonia santalensis,* a reconstruction of microsporophyll bearing two rows of synangia (After Sitholey and Bose).

trunk about two meters tall. The trunk is beset with rhomboidal scars left by the fallen leaves. At the summit, there is a crown of pinnate compund leaves like those of a *Cycas* plant. The leaves were once pinnate and have been separately described under the name *Ptilophyllum cutchense*. The branching is lateral and projects through the thick armour of leaf bases. Two kinds of lateral shoots can be recognised (*i*) the sterile or vegetative shoots; and (*ii*) fertile shoots that bore flowers. A distinct constriction is evident at the bases of lateral shoots. This had led to the belief that possibly some of them got detached and served as a means of vegetative propagation. The leaves are also of two kinds: (*i*) the simple scale-like and pointed leaves; and (*ii*) pinnately compound foliage leaves.The lateral branches bear only a few foliage leaves. The foliage leaves are once pinnate and the leaflets are attached to the rachis by means of a broad base to its upper side. The fertile branch of *W. sewardiana* has been shown to terminate in a female flower.

Anatomy

A cross section of the trunk of *W. sewardiana* revealed a rough circular outline. It had a wide parenchymatous cortex with secretory sacs. The pith was narrow and parenchymatous and possessed secretory sacs. The primary vasculature consisted of a ring of endarch, conjoint and collateral vascular bundles. The pith extended between the vascular bundles in the form of pith rays. The secondary wood formed an extensive cylinder around the primary xylem. It was traversed by both uniseriate and bi- or triseriate secondary medullary rays. The secondary wood was pycnoxylic, *i.e.*, compact and consisted of scalariform bordered pitted tracheids. The cambium was distinct. Multiseriated bordered pits were also present. They were found on radial walls. Leaf traces were multiple but not girdling.

Reproduction

The genus was unisexual and dioecious. The flowers as stated above were borne on lateral shoots arising in leaf axils. The floral branches bore simple scale leaves below and once pinnate green leaves near the apex. They served as bud scales. The flowers terminated these lateral floral shoots. Only female plants of *W. sewardiana* have been discovered so far. Every female flower was surrounded by a series of perianth-like bracts, whereas in male flowers such perianth lobes were not present.

Female Flowers

The female flowers in *W. sewardiana* and *W. gigas* (Fig 6.10) have distinct conical receptacle surrounded by simple perianth-like bracts, that had scales at their bases and hair-like ramenta above. The stalked ovules and interseminal scales were arranged in close spirals around the conical receptacle. The tip of the receptacle was naked. The ovules had short stalks that bear the body of the ovule which consisted of a single integument surrounding the nucellus. The integument was fused with the nucellus and was vascularised. The integument was prolonged beyond the surface layer, formed by enlarged tips of interseminal scales, and formed a characteristic micropylar beak. The nucellus had a prominent beak and a pollen chamber. The interseminal scales were interspersed among the ovules and were fleshy structures with expanded tips that formed a protective covering with small pores through which the micropyles slightly projected. The detailed structure of the ovule has not been studied.

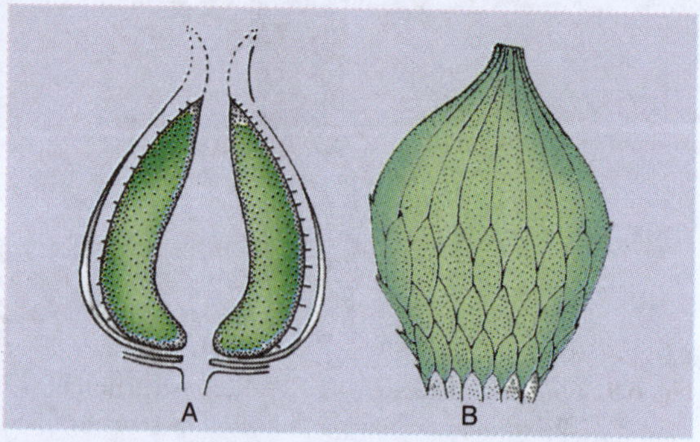

Fig. 6.10. (A—B) A. V.S. ovuliferous flower of *W. gigas* B. Floral bud of *W. gigas*.

Female flowers of *W. gigas, W. mexicana* and *W. scotica* have also been described from various horizons of the Mesozoic.

The other female fructifications referred to this family include *Bennetticarpus exiguus, B. tylotus, B. crossospermus, B. wettsteinii, Westersheimia and Vardekloeftia sulcata*. Presence of a vestigial cupule has been reported in *B. crossos permus* and *V. sulcata.* Presence of a cupule is regarded as an ancestral feature, but it cannot be said with certainty whether this cupule fused with the integument during the course of evolution or it could not develop.

At maturity, the ovule ripened into seeds protected by the inter seminal scales. There were two to six seeds in *V. sulcata* and many in *B. crossospermus*.In the former, they were large and about 7 mm long. The internal structure of the seed has recently been studied in *Williamsonia* by Sharma (1970). The seeds were dicotyledonous, stalked and endospermic.Sharma observed that the seed coat is differentiated into three layers; (*i*) the outer layer composed of tubular cells (comparable to cupule of Harris); (*ii*) a parenchymatous middle layer comparable to the actual integuments; and (*iii*) an inner vascularised and parenchymatous zone which probably represented the outer or peripheral part of the nucellus left over after embryo development.

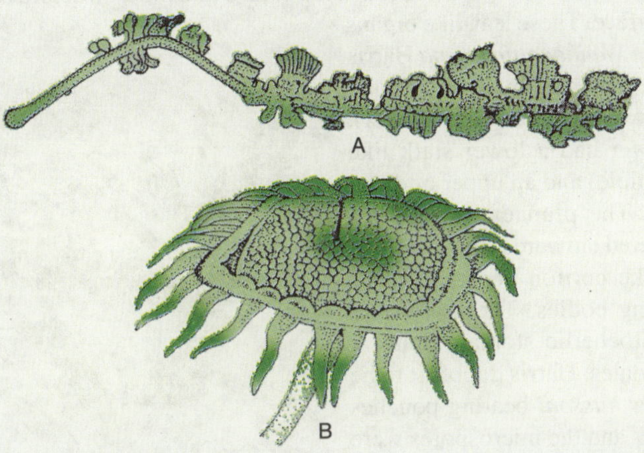

Fig. 6.11. A. Male flower of *Wonnacotia orispa*. B. Bisexual flower of *Sturiella langeri*. (From Andrews)

Westersheimia (1949) is a characteristic female flower described by Krausel (1949) from the Triassic of Austria, because it had pinnate megasporophylls whose ultimate segments bore ovules and interseminal scales.

Bisexual flowers, among this family, have been reported only in *Sturiella langeri* (Krausel, 1948). Discovered from the Triassic of Lunz, Austria, it is known from an axis 4 cm long that bore slender lateral branches that terminated in a sunflower-like hermaphrodite flower. The lateral branch flared into a cup-shaped receptacle that bore around its edges 25 to 30 ray-like lobes. A microsporangium was attached to the base of each lobe. In the centre of the cup were found ovulate structures that are not well preserved (Fig. 6.11, B).

Male Flowers

Williamsonia includes many species based on male flowers. These have mainly been discovered from the Jurassic. All of them consist of a whorl of varying number of microsporophylls united at the base to form a cup-like structure. In *W. spectabilis* (Fig. 6.9, D) the sporophylls branched pinnately, whereas in *W. whitibiensis* they were unbranched (Fig. 6.9, B). Sitholey and Bose (1953) described from the upper Gondwana of India the male flowers of *W. santalensis*. In this case the microsporophylls were bifid (Fig. 6.9, E). One-half is sterile, whereas the other half was fertile and bore two rows of finger-like synangia (Fig. 6.9, E). The synangia are arranged along the inner surface of the microsporophylls.

In *W. spectabilis* Nathorst (Fig. 6.9, D) the male flower measured about 9 cm across above and up to 3 cm in the region of the basal cup. There were about twelve segments which tend to bend out horizontally and then turn abruptly upwards, probably arching over the centre of the cup. Numerous slender branches arose from the inner side of the free ends of the microsporophyll. Each such branch bore rows of synangia.The male flowers were stalked.

In *W. whitibiensis* Nathorst (Fig. 6.9, B) the male flowers were sessile and the synangia were borne in two rows directly on the inner surfaces of the microsporophylls, *i.e.*, they were not branched.

The basal cup was comparatively shallow in *W. santalensis*. In this case, each microsporophyll forked into a blunt terminal portion and a narrow side branch which bore two rows of tapering finger-like appendages or synangia. Each such appendage contained two rows of chambers which enclosed microspores. The structure of synangia was similar to *Cycadeoiden dacotensis*. The microspores have not been found in them.

In *Bennettistemon ovatum,* the male flower consisted of an ovate foliar structure bearing separate sporangia on its upper surface. These leaf-like organs were about 7 mm. long. In *Wonnacottia crispa* Harris the male flowers were represented by a 12 cm long and 1.5 cm broad leaf-like structure (Fig. 6.11, A). This leaf-like male flower had a lower stalk like portion (resembling a petiole) and an upper or distal pinnately divided part. The pinnae were almost opposite and slightly tapered outwards (Fig. 6.11, A). The proximal petiole like portion bore numerous globose pollen-containing bodies. These leaf-like bodies contained syndetocheilic stomata that are characteristic of Bennettitales. Harris doubts it to be the leaf of *Anomozamites nitssoni* bearing pouches caused by gall mite attack and the microspores were carried into these pouches by these insects.

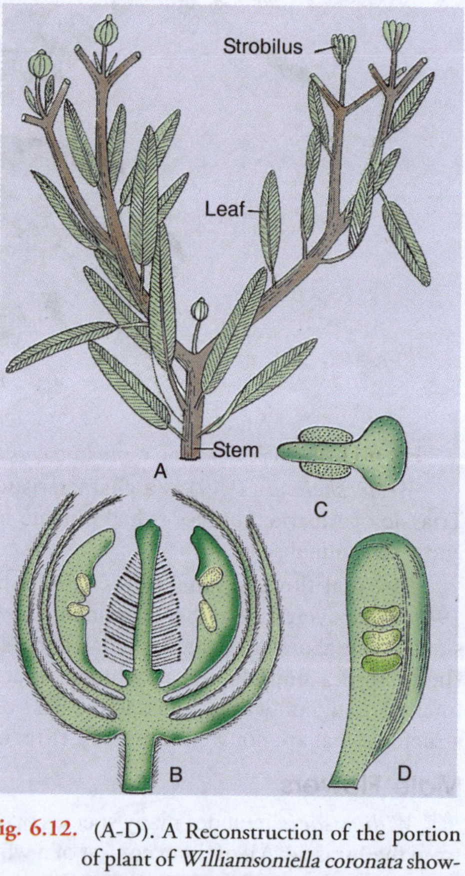

Fig. 6.12. (A-D). A Reconstruction of the portion of plant of *Williamsoniella coronata* showing branched stem, simple leaves and strobili, B. V.S. of bisexual flower of W. coronata. C—D. Microsporophylls. (A, after Thomas; B, after Harris)

FAMILY : WIELANDIELLACEAE

It includes the genera *Williamsoniella* and *Weilandiella*, discovered from the Jurassic of Yorkshire and from the Rheatic of Scania in Southern Sweden, respectively.

GENUS WILLIAMSONIELLA

It has 4 species all of whom were collected from the Jurassic of Yorkshire. *W. coronata* is the best studied species and consisted of a much branched stem bearing simple leaves. The stems appear to be dichotomously branched with a bisporangiate flower borne at the angle of the dichotomy (Fig. 6.12). It has also been interpreted that the flowers arise terminally on the stem and later two lateral branches arise from below, thus giving the appearance of a false or pseudodichotomy. Since these plants are available as fossil remains, it is not possible to study the exact mechanism of branching and hence it is difficult to say definitely whether branching is dichotomous or pseudodichotomous. The trunk was up to 1.5 metres long.

The leaves (Fig. 6.12) were entire, linear or lanceolate to narrowly ovate and petiolate. The lamina was simple, undivided and bore a strong midrib from which arose lateral veins. The venation was close. The leaves were borne separately along the axis. The leaf has been described under the generic name *Nilssoniopteris*. They were 3 cm wide and up to 20 cm long.

Stems in a cross section showed a conspicuous parenchymatous cortex and pith both of which were traversed by secondary sacs. There was a thin cylinder of xylem, the primary portion of which was endarch. The secondary wood was traversed by uniseriate or bi- to triseriate rays. The tracheids had scalariform bordered pits. The leaf traces were multiple but did not girdle the stem.

The hermaphrodite flower was borne terminally (if branching is supposed to be lateral) or in the angle of a dichotomy and had a 3.5 centimetres long stalk or the peduncle. Each flower bore a number of sterile perianth lobes (Fig. 6.12, B) that were beset with long hair and arose from the base of the floral receptacle. These perianth lobes enclosed the reproductive structures.

There was a ring of 12—14 microsporophylls that were free from each other and were attached to the base of the receptacle or the central column. Each microsporophyll looked like the pip of an orange fruit and was made up of parenchymatous tissue. It was fleshy. Each microsporophyll had finger-like outgrowths arising from its middle. These outgrowths partially enclosed two pairs of pollen capsules or synangia. Each capsule enclosed several sporangia that contained microspores. The capsule dehisced longitudinally along two lines into two valves. The microsprophylls were shed from the plant after the dissemination of the microspores, because isolated microsporophylls with empty bi-valved capsules have been collected.

The gynoecium consisted of a central conical receptacle that was broad at base and gradually tapered to narrow and capitate tip. The tip was sterile and naked. It was about 10 cm long. Borne around this central column were slender stalked 300 ovules and about 1,200 interseminal scales. They were spirally arranged. The ovules appeared as bulbous heads at the tips of stalks. The integument of the ovule was free from the nucellus.

GENUS WIELANDIELLA

The genus was described by Nathorst from the Rheactic of Scania, in Southern Sweden. The species *W. angustifolia* (Fig. 6.13) possessed dichotomously branched slender stems that usually measure 1.5 cm in diameter. The leaves in this species were pinnately compound (one pinnate) and were 8 cm long and had winged petioles. The dichotomous branching had been interpreted to be pseudodichotomous. The leaves arose in clusters at the point of forking of the stem and were usually classified as *Anomozamites minor*.

Fig. 6.13. Restoration of a part of the plant of *Wielandiella angustifolia* (After Nathorst).

The flowers were sessile and arose from the angles of the dichot omies (Fig. 6.13) or otherwise terminal if we regard the branching to be pseudo-dichotomous. The flowers were hermaphrodite and bore perianth-like leaves, microsporophylls and a central column bearing ovules (Fig. 6.14) and interseminal scales. The microsporophylls were not well preserved. The apical region of the ovule bearing, club-shaped central column or the receptacle was sessile.

Discussion

The affinities of the Bennettitales with cycads and angiosperms have been discussed in the introductory paragraph of this chapter. Rodin and Kapil (1969) while discussing the evolution of seed coats in *Gnetum* have vehemently opposed the concept of relationship between the seeds of the *Bennettitales* and the *Gnetales*. The presence of interseminal scales in the *Bennettitales* and their peculiar floral structure and some features of the stem and leaves do not show any close affinity between the two orders.

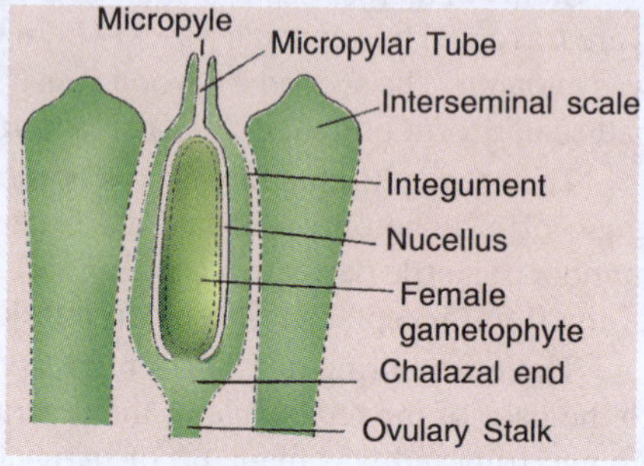

Fig. 6.14. Ovule and interseminal scales in *Wielandiella angustifolia.* (After Nathorst)

The *Bennettitales* have some characteristics in which they do not resemble any of the seed bearing plants: (*i*) the peculiar bisporangiate or hermaphrodite flower-like strobilii; (*ii*) synangial nature of the pollen-bearing organs, now called pollen-capsules; and (*iii*) presence of long stalked ovules and the interseminal scales.

Some authors consider that the *Cycadales* and the *Bennettitales* originated from a common ancestor among the pteridospermalean complex. There is no doubt that the stems of cycads and cycadeoids have some affinities with *Medullosaceae* (Pteridospermous family) but there is great difference in the reproductive structures of all these orders. There are many other anatomical features in which they differ from each other. It is not possible to trace a common origin for *Bennettitales* and *Cycadales* because these two orders also differ from each other. It has been proposed by Andrews that *Bennettitales* and *Cycadales* originated along two different and independent lines of evolution from a group other than the pteridosperms and which came into existence earlier. Progymnosperms of Beck (1970) can be regarded as ancestral to the gymnosperms.

Relationships of Cycadeoidales (Bennettitales) with other groups.

Members of Cycadeoidales show similarities and difference with Cycadaoles, Pteridospermales and Angiosperms. A brief account of these is as follows :

Cycadales : Cycadeoidales resembles with Cycas in (*i*) having morphologically similar fronds and exhibiting similar habit;

(*ii*) having manoxylic wood, monocolpate pollen grains, orthrotropous ovules and dicotyledenous embryo.

The points of differences between Cycadeoidales and Cycadales are :

(*i*) Syndetocheilic stomata in Cycadeoidales and haplocheilic Stomata in Cycadales,
(*ii*) Both the groups differ in anatomical details of the frond,
(*iii*) Girdle leaf traces are absent in Cycadeoidales while these are prominent in cycadales,
(*iv*) Cones are bisporangiate in Cycadeoidales and monosporaingate in Cycadales.
(*v*) Microsporophylls are arranged in whorls in Cycadeoidales and in Cycadales, microsprophylls are spirally arranged.
(*vi*) In Cycadeoidales, arrangement of stalked ovules alongwith interseminal scales is a characteristic feature while in Cycadales this is not found. Therefore Andrews (1961) was of the view that both the groups have evolved along two different independent lines.

With Pteridospermales. Cycadeoidales resemble Pteridospermales in many points such as

(*i*) presence of ramental hairs;
(*ii*) internal structure similar;
(*iii*) both having syndetocheilic stomata,
(*iv*) having direct leaf traces, and
(*v*) presence of microsporophyll bearing synangia and presence of cupules.

But the two groups have major differences as well. In Cycadeoidales, ovules are smaller and stalked while in Pteridospermales ovules are larger and sessile. Vascular supply to nucellar tissue is prominently exhibited in Cycadeoidales while this is absent in Pteridospermales.

It is therefore presumed by some authors that there were two lines of evolution from Cycadofilicales. Of these one line gave rise to the monosporangiate Cycads and the other to Cycadeoidales having both mono-and bisporangiate forms.

With Angiosperms. Few morphologists believe that angiosperms might have evolved from mesozoic Cycadeoidales because of the similarities in strobili and flower of Cycadeoidales and some angiosperms (like Ranales and Magnoliales). These two groups also resemble in having

(*i*) Spirally arranged hairy bracts around reproductive organs in Cycadeoidales comparable to the perianth of angiosperms,
(*ii*) Syndetocheilic stomata in Cycadeoidales and mesoparacylic stomata in Magnoliales,
(*iii*) Homoxylous wood without vessels in cycadeoidales and angiosperms like *Drimys, Trochodendron* and *Tetracentron,*
(*iv*) Dicotyledonous embryo in both Cycadeoidales and angiosperms.

But the Cycadeoidales differ from angiosperms in having (*i*) naked *Ovules;* (*ii*) large compound, stamens having internal synaniga as compared to simple stamens in angiosperms

(*iii*) no marginal cells in wood rays whereas in angiosperms prominent marginal cells are present in wood rays.

It was contended by Pant and Kidwai (1977) that Cycadeoidean flower may represent a parallel line of evolution to angiosperms, a contention substantiated by the fact that Cycadeoidales disappeared during cretaceous while angiosperms made their appearance by then.

QUESTIONS

ESSAY TYPE

1. Give an account of the strobili of Cycadeoidales. How do they differ from those of the Pentoxylales. **(Kanpur MSc., 1989, 1995)**
2. Discuss the Affinities of the Bennettitales. **(Kanpur MSc., 1984,1996,2004; Rohil Khand MSc, 2000)**
3. Compare the general Characters of Cycadeoidacea, and Williamsoniaceae. **(Kanpur MSc., 1985)**
4. Give a brief and general account of *Williamsonia.* **(Kanpur MSc., 1986; Calicut, 2003)**
5. Despite the nunerous similarities between Bennettitales and agiosperms they are not included among flowering plants. Explain. **(Kanpur, MSc., 1988)**
6. Describe the important features of Bennettitales . How would you discuss *Williamsonia* from *Williamsoniella?* **(Rohil Khand MSc, 1990)**
7. What is the justification for calling Bennettitales as living Cycads ? Discuss the affinities of this group? **(Rohil Khand MSc, 1993)**
8. Describe the strobili of Cycadeoidales and comment on their evolutionary significance. **(Calicut, 2001)**
9. Describe the salient features of *Williamsonia.* Give examples of two fossils reported for Bihar. **(Magadh, 1999)**
10. Enumerate the salient features of *Cycadeoidea.* Mention its affinities with other plant groups. **(Vinoba Bhawe, 1998)**

11. Give a concise account of fossil flora of Raj Mahal hills. **(Kanpur MSc., 2004)**
12. Discuss the evolutionary trends in the structural organisation of Bennettitalean fructifications. **(Gauhati, 2002)**
13. Give an account of *Williamsonia* and *Salhanianthus* flower. **(Punjab, 2003)**
14. Describe the reproductive organs of *Williamsonia.* **(Punjab, 2003)**
15. What is the name of Bennettitalean plant reconstructed by Prof. Birbal Sahni ? Give an illustrated account of various form genera of the plant. **(Calcutta, 2001)**
16. Give a detailed account of Bennettitales. **(Rohilkhand MSc., 2004)**

SHORT ANSWER TYPE

17. **Write short notes on**
    (*a*) *Williamsonia*
    (Kanpur MSc., 1994; Awadh MSc., 2000; Rohil Khand, 2001; Virkumar Singh, 1997; Raipur, 1996)
    (*b*) Fructifications of Williamsoniaceae **(Kanpur MSc 1984,)**
    (*c*) *Cycadeoidea* **(Kanpur MSc., 1998, Gauhati MSc., 2004)**
    (*d*) Bennettitales remains from India **(Awadh MSc., 1998)**
    (*e*) *Williamsoniella* **(Rohilkhand MSc., 1989)**
    (*f*) Affinities of Bennettitales
    (*g*) Male fructifications of *Williamsonia* and its occurrence in geological time **(Calcutta, 2002)**
18. **Differentiate between**
    (*i*) Leaf cuticles of Cycadales and Bennettitales **(Kanpur MSc., 1991)**
    (*ii*) *Caytonia* and *Williamsonia* **(Kanpur MSc., 1993, 1995)**
19. Assign *Williamsonia sewaridiana* to its systematic position and point features of special interest. **(Rohilkhand MSc., 1995)**
20. Compare with suitable diagrams :
    (*i*) *Cycadeoidea* and *Williamsonia* **(Allahabad MSc., 2001)**
21. Write the name of the male flower of *Williamsonia* **(Poorvanchal MSc., 2003)**

OBJECTIVE TYPE

22. **Select the correct answer:**
    (*i*) Bisporangiate cones are found in
    (*a*) *Lyginodendron*
    (*b*) *Cycadeoidea*
    (*c*) *Pinus*
    (*d*) *Ginkgo*
    (*ii*) *Williamsonia sewardiana* has been described by
    (*a*) H.N. Andrews
    (*b*) K.R. surange
    (*c*) Birbal Sahni (*d*) D.D. Pant
    (*iii*) Cycadeoidales resembles angiosperms in
    (*a*) having naked ovules
    (*b*) in large compound stamens
    (*c*) absence of marginal cells
    (*d*) having dicotyledonous embro
    (*iv*) *Cycadeoidia* differs from *Cycas* in having
    (*a*) morphologically similar fronds
    (*b*) having monoxylic wood
    (*c*) in having monocolpate stamens
    (*d*) in the presence of syndetochelic stomata
    (*v*) *Williamsonia* is represented by five species in India and all of these have been reported from
    (*a*) Raj Mahal hills
    (*b*) Shivalik hills
    (*c*) Vindhyam range
    (*d*) Nilgiris.

CHAPTER 7

Pentoxylales

PENTOXYLON FOSSIL

- Introduction
- Characteristic Features
- Reproductive Organs
- Affinities with Different Groups

Introduction

The name *Pentoxyleae* was proposed by Prof. Birbai Sahni in the year 1948 for a group of fossil plants discovered from the Jurassic horizon of the Rajmahal hills (Bihar) of India. These plants have been treated under a separate order by Sporne (1965). The order includes a number of stems (*Pentoxylon*, *Nipanioxylon*), leaves (*Nepaniophyllum*), pollen-bearing organs (*Sahnia*) and seed-bearing organs (*Carnoconites*). Sahni and Srivastava (1948) carried out extensive researches on the stems, leaves and seed-bearings organs. Vishnu Mittre described the pollen-bearings organs in 1953.

The members of this group are a unique combination of characters of *Bennettitales, Cycadales* and *Coniferales.* A few characters are peculiar to the group itself and hence it deserves a position as a separate order of the *Cycadopsida.* These have been found to occur in silicified petrification. Harris (1962) has also discovered them from New Zealand. He found only the seed-bearing organs.

Characteristic Features

The order is characterised by the following features :

1. The plants were probably shrubs or small trees. The habit is not known but the size and diameter of the stem, its branching and leaves reflect upon its shrubby or tree-like habit.
2. The branches were of two types : (*i*) long shoots and (*ii*) dwarf shoots.
3. The leaves were spirally arranged.
4. The stems were polystelic with 5 and 6 primary steles, hence the name Pentoxylae. Each stele has its own complete ring of cambium.
5. The development of secondary wood is exocentric.
6. The secondary medullary rays were uniseriate.
7. The leaves were simple, lanceolate and thick.
8. The venation was open, rarely reticulate.
9. The reproductive organs were unisexual. The female strobili looked like mulberries.
10. The male organs consist of a whorl of branched sporangiophores fused at the base into a disc. The sporangiophores bear microsporangia on its short branches.
11. The reproductive organs are borne terminally on short branches.
12. The tracheids have crowded, circular bordered pits on their radial walls.
13. The female reproductive organs lacked interseminal scales and consisted only of a thick central axis with ovules attached to it in a spiral manner.
14. The ovules were sessile.
15. The stomata were regarded by Sahni to be syndetocheilic by later Vishnu Mittre (1953) found that haplocheilic or cycadalean stomata also occur. So both the types were met with.

External Characters (Fig. 7.1)

Pentoxylon sahnii and *Nepanioxylon* are the two stem genera discovered from the Rajmahal hills in India. The stems were 3 mm to 2 cms across in the former species. It bore rhomboidal leaf scars and spirally arranged simple leaves. Short lateral shoots arose from the stem and the leaves were restricted to these shoots and were strap-shaped, simple, petiolate with entire margin and obtuse apex. There was distinct single mid-rib giving off lateral veins that ran outwards towards the margin and were parallel. The lateral veins were usually unbranched. Occasional dichotomies and reticulations have been noticed in rare cases. The reproductive organs were borne terminally on the short lateral shoots. The leaves were described under the name *Nipaniophyllum raoi.* They were later found attached to the shoots of *P. shanii*. These leaves were up to 7 cm long and one cm broad.

Anatomy (Fig. 7.1)

A cross section of the stem reveals five main primary steles, *i.e.*, the stem is polystelic. Each stele is a concentric structure having its own cambium that is uniformly active in young stems but in older stems, it produced more secondary tissue towards the centre so that the secondary wood became excentric. External to the cambium was primary phloem and primary xylem was in the form of an internal ring. The vascular strands run longitudinally and give off leaf traces. In addition to the 5 main strands there were 5 smaller strands alternating with them. According to Sahni, these were the strands of the lateral shoots. According to Vishnu Mittre, the number of primary strands varies at different levels along the length of the stem. Series of transverse sections cut at different levels revealed that there were three strands in the lower region of the stem and 5 in the middle, whereas the number increased to six at the top. In *Nipanoxylon guptai* there was a single vascular cylinder near the base of the stem whereas the number increases to 5 to 7 at a higher level and the strands again coalesce at apex to form a single vascular cylinder again.

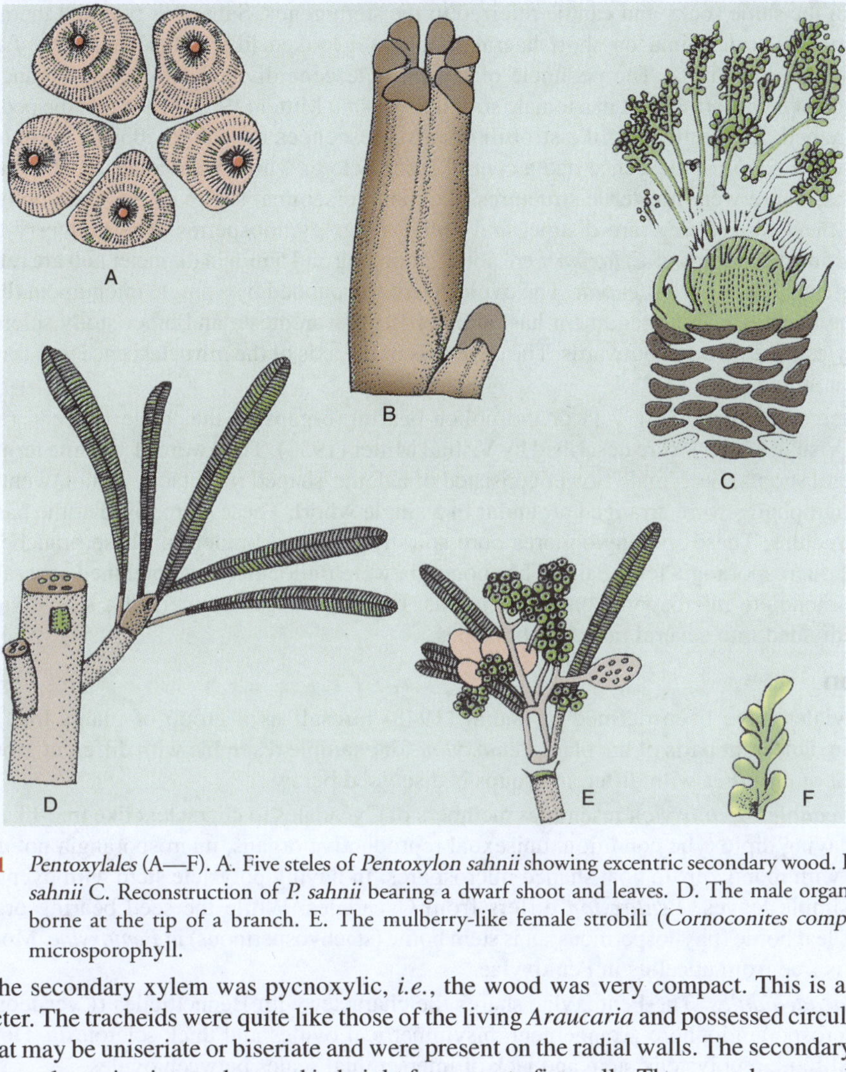

Fig. 7.1 *Pentoxylales* (A—F). A. Five steles of *Pentoxylon sahnii* showing excentric secondary wood. B. Trunk of *P. sahnii* C. Reconstruction of *P. sahnii* bearing a dwarf shoot and leaves. D. The male organs of *P. sahnii* borne at the tip of a branch. E. The mulberry-like female strobili (*Cornoconites compactum*). F. A microsporophyll.

The secondary xylem was pycnoxylic, *i.e.*, the wood was very compact. This is a coniferous character. The tracheids were quite like those of the living *Araucaria* and possessed circular bordered pits that may be uniseriate or biseriate and were present on the radial walls. The secondary medullary rays were also uniseriate and ranged in height from one to five cells. The secondary wood developed more towards the centre. The secondary wood shows growth rings but it cannot be definitely stated whether they represent annual rings.

The leaf traces are typically cycadalean because they posses vascular bundles with both centrifugal and centripetal xylem (diploxylic). Six strands entered each leaf base. They arose singly from the steles. Regarding stomata Sahni (1956) stated them to be syndetocheilic like those of the *Bennettitales*, but Vishnu Mittre (1957) observed them to be of both types. He also observed haplocheilic stomata on the leaf. So in this character *Pentoxylales* combine both Benettitalean and Cycadalean features.

Reproductive organs

Two species of seed-bearing organs have been described by Sahni and Srivastava from the Jurassic horizon of Rajmahal hills. These are *Cornoconites compactum* Srivastava and *C. laxum* Srivastava. These have not been found in organic connection with the stem genus *Pentoxylon* but they occur in the same rocks and can be referred to the stem genus. Sahni has pictured them in his reconstruction, to be terminal on short lateral shoots and looked like stalked mulberry fruits or annonaceous multiple fruits. The peduncle of the infructescene divided into several branches or pedicels. Each of these terminated in a female strobilus. Vishnu Mitter (1953) stated that the peduncles were unbranched. The mulberry-like strobili or infructescences measured about 1.8 cm long in *C. compactum*. Each infructescence had a central receptacle to which were attached about twenty sessile ovules. There were no sterile structures like the interseminal scales or megasporophylls. In this feature the *Pentoxylales* are distinct and unlike other gymnosperms. The mulberry-like or annonaceous infructescenes of *C. laxum* were about 3 cm long and 5 mm in diameter and are referable to the stem genus *Nipanoxylon guptai*. The ovules were surrounded by a single integument that was free from the nucellus. The integument has an outer fleshy sarcotesta and inner stony sclerotesta. The micropyles were directed outwards. The cone axis or the axis of the infructescence was traversed by 5 vascular strands.

The microsporangiate (Fig 7.1) or the pollen-bearing organs or the 'male flowers' (*Sahnia nipaniensis* Vishnu Mitter) were described by Vishnu Mitter (1953). They were also borne terminally on short lateral shoots. Each male flower consisted of a dome-shaped receptacle. About twenty-four microsporangiophores were arranged around it in a single whorl. These were fused at the base into a discoid structure. These sporangiophores bore spirally arranged branches. These branches bore pear-shaped microsporangia terminally. The sporangia were unilocular and contained several boat-shaped and monolete microspores or pollen grains. Each sporangium received a single vascular bundle that divided into several radiating branches.

Discussion

Pentoxylales have been defined by Sahni (1948) himself as a group of plants that defies classification. Different parts of the plant *Pentoxylon* for example resemble with different groups. A brief account of *affinities* with different groups is discussed below :

With Cycadales. Pentoxylon resembles members of Cycadales in characters like tree-like habit, leaf traces having diploxylic condition, unisexual reproductive organs, microsporangia not uniting in synangia with monocolpate boat shaped microspores. In having polystele stem with pycnoxyeic wood and simple leaves, *Pentoxylon* differs from Cycadales. While the seed bearing organ in *Cycadales* is leaf borne (phyllospermous), it is stem borne (stachyospermous) in *Pentoxylae.* Moreover integument is free from nucellus in Pentoxylae.

With Cacadeoidales. The Pentoxylae shares the character with Bennettitales (Cycadeoidales) are : the microsporangiophore arrangement, bisymmetrical ovules and thick sclerotesta. However Pentoxylae differs in polystelic stem and lack of interseminal scales between ovules.

With Medullosaccae. In having polystelic condition, *Pentoxylon* closely resembles with *Medullosa*. However in *Medullosa* the stele is centrifugally developed while in *Pentoxylon* it is centripetally developed. The reproductive structures too differ. While in *Pentoxylon*, these are stachy ospermous, in *Medullosa* these are phyllospermous.

With Conifers. Pentoxylae resemble conifers in having pycnoxylic wood with trachieds having bordered pits. In both the groups dimorphic shoots are present. However, in several other characters both differ.

With Angiosperms. The group Pentoxylae has been compared with *Pandanus* by Moeuse (1961). Both *Pentoxylon* and *Pandanus* are erect or suberect, dioecious plants with cylindrical stems bearing terminal tufts of strap shaped leaves and fleshy seeds.

The female infructescence with cone-like organisation and without any sterile appendages or megasporophylls is a character unique to Pentoxylales. The sporangiophores with spirally arranged branches bearing terminal and unilocular sporangia is also a character unique to Pentoxylales. These intresting features, in addition to those stated above, stamp the group of the Gymnosperms.

It would be interesting to learn that emblem of the Birbal Sahni Institute of Palaeobotany, Lucknow named after Prof. Birbal Sahni is based on *Pentoxylae* – the group instituted by Prof. Sahni. The internationally acclaimed journal *The Palaeobotanist* also carries the same emblem.

PROFESSOR BIRBAL SAHNI

Birbal Sahni

Prof. Sahni was born on the 14th of November, 1891 at Bhera, a small town in the Shahpur District of the West Punjab now in Pakistan. Birbal Sahni received his early education at Lahore. In the year 1911, he left for England to join Emmanuel College at Cambridge.

Birbal Sahni graduated from Cambridge in 1914 and soon settled down to research under the inspiring leadership and guidance of Professor Sir A.C. Seward. It was a relationship deeper and more exquisite than between a teacher and his disciple that stimulated him for a meaningful understanding of fossil plants, knowledge of living plants, geology and other related disciplines. For his researches on fossil plants he received the D.Sc. of the London University in 1919. Returning home in the same year, he served as professor of Botany for one year each at the Universities of Banaras and Punjab. In 1921 he took charge of the newly opened Botany Department of the Lucknow University, as its first Professor. Charisma of Professor Birbal Sahni as a teacher noted for brilliance, lucidity and illustrative skills became a legend and inspired a number of young minds to join the University of Lucknow. Despite his heavy teaching schedule and other preoccupations, he threw himself heart and soul into the work of organizing researches in palaeobotany and made enormous collections of fossils from various localities and geological periods. In recognition of his outstanding merits the Cambridge University conferred on him the degree of Sc.D. in the year 1929. In 1936 he was elected a Fellow of the Royal Society, London. It was Prof. Sahni who founded Institute of Palaeobotany at Lucknow in Sept. 1946. Later it was named Birbal Sahni Institute of Palaeobotany to recognize his contributions in the field of Palaeobotany particularly Gymnosperms.

The set of four stamps issued by the Department of Posts to commemorate the fiftieth anniversary of the Birbal Sahni Institute of Paleobotany in Lucknow, Uttar Pradesh, depict the richness and variety of plant fossils.

Williamsonia sewardiana - A model of the extinct plant Williamsonia sewardiana which thrived in Rajmahal, Bihar about 140 million years ago. This model is based on the reconstruction envisaged by Prof. Birbal Sahni.

Pentoxylon - An important discovery of Prof. Birbal Sahni is the extinct plant group named Pentoxylae from Nipania in Dumka district, Rajmahal Hills, Bihar (age 110-114 million years). Reconstruction of plant with leaves, stem, flowers.

Glossopteris - The tongue-shaped leaf Glossopteris, represents a unique group of extinct vascular plants (age : Permian, 250-280 million years). During this period India occupied a position south of equator close to South Pole as a part of a very large continent which included South America, Antarctica, Africa and Australia, called Gondwanaland. This vegetation was responsible for the precious coal reserves in peninsular India.

Birbalsahni divyadarshanii-Fossil of an enigmatic flower-like organ of the extinct plant named after eminent Indian Palaeobotanists — Prof. Birbal Sahni and Prof. Divya Darshan Pant, discovered from Hura Coalfield, Santhal Pargana, Bihar (age 250-280 million years).

Questions

ESSAY TYPE

1. Mention the characteristic features and affinities of Pentoxylales. **(Kanpur MSc., 1998)**
2. Briefly discuss Pentoxyleae. **(Kanpur MSc., 1986)**
3. Write an essay on Pentoxylales and discuss its affinities. **(Awadh MSc., 2000)**
4. Write detailed notes on Pentoxyleae. **(Rohilkhand MSc., 1992)**
5. Describe in detail anatomy of the *Pentoxylon* axis and compare the reproductive biology with that of Pteridospermales and Cycadeoidales. **(Rohilkhand MSc., 1995)**
6. Describe the structure of *Cornoconites* and discuss its affinities with Bennittitales.
7. Describe the systematic position of Pentoxylales.
8. Describe the anatomy of *Pentoxylon* stem. **(Calicut, 2001)**
9. Give an illustrated account of different form genera belonging to Pentoxyleae and comment upon the systematic position of the group. **(Poorvanchal MSc., 2001)**
10. Give an illustrated account of Pentoxylales. **(Kanpur MSc., 2004)**
11. Give concise account of flora of Rajmahal hills with special reference to those studied by Prof. Sahni.

SHORT ANSWER TYPE

12. **Write short notes on**
    (*i*) *Pentoxylon* **(Kanpur MSc., 1988, 1991, 1996; Rohilkhand MSc., 1997, 2000)**
    (*ii*) Pentoxylales **(Kanpur MSc., 1986, 1993, 2000; Rohilkhand MSc., 2002)**
    (*iii*) Discovery of Pentoxyleae **(Kanpur MSc., 1999)**
    (*iv*) Pentoxyleae **(Rohilkhand MSc., 1989)**
    (*iv*) Main features of Pentoxyleae **(Rohilkhand MSc., 1994)**
13. What do you know about *Pentoxylon* ?
14. Write a note on the structure and affinites of *Pentoxylon.*
15. Describe the male fructification of *Pentoxylon*.
16. Describe the structure of *Cornoconites.*
17. Comment on :
    (*a*) Pentoxyleae represents synthetic order. **(Awadh MSc., 1999)**
    (*b*) Pentoxyleae has characters of many groups.

OBJECTIVE TYPE

18. **Select the correct answers :**
    (*i*) The name Pentoxyleae was proposed by :
    (*a*) Vishnu Mittre
    (*b*) A.C. Seward
    (*c*) Birbal Sahni
    (*d*) D.D. Pant
    (*ii*) Since the plants of *Pentoxylon* have both long and dwarf shoots the group has affinites with
    (*a*) *Pinus* (*b*) *Cycas*
    (*c*) *Rhynia* (*d*) *Ephedra*
    (*iii*) The leaves of Pentoxyleae are known under the name
    (*a*) *Pentoxylon*
    (*b*) *Nipanioxylon*
    (*c*) *Nipaniophyllum*
    (*d*) All of the above
    (*iv*) The plants of Pentoxyleae was discovered from
    (*a*) Shivalik Hills
    (*b*) Rajmahal Hills
    (*c*) Nilgiri Hills
    (*d*) Vindhyan Hills
    (*v*) *Cornoconites compactum* represents the
    (*a*) Vegetative organs
    (*b*) Reproductive organs
    (*c*) Neither the reproductive organs or the vegetative organs
    (*d*) Both the vegetative and reproductive organs.
    (*vi*) A fossil plant with five steles is
    (*a*) *Cycas* (*b*) *Pentoxylon*
    (*c*) *Cycadesidea* (*d*) *Pinus*

CHAPTER 8

Cycadales

CYCAD SP.

- Cycadaceae : Characteristic Features
- Reproductive Characters
- Economic Importance of Cycads

Introduction

The order includes both living and extinct forms that originated in the Upper Triassic period of the early Mesozoic era, *i.e.*, about 200,000,000 years ago. The group reached its peak during the Mid-Mesozoic times and flourished during the Jurassic and Cretaceous periods and after that their decline started and in the present times there are only 10 genera with 107 to 117 species that are restricted to Australia, South Africa, Central America and Eastern Asia, including India. The living genera have fortunately attracted the attention of man and many of them are in cultivation and are, therefore, saved from the danger of extinction.

Their isolated occurance or disjunct distribution and fewer number of species speak of their antiquity. Their morphological features also speak of their ancient lineage. The extinct *Cycadales* are available from the mid-Jurassic Cayton Bay beds on the Yorkshire Coast. The fossils are referable to the order include leaves, seed-bearing organs and pollen-bearing organs that have been described under the generic names *Nilssonia*, *Bennia* and *Androstrobus,* respectively. *Pseudoctenus* is also a leaf referable to the *Cycadales*. *Bojuvia simplex* is a leaf ascribed to *Palaeocycas integer* (Florin, 1933) that was described by Florin from the Upper Jurassic of Sweden.

Classification

The Cycadales has been variously classified by different authors. Pilger and Melchoir included all the living genera in a single family *Cycadaceae* into five sub families : (*i*) *Cycadoideae*, (*ii*) *Stangeroideae*, (*iii*) *Bowenioideae*, (*iv*) *Dioonoideae*, and (*v*) *Zamioideae*. Sporne (1956) divided the order into two families :

(*i*) *Nilssoniaceae,* which includes the fossil forms.

(*ii*) *Cycadaceae* which includes one fossil genus (*Palaeocycas*) and 10 living genera *i.e., Cycas, Zamia, Dioon, Bowenia, Stangeria,* etc.

Bierhorst (1971) divided the living *Cycadales* into three families :

(*i*) *Cycadaceae :* includes *Cycas.*

(*ii*) *Stangeriaceae :* includes *Strangeria.*

(*iii*) *Zamiaceae :* includes eight genera, *e.g., Zamia, Lepidozamia, Macrozamia, Encephalortos, Dioon, Microcycas, Bowenia* and *Ceratozamia.*

Lyman Benson (1957) also includes all the extant cycads (10 genera) in a single family *Cycadaceae.*

In this text classification proposed by Sporne (1965) is followed.

FAMILY : CYCADACEAE

The *Cycadaceae* include 10 living and one fossil genera. The fossil genus is *Palaeocycas* whose species *P. niger* was described from Upper Triassic rocks of Sweden. This species includes the well known leaves that have been described under the same *Bjuvia simplex* and some structures bearing ovules. The living genera of the family include : *Cycas* with twenty species; *Microcycas* with one species; *Zamia* with thirty to forty species; *Lepidozamia* with two species; *Macrozamia* with 14 species; *Ceratozamia* with four species; *Encephalartos* with thirty species; *Dioon* with three to five species; *Bowenia* with two species and *Stangeria* with one species. So the total number of species varies between 107 to 119. Sporne (1965) mentions nine genera and 65 species. In 1990 Stevenson added a new genus *Chigua* to Cycadales. This is the only endemic cycad in South America. The genus differs from other cycads in the structures of leaflet & hexagonal peltate megasporophyll. Earlier this genus was considered as a species of genus *Zamia.*

Plate I. *Cycas circinalis.*

Distribution

The twenty species of *Cycas* are widely distributed and are encountered from Australia through the islands to Southern Japan, and then in the west to China, Madagascar and India. The other large

Fig. 8.1 (A—D). Cycadaceae A. Plant of *Zamia floridiana* showing habit and bearing a female cone; B. A male cone; C. A sperm; D. V.S. of female cone.

genus *Zamia* has also a wide distribution and is found in Florida, West Indies, Mexico, nothern parts of South America and along the West Coast of Chile. These two genera are geographically quite far apart. The genera *Macrozamia*, *Lepidozamia* and *Bowenia* are confined to Australia; *Stangeria* is restricted to south-east Africa; *Ceratozamia* and *Dioon* occur in Mexico; *Dioon* is also found in central America; one species of *Microcycas* is restricted to Cuba. Such a restricted and scattered distribution is suggestive of ancient lineage of the group.

CHARACTERISTIC FEATURES OF CYCADACEAE

The members of this family possesess a combination of characteristic features that make it a synthetic group of vascular plants. In certain respects its members resemble the ferns and the pteridosperms, in others they resemble the *Bennettitales, Pentoxylales, Cordaitales* and *Ginkgoales.*

Attempt has even been made to point out resemblances with angiosperms. A brief account of their diagnostic characteristics and variations will enable the student to form a general idea of the group. These will be discussed under separate headings :

Fig. 8.2. Cycadaceae. A plant of *Bowenia spectabilis.*

External Features

The members of this family give an external appearance simulating the palms. The stem is usually thick, cylindrical (Fig. 8.3), slow-growing, unbranched or occasionally adventitious branches may arise (Plate I, branched tree of *Cycas circinalis*), 10-18 metres tall [*Dioon spinulosum* (Fig. 8.3), *Microcycas calocoma* and *Macrozamia hopei*]. In some species of *Zamia* (Fig. 8.1) the stems are tuberous and nearly spherical, subterranean and inregularly branched. The stem in *Bowenia spectabilis* (Fig. 8.2) is almost spherical and tuberous, but is unbranched. The foliage leaves are spirally arranged and may be pinnate or bipinnate (*Bowenia*). The leaves have **Circinate vernation** in *Cycas* whereas it is **sub-circinate** in *Stangeria, Ceratozamia* and *Bowenia*. In all the other genera it is straight. In circinate ptyxis and frond-like large leaves that may be upto 3 meters long in *Cycas circinalis*, the cycads resemble the ferns. The smallest leaves among the family are those of *Zamia pygmaea* (5 centimetres long). The leaflets in *Cycas* and *Stangeria* have a distinct midrib. In *Cycas* there are no side veins whereas in *Stangeria* the lateral veins arise and branch dichotomously. The other genera are devoid of midribs and show open dichotomous venation. The leaves persist for 3-10 years and when they drop, their bases persist longer. In *Cycas* and *Dioon* the leaves are dimorphic, *i.e.*, within a single phyllotactic series there are zones of green and pinnate vegetative leaves and zones of brown **cataphylls** that are simple, hairy, 5 to 10 cms long and sharply pointed. They are also called scale leaves.

Fig. 8.3. *Dioon spinulosum.* A tree showing habit.

Fig. 8.4. Cycadaceae. A plant of *Stangeria paradoxa* showing its habit.

The trees are slow growing and a 6 to 7 feet tall plant of *Dioon* has been estimated to be one thousand years old. This is most probably due to their xerophytic habitat. The cycad stems have massive apices that do not possess **tunica** layer.

ANATOMY

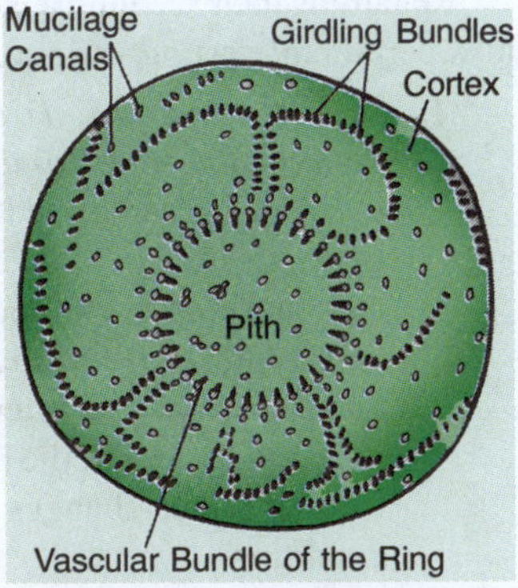

Fig. 8.5. T.S. stem of *Zamia* (outline) showing girdling leaf traces.

A cross section of the Cycad stems shows a roughly circular outline because of the persistent leaf bases. There is massive peripheral cortex and a massive central pith. Both are traversed by distinct mucilage canals. The endodermis and pericycle are not distinct. The primary vasculature consists of an interrupted ring of conjoint, collateral, open and endarch vascular bundles. The leaf traces originate singly from the vascular cylinder of the stem (cauline). Some traces originate quite close to point of attachment of the petiole, whereas others originate from distant points and completely girdle the stem (Fig. 8.5) cortex, where they branch and eventually enter the petiole. These girdling leaf traces are quite characteristics of the *Cycadaceae*. The leaf trace traverses half way around the stem almost horizontally and enters the base of the leaf almost opposite to its point of origin. It is joined by other traces during its course through the cortex so that the petiole receives a number of leaf traces. The arrangement of leaf trace bundles in the petiole may be in a horse-shoe-like pattern or like the word Omega. These leaf trace bundles are diploxylic, *i.e.*, they have both centripetal and centrifugal xylem.

The amount of secondary xylem is very small and this deficiency is made good by the persistent leaf bases whose sclerenchymatous tissues give sufficient strength to the stem and impart hardness to the otherwise soft wooded or **manoxylic** stem. In majority of the genera the secondary growth is **monoxylic**, *i.e.*, there is a single cambial ring that functions throughout the life of the plant; but in *Cycas*, and some species of *Macrozamia* and *Encephalarots* the wood is polyxylic, *i.e.*, it consists of successive rings of secondary wood formed by a succession of cambial rings developed external to the original ring. These rings originate in the secondary phloem of the youngest vascular cylinder. The secondary xylem has many rays, one or two cells wide and many cells high which are continuous to the pith. Broader rays are also present and through these the leaf traces pass outwards from their point of origin in the primary vasculature. In *Zamia* and *Stangeria* the tracheids in the secondary xylem have scalariform pits, whereas in all other genera they have circular bordered pits that are multiseriate and are present on the radial walls.

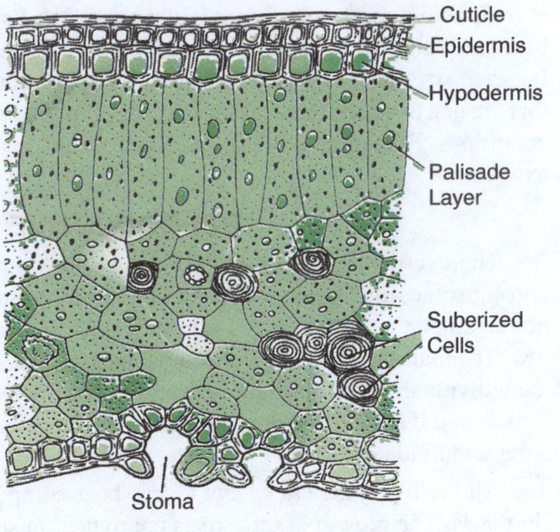

Fig. 8.6. V.S. leaf of *Dioon edule*.

The leaf epidermis (Figs. 8.6 and 8.7) is heavily cutinised and possesses sunken stomata that are **haplocheilic**. The vascular bundle of the midrib and of the leaf traces is diploxylic, *i.e.*, it possesses centripetal as well as centrifugal xylem, the former occurs in the form of a triangular area with a single protoxylem group. The centrifugal xylem is separated from it by a parenchymatous zone and consists of a semicircular area of tracheids associated with phloem and gives an idea that it is secondary in origin. Whether it is primary or secondary in origin is a debatable question.

Regarding their seedling anatomy the cycads show certain interesting variations :

1. In *Ceratozamia*, *Dioon edule*, *D. spinulosum* and *Zamia* the vascular cylinder in the cotyledonary node is a protostele. Later it becomes siphonostelic due to the appearance of pith. The cotyledonary traces are endarch first and later they become mesarch. In *Zamia* the trace is mesarch first and later changes into an exarch condition.
2. In *Microcycas* the stelar organisation in the seedling stage is an endarch siphonostele and later it changes to eustelic organisation. The cotyledonary trace is endarch first and later becomes mesarch or exarch.

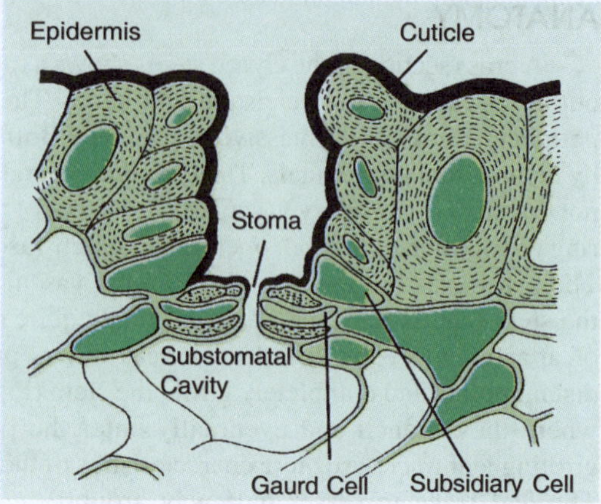

Fig. 8.7 *Dioon edule.* Lower epidermis showing detailed structure of a stoma.

The roots of cycads are very thick at their bases and are polyarch and possess up to 8 xylem bundles alternating with 8 phloem bundles. They become thinner towards the apex and the number of bundles goes on decreasing until a diarch condition is reached. The primary xylem is exarch. The older roots show accessory cambial rings that give rise to concentric cylinders of secondary wood. The root apex has common initial zone that gives rise to all tissues including the massive root cap.

Coralloid roots are present in almost all the living genera of cycads and contain an endophytic blue green alga belonging to the genus *Ansbaena*. It occurs in some cortical cells that are comparatively larger in size. These roots branch dichotomously and are negatively geotropic and possess root caps that are one to 4 cells thick. Such roots develop extensively when cycads are grown under mesophytic conditions. Branching of these roots is endogenous in all genera except in *Bowenia*, where they are exogenous.

REPRODUCTIVE CHARACTERS

The cycads are dioecious. Abraham and Mathew (1962) reported the occurrence of *X* and *Y* chromosomes in *Cycas pectinata*. They determine the sex of the plant. The female plants are homogametic and possess *XX*-chromosomes in addition to their normal number which is *twenty-two*. The males have *X* and *Y* chromosomes, that is, they are heteromorphic and determine the sex of the individual, *i.e.*, if during fertilization the gametes having *X* and *X* meet together the plants will be female and if female gametes meet a male gamete carrying *Y* chromosome the resulting plant will be a male individual.

All the living genera except Cycas bear compact female cones that terminate the female plant (Fig. 8.1). The growth of the axis is continued, in such plants (*Dioon*, *Zamia*, *Bowenia*, *etc.*), by the appearance of another bud near the base of the female cone. This bud grows and pushes the terminale female cone to one side. In *Cycas* the female plants possess spirally arranged megasporophylls that alternate with the cataphylls and the foliage leaves. The sequence of their appearance is : vegetative leaves → sporophylls → cataphylls → vegetative leaves and so on. Such an arrangement forms a loose megasporangiate zone consisting of leaf-like megasporophylls.This condition is regarded to be very primitive. In *Cycas revoluta* the megasporophylls have a highly pinnate sterile part and slender proximal part that bears sub-opposite pairs of ovules. There is a gradual reduction in the pinnate nature of the sterile part in the megasporophylls of other species until it is almost lost in *C. rumphii* and *C. circinalis.* In *Cycas* the female plants are monopodial as the apical geowth is continuous.

In other genera the megasporophylls are arranged compactly around a central axis to form a definite strobilus or the cone that is terminal (Fig. 8.1). Such genera are sympodial because the growth of the axis is continued by the appearance of another meristem at the base of the cone peduncle. Usually one female cone is borne at the apex, but sometimes two may appear due to the development of another meristem into a cone. The female cones in *Dioon spinulosum* may grow up to two feet tall when mature. In some cases they weigh up to 60 pounds. The female cones in *Macrozamia denisonii* are up to 70 centimetres long, whereas in *Zamia pygmea* they are only two centimetres long and are the smallest among the cycads. The megasporophylls in the cone bearing genera have a distal sterile laminate portion that is simple and a stalk-like portion that bears two or three ovules whose micropyles are directed towards the cone axis. The sporophylls are spirally arranged around the cone axis. In *Zamia* (Fig. 8.8 E), *Macrozamia* (Fig. 8.8 D) and *Ceratozamia* the sporophylls are peltate.

Plate II. Megasporophylls.
From left to right : *Cycas* (The big one), Others in 3 Rows—
Top : *Macrozamic, Dioon, Encephalartos*; Middle : *Stangeria*;
Bottom: *Lepidozamia, Zamia,* and *Ceratozamia*

Fig. 8.8. (A-E) Megasporophylls of cycads. A. *Cycas revoluta*; B. *C. circinalis*; C. *Dioon edule*; D. *Macrozamia*; E. *Zamia*.

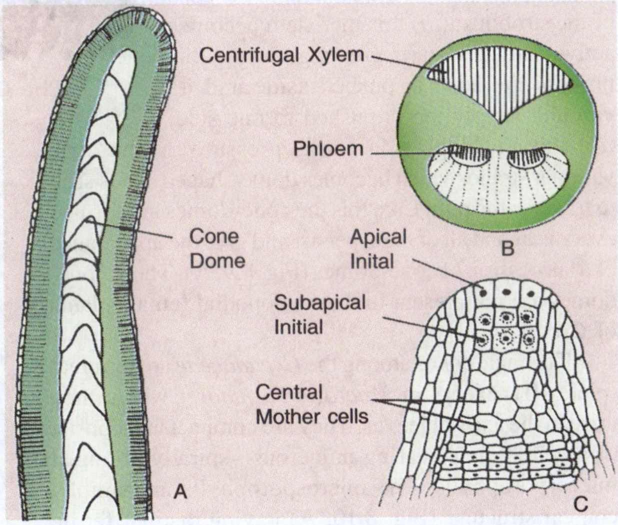

Fig. 8.9. (A-C) Cyadaceae. A. cone domes in Cycads; B. outline T.S. of vascular bundle of petiole of cycas to show centripetal and centrifugal xylems; C.V.S. apex of *Cycas* stem.

The *Cycas* megasporophylls is regarded to be the most primitive due to its leaf-like nature (Fig. 8.8 A, B). It is believed that during the course of evolution among the cycads its laminate distal part underwent gradual reduction accompanied by reduction in the number and size of the ovule. In the strobiliferous cycads the megasporophylls have 2 or 3 ovules and the distal part though laminate does not appear leaf-like and is never pectinate. This is the intermediate stage. The final stage, which is regarded as advanced, are the peltate sporophylls of *Zamia* (Fig. 8.8 E) and *Macrozamia* (Fig. 8.8 D) which bear two ovules. The ovules are foliar in origin and the cycads are **phyllospermous**. Another viewpoint is that peltate sporophylls with two ovules are primitive and the morphological series has progressed in an opposite direction leading to open female organs of *Cycas*. According to this view the peltate sporophylls are regarded as sporangiophores *instead* of sporophylls and the ovules are cauline in origin. The cycads according to this view are **stachyospermous**. The formation of **cone domes** (Fig. 8.9 A) is characteristic in the vascular supply of strobiliferous female cycadaceae. These cone-domes consist of convex or dome-shaped vascular strands that cross the pith at intervals. These can be seen in a longitudinal section of the stem below the insertion of the strobilus. The presence of these domes is an anatomical proof of sympodial nature of the stem. Their formation can be explained thus : (*i*) the cone is terminal and received its usual vascular supply and apical growth of the stem has ceased because its apex is used up in the formation of the strobilus; (*ii*) the stem grows further by the formation of a new meristem or bud that appears a little lateral to the base of the strobilus; (*iii*) this meristem becomes active and stem starts growing; during this process the cone is pushed aside and its vascular supply too is pushed to one side; (*iv*) this pushing aside exerts pressure on the cone vascular strand, which becomes dome-shaped and is seen to traverse the pith. Like this the cone-domes are formed every year and all of them persist and give the appearance of a succession of cone-domes (Fig. 8.9, A). These cone-domes are not present in the monopodial female plants of *Cycas*.

Plate III : A young male cone of *Cycas circinalis*.

The male cones among the *Cycadaceae* are terminal (plate III), except in *Macrozamia moorei* where they arise in the axils of leaves. They are compact and consist of a central axis bearing numerous, spirally arranged microsporophylls. The microsporophylls are simple, conical structure (Fig. 8.10. A) having distinct fertile and sterile portions. The sterile distal end is usually produced into single spinous projection in *Cycas*, *Dioon* and *Macrozamia* (Fig. 8.10, A, B, C). In *Ceratozamia mexicana* (Fig. 8.10, C) there are two such projections. There is no such sterile projection in *Zamia* (Fig. 8.10, D). The microporangia occur on the lower surface (abaxial) of the sporophylls and may either completely cover it or may occur in two zones. Their number may be over a thousand in *Cycas* and *Dioon* (Fig. 8.10 A, B) and slightly lesser in *Macrozamia* and *Ceratozamia* (Fig. 8.10, C, E). In *Zamia* there may be only 5 or 6

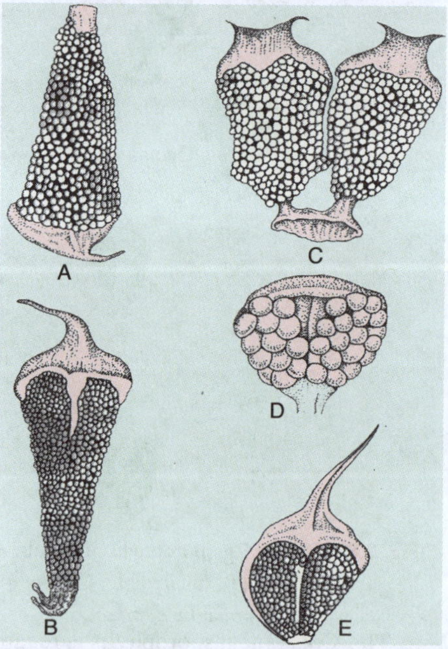

Fig. 8.10. (A-E). Microsporophylls of Cycadaceae; A. *Cycas circinalis*; B. *Dioon spinulosum*; C. *Ceratozamia mexicana*; D. *Zamia*; E. *Macrozamia*

microsporangia per microsporophyll in *Z. furfuracea*. In other species of *Zamia* (Fig. 8.10, D) the number is more (up to 50). The microsporangia appear to be arranged in a continuous mass but on closer examination they are found to be arranged in distinct groups of two or more. These groups are called **sori**. The sporangia of a given sorus have a common vascular supply. The sporangia dehisce by means of a longitudinal slit. Bierhorst (1971) observed bisporangiate synangia in *C. revoluta* and *C. circinalis*. They are fused externally and are unilocular.

The microsporangia have usually the same size, form and structure and are eusporangitiate in development (Fig. 8.11, A—E) and structure. A microsporangium has a massive wall made up of several layers of cells (Fig. 8.11 E—G) and most cells of the outer layer are involved in the dehiscence mechanism. The cells in the outer wall become thick (Fig. 8.11, F). A hypodermal cell on the abaxial side of the developing microsporophyll acts as the microsporongial initial (Fig. 8.11, A). It divides by a periclinal wall into an outer **primary wall cell** and an inner **primary sporogenous cell** (Fig. 8.11, B). The outer primary wall cell divides by periclinal and anticlinal walls to form a several layer thick sporangial wall (Fig. 8.11, C, D). The surrounding microsporophyll tissue may also take part in the formation of wall. The primary sprogenous cells divide repeatedly to form sporogenous tissue, the outer cells of which form a single layered tapetum. The rest round off and act as microspore mother cells which have 22 chromosomes in some species of *Cycas* (Shetty and Subramanyam, 1967; Rao, 1961; Abraham and Mathews, 1962). The tapetal cells disorganise to form the plasmodial fluid which nourishes the mother cells which grow and ultimately undergo meiosis to form four microspores of pollen grains each. The **microspores** are arranged in **tetrahedral** tetrads. The haploid chromosome number is 8 in *Stangeria* and *Ceratozamia*, 11 in *Cycas*, and 13 in *Microcycas* and 9 in others.

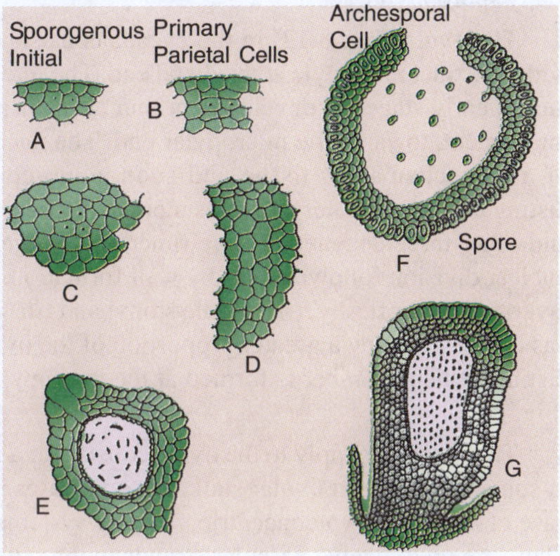

Fig. 8.11. (A—G). Stages in the development of microsporangium in *Stangeria paradoxa*. (For explanation see text) After Lang.

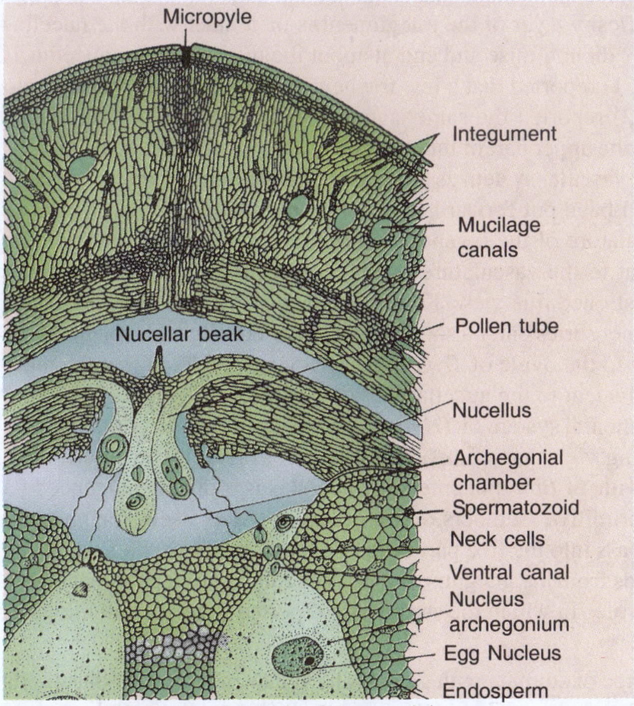

Fig. 8.12. V.S. Ovule of *Dioon edule*.

The vasculature of the central axis of a cycadaceous strobilus consists of discrete vascular bundles, which give off mesarch traces to the sporophylls. These traces may be single or double at the points of their origin from the vascular bundles of the central axis. Both these conditions may be found in a single strobilus. In the microsporangiate cones the single trace entering the microsporophyll divides into two. In the female plants of some genera this is the case with respect to the megasporophylls but in *Ceratozamia* the traces to megasporophylls divide further many times before entering the sporophyll and consist of many sporophyll traces.

The ovule (Fig. 8.12) in the cycads consists of a nucellus surrounded by a single integument with a distinct micropyle at the distal end. The integument is free for its major part from the nucellus during early stages in development, but at maturity only one-third of the nucellus is free from the integument, towards the micropylar end. The nucellus during early stages of development consists of a parenchymatous tissue and soon a sub-epidermal or deep seated megaspore mother cell distinguishes in it. Later it divides meiotically into 4 megaspores out of which one remains functional and other three degenerate. The functional megaspore grows into a female gametophyte by free nuclear division followed later by wall formation. The ovules are sessile. The integument has three layers : (*i*) outer fleshy; (*ii*) middle stony; and (*iii*) inner fleshy. At this stage the female gametophyte has come to occupy a greater proportion of the ovule and the nucellus is now a thin papery layer. A distinct pollen chamber is formed at the micropylar end, by the disintegration of the nucellar apex (Fig. 8.12).

The vascular supply to the ovule consists of a vascular strand that enters the base of the ovule or in some cases a short ovular stalk. In some cases it may fork befoer entering the ovule base. At the base of the ovule two concentric vascular systems are established : (*i*) the outer; and (*ii*) the inner. There are up to twelve vascular strands in the outer system. They originate by the branching of the main strand at the base of the ovule (chalazal end). These twelve strands remain unbranched during their longitudinal course upwards through the outer fleshy layer of the integument. They traverse right up to the micropylar end. The vascular strands of the inner system traverse upwards through that part of the ovule where the inner fleshy layer of the integument is in contact with the nucellus and the two are indistinguishable. They dichotomise and end at about the line where the nucellus is free from the integument. Shapiro (1951) reported that a few tracheids may extend beyond this into the nucellus in *Zamia*. Kershaw (1912) reported the same state in *Bowenia*. Chamberlain (1935) reported that in *Dioon* they extend into the upper part of inner fleshy layer of the integument. Various authors have tried to interpret the inner vascular system as either nucellar is nature or integumentary. Stopes (1904,1905) and Mattee (1904) have put forward a view that the inner vascular system of cycad ovule is equivalent to the vasculature of the lygenopterid ovule (*Lagenostoma lomaxi*) and the outer vascular system is equivalent to the vasculature of the cupule in *Lagenostoma*. Seward (1917) and Kershaw (1912) have questioned this view. Kershaw did it on the grounds that in the ovule of *Bowenia* the bundles of the inner series enter a short way into the nucellus. He regarded the cycad ovule to be closely comparable to the ovule of *Trigonocarpus*. He considered that the inner vascular system of cycad ovule is equivalent to the nucellar vascular system of *Trigonocarpus* and the outer vascular system to the integumental system of *Trigonocarpus ovule*. In terms of the telome concept these two apparently conflicting views are comparable and are, in fact, identical. The fact that the inner vascular system of the ovule of *Bowenia* enters the nucellus is probably evidence of a reduced condition since in the more primitive members of the order, in *Cycas* itself, the vascular strands are entirely integumental and pass into the free part of the integument above the level of the nucellus. The passage of vascular strands from the integument to the nucellus is perhaps comaparable to the situation in those angiosperm ovules in which the strands are reported to pass from the outer to the inner integument (Eames, 1961).

Shapiro (1951) reported the presence of stomata with guard cells and even with subsidiary cells on the outer surface of the nucellus in the ovules of *Zamia floridana* and *Z. integrifolia.* Substomatal chambers are not present. Guard cell like pairs have also been reported on the nucellar surface of ovules of *Encephalartos*, *Ceratozamia* and *Cycas* (Bierhorst, 1971). It is very interesting observation as it is quite surprising that what role the stomata had to perform in such an enclosed environment. These appear to be the vestiges of such structures that were present in the ancient ovules where nucellar surface was exposed as in *Genosperma*.

The young nucellus appears as a mound of parenchymatous cells representing the apex of the ovule. The integument arises sub-apically as a collar of tissue that grows up and eventually encloses the nucellus but leaving a small pore called the micropyle. The young nucellus has a rounded apex. Two to four cells below the apex and within the nucellar tissue, a megaspore mother cell is differentiated. As the development proceeds a linear tetrad of four haploid megaspores is produced as a result of meiosis (Fig. 8.13, C). Out of these, three cells lying towards the micropylar end of the nucellus disappear, and the fourth proximal cell enlarges and acts as the functional megaspore (Fig. 8.13, A, C). The megaspore enlarges (Fig. 8.13, D, E) and its nucleus and subsequent derivative nuclei divide until a sac of possibly 1,000 free nuclei is formed (Fig 8.13, F). Wall formation begins at the periphery (Fig. 8.13, G) and eventually the mass is entirely cellular (Fig. 8.13, G). This is theemale gametophyte. Wall formation is centripetal. As the female gametophyte develops within, the nucellus extends apically forming

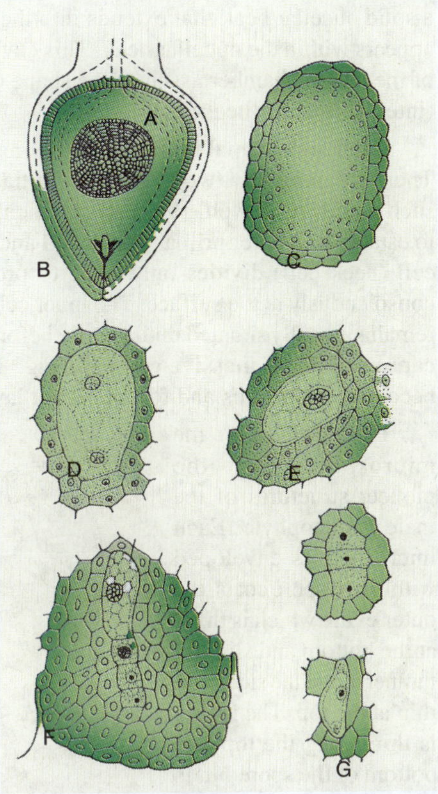

Fig. 8.13. (A-G) Megasporogenesis and development of female gametophyte in *Dioon edule* (D—G, after Chamberlain) and *Stangeria paradoxa* (A—C). For explanation see text.

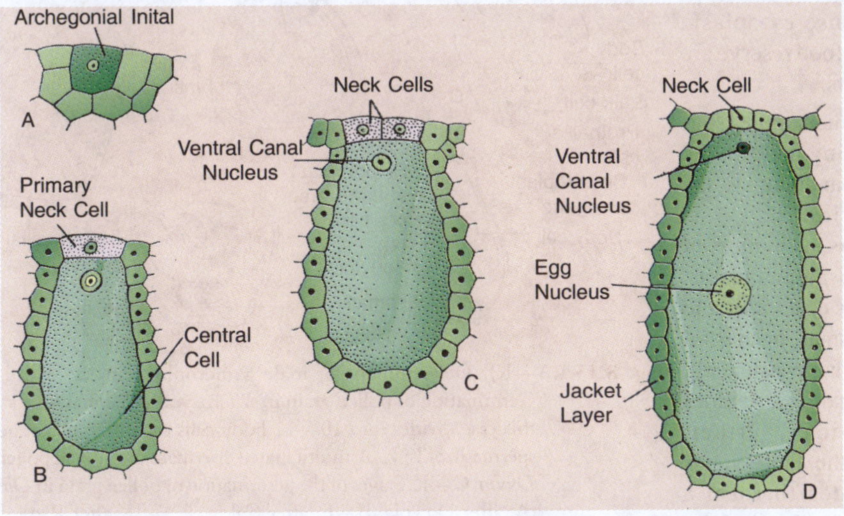

Fig. 8.14. Stages in the development of archegonium in the cycads.

a solid nucellar beak that extends into the micropyle. At a later stage a lysigenously formed cavity appears within the nucellar beak. This cavity gradually widens into a pollen chamber. Later the base of the pollen chamber assumes the shape of an inverted funnel with a slight projection in the centre (internal nucellar beak).

Later archegonial initials make their appearance at the micropylar end of the female prothallus. There are unusually two to ten such initials in the cycads, but in *Microcycas* there appear scores of such initials (Caldwell, 1907) all of which develop into archegonia. Each initial divides transversely to establish an outer primary **neck cell** and an **inner cell** (Fig. 8.14, B) or the **central cell.** The outer cell (neck cell) divides only once to produce two neck cells (Fig. 8.14, C), which later bulge conspicuously at the surface. The inner cell or the ventral cell enlarges occasionally and its nucleous remains apically situated until shortly before fertilisation, when it divides to produce the small **ventral canal cell** and a **massive egg cell** (Fig. 8.14 D). Cells of the nucellus surrounding the inner cell become conspicuous and form a jacket layer.

Pollen grains or the microspores are the pioneer structures of the male gametophyte. Each microspore is enveloped within two spore coats, the outer **exine** which is thick at the bottom and slightly thinner along the sides and thin at the top. The **intine** is thin along the top and bottom of the spore but is thicker along the sides of the microspore. The microspore has, therefore, a distinct upper and lower end. A big centrally located nucleus (Fig. 8.15.) dense cytoplasm and some food reserve are its contents.

The microspores start germinating within the microsporangium, *i.e.*, the germination is **precocious**, as in most of the heterosporous plants. Its nucleus divides first to cut off a small **pothallial cell** and a large sister cell. The prothallial cell does not divide further, whereas the large sister cell divides to form a **generative cell** and a **tube**

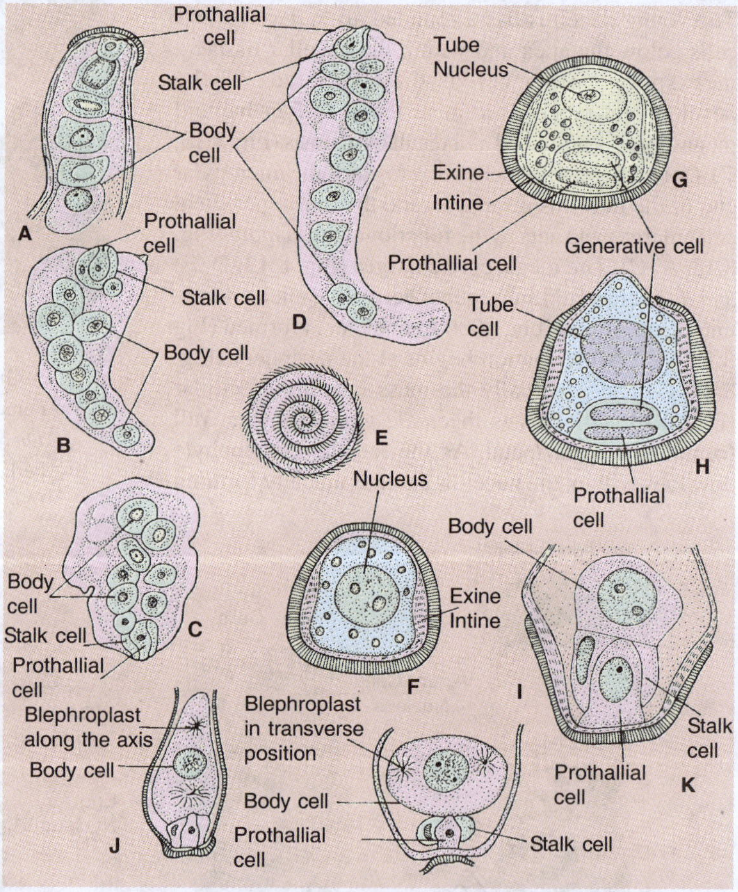

Fig. 8.15. (A—K). Development of male gametophyte in the cycads. A—D. Germination of pollen grain in *Microcycas calocoma*. The stalk cell in this case divides into 10—11 body cells each of which produces two spermatozoids; E. A multiciliated spermatozoid ; F. A pollen grain of *Dioon*; G—K. Stages of the germination of pollen grain in *Dioon edule*; (A, after Downie, B—E, after Caldwell, F—K after Rao).

nucleus (Fig. 8.15, G-H) which remains free within the gametophyte. At this three-celled stage the semi-germinated microspore is released from the microsporangium. After shedding most of its pollen the male cone increases to double its original length. Pollination is affected by wind. At the time of pollination the megasporophylls are loose enough so that the pollen grains can easily shift down to the micropyles of the ovules and get caught in the **pollination drop**. The drop dries and is pulled into the pollen chamber along with the entangled microspores. Sometimes fungal spores also get pulled in and they germinate there and result in complete destruction of the ovules. Kraus (1896), reported entomophily in *Macrozamia* and later Pearson (1906). Rattaray (1913) and Marloth (1914) reported it in *Encephalartos*. Tang (1987) reported insect (beetles) pollination in *Zamia pumila* while similar role of Beetle is also discussed in *Zamia furfuracea* (Norstog *et al*, 1986). Amphiphilous pollination is reported in *Bowenia serrulata*, where weevil *Tranes subopaca* transfer pollen to female cone (Wilson 1993). The semi-germinated, 3-celled microspores germinate now on the nucellus. Each pollen grain gives out a pollen tube that penetrates through the nucellar tissue along the sides of the pollen chamber. The pollen tube usually branches in the nucellar tissue. The other end of the gametophyte enlarges and remains protruding into the pollen chamber. The tube nucleus usually moves into the pollen tube except in a species of *Zamia* where it remains in the main body. Eighteen days after pollination the generative cell divides to form the **stalk cell** and the **body cell** (Fig. 8.15, I). The stalk cell is smaller and is in contact with the prothallus cell. Prothallus cell in the cycads is persistent. The body cell divides to form two spermatozoids that are multiflagellate. The stalk cell takes no further part, except in *Microcycas* (Fig. 8.15, A-D) where it divides repeatedly to produce 10 to 11 **spermatogenous cells** or **body cells** each of which produces two spermatozoids. The sperms are massive and perhaps the largest in the plant kingdom measuring 200 by 275μ in *Dioon*. Each **spermatozoid** bears ciliated band that originates from the **centriole** and bears hundreds of flagella (Fig. 8.15, E).

Detailed structure of spermatozoid

Norstog (1967, 1977, 1974, 1990) made significant contribution in the study of detailed structure of spermatozoids/sperms in cycads. According to the literatures available *Chigua* (earlier *Zamia chigua*) has largest spermatozoid of about 400μm in size. The spermatozoids in cycads are without cell wall but contains a prominent nucleus. The anterior half of the body is covered with six flagellated spiral bands except the first turn of band which is without flagella (Fig. 8.16). The length of flagella on these bands are of about 30 μm in length. These flagellar bands are formed by depressions in plasmamembrane and cytoplasm. The internal structure of flagella has typical 9 + 2 arrangement in the middle portion of it. Tip portion & basal body does not follow such arrangement.

The spiral band on the body of spermatozoids consists of 3-distinct regions. The outermost electron dense layer is followed by the intermediate granular layer. The most complex is the inner most layer called Vierergruppe. The Vierergruppe of *Zamia* is similar to the spermatozoids of *Marchantia* (Carothers and Kreitner, 1967), *Pteridium* (Tourte and Hurel-Py, 1967, Duckett and Bell, 1971), and *Equisetum* (Duckett and Bell, 1969 ; Duckett, 1973).

It is also similar to the Driergruppe of spermatozoids of *Polytrichum* (Heitz, 1960, Paolillo *et al*, 1968). The extensive system of microtubules associated with *Vierergruppe in Zamia* is also similar to the spline of *Marchantia's* spermatozoids, though number of microtubles are extremely high in case of *Zamia* (60,000) as compared to *Marchantia* (only 17).

The cytoplasm of spermatozoids are composed of granular endoplasmic reticulum and mitochondria which are dispersed in the cytoplasm. The anterior tip of the spermatozoid is occupied by an electron dense cap. The flagella arise on the body of the spermatozoid through a shallow channel formed by the depression of cytoplasm and plasma membrane. Their basal bodies are embedded in the dense layer. The first formed spermatozoids are better organised and more active whereas last formed spermatozoids are slow and less organized. Thus spermatozoids exhibit a developmental gradient in mobility and complexity. These spermatozoids are of considerable importance because of their phylogenetic implications and morphology.

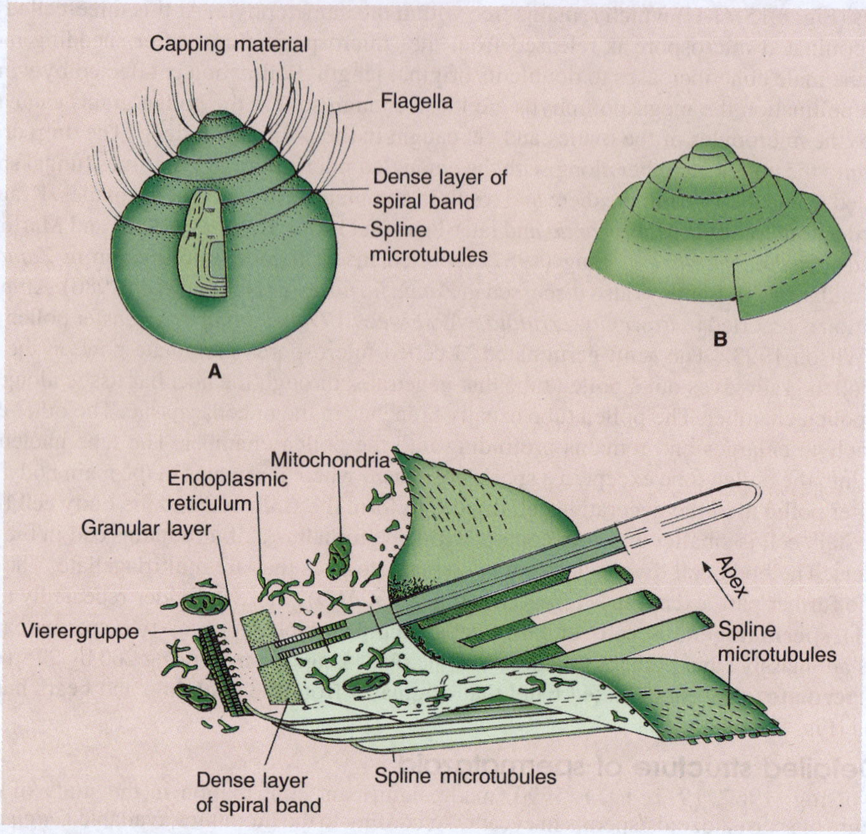

Fig. 8.16. (A-C) Detailed structure of the spermatozoid of *Zamia* A. Reconstruction of sperm with prominent ciliate bands (dense layer); B. Diagrammatic sketch showing spiral nature of the ciliate band. C. Diagrammatic representation of the flagellar apparatus.

As the above changes are taking place the tissue of the nucellus that separates the pollen chamber from the female gametophyte breaks down and disappears (Fig. 8.12). The main body of the microspore containing the spermatozoids, stalk cell and the prothallial cell hangs down in the cavity thus created (Fig. 8.12). The spermatozoids break the wall of the body cell and ultimately the pollen grain wall and are discharged into the tip of the female gametophyte. This process is reported to take place in $1\frac{1}{2}$ **hours.** Bierhorst (1971) noted that in *Zamia* it lasts for twenty to thirty minutes. As the sperms wriggle out of the pollen tube and fall on the female gametophyte a drop of liquid with an osmotic pressure equivalent to that of 30% sucrose solution is also spilled along with them. As this liquid touches the archegonial neck, it lowers its turgidity and results in the opening of the necks. The sperm is then violently pulled into the archegonium. In doing so it leaves its ciliated band and surface membrane behind. The ventral canal nucleus usually degenetrates immediately before fertilisation so that the naked male nucleus fuses with the oosphere to form the diploid **oospore**. Lawson (1926) reported that archegonia in *Bowenia serrulata* secrete drops of fluid into the archegonial chamber (concavity at the micropylar end of the female gametophyte). Fluid is also discharged by the pollen tubes. In *Encephalartos villosus* the ventral canal nucleus does not degenerate but starts enlarging and moves towards the oospore or the egg. Sedgwick (1924) was of the opinion that it fertilised the egg just as is the case in some conifers (*Picea*) and *Ginkgo*.

Embryogeny (Figs. 8.18) in cycads follows an almost similar pattern and the zygote nucleus enlarges in the archegonial venter and undergoes repeated free nuclear division without wall formation.

As many as 1,000 free nuclei develop in *Dioon* and about 64 in *Bowenia*. Divisions up to the formation of 256 free nuclei are regular and after that irregularities set in and counting is difficult. Chamberlain was of the opinion that large size of the oospore or the zygote stood in way of wall formation because the large mass of protoplasm hindered cytokinesis. Such an interpretation is not justified because large eggs and large cells elsewhere among seed plants do not show free nuclear divisions. Wall formation starts at the base and proceeds towards the neck of the archegonium. In *Cycas*, *Macrozamia* and *Encephalartos* the whole of the embryo becomes cellular, whereas in other genera (*Stangeria*, *Zamia*, *Bowenial* etc.) only the basal region becomes cellular. Cells in the upper part of the cellular tissue elongate and form the suspensor. The basal cellular mass is pushed by the enlarging suspensor cells into gametophytic tissue. The basal cells then develop further into the various parts of the young embryo. Two pronounced tissue bulges representing the two cotyledons become evident at a later stage (Fig. 8.18, H). The root apex becomes organised at the opposite end toward the archegonial neck. The tissue between the root apex and

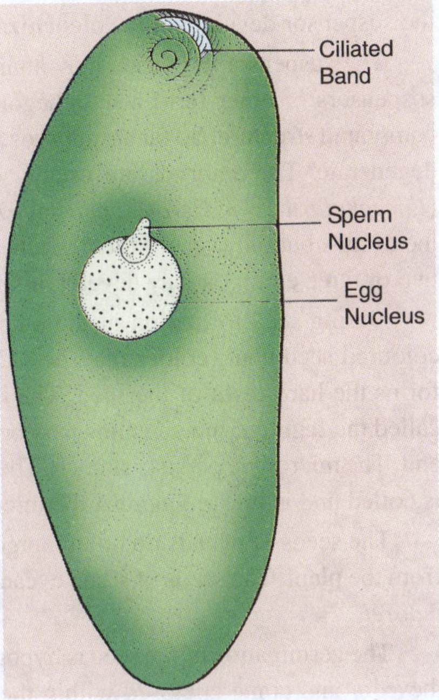

Fig. 8.17 Fertilization in *Stangeria paradoxa*.

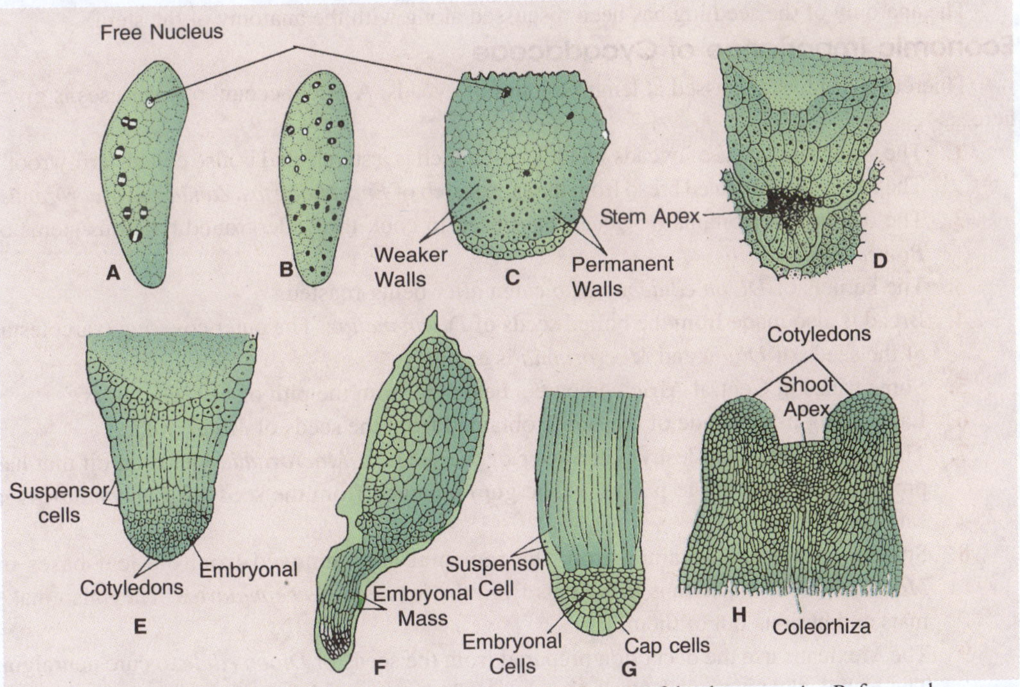

Fig. 8.18. (A—B). Embryogeny of *Zamia umbrosa* illustrating early stags of development. A—B. free nuclear stages; C. segmentation has started ; D. a later stage showing differentiation into cap, stem apex, suspensor and the buffer region. (After Bryan, 1952); (E-H) Embryogeny of the Cycads; E. *Zamia umbrosa*, a completely developed proembryo; F. *Macrozamia Spirais*. Proembryo with suspensor; G. *Bowenia serrulata*. developing embryo with suspensor; *Dioon edule*. L.S. embryo showing two developing cotyledons. (E.G after Lawson; H after chamberlam

the suspensor develops into **coleorhiza**. In *Ceratozamia* there is only one cotyledon.

The suspensor in cycads may attain a length of up to eight centimeters. It remains coiled. The suspensors of other fertilized archegonia may come together and coil about each other to form a compound structure. So far as embryos are concerned only one embryo reaches maturity, while others degenerate. The embryos are usually dicotyledonous, but the number may vary from one to six (*Encephalartos* = 3; *Ceratozamia* = 1). As several archegonia are fertilised there is **polyembryony** in the cycads but only one of them reaches maturity. Baird (1939) reported that in *Macrozamia reidlei* two or three embryos may develop in the same archegonium.

During seed formation of the outer fleshy layer of integument becomes the outer beautifully coloured seed coat (colour may be red, orange or various combinations of both); the stony layer forms the hard **testa** of the seed. The inner fleshy layer may persist in the form of a papery layer called the **tegmen**; the nucellus may be wholly absent or may form a nucellar cap at the micropylar end. The micropyle persists as such. The embryo extends the whole length of the seed. The suspensor is coiled and is pressed against the micropylar end.

The seeds of cycads do not undergo any resisting period and germinate immediately on falling from the plant. The seeds of some cycads may retain their vitality for a year or two *e.g.*, *Macrozamia moorei*.

The germination of the seed is hypogeal. The cotyledons mainly act as food absorbers. Ultimately they dry and come enclosed within the testa of the seed that remains attached to the seedling for about a year or even more. Only one leaf appears first.

The anatomy of the seedling has been discussed along with the anatomy of the stem.

Economic Importance of Cycadaceae

Thieret (1958) has discussed at length the use of cycads. A brief account of their uses is given here.

1. The seeds and stems of cycads yield 'sago' which is a starch and is also called "arrowroot". The Hottentots obtained bread from the sago starch of *Encephalartos*, *Zamia* and *Macrozamia*.
2. The original inhabitants of Queensland used to cook the underground tuberous stems of *Bowenia spectabilis*.
3. The kernels of *Dioon edule* are also eaten after being roasted.
4. Bread is also made from the boiled seeds of *Dioon mejiae*. The outer covering (sarcotesta) of the seeds of *Dioon* and *Macrozamia* is eaten.
5. Some tribes in Central Africa prepare a beverage from the pith of *Encephalartos*.
6. Laundrying use is made of the starch obtained from the seeds of *Macrozamia*.
7. The orange coloured fleshy outer layer of the seeds of *Macrozamia* yields an oil that has properties similar to the palm oil. The gum obtained from the seeds of *Encephalartos* is eaten by the natives.
8. Stuffing fibres are obtained from the ramental hair removed from the leaf bases of *Macrozamia*. The Africans make hats from the leaves of *Encephalartos*. They also make mats and baskets out of them.
9. The Mexicans use the decoction prepared from the seeds of *Dioon edule* to cure neuralgia. The leaves of *Macrozamia* are poisonous and may cause rickets if eaten continuously by animals.

Phylogeny

The consensus of opinion favours the origin of *Cycadales* from some members of the *Pteridospermales*. Both these orders resembles each other in many respects. The points of resemblance can be enumerated as below :

1. Presence of centripetal xylem in the foliar strands or even in the strands entering the strobilus stalks of cycads are similar to those present in the vegetative organs of the pteridosperms.
2. The ovules of cycads resemble those of some pteridosperms, *e.g.*, *Trigonocarpus* because they possess two systems of vascular strands. Stopes (1905) envisaged that if the two integuments of Medullosan ovules fuse, they would resemble those of cycads.
3. The megasporophylls of *Cycas* are leaf-like structures and remind us of the leaf borne ovules of pteridosperms. In *Calamatotheca* the leaves are sterile at their tips and bear ovules on their proximal portions.
4. The megasporophylls in *Cycas* are spirally arranged and do not form a compact cone and resemble the pteridosperms in this respect.
5. The spermatozoids are motile and multiciliate in both the groups.
6. The wood is manoxylic in both the orders.
7. Presence of mucilage canals in the pith and the cortex of the stem in both the orders.
8. Presence of extensive cortex in the *Cycadales* and in some Medullosan pteridosperms like *Medullosa* and *Sutcliffia*.

Based on these similarities some authors suggested the origin of *Cycadales* from the pteridosperms. These views are given below.

Worsdell (1896, 1898, 1900 & 1906) suggested the origin of *Cycadales* from Medullosaceae. His view is based on the detailed anatomical studies of the cycads. He beleived that the cycads descended from the *Medullosaceae* whose stems possessed a medullated vascular cylinder, *e.g.*, *M. stellata.* This was achieved by a progressive loss of centripetal parts of stelar rings and complete loss of central "star rings." The tertiary arcs of inverted vascular bundles in cycads like *Bowenia* and *Macrozamia* are vestiges of such centripetal arcs found in the *Medullosaceae*. Presence of two or more smaller steles around the large central stele in *Cycas siamense* were regarded by Worsdell to stimulate such an arrangement in *Medullosa porosa*. Successive rings of secondary wood found in most of the cycads are also regarded as a proof of Medullosan ancestry. Stewart and Delevoryas (1965) have also supported this view.

Scott (1909, 1923) believed that the *cycad* eustele originated from the monostelic *Lyginopteridaceae*. Arnold (1953) has supported this view.

Beck (1970) and Namboodri and Beck (1968) also traced the origin of gymnospermous eustele from the protostelic ancestors by the longitudinal splitting of the protostele (see Chapter I for details). This ancestor is believed to be a member of **Progymnosperms**.

De Fraine (1912) suggested that forms like *Sutcliffia* gave rise to *Lygenopteridaceae* and *Cycadales* on one hand and polystelic forms like *Medullosaceae* along another line of evolution. She suggested that *Sutcliffia* resembles both monostelic and polystelic pteridosperms. This stem has a large central protostele surrounded by many small steles. In possessing the central protostele it resembles monostelic forms like *Heterangium* and the presence of subsidiary steles brings it closer to the polystelic *Medullosaceae*. The monostelic cycads originated from forms like *Sutcliffia* by the medullation of the central protostele and modification of the smaller peripheral stele into cortical accessory bundles found in many cycads. This theory was also supported by Bancroft (1914).

Taylor (1969) takes into account the manoxylic nature of the wood, presence of mucilage canals, persistent cortex, similarity in the structure or the ovule and presence of centripetal xylem in both the cycads and the pteridosperms, as clues to the origin of *Cycadales* from some Palaeozoic pteridospermous group of plants.

Mamay (1969) described lower Permian megasporophylls that were convincingly cycadalean. Each such organ bore two rows of ovules on its proximal end and a laminate structure distally. They were more leaf-like and led Mamay to propose a Palaeozoic origin for the cycads.

These views have been challenged by some authors becaused of some striking difference in the male organs and vegetative organs of the two groups. Pant and Mehra (1962) have suggested, "the absence of intermediate types of male fructifications between the two groups is another, difficulty, but this presents no serious problem since the microsporophyll of the cycads is admittedly homologous with their megasporophyll. Moreover the sporophyll traces in cycads are all direct and if, as is usual, the reproductive parts are to be regarded as more primitive, the girdle traces of the vegetative leaves may be taken to be a later achievement."

Mamay's (1969) discovery of *Cycas* like megasporophyll from the Palaeozoic has, however, lent much support to the view that the cycads have a Palaeozoic ancestry, but whether the ancestors were really pteridosperms of some other more primitive group of plants *e.g.*, the progymnosperms cannot be definitely stated.

Cycads also resemble othe plant groups (both fossil and living) and it appears to the author that their resembles with pteridosperms have been over-emphasised. They resemble the *Bennettitales* in the following respects :

1. Their similar tree-like habit and conical, stumpy and barrel-shaped stems.
2. The leaves are pinnate compound. There is striking resembles between the fronds of *Bowenia* and *Encephalartos* with those of the genus *Cycadeoidea*.
3. The secondary wood in both is **manoxylic**.
4. They have extensive cortex and large pith.
5. Presence of two cotyledons in the embryo.
6. Presence of terminal strobili in *Williamsoinella.* In *Williamsoniella* and *Wielandiella,* strobili are regarded as terminal by some modern authors (Bierhorst, 1971)
7. Presence of unisexual fructifications in some *Bennettitales e.g.*, *Williamsonia.* Like cycads the plants of *Williamsonia* are dioecious.
8. Presence of bisporangiate synangia in some species of *Cycas*. Bierhorst (1971) noticed external fusion of sporangia in the sporophylls of *Cycas circinalis*. This feature reminds us of the *Bennettitales*.

There are, however many differences, between the two groups : (*i*) presence of flower-like reproductive organs, (*ii*) presence of bisporangiate reproductive structures, (*iii*) Syndetocheilic stomata in *Bennettitales* and (*iv*) absence of girdling leaf traces.

The Cycadales resembles the *Pentoxylales* in the following respects :

1. Their tree-like habit.
2. The reproductive organs are unisexual.
3. The tracheids have crowded, bordered pits on their radial walls.
4. The *Pentoxylales* bore both syndetocheilic and haplocheilic stomata.
5. The leaf traces possess both centripetal ans centrifugal xylem.
6. The microspores are monolete.
7. The microsporangia occur singly *i.e.*, they are not fused to form synangia.

There are, however many differences : (*i*) *Pentoxylales* have pycnoxylic wood, (*ii*) their polystelic stems, (*iii*) organisation of the female strobilus, (*iv*) organisation of the male strobilus, (*v*) the leaves are simple, and (*vi*) the ovules have integuments free from the nucellus.

The *Cycadales* resemble the *Cordaitales* in the follwoing respects :

1. Their tree-like habit.
2. The stems possess large pith and endarch primary vascular bundles.
3. The leaves possess a lot of sclerenchyma.
4. The spermatozoids are motile.
5. Structure of the ovules and seeds.

The resemblances of cycads with these fossil groups, however, lead us to one conclusion that they are "living fossils."

The cycads also resemble the *Ginkgolales* in the following respects :

1. The primitive organisation of the apical meristem of the stem.
2. The female gametophytes are extensively developed and have archegonia with large venters and large eggs.
3. The spermatozoids are multiciliate.
4. The pollen tube serves only the haustorial function of absorbing food.
5. The embryos have two cotyledons and there is abundant free nuclear division in the proembryo.

The Cycads also resemble some dicotyledonous angiosperms (*Liriodendron*).

ECONOMIC IMPORTANCE OF CYCADS

The economic importance of this primitive group of plants will be discussed under the following headings :

1. Cycads as Food plants
2. Cycads as Gum yielders
3. Cycads as Medicinal plants
4. Cycads as Oil yielding plants
5. Cycas Leaf Industry
6. Cycads as Poisonous plants
7. Miscellaneous uses of Cycads.

1. Cycads as Food plants : The stems and the seeds of various species of Cycads yield a starch commonly called Sago. The term arrowroot is also used for this starch. There two terms are confusing and are loosely used. Starches obtained from the stem and seeds of Cycas are called **Sago** by some authors whereas starches obtained from these two organs from *Zamia* and *Macrozamia* are called arrowroot. The terms are loosely applied and there is no rigidity about it.

In actual practice the term 'Sago' should be used for starches obtained from the stems and seeds of Cycads. This is, because 'arrowroot' is a starch commonly obtained from the tuberous rhizomes of *Maranta arundinacea* (an angiosperm belonging to family Marantaceae). East India 'arrowroot' is obtained from the tuberous rhizomes of *Curcuma* of *angustifolia*. Talema 'arrowroot' is obtained from the rhizomes of *Canna edulis*. Brazilian 'arrowroot' is obtained from the swollen roots of *Manihot esculenta*. British 'arrowroot' is the potato starch.

Similarly, the term 'Sago' is commonly used for the starch obtained from the stems of some palms such as *Metroxylon*, *Caryota* and *Corypha*. Starch obtained from the stems of Cycads is also called 'Sago'. So in this text the term 'Sago' is used for starch obtained from the stems of cycads whereas the term "seed starch" is used for the starch obtained from the seeds of cycads.

The aborigines in Queensland cook the subterranean stems of *Bowenia spectabilis* as vegetables. The young leaves of *Cycas circinalis, C. pectinata, C. rumphii* and *C. siamensis* are also cooked as vegetables by Malayans, Filipinos, Assamese and Indonesians, Kinch (1883) mentions the use of young leaves of C. revoluta as food in Japan. Extensive or prolonged use of young leaves or *C. rumphii* is known to cause rheumatism in man (Ochse, 1925).

The endosperm or the kernel of the seeds of *C. circinalis* and *C. rumphii* contain 20 to 30% of starch. It is made into flour or starch after various treatments to remove poisonous contents. In the Andamans the cycas seeds are eaten after cooking. In New Caledonia, Aru and Key Islands the seeds are raosted before use. In Fiji the seeds of *C. circinalis* are boiled before use. In Ceylon the cycad seed kernels are cut into slices, dried and ground into a meal. A dough made from this meal is then baked into cakes. In Jawa and Prince's Islands, the seed kernels are soaked in frequent changes of water over several days before they are made into cakes. They may also be made into cakes or served as extender with rice. Cakes prepared from seed kernels were also used by aborigines of Guam. Seed starch obtained from seeds of *C. media* was used as food by aborigines of Australia. In India also starch is prepared from ripe or even unripe seeds of *Cycas*.

The starchy pith of *C. circinalis* and *C. rumphii* and cortex of *C. rumphii* and some other species are eaten as food after cooking. They yield a starch called 'Sago' from the stem. It is used as food in several areas of southern and south-eastern Asia and Oceania, including New Caledonia, Timor, Indo-China, Malaya, India, Burma, Fiji, Ceylon, Madagascar and the Solomons. Sago is extracted from the cycas plants before fruiting because fruiting depletes the store of starch in the stem. The best time for extraction of starch is just before a flush of new leaves. Some plants of cycas start yielding starch at the age of seven years.

The root nodules of *C. revoluta* have also been reported to yield starch. They contain about 18% starch.

The seeds and stems of *Dioon edule*, *D. spinulosum* and *D. mejiae* also yield sufficient "seed starch" and 'Sago' or stem starch.

The stem pith in *Encephalartos* has been reported to be made use of in preparing a "Kaffir bread". The Dutch colonists of South Africa used this preparation in plenty.

The starch is obtained from the species of *Encephalartos* such as *E. transvenosus*, *E. ferose* and *E. hilderbrandtii*.

The seed starch is also obtained from seeds of *Macrozamia reidlei, M. dyeri, M.spiralis* and *M. miquelii*. Stem starch or 'Sago' is obtained from the stems of *M. reidlei, M. spiralis* and *M. fraseri*. Commercial adhesive pastes are also prepared from residual starches of some species of *Macrozamia*. The starchy interior of the stem of several species of *Macrozamia* can be ground or grated and boiled in water and used as food for pigs, poultry and calves.

The fleshy testa of the seeds of *Stangeria paradoxa* is eaten by the baboons.

2. Cycads as Gum Yielders : Gum is secreted by the injured megasporophylls (plate III) of *Cycas revoluta, C. circinalis, C. rumphii, C. siamensis* and *C. thuarsii.* The gum also exudes from wounds on stem and leaves. The gum is contained in the 'gum canals'. When placed in water the cycas gum begins to swell immediately. By the end of several days, it expands to many times its original size and becomes absolutely colourless. Cycas gum has been used medicinally to produce rapid suppuriation of malignant ulcers. It is also a good antidote for snake and insect bites. It is also used as an adhesive. The petiole of *Dioon edule* is also said to yield gum. Some species of *Encephalartos* such as *E. lemarinelianus* and *E. hildebrandtii* yield gum from their cones and leaf petioles. It is eaten by Kaffir children and birds. Species of *Macrozamia* yield gum from their wounded cones, stems and leaf bases (*M. spiralis*).

3. Medicinal uses of cycads : In Assam the pounded stem of *Cycas pectinata* is used as a hair wash for diseased hair roots. In Cambodia the mucilaginous terminal bud of *C. circinalis*, crushed in rice, is used in dressing of ulcerated wounds, of swollen glands, and of boils. The juice of young *Cycas* leaves is supposedly good for flatulence and vomiting blood. The abundant pollen of *C. rumphii* is said to be strongly narcotic, as are the male cone scales of this species and *C. circinalis* that are commonly sold as an **anodyne** in bazars in India. The female cone, reduced to a poultice and applied to the lions, is reported to allay nephritic pains. **Cycas** seeds, pounded to a paste in coconut oil, are used for sores, swellings, wounds, boils, and various skin complaints in parts of southern Aisa, Indonesia, and Philippines. The decoction of young seeds of **Cycas** in water are emetic and stomach purifying. Cycas seed flour in Ceylon is esteemed highly as a remedy in bowel complaints and haemorrhoids, for which purpose the flour is boiled and eaten. A liquid decoction of the seeds of *Dioon edule* is said to be employed in Mexico for neuralgia.

4. Cycads as Oil yielding Plants : Oils from two species of cycads have been studied. The fleshy layer of the seeds of *Macrozamia reidlei* contains a bright orange oil (28.2%) whose physical and chemical contents were found to resemble those of palm oil.

In his analysis of the seeds of *Cycas revoluta*, Peckolt found the fleshy testa to contain 3.89% of an orange-yellow oil. Much more recently, analysis of sotetsu seeds revealed from 20.44—23.37% of oil, depending upon the method of extraction. Oil from the seeds of *Cycas revoluta* is said to have been used on Okinawa during the food crisis of World War II.

5. Cycads as Nitrogen Fixers

Many and possibly all cycads have, in addition to the primary tap root, numerous small secondary roots, some of which form the so-called root tubercles, root nodules, or coral-like roots. Reinke studied these structures and showed the presence of an alga, Anadaena, in them. The presence of bacteria in the nodules was demonstrated by Schneider who considered the alga, bacteria, and cycad to exist in a symbiotic relationship. Life concluded that the nodules of cycads may have at least two functions ; (1) that of aerating, and (2) that of assisting in nitrogen assimilation, In all bacterial cultures prepared by Bottomley from the algal zone of cycad nodules, two nitrogen-fixing bacteria, Pseudomonas radicicola and Azotobacter were associated.

That the roots of *Cycas revoluta*, with the assistance of *Anabaena*, are able to use the free nitrogen of the air was the conclusion reached by Yosimura. More recently, Douin showed that cycad nodules with *Anabaena* contain more nitrogen than normal roots and that the alga is able to fix atmospheric nitrogen.

In addition to the papers cited above, several others have contributed to our knowledge of the root nodules of cycads. Photographs of nodules of *Cycas* and *Encephalartos* are reproduced in Kellerman.

While it is true that the cycads have no agricultural significance as nitrogen gatherers, their importance in the nitrogen Cycle is a present unknown. Surely in areas where they are abundant they have played a role in the maintenance of the nitrogen supply.

6. Cycas Leaf Industry : Dried leaves of *Cycas revoluta* are exported in large quantities from the Ryukyu Islands for use as decorative material. This little-known industry is perhaps about 50 years old. A 1912 publication of the British Foreign Office refers to it as "recently developed". About 15 people are employed by the industry in Japan and the Ryukyus at the height of the season.

Cycas leaves are collected, starting at the end of July, in several islands of the Ryukyu group such as Amami-Oshima, Oki-erabu-shima, Yoron-juma, and Okinawa. There is no attempt to cultivate cycads for leaf harvest. The supply of wild plants is quite adequate although, at least in Okinawa, it is being reduced by the expansion of agriculture. After being cut from the plant and sorted into several size classes, cycas leaves are ties in bundles of five each and hung in a shady place to dry for about two months. They are then packed in bales that measure about 5 × 4 × 3 feet. The number of leaves per bale varies, depending upon the size of the leaves therein. Bales of 8-12 to 12-inch leaves contain about 26,000 leaves, while those of the 32- to 36-inch contain only 5000. About 1000 bales of cycas leaves are exported every year. They are sent from the Ryukyus to at least six brokers in Kobe, Yokohama and Tokyo (Plate IV). From there the leaves are shipped to overseas market chiefly the United States, Germany, and Switzerland.

An estimated 3,000,000 *cycas* leaves with a gross value of about $30,000, are imported annually into the United States (Plate IV). The price to the importers has increased about seven-fold since before World War II. The several processors here preserve and hand-paint the leaves. A typical prserving procedure involves oiling the leaves in water for a short time and then soaking them for several days in a preserving solution. A large precentage of prepared cycas leaves is used in the manufacture of wreaths and other floral decorations. Some leaves are used to make artificial palm tree for window displays.

7. Cycads as Horticultural Plants :— Many speceis of cycads, representing most genera, are cultivated as ornamentals. They are farily easily grown and are indeed handsome plants. In the tropics and subtropics the world over, cycads are used in out-of-door plantings. In cooler regions they are familiar in conservations where they usually seem quite well adapated to their artificial surroundings.

8. Cycads as Fiber Plants :— The fine, pulu-like surface fibers— called "palm wool" —found at the base of the leaves of several species of *Macrozamia* and *Cycas* are used for stuffing pillows and mattresses. Surface fibers from the leaves of *Cycas revoluta* and *C. circinalis* have been made into cloth. Before rubber was known in the Amami Islands, girls made balls by winding waste silk tightly on a core of "floss" from cycad leaves.

The leaves of various cycads (*Cycas*, *Encephalartos*) are plaited or woven into hats, baskets, mats, and fences, and are utilized for thatch and brooms. Cordage and twine are said to have been made from *Cycas* "bark" ("ecorce") and from structural fibers of *Cycas* leaves. During the Exposition Universelle at Paris in 1867, "fibres du-tronc" of *Cycas circinalis* were shown. The stem of *Macrozamia spiralis* has been found to be unsuitable as a source of paper-making fiber.

9. Miscellaneous :— At one time—but apparently not at present so far as I have been able to determine—the wood of *Cycas revoluta* was used in Japan to make small articles such as boxes, plates and bottle stands. Rein's *The Industries of Japan* mentions "the beautifully spotted but very light and porous wood" of sotetsu being utilized in the woodworking industry of the Hakone Mountains near Yokohama. Stems of *C. rumphii* are, in Indonesia, used to build small houses. Maiden says of the stem of *C. media*: " Of no use for timber purposes, but perhaps it might be useful to a limited extent for rustic work".

The dry, stony seed-coats of some of the cycads—*e.g.*, *Dioon spinulosum* and *Cycas circinalis*— are sometimes used as playthings or whistles by children. In parts of India, empty seeds of *Cycas circinalis* are used as snuff boxes, and, in Australia, empty seeds of *Macrozamia denisoni,* cut in two and provided with a hinge and clasp, are used as match boxes.

The fresh or dried leaves of cycads serve as decorative material and are, in many areas, used in religious ceremonies, *e.g.*, Palm Sunday processions as in Australia, the Philipines (46, 50), and Goa, and in native rituals as in the New Hebrides and the Solomons. Around Dresden, some years ago, entire hot houses were devoted to the growing of *Cycas revoluta* for its fronds that were used in funeral processions. Maumene wrote on the production of *Cycas* leaves for decorative purposes.

QUESTIONS

ESSAY TYPE

1. Give an explanatory account of seedling anatomy of Cycads. **(Kanpur MSc., 1985)**
2. Describe the characteristic features of a member of Cycadales.
3. Give an account of reproductive character of Cycadales.
4. With the help of diagrams, give an account of development of archegonia in Cycads.
5. Describe the economic importance of *Cycas*.
6. Describe the living members of Cycadales. What is their distribution worldwide ?
7. Describe the secondary growth in Cycadales.
8. With the help of diagrams, describe the megasporogenesis in Cycads.
9. Give a brief account of development of male gametophytes in Cycads.
10. Describe detailed structure of spermatozoid in *Zamia*.

SHORT ANSWERS TYPE

11. **Write short notes on**
    (*a*) Estimation of age of Cycads **(Rohilkhand MSc., 1998)**
    (*b*) Characteristic features of Cycads
    (*c*) Phylogeny of Cycadales **(Rohilkhand MSc., 2002)**
    (*d*) Leaf cuticles of Cycadales
    (*e*) Spermatozoid structure in *Zamia*
    (*f*) Fern like characters of Cycads
    (*g*) Distribution of Cycads in India **(Kanpur, 2002)**
    (*h*) Economic importance of Cycads.
12. What is circinate vernation ?
13. In which genera of Cycadales, the leaves are dimorphic ?
14. Where are haplocheilic stomata found ?
15. What is the chromosome number in Cycads ?
16. Draw neat and labelled diagrams of :
    (*a*) T.S. leaf of *Zamia*
    (*b*) A male plant of *Zamia*
    (*c*) Microsporophylls of *Ceratozamia*, *Zamia* and *Macrozamia*
17. Describe the medicinal properties of Cycads.

OBJECTIVE TYPE

18. **Select the correct answer :**

(*i*) Sago is obtained from
(*a*) *Cycas circinalis*
(*b*) *Zamia*
(*c*) *Dioon*
(*d*) *Macrozamia*

(*ii*) Circinate vernation is common to both
(*a*) *Cycads* and *Pinus*
(*b*) Cycads and Ferns
(*c*) Cycads and Horsetails
(*d*) Horsetails and Ferns.

(*iii*) Manoxylic stem is characteristic features of
(*a*) *Cycads* (*b*) Conifers
(*c*) Ferns (*d*) Angiosperms

(*iv*) When the epidermis is heavily cutinised and have sunken stomata these are known as
(*a*) Haplocheilic
(*b*) Spirocheilic
(*c*) Heterokantae
(*d*) Diplophyllic

(*v*) The longest female cone is present in
(*a*) *Cycas*
(*b*) *Zamia*
(*c*) *Ceratozamia*
(*d*) *Macrozamia denisonii*

(*vi*) The heaviest female cone is found in
(*a*) *Cycas circinalis*
(*b*) Dioon *spinulosum*
(*c*) *Macrozamia demisonii*
(*d*) *Zamia pygmala*

(*vii*) The smallest female cones are present in
(*a*) *Cycas circinalis*
(b) *Dioon spinulosum*
(*c*) *Macrozamia denisonii*
(*d*) *Zamia pygmaea*

(*viii*) Which of the following belongs to Cycads ?
(*a*) *Taxus* (*b*) *Zamia*
(*c*) *Ginkgo* (*d*) *Araucaria*

CHAPTER 9

Cycadopsida (contd.)

- Cycadopsida Charaders Distribution and General
- Morphology of Cycadopsida
- Secondary Growth
- Development of male and Female Gametophyle
- Embryogeny
- Economic Importance of Cycads
- Fanuly Nilssoniaceae

Order : **Cycadales**
Family : **Cycadaceae**
Type : **Cycas Linn.**

Distribution and General Characters of Common Indian Species

This chapter is devoted to the life history of *Cycas* the only representative of the family met within India. Only six species of *Cycas* grow wild in India. These are :

1. *Cycas circinalis* L. (plate I). It is widely distributed in the dry deciduous forests of Southern India. Sago is extracted from its stem and seeds. It is

CYCAS RUMPHII

the type species of the genus and the specific name *circinalis* is derived from the circinate ptyxis of the leaves in the young state. In Hindi language it is known as *Jangli madanmast Ka Phool* (phul). The cylindrical trunk of this medium sized tree is covered with persistent leaf bases and bears a crown of once pinnate compound leaves (upto 3 metres long) at its summit. It grows upto 4 metres high and is usually unbranched, but old trees show branching (see plate 1). The petioles of the leaves possess spines at their bases. The leaflets are not revolute and have smooth margins with acute apex and linear lanceolate shape. The upper surface of the leaflets is glaucous. The male plants bear cylindric-ovoid male cones that may be upto $1\frac{1}{2}$ feet long. The megasporophylls have a rhomboidal and slightly dentate blade with a stalk bearing upto 12 ovules on it. It is also cultivated in the gardens throughout the country as an ornamental.

2. *Cycas beddomei* Dyer is found wild in Cuddapah district of Andhra Pradesh. Its local name in Andhra is *Per ita*. As compared to *C. circinalis* this species has a dwarf trunk upto 40 cm long. The leaves have leaflets that are strongly revolute. There are no spines on the petiole but it is covered at its base with tufted hair. The male cones are short peduncled oblong-ovoid in shape and reach a length of 33 cms. The megasporophylls have an ovate-lanceolate blade with long linear teeth. The stalk bears 2—4 ovules.

3. *Cycas pectinata* Griff is found in several parts of Eastern India and grows upto 3.5 metres tall. The stem bears a crown of recurved leaves with flat, narrow and linear leaflets. A few spines are present at the base of the petiole. The male cones are shortly peduncled, cylindric-ovoid and upto 40 cm long. The megasporophylls have pectinate leaf blades and bear 4-6 glabrous ovules on their stalks. Its common name in Nepal is *Thakal*.

4. *C. rumphii* Miq. (Plate III). The stem is often branched and may grow as high as *C. circinalis*. The leaflets are dark green and shining with flat blades. The leaves are upto two metres long. The almost ellipsoidal and short peduncled male strobili grow as long as those of *C. circinalis*. The megasporophylls have an ovate to lanceolate blade with a long stalk bearing 6—10 ovules. The blade is toothed above and entire below. This species is found in Andaman and Nicobar Islands, Ceylon, Cocosiislands.

5. *C. siamensis* Miq. is found in Burma, Cochin-China and Siam. Burkill (1933) considered it to be a variety of *C. pectinata* with which it resembles very much. It differs in possessing smaller leaves, smaller megasporophylls and is also short in stature. It is not an Indian species but is commonly cultivated.

6. *C. revoluta* Thunb. (Plate II) It is a Japanese species that is widely cultivated as an ornamental plant in India. It is vegetatively propagated by means of bulbils. Only female plants grow in India. The leaflets have a revolute margin and the megasporophylls have a pinnate blade covered with extensive yellowish-brown tomentum. The ovules are large, tomentose and number 4—6 per megasporophyll. Its local name in Japan is *Tosso*.

The total number of species according to Bierhorst (1971) is twenty. Some other common species include : *C. media, C. normambyana, C. madagascriensis, C. taiwaniana, C. nathorstii, C. thouarsii, C. pupuana, C. cairnsiana, C. seemannii, C. riuminiana* etc.

The account given below refers mostly to *Cycas revoluta* and *C. circinalis*.

Habit and Habitat

The plants are short or medium sized trees which look like palms. The trunk is covered with persistent leaf bases. The stem is usually unbranched but branched old stems are not rare. The male plants have sympodial axis whereas in the female plants the axis is monopodial. Regarding their habitat the plants grow under xerophytic conditions like the exposed slopes of hills and other sunny places where water is scarce. They grow well under cultivated conditions.

General Morphology

The plant, which belongs to the sporophytic generation, is differ entiated into root, stem and leaves. The stem in the young plants, is stout, almost tuberous and subterranean, the top bear ing a crown of spirally arranged leaves. The growth is very slow but it lives to a great

Plate I : Plants of *Cycas circinalis*

Plate II : *Cycas revoluta.*

age. In the course of time the stem builds up into an aerial thick columnar normally unbranched trunk from the summit of which arises a crown of large fern-like leaves (Plate III). It resembles in its appearance the tree ferns or some palms. The older parts of the stem are covered with an armour of persistent woody leaf bases. They closely cover the stem and fortify it. The stem bears adventitious buds at the base. They are large and are covered with scales. They are often called the **bulbils** and serve as a means of vegetative propagation (Plate IV). The leaves are of two kinds :— They are arranged in close spiral succession alternating with each other. The alternating clusters of persistent

Plate III : *C. revoluta; B, Circinalis.*

leaf bases of foliage and scale leaves are of different thickness and give the stem a rugged appearance. The scale leaves are brown and are far more numerous than the foliage leaves. They are persistent and protective in function. They cover the stem apex and offer protection to the young foliage leaves. The foliage leaves are large, compound and pinnately divided into many leaflets (Fig. 9.1).

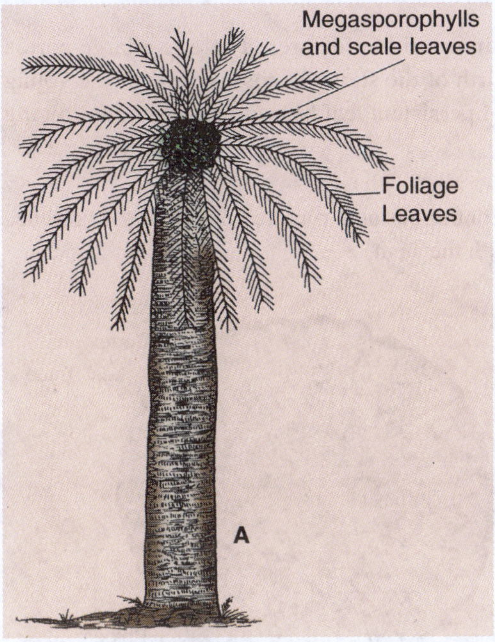

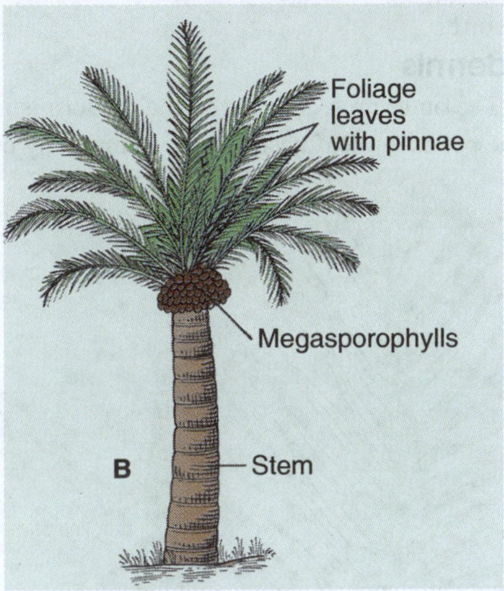

Fig. 9.1. External features of Cycas. *A, C. revoluta; B, C. circinalis.* Note the position of megas porophylls

They are pari-pinnate and the pinnae are closely set on the rachis and are sessile. Some of the lower pinnae may be rather tough and leathery in texture showing their xerophytic nature. The *Cycas* leaves are extensively used for floral decorations. The pinnae are linear in outline with a narrowed structure. The margin is curved downwards and inwards. The revolute margin of the leaflets has given the plant its specific name (*C. revoluta*). Each leaflet is traversed by a single mid-rib. Like the ferns the rachis is incurled and the leaflets are inrolled in the bud condition. The ptyxis is thus of **circinate** type, a new cluster of leaves is formed every year or every second year. In the interval, the succession is kept up by the production of scale leaves.

The primary root persists and forms a **tap-root system**. The main tap root may be long but usually it is short and thick. Some of the lateral roots give out branches which become apo-geotropic, grow vertically upward just below the ground level. They branch repeatedly to form dwarfed dichotomously branched coral-like masses. They are called the **coralloid roots** or **corallorhiza.** They become inhabited by a blue green alga, *Anabaena cycadacearum*. The surface of the coralloid roots is beset with lenticel-like apertures which suggest that they are respiratory in function.

Plate IV : *Cycas revoluta* young leaves.

Pant and Das (1990) have reported the presence of non-coralloid aerial roots in *C. revoluta* and *C. rumphii*. These are positively geotropic adventitious roots and grow out from the lower *sides* of young and old bulbils or leafbases while still attached to that plant. Occasionally these roots are branched. Anatomically these roots are di-to polyarch with wide cortex, mixed parenchymatous pith and scattered xylem elements.

Anatomy of the stem

The stem anatomy is simple and shows certain primitive features. There is a large cortex and pith but a small wood as compared with the girth of the stem. A cross section of the young stem is irregular in outline owing to the presence of persistent leaf bases. It shows the following structures :—

Epidermis

The outer protective layer, the epidermis is discontinuous and ruptured owing to the presence of the persistent leaf bases which completely ensheath the stem.

Fig. 9.2. Leaf of *Cycas revoluta* (A) and its leaflet (B).

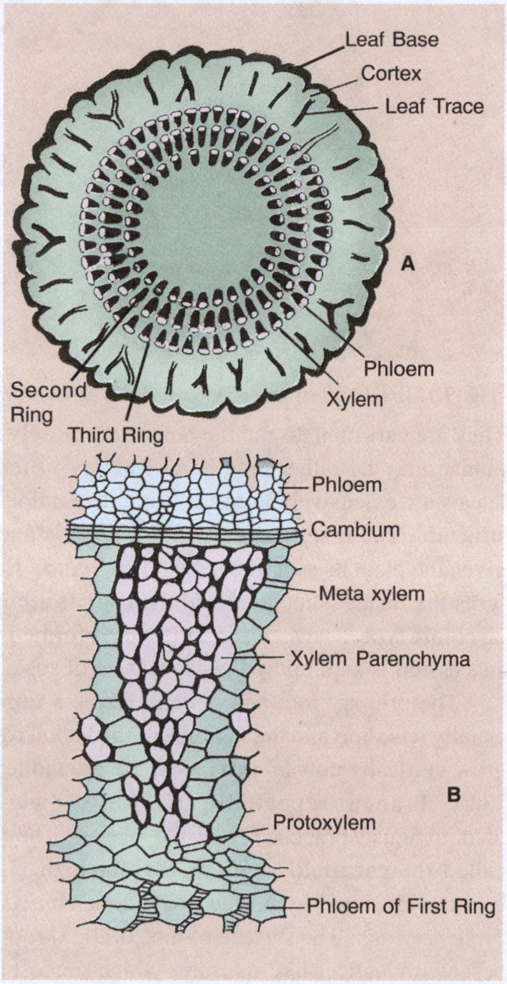

Fig. 9.3. Anatomy of stem of *Cycas revoluta*. A. outline diagram of T.S. stem; B. T.S. of a part of vascular bundle of first ring.

Cortex

It forms the greater bulk even in old stems and consists of thin-walled, parenchymatous cells densely filled with starch grains. Numerous mucilage canals are found in this region. They are connected with the mucilage canals of the pith by medullary rays. The occurrence of peculiar curving leaf traces, the girdles in the cortex, is a striking feature of the stem anatomy of *Cycas*. We shall refer to them in detail afterwards.

Endodermis, pericycle and a narrow vascular cylinder

Endodermis and pericycle are not distinct. The vascular cylinder consists of a primary ring of numerous small closely set vascu lar bundles. The bundles are conjoint, open, collateral and endarch. The xylem consists only of **tracheids** (Fig. 9.3 A) and xylem parenchyma. There are no wood **vessels.** The protoxylem has tracheids with spiral thickenings and the metaxylem with scalariform thickenings and bordered pits in several alternating rows on the radial walls. The phloem contains **sieve tubes** with sieve plates on the oblique end walls, and **phloem parenchyma** (Fig. 9.3 B). There are no companian cells. The primary cambium lies between the phloem and the xylem. It remains active only for a short time, but is succeeded by another, the **secondary cambium** which may arise in the pericycle or the cortex. It also dies. In this way several rings of cambium may be formed.

Medullary rays

There are well developed **medullary** rays between the vascular bundles. They are fairly broad and deep.

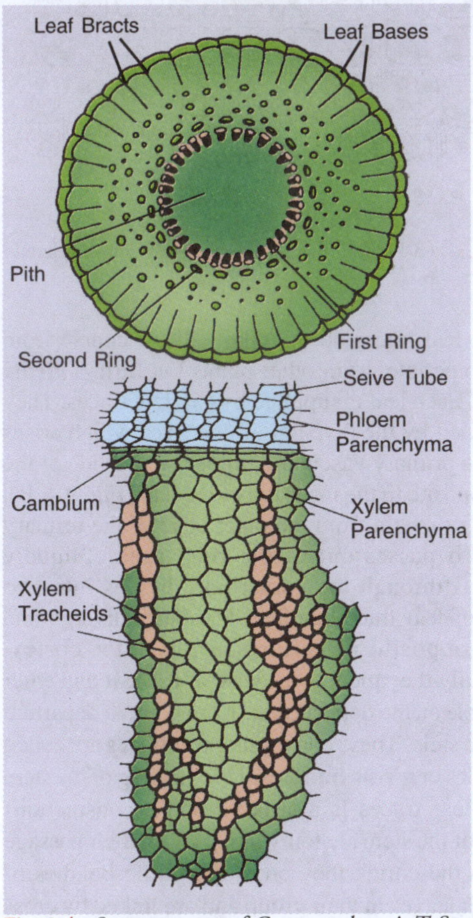

Fig. 9.4. Stem anatomy of *Cycas revoluta*. A. T. S. of stem showing 3 rings of vascular bundles; B, A vascular bundle of secondary ring.

Pith

The pith is large and consists of parenchymatous cells rich in starch. It is traversed by numerous mucilage canals.

Leaf Traces

The vascular supply to the leaves in *Cycas* is peculiar and interesting. The noteworthy feature is

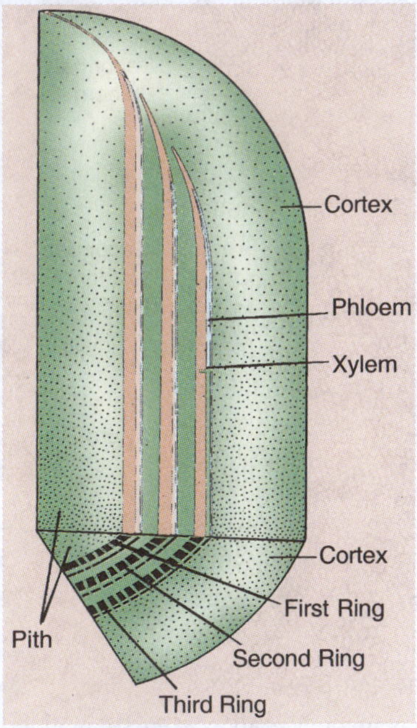

Fig. 9.5. Diagrammatic radial and transverse aspects of stem of *Cycas revoluta* showing concentric rings of vascular bundles.

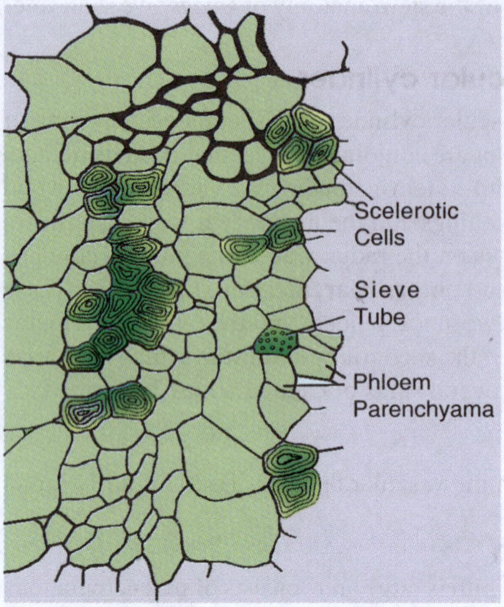

Fig. 9.6 Portion of vascular bundle of stem showing the details of phloem in a T.S.

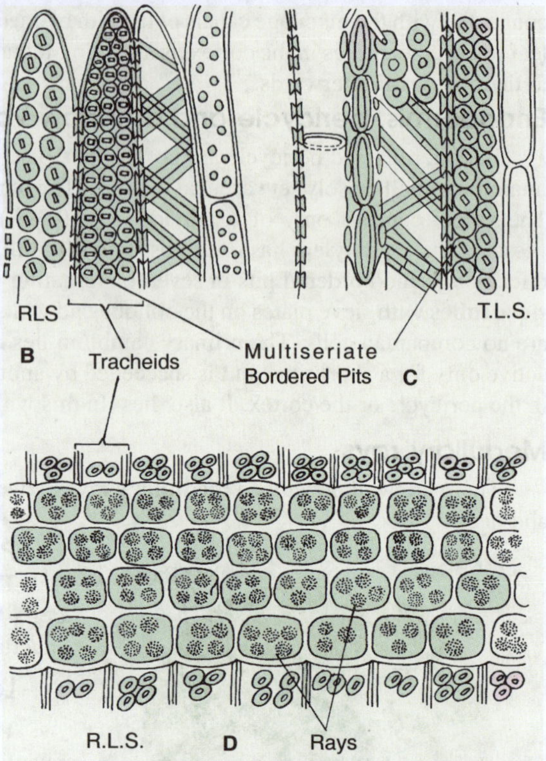

Fig. 9.7 (B-D) L.S. Cycas old stem showing bordered pits.

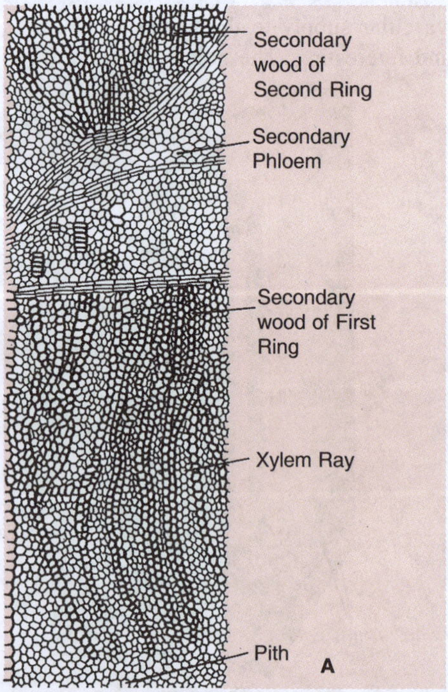

Fig. 9.7A. Cycas T.S. of a portion of an old stem showing two rings of secondary growth.

that the leaf traces do not take a direct course from steles to petiole; as in other plants but form a **girdle system.** Each leaf is supplied by two leaf traces. They are formed by the division of one trace which arises from the primary vascular cylinder on the side of the stem opposite to the insertion of the leaf (Fig. 9.4, B). It takes its origin from the inner edge of the primary xylem. It passes out of the ring, runs obliquely outwards through the cortex and divides into two halves of leaf trace bundles. The latter diverge and run in opposite directions in the outer cortex. Eventually they meet at the base of the leaf and enter the petiole on the opposite side from point of departure from the stele. These converging leaf traces are called the girdles or **girdle bundles.** This girdling of the stem by the leaf traces is a striking and a conspicuous feature of the stem anatomy of *Cycas*. In their passage through the cortex they cross the girdle bundles of the other leaves in their group and are linked by cross connections with each other, with other leaf trace bundles of other leaves in the same group and with the **radial traces** arising from other points around the vascular ring. The two girdle bundles enter the leaf base and divide repeatedly to form a large number of bundles which are arranged in the horse-shoe (omega) pattern characteristic of the petiole (Fig. 9.7). In the seedling and the cones the girdles are absent. The leaf trace bundles take a direct course from the stele of their destination. The girdles function as supporting structures to the trunks of old plants. The girdle bundles (Fig. 9.7) are collateral with xylem facing inwards. The xylem elements are scalariform as in **ferns** and are centrifugal. The protoxylem is endarch. During their passage up in the petiole, a centripetal mass of xylem appears on the other side of the protoxylem and the centrifugal xylem gradually diminishes. In this way the bundle structure is modified. It consists of a mass of centripetal xylem and a remnant of centrifugal xylem (Fig. 9.8). The latter is separated from the protoxylem by parenchymatous cells. Such bundles are called mesarch. They are characteristic of the ferns. It is a primitive character and indicates the antiquity of the *cycadales*. Presence of two xylem types gives them the name **diploxylic bundles**.

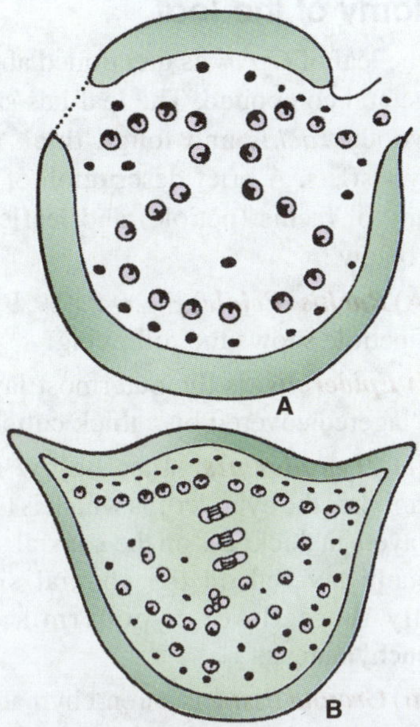

Fig. 9.8. Outline diagrams of petiole of Cycas as seen in a transverse section; they show the omega-like or horse-shoe-like arrangement of the vascular bundles. A. Section cut through the middle region of the petiole. B. Section cut through the base of the petiole.

Secondary growth (Fig. 9.7)

Secondary thickening takes place in the same manner as in the dicotyledonous stem but is not so vigorous. The ring of vascular bundles slowly increases in size radially by means of cambium which forms secondary xylem to the inside and secondary phloem to the outside. As a result a continuous vascular ring is formed. The medullary rays become narrow. The primary cambium then dies and is succeeded by a secondary cambium formed outside the secondary zone derived from the primary cambium. It is formed in the pericycle or the cortex. It forms the second ring of wood concentric with the first. The secondary cambium also perishes and is superseded by a third. It arises outside the second vascular ring. In this way several rings of secondary wood may be formed. The xylem and phloem elements of the rings are normally orientated. Areas of the inverted bundles may occur here and there between the normal rings.

With the formation of the secondary vascular tissues, the outer tissues of the stem are subjected to tangential pressure. Eventually they rupture, but before this take places a special protective tissue the corky periderm is formed outside the cortex by the activity of a special cambium which arises in the outer cortex immediately below the epidermis. This cambium is a secondary meristem. It is called the **phellogen** or **cork cambium.** By repeated division of the phellogen cells a tissue is formed to the exterior. It is the **cork.** It forms the stem surface in the lower part of the very old stem replacing the outer armour of leaf bases.

The anatomy of the stem of *Cycas* presents certain interesting features. They are, the large pith and cortex traversed by numerous mucilage canals, thousands of girdle bundles in the cortex, scanty wood region, short lived primary cambium with well developed broad medullary rays, and several rings of secondary vascular tissue formed by successive rings of cambium formed independently in the cortex or pericycle regions The wood formed is known as *manoxylic*.

Anatomy of the leaf

The leaf of *Cycas* as mentioned above is large and compound. The leaf has stout and woody *rachis* and tough thick and leathry lesflets. A brief description of the anatomy of rachis (petiole) and leaflet is given below :

(A) ***Rachis (Peiole)*** Fig. 9.8, 9.9, 9.10. T.S of petiole shows the following:

(i) Epidermis is the outermost layer, single lagered covered by a thick cuticle.

(ii) Hypodermis. Just below the epidermis lies the hypodermis which is two-three layers in thickness on the adaxial side and many layered on the abaxial side. Usually the cells of hypodermis are chlorenchymatous.

(iii) Ground tissue is paren-chymatous and follows the hypodermis.

(iv) Vascular bundles arranged in inverted omega shaped (Ω) manner is the characteristic feature. Vascular bundles vary in their structure at the base, centre and apex. Each vascular bundle, ensheathed within a single large sclerencymatous sheath is open and collateral. A single layered endodermis and one or many larged pericycle surround the vascular bundle.

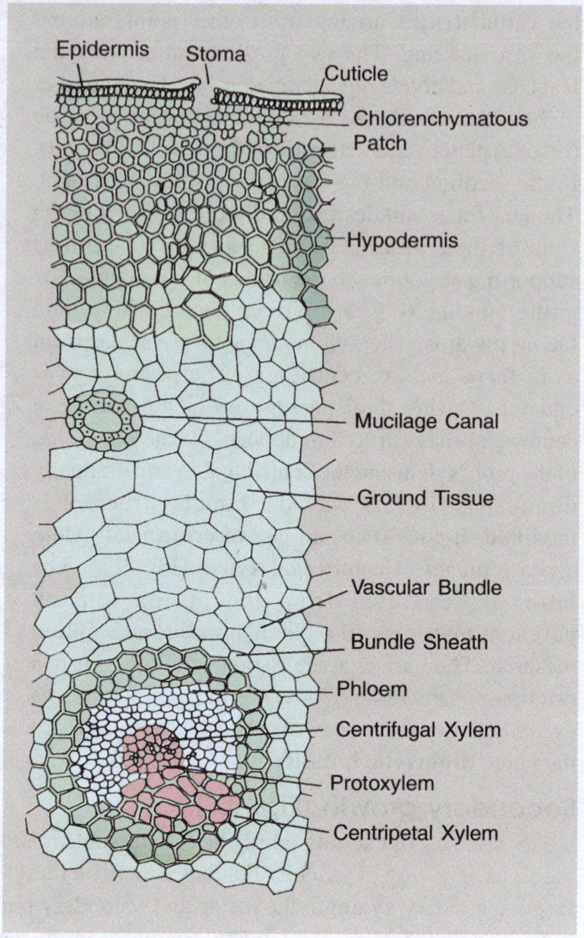

Fig. 9.9. T.S Petiole of *Cycas revoluta* showing one vascular and a portion of other tissues in detail.

At the base the vascular bundle is endarch *i.e.* protoxylem remains towards the centre, the xylem develops centrifugally, the phloem lies towards the abaxial side. The cambium is present on the upper surface of the metaxylem. During the course of upward movement of the bundles, the metaxylem appear opposite the protoxylem and constitute the centrally formed metaxylem. In this way metaxylem is present on both the sides of protoxylem and the bundles are mesarch. As a result the cambium occupies a position on the inner side. In further upward course, the centripetally developed metaxylem increases in quantity, the centrifugal metaxylem decreases in quantity and only two patches of metaxylem remain, each consisting of only a few tracheids. The phloem during this change remains on the abaxial side in this way from the base to the apex of rachis, the xylem elements change from endarch to exarch in the vascular bundles.

(B). ***Leaflet*** (Fig. 9.11, 9.12) A vertical section of the leaflets in the region of midrib shows the following tissues:

1. Cuticle. The upper surface of the leaf is covered with a thick cuticle. It serves to check excessive transpiration.

2. Epidermis. It forms the outermost cellular layer on both the surfaces of the leaflet. It is protective in function and consists of a single layer of thick-walled parenchymatous cells. The upper epidermis is continuous but the lower is punctured here and there with pits or stomata. The stomata which are confined only to the lower epidermis, are sunken and lodged at the bottom of these pits with overraching rims. (Fig. 9.11.)

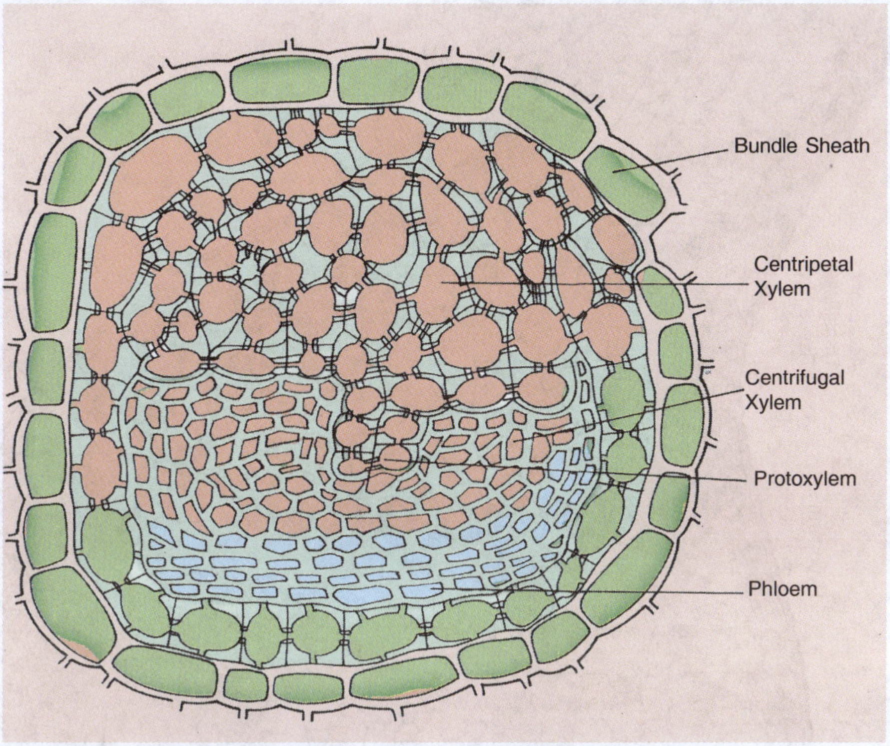

Fig. 9.10. A vascular bundle of petiole of *Cycas revoluta* showing detailed structure.

3. Hypodermis. It forms a single layer of thickened cells on both sides immediately below the epidermis. The cells of the upper hypodermis are highly thickened. It checks excessive transpiration and serves as a heat screen preventing over heating.

4. Mesophyll Tissue. It lies on both sides, between the upper and lower epidermis. It is differentiated into a well developed **palisade layer** on its upper side immediately below the upper hypodermis. It functions as the main photosynthetic tissue. Its elongated coloumnar cells contain abundant chloroplasts. The lower part of the mesophyll which lies within the lower hypodermis consists of loosely arranged parenchymatous cells with scanty inter cellular spaces between them. The cells contain chloroplasts. It is called the **spongy mesophyll.** It functions, as aerating and supplementary assimilatory tissue.

5. Transfusion tissue. On either side of the midrib between the palisade and spongy layers is a zone, 3 cells thick, of elongated colourless cells which run parallel to the leaf surface from the midrib to the margin. It constitutes the **accessory transfusion tissue.** The development of this tissue in the leaflet of *Cycas* serves to compensate for the unbranched condition of the midrib as it probably serves as a lateral conducting channel for water.

6. Vascular Bundle. It consists of xylem facing the upper surface and phloem facing the lower surface. The latter consists of sieve tubes and phloem parenchyma. It is in the form of an arc. The xylem consists of a mass of tracheids which are centripetal. There is a small patch of centrifugal metaxylem separated from the protoxylem by parenchymatous cells. The protoxylem thus occupies the centre of the bundle. The bundle is, therefore, mesarch in character.

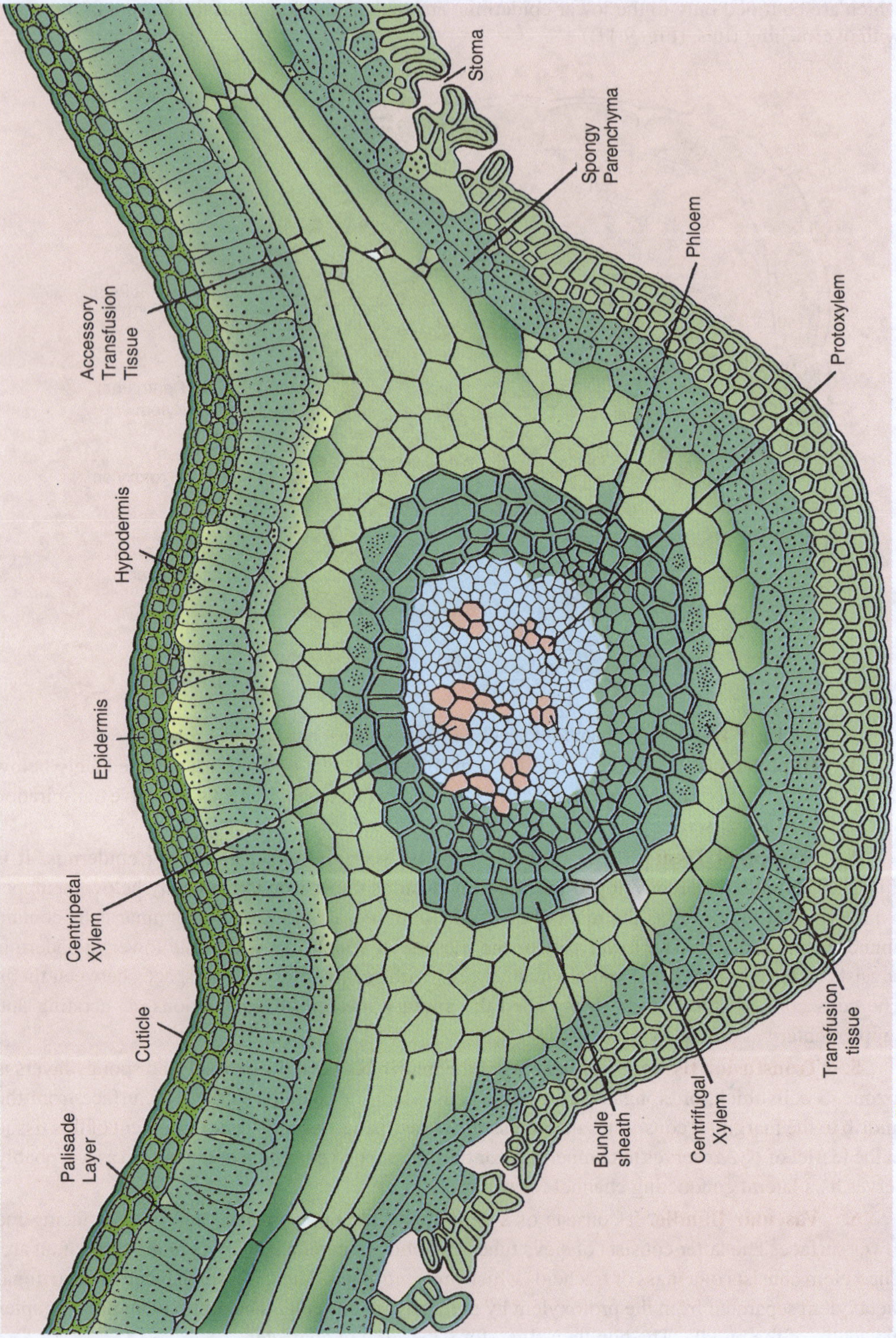

Fig. 9.11. Anatomy of leaf, T.S. of *Cycas circinalis* showing detailed internal structure.

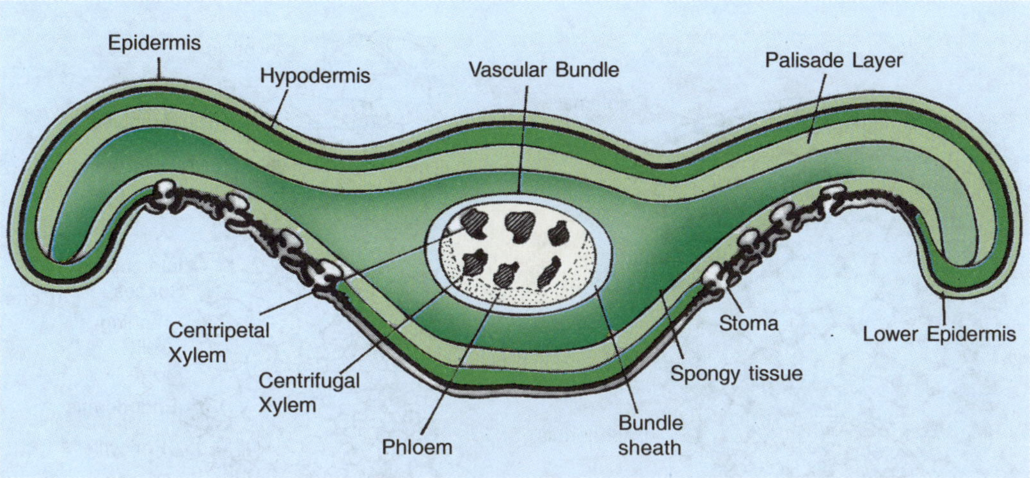

Fig. 9.12. Outline diagrams of T.S. leaf of *C. revoluta*.

7. **Primary transfusion Tissue.** It occurs on the sides of the xylem mass and consists of tracheidal cells with bordered pits on the walls. It is connected on either side with the accessory transfusion tissue.

Xerophytic Adaptation of the Cycas leaf

The tough and leathery texture and other structures such as strongly cutinized thick-walled epidermis, highly thickened hypodermis on both sides of the leaflet, sunken stomata restricted only to the lower surface, unbranched midrib and occurrence of primary and secondary transfusion tissues, all point to the fact that *Cycas* leaf is a strongly xerophytic object.

Anatomy of the root

1. Primary root (Fig. 9.13). It is identical to that of dicotyledonous root. There is the single layered **epiblema** or **exodermis** surrounding a large thin-walled parenchymatous **cortex.** The cortex is delimited by single layered **endodermis** consisting of four-sided slightly longitudinally elongated cells. Next is the single layered pericycle. It forms the outer layer of the stele which is **tetrarch,** but **diarch** in the youngest parts. There is pith in the centre. The xylem bundles are exarch with protoxylem towards the outside and metaxylem towards the centre. The phloem bundles are equal in number and alternate with the protoxylem elements. The metaxylem elements may fuse or meet in the centre. In this case there is no pith.

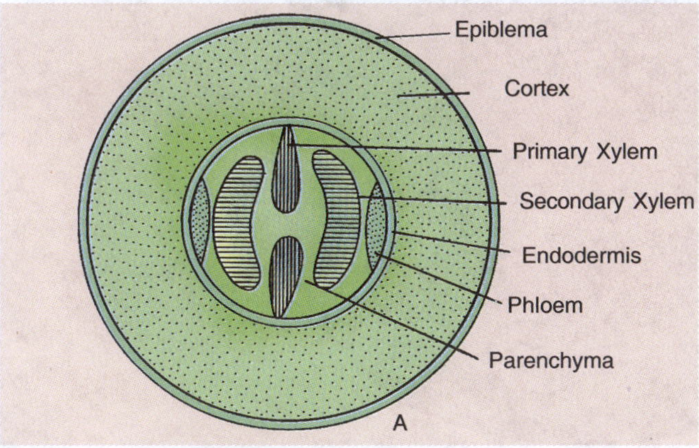

Fig. 9.13. (A) T.S normal root of *Cycas*. (diagrammatic).

Secondary growth (Fig. 9.12) takes place in the older portions. It takes place early and is more or less irregular. Several concentric rings of vascular bundles as well as the periderm develop, in old roots. The tip is protected by a root cap developed from the **periblemma.** There being no true **calyptrogen.**

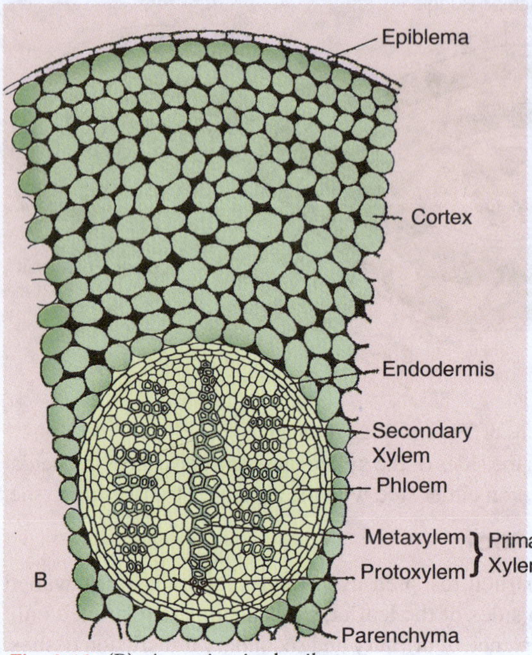

Fig. 9.13 (B), A portion in detail.

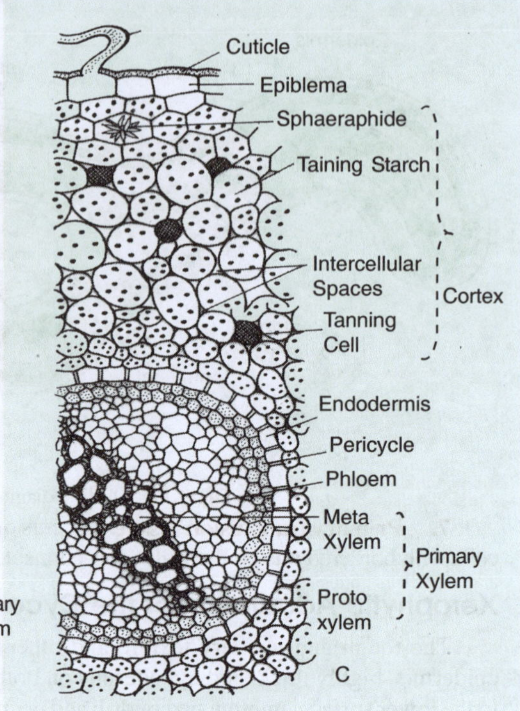

Fig. 9.13. (C) T.S. Primary root of *Cycas revoluta.* It is diarch.

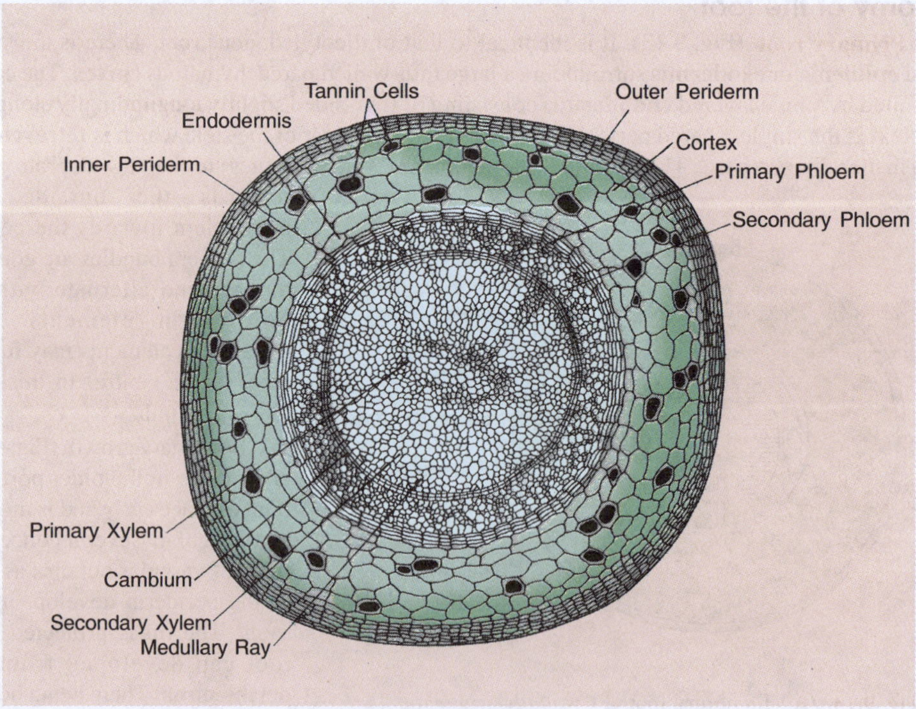

Fig. 9.14. T.S. root of *C. revoluta* showing secondary growth.

2. **Coralloid Roots** (Fig. 9.15). The structure is similar to that of the primary root. In addition, there is a conspicuous broad blue green zone in the middle cortex. This is the **algal zone.** It lies midway between the vascular bundle and the epiblemma. The cortical cells in this region get disorganised and are inhabited by blue green algae or cynobacteria such as *Nostoc* and *Anabaena cycadacearum* and *Calothrix*

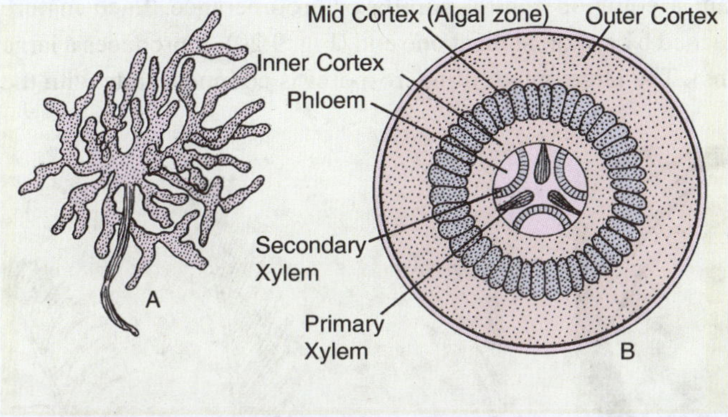

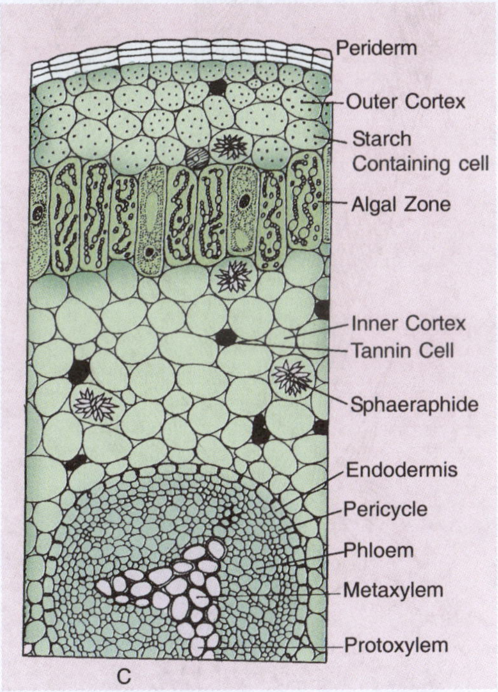

Fig. 9.15. (A-C). Coralloid roots of *Cycas.* A. As seen externally, B.T.S diagrammatic, C, A portion of B in detail.

There is difference of opinion about the nitrogen fixing ability of these roots as suggested by Life (1901). However recent reports suggest that cynobacteria fix-nitrogen and promote the growth of host plants.

Reproduction

1. **Vegetative.** *Cycas revoluta* is extensively propagated vegetatively by means of adventitious buds or bulbils (Fig. 9.16) developed in the basal parts of the stem. The male plants of *C. revoluta* are not found in Northern India. Hence no seeds of this species are available in this part of the country. The female plants are raised entirely by means of the bulbils. The female plants of *C. circinalis* are relatively more common.

Sexual Reproduction

All living species of *Cycas* are dioecious as the male and the female structures occur on separate plants. The **microsporophylls** are aggregated into large compact male strobili or cones. **The megasporophylls** are loosely arranged. They do not form a true cone.

Male Strobilus (plates V—X)

The male or staminate cones are borne singly and terminally on the main stem. The apical growth of the stem is continued by the formation of the axillary bud at the base of the cone. It becomes the new stem apex. As it grows the male cone is displaced to one side (Fig. 9.18 A). The stem of the male plant of *Cycas* is, therefore, a **sympodium.**

The male cone is woody in texture. It is a compact, long, spindle-shaped and blunt-ended object 40—50 cm (Fig. 9.17). It consists of a central woody axis bearing microsporophylls (stamens) in a close and compact (close fitting) spiral. They develop in acropetalous succession. The sporophylls at the apex, and the base of the cone may remain sterile. Others are fertile. Each microsporophyll of *C. circinalis* is a woody flattened wedge-shaped structure consisting of a narrow lower portion broadened into a flat sterile disc above (Fig. 9.18 B—D). It bears several hundred microsporangia (pollen sacs) on its lower or abaxial surface (Fig. 9.17 B). The microsporangia are arranged in clusters of three to six. Each cluster or group of microsporangia is called **sorus** (Fig. 9.19). The microsporangia in a sorus arise from a common central papilla or protuberance. Each mature microsporangium is an oval body attached by a short stalk at one end (Fig. 9.20). It produces a large number of **microspores** (pollen grains). The large output of microspores is commensurate with the primitive nature of *Cycas*.

Plate V : A very young male cone of *Cycas circinalis.*

Fig. 9.16. A bulbil of *Cycas circinalis.*

Development of the microsporangium (Fig. 9.21)

Each microsporangium develops from a single sporangium initial which is hypodermal in origin (as in the angiosperms) and not superficial. The sporangial initial arises in the hypodermis of the microsporophyll (Fig. 9.21 A). It divides by a periclinal wall into an outer **primary wall cell** and inner **archesporial cell** (Fig. 9.21B). The primary wall cells by repeated periclinal and anticlinal divisions form a sporangial wall several layers (1—6) thick beneath the epidermis (Fig. 9.21 C). Some botanists designate the sporangium wall cells as **endothecium** and the epidermis external to the sporangium as **exothecium**. The cells of the epidermis external to the sporangium (Exothecium)

Plate VI : A young male cone of *Cycas*.

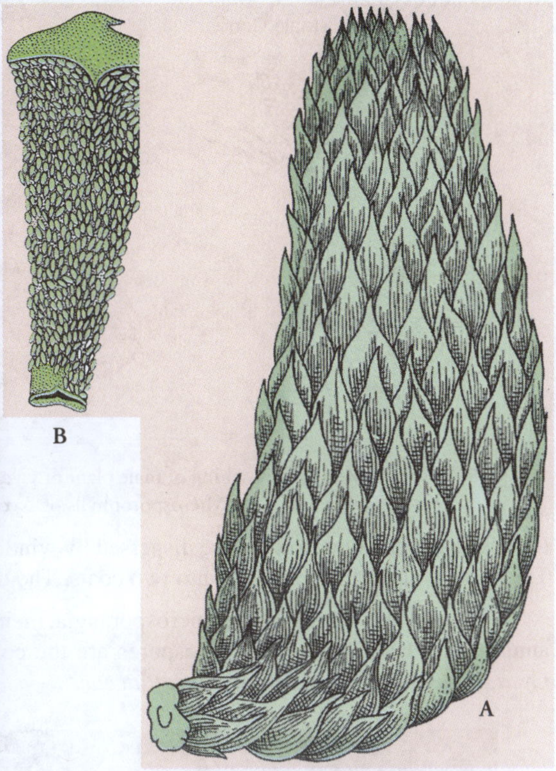

Fig. 9.17. (A-B) *Cycas circinalis*. A *Male cone* B. *Microsporophyll*

become thickened with the exception of two rows of cells along the abaxial route. They become elongated and mark the line of dehiscence. The archesporial cell undergoes divisions to form a mass of sporogenous cells. The sporogenous cells at the periphery of the mass form a single investing layer, the tapetum (Fig. 9.21 C). The cells of the last cell generation of the sporogenous mass consti tute the spore mother cells. They separate from one another, enlarge, round off (Fig. 9.21 D) and begin to divide. The tapetal cells along with wall cells break down providing nourishment for the developing spore mother cells. Each spore mother cell becomes a four-lobed structure. Its nucleus divides twice. The nuclear division is meiotic. Each haploid daughter nucleus migrates to one of the lobes of the spore mother cell. The lobes become separated by thickened partitions. A single microspore develops within each lobe.

Dehiscence of the Sporangium (Fig. 9.17)

It is longitudinal. When the microspores reach maturity, the male cone elongates considerably and rapidly. The scales separate from one another (Plate X) so that the sporangia are exposed. The sporangia lose water and with the loss of water from its cells the exothecium shrinks. As a result the wall splits open on the abaxial side along the elongated cells marking the line of dehiscence. The sporangia thus open by a slit on its outer face. The spores fall out. The slits are radial to the centre of

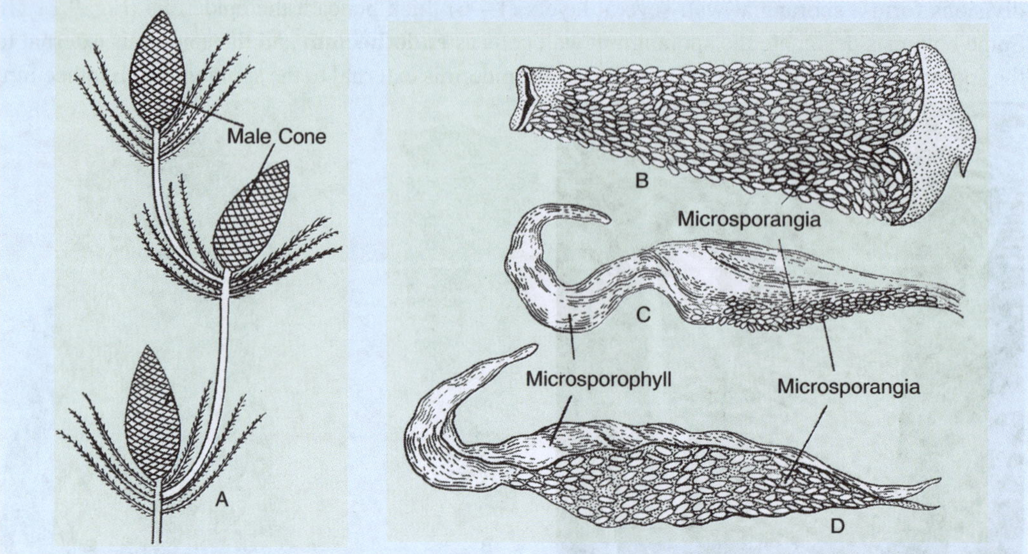

Fig. 9.18. (A-D). A. Mode of branching of male plant of *Cycas circinalis*. B. Microsporophyll of *Cycas circinalis* as seen from above. C—D. Microsporophylls of *Cycas circinalis* as seen in lateral view.

the rows. The liberated spores are dispersed by wind. Each spore is thickened at the opposite poles. The spore wall is differentiated into two coats. The outer is called the **exine** and the inner **intine.**

The soral arrangment of the microsporangia, their longitudinal dehiscence, absence of organised annulus and large output of microspores are the common primitive features which serve to link *Cycadales* with the fern family *Marattiaceae*.

Plate VII : Mature male cone of *Cycas circinalis.*

Female or Ovulate Strobilus (Plates XI and XII)

Plate VIII. *Cycas circinalis* : emerging male cones.

In *Cycas* there is no true and compact or properly organised female cone. There is a crown or a rossette of megasporophylls arising spirally in acropetalous succession at the apex of the stem in between a cluster of foliage and scale leaves. They are far more numerous than the foliage leaves and are formed each year between successive crowns of foliage leaves. They are loosely arranged and thus form a loose female strobilus (Plate XI). The growing point of the stem is unaffected by their development and continues its growth directly through the loose strobilus. The stem of the female plant is, therefore, **monopodial**.

Megasporophyll

The megasporophylls of *Cycas revoluta* (Fig. 9.22) are large ranging from 6—8 inches in length. Each megasporophyll is divided into an upper broad leafy portion and a lower (Fig. 9.22) stalk like portion. The broad leafy portion is lobed rather pinnate, tapering to a point. The megasporophylls, therefore, show a marked resemblance to foliage leaves but are much reduced. The ovules are borne laterally on the

Plate IX : Closer view of male cones of *Cycas circinalis.*

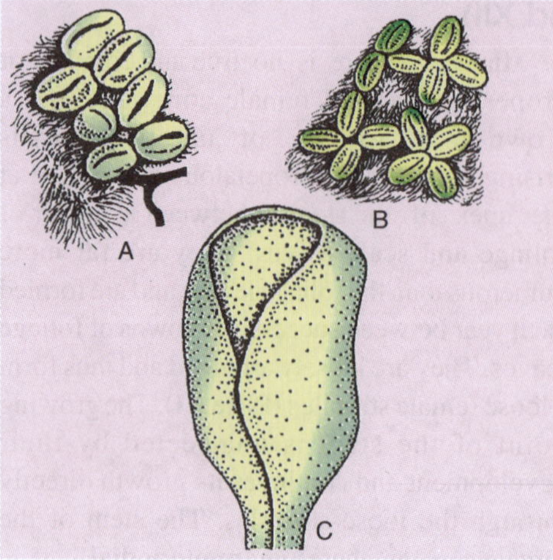

Fig. 9.19. (A—C).A sorus with sporangia that have shed their spores. B.A sorus with sporangia showing dehisence slits. C.A mature and a dehisced sporangium.

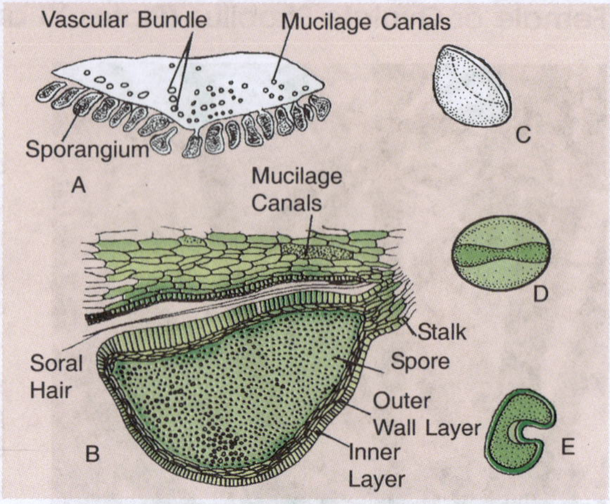

Fig. 9.20. (A-E). A. T.S. microsporophyll showing sporangia on its abaxial side. B Enlarged view of a portion of A showing highly magnified and longitudinally cut mature microsporangium. C—E. Mature microspores after being shed (After Woodehouse).

lower stalk-like portion. They occupy the positions of the lower pinnae on this part of the organ and vary in number from 2—10 per megasporophyll. The megaporophylls as well as the young ovules

Plate X : A female plant of *Cycas revoluta* bearing megasporophylls.

are densely covered with soft brownish wooly hairs but these are lost from the ovules as they reach towards maturity, so that the ripe ones are orange red in colour. The broad pinnate portion of the megasporophyll of *C. circinalis* (Fig. 9.21, C) is further reduced and shows only traces of a pinnate structure. The number of ovules is also reduced. In still other, species of *Cycas* the megasporophyll has only a serrate margin and further reduction in number of ovules to a single pair (Fig. 9.21 A). In fact a series of types showing progressive and gradual reduction in the foliar nature of the

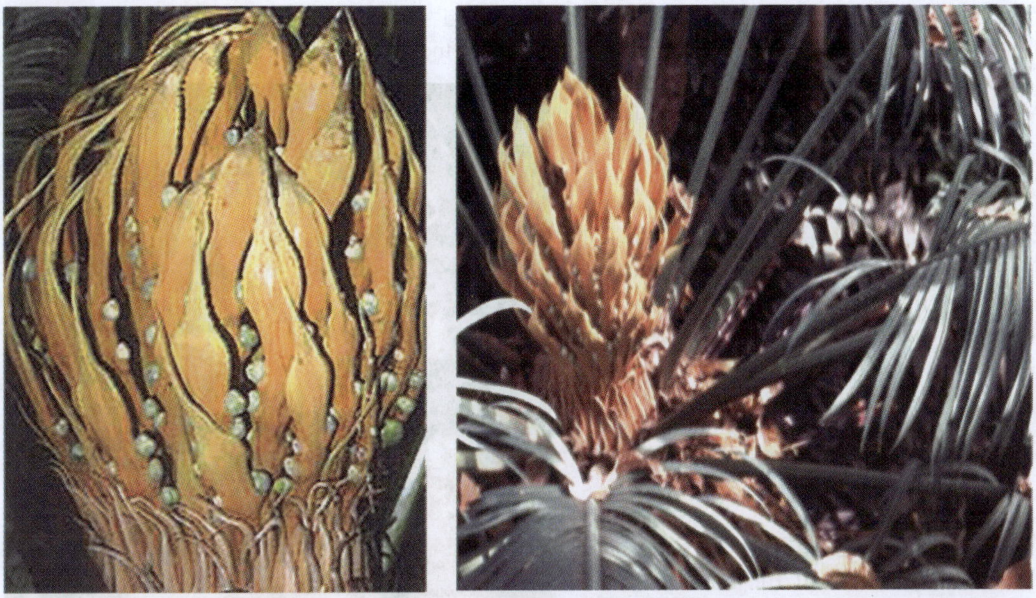

Plate XI : *Cycas circinalis.* a female plant bearing megasporophylls.

megasporophyll can be arranged among the members of the Cycadales. It starts Structure of the mature ovule (Figs. 9.23 and 9.24)

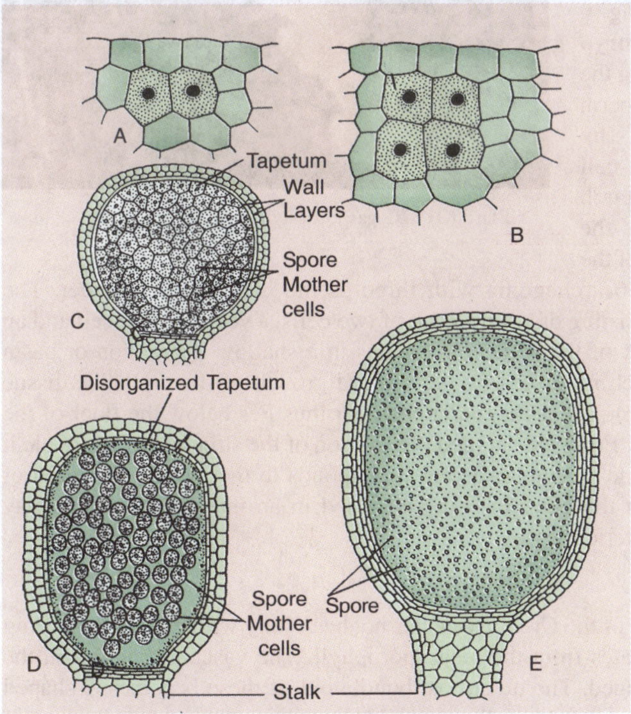

Fig. 9.21. (A-E). Stages in the development of microsporangium of *Cycas revoluta*. For explanation see text.

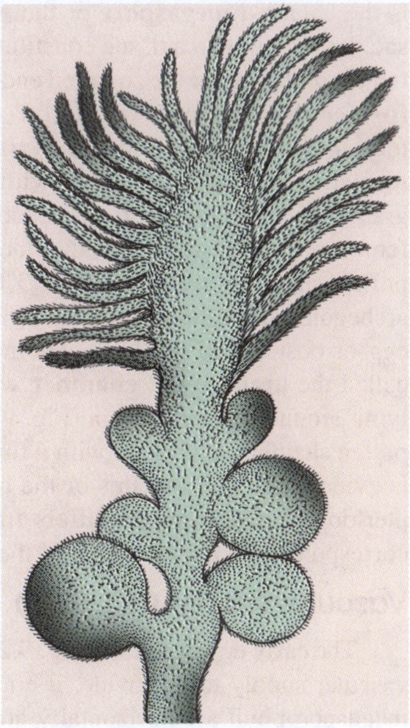

Fig. 9.22. Megasporophyll of *C. revoluta.*

The ovules are large, erect (orthotropous) and slightly flattened in the plane of the megasporophyll. Each ovule is enclosed in a thick **coat,** the **integument.** In *Cycas* the integument is differentiated into three layers—the middle stony layer with fleshy layers both inside and outside it. The outer fleshy layer is of orange red colour and the stony layer is light yellow.

Plate XII : Megasporophyll of *C. revoluta.*

The inner fleshy layer is used up as the female gametophyte matures and is left as a dry membranous covering or layer united to the inner tissue (nucellus) except near the apex. Owing to the presence of the stony layer it is hard to cut the ovule. The integument does not completely invest the ovule. The tissue within the integument is called the **nucellus.** It constitutes the body of the ovule. The apex of the nucellus grows up into a beak-like outgrowth, the **nucellar beak** which projects into the micropyle. Some of the cells of the nucellar beak disorganise to form flash-like hollow cavity. It is the **pollen chamber.** It plays an important role in the early stages of fertilization. The micropyle leads into the pollen chamber. The pollen grains germinate here. Within the nucellus, almost filling it, is the enlarged **megaspore** or the **embryo sac.** It is filled with a tissue constituting the female prothallus or the so called endosperm formed by repeated division of the megaspore nucleus followed by free cell formation. Its cells are filled with starch grains. Towards the micropylar end of the female prothallus just below the floor of the pollen chamber are developed 3 to 6 archegonia with three as the commonest number. The archegonium is a simple structure consisting of a short neck of two cells, a ventral canal cell and an egg or oosphere (Fig. 9.26). The neck of the archegonia opens in a shallow depression or basin called the **archegonial chamber** which is formed by the upward growth of the prothallus tissue lying around the archegonia (Fig. 9.26). The archegonial chamber thus lies below the floor of the pollen chamber. It is filled with a fluid. From the detailed description of the structure of the ovule it is evident that the nucellus of the ovule of gymnosperms corresponds to the megasporangium of pteridophytes. The ovule differs from the latter in being enclosed in an integument which may correspond with the indusium of the pteridophytes.

Vascular supply (Fig. 9.25)

The early development (Fig. 9.24) of the *Cycas* ovule has not been fully worked out. Regarding vascular supply to the ovule, it emanates from the megasporophyll. The vascular bundles in the megasporophyll are horizontally arranged. The outermost bundle out of these becomes C-shaped first and then assumes a concentric form. It later divides into 4 or more vascular bundles out of which one bundle springs towards the centre. The others are arranged in a ring. The central bundle

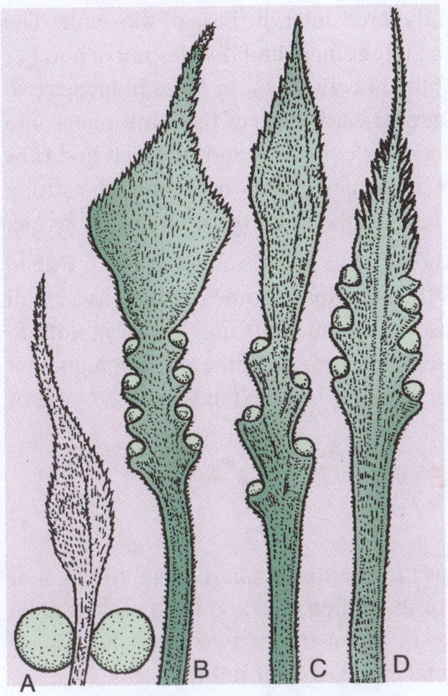

Fig. 9.23. (A-D) Megasporophylls of various species of *Cycas*. A. *C. normambyana* B. *C. rumphii* C.*C. circinalis* D.*C. thourasii*

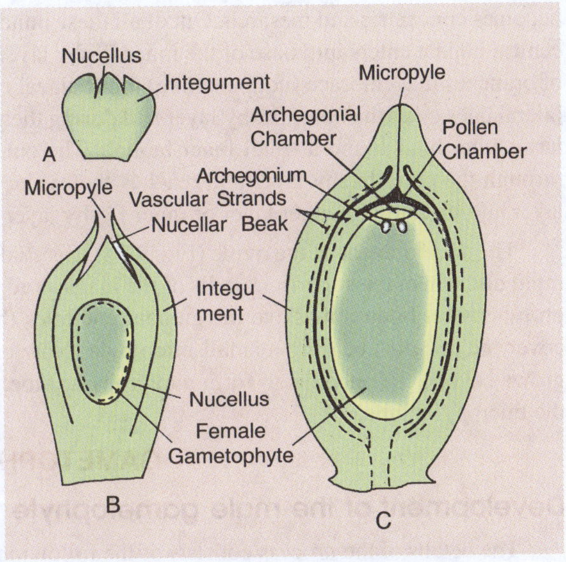

Fig. 9.24. (A-C) Stages in the development of the ovule *Cycas* (After De Silva and Tambiah)

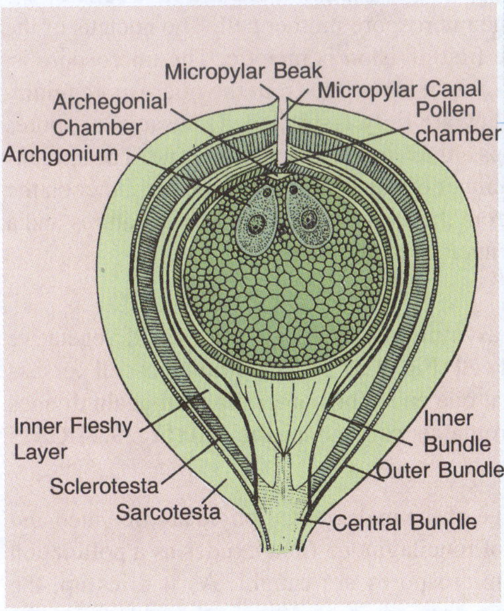

Fig. 9.25. *Cycas revoluta.* V.S. of ovule showing internal structure before fertilisation and its vascular supply.

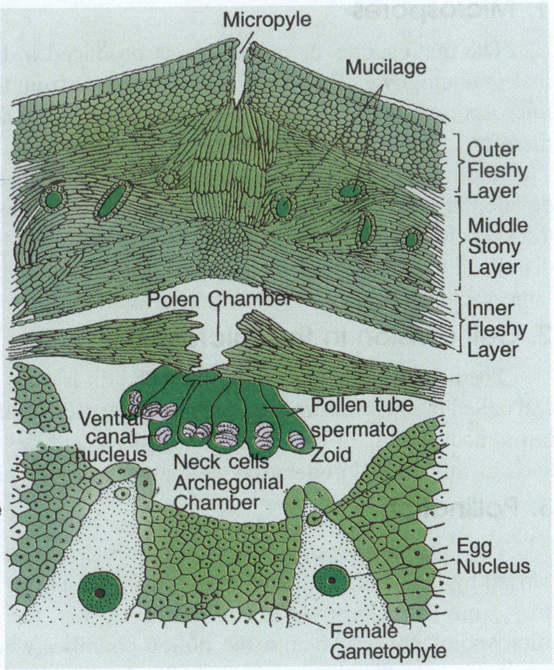

Fig. 9.26. V.S. mature ovule of *Cycas revoluta* showing detailed internal structure (After Swamy).

becomes concentric and mesarch. Out of all these bundles only three enter the base of the ovule. The central bundle enters into base of the inner fleshy layer of the integument and divides into a number of branches that embrace (not penetrate) the chalazal end of the nucellus. Out of these branches two lateral ones enter the outer fleshy layer and during their course upwards each of them bifurcates into large outer branch and a small inner branch. The outer branch is collateral and mesarch and runs through the outer fleshy layer and reaches the ovular apex. the inner branch penetrates the strong layer and supplies the inner layer of inner fleshy layer and reaches the micropylar end of the ovule.

The development of the ovule (Fig. 9.24) was studied by De Silva and Tambiah in 1952. Due to rapid division of the superficial cells of the megasporophyll the nucellus becomes evident as a small protuberance. Soon a massive integument encircles the base of the nucellus and grows rapidly to cover the nucellus bearing a small micropyle. Now onwards they grow together. The integument grows beyond the nucellus to form a micropylar tube. Later nucellar beak extends into the base of the micropylar tube.

GAMETOPHYTE

Development of the male gametophyte

The details of the process consists of the following steps : formation of microspore, division of microspore, formation of stalk and body cell, development of the pollen tube and formation of male cells or gametes. These developmental changes may be grouped under two main heads :— Developmental changes before pollination and developmental changes after pollination.

A. Developmental changes before pollination

1. Microspores

The microspores or pollen grains produced in the microsporangium are the first cells of the male gametophyte. They are formed in tetrads from the microspore mother cell. The nucleus of the microspore mother cell undergoes two divisions. The first division is meiotic. The microspore is, therefore, a haploid structure. It begins to germinate before it is set free from the microsporangium. The microspores are almost spherical in shape after they are fully fashioned. Thousands of spores develop in each microsporangium. Each spore develops a thick exine and a thin intine. The exine is thickest on the proximal side and becomes gradually thinner on the lateral sides and is thinner on the distal end of the spores. The starch grains disappear from the cytoplasm as the exine matures and a large vacuole appears on one side (Fig. 9.27, E). The nucleus is on one side.

2. Cell division in the microspore

The microspore nucleus divides. The division is asymmetrical and cuts off a small vegetative cell called the **prothallus cell,** and a large **antheridial cell** (Fig. 9.27 F). The prothallus cell persists throughout the life of the male gametophyte and lies at one end. The antheridial cell again divides and cuts off a small **generative cell** attached to the prothallus cell and a **tube cell** (Fig. 9.27 G.)

3. Pollination

The light, dry microspores at this three-celled stage of the male gemetophyte are liberated and carried by the wind, to the ovule. At this time a drop of mucilaginous fluid exudes as a pollination drop from the micropyle of the ovule. In this, the microspores are caught. As it dries up, the microspores are drawn into the pollen chamber where they are sealed by further drying of the mucilaginous exudation.

B. Developmental changes after pollination (Fig. 9.27 H—L)

4. Formation of stalk and body cell

The small, three-celled male gametophyte enclosed within the microspore wall resumes growth about a week after pollination. The generative cell now divides to form **the stalk cell,** which takes no further part in the development, and a **body cell** (Fig. 9.27 H). The two lie side by side.

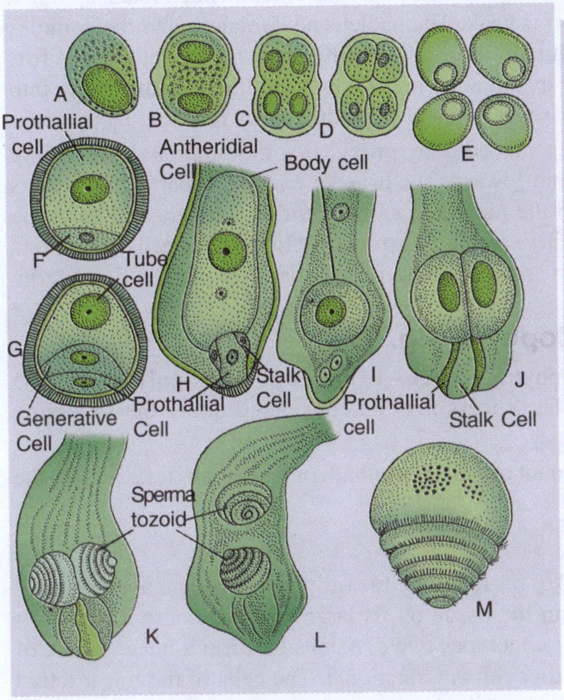

Fig. 9.27. (A—M). Microsporogenesis and development of male gametophyte in *Cycas revoluta* (After Rao). A—D Formation of micropores from the mother cell. E. young microspores. F—G. Two-celled and three-celled stages. H—L. Later stages of germination. M.A spermatozoid.

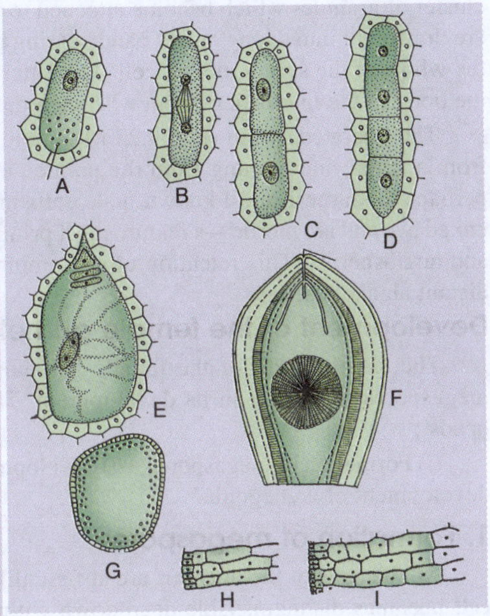

Fig. 9.28. (A—I). Stage in the development of the female gametophyte in *Cycas*, A-D. Stages in the formation of a megaspore tetrad that is linear. E. The lower cell of tetrad has enlarged as a megaspore. F. shows free nuclear division. G. shows cellular gametophyte. H. to I. illustrate the development of endosperm (A—E; after De Silva and Tambiah; H. I. after Chamberlain)

5. Development of the pollen tube

Soon after pollination the microspore absorbs moisture from the pollen chamber and swells. The exine bursts at its thin-walled apex. The tube cell stretches out through the break in the microspore wall in the form of a slender tube called the **pollen tube.** As the pollen tube elongates it grows downward through the nucellus (Fig. 9.26). Its action and function are entirely different from the pollen tube of angiosperms and conifers. It is not an agent of fertilization as it is in the angiosperms and conifers. It simply serves as a nutritive haustorium. It eats its way downwards gradually digesting and breaking down the tissue constituting the floor of the pollen chamber. The pollen tube now hangs freely in a cavity which is partly a pollen chamber and partly an archegonial chamber (Fig. 9.26). No further development of the male gametophyte takes place until the archegonia are mature. This interval between pollination and fertilization is about 4 months.

6. Formation of the male cells or gametes

Just before fertilization the body cell enlarges considerably (Fig. 9.27, H—I), The blepharoplasts, one at each pole of the body nucleus, appear (Fig. 9.27, I). The blepharoplasts consist of a conspicuous centre surrounded by a number of radiating fibrils. The body cell now divides, each half (daughter cell) carrying one blepharoplast (Fig. 9.27, J). The two daughter cells function as sperm mother cells, each metamorphosing into an antherozoid or a sperm (Fig. 9.27, K—L). The nucleus of each sperm mother cell enlarges and grows out into a beak. Its blepharoplast meanwhile breaks up into a cluster of granules which become attached to this beak. The nucleus now rotates, and the granules are drawn out into a long coiled band bearing hundreds of cilia. At this stage the nucleus nearly fills the whole of the sperm mother cell. The latter eventually breaks down and liberates the sperm into the body of the microspore from whence it passes into the pollen tube finally reaching its tip.

The liberated sperm (Fig. 9.27 M) is a top shaped body provided with numerous cilia arising from a spiral line running from the pointed end towards the broader end. The sperm of *Cycas* is perhaps the largest of all known male cells in the plant and animal world. They are visible to the naked eye and are motile—a feature most primitive and in sharp contrast to the male cells of conifers and angiosperms. This retention of swimming sperms in *Cycas* is a character handed down from distant algal ancestors.

Development of the female gametophyte (Fig. 9.28)

The development of the female gametophyte of *Cycas* is **monosporic** as only one of the megaspores takes part in its development. The process may be considered under the following heads :

(*i*) Formation of megaspores; (*ii*) development of female prothalli or gametophytes; and (*iii*) the development of archegonia.

1. Formation of megaspores

All the cells of the nucellus are at first alike. By the time it is fully developed an **archesporial cell** becomes distinguishable deep down within the tissue by its large size and dense contents. It functions as the **megaspore mother cell** and it undergoes two divisions to form a linear tetrad, of four cells. The first division of megaspore mother cell is reductional. The cells of the linear tetrad which are the potential megaspores are therefore, haploid. Usually only one cells of the linear tetrad, the lowest develops further. The upper three cells of the row abort. It is, sometimes, called the **embryo sac cell.** It is the first cell of the female gametophyte. It grows at the expense of the three abortive cells and the other adjoining cells. As it increases in size the surrounding cells of the nucellus take on the character of spongy tissue. This functional megaspore develops into the female gametophyte.

2. Development of the Female Prothallus

The nucleus of the enlarged megaspore enters upon rapid mitotic division forming a large number, about 1000 or more small nuclei by free nuclear divisions. They lie embedded in the cytoplasm of the megaspore. No cell walls are laid between them. A central vacuole then appears in the megaspore. As the vacuole grows in size the nuclei embedded in a thin layer of cytoplasm are forced to form a peripheral layer within the megaspore wall. Cytoplasm aggregates round these nuclei and finally cell walls are laid down between them. As a result a parietal tissue is developed in the megaspore. The cells of the parietal tissue grow towards the centre and eventually fill up the whole space (central vacuole or cavity) inside the megaspore or embryo sac. The tissue thus formed within the megaspore wall is the female **gametophyte** or **prothallus.** All its cells, except the outermost layer of small cells become filled with starch grains. The megaspore wall thickens and persists around the prothallus. The mature female prothallus is enclosed in the nucellus surrounded by the integument. It has no chlorophyll nor are the rhizoids developed.

3. Development of Archegonia (Fig. 9.29)

Immediately after pollination some of the cells at the periphery of the micropylar end of the female prothallus begin to increase in size. They are the **archegonial initials.** Usually 3 to 6 of them develop into archegonia, others perish. The functional archegonial initial (Fig. 9.29A) divides transversely into an outer **primary neck cell** and an inner **central cell** (Fig. 9.29 B). The primary neck cell by a vertical division forms the short neck of the archegonium which consists of two cells only (Fig. 9.29 C). There are no **neck canal cells.** The central cell increases in size and undergoes one asymmetrical division to form a **small ventral canal cell** and a large **egg cell** or the **oosphere** (Fig. 9.29, D). The ventral canal cell soon disorganizes. The egg or the oosphere enlarges. It is nourished by the adjacent prothallial cells through pits developed in the oosphere wall. The *Cycas* oosphere and its nucleus are the largest in the plant kingdom. The tissue of the female prothallus around the archegonia grows more vigorously than the rest and forms a depression, the **archegonial chamber,** into which open the necks of the archegonia (Fig. 9.26).

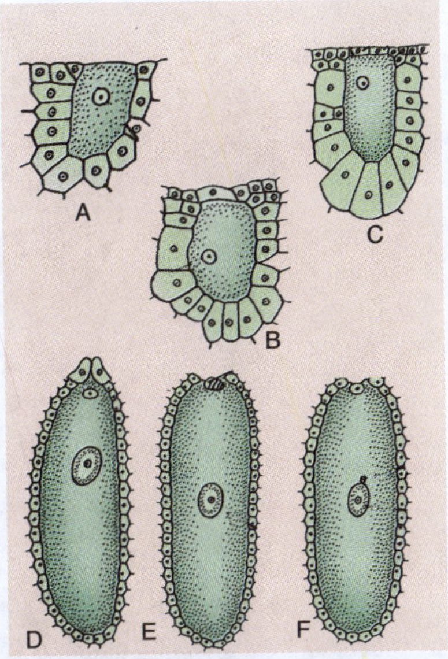

Fig. 9.29. (A—D). Stages in the development of an archegonium, E—F. Stages in the process of fertilisation. (After Tambiah and De Silva)

Fertilization (Fig. 9.29, E, F)

The end of the pollen tube containing the sperms and the tube nucleus grows downwards piercing the tissue of the pollen chamber which it gradually digests and disorganises. Now it hangs in the cavity formed partly by the pollen chamber and partly by the archegonial chamber (Fig. 9.26). It is swollen and highly turgid. Eventually it bursts and discharges its contents into the archegonial chamber. The sperms swim about in the fluid within the archegonial chamber. One of them swims to an archegonium, enters it and makes its way down the egg (Fig. 9.29, F). As the sperm is heading towards the egg its cytoplasmic membrane along with cilia is stripped off. Fertilization is accomplished by the naked male nucleus approaching and fusing with the female nucleus.

EMBRYOGENY (Fig. 9.30)

(*i*) Pro-embryo stage

The fertilized egg or the oospore increases in size (Fig. 9.30A).Its large nucleus by repeated free nuclear division gives rise to hundreds of free nuclei (Fig. 9.30C). They lie scattered through out the cytoplasm of the huge oospore. Eventually the cell walls are laid between the nuclei throughout the oospore. The central cells then break down resulting in the formation of a large central cavity surrounded by a layer or two of cells. This is the pro-embryo stage (Fig. 9.30, D—E). It consists of an embryonal mass of cells without any differentiation. The cells are more numerous at the base of the pro-embryo as the early free nuclei divide continuously for a longer period at the base of the oo spore.

(*ii*) Formation of the suspensor

Differentiation of cells starts at the basal region of the proembryo. The cells just behind the apex elongate considerably pushing the group of cells in front of them at the tip out of the arche gonium into the nutritive tissue or the female prothallus. These elongated or suspensor cells continue to elongate till they form long, tortuous, spirally coiled structures (Fig. 9.30, F—H). It is the **suspensor**. It may be several centimetres long. It serves to thrust the embryo into the nutritive prothallus tissue.

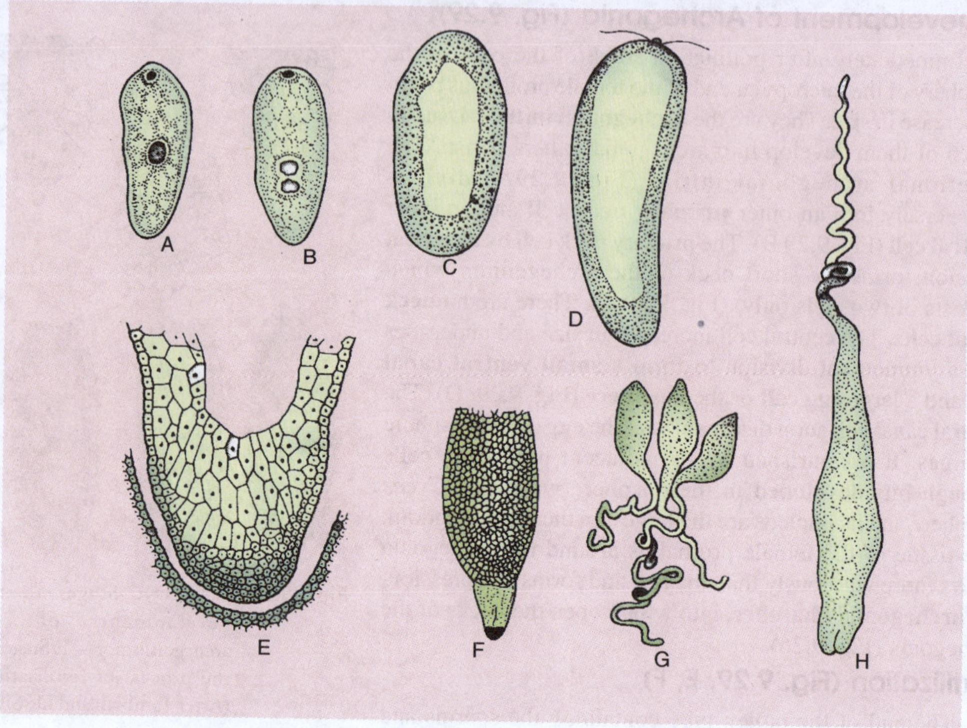

Fig. 9.30. (A—H). Embryogeny in *Cycas*. A. The oospore. B. First division of the oospore. C. Showing large number of free nuclei around a central vacuole. D. Young pro-embryo showing segmentation. E. It shows apical end of the proembryo with cell walls laid down completely. F. A young embryo. G. It shows three embryos, with long and coiled suspensors. H. An embryo with two cotyledons. (A—D, G after Swamy; E after De Silva and Tambiah; F. after Treub; H. after Schuster).

(*iii*) Differentiation of the embryo (Fig. 9.31)

The cells which are carried in front of the suspensor constitute the embryo proper. They are gradually differentiated into two cotyledons, plumule and the radicle embedded in a **coleorhiza.** The embryo is perfectly straight. Its development is slow. More than one archegonium may be fertilized in the ovule. The oospores of these develop into the pro-embryo stage. Each proembryo may produce a suspensor. All the suspensors twist together to form a compound suspensor. It supports only one embryo which reaches maturity. All others perish.

Seed formation (Fig. 9.32, A—B)

The integument hardens to form the seed coat. Its outer fleshy layer dries up and adheres firmly to the stony middle layer forming a hard type of seed. The inner fleshy layer had already shrivelled. The nucellus is completely crushed and utilized by the developing embryo. Within the seed coat lies the female prothallus. It is fleshy and functions as the food storage region of the seed. It is called the **endosperm.** Lying embedded in the middle of the endosperm is the straight embryo which extends the sole length of the seed. The suspensor is pushed back and packed against the micropylar end. At the radicle end of the embryo is developed a pad of tissue, the **coleorhiza.** It protects the delicate tip of the radicle against injury. It bursts at the time of germination.

Germination of the seed (Fig. 9.32, C—E)

It remains viable for not more than a few months. This is a primitive condition as compared with the highly persistent seeds of many of the angiosperms. During germination the seed absorbs water and the contained embryo resumes growth. The expansion of the embryo breaks open the hard shell of the seed. The coleorhiza protrudes and turns papery. It is then pierced by the emerging radicle which grows down and forms the tap root. The cotyledons remain in the seed coat. Their stalks elongate to carry the plumule out of the seed. The plumule grows erect. The growing plumule, at first, bears only scale leaves to pinnate foliage leaves with a few pinnae. The cotyledons remain underground (hypogeal) enclosed in the female prothallus or endosperm. They remain active and absorb food from the endosperm for the growing embryo. When the endosperm is completely used up and finished the cotyledons also dry up.

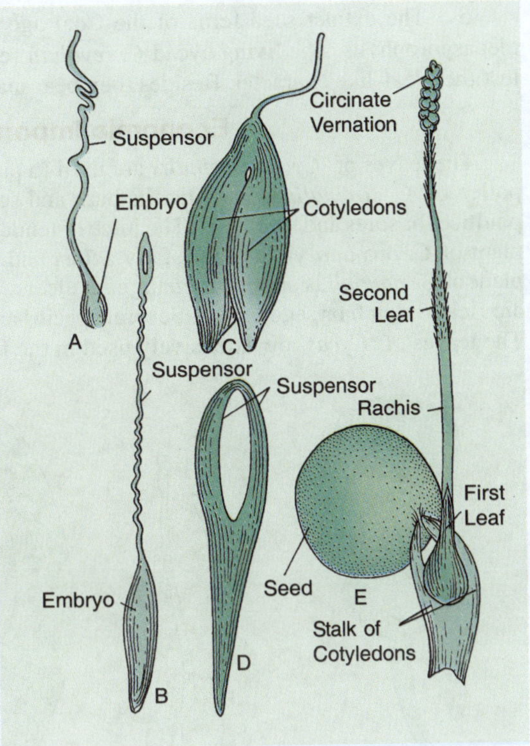

Fig. 9.31. (A-E). *Cycas*. Embryogeny. A and B, embryos with long and coiled suspensors; C, an embryo with two cotyledons; D, young embryo with two suspensors; E, young seedling.

Conclusion

The Cycads are obviously primitive plants. They in many respects bear resemblance to that group of remarkable plants which are clearly intermediate between the ferns and the typical Gymnosperms. This group is known as the **cycad ferns** or **Cycadofilicales.** The latter are the earliest seed bearing plants and are fern-like in their appearance. Many of these, which were formerly thought to be ferns from their foliage, are now known to be primitive gymnosperms, because well-developed seeds have been found attached to their leaves. Such plants were abundant in the Coal period and are now extinct. The living Cycads resemble these plants in many respects and hence are often called the 'living fossils'. The primitive features common between the living cycads and the fossil cycadofilicales on one hand and the ferns on the other hand are as follows :

1. The living cycads and the fossil cycad ferns have a fern like habit. Many of these resemble true ferns. Their stems are covered with an armour of persistent leaf bases. Their foliage is definitely fern-like. They have a circinate ptyxis as in the ferns.

2. Their stems possess large pith and cortex and a well developed fern-like primary wood consisting of tracheids and xylem parenchyma only. There are no true vessels as in ferns. The phloem has sieve tubes and phloem parenchyma. Like ferns the companion cells are absent.

N.B. The presence of secondary wood the living cycads share only with some of the cycad ferns. It is entirely absent in the ferns.

3. As in the ferns the megasporophylls are leaf-like.

4. In both the cycads and fern cycads the ovules are borne laterally on the megasporophylls. Each ovule consists of a megasporangium surrounded by an integument. The fern megasporangium differs from the cycad ovule in the absence of the integument. It corresponds to the nucellus.

5. Like ferns the sperms are multiciliate and motile. The gymnosperms with motile sperms form a connecting link between ferns and more modern gymnosperms in which the sperms are not motile.

6. The extinct seed ferns of the Coal age bear their ovules along the margins of leaf-like megasporophylls. The living cycad *C. revoluta* retains that feature. The megasporophylls have not lost their leaf-like character. Besides they bear marginal ovules.

Economic Importance of Cycads

The leaves of *Cycas circinalis* are used to prepare hats, baskets and mats, by the natives. The pollen of *C. circinalis* is narcotic. Its bark and seeds ground to paste with coconut oil are used as poultice for sores and swellings. The juice of tender leaves is useful for flatulence and vomiting. The plant of *C. revoluta* yields sago, fatty acids, and a colouring matter. The resin obtained from the plant of *C. rumphii* is applied to malignant ulcers. The pith and cortical cells of *Cycas revoluta* stem are rich in starch or sago grains. Several species are extensively cultivated for decorative purposes. The leaves of '*Cycas*' are extensively used in the floral decorations.

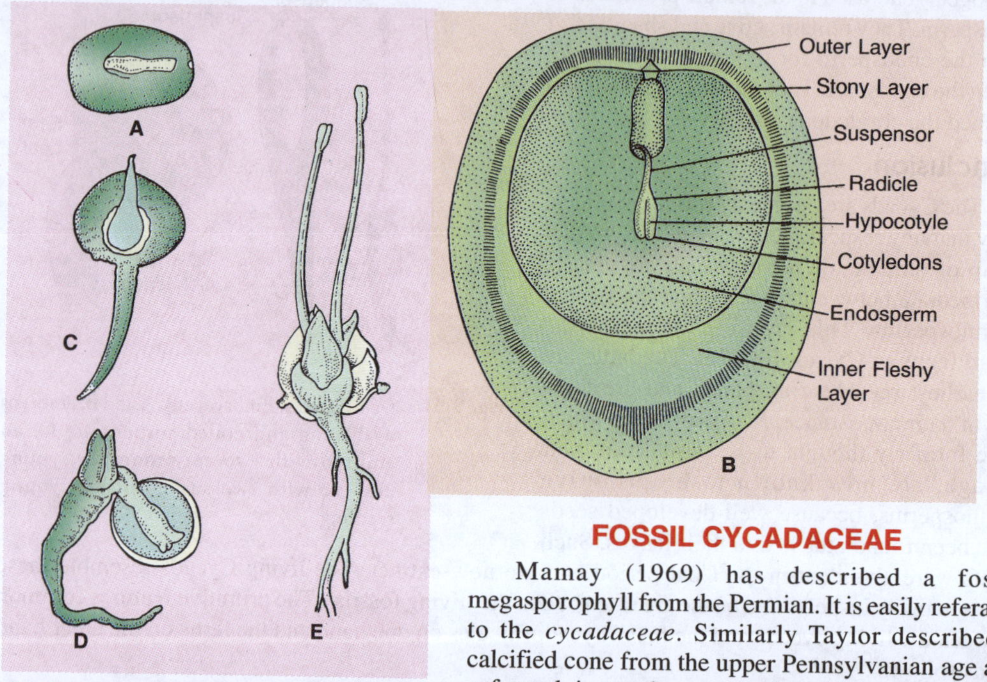

Fig. 9.32. (A-E). Seed and its germination in *Cycas cercinalis.* A—B. L.S. through mature seed showing embryo and endosperm. C—E. Stages in germination. (A—B

FOSSIL CYCADACEAE

Mamay (1969) has described a fossil megasporophyll from the Permian. It is easily referable to the *cycadaceae*. Similarly Taylor described a calcified cone from the upper Pennsylvanian age and referred it to the cycadales. This cone bears appendages in a spiral sequence and each bears on its lower surface 8—10 elongate microsporangia arranged in a single row. These findings suggest that *cycadaceae* may have come into existence during the Palaeozoic era.

Yokoyama (1911) reported a species of *Cycas* from the early Tertiary of Japan and named it as *C. fujiana*. He described its leaves which resembled those of *C. revoluta*. Some other leaves were also described from the Mesozoic era and only *Cycaditescteis* described by Harris (1950) from Yorkshire and *C. blomquistii* described by Florin (1933) are referable to the genus *Cycas*.

Palaeocycas integer (Nathorst) Florin was described from the upper Triassic of Sweden (Fig. 9.30, A). This generic name has been given to the structures that bore seeds. Florin (1933) described its leaves under the generic name *Bjuvia* and described them under the specific name *B. simplex*.

Plate XIII : *C. circinalis* : Cluster of mature seeds attached with sporophyll

Plate XIV : Large seed of *C. circinalis*

These leaves were simple and measured about one metre in length and 20 cm. broad. They looked like the banana leaves. The stems were not available. The seed-bearing organs or megasporophylls

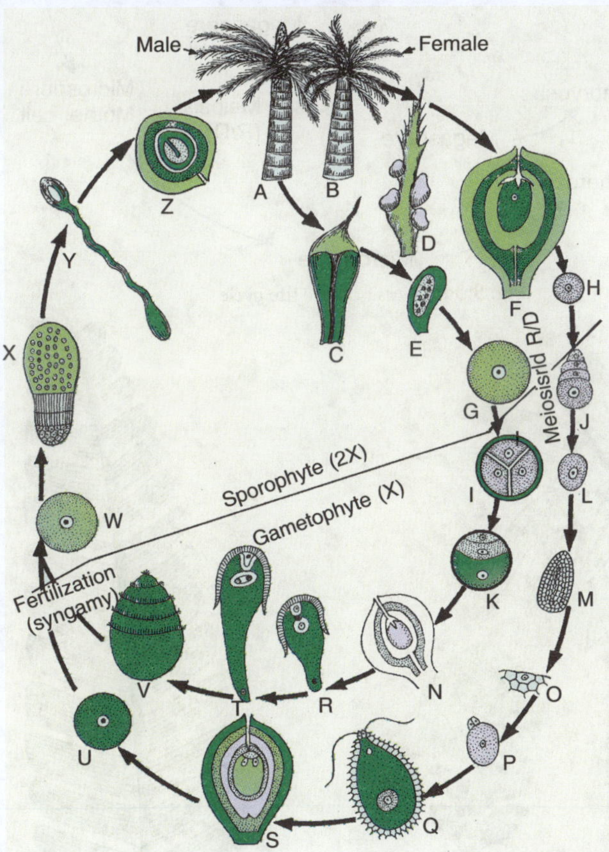

Fig. 9.33. *Cycas* sp. Diagrammatic life-cycle; A, male plant; B, female plant; C. microsporophyll; D, female strobilus; E, microsporangium; F, ovule; G, microspore mother cell; H, megasopore mother cell; I, microspore tetrad; J, megaspore linear tetrad; K, male gametophyte; L, megaspore; M, female gametophyte; N,ovule; O, archegoinal initial; P, young archegonium; Q, mature archegonium; R, germinating pollen; S, germinating pollen in pollen chamber; T, germinating pollen; U, egg; V, spermatozoid; W, oospore; X, young embryo; Y embryo; Z, seed.

resembled those of modern *Cycas* and had a laminate upper part with a cylindrical stalk bearing four seeds (Fig. 9.35, B). Only scars of seeds were seen and it was supposed by Florin that the four scars bore the seeds. Florin reconstructed the whole plant from her imagination and showed it to

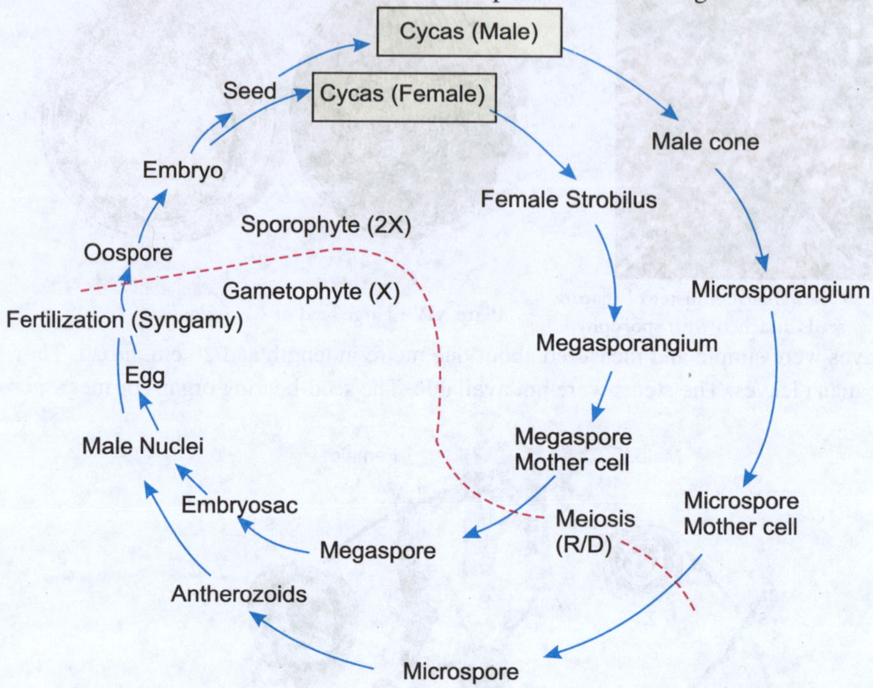

Fig. 9.34. *Cycas* graphic life cycle

Fig. 9.35. Fossil *Cycadaceae*. A. reconstruction of the plant of *Palaeocycas integer* by Florin. B. a megasporophyll of *Palaeocycas niger*.

Fig. 9.36. *Nilssoniaceae*. Leaf of *Nilsonia compta* (= *Baenia gracilis*) (After Andrews).

bear a crown of leaves on a cylindrical axis (Fig. 9.35, A). The megasporophylls were shown loosely arranged in the centre of the crown of leaves. The leaves (*B. simplex*) and the seed-bearing organs (*P. integer*) belong to one and the same plant is evident from their similar cuticular, stomatal and epidermal structures. The entire reconstruction by Florin is merely a guess-work. The microsporan giate organs are unknown.

Another genus *Cycadospadix hannocquei* (Panel) Schimper is another *Cycas* like megasporophyll described from the Jurassic age of the Mesozoic era.

Diooniticarpidium pennaeforme is another pinnate megasporophyll described from the Triassic beds of Germany by Lilienstern (1928) and Kraussel (1949).

Family NILSSONIACEAE (Figs. 9.36 & 9.37)

Sporne (1965) recognised this family as including certain leaf-bearing (*Nilssonia pseudoctenis*), seed-bearing (*Baenia*) and pollen-bearing (*Androstrobus*) organs.

Harris (1961) described two species of the seed-bearing genus *Baenia* from the Jurassic of Yorkshire. The genus had peltate sporophylls that bore two ovules on the underside of the peltate disc. The sporophylls were spirally arranged around a central axis. The integument in the ovules was fused with the nucellus except at its upper part. The single integument was differentiated into three layers usually for the *Cycadaceae*. An evidence of vascular traces running along the integument upto the point where it was free from the nucellus was also available. *B. gracilis* and *B. mamayi* are the two species described by Harris.

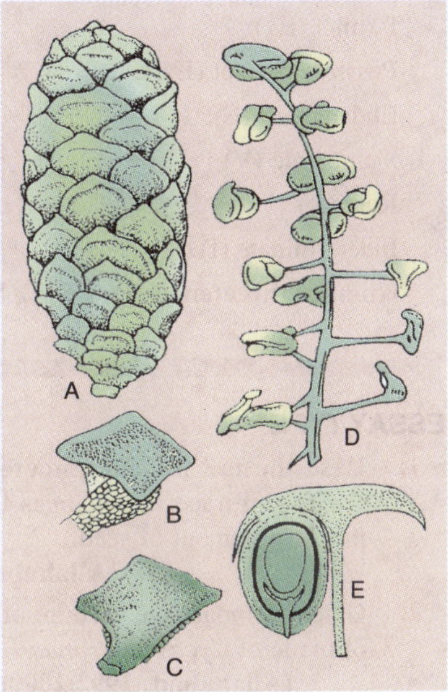

Fig. 9.37. Fossil cycads. A. Male cone of *Baenia gracilis* B. C. The microsporophylls. D. Female cone of *Baenia gracilis.* E. L. S. of megasporophyll with attached ovule. (After Harris).

Thomas and Harris (1960) discovered the pollen grains of male cones known as *Androstrobus mamis* from the micropyles of ovules of *Baenia gracilis*. *A. mannis* and *A. wonnacottii* are small male strobili that are very compact and measure about 5 cm long and 2 cm broad. The microsporophylls are peltate and arranged spirally around a central axis. Numerous finger-like microsporangia are scattered along their abaxial sides. *A. mannis* belongs to the ovules attached to the latter fructification. Morevoer their stomatal structure is also similar. The leaf *Nilssonia compta* (Fig. 9.36) was also attributed by Harris (1941) to these fructifications for similar structures of their stomata and their occurrence in the same rocks. *N. compta* is a simple leaf 40 cm long and 9 cm wide, shortly petiolate, linear (Fig. 9.36) or subulate with entire margin. The lamina is irregularly cut and has parallel veins that arise at right angles from the midrib. The stomata are haplocheilic with straight walled epidermal cells. Stomata occur on the lower epidermis.

Baenis mamayi is another seed-bearing species whose megasporophylls slightly bifurcate at the top and are not expanded like those of *B. gracilis*. *Androstrobus wonnacottii* and *Nilssonia tenuinervis* are the male cones and the leaves that can be associated with it, respectively.

Harris has reconstructed the plant of *B. mamayi* depending merely upon speculation. He arrived at his conclusion by imagining that the lax male and female cones must be borne upon an aerial axis and this stem should be stout because the leaf bases of *N. tenuinervis* are broad.

Pseudoctenis and *Ctenis* are other leaf genera of *Cycadales*. Both are unipinnate with parallel venation which in the latter have cross-connections. The stomata are haplocheilic and the epidermal cells are wavy. *Pseudoctenis lanei* is associated with *Androstrobus prisma*.

1997 IUCN Red List of Threatened Plants

Family Cycadaceae

Madagascar; Eastern and Southeast Asia; Indomalaysia; Australia; Polynesia.

Number of genera : 1

Number of species : : 35

54.3% of Total Species in Cycadaceae are threatened :

IUCN category	Taxa	Species
Extinct (Ex):	1	1
Possibly extinct (Ex/E) :	—	—
Endangered (E):	5	5
Vulnerable (V) :	10	10
Rare (R) :	4	4
Indeterminate (I) :	–	–
Number threatened (Ex / E, E, V, R, I)	19	19

Globally threatened Cycadaceae

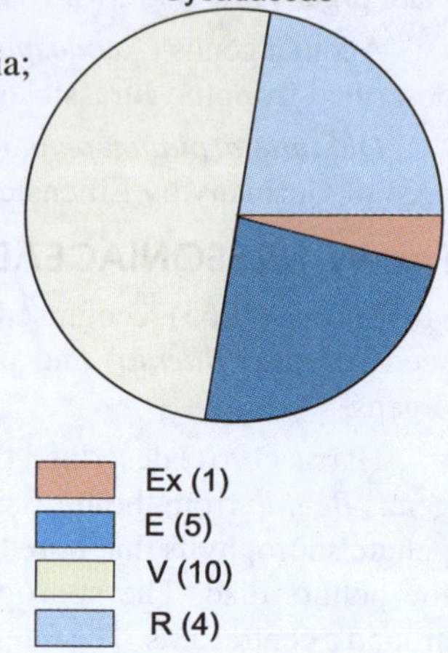

Questions

ESSAY TYPE

1. Describe the female gametophyte of *Cycas* and trace the changes leading to the development of seed. **(Allahabad, 1993)**
2. Give a comparative account of structure of ovule of *Cycas* and *Pinus*. **(Allahabad, 1992,2000; Kanpur 2002; Lucknow, 1997; Gorakhpur, 1995; Kerala, 1998; Guru Nanak Dev, 1992)**
3. Mention the systematic position of *Cycas* and *Pinus* and point out the differences in their wood structure. **(Allahabad, 1993)**
4. Give a comparative account of the female reproductive organs in *Cycas* and *Pinus*. **(Allahabad, 1994)**
5. Describe the internal structure of stem of *Cycas* and *Pinus* before and after secondary growth. **(Allahabad, 1995, 2000; Gorakhpur, 1997)**
6. Give a comparative account of male gametophyte of *Cycas* and *Pinus*. **(Allahabad, 1996; Gorakhpur 2003; Kerela, 1996; Vir Kumar Singh, 1992)**
7. Describe the development of embryos in *Cycas* and *Pinus*. **(Allahabad, 1997)**
8. With the help of only labelled diagrams show the salient features of the life cycle of *Cycas*. **(Allahabad, 1998)**
9. Discuss critically
   (*a*) The plants of *Cycas* are indeed ancient and archaic and are also called as living fossils of cycadaceae.
   (*b*) Some authors believe that the presence of transfusion tissues in *Cycas* leaflets merely represents the vestigial side veins. **(Allahabad, 1999)**
10. Compare male and female gemetophytes, of *Cycas* and *Pinus*. **(Kanpur, 1996)**
11. Where does *Cycas* occur ? What are its Indian species? Give fern like characters of *Cycas*. **(Kanpur, 1998)**

12. Describe the post pollination development of male gametophyte in *Cycas*. **(Kanpur, 2002)**

13. Explain the development of female gametopyte in *Cycas*. **(Kanpur, 2003)**

14. Describe the megasporogenesis, megagametogensis and fertilisation in *Cycas*. **(Agra, 1992)**

15. **Write detailed notes :**

    (*a*) Coralloid root of *Cycas* **(Agra, 1995)**

    (*b*) Male gametophyte of *Cycas* **(Agra, 1995; Lucknow, 2001)**

    (*c*) Anatomy of Rachis of *Cycas* **(Kanpur MSc, 1986)**

    (*d*) Pollination in *Cycas* **(Lucknow, 2000)**

    (*e*) Female gametophyte of *Cycas* **(Lucknow, 2001)**

    (*f*) External morphology of *Cycas* **(Kerla, 1998)**

16. Give an account of the distribution of *Cycas* in India. **(Rohilkhand, 1994)**

17. Describe the male gametophyte of *Cycas*. **(Rohilkhand, 2003; Kumaon, 1996, 1999; Awadh, 1995, 1997, 2000; Kerla, 2001)**

18. Describe mechanism of pollination and fertilisation in *Cycas*. Why is it called a living fossil? **(Awadh, 1999)**

19. Give an illustrated account of the internal structure of the young and old stem of *Cycas*. **(Lucknow, 1993,2001; Allahabad, 1999)**

20. (A) Describe secondary growth in the stem of *Cycas*. **(Lucknow, 1997, 1998, 2003; Bundel Khand, 1989)**

    (B) Give a detailed account on the anatomy of stem (old and young) and reproduction in *Cycas* with labelled diagrams. **(Rohilkhand, 2001)**

21. Describe the development of embryo in *Cycas*. **(Poorvanchal, 1998; Mahatma Gandhi, 1996)**

22. Give a comparative account of the root anatomy in *Cycas* and *Pinus*. **(Gorakhpur, 1998)**

23. Describe the mechanism of pollination, fertilisation, development of embryo and seed formation in *Cycas*. **(Punjab, 1992)**

24. Describe the female cone of *Cycas* with a note on the development of archegonium. **(Kerala, 1997)**

25. Give an account of microgametogenesis is in *Cycas*. **(Kerela, 1997)**

26. Give an account of mature ovule and embryo development in *Cycas*. **(Mahama Gandhi, 1998)**

27. Describe internal structure of *Cycas* leaflets and point out the xerophytic adaptations in it. **(Mahatma Gandhi, 1997)**

28. Give an account of the microsporangia and development of male gametophyte of *Cycas*. **(Madras, 1998)**

29. Describe the events taking place inside the ovule from the time of appearance of archegonium up to the formation of mature archegonium. **(Vikram, 1999)**

30. Give an illustrated account of reproduction in *Cycas*. **(Karnataka, 1999; Agra, 1998)**

31. With the help of labelled diagrams only depict the life cycle of *Cycas*. **(Kanpur, 1998)**

34. Give a brief account of fossil cycads.

35. Describe the fossil members of family Nilssoniaceae.

36. With the help of diagram, briefly discribe the fossil members of family Cycadaceae.

SHORT ANSWER TYPE

37. Point out differences between the following pairs:

    (*i*) Normal and Coralloid roots of *Cycas* **(Allahabad 1992, 1996, 2000; Rohil Khand MSc, 1990; Lucknow 1996,1997, 2000; Poorvanchal, 1996)**

    (*ii*) Microspores of *Cycas* and *Pinus* **(Allahabad, 1992, Rohil Khand, 1995, Gorakhpur, 1995)** (*iii*) Manoxylic and pycnoxylic wood **(Allahabad, 1992, Kanpur M.Sc., 1984, 1995, Lucknow, 1995, 1996, 2000, 2001; Gorakhpur, 1990, 1992, 1995, 1996, 1998; Himachal Pradesh, 2000; Poorvanchal M.Sc., 2002)**

(*iv*) Microsporophylls of *Cycas* and *Pinus* **(Allahabad, 1993)**

(*v*) Pollengrains of *Cycas* and *Pinus* **(Allahabad, 1994, 1996, 2000; Lucknow, 1997)**

(*vi*) Mesophylls of *Cycas* and *Pinus* **(Allahabad, 1994)**

(*vii*) Certripetal and centrifugal xylen **(Allahabad, 2002; RohilKhand, M.Sc., 1990; Lucknow, 1996, 2000; Gorakhpur,1995)**

(*ix*) Male prothallus of *Cycas* and *Micro Cycas* **(Kanpur, M.Sc, 1993; Rohil Khand, M.Sc, 1993)**

(*x*) Archegonium of *Cycas* and *Pinus* **(Rohil Khand M.Sc 1993)**

(*xi*) Pre-pollination changes in the male gametophyte of *Cycas* and *Pinus* **(Rohil Khand, 1993)**

(*xii*) Megasporophylls of *Cycas* and *Pinus* **(Rohil Khand, 1995;)**

(*xiii*) *Palaeocycas* and *Cycas*

(*xiv*) Cycadaceae and Nilssoniaceae

38. Draw well labelled diagrams to show the structure of the following:

(*a*) T.S young stem of *Cycas* **(Allahabad, 1992; Meerut, 1999; Kumaon, 2001, 2003)**

(*b*) T.S Coralloid Root of *Cycas* **(Allahabad, 1993, 1994, 1998; Kanpur 1996,1998; Rohil Khand, 2000, 2003; Awadh, 1993,1996; Merrut, 1998, 2000; Lucknow, 1993, 2000; Poorvanchal, 1996,1999,2002; Gorakhpur, 1995 1999, 2003; Punjab 1992; Kerala, 2000; Agra, 1998)**

(*c*) T.S leaflet of *Cycas* **(Allahabad, 1993, 1995, 2000, Kanpur, 1994, 1995, 1997; Agra, 1993; Rohil Khand, 1995; 2000; 1997; Awadh, 1991,1994,1996; Poorvanchal, 2001; Gorakhpur, 1999; Lucknow, 2003 Virkumar Singh,1992)**

(*d*) T.S Rachis of *Cycas* at middle region **(Allahabad, 1994, 1996, 2000; Kumaon, 2000; Awadh, 1995, 1997; Merrut, 2000; Lucknow, 1996; Gorakhpur, 1992, 1995, 1997)**

(*e*) L.S Ovule of *Cycas* **(Allahabad, 1995, 1996, 1998; Rohil Khand, 1994, 1997; Awadh, 1993, 1998, 2000, 2003; Merrut, 1999; Lucknow, 1998, 2000; Poorvanchal, 1997; Gorakhpur, 1990, 1993, 1997, 1999, 2003; Punjab 1998)**

(*f*) Microsporophyll of *Cycas* **(Allahabad, 1997; Gorakhpur, 1998)**

(*g*) L.S Ovule of *Cycas* at the time of fertilisation **(Kanpur, 1994; Lucknow, 1999; Gorakhpur, 2003)**

(*h*) Male cone of *Cycas* (Awadh, 1998)

(*i*) T.S. Leaflet of *Cycas revoluta* **(Lucknow, 1997, 1999, 2001; Gorakhpur, 1996)**

(*j*) T.S. Normal Root of *Cycas* **(Poorvanchal, 1996; Gorakhpur, 1990)**

(*k*) L.S. Mature seed of *Cycas* **(Gorakhpur, 1992)**

(*l*) Apogeotropic roots of *Cycas* **(Gorakhpur, 1993)**

(*m*) T.S. microsporophyte of *Cycas* **(Gorakhpur, 1994; Agra, 1999)**

(*n*) T.S. old stem of *Cycas* **(Gorakhpur, 1994)**

(*o*) Petiole of *Cycas* **(Gorakhpur, 1996)**

(*p*) Coralloid root of *Cycas* **(Meerut, 1998; Jawaji 1998; Jodhpur, 1999; Bundelkhand, 1999; Kumaon, 2002)** (*q*) Reconstruction of *Paleoecycas integer*

(*r*) leaf of *Nilsoni compta*

39. Write short notes on the following :

(*i*) Transfusion tissue
(Allahabad, 1992, **1996; Kanpur 1996; Awadh, 1994, 1997; Lucknow, 1993, 1997, 1999, 2001; Poorvanchal, 1992, 2002; Gorakhpur 1994, 1998; Lucknow, 1993, 1997, 1990, 2001; Gorakhpur, 1994, 1998; Bhagalpur,1991; Magadh, 1994; Calicut, 1997; Bangalore, 2001; Chhapra, 1999)**

(*ii*) Microsporphyll of *Cycas* **(Allahabad, 1992; Awadh, 2000; Lucknow, 1993, 2001; Poorvanchal, 1999, 2000, 2003; Kerala, 2000; Bangalore, 2001; Rohil Khand, 2002)**

(*iii*) Pollination and fertilisation in *Cycas* **(Allahabad, 1993, 1997, 1998; Kanpur, 1995; Awadh, 1994; Gorakhpur, 1992, 1994, 1998)**

(*iv*) Fern like characters of *Cycas* **(Allahabad, 1995, 1997; Kanpur MSc, 1999; Poorvanchal; 1998; Kerla, 1996)**

(*v*) Habit and Habitat of *Cycas* and *Pinus* **(Allahabad, 1998)**

(*vi*) Monoxylic and Polyxylic stems of *Cycas* **(Allahabad, 1999,2001)**

(*vii*) Antherozoid of *Cycas* **(Kanpur, 1996)**

(*viii*) Male cone of *Cycas* **(Kanpur,1997; Gorakhpur, 1990, 1993; Meerut, 2001)**

(*ix*) Secondary xylem of *Cycas* and *Pinus* **(Kanpur, 2003)**

(*x*) Embryo and seed formation in *Cycas* **(Agra, 1993)**

(*xi*) Economic importance of *Cycas* **(Agra, 1995; Gorakhpur, 1991)**

(*xii*) Centripetal and centrifugal xylem in *Cycas* **(Kanpur MSc 1999; Lucknow, 1999)**

(*xiii*) Diploxylic condition in *Cycas* **(Rohilkhand, 1995)**

(*xiv*) Megasporophyll of *Cycas* **(Rohilkhand, 1999; Kumoon, 1996; Meerut, 2000; Lucknow, 1996; Poorvanchal, 1997, 2001; Gorakhpur, 1991, 1993, 1995; Kerala, 1997; Bhopal, 1997)**

(*xv*) Coralloid Root of *Cycas* **(Rohilkhand, 1999; Kumaon, 1996; Awadh, 1999; Lucknow, 1998; Poorvanchal, 1998, 2001; Mahatma Gandhi, 1997)**

(*xvi*) Distribution of *Cycas* in India **(Rohilkhand, 2000; Poorvanchal, 2002; Himanchal Pradesh, 2000)**

(*xvii*) Ovule of *Cycas* **(Kumaon, 1997,1998; Lucknow, 1995; Kerala, 1996, MahamtaGandhi, 1997)**

(*xviii*) Male *Cycas* are rare **(Rohilkhand MSc. 1992, 1995)**

(*xix*) Vascular bundle of *Cycas***(Awadh, 1991)**

(xx) Pollen grain in *Cycas* **(Awadh, 1991)**

(*xxi*) *Cycas* is living fossil **(Meerut, 1998; Lucknow, 2001)**

(*xxii*) Primitive and Advanced characters of *Cycas* **(Meerut, 1999)**

(*xxiii*) Rachis of *Cycas* **(Lucknow, 1994)**

(*xxiv*) *Cycas revoluta* leaflet **(Lucknow, 1998)**

(*xxv*) Archegonum in *Cycas* **(Lucknow,1999)**

(*xxvi*) Leaflets of *Cycas* **(Lucknow, 1999; Poorvanchal, 1996; Kerala, 1998)**

(*xxvii*) Apogeotropic roots **(Gorakhpur, 1992)**

(*xxviii*) Centripetal xylem **(Lucknow 2003)**

(*xxix*) Internal structure of coralloid root of *Cycas* **(Kerala, 1996)**

(*xxx*) Male gametophyte and sperms of *Cycas* **(Mahatma Gandhi, 1998)**

(*xxxi*) Anatomy of petiole of *Cycas* **(Chhapra, 1999)**

(*xxxii*) Fossil cycadaceae

(*xxxiii*) Nilssoniaceae

40. With the help of labelled diagrams only differentiate between :

(*a*) Vascular bundles of Stem and petiole of *Cycas* **(Meerut, 1997)**

(*b*) Normal and coralloid roots of *Cycas* **(Meerut, 1997; Agra, 1999; Kanpur,1998,1999)** (*c*) Roots of *Cycas* and *Pinus* **(Gorakhpur,1992)**

(*d*) Megasprophylls of *Cycas revoluta* and *C. cercinalus* **(Gorakhpur, 2000)**

41. In which shape the Vascular bundles in the rachis are arranged? **(Gorakhpur, 1994)**
42. Give the characteristic features of the microspores of *Cycas.* **(Gorakhpur, 1997)**
43. Mention four fern like characters of *Cycas.* **(Gorakhpur, 2003)**
44. (*a*) Describe the structure of seed of *Cycas* in detail. **(Punjab, 1992)**
    (*b*) Discuss the economic importance of *Cycas.* **(Punjab, 1992)**
45. What are the primitive characters of *Cycas.* **(Kerala, 1996)**
46. What is the ploidy level of the endosperm in a diploid *Cycas* plant? **(Kerala, 1996)**
47. Why is the *Cycas* called a palmfern? **(Punjab, 1996, 1999)**
48. What is the function of pollen tubes in *Cycas*? **(Mahatma Gandhi, 1998; Madras, 1998)**
49. Name two common species of *Cycas.*
50. Why is *Cycas* called a living fossil ? Draw L.S. seed of *Cycas.* **(Himanchal Pradesh, 2000)**
51. Which gymnosperm is known as living fossil?
52. Mention two characters common to *Cycas* and *Marsilea.* **(Allahabad, 1995; Lucknow, 1996; Poorvanchal, 2000; Gorakhpur, 1990, 1993)**
53. How is epidermis of stem of *Equisetum* different from that of *Cycas* leaflet? **(Allahabad, 1995)**
54. Name the plant from which sago is obtained. **(Allahabad, 1996; Lucknow, 1996; Gorakhpur, 1995)**
55. Mention the function of pollen tube in *Cycas.* **(Allahabad,1998)**
56. Amongst *Cycas* and *Pinus*, whose ovules are larger in size? **(Allahabad, 1998)**
57. Although the trees of *Cycas* may reach to a height of several metres, yet they do not producetimber why? **(Allahabad, 1999; Gorakhpur, 1999)**
58. What is toxic about *Cycas* plant? **(Allahabad,1999)**
59. Name the alkaloid which is found in *Cycas* plant. **(Allahabad, 2000; Gorakhpur, 2003)**
60. Give two points of resemblance between *Cycas* and *Pinus.* **(Allahabad, 2000)**
61. What is the botanical name of sagopalm ? **(Allahabad, 2001; Kanpur, 1997, Poorvanchal, 1997)**
62. When and by whom the spermatozoids, of *Cycas revoluta* were discoverd ? **(Allahabad, 2001)**
63. Which gymnosperm has ciliated sperms ? **(Kanpur, 1995)**
64. Name a gymnosperm which has algal colony in its roots. **(Kanpur, 1995; Meerut, 1997)**
65. Write the systematic position of *Cycas.* **(Knapur, 1995)**
66. The algae found in corolloid root of *Cycas* belong to which class ? **(Kanpur, 1996)**
67. Where are the male cones of *Cycas* located? **(Kanpur, 1996)**
68. Small amount of sago is obtained from which plant ? **(Kanpur, 1998)**
69. Which plant has manoxylic wood? **(Kanpur, 1998)**
70. With the help of suitable diagrams ? describe the structure of sperm of *Cycas revoluta.* **(Kanpur, 2001)**
71. Give on account of the structure and function of transfusion tissues. **(Kanpur, 2001)**
72. Name five Indian species of *Cycas* and their distribution. **(Agra, 1991)**
73. What is the significance of leaf like appearance of megasprophyll of *Cycas?* **(Kanpur MSc, 1989)**
74. Differentiate between the leaflet of *Cycas circinalis* and C. *revoluta* on the basis of two characters. **(Lucknow, 1996)**
75. Describe sperm of *Cycas.* **(Poorvanchal, 1998)**

76. Can *Cycas* plants be grown without coralloid roots? **(Poorvanchal, 1999)**

77. Name the term used for tracheid like cells occurring outside the vascular bundles of *Cycas*. **(Poorvanchal, 2000; Gorakhpur, 1995)**

78. Give comparative account of the following:

(*a*) Secondary growth in stem of *Cycas* and *Pinus*. **(Rohilkhand, 1993; Gorakhpur, 1991)**

(*b*) Megasporophyll of *Cycas* and *Pinus* **(Rohilkhand, 1993, 1994)**

(*c*) Foliage leaves of *Cycas* and *Pinus* **(Rohilkhand, 1994)**

(*d*) Leaf of *Cycas* and *Pinus* needle **(Rohilkhand, 1994)**

(*e*) Fertilisation in *Cycas* and *Pinus* **(Meerut, 1997)**

(*f*) Roots of *Cycas* and *Pinus* **(Lucknow, 1998)**

(*g*) Male cone of *Cycas* and *Pinus* **(Lucknow, 1998)**

(*h*) Misospores of *Cycas* and *Pinus* **(Gorakhpur, 1991)**

(*i*) Coralloid root of *Cycas* and mycorrhizal roots of *Pinus* **(Kumaon, 2004)**

79. Give an account of structure of ovule in *Cycas*. **(Poorvanchal, 2002)**

80. What could be the best function attributed in transfusion tissues in *Cycas ?* **(Poorvanchal, 2003; Gorakhpur, 1996)**

81. List the characters common to *Cycas* and ferns. **(Gorakhpur, 1991, 1997)**

82. Are the coralloid root of *Cycas* useful for aeration of nitrification ? **(Gorakhpur, 1991)**

83. What are the resamblances between *Cycas* and ferns? **(Gorakhpur, 1992)**

84. What type of leaf traces are found in *Cycas* ? **(Gorakhpur, 1994)**

85. Mention the name of alga found in old roots of *Cycas* **(Gorakhpur, 1994)**

86. Mention *four features* of importance of the following :

(*a*) Endosperm in *Cycas* **(Mahatma Gandhi, 1998)**

(*b*) Sperms of *Cycas* **(Mahatma Gandhi, 1997)**

(*c*) Function of endosperm in *Cycas* **(Calicut, 1998)**

(*d*) *Cycas ovule* **(Calicut, 1999)**

(*e*) Roots of *Cycas* **(Calicut, 1997)**

(*f*) Leaflets of *Cycas*

(*g*) Rachis of *Cycas*

(*h*) Stem of *Cycas*

(*i*) Ovule of *Cycas*

(*j*) *Cycas* plant

(*k*) *Palaeocycas*

(*l*) *Nilssonia compta*

Objective type

87. Select the correct answer :

(*i*) Which of the following structure in a mature ovule of *Cycas* is diploid ?
(*a*) Integuments (*b*) Nucellus
(*c*) Archegonia (*d*) Endosperm

(*ii*) Companion cells are absent in the phloem of
(*a*) *Cycas* (*b*) *Pinus*
(*c*) *Gnetum* (*d*) All the three.

(*iii*) Neck canal cells are found in the archegonia of
(*a*) *Cycas* (*b*) *Pinus*
(*c*) *Ephedra* (*d*) All the three.

(*iv*) Both centripetal and centrifugal xylem are found in
(*a*) *Cycas* (*b*) *Ephedra*
(*c*) *Pinus* (*d*) None

(*v*) Male gametophyte in *Cycas* is
(*a*) 2 celled (*b*) 4 celled
(*c*) 5 celled (*d*) 6 celled.

(*vi*) Coralloid roots of *Cycas* are
(*a*) Chemotropic (*b*) Phototropic
(*c*) Apogeotropic (*d*) Positive geotropic

(*vii*) Algal zone is characteristic of
(*a*) Roots of *Pinus*

(*b*) Stem of *Cycas*
(*c*) Normal Root of *Cycas*
(*d*) Coralloid Roots of *Cycas*

(*viii*) *Cycas revoluta* is popularly known as
(*a*) Date palm (*b*) Sea palm
(*c*) Sago palm (*d*) Royal palm

(*ix*) The function of transfusion tissues in *Cycas* leaflets is
(*a*) Mechanical (*b*) Photosynthetic
(*c*) Conduction (*d*) Storage

(*x*) The male gametes of *Cycas* are
(*a*) Large and have numerous spirally ar ranged cilia
(*b*) Small and have no cilia
(*c*) Large and non-ciliate
(*d*) Small and non-motile.

(*xi*) Blue green algae present in coralloid root of *Cycas* are
(*a*) *Nostoc* and *Oscillatoria*
(*b*) *Anabaena* and *Nostoc*
(*c*) *Chlorella* and *Spirogyra*
(*d*) *Scytonema* and *Gloeotrichia*.

(*xii*) Normal Roots of *Cycas* is
(*a*) Positively geotropic
(*b*) Negatively geotropic
(*c*) Positively phototropic
(*d*) All of the above.

(*xiii*) How many species of *Cycas* are found in India ?
(*a*) 16 (*b*) 10
(*c*) 08 (*d*) 06

(*xiv*) Spermatozoids large enough to be seen by naked eye are produced in
(*a*) *Pinus roxburghii* (*b*) *P. wallichiana*
(*c*) *Cycas revoluta* (*d*) *Gnetumula.*

(*xv*) The anatomical features of *Cycas* leaflet indicate that *Cycas* is a
(*a*) Xerophyte (*b*) Nyorophyte
(*c*) Mesophyte (*d*) Halophyte.

(*xvi*) Which of the following is not the characteristic feature of *Cycas* ?
(*a*) Circinate vernation of foliage
(*b*) Presence of arm parenchyma
(*c*) Presence of motile sperms
(*d*) Absence of vessels in xylem.

(*xvii*) The wood of *Cycas* consists of
(*a*) Tracheids only
(*b*) Vessels only
(*c*) Equal number of tracheids and vessels
(*d*) More of tracheids and less of vessels

(*xviii*) In *Cycas* the pollen grains are shed at the
(*a*) 4-stage celled stage
(*b*) 3-celled stage
(*c*) 2-celled stage
(*d*) 1-celled stage.

(*xix*) In which of the following sexual reproduction is possible without water ?
(*a*) *Ulothrix* (*b*) *Funaria*
(*c*) *Dryopteris* (*d*) *Cycas*

(*xx*) In *Cycas*
(*a*) Male and female strobili occur on separate plants
(*b*) Male and female strobili occur on same plant
(*c*) One cone contain both microsporangia and ovules
(*d*) The same sporophyll bears both male and female reproductive organs.

(*xxi*) The male cone of *Cycas* lacks
(*a*) Microspores
(*b*) Microsporophyll
(*c*) Micro sporangium
(*d*) Nucellus.

(*xxii*) The tracheids of *Cycas* are
(*a*) With uniseriate borderd pits
(*b*) With biseriate bordered pits
(*c*) With multiseriate bordered pits
(*d*) Without bordered pits

(*xxiii*) Endosperm in *Cycas* is formed
(*a*) At the time of fertilisation
(*b*) Before fertilisation
(*c*) After fertilisation
(*d*) Along with the development of *embryo*.

(*xxiv*) *Cycas* ovule is
- (*a*) Anatropous
- (*b*) Othrotropous
- (*c*) Hemi anatropous
- (*d*) Campylotropous

(*xxv*) The pollination takes place through
- (*a*) Water (*b*) Wind
- (*c*) Bat (*d*) Man.

(*xxvi*) What type of plant is *Cycas ?*
- (*a*) Heterocious (*b*) Hetrothallic
- (*c*) Heterophic (*d*) None.

(*xxvii*) What is the name of algae found in Coralloid root of *Cycas* ?
- (*a*) *Nostoc* (*b*) *Anabaena*
- (*c*) *Oscillatoria* (*d*) All of them.

(*xxviii*) Diploxylic condition is found in which part of *Cycas* ?
- (*a*) Leaflet (*b*) Normal root
- (*c*) Stem (*d*) Coralloid root.

(*xxix*) *Cycas* plant is
- (*a*) Monoecious
- (*b*) Diocious
- (*c*) Monoecious and Dioecous
- (*d*) None.

88. Fill in the blanks :

(*i*) Coralloid root of *Cycas* contain an endophytic .... named ....

(*ii*) Transfusion tissue and Diploxylic condition is found in ....

(*iii*) Negatively geotropic roots are found in ....

(*iv*) Botanical name of Sago palm is ....

(*v*) Coralloid roots differ from normal roots of *Cycas* in having .... in the cortex.

(*vi*) The starch called Sago is obtained from ....

(*vii*) Nitrogen fixing organisms are present in the .... of ....

(*viii*) The anatomical features of *Cycas* leaflet indicate that it is ....

(*ix*) At the under surface of microsporophyll of *Cycas* sori are found, each sorus having ....

(*x*) *Cycas circinalis* has .... shaped megasporophylls.

(*xi*) Cleavage polyembryony is not met with during the development of embryo in ....

(*xii*) The Gymnosperm in which you find coralloid roots is ....

(*xiii*) Vascular bundles in *Cycas* rachis are arranged in .... shape manner.

(*xiv*) Revolute margins of the leaflet is present in ....

89. State whether the following statement is true (T) or false (F):

(*a*) Coralloid root of *Cycas* inhabit blue green algae.

(*b*) Coralloid roots are positively geotropic.

(*c*) Leaflets show xerophytic characters.

(*d*) *Cycas revoluta* is named because the margins of the leaflets are not folded inside.

(*e*) *Cycas* plant is ------ like in appearance.

(*f*) The plant of *Cycas* show only gymnospem like characters.

(*g*) Fruits are formed in *Cycas*.

(*h*) Endosperm is not formed in *Cycas*.

(*i*) Secondary growth is present in *Cycas* stems.

(*j*) *Cycas* ovule in anatropous.

(*k*) The pollination in *Cycas* takes place through water.

(*l*) *Cycas* is popularly known as Sago palm.

(*m*) The wood of *Cycas* is manoxylic and is made up of tracheids only.

(*n*) Spermatozoids in *Cycas* are small.

CHAPTER 10

Coniferopsida

Order : Cordaitales

- Order Cordaitales-Introduction
- Diagnostic Characters
- Family-Cordaitaceae
- Family-Eristophytaceae
- Affinities of the Cordaitales

WOODY STEM OF CARDAITES

The **Coniferopsida** include much more massive, large and dominant plants that, in the present times, occupy eight percent of the earth's surface. They came into existence during the Carboniferous period of Palaeozoic era. The **Cordaitales** even lived in the Devonian along with **Pteridospermales**. This clearly brings home to us that neither of the **Cycadopsida** and **Coniferopsida** is the ancestor of the other, because both are of same antiquity. This lends support to the view that gymnosperms have monophyletic origin. It is further strengthened by data available from the study of morphological and anatomical characteristics of stem, leaf and the seed. It has been discussed in detail in the first chapter.

The **Coniferopsida** differ from the **Cycadopsida** in possessing (*i*) simple leaves; (*ii*) narrow pith; (*iii*) narrow cortex; (*iv*) active cambium; (*v*) extensive development of secondary wood; (*vi*) pycnoxylic wood; and (*vii*) bilaterally symmetrical ovule and seeds.

The **Coniferopsida** have been classified into 4 orders by Sporne (1965). This classification is followed in this text. These orders are : (*i*) **Cordaitales**; (*ii*) **Ginkgoales**; (*iii*) **Coniferales** and (*iv*) **Taxales**. Of this, Cordaitales is represented only by fossil plants. The others include both the fossil and living genera. These orders will be briefly discussed here.

Order : CORDAITALES

Introduction

The **Cordaitales** represent the most ancient and primitive order of the class **Coniferopsida**. They appeared in the Devonian and were at their peak of development during the carboniferous period of the Palaeozoic era. They were the only representatives of the **Coniferopsida** in that age and were succeeded by the **Ginkgoales** and the **Coniferales** in the Mesozoic era. The **Ginkgoales** declined as evolution proceeded and are now represented only by a single living species (*Ginkgo biloba*), whereas the **Coniferales** went on spreading and dominating the world and are now the most dominant gymnosperms. The **Ginkgoales** and **Coniferales** have both independent connections with the **Cordaitales**. The fossil conifers are represented by three families : *Lebachiaceae*, *Voltziaceae* and *Palissyaceae*. The *Lebachiaceae* originated towards the close of the Carboniferous period and disappeared in the Middle Permian period of the Palaeozoic and at the same time the *Voltziaceae* started appearing and continued up to the Jurassic (Mesozoic). The Permian which witnessed the fall of **Cordaitales** and *Lebachiaceae* also saw the rise of the **Ginkgoales**. During the Triassic and the Jurassic periods of the Mesozoic when the **Ginkgoales** were at their best, the *Voltziaceae* faded out of existence and the other coniferous family *Palissyaceae* and some modern conifers started appearing on the horizon. The **Ginkgoales** and the *Palissyaceae* faced decline during the Cretaceous period. The **Ginkgoales**, however, continued and have survived the test of time by leaving at least one living representative, for us to see and study. This is the famous Maiden Hair Tree, *Ginkgo biloba*. To this day *Ginkgo* shares many characters with the *Cordaites* and in the many respects a veritable cordaite—different no doubt from the Palaeozoic *Cordaites*, but a cordaite

Fig. 10.1. Reconstruction of *Dorycordaites* (After Scott)

nonetheless. It is an excellent example of a recent plant with general cordaitalean characters : (*i*) the motile male gametes; (*ii*) the double leaf trace; and (*iii*) the multinerved leaves. Passage of time has no doubt brought many changes, *e.g.*, (*i*) simple strobili, (*ii*) smaller pith; and (*iii*) uniseriate pitting. In the last two respects it resembles the pines and other conifers.

This evolutionary history of the **Coniferopsida** that began in the Permian (Palaeozoic) and passed through Mesozoic to the Coenozoic has also witnessed the evolution of the world we live in. During the Permian of the Palaeozoic the world had only one super-continent called the **Pangaea**. This super-continent started dismantling during the Permian and by the end of the Triassic it had splitted into two continents known as the Northern **Laurasia** and the Southern **Gondwana** Land. These continents started drifting during the Coenozoic (65 million years ago) and reached the position in which we observe them today. The significant feature of the Coenozoic was the collusion of Indian sub-continent with the undervalley of Asia (Tibet) thus creating the Himalayas. It took two hundred million years for India to travel 9000 km, which work out to an average of a little, less than 5 centimetres per year.

The Palaeozoic Cordaitales are an assemblage of fascinating plants that reveal the historical background of the most dominating group gymnosperms, that this day cover about 4,000,000 square miles of the earth's surface. These are the Conifers.

Classification

This fascinating group of gymnosperms was named after the early 19th century palaeobotanist called Corda. Grand Eury and Renault brought it into recognition in the year 1870. The order has been divided into three families : (*i*) *Cordaitaceae*, (*ii*) *Eristophytaceae* and (*iii*) *Poroxylaceae.*

The order is represented in India by 4 genera and 26 species. The genera include *Cordaicarpus* (4 sp.); *Dodoxylon* (14 sp.), *Noeggerathiopsis* (3 sp.) and *Samaropsis* (5 species). All were discovered from the Upper Gondwana.

Diagnostic Characters

The Cordaitales possess the following distinguishing features :

1. The plants were tall trees with slender and branched stems.
2. The leaves were simple, subulate or paddle-shaped.
3. The leaves were spirally arranged.
4. The venation of the leaves was parallel.
5. The secondary xylem was pycnoxylic.
6. The strobili were unisexual and compound structures, that consisted of a central axis bearing bracts that, in turn, bore fertile shoots bearing sterile and fertile appendages.
7. Megasporophylls bore four ovules.
8. The male appendages bore 4—6 microsporangia terminally.
9. The seeds were bilateral.
10. The stems were monopodial in habit.
11. Presence of an endospermic beak in some ovules.
12. The nucellus and the integument are free from each other. The nucellus is vascularised.

Family : CORDAITACEAE

External Characters

The members of this family formed the world's first great forests towards the close of Lower Carboniferous and included tall trees (upto at least 30 metres in height) that grew in dense stands. The stems attained a diameter of 65 centimetres or a little more. This diameter is comparatively small with respect to their height. This gave them the slender appearance. The branching was lateral and confined only to the top of the tree.

The leaves appeared on the branches and were spirally arranged, petiolate, simple, subulate, spathulate or grass-like with entire margins and obtuse or acute apices. The venation was parallel. The leaves showed considerable variation among the members of the family and were put under the type genus *Cordaites*. This name is now applied to the stem as well as to the whole plant. Based on the differences among the leaves of various *Cordaitaceae,* the genus *Cordaites* has been sub-divided into three sub-genera :

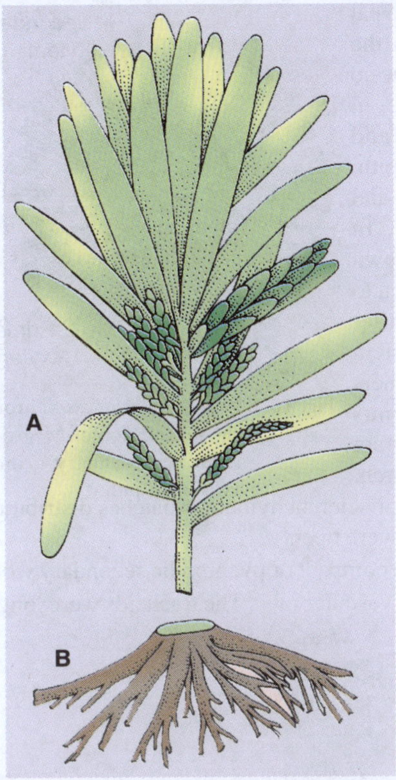

Fig. 10.2. (A-B) A reconstruction of *Cordaites* sp. A. It shows foliage bearing and inflorescence bearing branches. B. The Root system (After Grand, D. Eurry).

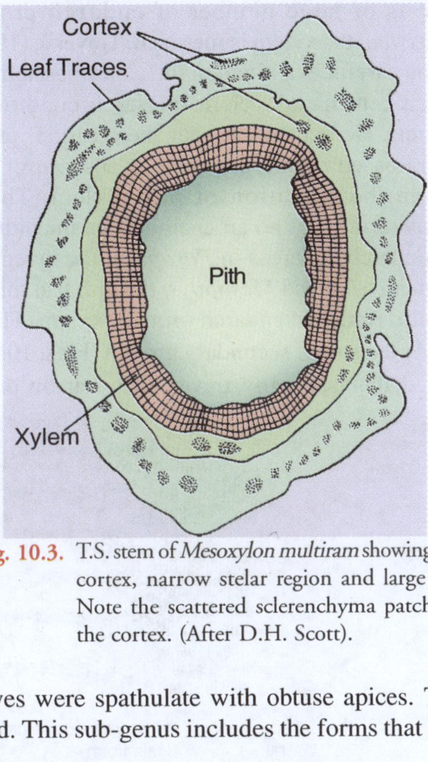

Fig. 10.3. T.S. stem of *Mesoxylon multiram* showing large cortex, narrow stelar region and large pith. Note the scattered sclerenchyma patches in the cortex. (After D.H. Scott).

Sub-genus : 1. **Eu-Cordaites** in which the leaves were spathulate with obtuse apices. They reached up to a metre in length and about 15 cm broad. This sub-genus includes the forms that have been fully investigated.

Sub-genus : 2. **Dory-Cordaites.** The leaves were lenceolate (Fig. 10.1) with pointed apices.

Sub-genus : 3. **Poa-Cordaites.** The leaves were grass-like and reached a length of 50 cms and about a centimetre wide.

The leaves were encountered in the form of compressions. The parallel venation was traceable to a series of dichotomies. The leaves were thick at their regions of attachment and then tapered proximally. Harms and Leisman (1961) described the leaf of *C. crassus* to be 4 mm thick at base, 1.1 mm thick near the middle and 0.8 mm thick at the apex.

The roots were also discovered and were named *Ameylon*. They arose from the base of the stem and were repeatedly branched.

ANATOMY

Stem

The cordaitalean stems have been described under several generic names such as *Mesoxylon*, *Cordaites*, *Metacordaites*, *Parapitys*, *Coenoxylon, Mesopitys, Cordaicladus* and *Artisia*. All of them, however, agree in their anatomy to the various species of *Cordaites* and *Mesoxylon*.

In *Cordaites* (Fig. 10.3) the stem has a large cortex with sclerenchymatous patches scattered amid the parenchyma. The secretory sacs were also present. The pith casts reveal it to be wide and extensive and peculiar in being discoid. The characteristic feature of the pith is the presence of large air chambers separated by diaphragms. The pith casts reveal the pith to be transversely septate at the margins but the petrified material reveals the diaphragms to be incomplete so that a transection shows two or three large air cavities, in addition to many smaller ones. The names *Artisia* and *Sternbergia* were given to the casts showing chambered pith. The primary xylem forms a narrow zone around the pith and consists of large number of **endarch** protoxylems. In *Mesoxylon,* the xylem is **mesarch**. Traverse (1950) has shown that on careful examination the *Cordaites*stem is also seen to have a centripetal xylem (mesarch condition) in smaller amounts. He does not favour the creation of separate stem genera based on the presence or absence of pith chambers and on the condition of protoxylem. The stem genus *Metacordaites* has no air chambers on the pith and the xylem was endarch, whereas in *Parapitys,* the xylem was mesarch. In *Coenoxylon* and *Mesopitys,* the pith had thick-walled or sclerenchymatous patches distributed in the pith. Both had endarch primary xylem. The bundles were open.

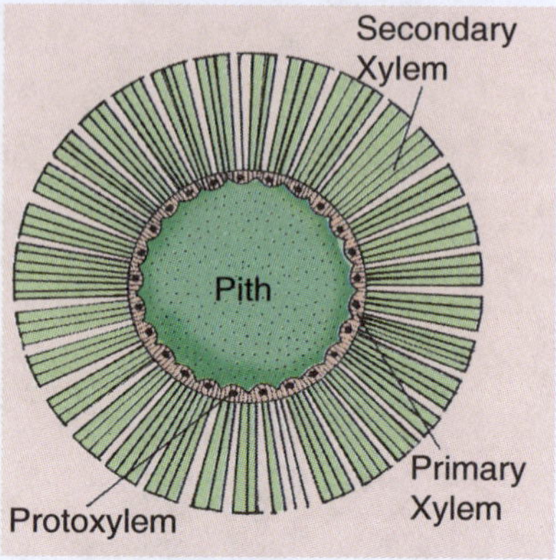

Fig. 10.4. Diagrammatic T.S. Through the stelar region of *Mesoxylon* showing secondary primary xylem and pith.

As a result of secondary growth (Figs. 10.4 and 10.5) compact or pycnoxylic secondary wood is formed, that is, usually, traversed by narrow or uniseriate vascular rays. The tracheids were long and

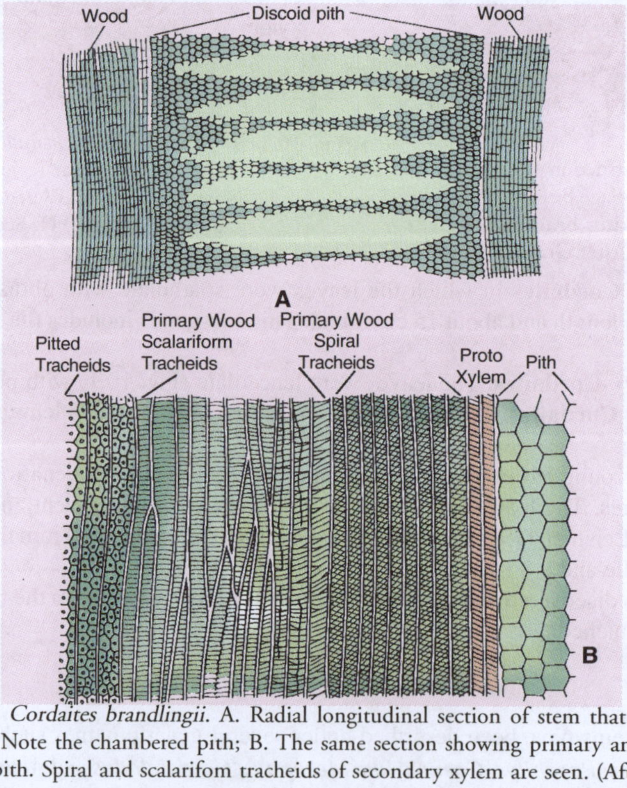

Fig. 10.5. (A-B). *Cordaites brandlingii.* A. Radial longitudinal section of stem that has undergone secondary growth. Note the chambered pith; B. The same section showing primary and secondary xylem, and a part of pith. Spiral and scalarifom tracheids of secondary xylem are seen. (After Scott).

narrow. The vascular pitting was primarily restricted to the radial walls. The bordered pits were hexagonal in outline (Fig. 10.5, B) and occurred usually in uniseriate rows. Biseriate, triseriate to alternate and crowding arrangements were not uncommon. Metaxylem also showed some scalariform pitting. It is believed that in *Mesoxylon* the leaf traces were mesarch and this condition extended downward to only some distance into the cauline system, but most of the cauline bundles had endarch xylem. In *Cordaites iowensis* the first formed secondary wood had bordered pits multiseriately arranged but in the later formed wood the arrangement was uniseriate. The wood resembles the modern coniferous family *Araucariaceae*. The large pith and extensive cortex are cycadalean characters.

The phloem, though not well preserved, showed sieve tubes and phloem parenchyma. Presence of phloem fibres is doubtful.

The stem named *Dodoxylon* (with 14 species from Gondwana in India) also belongs to Cordaitales. It was first named and described by Dawson from Canada and Eastern United States. Later many species were discovered from France, England, Saxony, Brazil and South Africa. Presence of chambered pith in the stem of some species confirms their inclusion in the order.

Leaf Traces

The cordaitean leaf traces were either single or double at their point of origin from the primary vasculature. In *Cordaites,* the strands were **endarch** whereas in *Mesoxylon* they were **mesarch**. During their course through the cortex they are sub-divided. Two traces supplied each leaf. Their branches through the cortex increase the number of strands entering each leaf base. The two traces arose from a single protoxylem group. The leaf traces have been linked to those of *Ginkgo*, but there is a difference. In *Ginkgo,* the two traces supplying a leaf originate from two protoxylem groups.

The branch traces were also double at the point of their origin and they fused together before entering the branch base.

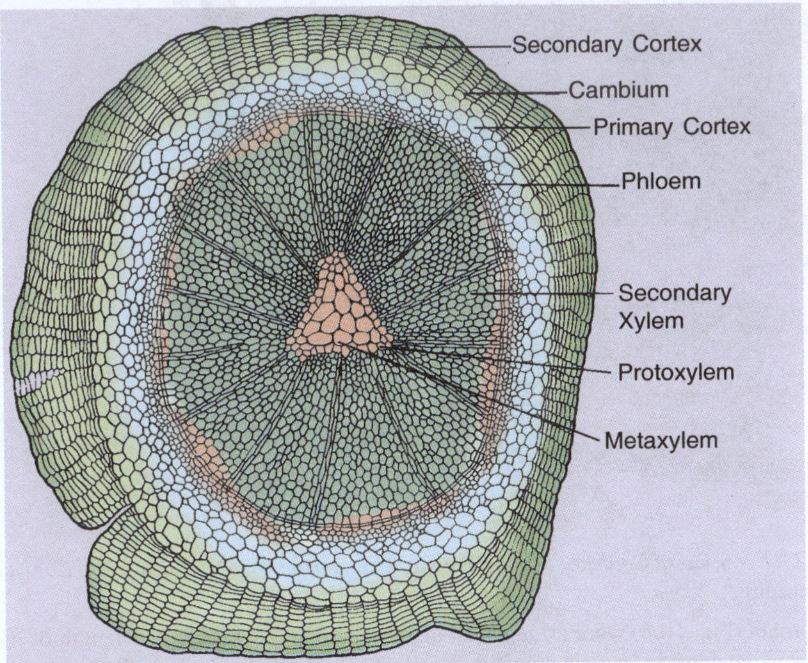

Fig. 10.6. T.S.root of *Cordaites* (*Ameylon radicans*) showing secondary growth (After Scott).

Root (Fig. 10.6)

The roots in cordaitales (*Ameylon*) were usually triarch, exarch and protostelic. Most of the petrified specimens also showed the presence of periderm around the secondary cylinder. Secondary wood in the root was extensive. Another root genus is *Premnoxylon*. It was also exarch with the number of xylem poles varying between two to seven or even greater, in various specimens. In this genus the root pith had sclerenchymatous patches and the secondary xylem formed a distinct ring from that of the primary xylem. The root cortex was divisible into an outer and an inner cortex. Periderm was present in the outer cortex. The inner cortex had some cells with dark contents and were infected with mycorrhizic fungi.

Leaf (Fig. 10.7)

The leaves reveal characteristic xerophytic internal structure. Leaves were well preserved and their anatomy has been well studied in some species of *Cordaites*. The leaves had a distinct epidermis on the upper and lower surfaces (bifacial). The epidermal cells were thick walled. The sclerenchymatous hypodermis occurred in the form of distinct patches on both sides in *C. angulostriatus* whereas in *C. rhobinervis* it formed a continuous layer interrupted by distinct sclerenchymatous **bundle-sheath extensions**. In *C. lingulatus* only sclerenchymatous bundle-sheath extensions were present. There was a clear distinction into palisade and spongy parenchyma in *C. lingulatus*. The vascular bundles were mesarch in all the species. The bundles were surrounded by a distinct sclerenchymatous bundle-sheath.

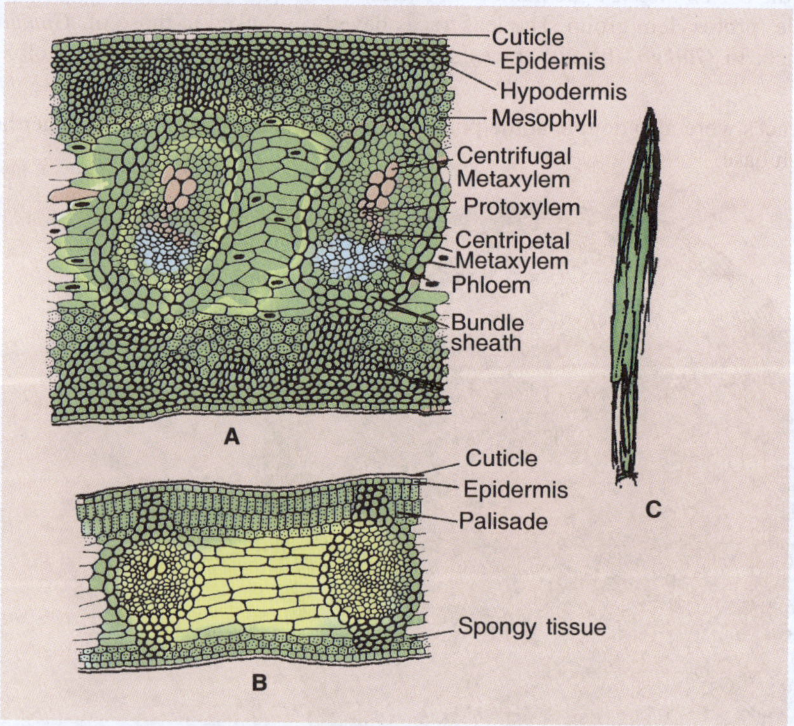

Fig. 10.7. (A-C) A.V.S. leaf of *Cordaites angulostriatus;* B. *C. lingulatus* and C. A single leaf of *Cordaites. sp.* (After Renault)

The stomata (Fig. 10.8) were arranged in distinct bands along the entire length of the leaf. The guard cells were surrounded by 4-6 subsidiary cells. *Cuticular* ridge and papillae were present in the subsidiary cells that were lying besides the guard cells. Two polar subsidiary cells in every stomata but on the ends of the guard cells.

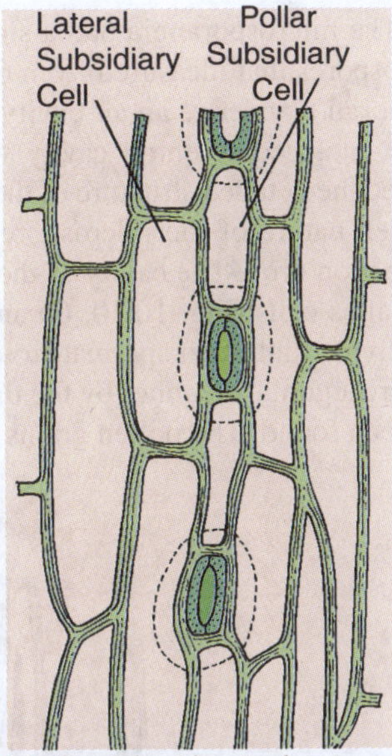

Fig. 10.8. Part of epidermis showing stomata on the epidermis of leaf of *Cordaites.* (After Florin).

Reproductive Organs

The fructifications of *Cordaites* are known under the form generic name *Cordaitanthus* (*Cordaianthus*). The fructifications occur among the leaves on the stem and are loose inflorescences or compound strobili. Each fructification was borne on a vegetative shoot and consisted of an axis which bore spirally arranged bracts with axillary reproductive shoots also designated dwarf shoots or short shoots. The compound strobili were unisexual *i.e.* they were either entirely megasporangiate or microsporangiate.

Male Strobilus (Fig. 10.9)

It is also known as microsporangiate short shoot and the three well studied forms include *Cordaitanthus concinnus*, *C. penjonii* and *C. saportanus*. Each dwarf shoot had a central axis with a large number of spirally arranged and uni-nerved scales. The scales at the proximal end of strobilus or dwarf shoot were sterile whereas those near the apex or the distal ones were fertile and possessed 4 (*C. saportanus*) or 6 (*C. penjonii*) or many microsporangia or pollen sacs. The microsporangia were elongate and finger-like structures each separately vascularised. These fertile and sterile appendages bore no axillary structures and are also termed as microsporophylls. The vascular strand that supplied the sporophylls or the appendages were mesarch and forked dichotomously near their apices to send a branch to the base of each microsporangium.

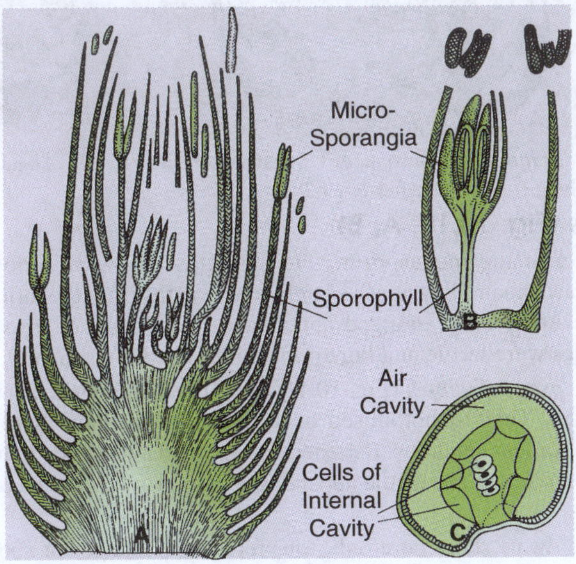

Fig. 10.9. (A—C). *Cordaitanthus.* Male reproductive organs. A. L.S. Male strobilus of C. *penjonii.* Note the microsporophylls and terminal microsporangia. (After Renault); B. Three microsporophylls of *C. penjonii.* The middle one bears four microsporangia with one of them dehisced (After Renault); C. Pollen grain of *Cordaitanthus* (After Florin).

The microsporangia had a single layered wall enclosing large number of pollen grains or the microspores that measured from 65 μ to 150 μ in diameter. The microspores (Fig. 10.9, C) had an equatorial bladder or an air cavity enclosed by the exine. This is the *saccus*. The inner wall or the intine enclosed a central cavity which contained cellular remains of microgametophytes. Florin studied the detailed structure of the microspores and the enclosed cells. Chamberlain considered the swollen nature of the microspores to be due to increase in size as a result of immersion in the pollination drop. The cavity of the germinated microspore is occupied by a row of peripheral male prothallus cells (Fig. 10.10, C) and a row of central spermatogenous cells. Each spermatogenous cell divided into two spermatozoids that were multiciliated. These were liberated into the trough-like archegonial chamber by the disintegration of the intervening nucellar tissue. Pollen tubes have not been found. The pollen grains were trilete on the proximal surface (Taylor, 1972).

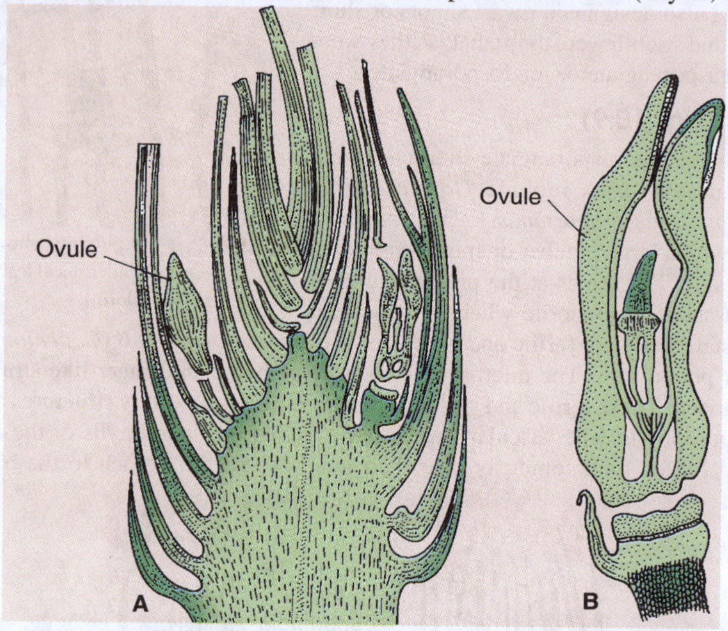

Fig 10.10. (A-B). *Cordaitanthus williamsonii.* A. L.S. female strobilus, B. L.S. of ovule. Note that the nucellus is free from the integument. (After Renault).

Female Strobilus (Fig. 10.10, A, B)

The organisation of the megasporangiate strobilus or dwarf shoot is similar to that of microsporangiate dwarf shoot. They arise as lateral outgrowths from the inflorescence axis and have each a central axis with spirally arranged appendages. The basal or proximal ones were sterile whereas the distal ones were fertile and bore one to three ovules terminally.

In *Cordaitanthus pseudofluitans,* (Fig. 10.11, B, C) the distal fertile appendages or sporophylls were greatly elongated. They dichotomised once or more times and bore two to more terminal ovules that were more or less pendulus, flattened and showed a tendency to be recurved towards the main axis of the inflorescence. The fertile appendages later grow more and project beyond the short shoot.

In *C. zeilleri,* the dwarf shoot bore only four fertile appendages or sporophylls or sometimes only one. The megasporophylls were unbranched and bore only one terminal erect ovule and remained short and concealed among the sterile appendages.

In compression-impression fossils the ovules are of platyspermic type. The structurally preserved ovule of *Cordaitanthus* sp. have been assigned to genera *Miterospermum* and *Cordaiocarpus.*

The ovules were bilaterally symmetrical and were enclosed within a single, but two parted integument that was completely free from the nucellus. A distinct pollen chamber was found containing germinated pollen grains. The ovules appeared heart-shaped from the broad side. The single median bundle entered the sporophyll and divided into three just below the ovule. From this point, the central vascular strand continued up to the base of the nucellus, whereas the two lateral ones enter the fusing lobes of the integument. A few examples will illustrate the structure of the cordaitalean ovules.

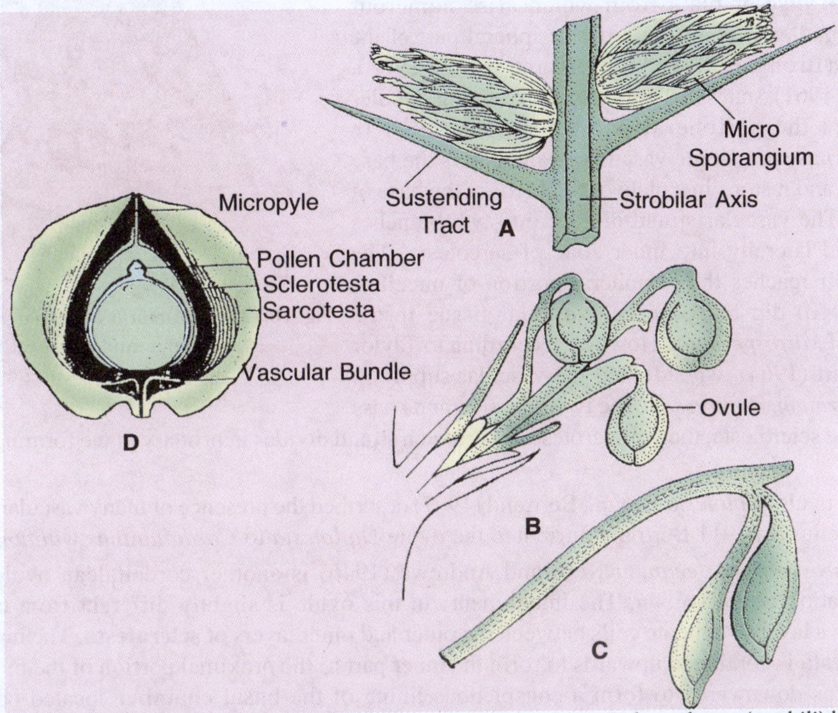

Fig. 10.11. (A-D). A. *Cordaitanthus concinnus.* A pair of microsporangiate short shoots (strobili) borne in the axils of bracts in the strobilar axis. (After Delevorayas). B. *Cordaitanthus pseudofluitans.* A megasporangiate short shoot, *C*. A single megasporophyll of *C. pseudofluitans* bearing a pair of ovules D. L. S of a cordaitean ovule (B—C, After Florin)

These ovules are bilaterally symmetrical and the integument is free from the nucellus and consists of two or three distinct layers of tissues : (*i*) **sarcotesta** or the outer fleshy layer; (*ii*) **sclerotesta** or the middle stony layer; and (*iii*) **endotesta** or the inner narrow and fleshy layer.

Cardaiocarpus spinatus Graham is the ovule described by Roth (1955) in detail. In this case (Fig. 10.11 D) the **sarcotesta** is composed of inner and outer zones : the cells of the outer zones are thin-walled and larger in size and are filled with a brownish substance that gives them an amber colour; the inner zone is composed of comparatively smaller cells. Likewise the **sclerotesta** is also made up of an outer layer composed of dark coloured, elongated and thick-walled cells that usually proliferate into the sarcotesta in the form of spines; and an inner layer which is made up of thick-walled cells that are not elongated. The endotesta is made up of thin-walled cells. The sarcotesta is fleshy, whereas the sclerotesta is hard and stony and is highly resistant to decay. The nucellus, in the mature ovules, appears as a thin band of tissue surrounding the female gametophyte which is well preserved in some specimens and consists of a parenchymatous mass of cells with two archegonia towards the micropylar end. At the apex of the nucellus is a small often conical pollen-chamber. The nucellus is free from the integument except at the base.

In the European specimens of *Cardaiocarpus,* Seward (1917) described the vascular supply to the ovule as consisting of a single central strand that enters the base of the ovule. It gives off two branches that pass up the inner layer of sarcotesta running in the principal plane of the ovule. The main bundle then passes through the sclerotesta and enters the base of the nucellus where it forms a **tracheidal disc** or plate from which arise numerous discrete bundles which run up the peripheral part of the nucellus (Brongniart, 1881; Bertrand 1907, 1908). Andrews (1961), states that only a pair of vascular bundles penetrates the peripheral part of the nucellus. In *Cardiocarpus* sp. a single vascular strand enters the base of ovules, and it stops in a plate of trachieds at the base of nucellus. The vascular strand divides into two branches that extend laterally into inner zone of sarcotesta. The second pair reaches the peripheral portion of nucellus. Arber (1910) did not observe vascular tissue in the nucellus of *Mitrospermum.* However, according to Taylor and Stewart (1964), who studied the vascular supply in *Miterospermum compressum,* the two lateral strands pass through the sclerotesta, then to sacrotesta where each strand divides in primary plane forming several bundles.

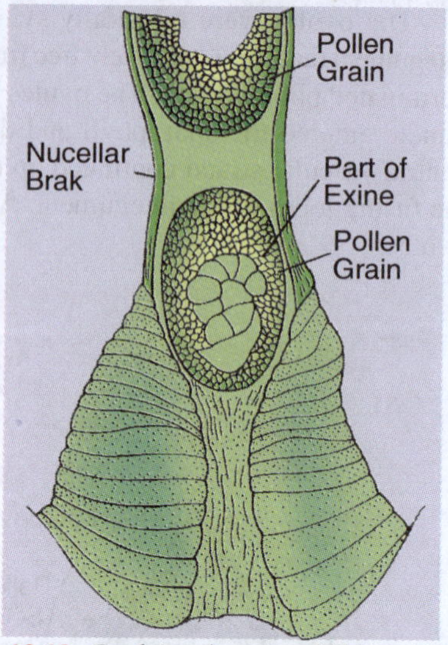

Fig. 10.12. *Cordaitanthus Grand d' Eurgi.* L.S. through nucellar beak of ovule showing two pollen grains. (After Renault)

In the ovule *Diplotesta* Brogn., Bertrand (1907) described the presence of many vascular bundles in the nucellus. In 1911 Bertrand attributed the ovule *Diplotesta* to *Cordaitanthus williamsonii.*

Kamarospermum leeanum Kern and Andrews (1946) is another cordaitalean ovule whose vascularisation is not known. The integument in this ovule is slightly different from others in possessing a layer of delicate cells between the outer and inner layers of **sclerotesta**. The inner layer of **sclerotesta** is continued upwards to form the inner part to the proximal portion of the micropyle. It then turns downwards to form a conspicuous lining of the **basal chamber** located below the **nucellar chamber**. Basal chamber is a cavity located below the nucellar chamber and is a characteristic of this ovule. This chamber is thought to be an air chamber that aided in the dissemination of the seed by water. The micropyle of this ovule is also characteristic in possessing two distinct zones: (*i*) a massive one directly above the nucellus; and (*ii*) a narrow and elongated terminal part that has the form of a gradually tapering and flattened funnel.

These ovules have no cupule.

Regarding the absence of vascularisation in the nucellus of *Cardaiocarpus spinatus*, three possible explanations have been put forth by David L. Smith (1964) :

1. It may be due to poor preservation because it resembles all other ovules in other respects.
2. The nucellus may be actually unvascularised.
3. Lack of vascularisation may be a derived state due to the loss of first integument with consequent loss of vascular tissue.

Brotzman, N.L. and J.T. Schabilion (1972) have discovered another Cordaitalean ovule from the coal balls of the Upper Des Moines series in Iowa. This is named as *Cardaiocarpus magnicellularis* Baxter and Roth. There is a well defined epidermis and fibres in the outer zone of sclerotesta and presence of nucellar pollen chamber.

These bilaterally symmetrical ovules have been considered to have evolved from an ancestral ovule with a radial symmetry. Its bilaterally symmetrical integument was derived by the fusion of two telomes. This probably evolved simultaneously with the aggregation of the ovules into a primitive strobilus.

The female gametophyte in most of the ovules encroaches upon the nucellus almost to the integument and also develops a characteristic **endosperm beak** which supports the nucellar beak and is called the **tent pole** as in *Ginkgo biloba*. The archegonial necks open into a distinct *Archegonial chamber* which surrounds the tent pole. Baxter (1964) even found starch grains in *Cordainocarpus spinatus* ovule each with a hilum.

Family : POROXYLACEAE

This family is represented by a single genus *Poroxylon* whose seeds are put under the form genus *Rhabdospermum*. This family is of Permocarboniferous origin. The genus *Poroxylon* is of singular interest because it shows certain characteristics in which it resembles the pteridosperms and others which necessitate its retention among the Cordaitales. It bridges the gap between these two orders. It resembles the pteriodosperms in the following respects :—

1. Poorly developed secondary wood.
2. The secondary xylem rays were 3 cells wide.
3. The primary xylem was exarch.
4. The pith was wide.
5. The cortex had bands of sclerenchyma like those of *Lyginopteris*.

It resembles the Cordaitales in the following respects :—

1. The leaves were simple and up to 1 metre long.
2. The cordate base of the seeds (*Rhabdospermum*).
3. The nucellus was vascularised.
4. The leaves were parallel veined.
5. Presence of a conical pollen chamber at the apex of the nucellus.
6. The ovule was bilaterally symmetrical.
7. The nucellus is free from the integument.

Family : ERISTOPHYTACEAE

It includes the form genera *Eristophyton*, *Endoxylon* and *Bilingea*. These genera were previously placed under the family *Calamopityaceae*, but Lacey (1953) pointed out that they have pycnoxylic wood and should therefore be included under the Cordaitales. Only stems of these genera are well known. The leaves and reproductive organs are not known and hence their taxonomic position remains unsettled.

In *Eristophyton* the primary vasculature consisted of five distinct vascular bundles arranged around the large and conspicuous pith. They were quite free from the secondary wood but during their course outward as leaf traces they came to touch the secondary cylinder. The strands were mesarch at this stage but during their course downwards they lost their centripetal wood and became endarch. In *Endoxylon* the strands remained endarch throughout. In *Bilingea* the pith was mixed and contained groups of tracheidal elements.

Their features suggest that like the *Poroxylaceae* this family also has certain pteridospermous and some cordaitalean characters.

AFFINITIES OF THE CORDAITALES

The Cordaitales are the oldest plants of the true Coniferophytes, and constitute an interesting group of fossil plants that reached a high state of development as early as the Palaeozoic era (Miller, 1977). These highly developed features include :—

1. The development of simple leaves with parallel venation.
2. The secondary wood was pycnoxylic and had uniseriate rays.
3. The development of definite strobilar arrangement of the reproductive structures.
4. The reproductive structures were unisexual.
5. The ovules and seeds were bilaterally symmetrical.

This group can also be regarded as an assemblage of fossil gymnosperms that combined the characters of most of the other fossil and living gymnosperms.

Resemblances with Cycads

They resemble the cycads in the following points :—

1. Their xerophytic habitat and aerial stems.
2. The stems had large pith.
3. The leaf had abundant sclerenchyma and showed xerophytic characters.
4. The ovule was surrounded by a single integument with three distinct layers.
5. Vasculature of the ovule is almost similar in both the groups.
6. Both have simple and large seeds.
7. The spermatozoids were motile.

These primitive characters seem to be inherited by both groups from a common ancestor.

They differ from the cycads in possessing pycnoxylic wood, simple leaves, leaves parallel veined, straight leaf traces and compound strobili.

Resemblances with Ginkgoales

The Cordaitales resemble the Ginkgoales in the following respects :—

1. Presence of double leaf traces in both, but there is a little difference regarding the origin of leaf traces. In the Cordaitales the two traces originate from the same protoxylem group, whereas in the Ginkgoales they originate from separate groups of protoxylem.
2. Presence of motile sperms to zoids.
3. Presence of an endospermic beak in the mature ovules.

Resemblances with Pteridosperms

The Cordaitales resemble the pteridosperms in the following features :—

1. Both have stems with wide pith.
2. The tracheids have multiseriate pitting in many cordaitales.
3. The vascularisation of the ovule and its general structure is similar in both the groups and many intermediate ovule genera bind them together (see Chapter 1), *e.g.*, *Eosperma oxroadensis* Barnard (1959).
4. The xylem is centripetally developed in the primary wood.
5. Both have double leaf traces.
6. The internal structure of the leaf is also similar in the pteridosperms and the cordaitales.
7. There is an evidence that in both the groups fertilisation was accomplished by motile spermatozoids (Arnold, 1949), but this can be regarded as merely a primitive condition.
8. Their seeds lack embryos. Emberger (1944) regarded it to be an important point linking the two groups. Arnold (1948) regards it to be coincidental.

The *Poroxylaceae*, however, show many points of resemblance with the pteridosperms (discussed before), otherwise the cordaitales, as a whole, exhibit more differences rather than resemblances with the petridospermales.

Resemblances with the Coniferales

The points of resemblances with the Coniferales can be listed as below :—

1. Both include tall trees that are branched.
2. The leaves are simple and parallel veined.
3. Presence of sclerenchymatous hypodermis in the leaves of both the orders.
4. The wood is pycnoxylic in both.
5. They resemble the Araucariaceae in possessing multiseriate bordered pits on the tracheidal walls and absence of resin ducts.
6. Florence (1951) has pointed out homologies between the compound strobili of Cordaitales and the cones of *Abietaceae* (whose genera have now been placed under *Pinaceae*). Such a link is supported by some Palaeozoic coniferous genera, *e.g.*, *Lebachia*, *Pseudovoltzia* and *Ernestiodendron* that provide connecting links.
7. The ovules are bilaterally symmetrical and have two parted integuments.

Resemblances with Ephedra

The Cordaitales have also been compared with the Gnetalian member *Ephedra* (put by some authors under a separate order Ephedrales). Eames (1952) pointed out resemblances between the microsporangiate and megasporangiate short shoots of Cordaitales with *Ephedra*:

1. In *Cordaitanthus* and some species of *Ephedra,* the microsporophylls are laminate or flattened.
2. Both bear sporangia at the apex of the sporophylls.
3. The number of sporangia at each tip varies from 1 to 6 in both the groups.
4. In some species of *Cordaitanthus,* the number of ovules is reduced from three to two to one. Such species have comparatively short megasporophylls that do not extend beyond the sterile appendages. This shortened megasporophyll with one terminal ovule can be compared to a still shorter axis bearing a solitary ovule in *Ephedra.*There is an evidence among the genus *Cordaitanthus* that reduction in the number of ovules is accompanied by reduction in the size of the appendage (sporophyll) bearing them.
5. The leaves have parallel venation.
6. Each leaf is supplied with two leaf traces.
7. Pycnoxylic secondary wood.
8. The tracheids are long and narrow with tapering ends.
9. The stomata are haplocheilic.

To add to the above list, it may be pointed out here that the microsporangiate fructifications of *Ephedra* can be seen as compound strobili bearing a series of decussately arranged, broad and cupped bracts. In the axil of each bract is the simple strobilus consisting of a pair of basally attached and fused bracts or perianth parts enclosing the one compounded sporangiophore (*E. antisyphilitica*) or in some species two or occasionally more sporangiophores with or without protruding axial remnant between them (*E. distachia*, *E. intermedia*). Each terminal sporangium is supplied by its own vein ending, which is a terminus of a dichotomous tree.

The megasporangiate fructification of *Ephedra* also seems to retain the compound strobilar organisation. The megasporangiate strobilus arises in the axil of a bract and bears a short axis to which are attached usually two or three pairs of decussate bracts and one of two terminal ovules.

The facts stated above, according to Eames (1952), present the most convincing case for the derivation of *Ephedra* from near the *Cordaitales*.

QUESTIONS

ESSAY TYPE

1. Give an account of morphology and systematic position of Cordaitales. **(Awadh, M.Sc., 2000)**
2. Give an illustrated account of the order Cordaitales. Discuss its affinities. **(Kanpur, M.Sc., 1986)**
3. Discuss the affinities of Cordaitales. **(Kanpur, MSc., 1989, 2002)**
4. What are Cordaitales? Write a brief account of fructifications and anatomy of vegetative parts of Cordaitales. **(Rohilkhand, M.Sc., 2003)**
5. Describe the fructifications of Cordaitales. **(Kanpur, M.Sc., 2004)**

SHORT ANSWER TYPE

6. **Write notes on :**
   (*a*) Anatomy of Cordaitalean woods **(Allahabad, M.Sc., 2001)**
   (*b*) Spore producing organs in Cordaites **(Kanpur, M.Sc., 1999, 2001)**
   (*c*) Cordaitales **(Kanpur, M.Sc., 1992, Gauhati, 2004; Garhwal, M.Sc., 2004)**
   (*d*) Compare Cycadopsida with Coniferopsida
7. **Describe briefly the following :**
   (*a*) *Cordaites* **(Rohilkhand, M.Sc., 1999, 2003; Kanpur, M.Sc., 1996, 1994)**
   (*b*) *Cordainthus* **(Rohilkhand, M.Sc., 1994)**
   (*c*) *Cordaiocarpus* **(Rohilkhand, M.Sc., 1996, 2001)**
   (*d*) Morphology of ovule bearing organs in *Cordaites* **(Gauhati, M.Sc., 2003)**
8. **Draw suitable sketches to illustrate the following :**
   (*a*) Male organs of Cordaitales **(Kanpur, M.Sc., 1985)**
   (*b*) Ovule of *Cordaites*

OBJECTIVE TYPE

9. **Select the correct answer:**

(*i*) The Coniferopsida has been divided into;
(*a*) One order (*b*) Two order
(*c*) Three order (*d*) Four orders

(*ii*) The Coniferopsida appeared for the first time in the
(*a*) Permian (*b*) Cambrian
(*c*) Mesozoic (*d*) Coenozoic

(*iii*) In *Cordaites* the xylem is
(*a*) Exarch (*b*) Mesarch
(*c*) Endarch (*d*) Absent

(*iv*) The roots in Cordaitaceae were usually triarch and
(*a*) Exarch (*b*) Mesarch
(*c*) Endarch (*d*) Absent

(*v*) The secondary xylem in Cordaitales is
(*a*) Manoxylic (*b*) Pycnoxylic
(*c*) Monoxylic (*d*) Absent

CHAPTER 11

Coniferopsida – Ginkgoales

A TREE OF GINKGO BILOBA

- Ginkgoales Characteristic Features
- Ginkgo-External Characters
- Microsporangiate Strobilius
- Megasporangiate Strobilus
- Gametophytic Generation
- Embryogeny
- Economic Importance of Ginkgoales
- Fossil Members of Ginkgoaceae

Introduction

This remarkable order of great antiquity is represented in the present age by a sole survivor *Ginkgo biloba,* which by many palaeobotanists is regarded as a "living fossil". It occurred as far back as Triassic period of the Mesozoic age (about 200,000,000 years ago). The order as a whole consists of 16 genera and many species, all of which except one are extinct. It came into existance during the Permian and achieved worldwide distribution and luxuriance during the existence during the Triassic and Jurassic periods of Mesozoic age. It gradually thinned out and started fading out of existence during the Cretaceous and onwards into the Coenozoic. The sole living respresentative, *G. biloba,* owes its existence and has spread in the modern period due to the interest man has taken in this plant. It has been protected., worshipped and cultivated by man. Still it grows in a wild state in a small and inaccessible area in Southern China (Li, 1956). It has also been reported as a native plant in the province of Chekiang in eastern China. This plant has vegetative and anatomical characters that are purely coniferophytic, but in many reproductive features it resembles the modern cycads; it also resembles the extinct Cordaitales and even Cycadeoids (see Delevoryas, 1972) in some features. This synthesis of characteristics from various extant and extinct groups in one living plant certainly makes it a treasure for the botanists and stamps it as a "living fossil". Foster and Gifford (1959) state, "such similarities appear as the result of the retention, by both Cycads and *Ginkgo,* of certain very ancient patterns of reproduction which probably were shared by many of the Palaeozoic and Mesozoic gymnosperms". If these observations are correct, then it follows that the seed must have evolved the ultimate common ancestor of the *Cycads* and *Ginkgo,* for the common reproductive features of the two groups, are those associated with the seed habit itself.

Plate I : *Ginkgo Maidenhair Tree*

Characteristic Features

The living *Ginkgo biloba* reveals the following characteristics :

1. The tall trees give a stately appearance and have an excurrent habit.
2. The roots penetrate deep into the substratum and constitute a tap root system.
3. The leaves are fanshaped and deciduous.
4. The venation is open dichotomous.
5. The leaves arise singly along the terminal branches and bear in their axils buds that grow into characteristic short shoots, which also bear a dense cluster of leaves at their apices.
6. The plant is dioecious and slow growing.
7. The wood is pycnoxylic.
8. The catkin-like inflorescences that bear microsorangiophores with two to twelve pendent microsporangia arise in the axils of leaves on dwarf shoots.
9. The ovules arise in groups from the apices of axillary branches.
10. Presence of an endosperm **tent pole** is characteristic of this living plant and brings it akin to the cordaitales.
11. The mature seeds are large, fleshy and about the size of small apricot. The integument is composed of an outer orange coloured fleshy portion and an inner hard and stony layer. The fleshy coat is rich in butyric acid and when crushed emits an odour that is unpleasant. The seeds are produced in large numbers and tend to hang on for a month or two after the leaves are shed. A female tree in this late fall attire gives a beautiful look.
12. The spermatozoids are motile and bear a spiral band of cilia.

Classification

Sporne (1965) classifies Ginkgoales into two families :

1. **Trichopityaceae** including *Trichopitys.* It is an extinct family.
2. *Ginkgoaceae* including both extinct and one living memeber. The extinct genera include : *Baiera, Sphenobaiera, Arctobaiera, Ginkgoites, Stephanophyllum, Dichophyllum, Saportaea, Windwardia, Eretmophyllum, Ginkgodeum* and *Ginkgo* (includes both living and fossil species)

GINKGO Linn.

This name was brought into the botanical nomenclature by Carlous Linnaeus in 1771 about two hundred and thirty four years ago. Some authors (Moule, 1944; Pulle, 1946; Widder, 1948; and Thommen, 1939) have questioned the validity of the term *Ginkgo.* They propose *Ginkyo* to be the correct name according to the Chinese description of its characteristics. Kaempfer was the European who came across this tree in 1690 in Japan.

Fig. 11.1 *Ginkgo biloba*. A. Long shoot bearing leaves. B. A long shoot bearing dwarf shoots.

The living species *G. biloba* originated in the Triassic period of the Mesozoic era and since then it has continued to live unchanged and undeterred by the innumerable environmental hazards, concealing in its stately stature the history of the immeasurable past, that counts down to about two hundred million years ago. It can safely be regarded as the oldest species of vascular plants that is in existence; but not the oldest genus because other vascular plants like *Lycopodium* and *Selaginella* are much older. It is also known as the Maidenhair Tree in common European language.

Distribution

It grows wild in a small, restricted and an almost inaccessible region in South-eastern China, and in the province of Chekiang in eastern China. Some authors doubt them to be re-establishments from cultivation or escapes from gardens. *Ginkgo biloba* is known to have been in wide cultivation for many centuries in Chinese gardens and we are indebted to those ancient horticulturists for its present-day existence. Western travellers first came across *Ginkgo biloba* in eastern China and intoduced it to Europe in the early eighteenth century. Later it was introduced into the united states, where it is being cultivated successfully as a most valuable shade tree. One of the *Ginkgo's* greatest properties is its immunity to serious diseases, having apparently outlived any that may have bothered it in the past. The variations exhibited by *Ginkgo* in its vegetative and reproductive parts clearly manifest that it is not as genetically stagnated as its antiquity might lead us to suspect.

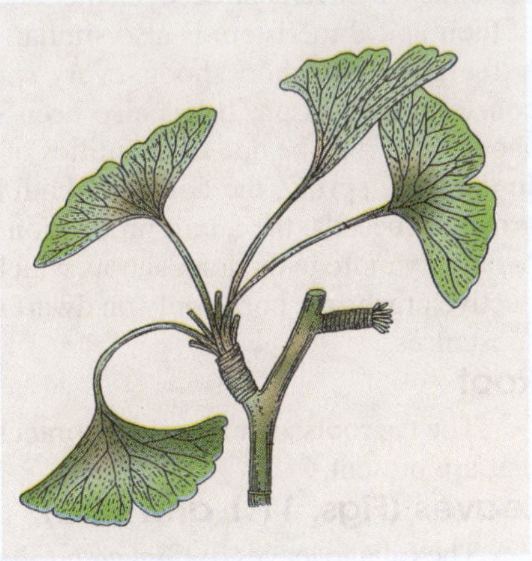

Fig. 11.2. A dwarf shoot bearing leaves that are characteristically lobed.

The tree has also been sucessfully cultivated in India and grow well under mild climate. The plant is hardy to about – 30°F, but the reproductive cycle is rarely complete at such a lower temperature range. It cannot either tolerate hot climate and slow down in growth.

Habit and Habitat

Ginkgo biloba is tall tree that may grow up to 100 feet tall with excurrent habit but may achieve irregular form in older trees. The young plants grow erect and straight for the first twenty to thirty years with a strangely spindly growth habit. The lower branches gradually begin to spread up, a few of them becoming quite massive in the course of a century. The leaves are deciduous. This slow growing tree is considered to be one of the best shade trees for future generations. It can grow along the hill slopes or even in the plains.

EXTERNAL CHARACTERS

Stem

The stem is erect, tall, slow growing and branched. The branching is lateral. There are two types of branches :

1. *Long Shoots* (Fig. 11.1, A) : They arise from the main stem and grow rapidly and may again bear long shoots and dwarf shoots. The long shoots bear green leaves that arise singly from a node and are spirally arranged with a bud in their axils. They bear deeply lobed foliage leaves.

2. *Dwarf Shoots* or *Spur Shoots* (Figs. 11.1, B, 11.2) : They develop from the axils of foliage leaves on the long shoot and grow very slowly producing each year a small crown of foliage leaves and the reproductive structures. The foliage leaves borne on them are not deeply lobed and are more or less entire. They simulate the pinnae of the fern *Adiantum.* The dwarf shoots may persist as such for years. They differ from the long shoots in their short length, in anatomical details and also in being structures that bear reproductive organs.

The long shoots and the dwarf shoots are interconvertible. The dwarf shoots have been observed to grow into long shoots if the apical bud of the subtending long shoot is removed, but cases are not rare when the dwarf shoots grow as long shoots without removing the apices of adjoining long shoots. Similarly long shoots have been observed to grow as dwarf shoots for a year or two and then revert to long shoot growth phase. Sometimes after an active growth of one or two years, the long shoot may permanently enter dwarf shoot phase. These observations suggest that there is no fundamental difference between the shoot apical meristem of both kinds of branches. The organisation of their apical meristem is also similar. Cunckel and Wetmore (1946) describe that, all the buds on a tree grow into short shoots every spring and some of them change into long shoots during the course of the season. It has also been suggested that differences in the growth pattern of the two shoot types may be due to quantities of auxins produced in the apical meristems (Bierhorst, 1971). In the early spring, the apices of both kinds of shoots produce similar quanities of auxins and as growth proceeds, the auxin production increases. At a defite stage, the auxin poroduction is comparatively more in the long shoots which grow more and thus increase in length. The fact that reproductive organs are borne only on dwarf shoots seems to be related to the production of some specific substances.

Root

The tap roots are extensively branched and grow quite deep into the soil. The root caps and root hair are present.

Leaves (Figs. 11.1 and 11.2)

The foliage leaves of *Ginkgo biloba* have a characteristic shape, which distinguish the tree from all other gymnosperms. They are simple, large and fan-shaped with expanded apex and narrow base. They are petiolate and are highly variable. When the leaf emerges from the bud, the two lobes of the leaf are folded in towards each other, but it becomes flattened out later. In the early stages, the leaf is pale yellow-green in colour, but it becomes typical dark green about half way through its development. During this colour change, it becomes more fan-shaped and both the lobes of the leaf increase in area. They are deeply bilobed on the long shoots and on the seedling and may also show secondary lobing. On the dwarf shoots, they are not deeply lobed and may be almost entire or sinuate. They have an open dichotomous venation, there being no interconnecting cross veins. The leaves usually fall during the autumn and turn beautiful golden-yellow before leaf fall. Their venation is straight like those of other conifers. Another feature of the leaf, besides leaf shape, that distinguishes the Ginkgoales from other conifers is the shape of epidermal cells. The epidermal cells between the veins are polygonal in outline whereas over the veins they are rectangular in shape. The walls of epidermal cells are wavy. The stomata are restricted to the lower epidermis and are surronded by 4-6 subsidiary cells that form a circle around it and have characteristic finger-like projections overarching the guard cells. The stomata occur in between the veins and are arranged in broad bands or ill-defined rows. Kanis and Karstens (1963) have described the presence of stomata on the upper surface of leaves in the long shoots of male plants. Two vascular bundles enter at the base of leaf and undergo repeated dichotomy.

Foster (1938) studied the shoot apices of *Ginkgo biloba.* A shoot apex consist of a few superficial **apical initials.** Below them is the **sub-apical zone** consisting of a group of **central mother cells.** Around the proximal part of this zone is a zone of **rib meristem.** There is no distinct **tunica region** because the superficial cells usually undergo a periclinal division. The rate of division in sub-meristem cells determine whether the branch will be a dwarf shoot or a long shoot.

ANATOMICAL CHARACTERS

The anatomical details of the species also present certain interesting features that are discussed below :

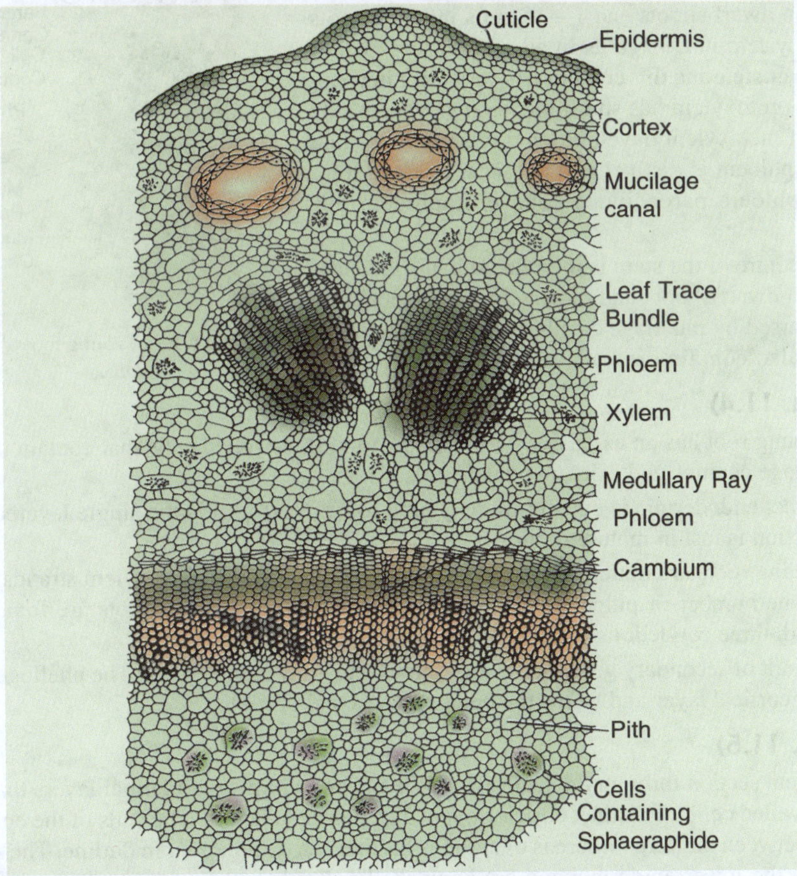

Fig. 11.3 T.S. young stem (long shoot) of *Ginkgo biloba.* Note the leaf trace bundles in the cortex.

Stem (Fig. 11.3)

A cross section of the young stem presents a roughly circular and wavy outline. There are some variations in the internal structure of a dwarf shoot and the long shoot.

The outermost layer of the stem is the epidermis which is quite distinct, in the young stems that have not undergone secondary thickening. It is made up of a single layer of brick-shaped cells and is covered with a thick cuticle (Fig. 11.3). In older stems, the epidermis is replaced by periderm which originates in the cortex.

Next to the epidermis is the cortex, which is quite extensive in the dwarf shoots and comparatively narrow in the long shoots. The outer cortical cells contain crystals of calcium oxalate (Fig. 11.3) and tannin filled cells. Large resin canals (Fig. 11.3) are found scattered in the inner coritcal layers.

There are no distinct endodermis and pericycle.

The very young stem consists of a number of discrete primary vascular bundles arranged in a ring. These bundles are conjoint, collateral, open and endarch. They run logitudinally throughout the stem and branch sympodially to give off leaf traces. Each leaf receives two leaf traces that originate from two different primary cauline strands (Gunkel and Wetmore, 1946). These leaf traces arise by tangential division of the cauline strand. Soon secondary growth starts and the vascular ring becomes complete and appears to be siphonostelic. The meduallary rays are uniseriate and 1—5 cells high in dwarf shoots and 1—15 cells in long shoots. The stelar system in *Ginkgo* as in all other living gymnosperms, is a **eustele** and this condition is very clear in young stems. The protoxylem has spiral thickenings whereas the tracheids of metaxylem have bordered pits on their radial walls. The phloem is distinct and consists of sieve tubes and little phloem parenchyma. The protophloem gets crushed.

In the centre of the stem is the pith which is large and extensive in dwarf shoots and narrow in long shoots. The pith is traversed by mucilage canals and its parenchymatous cells also contain crystals of calcium oxalate.

Fig. 11.4 T.S. young root of *Ginkgo biloba.*

Root (Fig. 11.4)

The young root has an extensive cortex made up of thin-walled cells that contain tannin filled cells, mucilage cavities and calcium oxalate crystals.

A distinct endodermis is evident in the young roots. Next to it is the single layered pericycle. This distinction is lost in mature roots.

The young roots are diarch or triarch *i.e.,* there may be two or three xylem strands alternating with the same number of phloem stands. The xylem is exarch. Triarch roots are found in young seedling with three cotyledons.

As a result of secondary growth, the root shows indistinct annual rings. The phellogen develops in the outer cortical layer and froms the outer corky layers of root.

Leaf (Fig. 11.5)

A vertical section through the leaf reveals the upper and lower epidermal layers that are made up of thin-walled cells. A distinct cuticle covers the epidermal layers. The cells of the epidermis are polygonal between the veins whereas over the veins they are rectangular in outline. The stomata are restricted to the lower epidermis and are haplocheilic. Kanis and Karstens (1963) reported some stomata in the upper epidermis of leaves of long shoots in male plants. The guard cells of the stomata are surrounded by a ring of 4—6 or 7 accessory cells whose surface projects into finger-like outgrowths that overarch and guard cells. The cells of the lower epidermis have slightly wavy outer walls. The stomata are arranged in broad but indistinct bands between the veins.

Between the two epidermal layers is the mesophyll tissue (Fig. 11.5) that shows no distinction into palisade and spongy parenchyma except in old and mature leaves of some long shoots. The mesophyll cells contain abundant chloroplasts and are loosely arranged so as to enclose small air spaces. Mucilage ducts also traverse the mesophyll. Tannin-filled and crystal-containing cells also occur in this region.

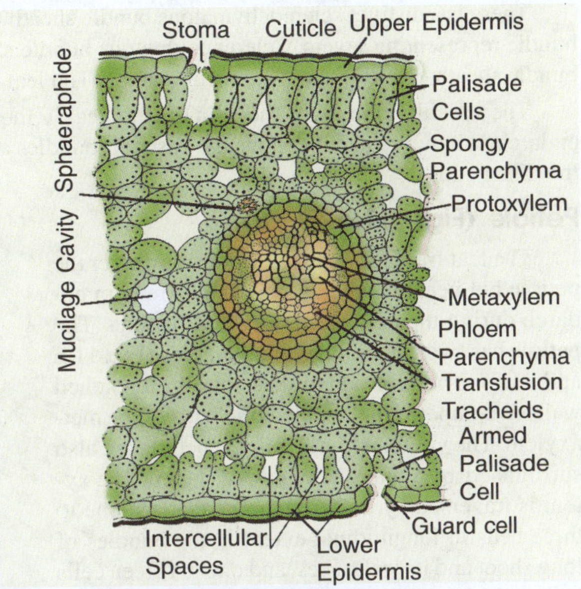

Fig. 11.5. T.S. leaf of *Ginkgo biloba*.

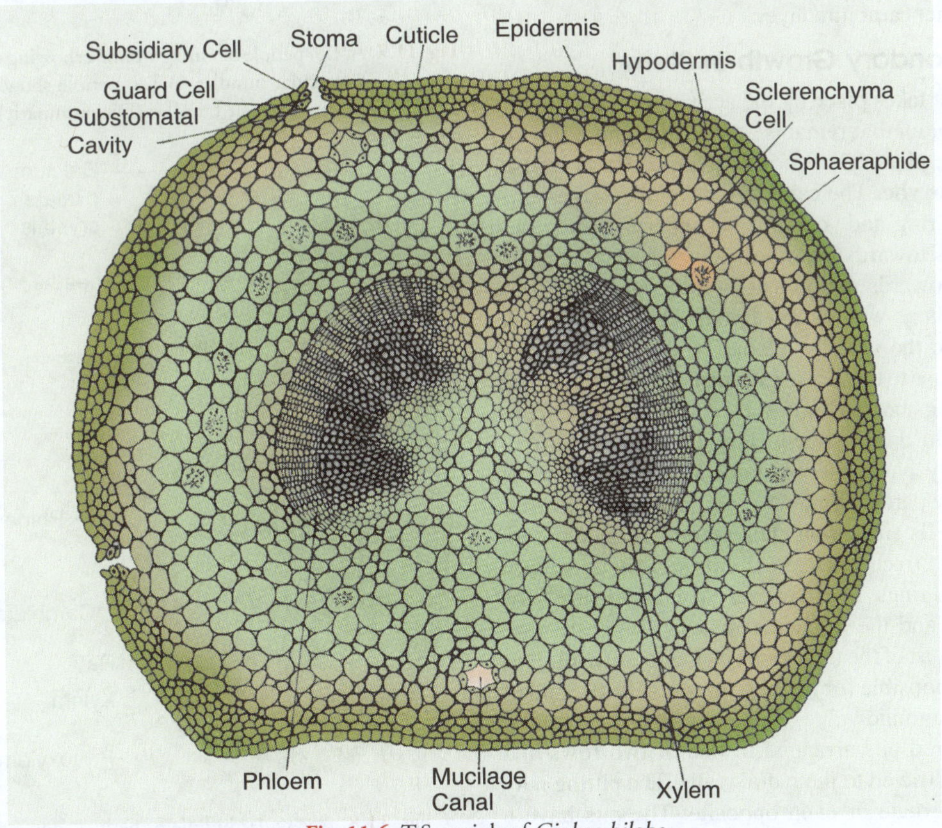

Fig. 11.6. T.S. petiole of *Ginkgo biloba*.

There is a distinct sclerenchymatous bundle sheath (only in mature leaves) around each vascular bundle representing a vein. Sclerenchymatous bundle sheath is not developed in young leaves. The bundle sheath extensions are parenchymatous. The leaf otherwise lacks sclerenchymatous tissue.

The vascular bundle of the vein has a weakly mesarch xylem. Majority of the bundles show endarch xylem. In the cotyledons the vascular bundles are mesarch at the base but become exarch at the distal ends.

Petiole (Figs. 11.6—11.8)

The leaf traces are endarch when they enter the petiole but become weakly mesarch or remain endarch during their course through the lamina. The petiole has two endarch vascular bundles (Fig. 11.6 and 11.7). The protoxylem has spirally thickened walls as compared to the pitted tracheids of the metaxylem. The vascular bundles in the petiole are also surrounded by a sclerenchymatous sheath. The xylem is traversed by uniseriate rays, that are one to three cells in longitudinal extent in the bundles of long shoot and its leaf traces, and one to fifteen cells high in the bundles of dwarf shoot and its leaf traces. The medullary rays of xylem are continuous with those of the phloem. The leaf trace bundles have a distinct cambium layer.

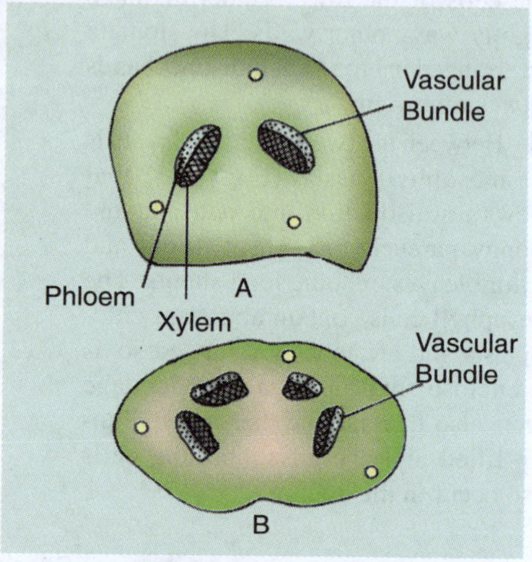

Fig. 11.7. A.T.S. petiole of *Ginkgo-biloba* showing the vascular bundles. B.T.S. petiole showing four vascular bundles (Diagrammatic).

Secondary Growth in Stem

It takes place by the activity of a single ring of cambium that remains active throughout the life of the plant. The wood is **pycnoxylic** as well as **monoxylic.** The cambium develops into a complete ring and cuts off secondary phloem elements towards the outer side and secondary xylem towards the pith. It is made up of fusiform initials as well as ray initials. The former give rise to the vascular elements whereas the latter give rise to uniseriate rays that are 1—5 cells high in long shoots (Fig. 11.11, A) and up to 15 cells high in dwarf shoots (Fig. 11.11, B). The secondary pholem consists of sieve elements and scanty parenchyma. Pholem fibres and companion cells are absent. The secondary xylem consists of tracheids and exhibits weahly developed annual rings. The tracheids have moderately thick walls and the tracheidal overlap is inextensive and most of the tracheids end the same level. This is responsible for making the wood brittle and of no economic value. The tracheids have circular bordered pits arranged in one or two rows and are restricted to the radial walls. The pitting may be alternate or even opposite. The pits have a distinct torus and bars of Sanio are also present

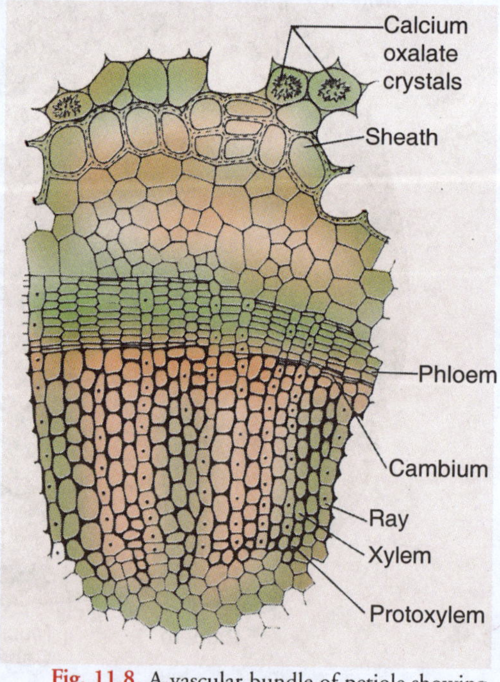

Fig. 11.8. A vascular bundle of petiole showing detailed structure.

(Fig. 11.10). The bars of Sanio are thickened lines that separate the pits (Fig. 11.10) and are present in the lumen of the tracheids. They do not cross the lumen or cavity of the tracheidal cell. There also occur trabeculae of Sanio that cross the tracheidal lumen and can, therefore be easily distinguishable from the bars. The tracheids are of moderate length as compared to very long coniferous tracheids.

Cork cambium appears in the outer cortex (Fig. 11.9) and forms a well developed periderm that replaces the epidermis. This cambium also forms secondary cortex towards the inner side. Mucilage canals are absent from the secondary cortex, but tannin cells and calcium oxalate crystals are present in smaller amounts.

REPRODUCTION

There are different male and female trees *i.e., Ginkgo biloba* is dioecious like the cycads. It is very difficult to differentiate between the young male and female trees. This distinction is only

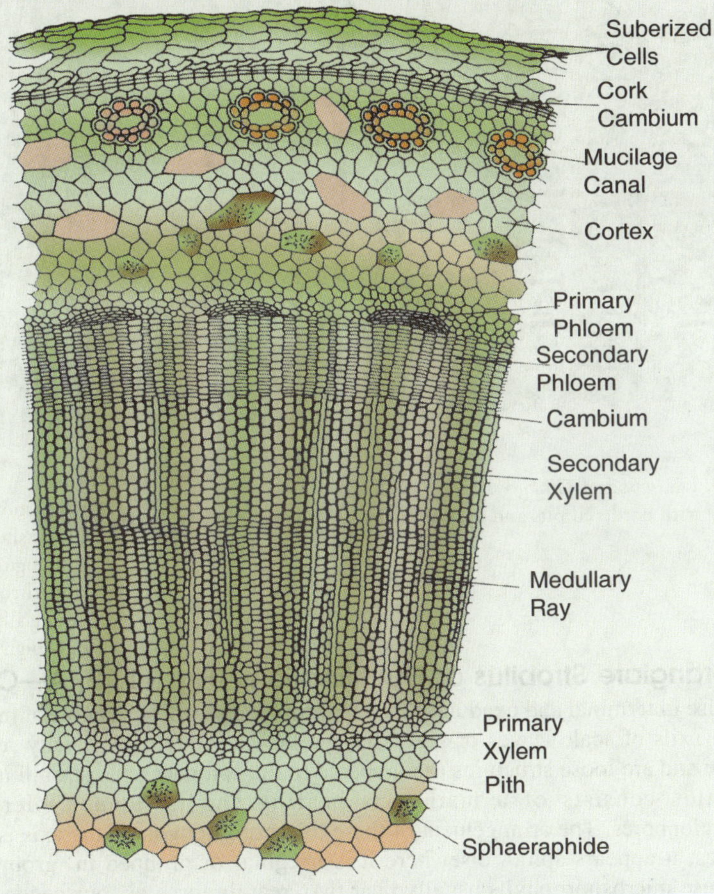

Fig. 11.9. T.S. old stem of *Ginkgo biloba* showing secondary growth.

possible when they bear male and female fructifications. Lee (1954) and Pollock (1956) are of the opinion that it is possible to differentiate between the chromosome complements of the male and the female trees. Lee (1954) described the presence of sex chromosomes in the female plant and XY complement in the male plant. The existence of sex chromosome has been denied by some workers. In case smears of stem tip and leaf primordia help us to distinguish the sex by studying chromosome morphology, it can be possible to avoid planting female plants in parks and streets because of the nasty odour produced by the ripe ovules and seeds. Both the microsporangiate and megasporangiate fructications arise in the axils of scales or foliage leaves and are found only at the apices of dwarf of spur shoots.

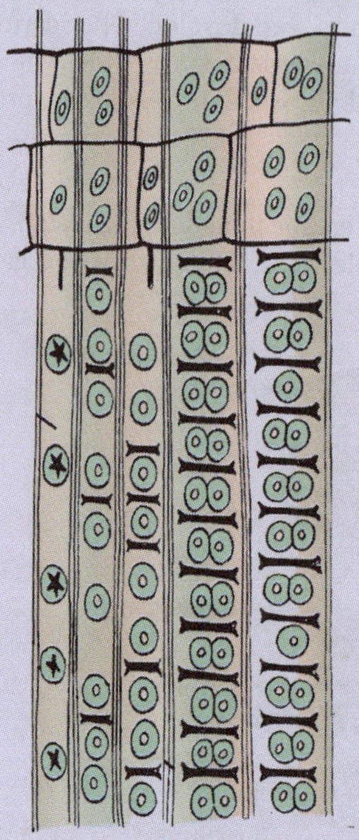

Fig.11.10. L.S. wood of *Ginkgo* showing tracheids with bordered pits and bars of Sainio.

Fig. 11.11. (A-B) A. Longitudinal tangential section of shoot of *Ginkgo biloba* showing medullary rays; B. Longitudinal tangential section through dwarf shoot of Ginkgo biloba showing 1-15 cells high medullary.

Microsporangiate Strobilus or Fructifications (Fig. 11.12, A—C)

They arise in terminal and pendulus clusters from the tips of spur-shoots of male trees. They are borne in the axils of scale leaves or green foliage leaves. They superficially resemble the catkin inflorescence and are loose structures as compared to the compact male strobili in the cycads. Each microstrobilus consists of a main axis bearing small, slender microsporophylls or microsporangiophores. The arrangement of these microsporophylls on the axis is very irregular. At certain places, it appears spiral, elsewhere it is irregular or grouped in groups of two or three together. These microsporophylls usually bear two, rearely three or four pendent microsporangia terminally. The microsporangiophores or sporophylls have a characteristic terminal ridge of tissue to which the sporangia are actually attached. Some morphologists are of the view that the terminal ridge is an abortive microsporangium and subscribe to the view that the sporangiophore of *Ginkgo* is derived by way of reduction from the sporangiophores of some fossil genera that had many microsporangia *e.g., Baiera.* This hump–like knob or the ridge has two variably developed horns each with a mucilage cavity. Lines of dehiscence of adjacent sporangia oppose each other and as they dehisce by generalised surface cell shrinkage, they are pulled apart by the shrinkage of two mucilage cavities located below each horn.

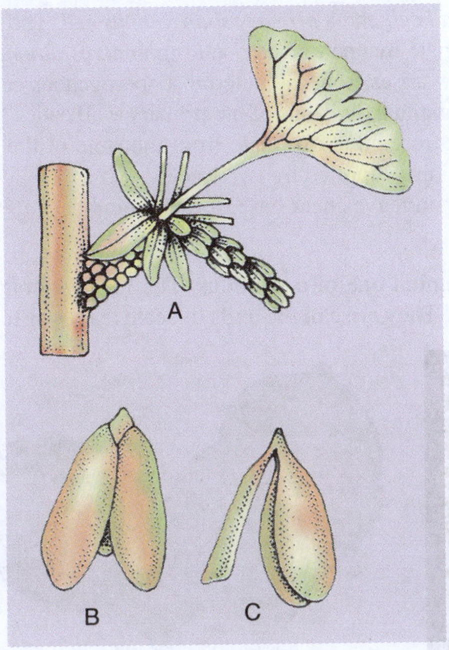

Fig. 11.12 (A—C). A Dwarf shoot or a spur shoot bearing terminal clusters of microsporangiate strobili; B—C. Microsporophylls bearing two microsporangia. Note the hump-like knob or ridge.

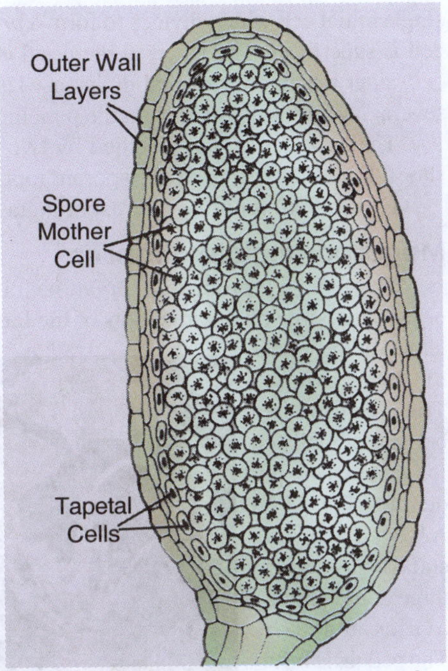

Fig. 11.13. Structure of sporangium of *Ginkgo biloba*.

Structure of Sporangium (Fig. 11.13)

The sporangia are tubular structures. The sporangial wall is many layered, but the cellular distintegration starting at the periphery of the sporogenous tissue ultimately involves all layers but the outermost layer. The outermost layer of the sporogenous cells differentiates into a single layered tapetum. The sporogenous cells undergo meiosis to form many tetrads of haploid microspores that later separate into spherical microspores.

Development of Microsporangium (Fig. 11.14, A—E)

The development of the microsporangium is *eusporangiate*. The microsporangium arises from a

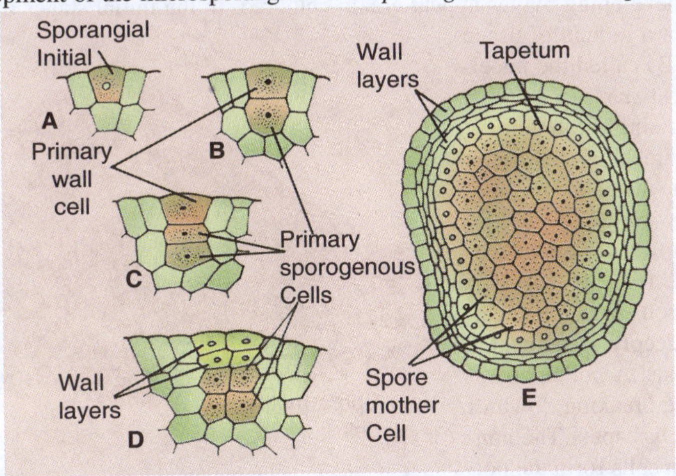

Fig. 11.14. (A-E) Development of microsporangium in *Ginkgo biloba*.

single initial cell which divides to form a *primary wall cell* and a *primary sporogenous cell.* The initial cell is superficial. The latter, in the usual eusporangiate manner, gives rise to the *microspore mother cells* after undergoing several divisions. The *tapetum* arises from the outermost sporogenous cells. A several layered wall arises from the periclinal and anticlinal divisions of hte primary wall cell.

The strobilar axis is supplied by two vascular traces that originate from the stele of the short shoot. Each sporophyll or the sporangiophore is in turn supplied by two traces.

Fujii (1896) reported microsporangia growing on leaves near base of the lamina.

Megasporangiate Strobilus

The female reproductive organ has little to remind one of a strobilus. The megas-porangiate organs are borne on dwarf shoots of the female tree. They arise in the axils of scale leaves or foliage leaves at the apices of dwarf shoots. The ovules are borne terminally on long stalks or peduncles (Fig. 11.15 A&B). Each long stalk bears two or three or rarely more ovules at its tip. The peduncle bifurcates at its tip and each branch bears sessile ovules, out of these only one matures and the others abort. Each young ovule is encircled by a collar at its base. This collar is regarded by Chamberlain as a megasporophyll. Four vascular traces supply the peduncle bearing two ovules. In case there are three ovules the number of vascular traces increases to six. The number of traces doubles with the increase in the number of ovules. The petiole in the leaf has two traces whereas the peduncle has 4 or 6 or more lends support to the view that it is a stem. Ovules arise on vegetative leaves (Pilger, (1 926) and sometimes the ovule bearing peduncle may become partly laminate.

Plate II : A portion of *Ginkgo*, plant with ovulate short shoot.

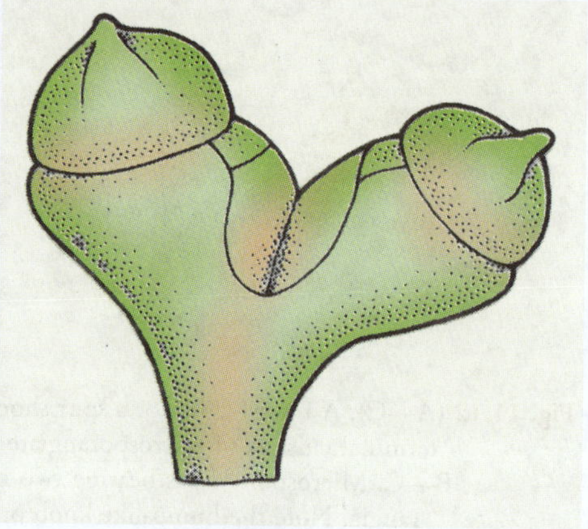

Fig. 11.15. A stalk bearing two ovules.

The young or developing ovules appear cream coloured. In the initial stages of development, the ovule consists of a mound of tissue (Figs.11.16, 11.17, B) called the nucellus. The development starts in April. It is surrounded by a single integument only at its base (Fig. 11.17, B). After about a month's time the naked nucellar mass, which is only partly covered by the integument, shows a distinct *megaspore mother cell* (Fig. 11.16). It is situated only a few cells from the nucellar tip but bcomes deeply seated as the nucellar apex expands. A pollen chamber is formed by the breaking down of the cells at the nucellar apex. The contents of these broken cells form the pollination drop. The integument has by

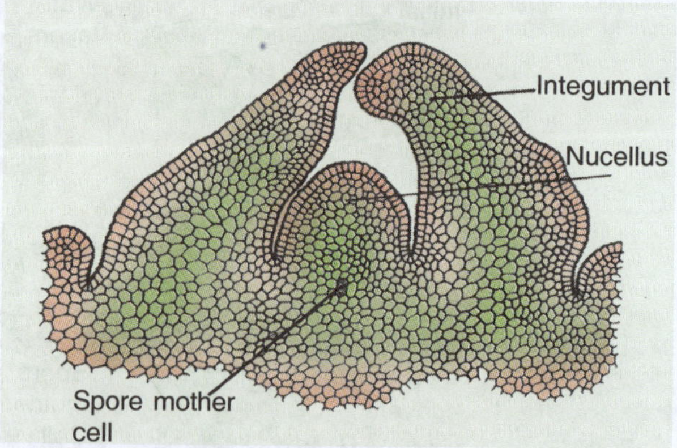

Fig. 11.16. L.S. Young ovule of *Ginkgo biloba*.

now grown rapidly around the nucellus leaving a small pore at the tip. This pore is the micropyle. The pollination drop exudes through the micropyle and catches the pollen grains that are drawn into the pollen chamber. The pollination drop dries and seals the micropyle. During all these changes, the megaspore mother cells had undergone meiosis to form a linear tetrad of four megaspores. Sometimes, due to the failure of one cell to divide further, only a row of three megaspores is formed. Out of these only one is functional. The diploid number of chromosomes as counted by Lee (1954) and Pollock (1956) is $2n = 24$. Miss Carothers (1907) counted $n = 8$, whereas Sprecher (1907) estimated it to be between 7—10. The young ovules are almost green or greenish yellow in colour. A collar-like structure also develops around the base of the ovule. It is more prominent in young ovules (Fig. 11.17C).

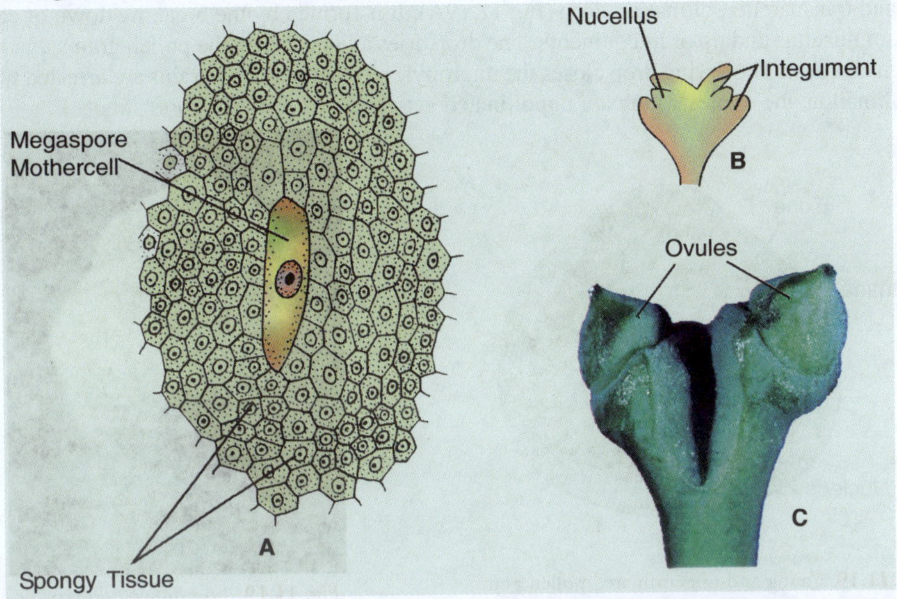

Fig. 11.17. (A-B). *Ginkgo biloba.* A. Nucellus with megaspore mother cell; B. Young ovule; C. A pair of ovules.

Usually only one out of a pair of ovules develops to maturity, the other aborts at an early stage. As the ovule matures, the integument becomes thick and can be easily distinguished to have three layers : (*i*) the outer greyish green and fleshy layer, (*ii*) the narrow stony middle layer, and (*iii*) inner fleshy layer well supplied with a watery substance. The integument is at first free from the nucellus except at the base but later due to intercalary growth and the consequent enlargement of the chalazal end the integument becomes fused with the nucellus except at the apex. The nucellus has a prominent apical beak. The apical part of the micropyle is generally two lipped—a feature rather general throughout the conifers. The mature ovule is about spherical in outline. The outer fleshy layer contains lactic acid that emits a musky odour. The vascular supply of the ovule consists of two

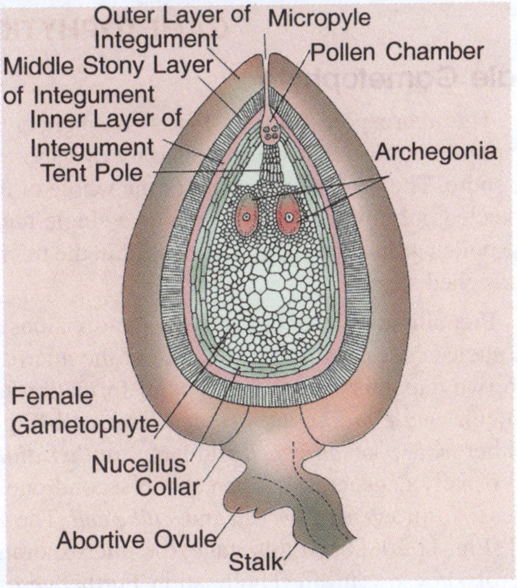

Fig. 11.18. L.S. ovule of *Ginkgo biloba.*

vascular strands that enter the fleshy layer. Sometimes, when the stony layer is three sided, a third vascular strand also enters the inner fleshy layer. The outer fleshy layer is entirely devoid of vascular trace.

The megaspore develops into the female gametophyte that is cellular at maturity and contains two or three archegonia at its micropylar end.

Pollination

It is anemophilous *i.e.,* pollination is affected by wind. In the Northern Hemisphere, the pollen grains are released during April or May when the ovules are quite exposed and the leaves are partly expanded. The semi-germinated pollen grains fall on the micropyle, where they are caught in the exuded and transparent pollination drop (Fig 11.19A). It is formed by the breaking down of cells at the apex of nucellus and tip of integuments. The drop dries and pushes in the pollen grains that fall in the pollen chamber. The drying drop closes the micropyle so that the pollen grains are arrested within. After pollination, the ovules which are unpollinated get abscissed from the short shoot.

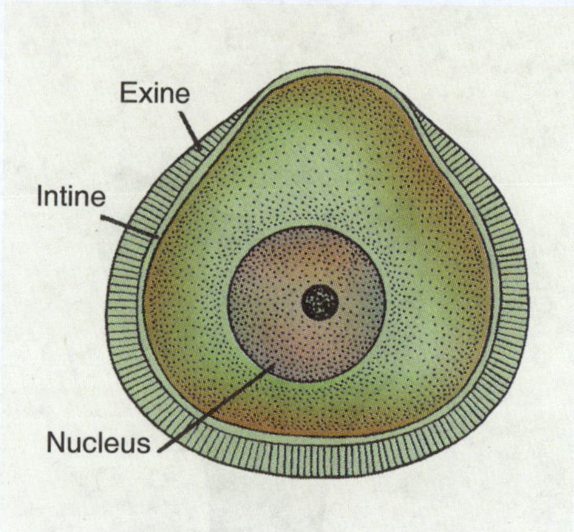

Fig. 11.19. Young and ungerminated pollen grain of *Ginkgo biloba.*

Fig. 11.19. A *Ginkgo biloba* ovule with pollination droplet :

GAMETOPHYTIC GENERATION

Male Gametophyte

The microspore (Fig. 11.19) is the pioneer structure of the microgametophyte. It is almost circular in outline and is surrounded by two spore coats. This inner or the intine completely covers the spore. The exine does not cover the whole of the spore and is lacking at the apical end, which is protected only by the intine. There is a single nucleus surrounded by dense cytoplasmic contents. The pollen grains start germinating within the microsporangium (precocious) and are released in the four-celled stage.

Prepollination changes include the divisions of the microspore nucleus. Before this division, the nucleus migrates from the centre of the microspore to the other end, where the exine is lacking, into two daughter nuclei. Later, wall formation takes place resulting in the formation of a small lenticular *male prothallial cell* and large cell (Fig. 11.20, B). The large cell divides again to from another *male prothallus cell* and larger *antheridial cell* (Fig. 11.20, C). The first formed prothallial cell usually degenerates, whereas the second one persists (Fig. 11.20, D). The antheridial cell divides to form a *generative cell* and *a tube cell.* The generative cell is in contact with second prothallial cell (Fig. 11.20, D). At this stage, the microsporangium dehisces and liberates the semigerminated microspores so as to effect pollination. Further germination takes place in the pollen chamber throughout the summer. Upon reaching the pollen chamber, the intine protrudes out slightly and fixes the semigerminated pollen grain to the tissue of the nucellus. This protruded intine acts as a haustorium and absorbs food for the germinating spore.

Other post-pollination changes include the division of the generative cell into a **stalk cell** or a **sterile cell** and a **body cell** or **spermatogenous cell.** (Fig. 11.20, E). The division of the generative cells is anticlinal. During later development, the sterile cell or stalk cell surrounds the second male prothallus cell (Favre-Duchartre, 1956). In cycads, the male prothallus cell protrudes into the stalk cell. The body cell divides to form two motile spermatozoids (Fig. 11.28, F, G). Blepharoplasts appear before nuclear division and develop into ciliated bands that are attached to the nuclear tips which form a small beak each. The spiral band gets wound around the tip of the nucleus to form few spiral turns. The pollen tube grows through the nucellus and may branch . The multiciliated sperms in *Ginkgo* are not separated by any wall within the body cell (in cycads, a wall is present). Schimamura (1937) reported the top-shaped spermatozoid to measure 90μ in length and 72μ in breadth.

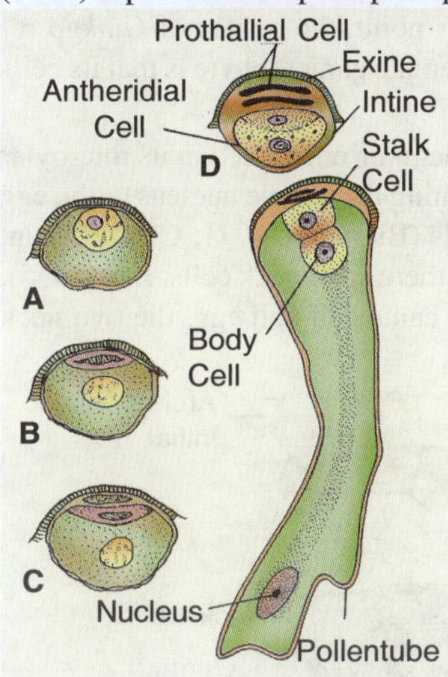

Fig. 11.20. (A-D). Microspore and its germination in *Ginkgo biloba.* Only earlier stages are shown.

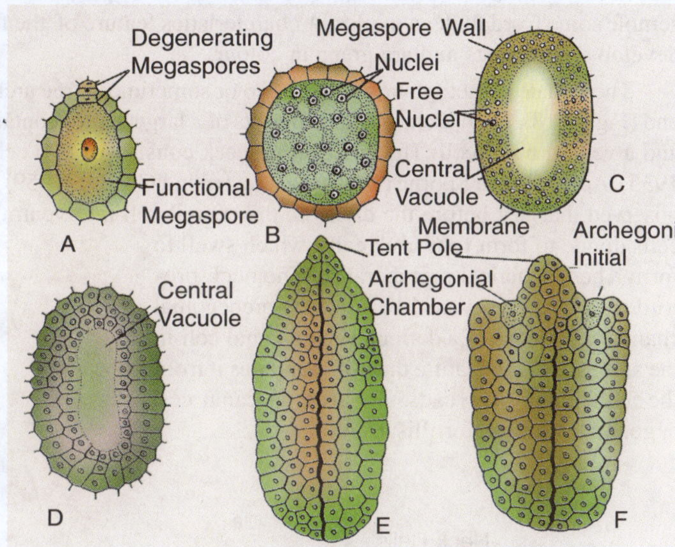

Fig. 11.21. (A—F) Development of female gametophyte in *Ginkgo biloba* (modified from Lee, 1955).

Tulecke (1957) cultured pol1len grains of *Ginkgo* on synthetic media and found cellular proliferation of the male prothallus cell and the stalk cell. Morphological significance of such a proliferation is not clearly understood.

Female Gametophyte

The megaspore (Fig. 11.17, A) is the pioneer structure of the female gametophyte. It enlarges considerably and assumes an almost spherical shape (Fig. 11.21, A). These changes start after the pollination has been effected. The nucleus undergoes repeated free nuclear divisions untill 256 free nuclei are formed. Lee (1955) obsrved that the number of free nuclei in the female gametophyte is far more than 256 and he estimated it to be more than one thousand. Such a condition is similar to the Cycadales (*Dioon*). These nuclei are embedded in a thin peripheral cytoplasmic layer that surrounds a central vacuole which later starts diminishing (Fig. 11.21, C). These changes take about two months. At the close of free nuclear division, the peripheral cytoplasm in which they are embedded secretes a thin membrane around it (Fig. 11.21, C), just internal to the megaspore wall.The cell walls, according to Chamberlain (1935) start appearing perpendicular to this membrane. The wall formation is centripetal *i.e,* it starts from the periphery and proceeds towards the centre (Fig. 11.21, D). During wall formation, the outermost cells are completely surrounded by cell wall but the innermost cell has no wall toward its inner side (Fig. 11.21, D) that faces the centre of the megaspore. This side is later covered by a wall when the wall formation comes to a stop. This means the inner walls of all inner cells are secreted by the cells themselves. The wall formation proceeds towards the centre from all sides. The cells in the central region of the the gametophyte are not joined by common wall leading to the formation of definite structure, along which the gametophyte can be spilt into two (Fig. 11.21, E). The developing gametophyte encroaches upon the surrounding nucellar tissue, which is reduced to a thin papery later. A characteristic beak-like protuberance develops (Fig. 11.21, A) from the micropylar end of the female gametophyte and touches the base of pollen chamber. Surrounding this beak develops a circular crevice (Fig. 11.21, F) into which the archegonial necks open. This protuberance is called the **tent pole.** In this point, the ovules of *Ginkgo* resemble some Cordaitalean ovules. A charcteristics feature of the female gametophyte is that its cells develop chloroplasts and are green in colour.

The female gametophyte contains two or sometimes three archegonia embedded in its microylar end (Fig.11.18). Each archegonium consists of a large venter containing the female nucleus or the egg and a ventral canal cell. The archegonial neck consists of two cells (Bierhorst, 1971; Chamberlain, 1935). According to Sporne (1965), and Foster and Gifford (1959), there are 4 neck cells. Lee (1955), observed that just before the division, the central cell into ventral canal cell and egg, the two neck cells divide to form four neck cells, which swell to form a hemispherical protuberance. The neck protrudes into the crevice-like circular archegonial chamber. Presence of a definite ventral canal cell in the archegonia is a feature that differentiates it from the archegonia of cycads. The ventral canal cell degenerates before fertillisation.

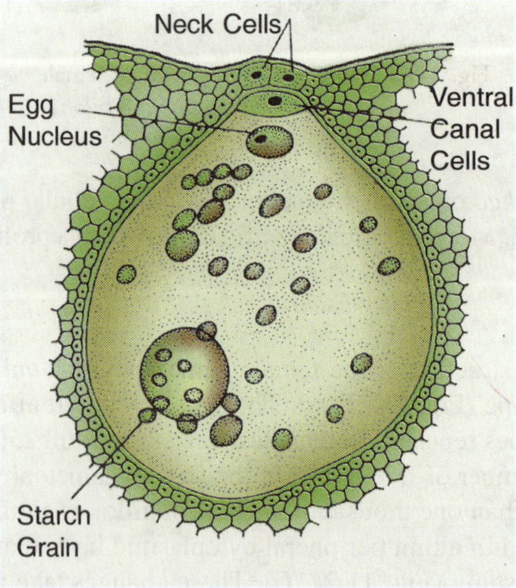

Fig. 11.22. Archegonium of *Ginkgo biloba.*

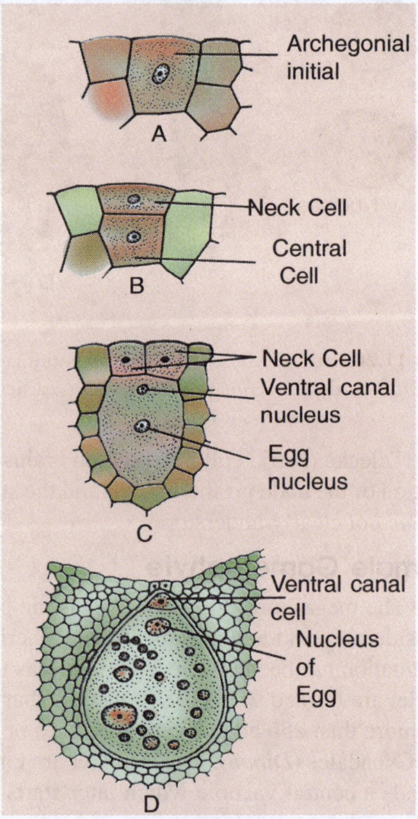

Fig. 11.23. (A-D) Development of archegonium in *Ginkgo.*

Fertilisation

At the time of fertilisation, the archegonial chamber in *Ginkgo* is filled with a liquid (in cycads the chamber is moist and contains no free liquid). When the pollen tubes discharge their sperms into the archegonial chamber, through the opening near the prothallial end of the pollen tubes movements of the sperms into the archegonia follow. The sperms squeeze their way through the archegonial neck and become considerably stretched without any injury. During their journey towards the female nucleus, the sperms are stripped off their flagella and cytoplasmic membrance. The naked male nuclei move towards the egg and one of them is able to fuse with it and effect fertilisation or syngamy. Lee (1955) describes that the egg cell forms a small projection which extends between the neck cells. The sperms get attached to this projection, which retreats leaving a path through the neck. Only the head of the spermatozoid which includes the chromatin, penetrates the egg cytoplasm, where it forms a globular mass. Simultaneously, the chromatin in the egg nuclear forms another group of granules. The union of chromatin granules of male nucleus and those of egg nucleus is the real fertilisation process. Banerjee and Redforth (1966) reported that the amount of total amino acids increases during fertilisation and again decreases a month later. During embryo development, the amino acids again increase in amount to be maximum during the next two months, when the apical meristem is organised. The amino acids proline and aspartic acid are utilised before fertilisation and during the establishment of polarity of zygote. Fertilisation may occur when the ovules are on the tree or after they have fallen to the ground (Eames, 1955).

EMBRYOGENY (Fig. 11.24 A—D)

Lee (1934) and Lee and Shen succeeded in culturing *Ginkgo* embryos by adding to the medium and extract of *Ginkgo* endosperm. Steward and Caplin (1952) also obtained extracts from *Ginkgo* endosperm for promoting embryogenesis *in vitro* cultures. This extract contains some amino acids and other substances that are necessary to promote nuclear division in the zygote. The work of Banerjee and Radforth (1966) also supports such an inference.

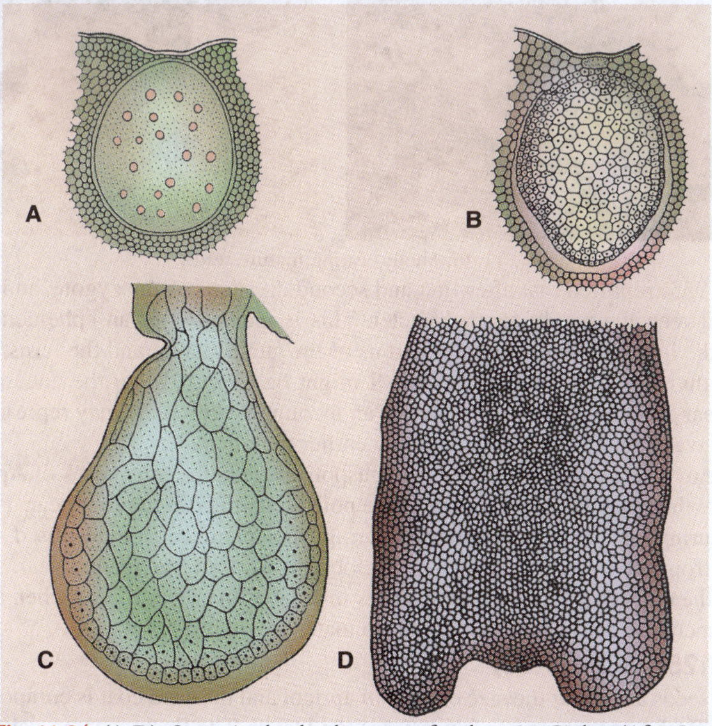

Fig. 11.24. (A-D). Stages in the development of embryo in *Ginkgo* (After Lyon).

The zygote enlarges in the archegonial venter and its nucleus divides repeatedly to form 256 free nuclei (Fig. 11.24, A) that are evenly distributed throughout the cytoplasm, there being no central vacuole. Lee (1955) reported the formation of a vacuole at the lower parts of the proembryo. Later as a result of centripetal wall formation, the developing embryo becomes cellular throughout (Fig. 11.24, B). The cells in the micropylar end of the embryo enlarge much more than those at the chalazal end (Fig. 11.24, C). The former grow into an indistinguishable suspensor that grows about 1 millimetre out of the neck of archegonium and curves over and generally touches the gametophytic beak or the tent pole. Lee (1955) reported an abnormal condition of occurrence of nuclei of various sizes in the free nuclear embryo. Soon active divisions, at the chalazal end of the embryo, form a distinct meristem. The shoot apex and the two cotyledons develop from the apical meristem (Fig. 11.24, D). Ball (1956) reported that the root or radicle develops from an apical meristem at the micropylar end. In the ripe seeds the embryo, in addition to the two cotyledons (Sometimes even three) develops several foliar primordia that along with the shoot apex constitute the first apical bud of the seedling.

Fig. 11.25. Shoot-bearing mature seeds.

Lee, C.L. (1955) reported that after first and second divisions of the zygote, an incomplete cell plate appears between the newly formed nuclei. This is indicative of an ephemeral type of wall formation after the initial divisions that remind us of the angiosperms and the ferns, where there is no free nuclear division. Lee interpreted that "I might be expected that the ancestors of *Ginkgo* lacked free nuclear division, and the presence of an incomplete cell plate may represent a vestige of multicellular cleavage found in phylogenetically earlier members."

In *Paris*, the ovules are pollinated at the megaspore stage in the third week of April. The pollen initials germinate three weeks after landing in the pollen chamber in the first year. The archegonial venter develop during June. Fertilisation begins during the September. Both fertilised and unfertilised ovules are shed from the tree irregularly from October to April of the Second year.

As soon as the embryo matures, it germinates in June of the second year when adequate water and suitable temperature are present. The embryo matures without dormancy.

SEED (Fig. 11.25A & B; 11.26)

The mature seeds are about the size of a small apricot and the seed coat is composed of an outer orange coloured fleshy portion and an inner hard stony layer. The fleshy coat is rich in butyric acid and when crushed, emits an unpleasant odour like that of a rancid butter. Next to the strong layer is the papery inner layer made up of crushed nucellus and inner fleshy layer of the integument. In the centre is dicotyledonous embryo surrounded by abundant endosperm. The pollen chamber and tent pole have disappeared. The micropyle is visible as a minute opening. The reduced collar is also visible at the base.

The germination of the seed is hypogeal (Fig. 11.27). During seed germination, the primary root emerges early and penetrates the soil as a result of active cell division at the tip of the hypocotyl. The cotyledons remain within the seed coat, where they absorb food from the surrounding gametophyte and pass it on to the developing seedling axis. The first leaves that unfold are typically bilobed. Later the stem tip grows and more leaves unfold. The remains of the seed remain attached to the base of the seedling stem for one or two years.

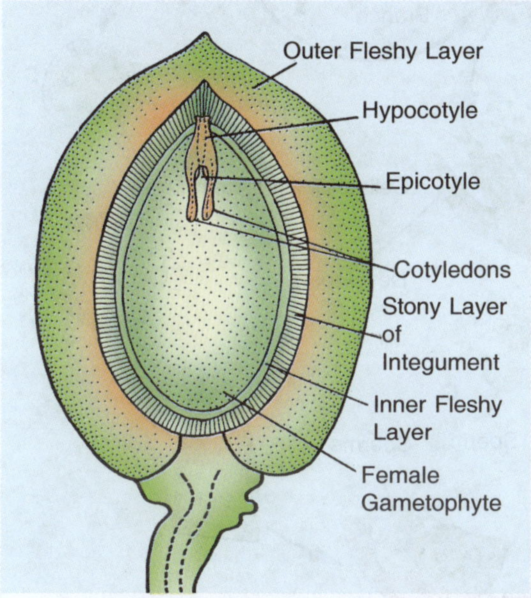

Fig. 11.26. L.S.seed of *Ginkgo*.

Fig. 11.27. Young seedlings of *Ginkgo biloba*.

A noteworthy feature is the development of the embryo continuously from the time of fertilisation until germination of seed. Dogra (1992) mentions that major part of the growth of embryo takes place when the seed is detached from the tree and the seed is lying on the ground.

Usually the reproductive cycle of *Ginkgo* is completed within two years.

ECONOMIC IMPORTANCE

Ginkgo is commonly grown as an ornamental tree in the gardens all over the world. In the United States, it is also grown as a shade tree. In China and Japan, the tree is worshiped.

The endopserm of the roasted seeds is edible. Too much of the seeds, if eaten, may prove fatal. The tree has a remarkable property of resisting fungal and bacterial attacks. Its wood is useless commercially.

The male trees are usually preferred in the city gardens and as shade trees because the ripe seeds of female trees emit an unpleasant odour. The tree is known in Hindi as *Bal Kunwari.*

Ginkgo Biloba seeds.

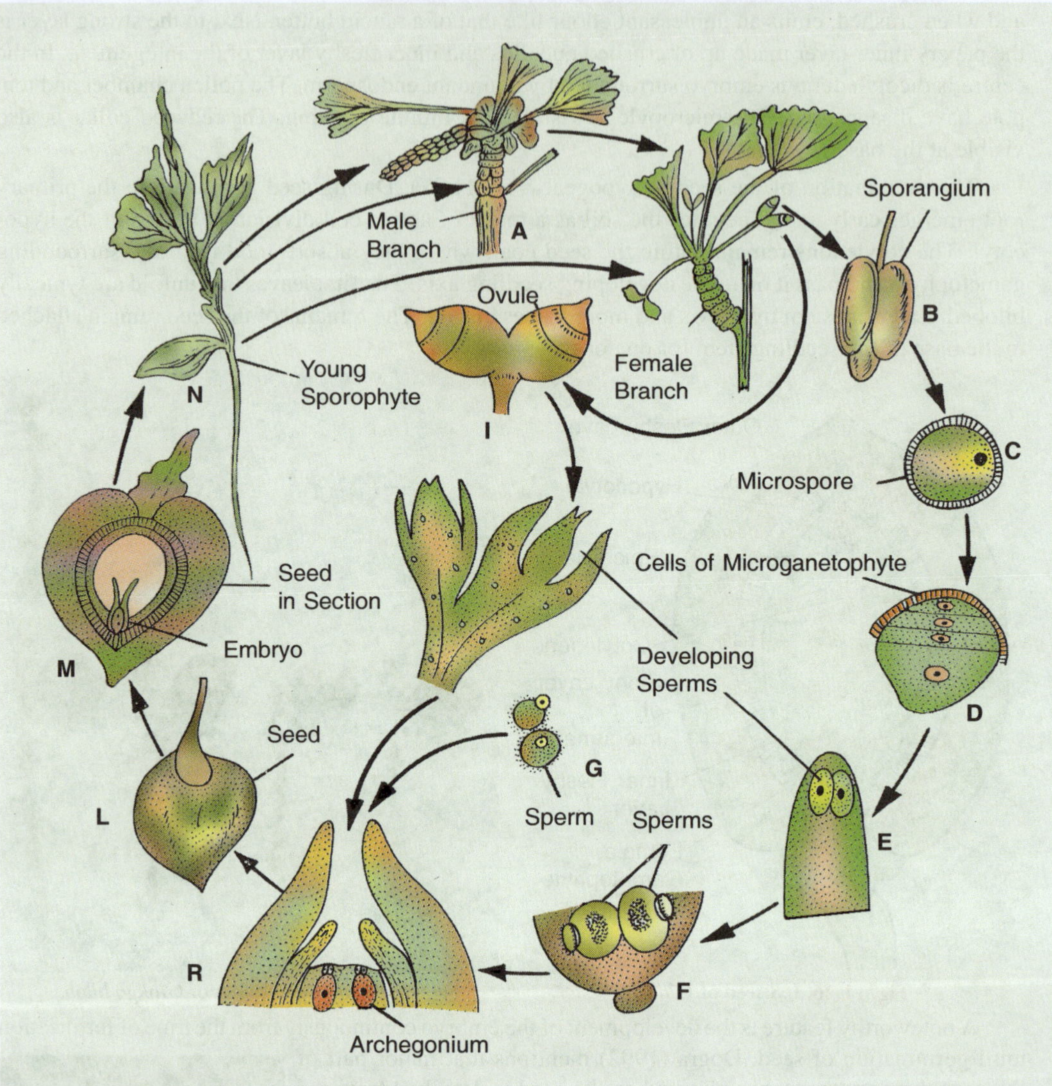

Fig. 11.28. Life cycle of *Ginkgo biloba*

PHYLOGENY

The *Ginkgo biloba* is the only surviving species of this ancient order has certain characteristics peculiar to itself. These include :

1. The fan-shaped and bilobed leaves with dichotomous venation.
2. The presence of a collar below the ovules. The ovules are borne on stalk-like peduncles and there are no structures likened to the megasporophylls.
3. Presence of a hump-like outgrowth at the tips of microsporangiophores.
4. The female gametophyte is produced into a tent pole like structure at its apex.
5. The embryo lacks a well developed suspensor.

Ginkgo resembles Cordaitales in certain respects. These points have been discussed in the earlier chapter.

The other groups with which it shares certain characteristics include Pteriodspermales, Filicales, Cycadales and Coniferales.

Resemblances with Filicales

Ginkgo resembles ferns in possessing : (*i*) dichotomously veined leaves; (*ii*) motile and many ciliated spermatozoids; and (*iii*) a distinct ventral cell in the archegonia.

The primary xylem of *Ginkgo* is very specialised type and resembles in certain respects with the Ophioglossales. *Ophioglossum* is the only fern that exhibits a coniferophyte transition from helical type of tracheids to the ultimate type with bordered pits. In this respec, it resembles *Ginkgo.*

Bierhorst (1971) while discussing the primary and secondary xylem elements in coniferophytes and some ferns like *Ophioglossum* and *Botrychium* states "The similarities between the ophioglossaceae and the coniferophytes in primary xylem structure, secondary xylem and periderm in *Botrychium* is anomalous in view of their divergent kinds of life cycles. It may be interpreted as evidence of an ultimate inter-relationship going all the way back to a common nonseed bearing ancestor such as an aneurophyte."

Resemblances with Pteridospermales

A few points of resemblances with the seed ferns are worth mentioning. These include :

(*i*) the abnormal occurrence of microsporangia and ovules of *Ginkgo* on the leaves; (*ii*) the presence of collar at the base of the ovule in *Ginkgo* is compared to the cupule of pteridospermous ovules; (*iii*) presence of a leaf gap in the stem; (*iv*) the wedge-shaped leaf with dichotomous venation can be compared with wedge-shaped pinnules of some pteridosperms. The seed of *Ginkgo* resembles some pteridosperms in size, presence of a distinct pollen chamber and in the nature of a massive fleshy layer.

Resemblances with Cycadales

Ginkgo bears some resemblances with the Cycadales in so far as their reproductive structures are concerned. These can be listed as under :

1. The pollen tubes are haustorial in function.
2. The spermatozoids are large and multiflagellated.
3. Presence of a pollen chamber and a distinct nuclear beak.
4. The female gametophyte is large and massive.
5. The large egg and the venter of the archegonia.
6. The longer period of free nuclear division in the proembryo.
7. The embryo is endoscopic.
8. The embryo has two cotyledons.
9. The germination is hypogeal and the cotyledons remain embedded between the endosperm tissue.
10. Thick and massive integument of the seed and the persistent outer fleshy layer.

Many points of difference between the reproductive structures of the two orders can be listed *e.g.,* (*i*) organisation of male and female fructifications; (*ii*) presence of a collar below the ovule; (*iii*) vascular supply to the ovule; (*iv*) the integument is free from the nucellus in young ovules; (*v*) presence of a tent pole; (*vi*) the female gametophyte is green in colour; and (*vii*) absence of suspensor in the embryo. Moreover the points in which the reproductive structure resembles those of cycads are also shared by other conifers. Only certain features of the male gametophyte and motile sperms are shared exclusively by *Ginkgo* and cycads. Regarding vegetative features, the *Ginkgoales* differ from Cycadales almost exclusively even during the development of male gametophyte. *Ginkgo* shows slight differences from the cycads *e.g.,* in *Ginkgo,* the generative cell divides anticlinally into two spermatozoids whereas in cycads, it divides periclinally (*Cycads revoluta* is, however, an exception). In *Ginkgo,* the stalk cell surrounds the prothallial cell during later stages of development whereas in cycads, the prothallial cell protrudes into the stalk cell. The differences in vegetative organisation include. : *Ginkgo* has a cone-like excurrent habit with branched stems and simple leaves whereas in cycads, the stem is unbranched with a tuft of compund leaves at its summit. In *Ginkgo,* the venation is not circinate which is a characteristic of the cycads. Large cortex, large pith, sluggish secondary growth and large size of the strobili also distinguish cycads from the *Ginkgo.* The microsporangia in cycads are arranged in distinct sori on the abaxial side of the microsporophylls, whereas in *Ginkgo,* microsporangia are terminal. The ovules in *Ginkgo* are also terminal whereas in cycads they are lateral in position (*Cycas*) on adaxial side of sporophylls as in *Dioon, Stangeria* and others. The microsporangiate strobili are catkin-like and lax as compared to compact strobili in cycads.

The similartities in the reproductive structures of *Ginkgo* and the cycads, according to Foster and Gifford (1959), "represent the persistence of certain of the generalised characters shared by many of, he ancient gymnosperms, particularly the pteridosperms. If this view is correct, the living cycads and *Ginkgo* would represent two terminal and independent lines of evolution. From the standpoint of vegetative morphology, the cycads may have evolved, together with Bennettitales, from a common pteridosperm stock—both are cycadophytes. On the other hand, the anatomy and organography of *Ginkgo* include, typical coniferophyte characteristics, and this relict plant may have developed, along with the Palaeozoic Cordaitales and the extinct confinct conifers, from a separate, but at present unknown, group in the seed ferns."

Chamberlain (1935) stated, "Similarly, it seems quite possible that both Cordaitales and Ginkgoales have a common ancestry. If reproductive organs could be found associated with the Ginkgoales—like leaves of the Carboniferous, the problem would be simplified ." He also suggested that the fern-like leaves of Ginkgoales suggest their origin from some heterosporous group of the Filicales. The motile sperms and resemblances between tracheidal transformation of *Ginkgo* and some ferns like *Ophioglossum* also suggest a remote origin from forms like *Aneurophytes* (Bierhorst, 1971).

Resemblances with Coniferales

Ginkgo can be viewed as a well defined coniferophyte because of many points in which it resembles the coniferales. These points of resemblances can be listed as below :

1. The cone-like appearance of the tree.
2. The stem is monopodial and is extensively branched.
3. Two types of branches *e.g.,* long shoots and dwarf shoots.
4. Presence of two types of leaves *e.g.,* green leaves and scale leaves.
5. The leaves are simple.
6. The wood is pycnoxylic.
7. The pitch and the cortex are narrow.
8. There is extensive development of secondary wood. The tracheids have circular bordered pits arranged uniseriately along the radial walls. The rays are uniseriate.
9. The leaves have sunken stomata.
10. Presence of two ear-like structures laterally on the microspores of *Ginkgo.* They resemble the winged pollen grains in *Pinus* and other conifers.
11. Longitudinal dehiscence of the microsporaniga.
12. The integument in *Ginkgo* ovule is two parted as in many coniferophytes.

13. The ovules in *Ginkgo* is sessile like those of conifers and is likewise surrounded by a single integument that can be differentiated into three similar layers. There is, however, difference in the degree of development.
14. The pollen chamber of the ovule in *Ginkgo* differs only relatively from the apical concavities formed also by cell breakdown, which occured in many conifers and taxads.

Ginkgo has all the characteristics that can enable us to view it as a well defined coniferophyte, that retains certain features of the male gametophyte and sperms (motile) which probably existed in ancestral forms common to conifers and taxads.

Florin also concludes that the Cordaitales, Ginkgoales, Coniferales, and Taxales all belong to the same natural group named as Coniferopsida. Florin regards them to constitute parallel lines of evolution which separated from each other in the Upper Devonian or Carboniferous times. They all originated from an unknown (at present) common ancestor and later followed their own lines of evolution. Such ancestors can, however, be traced among the Progymnosperms of Beck.

FOSSIL MEMBERS OF GINKGOACEAE

The fossil remains of several members belonging to the *Ginkgoaceae* have been found, but unfortunately very little is known about their reproductive structures. Majority of them are known in the form of leaves and have been found from the late Palaeozoic (Permean rocks) and Mesozoic age. The leaves belonging to fossils that date back to Palaeozoic and Mesozoic range in their external morphology from deeply dissected to gradually less dissected ones and ultimately entire leaves of the modern Ginkgo. Andrews (1961) considers these fossil genera under three headings:

A. Some Palaeozoic fossils that are suggetive of being ancestral ginkgoales

These include the genera *Saportaea* (*S. nervosa*) described by Halle from the Permian of Shansi in China. Ii is essentially a leaf resembling the modern *Ginkgo* in shape and venation. This fossil leaf (Fig. 11.29, A) has a long petiole (5 cm long and 4 mm board). The lamina or the leaf blade which is roughly wedge-shaped is 14 cm broad and has sinuate upper margin *Ginkgodium nathorsti* Yokoyama and *Dichophyllum morrei* Elias are the other two general that belong to this category. The leaves in the former species were spoon shaped, shortly petioled and have a medium sinus. They had a length of about 6.6 cm and breadth 2 cm. *D. moorei* from the upper Carboniferous consists of a peculiar branching pattern without any distinction into the main axis and the leaves. The specimens exhibit a sequence of repeated equal dichotomies that terminate in linear ultimate lobes. A distinct cuticle has also been detected from the specimens. Seeds with two distinct horn-like projections at the micropylar end were associated with fossil. Andrews regards it to be a, "preginkgophyte in which the characters of the group are becoming established."

B. Mesozoic and Tertiary fossils that seem closely related to the living *Ginkgo biloba*

Some of these belong to the genus *Ginkgo* and those from the United States have been assigned to three species of the genus by Brown. These include :

1. *G. adiantoides* (Unger) Heer (Fig. 11.29, C) had kidney-shaped leaves with distinct notch at the apex. It has been encountered in North America from Alaska to South Dakota from Eocene rocks belonging to the Oligocene and Miocene ages. The species upto Pleiocene was restricted to Asia and Western Europe and since then only to Eastern Asia.

2. *G. digitata* (Brongniart) Heer (Fig. 11.29, D) had deeply dissected wedge-shaped lamina and has been reported from the Jurassic rocks of Europe, Australia, Japan and Asia.

3. *G. lamariensis* Ward. (Fig. 11.29, B) from the late Mesozoic had a wedge-shaped and slightly notched blade.

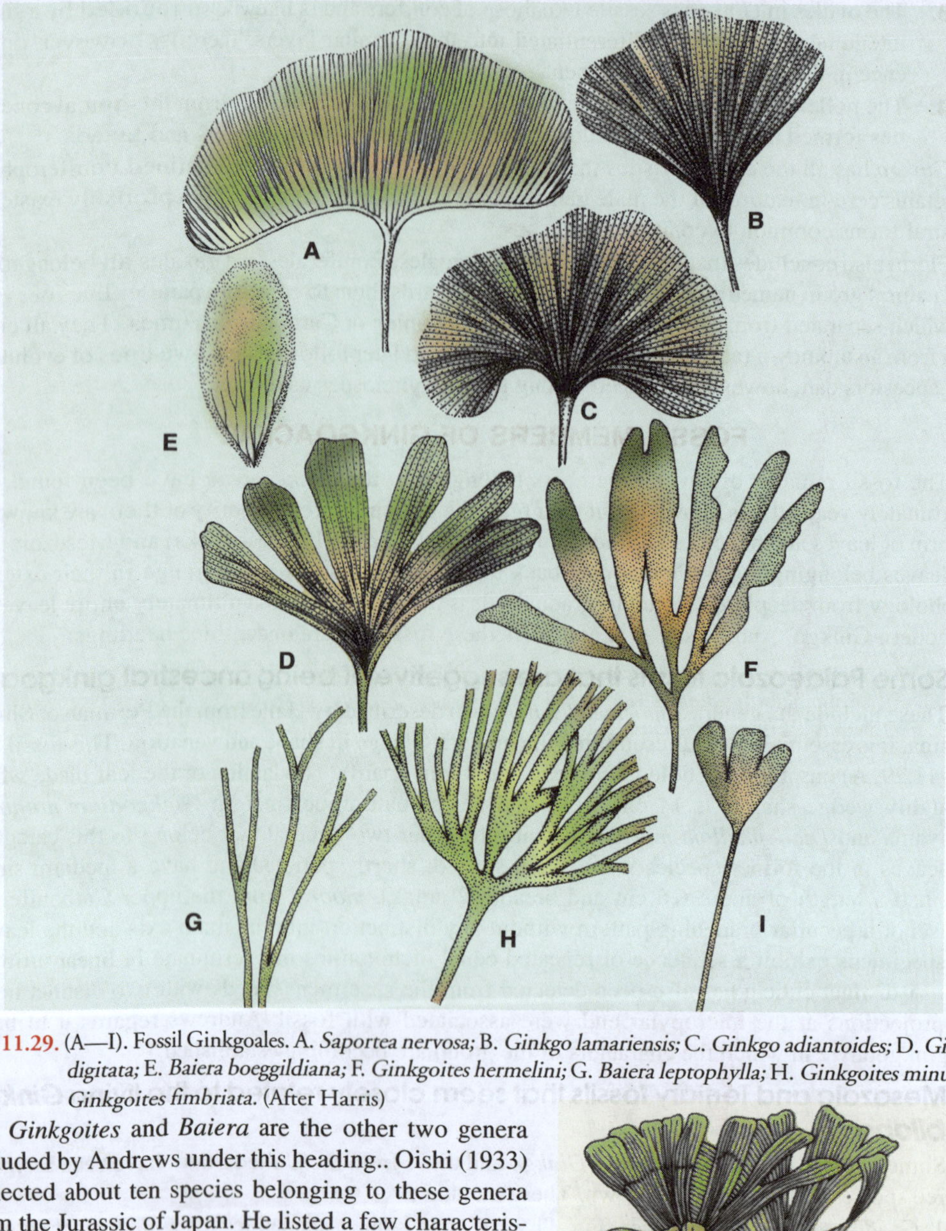

Fig. 11.29. (A—I). Fossil Ginkgoales. A. *Saportea nervosa;* B. *Ginkgo lamariensis;* C. *Ginkgo adiantoides;* D. *Ginkgo digitata;* E. *Baiera boeggildiana;* F. *Ginkgoites hermelini;* G. *Baiera leptophylla;* H. *Ginkgoites minuta;* I. *Ginkgoites fimbriata.* (After Harris)

Ginkgoites and *Baiera* are the other two genera included by Andrews under this heading.. Oishi (1933) collected about ten species belonging to these genera from the Jurassic of Japan. He listed a few characteristic features fo these genera. These are : (*i*) the stoma were found both on the upper and lower epidermises but more on lower epidermis; (*ii*) guard cells were sunken below the surface; and (*iii*) the subsidiary cells that number five to seven are arranged in a circular pattern and overarch the guard cells. Some species of *Ginkgoites* described from Greenland include :

Fig. 11.30. *Ginkgoites lunzensis* (After Kraussel)

1. *Ginkgoites fimbriata* Harris (Fig. 11.29, I). was petiolate leaf with petiole upto 4.5 cm long and small

d had dichotomous venation.

riera flettii. Florin (Fig. 11.31. A,B) also had strap-shaped, linear and sessile leaves argins and obtuse apex. The leaves were smaller than *Windwardia* and measured 6 cm ut 4 mm broad. Some were bilobed whereas others were entire. These were also borne short shoots. The venation was open dichotomous.

TRICHOPITYACEAE (Fig. 11.32, A-C)

amily is represented by a single genus *Trichopitys. T. heteromorpha* was described in the 5 from the Lower Permian of southern France by Saporta. Florin (1949) studied the species d confirmed its inclusion in the Ginkgoales and also regarded it as the oldest member of the Due to its differences from other fossil members of the *Ginkgoaceae,* it was put under sepa- mily. Florin reconstructed the plant from fragments about 8 mm in diameter and bore spirally ged leaves (Fig. 11.32, A) and fertile ovulate shoots. The leaves were narrow, terete, entire and tomously branched. They appeared circular in a transection and branched in several planes. fertile or ovuliferous trusses or branches were found in the axils of leaves. Florin described m as "sporangiophoric complexes", and were found on long shoots. Each such structure con- ted of an axis that branched repeatedly in a dichotomous manner. Each branch bore a single rminal ovule. The ovule was recurved or inverted. Each such sporangiophoric system bore com- nonly 4—6 ovules but upto 20 ovules have also been reported. Florin is of the opinion that these reproductive structures that are borne on long shoots in the Permian *Trichopitys* were gradually

Fig. 11.32. *Fossil Trichopityaceae.* A. *Trichopitys heteromorphia*, a shoot fragment with leaves bearing seed branches; B. single ovuliferous appendage of *Trichopitys*; C. Ovule of *Trichopitys.* (After Florin)

lamina with a lobed apex. The guard cells were surroun... ranged in a circular pattern and had very thick outer sur...

2. *G. acosima* Harris had long petioled wedge-shape... was again sub-divided into three smaller lobes. The smaller

3. *G. minuta* (Nathorst) Harris (Fig. 11.29, H) was also a wedge-shaped leaf with a distinct petiole and finely dissected lamina. Presence of petiole distinguishes it from the leaves of *Baiera.*

4. *G. hermelini* (Hartz) Harris (Fig. 11.29, F) was a petiolate and wedge-shaped leaf divided into six or more elliptical lobes. The 4-6 subsidiary cells possessed papilla-like outgrowth that instead of overarching the stoma grow straight upwards. The stomata has sunken guard cells.

5. *G. obovata* (Nathorst) Seward had entire lamina.

6. *G. lunzensis* (Fig. 11.30) also had petiolate and wedge-shaped leaves and lobed lamina. These were discovered on dwarf shoots. The leaves had sunken stomata surrounded by 4-7 subsidiary cells whose papillae overarched the guard cells.

Ginkgoites crassipes, G. lobata and *G. feistmantelli* are the Indian species of the genus discovered from the Gondwana land.

Fig. 11.31. Fossil Ginkgoales. A-B. leaves of *Arctobaiera flettii.* C. leaf of *Sphenobaiera horntiana.*

The common species of the genus *Baiera* include *B. spectabilis, B. leptophylla* (Fig.11.27, G.) *B. boeggildiana* (Fig. 11.29, E) and *B. munsteriana.* An unidentified species has also been reported form the Gondwana of India. The leaves of *Baiera* were like those of *Ginkgo* and *Ginkoites* but had very short petioles or were sessile. The Jurassic *B. munstriana* had catkin-like structures assoicated with it. These were microsporangiate strobili that had distinct axis with loosely arranged sporangiophores that bore 10-12 terminal sporangia that were pendant. A distinct hump-like that of modern *Ginkgo* was also recoginsable. The almost sessile leaves of *B. boeggildiana* (Fig. 11.29, E) had 3—4 cms long lamina with two obtuse apical lobes and dichotomous venation. The stomata were sunken in a pit. The guard cells were surrounded by subsidiary cells with finger-like projections that overarched the stomatal aperture.

C. Some Ginkgoaceae from Mesozoic and Palaeozic eras

These include the following genera :

1. *Sphenobaiera* (Fig. 11.31, C). It extended from Lower Permian to Lower Cretaceous periods. In *S. furcata* (Heer) Florin, both short shoots and long shoots bearing sessile and repeatedly forked leaves have been found. The microsporangiate organs have been found in organic connection on the short shoots. These consisted of a central axis that divided into branches. Each branch bifurcated once or twice and bore terminal clusters of three to five sporangia. The leaves of *S. horniana* (Fig.11.31, A) were strap-shaped and dichotomously branched.

2. *Windwardia crookalli.* Florin was described from the high Arctic Jurassic rocks of Franz Joseph Land. The leaves were long, linear, sessile and starp-shaped. They had entire margins and truncate apex and measured about 12 cm long and 5 mm broad. The leaves were borne in clusters on

transferred to the short shoots, during the course of evolution. This is borne out by the fact that in the living *Ginkgo,* the long shoots and short shoots are interconvertible and sometimes the long shoot behaves as short shoot for one season and again starts growing as a long shoot. Regarding the number of ovules, there are only one or two on a single truss in *Ginkgo* but abnormal cases with initiation of more than two ovules on a single truss are on record. These observations suggest that *Trichopitys* is very much a member of the Ginkgoales.

QUESTIONS

ESSAY TYPE

1. Draw L.S. of ovules of *Ginkgo* and *Ephedra*. Describe the development of female gametophyte in any gymnosperm. **(Garhwal MSc, 2004)**
2. Give illustrated account of Ginkgoales and compare it with Gnetales. **(Rohilkhand MSc, 2000)**
3. *Ginkgo* is a living fossil ? Why ? Describe the distinguishing features and interrelationships of *Ginkgoales*. **(Rohilkhand MSc, 1991)**
4. Describe important features of *Ginkgoales* and state why *Ginkgo-biloba* is called a living fossil. **(Awadh MSc, 1999)**
5. *Ginkgo* is a living cordaite. Discuss this statement. **(Kanpur MSc, 1995, 1984; Vinoba Bhave, 1998)**
6. Describe the characteristic features of *Ginkgo* and discuss its affinities. **(Kanpur, 1998)**
7. Discuss the affinities of *Ginkgoales*. **(Kanpur MSc, 1985, 1989)**
8. Write in brief life history of *Ginkgo biloba* giving suitable sketches. **(Kanpur MSc, 1986)**
9. Describe the morphology and anatomy of the vegetative part of the sporophyte.
10. Describe the fossil member of *Ginkgoales*.

SHORT ANSWER TYPE

11. **Write short notes on :**
    (*i*) L.S. ovule of *Ginkgo*. **(Garhwal MSc, 2004)**
    (*ii*) Morphology of the female strobilus in *Ginkgo*. **(Gauhati MSc, 2003)**
    (*iii*) *Ginkgo* is a living fossil. **(Rohilkhand MSc, 1995, 2001, 2004; Kanpur MSc, 1999, 2000)**
    (*iv*) Short shoot in *Ginkgo*. **(Rohilkhand M.Sc, 1996)**
    (*v*) *Ginkgo*. **(Rohilkhand MSc, 1999; Awadh MSc, 2001; Kanpur M.Sc, 1996)**
    (*vi*) Fossil members of Ginkgoales. **(Rohilkhand MSc, 2002)**
    (*vii*) Morphology of Ginkgoales. **(Rohilkhand MSc, 2003)**
    (*viii*) Phylogeny and interrelationships of Ginkgoales. **(Kanpur MSc, 1993)**
    (*ix*) Living fossil. **(Kanpur MSc, 1994)**
    (*x*) Morphology of seed bearing part in *Ginkgo*. **(Kanpur MSc, 2000)**
    (*xi*) Anatony of the stem.
12. Name a living fossil in gymnosperms. **(Gurunanak Deo, 2004)**
13. Give a morphological note on *Ginkgo*. **(Gurunanak Deo, 2004)**
14. Compare ovules of *Cycas* and *Ginkgo*. **(Allahabad M.Sc, 2001)**
15. **Diffrentiate between :**
    (*i*) Ovules of *Taxus* and *Ginkgo*. **(Kanpur MSc. 1988, 1992, 1994)**
    (*ii*) Male cone of *Cycas* & *Ginkgo*.
    (*iii*) Roots of *Cycas* and *Ginkgo*.
16. Mention important features of Ginkgoales and mention why it is called a living fossil ? **(Kanpur M.Sc, 2001)**
17. Write in brief with suitable sketch about the female flower of *Ginkgo*. **(Kanpur M.Sc, 2001)**
18. Justify that *Ginkgo* is a living fossil. **(Kanpur M.Sc, 2003)**

OBJECTIVE TYPE

19. Fill in the blanks :

(*a*) Venation in *Ginkgo biloba* leaf is ______________ .

(*b*) ______________________ is a living fossil.

(*c*) *Ginkgo biloba* is cultivated as ____________________ .

(*d*) The wood of *Ginkgo* is ________________ .

(*e*) Bal Kunwari is the name given to the genus ________________ .

20. Select the correct answer :

(*i*) Which one of the following is called a living fossil ?

(*a*) *Gnetum* (*b*) *Ginkgo biloba*

(*c*) *Taxus* (*d*) None of the above.

(*ii*) Which plant is known as living fossil?

(*a*) *Cordaites* (*b*) *Pentoxylon*

(*c*) *Ginkgo* (*d*) *Gnetum.*

(*iii*) The pollination in *Ginkgo* is

(*a*) Hydrophilous

(*b*) Zoophilous

(*c*) Anemophious

(*d*) All of the above.

(*iv*) In which of the following tap root is present ?

(*a*) *Lycopodium* (*b*) *Selaginella*

(*c*) *Marsilea* (*d*) *Ginkgo.*

(*v*) The pollen grains in *Ginkgo* are

(*a*) almost circular and uninucleate

(*b*) almost circular and binucleate

(*c*) oval and multinucleate

(*d*) oval and uninucleate.

(*vi*) Maiden Hair Tree is the name given to

(*a*) *Cycas* (*b*) *Ginkgo*

(*c*) *Pinus* (*d*) *Ephedra.*

CHAPTER **12**

Coniferopsida - Coniferales

- Cycadospsida : Economic Importance
- Primitive Conifers
- Family : Pinaceae : Reproductive Structures
- Family : Taxodiaceae
- Family : Podocarpaceae
- Family : Cephalotaxaceae
- Family : Araucariaceae
- Family : Lebachiaceae
- Family : Voltzaiceae
- Family : Palissyaceae
- Economic Importance

GENERAL ACCOUNT AND ECONOMIC IMPORTANCE

Introduction

The living conifers include 52 genera and 570 species and are distributed mainly in temper ate regions or in the mountains of the tropics. They are tall, elegant and graceful trees with foliage and branches presenting a magnificent cone-like appearance. All are woody perennials. They range in size from the tallest trees of californian *Sequoia sempervirens* and *Sequoia gigantea* that reach a height of 100 metres or more and a diameter of 5-11 metres (the former is as old as 100 years whereas the latter lives upto 4,000 years) to species of *Dacridium* that are only a few inches tall. Some species of *Juniperus e.g. , J. horizontalis* grows completely prostrate. *Taxodium mucronatum* from Mexico has lived for 2,000 years and has achieved a girth of 15 metres. In between these extremes of size are included *Pseudotsuga taxifolia* a native of New World that grows 150-200 ft. tall, 3-6 ft. diameter, *Abies grandis* of Vancouver and *A balsamea* of New Foundland that grow from 80 ft to 200 ft tall and 2-8 ft in diameter, *Abies pindrow* and *A.. spectabilis* grow in India at 8,000 to 14,000 ft above sea level , reach a height of 105 ft. and a diameter of 4 ft, maority of the species of *Pinus* grow upto 120 ft tall and vary in diameter from 2-5 ft, *Cedrus deodara* grows in India upto12,000 feet above sea level and reaches a height of 175 ft and a diameter of 6 ft, *Tsuga canadensis, Phyllocladus trichomanoides, Cupressus, Juniperus,* and *Taxus* grow from 40-80 feet tall and vary in diameter from one to three feet. *Juniperus communis* grows in the Western Himalayas upto a height of 4,250 metres and attains a height of 0.6 metre only. *Juniperus wallichiana* from the same locality has two spreading branches 30 centimetres to 1.5 metres high. *J.wallichiana* grows upto 18 metres high in the East Himalayan dry Juniper forests J. Macropoda grows 5 – 8 meters high in the west Himalayan dry juniper forests. Similarly *Cupressus torulosa* that forms open forests in the outer and middle ranges of Himalayas from Chamba (H.P.) to the Aka Hills (NEFA) at 1,800 to 2,800 metres, varies in height from 10 metres to 45 metres. They usually grow on limestone rocks, particularly on screes of limestone shales. *Abies delavayi* forms forests on the northern slopes of Piri between 2,750 -3,350 metres and grows upto 20 metres tall with a girth of 2 metres. It is a Chinese species but occurs in remarkably pure stands in this locality in the Eastern Himalayas. *Tsuga dumosa* from Darjelling division forests has been seen to measure 6 metres in girth. It grows along with *Abies densa* upto 3,500 metres above sea level. *Larix griffithiana* grows abundantly in the Eastern Himalayas from Sikkim and through Chamba valley of Tibet to Bhutan and Mishmi Hills of NEFA, usually above 3,000 metres. It grows 10-17 metres high and 1-2 ft in girth and is mixed with other conifers like *Pinus wallichiana, Abies densa, Picea spinulosa, Tsuga dumosa* and *Junipers wallachiana. Picea smithiana* grows along with *Cedrus deodara* or even above the deodar forests in Western and Northern Himalayas and may reaches upto 50 meters in height and a girth of 2-4 ft. It grows in central and eastern Himalayas, including North-West Kumaon westwards through Jaunsar, jammu and Kashmir and reaches an altitude of up to , 3,000 meters. *Pinus gerardiana* grows in the inner dry valleys of the Western Himalayas including the Kishtwar valley but not Kashmir. It grows upto 25 meters tall and 2-3 ft in girth. Its altitudinal range is 1,600 meters to 2,440 meters. *Pinus roxburghii* forms extensive forests along steep slopes in the western Himalayas at 1,500 meters to 2,000 meters and grows in height from 23-35 meters. *P. insularis* forms forests in the Khasi hills, Naga hills and in Manipur (Eastern Himalayas) at 800 to 1,600 meters and grows upto a height of 25 meters. *Pinus wallichiana* grows both in Western and Eastern Himalayas at an altitude ranging between 1,800 meters to 3,600 meters. It grows as tall as other species of *Pinus* in India. *Podocarpus neriifolnis* and *Cupressus funiberis* also grow in the central Himalayas at an altitudinal range of 800-950 meters above sea level. They are medium sized trees. *Cryptomeria japonica* is a small japanese tree that forms pure stands in Khasi and Jaintia hill in Eastern Himalayas. *Cephalotaxus griffithii, C. mannii, & Biota orientalis* also grow in Eastern Himalayas and are small trees that grow at altitudes ranging between 600

Plate I : A conifer Picea

meters and 2000 meters above sea level. *Cryptomeria* grows at higher altitudes (2,000 to 4,500 meters). The leaves in the conifers vary from scale leaves to linear to broad types and show xerophytic characteristics. Billings and Thompson (1957) report a tree of *Pinus aristata* in California that is about 4000 years old.

Plate II : Pinus

The conifers are branching woody plants that have usually two types of branches : (*i*) the dwarf shoots, and (*ii*) long shoots. Similarly there are two types of leaves : (*i*) scale leaves; and (*ii*) green of foliage leaves. The root system is an extensively branched tap root system which commonly infected with mycorrhizic fungus. The mycorrhiza may be ectotrophic or endotrophic (*Araucariaceae*). The roots in Podocarpaceae possess root nodules. Pneumatophores have been reported in *Taxodium distichum.* There is a high degree of histological uniformity expressed in the xylem, the phloem, and the cortical tissues. The secondary wood in pycnoxylic with uniseriate or biseriate (early multiseriate) secondary medullary rays that are often heterogenous. The tracheids bear circular bordered pits (uniseriate, rarely bi- or multiseriate) or their radial walls. Resin canals are found in the cortex, pith and even in the xylem of leaves, stems and roots. The leaves may be spirally arranged or may be opposite or whorled. The reproductive organs are unisexual and form compact cones or strobili. The female cones are compound structures with a central axis bearing spirally arranged bract scales that are fused with or subtend a single ovuliferous scale. The ovuliferous scale bears many or two or rarely one ovules. The ovuliferous scale is also regarded (Florin) as the secondary fertile axis with fertile and sterile appendages. The male cones are compact and simple structures with a central axis having spirally arranged microsporophylls. Each microsporophyll bears many or two microsporangia containing winged (*Pinus, Abies*) or wingless pollen grains. The spermatozoids are non-motile. The embryo has two to many cotyledons. Polyembryony is of common occurrence. The seeds are endospermic.

Fossil History

The living conifers are believed to have been differentiated by the Jurassic period of the Mesozoic era. The *Pinaceae* and the *Araucariaceae* are believed to be slightly older. The *Cephalotaxaceae* and the *Cupressaceae* probably came into existence in the Upper Jurassic or the Lower Cretaceous periods. With the recent discovery of *Metasequoia glyptostroboides* from China as a living representative of a genus that became 20,000,000 years ago is of immense scientific value. It is the oldest extant conifer living uptodate. This family came into existence somewhere in the upper Cretaceous of the Mesozoic. A careful study of the comparative morphology of the living and fossil types suggests that the modern coniferous families originated independently from a group of extinct conifers called the **transition conifers** by Florin. These have been put under the family *Voltziaceae* that originated in the Upper Permian period of Palaeozoic era. The distribution of the modern conifers clearly suggests their being an ancient and a relict group. The various families are, no doubt, widely distributed but their genera are of limited distribution and indicate their great antiquity, which is completely borne out by the fossil record. The earliest conifers are included in the family *Lebachiaceae* that originated in the Early Permian and were in existence before the Coredaitales became extinct. The conifers reached their climax in the mid-Mesozoic times and have been on the decline ever since. Andrews (1961) remarks, "It seems to me safe to assert that the coniferales, as a major group of plants, exceeds all others in the combination of the length of time that it has existed and its dominance over vast areas of the earth's surface. The oldest members of the order (from the liver Permian) flourished approximately 200 million years ago; their evolution appears to have been rapid during the late Palaeozoic and early Mesozoic with relatively minor deviations since that time. Until men began cutting away the coniferous forests for their choice lumber, they continued to compete favourably with other groups." Andrews is of the opinion that the conifers and the cordaites originated along independent lines of evolution ,as seed plants, did not have common ancestors. Florin (1949, pp. 101–102) also hinted at such a conclusion much earlier. He regards Ginkgoales, Cordaitales Coniferales and Taxales as belonging to the same class (Coniferopsida) but differs from the view that they had a common ancestor. He states, "but they constitute parallel evolutionary lines which probably were already separated from each other in Upper Devonian or lower Carboniferous times. At all events, a clear differentiation can be seen as far back as the fossil records go."

Primitive Conifers

The living conifers show a mingling of both primitive and advanced features. There is no living family that portrays all primitive features in any of its members. There is yet no agreement as to which characters should be definitely regarded as primitive and others as advanced. Greguss (1955), Buchholtz (1948), Thomson (1940) and Sporne (1965) have tried to define the primitive and advanced features among conifers. The primitive characters about which there is general agreement are listed below;

1. The shoot apex does not show any distinction into tunica and corpus layers.
2. Presence of central mother cells in the shoot apex.
3. The tracheids in the wood have multiseriate pitting.
4. The medulary rays are homogenous and have no resin ducts. They have no pits on radials as well as tangential walls. The heterogenous rays according to fossil evidence appeared late and are hence regarded as advance.
5. The female cone is well developed and larger in size.
6. The bract scale is free from the ovuliferous scale.
7. The pollen grains are winged.
8. The male gametophyte has several male prothallial cells enclosed within the microspore.
9. The number of free nuclear divisions is large in the early embryogeny and this condition persists for a longer period.
10. There is no cleavage polyembryony in primitive forms. Buchholtz (1918) regarded cleavage polyembryony as a prinitive condition. Thomson (1940), however, regraded clearage Polyembryony as an advanced character.
11. The number of chromosomes is high in primitive forms.
12. The spiral arrangement of the leaves is primitive as compared to the opposite and whorled arrangements.
13. The haplocheilic type of stomata are primitive as compared to the syndetocheilic type.
14. The broad multiveined leaves as those of *Agathis* and *Podocarpus* are regarded as primitive and more simple veined types of leaves as advanced (Laubenfels, 1953, however regards broad leaves as advanced).

According to the list given above the *Podocarpaceae* seem to be the most primitive members of the coniferales. The *Cupressaceae* and the *Taxodiaceae* are the most advanced families, whereas the *Pinaceae, Cephalotaxaceae* and the *Araucariaceae* take up the intermediate position. So far as characters of wood and microgametophyte are concerned the *Araucariaceae* are primitive. They have microgametophytes with numerous male prothallial cells, but some authors consider this condition to have developed secondarily in *Araucaria* and *Agathis*. To sum up the conifers are advanced gymnosperms that show most complex organisation of any genus in its growth phenomena, in the organisation of its female cones, in the structure of its wood, medullary rays, ray tracheids and resin canals, in its embryology and in its symbiotic relationship with fungi.

Plate III : Sequia Sp.

Classification

Coulter and Chamberlain (1917) recognised only two families in the conifers (*i*) *Pinaceae*, and (*ii*) *Taxaceae*. Earlier Sinnot (1913) included the *Taxaceae* along with *Podocarpaceae*. Pilger 1926 splitted coniferales into seven families:

1. *Taxaceae* (including *Torreya*, *Taxus* and *Austrotaxus*).
2. *Podocarpaceae including Pherosphaera, Microcachrys, Saxogothea, Dacrydium, Acmopyle, Podocarpus, Phyllocladus).*
3. *Araucariaceae (including Agathis and Araucaria).*
4. *Cephalotaxaceae (including Amenotaxus and Cephalotaxus).*
5. *Pinaceae (including Abies, Keteleria, Pseudotsuga, Tsuga, Picea, Pseudolarix, Larix, Pinus).*
6. *Taxodiaceae (including Sciadopitys, Sequoia, Glyptostrobus, Taxodium, Cryptomeria, Athrotaxus, Taiwania, Cunninghamia).*
7. *Cupreassaceae (including Actinostrobus, Callitris, Tetraclinis, Callitropis, Widdringtonia, Fitzroya, Diselmna, Thujopsis, Libocedrus, Fokienia, Cupressus, Chamaeyparus, Arceuthos, Juniperus).*

Buchhaltz (1933) recognised two subordinate groups among the coniferales : (A) **Phanerostrobilares** that have a well developed female cone with ovulate or ovuliferous scales bearing ovules on their adaxial surface. This group includes the families *Pinaceae, Taxodiaceae, Cupressaceae, and Araucariaceae;* and (B) **Aphanostrobilares** that include families in which there are no well defined female cones and the ovules are exposed. This group includes the families *Podocarpaceae, Taxaceae and Cephalotaxaceae*.

Chamberlain (1935) recognised only six families. He did not recognise Cephalotaxaceae and included *Cephalotaxus in Taxaceae*.

Pulle (1940) recognised *Taxales* as the separate order and included the *Taxaceae* and the *Cephalotaxaceae* in this order.

Zimmermann Takhtajan and Cronquist (1959) also recognised Taxales as a distinct order of the same rank as coniferales.

Rudolf Florin (1949) also recognised Taxales as a distinct order.

Lyman Benson (1957) divided the conifers into three orders- Ginkgoales, Pinales, and Taxales. He included the family *Taxaceae* in this order. The Pinales include six families (see chapter 1).

D.D. Pant (1957) includes orders Taxales in a separate class Taxopsida (see Chapter 1).

Andrews (1961) includes *Taxaceae* as a family of coniferales. He has followed Pilger (1926).

Sporne (1965) classifies coniferales into nine families as follows:

1. *Lebachiaceae :* it is an extinct family and includes the general *Lebachia, Ernestiodendron, Walchia, Walchiostrobus, Carpentieria, Buriadia.*
2. *Voltziaceae* : it is an extinct family and includes the genera *Psudovoltzia, Voltziopsis, Ullmania.*
3. *Palissyaceae* : it is an extinct family and includes the genera *Palissya* and *Stachyotaxus.*
4. *Pinaceae* : it is a modern family of living genera and includes *Abies, Pseudotsuga, Picea, larix, Cedrus, Tsuga and Pinus.*
5. *Taxodiaceae* : includes the living genera *Sequoia, Sequoiadendron, Metasequoia Taxodium, Cryptomeria, Sciadopitys, Athrotaxis.*
6. *Cupressaceae* : includes the living genera *Cupressus, Chamaecyparis, Thuja, Juniperus, Callitris, libocedrus, Papuacedrus.* The monotypic genus Biota also belongs here.
7. *Podocarpaceae* : includes *Phyllocladus, Dacrydium, Podocarpus, Saxegothaea, Microcachrys, Microstrobos, Acmopyle.*
8. *Cephalotaxaceae* : include *Cephalotaxus.*
9. *Araucariaceae*: includes *Araucaria* and *Agathis.*

Sporne also raises *Taxaceae* to the rank of a separate order Taxales.

Gangulee (1968, 2002) added two more extinct families to the list of Sporne's classification thus raising the number of coniferous families to eleven. These two families include: (*i*) *Cheirolepidaceae* and (*ii*) *Protopinaceae.* Of these five families namely Labachiaceal, Voltziaceae, Pallissyaceae Cheirolepidaceae and Protopinaceae are represented by only fossil members. The remaining six families pinaceae Taxodiaceae, Cupressaceae, Podocarpaceae, Cephalotaxaceae and Araucariaceal include living genera.

Bierhorst (1971) also recognises seven families and includes Taxaceae as a family of the order coniferales. He includes five genera in this family: *Amentotaxus* , *Austrotaxus, Pseudotaxus, Taxus* and *Torreya*.

The classification schemes discussed above are based on the following characters:

1. The arrangement of leaves on the stem and branches is spiral, opposite or whorled.
2. The arrangement of bract scales and ovuliferous scales on the cone axis.
3. The arrangement of microsporophylls on male cone axis.
4. Degree of fusion between the bract scale and the ovuliferous scale.
5. The number of ovules on each ovuliferous scale.
6. The number of microsporangia on each microsporophyll.
7. Arrangement of sporophylls to form a distinct cone or they occur singly in leaf axils.

These characteristics are not sufficient to bring out a natural scheme of classification. Other criteria like anatomy, embryology, and cytological detail like chromosome numbers and morphology should also be taken into consideration to erect a natural system of classification. Later Mohd. Yunus (1972) has tried successfully to employ anatomical characters of leaves of Indian conifers to classify them.

The following table 12.1 portrays the summary of important morphological features of the coniferales. The number of species is given in brackets after the genera.

Leaves of Conifers

The table 12.1 clearly illustrates the shape, kind and arrangement of the leaves in all the living coniferous genera but it does not illustrate certain aspects of morphology which will be disscussed in detail below.

Table 12.1 indicating Morphological features of Conifers

Family	*Genera and No. of species*	*Distribution*	*Shape of leaf*	*Phyllotaxy*	*Number of sporangia per Microsporophyll*	*Ovules per ovuliferous scale*	*Number of cotyledons*
Pinaceae	*Abies* (50)	North temperate regions, Central America	Needle like	Spiral	Two	Two	Five to seven
	Cedrus (4)	India, Cyprus, Algiers, Middle East.	Needle like	Spiral	Two	Two	Nine to ten
	Picea (50)	India, North temperate	Needle like	Spiral	Two	Two	Four to fifteen
	Pinus (90–100)	India, Burma, North temperate regions.	Needle like as well as scale like	Spiral	Two	Two	Four to fifteen
	Cathaya (2)	China	Linear	Spiral	Two	Two	Three to four
	Keteleria (4–8)	Indochna	Linear as well as broad	Spiral	Two	Two	Two
	Larix (10–12)	Europe, North Asia, North America, India.	Needle like	Spiral	Two	Two	Six
	Pseudolarix (2)	China	Needle like	Spiral	Two	Two	Five to seven
	Pseudotsuga(7)	East Asia, and Westen North America	Linear	Spiral	Two	Two	Six to twelve
	Tsuga (15)	East Asia, India and North America	Scale like and broad	Spiral	Two	Two	Three to six
Taxodiaceae	*Athrotaxis* (3)	Tasmania, Australia	Scale like and broad	Spiral	Two	Three to six	—
	Cryptomeria (1)	Japan, India	Scale like	Spiral	Five to seven	Three to six	—
	Cunninghamia (3)	Formosa, China	Broad	Spiral	Two to four	Two to three	Two
	Glyptostrobus (1)	China	Linear or broad	Spiral	Four to seven	Two	—
	Metasequoia (1)	China	Linear	Opposite	—	Five to eight	—
	Sciadopitys (1)	Japan	Scale like and Linear	Spiral	Two	Five to nine	—
	Sequoia (1)	California	Linear	Spiral	Three to four	Three to seven	Two
	Sequoiadendron (1)	California	Scale like or Needles	Spiral	Three to six	Five to eight	Three to five
	Taiwania (3)	China, Formosa	Scale leaves	Spiral	—	Two	—
	Taxodium (3)	Mexico & S.E. , U.S. A.	Linear	Spiral	Four to seven	Two	—
Araucariaceae	*Agathis* (20)	Indochina, West Malaysia to New Zealand	Broad	Spiral	Five to fifteen	One	Two

Family	Genera and No. of species	Distribution	Shape of leaf	Phyllotaxy	Number of sporangia per Microsporophyll	Ovules per ovuliferous scale	Number of cotyledons
	Araucaria (18)	Australia, New Zealand, Chile, Brazil , North Caleonia	Broad	Spiral	Five to fifteen	One or two	Two
Cepha-lotaxaceae	*Cephalotaxus* (4–7)	East Asia	Linear	Opposite	Three to eight	Two	Two
Cupressaceae	*Actinostrobus* (2)	South West Australia	Scale like	Spiral	—	One to two	—
	Austrocedrus (1)	Temperate South America	Scale like	Opposite	—	One to two	—
	Callitris (16)	Australia, North Caledonia	Scale like	Opposite	—	Numerous	—
	Calocedrus (3)	S.E. Asia , Formosa, N. America	Scale like	Opposite	Two to four	Two	—
	Chamaecyparus (7)	Japan, Formosa, North America	Scale like	Opposite	Two to five	Two to five	Two
	Cupressus (16–20)	Asia, India, Sahara, Mediterranean, North America	Scale like	Spiral	Two to six	Six to twenty	Two to five
	Diselma (1)	Tasmania	Scale like	Opposite to whorled	—	Two	—
	Fitzroya (1)	Chile	Scale like	Whorled	—	One to six	Two
	Fokienia (1–3)	Indochina, China	Scale like	Whorled	Two to five	Two	—
	Juniperus (60)	Northern Hemisphere India	Scale like	Spiral	Three to six	One to three	Two to six
	Libocedruer (5)	North Caledonia, New Zealand	Scale like	Opposite	Three to five	One to two	—
	Neocallitropsis (1)	N. Caledonia	Scale like or linear	Whorled	Many	Two to five	Two
	Popuacedrus (3)	New Guinea	Scale like	Opposite	Many	Two	—
	Pilgerodendron (1)	Chile	Scale like	Opposite	Four to eight	One to two	—
	Tetraclinis (1)	Spain, Malta	Scale like	Opposite	—	—	—
	Thuja (5)	China, Japan, North America, India	Scale like	Opposite	Two to five	Two to three Four to eight	—
	Thujopsis (1)	Japan	Scale like	Opposite	Two to five	Three to five	—
	Widdringtonia (5)	Tropica & South Africa	Scale like	Opposite	Three to five	Five or more	Two
	Acmophyle (3)	Fiji & New Caledonia	Linear	Spiral	Two	One	Two
Podocarpaceae	*Dacridium* (20–25)	New Zealand, Chile, Indomalayasia	Linear and Scale like	Spiral	Two	One	Two
	Microcachrys (1)	Tasmania	Scale like	Spiral	Two	One	Two
	Microstrobus (2)	Australia, Tasmania	Scale like	Spiral	Two	One	Two
	Phyllocladus (7)	New Zealand, Tasmania Philippines	Linear and Scale like	Spiral	Two	One	Two
	Podocarpus (100)	Tropics, North to Himalayas in India; Japan	Scale like to Broad	Spiral	Two	One	Two
	Saxagothea (1)	Andes of Argentina	Linear	Spiral	Two	Two	Two

The leaves in conifers are either scale-like or have well developed laminae. They are as a rule simple, and possess a single midrib in majority of the taxa. In *Podocarpus* and *Agathis* the leaves are

Fig. 12.1. Leaves of conifers. A. Trifoliar spur of *Pinus roxburghii*; B. A branch of *Picea* with needle like leaves arranged spirally on the stem; C. Single leaf of *Picea*; D. Leaves of *Araucaria bidwilli*; E. Branch of *Sequoia sempervirens* bearing leaves; F. Branch of *Cupressus sempervirens* bearng closely adpressed leaves; the leaves are scale-like; G. A branchlet of F. enlarged; H. Branch of *Cedrus deodara* bearing leaves.

broad and may be as long as 30 cm and 17 cm respectively, and possess an open and dichotomous venation. Mostly the leaves are linear and needle like (Fig. 12.1 A, B, C, H). Mostly the conifers are evergreen and in *Araucaria* (Fig. 12.1, D) foliage leaves remain green and persist for ten to fifteen years. In *Pinus* the needles on spur shoots fall along with the spur and are deciduous. A few genera are, however, deciduous and experience leaf fall every autumn *e.g. Taxodium, Metasequoia, Pseudolarix* and *Larix*. Due to their small size and a single vein the leaves of conifers are regarded as microphyllous, but if we believe that the leaf traces leave leaf gaps in the stem stele then the conifier leaves are not microphyllous, because according to this view the leaves are megaphyllous (megaphyllous leaves are anatomically defined as those whose leaf traces leave a leaf gap in the cauline stele). As is clear from the table the *Cupressaceae* are distinct from other conifers by the whorled and opposite phyllotaxy of the leaves that always appear on long shoots. Majority of conifers have an alternate or spiral leaf arrangement. In most of the *Pinaceae* (*Pinus, Picea, Abies* etc.) the green needle like leaves are borne or spur-shoots or dwarf shoots along (Fig. 12.1, A). They have no axillary buds and branches usually appear from the buds in the axils of scale leaves. *Cedrus, Larix* and *Pseudolarix* (*Pinaceae*) bear foliage leaves on both the spurshoots and the long shoots. The spur shoots in *Pinus* are characteristic and bear a variable number of foliage leaves or needles at their tips and during the time of shedding the whole spur is shed. The spur arises from an axillary bud of a scale leaf. It has a small $\left(\frac{1}{2} - 1 \text{ long}\right)$ dwarf shoot or the stem that bears a few bud scales or imbricated scale leaves and terminates in one (*Pinus monophylla*), two (*P. sylvestris*), three (*P. roxburghii*) or five (*P. wallichiana*) needles. The spur shoot has a definite shoot apex with inactivated cells, that ultimately collapse and die (Sacher, 1955). In *Pinus* the seedling bears spirally arranged needles. The first dwarf shoots arise in the axils of these green leaves, but the subsequent ones will arise only in the axils of brown scale leaves. The scale leaves in some genera (*Thuja, Cupressus, Biota*) are green in colour and are closely apposed to the stem surface (Fig. 12.1, F, G). The *Cupressaceae* are also characteristic in exhibiting heterophylly *e.g.* in the seedling stage the young plants bear needle-like leaves but as the sporophyte grows the leaves become scale-like and are closely oppressed to the branches. The leaves in the living conifers have been grouped into four distinct types by Laubenfels (1953). These types are :

1. The needle-like leaves (Fig. 12.1 A—C) that are traversed by a single midrib and appear three-sided or four-sided in a transection. This form was

Fig. 12.2. *Agathis alba.* A. A branch bearing multinerved leaves; B. A branch with male cones; C. A female cone.

very common among the fossil conifers and in the living forms it is found in many genera of the families *Pinaceae, Podocarpaceae, Araucariaceae* and *Taxodiaceae*.

2. The leaves are bifacial, uni-nerved and lanceolate or linear in shape (Fig. 12.1, D). This type is very common among living confiers (see table).
3. The scale-like leaves of photosynthetic and non-photosynthetic type. The former occur in *Cupressaceae* (Fig. 12.1, F), whereas the latter are encountered in many genera.
4. The leaves are broad and multiveined *e.g. Agathis* (Fig. 12.2) *Podocarpus* and some species of *Araucaria*.

Florin (1950, 1951) made an extensive study of the fossil conifers that existed in the paleozoic era and found that they possessed two types of foliage leaves :

1. Dichotomously forked leaves that were supplied by single leaf trace. The venation was also dichotomous *e.g. Carpentieria frondosa*.
2. Needle-like or scale-like leaves that were innervated by a single, unbranched mid-rib. Both these type of leaves occurred on the branches of some members of the *Lebachiaceae e.g., Lebachia* and *Ernestiodendron* (Fig. 12.2). The bifurcated leaves were found on the main axis and older branches, whereas the simple leaves were restricted to the super younger branches.

Anatomy of Leaf

The leaves in conifers have a well defined epidermis covered with a thick cuticle. The cell walls are also thick. In *Sciadonitus* the epidermal cells bordering the stomata are produced into strongly cutinised projections (Fig. 12.3), which are absent elsewhere. The stomata are sunken and are generally disposed in parallel rows forming vertical bands on either side of the midrib and are of haplocheilic type as the guard cells originate from a common mother cell and certain of the neighbouring epidermal cells become modified as subsidiary cells. In *Cedrus, Cupressus, Thuja* and *Cryptomeria* the stomata have no regular arrangement. In the *Pinaceae* that have needle like leaves the stomata are found on all sides of the leaves. Such a stomatal distribution is called *amphistomatic* (Fig. 12.4). In *Abies, Cephalotaxus, Podocarpus, Tsuga* and some other genera the stomata are found only along the abaxial surface. Such an arrangement is called *hypostomatic*. In *Cupressus, Thuja* and *Juniperus* the stomata are found only on the adaxical epidermis and are called *hyperstomatic*.

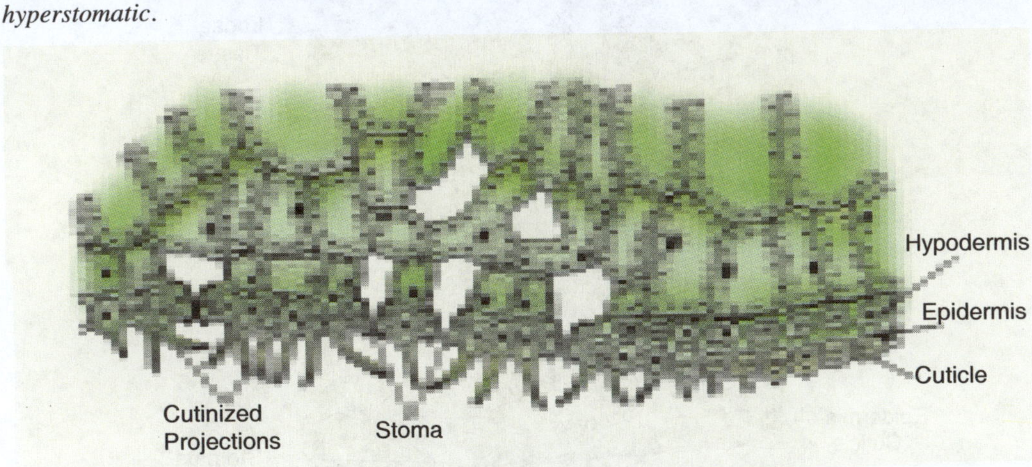

Fig. 12.3. *Sciadopitys.* Portion of stomatal region. The epidermal cells have strongly cutinised projections.

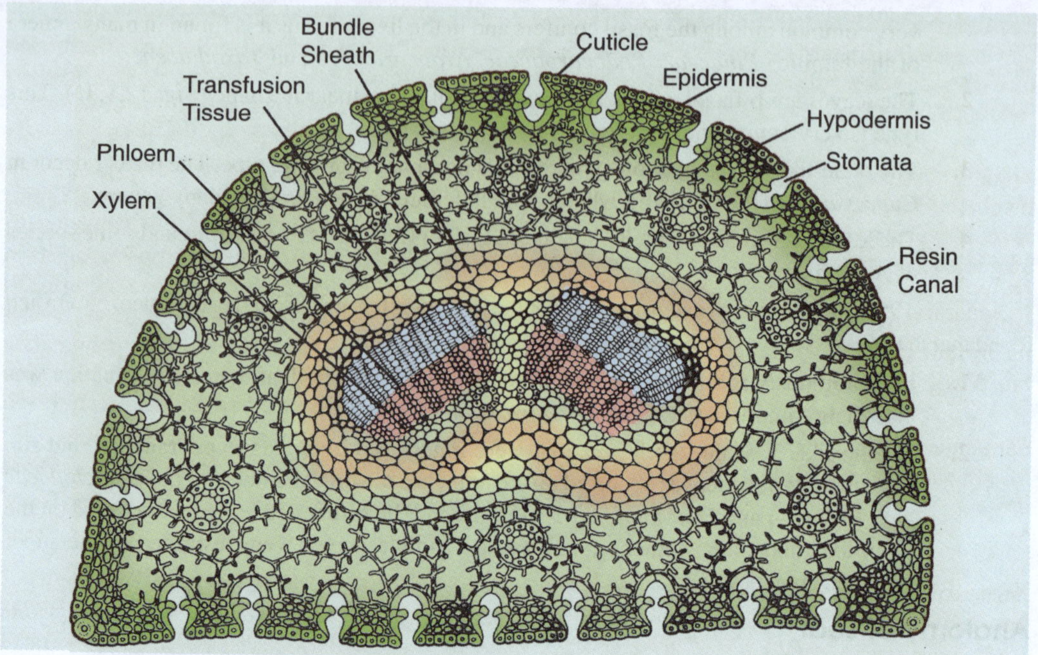

Fig. 12.4. T.S. leaf of *Pinus sylvestris.*

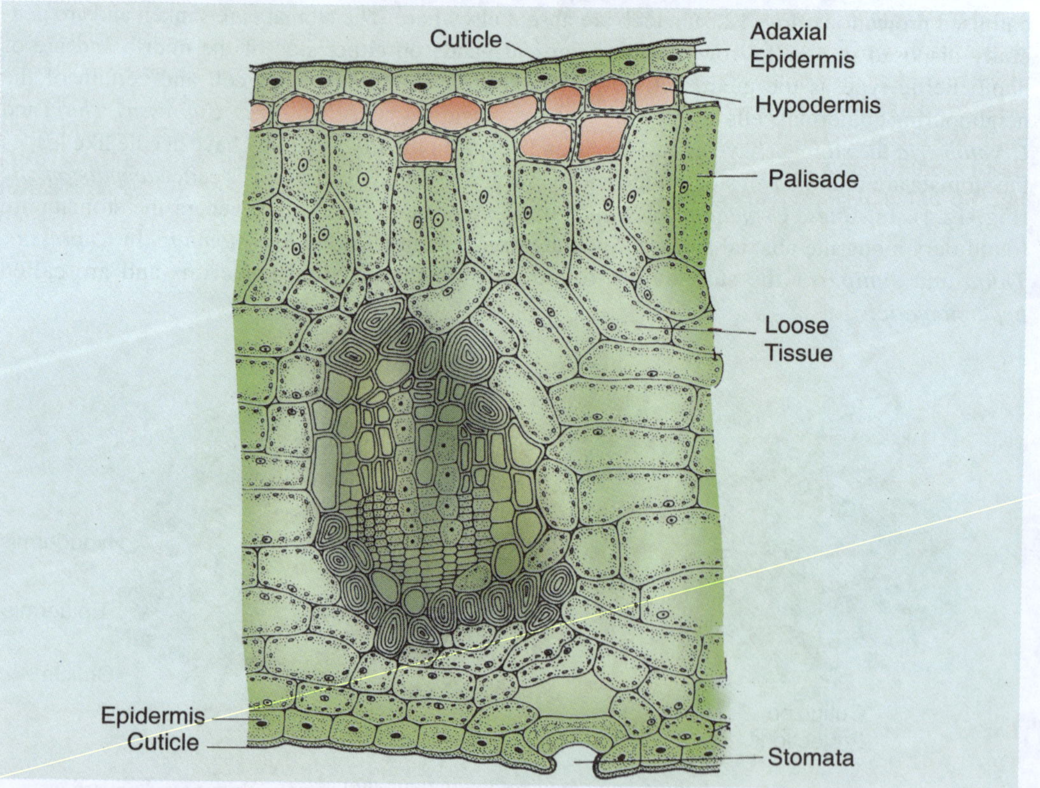

Fig. 12.5. T.S. of two fused leaves of *Sciadopitys verticillata.*

The hypodermis is absent in the leaves of *Cephalotaxus* and *Tsuga*. In *Pinus roxburghii* and *P. merkusii* it occurs in distinct patches that are well developed at the corners. In *Abies, Cedrus Picea, Juniperus, Cryptomeria* and *Agathis* (Fig. 12.5) the sclerenchyma is of uniform occurrence.

In *Agathis* (Fig. 12.5) and *Podocarpus* there is a well developed palisade tissue and a spongy parenchyma (Fig. 12.5). In others there is no such distinction. In *Pinus* the cells of mesophyll tissue show infoldings. The mesophyll tissue in *Sciadopitys verticillata* possesses branching sclereids (Fig. 12.6). The mesophyll in *Crytomeria* shows well developed air spaces. A well developed endodermis is present around the vascular bundle. In *Cryptomeria* (Fig. 12.7) and *Cunninghamia* (Fig. 12.9) there is a big resin canal above the vascular bundle. The distribution of resin canals is variable in various genera and species.

The leaf trace is double in the living *Pinaceae* and single in the *Cupressaceae*. The traces in the cotyledons are single in *Pinaceae, Cupressaceae* and *Taxodiaceae*, but they are double in all the *Araucariaceae*.

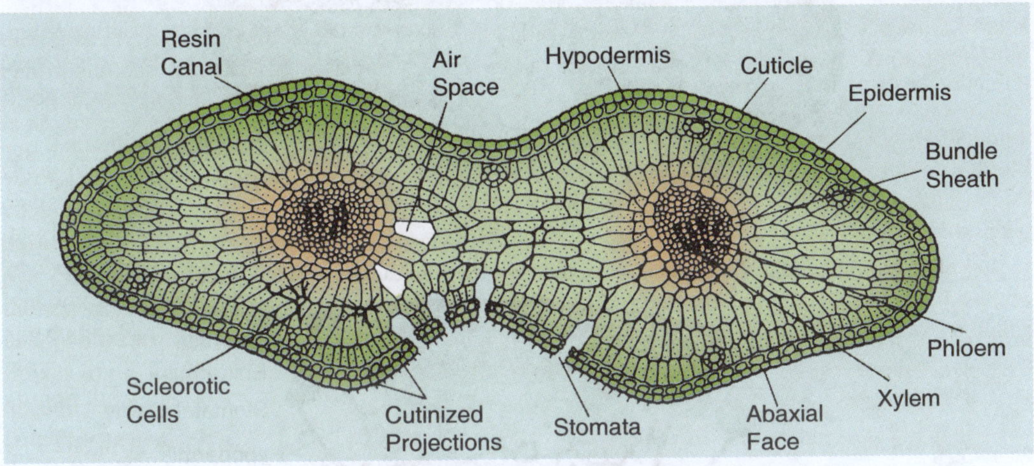

Fig. 12.6. T.S. leaf of *Agathis robusta*.

Transfusion tissue is a characteristic feature of the conifer leaves. It consists of two types of cells in the leaves of *Pinus* (Figs. 12.4 and 12.8).

1. Transfusion tracheids that are dead cells having thick and lignified walls and circular bordered pits.

2. Transfusion parenchyma cells that are living and contain tannin-like contents.

Both these types of cells occur within vascular region and are enclosed by the endodermis. They are in direct contact with the endodermis and form a complicated system of interconnecting cell types. Its main function is to conduct water and food materials between vascular bundle and the surrounding mesophyll tissue. Sacher (1953) studied the development of this tissue in *Pinus lambertiana*. He found that its cells originate from the outer border of the *procambial strands* in the young leaf and the maturation process is centripetal. In other conifers the transfusion tissue is made up of only transfusion tracheids. In the *Cupressaceae* this tissue forms two lateral wing-like extensions that arise from the sides of the vascular bundles. In the leaves of *Araucaria* the transfusion tracheids competely surround the xylem of the vascular bundle in the lower regions of the leaf. In the upper portions of the leaf it extends a little up to enclose a part of the phloem too. The thickenings

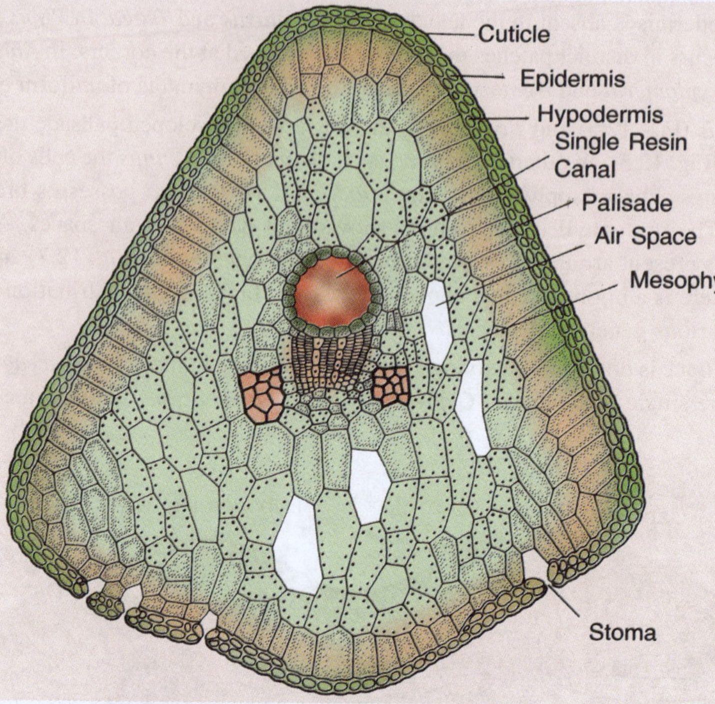

Fig. 12.7. T.S. leaf of *Cryptomeria japonica.* Note the resin canal next to the vascular bundle.

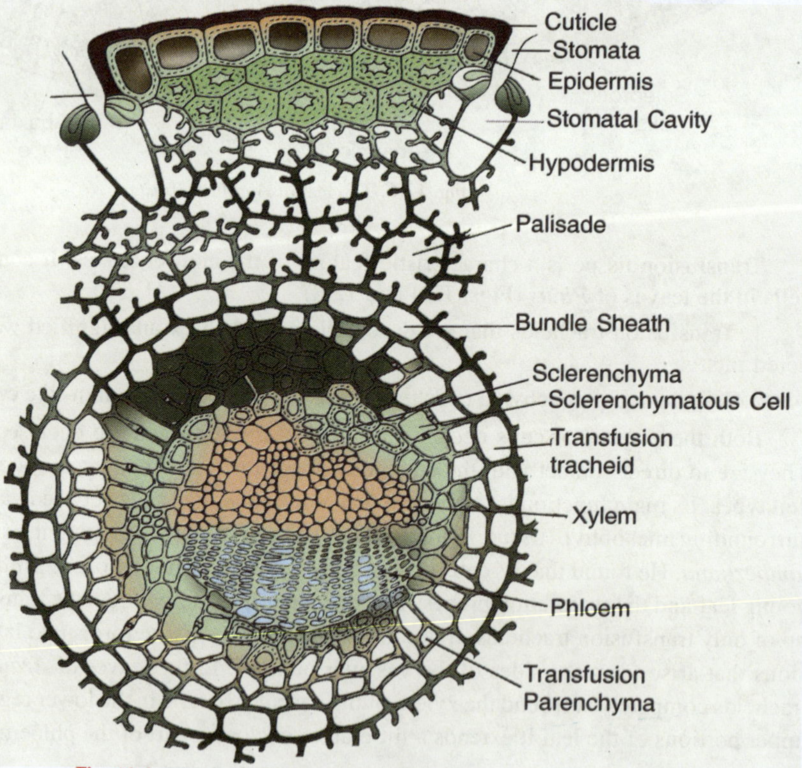

Fig. 12.8. T.S. leaf of *Pinus monophylla.*

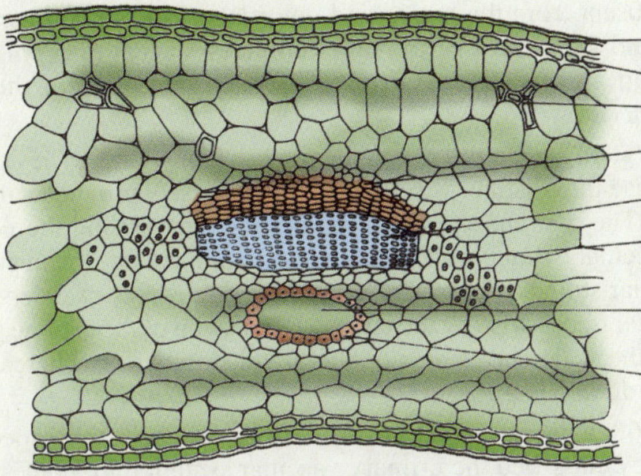

Fig. 12.9. T.S. leaf of *Cunninghamia sinensis.* Note the single resin canal next to the vascular bundle.

of transfusion tracheids may also reticulate and many transitional stages between reticulate and pitted types occur *e.g.* in *Cupressaceae* (Al-Sherifi, 1952).

Stem Anatomy

Development of the shoot apices of the conifers is substantially similar to that in *Ginkgo* and most other gymnosperms although there is quite a difference within the coniferales in seasonal distribution of growth. *Sequoia* has shoot apex that exhibits growth actively over extremely long periods of slow but continous growth, while in others, such as *Pinus, Pseudotsuga* and *Torreya* there is a definite phase of annual shoot development which takes place with a specific seasonal rhythm. In the apex there are zones with differing degrees of mitotic activity, different directions of growth, and difference in cellular constitution. These zones vary during the growing seasons and among. The different species. In the *pinaceae* the apical meristems show nuclear distinction between *tunica* and *corpus* and there is an initial group of apical cells that divide anticlinally and periclinally to give rise to lateral derivatives and to the central mother cell zone below. In the centre of this zone there is a gradual differentiation of cells with meristematic properties that give rise to a peripheral zone from which arise, in part leaf primordia, cortex, procambium, and epidermis. The central mother cells also give rise basally to the *rib meristem* which pro-

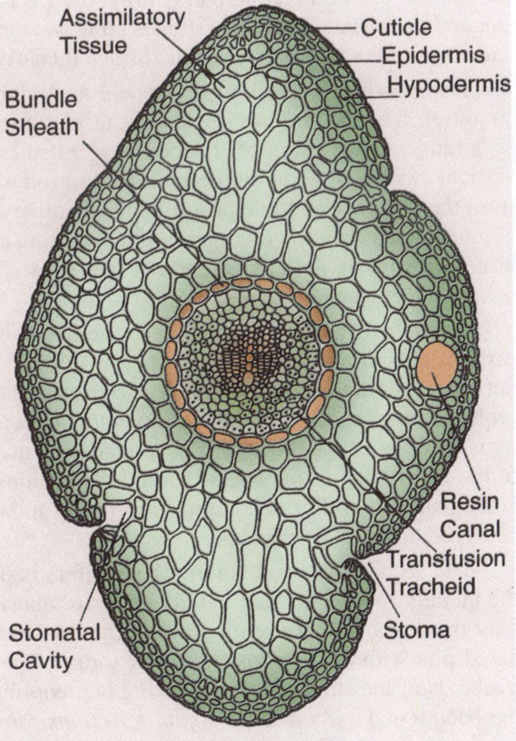

Fig. 12.10. T.S. leaf of *Picea excelsa.* Note the single resin canal.

duces the pith. In the *Araucariaceae* the shoot apex shows clear distinction into tunica and corpus. The periclinal divisions are absent from the tunica and outer layers of the corpus. Central mother cells are absent from *pinus* and from most of the other conifers, except *Sequoia, Araucaria* and *Pseudotsuga*. The tunica and corpus organisation is also lacking among the *Taxodiaceae, Cuprassaceae* and *Podocarpaceae*.

The peripheral zone of the stem apex gives rise to the provascular strands. They differentiate into the vascular tissues that organise into distinct vascualr strands or vascular bundles that are conjoint, collateral, open and endarch. These strands form an interrupted ring around the central pith. The conifers have very massive vascular cylinders and small areas of pith cortex. Resin passage occurs in the cortex and pith. The vascular strands of conifers show varying degrees of independence in their vertical course upwards. The leaf traces arise by the tangential division of the cauline strands and may be single or double. In *Pinus* the scale leaves are supplied by a single trace whereas the photosynthetic leaves are supplied by a double leaf traces.

Namboodri and Beck (1968, a, b, c) studied the primary vasculature of 30 species of conifers belonging to five families. They classified the primary vascular system into 3 types :

Type 1. In this case the leaf arrangement is spiral and the cauline strands were vertical and sinuous in their upward course. They divide tangentially to give rise to the leaf traces all along one side and at points where two of the vertical strands closely approached each other. This type is common among the *Pinaceae, Araucariaceae, Cephalotaxaceae, Podocarpaceae* and *Taxodiaceae*.

Type 2. This type is similar to type 1 but the leaf traces arose from both the sides of each vertical strand. It is found in some members of *Cupressaceae, Podocarpaceae* and *Taxodiaceae*.

Type 3. It is also characteristic of *Cupressaceae, Podocarpaceae* and *Taxodiaceae*. In this case the cauline strands occur in pairs during their vertical course upwards and give off two strands by a tangential division. These two strands unite and then traverse between the two vertical cauline strands and constitute a leaf trace. This leaf trace then diverges radially to the leaves.

Namboodri and Beck regard type 3 as derived from types 1 or two and hence is an advanced condition. They regard the single trace as a primitive condition. They related the origin of leaf trace by a tangential division of Indepedent vertical cavline strands to the condition where indepedent vertical cavline strands divide radially to produce a leaf trace. They further presented evidences from the anatomy of some primitive paleozoic gymnosperms that the eustelic condition of modern gymnosperms arose by the longitudinal splitting of the protostele and not by the medullation of the protostele and by the appearance of leaf gaps. This view is quite opposed to that of Jeffery (For details see chapter I).

Another feature worthy of note among the conifers is that the cambial activity appears very early in the ontogeny, and usually nearly all the xylem elements are radially aligned from the pith outwardly. A few small bundles of the ring arise later, entirely from the cambium and contain typical protoxylem elements. As the stem matures the cambium forms a complete ring and by tangential division of its cells gives rise to secondary xylem and secondary phloem, which form the great bulk of the coniferous stem. The structural similarities of the secondary xylem and secondary phloem among all the families of conifers are striking and more important than the differences that appear to be trivial and less significant.

The secondary wood among the conifers is pycnoxylic and is of homogenous texture, except the medullary rays that are heterogenous in many genera especially in *Pinus*. The xylem consists only tracheids, vessels being absent. The tracheids are long and narrow and possess circular bordered pits with distinct *tori* (singular torus). The pits appear on radial walls except in fibre-like tracheids of the summer wood. Pitting is predominantly uniseriate but biseriate or grouped pitting has been found in *Araucaria, Agathis, Actinostrobus, Microbiota, Cedrus* and *Keteleria* (this condition is primitive). The medullary rays are almost entirely uniserite (one cell wide) and only few cells high. Ray tracheids or dead cells with bordered pits develop within the medullary rays of certain *Pinaceae* and *Cupressaceae*. These cells are vertically extended and appear to be a part of the vertical system of elongated cells. Vertical parenchyma is scarce or altogether absent in the conifers. The secondary xylem of many *Pinaceae* contains resin ducts. They are also found in a few *Taxodiaceae*. Bars of Sainio is a characteristic feature of the coniferous wood. In the secondary wood of Thuja, Bannon (1958) described perforations in the tracheids. His figures illustrate these openings to be between adjacent tracheids. Bierhorst (1971) regards them "H-shaped tracheids in which the vertical arms were adjacent to each other."

The secondary phloem consists of elongated **sieve elements** or **sieve cells** and undifferentiated parenchyma oriented vertically besides the sieve cells and horizontally in the rays. The associated parenchyma cells are not companion cells in the angiosperm sense but may have a similar physiological association. These cells have larger nuclei and abundant starch. The long sieve cells have small, rounded and generally separated **sieve areas** that are confined to the radial walls. There are no companion cells. In *Pinus* two kinds of phloem rays are found, one dying with the rest of the phloem and the other starch filled cells that live even after the death of the phloem. Distinct crystalliferous cells are also associated with the phloem of some conifers (*e.g. Pinus*). Some conifers also contain many thick-walled fibres and sclereids in the secondary phloem. The original cambial ring remains active for hundreds of years or in rare cases for thousands of years (*e.g. Sequoia, Pinus aristata*) and results in the formation of stems with enormous diameters. The annual rings are quite distinct and represent yearly increments in the secondary xylem. False annual rings may also result due to abnormal cambial activity.

A distinct cork cambium arises in the outer cortex and its cells divide horizontally to cut off cork cells or *phellem cells* outwardly and secondary cortex or *pheloderm* cells within. Due to constant increase in the girth of the secondary wood and phloem the outer cork layers are sloughed off and new ones arise by the activity of a freshly developed cork cambium in deep seated cortical cells and ultimately the cork cambium may arise in the secondary phloem. So in older stems the bark of the tree consists of the **periderm** (cork layers, cork cambium, secondary cortex) and a part of nonfunctional secondary phloem. A term **rhytidome** is used to designate the firssured or wrinkled and furrowed outer portion of the bark. The rhytidome is lacking in smooth barked trees. In sequoias the bark may become as thick as two feet and forms an excellent insulating tissue from the heat of forest fires for the tender cambial layer and functional phloem cells within. This is the main reason why such trees live long and successfully withstand the ravages of forest fires during their thousands of years of existence.

Anatomy of the Root

In general the tissue organisation in the root tips of gymnosperms is not as distinct and as well differentiated as in the angiosperms *i.e.*, there is no clear demarcation between *protoderm, ground meristem* and *procambium.* Studies on the root tip organisation of *Abies, Larix* and *Pseudotsuga* show that at the root apex a group of **apical initial cells** undergoes repeated divisions to contribute cells to the **root cap zone** and to the **stelar mother cell zone**. The initials also divide to form several vertical rows of cells that form the core of the root cap. These cells of the core divide to give rise to the cortex and to the lateral cells of the root cap. So cortex and root cap have a comon origin and that is why there is no well defined dermatogen and periblem in the root tip. The cells of the stelar mother cell zone divide transversely to produce procambial cells that start elongating immediately and undergo numerous logitudinal divisions to form a distinct procambial zone about 100 μ from the apical initials.

The primary root has usually a **diarch** stele but **triarch** and **tetrarch** conditions are also common. In *Taxodium imbricatum* the usual number is five and the roots are **pentarch**. The primary xylem is exarch and in some cases (*Pinus*) the protoxlem is *Y*-shaped with a resin duct in the arms of Y. In *Abies balsamea* there is a single resin duct in the metaxylem. The pericycle and endodermis are usually clear. The primary cambium is very active and starts dividing early during the ontogen of the root to produce a secondary vascular tissue. The endodermal cells possess casparian strips. The cortex is distinct. As a result of secondary growth continuous layers of secondary xylem and phloem are formed and the old roots look like stems in their internal organisation. The annual rings are not very prominent. A distinct periderm is formed by the activity of the cork cambium.

FAMILY : PINACEAE

The family Pinaceae (Abietaceae) with 9 genera and 145 species is the predominant conifer of temprate to coder zones. The bracts and the ovuliferous scales are distinct in the female shoot.

The family has been classified into three sub families as follows :

Sub family : *Abietoideae*

Genus : 1. *Abies* (Fir tree) with 40 Sp; grows abundantly in Northern hemisphere mountains. *A. balsamea* yields canada Balsam.

2. *Picea* (Spuce) with 40 spp; northern hemisphere mountains.

3. *Keeteleria* with 3–4 spp; restricted to China.

4. *Tsuga* with 14spp; spread from himalayas to Japan, Northern America.

5. *Pseudotsuga* with 7 spp grows in East Asia, Pacific, North America.

Sub-family : *Laricoideae*

Genera : 6. *Larix* (Larch tree) with 10sp. grows in North Hemisphere,

7. *Cedrus* (Cedar tree) with four sp. grows in Mediterranean and Himalyas.

8. *Pseudolarix* with one spp. restricted to China.

Sub-family : *Pinoideae*

Genus : 9. *Pinus* with 90 spp. Dominant conifer in Northern Hemisphere, and also sunda Islands.

REPRODUCTIVE STRUCTURES

The reproductive structures in the Coniferales show a greater amount of variations and here these will be discussed separately in the individual families.

The details are given here.

Megasporangiate Cones

These are composed of a number of spirally arranged **bracts** in the axils of which are produced undivided **ovuliferous scales**. The ovuliferous scales bear two ovules each, on their adaxial or upper sides. The bract scales in *Larix* and *Pseudotsuga* are trifurcate at their apices. The bract scale develops first and the ovuliferous scale or the cone scale arises later in its axil. During pollination either the former or the latter projects beyond and grows more than the other. In *Pseudotsuga* and *Larix* the bract scales grow very long and completely cover or hide from view the ovuliferous scales. The bract scales may be free or fused to varying degrees with the ovuliferous scales. The ovules are sessile and inverted with their micropyles directed towards the cone axis.

Megasporangium and Megasporogenesis

The ovule has a single integument (unitegnic) that has the usual three layers (outer fleshy, inner fleshy, and middle stony). The integument is two lipped at the micropyle and is fused to the nucellus for a part of its length except most of the upper portion. The nucellus is thick and well developed (carassinucellate). The nucellus is also called the **megasporangium**. In *Larix decidua* the micropyle develops into a stigma-like outgrowth that catches the pollen grains.

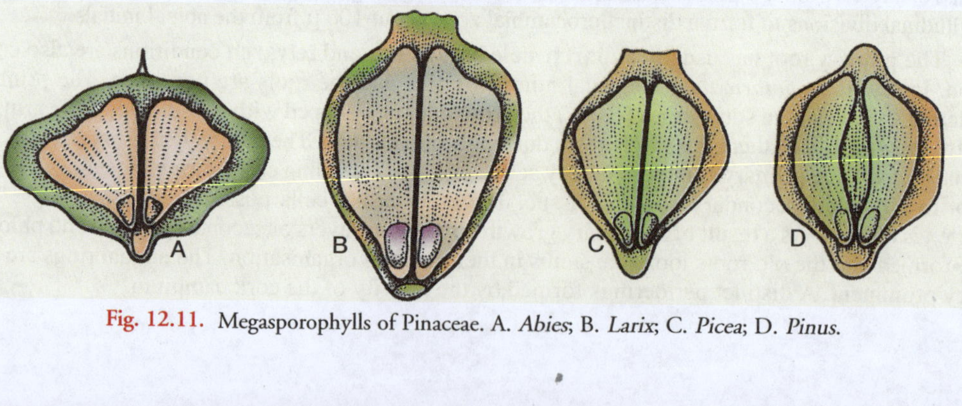

Fig. 12.11. Megasporophylls of Pinaceae. A. *Abies*; B. *Larix*; C. *Picea*; D. *Pinus*.

In majority of the *Pinaceae* (*Pinus wallichiana, Cedrus deodara*, etc.) a deep seated cell in the nucellus tissue differentiates as a **megaspore mother cell**. Konar (1960) observed for the first time that in *P. roxburghii* a hypodermal **archesporial cell** differentiates and divides immediately into an outer **primary parietal** cell and an inner **sporogenous cell**. The sporogenous cell functions as the megaspore mother cell. It undergoes reduction division to form three or four haploid megaspores. In *Pinus roxburghii, Larix decidua* and *Keteleeria evelyniana* the mother cell forms a linear tetrad of haploid megaspores. The one towards the chalazal end is functional, whereas the upper three degenerate.

Female Gametophyte

In *Keteleeria* (Wang, 1948) the functional megaspore enlarges and without undergoing any resting period, its nucleus divides uninterrupted into 512 free nuclei. In *Cedrus deodara* (Roy Chowdhry, 1961) the megaspore undergoes a resting period of 4 months before starting free nuclear division. In *Pinus roxburghii* and *P. wallichiana* the megaspore nucleus divides into a few free nuclei and then undergoes a resting period. The number of free nuclei in *P. roxburghii* is 2,500. The functional megaspore in *Keteleeria evelyniana* has been observed to possess a two layered membrane (inner endospore and an outer exospore which is 5 μ thick). It grows along with the megaspore. During free nuclear stage the megaspore is surrounded by a distinct spongy tissue (surrounding cells of nucellus that become distinct by their dense contents and conspicuous nuclei) which disappears after the female gametophyte has completely developed. Roy Chowdhry (1961) observed that in *Cedrus deodara* this tissue persists upto pro-embryo stage and starts degenerating after that. After the free nuclear division wall formation starts. It is centripetal and may be followed by transverse septa (*Cedrus deodara*, Roy Chowdhry, 1961) or by alveoli (*Pinus*). Konar (1959) observed that sometimes in *P. roxburghii* the megaspore fails to develop and the cells of the surrounding spongy tissue look like megaspores and many undergo free nuclear division, but they have not been seen to develop into functional gametophytes. In *Tsuga canadensis* Sterling (1948) observed two megagametophytes enclosed within one megaspore membrane. He recorded such a situation in 1% of the ovules.

The archegonia develop from the superficial cell of the female gametophyte, at the micropylar end and their number varies in the different genera : (*i*) 1–5 in *Larix decidua* (Schopf, 1943); (*ii*) 3–5 in *Cedrus deodara* (Roy Chowdhry, 1961); (*iii*) 2–4 in *P. roxburghii* (Konar, 1960); (*iv*) 1–2 in *P. wallichiana* (Konar and Ramchandani 1958); and (*v*) 7–12 in *Keteleeria* (Sughiara, 1943 and Wang 1948). Roy Chowdhry (1961) observed archegonia on the chalazal end of the female gametophyte in *Cedrus deodara*. Sterling (1948) observed lateral archegonia in *Tsuga canadensis*. At maturity the neighbouring gametophyhtic tissue grows upwards to form a sunken area called the **archegonial chamber**. The archegonium consists a short neck and a large venter. The neck is composed of varying number of **neck cells** but does not enclose a **neck canal**. Sughiara (1943) and Wang (1948) reported a neck of two to four tiers of 4 cells each in *Keteleeria*. In *Pinus roxburghii* and *P. wallichiana* Konar (1961) and Konar and Ram Chandani (1958) reported a single tier of 4 cells. Same is the case in *Larix* and many other *Pinaceae*. Roy Chowdhry (1961) reported them to vary in numberand without any definite arrangement in *Cedrus deodara*. The venter has a small ventral canal cell that may degenerate before fertilization or may persist for some time after fertilization. Allen (1943) reported that the egg is fertilized by the ventral canal nucleus, but further development was not noticed. In *Cedrus* and *Keteleeria* the ventral canal nucleus and egg nucleus have been seed to lie side by side.

At maturity cytoplasm of the egg develops fibrils that radiate outwards from the nucleus towards the periphery. A number of **proteid vacuoles** or **para-nuclei** also appear in cytoplasm of the egg and they have been observed to increase in size along with the increase in size of the egg and reach maximum enlargement before fertilization. They have a nutritive role to play and start dwindling in size and number as the zygote starts dividing. They disappear completely when the suspensor starts elongating (Takao, 1959). Vazart (1958) observed a considerable increase in the ribonucleo proteins in the ventre cell cytoplasm during the differentiation of ventral canal cell and migration of the female nucleus towards the centre of the ventre in the archegonia of *pinus sylvestris* and *P. nigra*. DNA has also been deteted in the central cell before its division into the ventral canal cell and the egg cell. The para-nuclei in *Pinus thunbergii* have been shown (Tako, 1959) to contain proteins, polysaccharides and nucleoproteins.

The female cones take the place of long shoots and take two years to complete their development.

Microsporangium and Microsporogenesis

The microsporangiate cones bear a spiral sequence of microsporophylls, each with a variably developed sterile distal flap and a pair of elongate microsporangia that usually dehisce by a longitudinal split. Rarely oblique and transverse dehiscence has also been observed. The male cones take the place of short shoots or spurs and are aggregated together in *Pinus*.

The structure of the microsporangium presents no major variations among the *Pinaceae*. The mature sporangium has a wall made up of two to six layers of cells. Tapetum is present. In *Pinus roxburghii* it originates from the innermost wall layer (Konar, 1961) and becomes binucleate at maturity. The tapetum and all the wall layers, except the outer layer or the epidermis, degenerate during meiosis and maturation of the microspores. The epidermal cells undergo significant changes and develop tanniferous contents. Their walls develop bands of thickenings and thick cuticle develops on the outer tangential walls.

The microspores in *Pinaceae* possess two wings. The genus *Tsuga* is an exception beause it has wingless pollen grains in all its species except *T. martensiana*. Carr. Martens and his colleagues (1967) studied the wall structure in the pollen grains of *P. sylvestris*. The young microspore has a three-layered *exine* (outer *ectine*, middle *mesine* and inner *endine*) and an intine whose outer layer is well developed whereas the inner layer does not develop in young grains but in mature grains it is recognisable as a pectocellulosic layer. The microspores are uni-aperturate in *Abies pindrow* and *Larix griffithiana*. The pollen grains start germinating within the microsporangium and are shed either at 4-celled stage (*Pinus, Pseudotsuga, Keteleeria davadiana*) or 5-celled stage (*Abies, Cedrus, Picea, Pseudolarix, Larix and Keteleeria evelyniana*). These changes constitute pre-pollination changes and are described below.

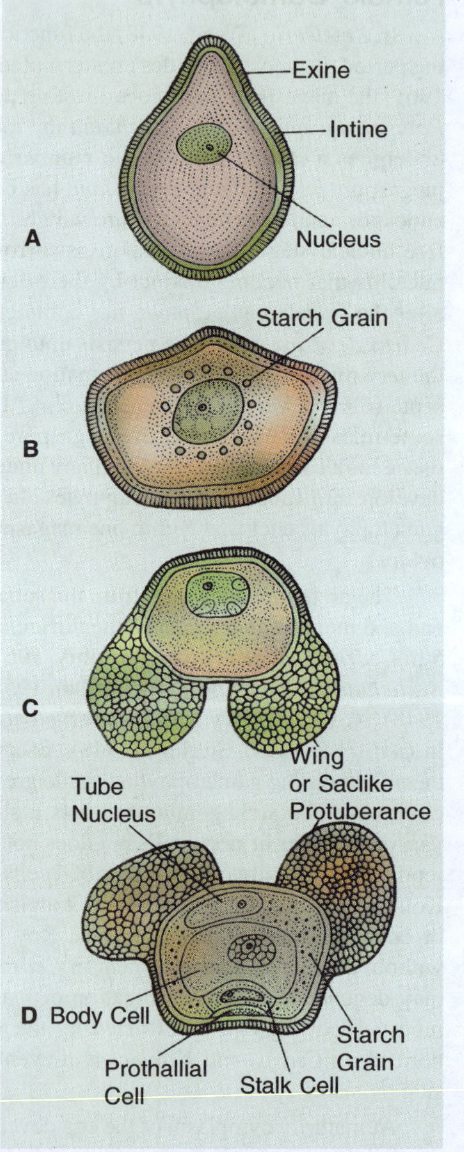

Fig. 12.12. (A-D) Pollen grains of Conifers. A. *Cryptomeria japonica*; B. *Araucaria cunninghamii*; C. *Pinus*; D. *Abies*.

Male Gametophyte

Before pollination the nucleus of the microspore divides twice by pericinal walls to cut off two **prothallial cells** and an **antheridial cell**. The latter divides by a perichial wall to form a small **generative cell** and a large tube cell. In *Pinus, Pseudotsuga* and *Keteleeria davadiana* the semigerminated pollen grains are released from the microsporangium and further germination takes place on the nucellus. In *Abies, Cedrus, Picea, Larix, Pseudolarix* and *Keteleeria evelyniana* the generative cell divides periclinally to form a **stalk cell** (or **sterile cell**) and a **body cell** (or **spermatogenous cell**) so that the pollen is released in a five-celled stage.

Pollination

It is anemophilous. The nucellus secrets a **pollination drop** which contains sucrose, glucose and fructose (McWilliams, 1958). The secretion fills the micropylar tube upto the rim and usually takes place at night. In *pinus* the winged pollen orientates itself in the pollination drop in such a manner that its germinal furrow is in contact with the nucellus. In *Pseudolarix* and *Larix* the pollen is not deposited in the nucellus but is in touch with the lateral sides. The pollination drop dries and the semigerminated pollen is retracted back to the nucellar region or the lateral sides of the micropyle tube. Doyle and Kave (1943) report that there is no pollination drop in *Abies* and the pollen is deposited in the funnel-like integumentary expansion around the micropyle. The nucellus grows and comes in contact with pollen grains.

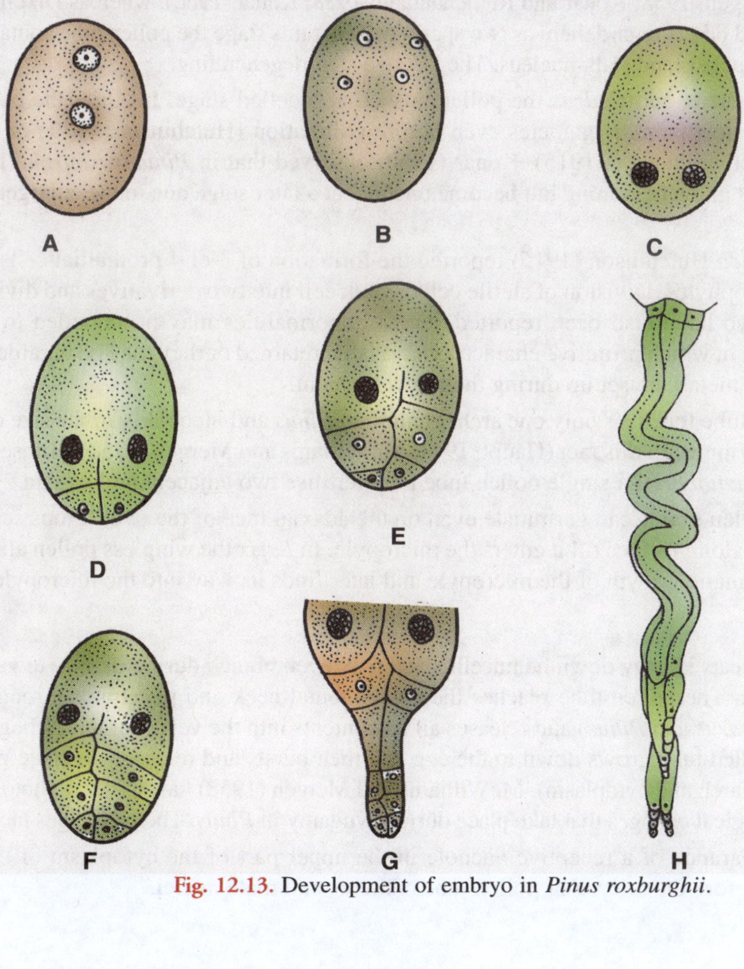

Fig. 12.13. Development of embryo in *Pinus roxburghii*.

Post-pollination Changes

The seimigerminated microspores rest on the nucellus before starting to germinate again. The period of dormancy is variable. It is 9 months in *Cedrus* (Roy Chowdhry, 1961), 3 months in *Keteleeria evelyniana* (Wang, 1948), 3 weeks in *Pseudotsuga* (Allen, 1942) and a few days in *Picea*. The period between pollination to fertilization also varies from a few days to a year. In *Pinus* it is one year.

After pollination the **generative cell** divides sometimes between the fall of the year of pollination and the following spring. It divides periclinally into the stalk cell (sterile cell) and the body cell (spermatogenous cell). The pollen tube also appears by the growth of intine through the germ furrow. The tube nucleus migrates into the pollen tube soon after pollination and later the stalk cell and the body cell also migrate into the tube. Sethi (1928) and Haupt (1941) claim that the stalk cell persists as a cell throughout the development, whereas Coulter (1897), Blackman (1898) and others claim that only its nucleus migrates into the pollen tube. None of these authors has made explicit the meaning of the term 'cell' in the coniferous microgametophyte. It is rather not possible to say whether the workers refer to the cell as a structure having cell wall or only cell membrane or only it is surrounded by a thin layer of cytoplasm that stains deeply. About a week before fertilization the body cell divides into two non-motile male gametes. The male gametes are unequal in size in majority of the conifers but Ray Chowdhry (1961) observed equal male gametes in *Cedrus deodara*. Some authors regard the two male gametes as two nuclei of a binucleate sperm cell (Sethi, 1928; Mathews, 1932; Haupt, 1941; McWilliams and Mergen, 1958; Konar and Ramchandani, 1958; Konar, 1960), whereas Dixon (1894) and Hirase (1918) and others regard them as two sperm cells. At this stage the pollen tube contains two male gametes, tube nucleus and stalk nucleus. The last two start degenerating.

In *Abies, Picia, Larix* and *Cedrus* the pollen is shed at 5 celled stage. In *Abies* the body cell sometimes divides into two male gametes even before pollination (Hutchinson, 1914) or shortly after the pollination (Hutchinson, 1915). Konar (1960) observed that in *Pinus roxburghii* the two male nuclei are equal in the beginning but become unequal at a later stage due to the enlargement of one of them.

In *Abies balsamea* Hutchinson (1915) reported the formation of 3 or 4 prothallial cells and of bi-antheridial gametophytes. Division of sterile cell or stalk cell into two derivatives and division of tube nucleus into two have also been reported. These abnormalities may be regarded to reflect stages of past history in which primitive characters have been retained or they may be regarded as an outcome of aberrant metabolic set up during their development.

A single pollen tube fertilizes only one archegonium in *Pinus* and inequality in the size of male gametes has been attributed to this fact (Haupt, 1941; Mcwilliams and Mergen, 1958). Hirase (1918) observed that in *P. thunnbergii* a single pollen tube may fertilise two adjacent archegonia.

In *Tsuga* the pollen grains can germinate even on the adaxial face of the ovuliferous scale, the pollen tube growing along its face till it enters the micropyle. In *Larix* the wingless pollen alights on the cushion-like stigmatic growth of the micropyle and later finds its way into the micropyle.

Fertilization

The pollen tube eats its way down he nucellus and may even branch during its way towards the archegonial chamber. The pollen tube reaches the archeogonial neck and penetrates through it by crushing its cells (*Cedrus* and *Pinus*) and releases all its contents into the venter of the archegonium. In *Keteleeria* the pollen tube grows down to the egg and then bursts and releases the male gametes and other contents (starch and cytoplasm). McWilliams and Mergen (1958) have given a photographic account of the cytological changes that take place during syngamy in *Pinus*. These changes include :

1. The appearance of a receptive vacuole in the upper part of the cytoplasm of the egg, shortly before the release of pollen tube contents and male nuclei.

2. The migration of functional male nucleus towards the egg cytoplasm.
3. The male nucleus approaches the receptive vacuole and sinks into it.
4. The fusion of the male and the female nuclei. They first lie in pairs and are separated by their double membranes.
5. The dissolution of the double membranes separating the paired male and female nuclei.
6. The chromatin material in each nucleus condenses and ultimately distinguished into short and thick chromosomes.
7. The establishment of a diploid nuclear condition of the zygote.

Embryogeny

The zygote enlarges and its diploid nucleus divides twice to form four free nuclei that migrate to the chalazal end of the archegonium. In *Pinus* (Fig. 12.13, A—H) the four nuclei divide again and arrange themselves in two tiers of four nuclei each. The lower one is enclosed by walls all around whereas the upper one is free from walls at their distal end. These two tiers divide to form four tiers of four cells each. These are from below upwards (*i*) embryonal tier, (*ii*) suspensor tier, (*iii*) rosette tier and (*iv*) upper tier. The cells of the upper tier have no walls towards their upper side and are also called open cells. Their nuclei slowly degenerate. The rosette tier of *Pinaceae* behaves as suspensor tier in other coniferous families. The cells of the suspensor tier elongate considerably and push the terminal embryonal cells deep into the endosperm tissue outside the archegonial venter. The four embryonal cells divide transversely into upper 4 cells that act as secondary suspensor cells and lower four cells. The primary and secondary suspensor cells elongate considerably and ultimately split apart longitudinally into 4 parts each carrying an embryonal cell at its tip. This splitting results in **cleavage polyembryony** which is in addition to the **simple polyembryony** resulting from the fertilisation of eggs in other archegonia of the female gametophyte. Sometimes the cells of the rosette tier also develop into embryos and this is called **rosette polyembryony**. **Suspensor polyembryony** has been reported in *Keteleeria* and *Pseudolarix*. *Pseudotsuga, Pseudolarix* and *Picea* do not show cleavage polyembryony. In *Larix* cleavage is delayed and Schopf (1943) inerpreted it as 'late cleavage polyembrony'. In *Cedrus* and *Keteleeria* too cleavage is delayed. Cleavage polyembryony in *Abies* is not frequent.

In *Pseudotsuga* (Allen, 1943, 1946) out of the four free nuclei in the pro-embryo two migrate to the micropylar end and two to the chalazal end of the archegonial venter. These nuclei divide again to form two groups of four nuclei each. The upper (micropylar) groups of four nuclei slowly degenerate whereas the lower nuclei divide into eight nuclei that arrange in two tiers of four each. The four nuclei of the lower tier divide again forming three tiers of four each. Wall formation takes place and a three-tiered proembryo is formed.

Detailed account of development of embryo from the embryonal cells has been described by Buchholz (1945, 1946), in *P. ponderosa* by Schopf (1943) in *Larix*; by Wang (1947) in *Keteleeria*; and by Allen (1946) in *Pseudotsuga*. Buchholz described eight stages that complete the embryo development. Schopf described 4 stages in *Larix*, which he named as prostage, meta-stage, ana-stage and telo-stage.

Of all the embryos differentiated by various types of polyembryony only one reaches maturity. Buchholz found that in *P. ponderosa* that developing embryos in the seed are in a state of extensive competition for survival within the first five or six weeks. The embryo that has to survive shows active growth and differentiates all its organs within 10–12 days. The seeds ripen in a few weeks time and the volume of the embryo increases during cotyledon initiation and differentation of shoot apex. This increase in volume proportionate to the volume of the endosperm and the seed number of cotyledons varies in the different genera (see table in the beginning of this chapter).

After seed formation the female cones turn brown in colour and become woody, the cone axis grows a little so that the tight scales separate and become loose and allow the seeds to escape. The seeds develop wings that are formed from the adaxial surface of the ovuliferous scale. The wing is not a part of the seed. In *Picea* the mature female cones become pendulous thus facilitating seed dispersal. The cones may persist after the seeds have been shed or may fall down. In *Abies* the cone axis persists on the tree but the scales fall down. In *Abies, Picea, Tsuga* and *Pseudotsuga* there are no dwarf and long shoots. The falling leaves in *Abies* leave characteristic disc-like scars on the skin whereas in *Picea* peg-like projections develop at places from where leaves have fallen.

The *Pinaceae* have 12 as the basic haploid chromosome number except in *Pseudotsuga* that has 13.

FAMILY : TAXODIACEAE

The family includes 10 genera and eighteen species. Six genera are monotypic whereas the other 4 have three species each (see table). Some members of this family *Sequoia gigantea* and *S. sempervirens* are the largest and oldest trees of the world. The list of genera and number of species along with their distribution is given in table 1 in this chapter. *Metasequoia glyptostroboides* is a member of the family discovered from China by T. Wang in 1944. Previously the genus was known only as a fossil from Pliocene deposits. This living species is the oldest living tree.

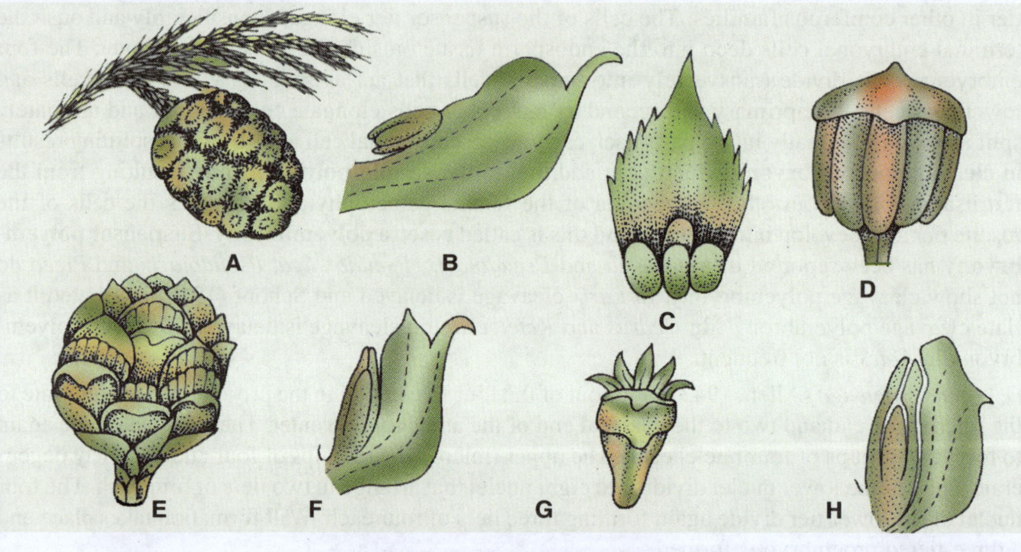

Fig. 12.14. *(A—G). Taxodiaceae.* A. Shoot bearing female cone of *Sequoiadendron giganteum* (After Eichler); B. L.S. cone scale complex of *Cryptomeria Japonica* (After Hirmer); C. microsporophyll of *Sequoiadendron gigantenum* (After Eichler); D. Megasporophyll *Sequoia sempervirens* with ovules (After Eichler); E. Female cone of *Taxodium distichum* (After Eichler); F.L.S, megasporophyll of *Taxodium* (After Hirmer); G. Megasporophyll complex of *Cryptomeria japonica* (After Goebel).

The family is characterised by : (*i*) The trees are monoecious that vary in size from small to gigantic size (upto 340 ft high and 825 ft. diam); (*ii*) the leaves are spirally arranged or rarely opposite (see table); (*iii*) the leaves may be scale-like or broad; (see table for examples); (*iv*) *Taxodium, Metasequoia, Sequoia, Cryptomeria, Glyptostrobus* and *Athrotaxis* produce two types of branchlets, one type is persistent and other deciduous and falls along with the leaves; (*v*) the strobili are small and retain their cone-like appearance; (*vi*) bracts and ovuliferous scales are free when young, but become fused in mature cones; (*vii*) the number of ovules varies from 2 to 9 on each ovuliferous scale; (*viii*) microsporangia also vary in number from 2 to 9 on each microsporophyll; (*ix*) the pollen grains are wingless; (*x*) the number of cotyledons varies from 2 to 5; (*xi*) the male gametophytes lack prothallial cell; (*xii*) the female cones are solitary and terminal; (*xiii*) the male gametes are equal in size; and (*xiv*) the archegonia occur in closed groups on archegonial complexes.

In *Taxodium* there are two types of branchlets : (*i*) persistent branchlets with axillary buds that are produced at the end of year's growth; and (*ii*) deciduous branclets that are devoid of axillary buds and are shed along with the leaves. *Metasequoia* also bears deciduous branchlets that are shed along with the leaves. In *Sequoia* the lateral branchlets bear linear leaves that are about 2 cm long whereas the long shoots bear lanceolate leaves. The deciduous branchlets in *Cryptomeria japonica* bear awl-shaped leaves. In *Sciadopitys verticillata* the photosynthetic organs arise in pseudo-whorls of 10 to 30 leaf-like structures that resemble a pair of pine needles fused throughout their length and about 12 cm long. These structures arise in the axils of scale leaves and have been described as cladodes. *Athrotoxis* also bears deciduous ultimate branches.

Plate IV : *Sequoia gigantea*

Microsporangia, microsporogenesis and Microgametophytes

The microsporangiate cones are frequently grouped on tasseloid brances as in *Taxodium* and resemble pinaceous cones in their general morphology. The number of microsporangia per sporophyll varies from two to seven or nine. The microsporangia have a three layered thick wall with tapetum as the innermost layer. A mature sporangium has a single wall layer whose walls develop fibrous thickenings. It acts as endothecium (Vasil and Sahni, 1964). The sporangium develops from a group of hypodermal meristematic initials that appear on the abaxial side of the microsporophyll. These are designated as archesporial cells and divide periclinally into outer parietal layer and an inner sporogenous layer. The cells of parietal layer divide by periclinal walls to form a three-layered sporangial wall, the innermost functioning as tapetum. The sporogenous cells divide in all directions to form a growth of microspore mother cells that undergo reduction division to form tetrads of microspores. The tapetum and inner wall layers degenerate during spore formation and maturation. The microspores are uni-aperturate. In *Cryptomeria japonica* the apertural area is marked by a papillate projection. They are usually spheroidal and devoid of wings. The exine is upto 1 μ thick and psilate in *Cryptomeria*. The intine is two-layered in *Athrotaxis cupressoides* (Elliot, 1951) and three-layered in A. *selaginoides* (Brennan and Doyle, 1956). These layers are exo-intine, moeso-intine and endintine. Elliot (1951) reported that a germ pore is absent in the pollen grains of *Athrotaxis*.

The microspore nucleus divides before pollination into a **generative cell** and a **tube cell**. There are no prothallial cells in *Taxodiaceae*. At this stage the grains are released and pollination takes place landing the semigerminated grains on nucellus. Such two celled pollen grains have been reported in *Cryptomeria* (Lawson, 1904). Singh and Jyotsna (1963) reported that the pollen grains in *Cryptomeria* (Fig. 12.12, A) are shed in a uninucleate state. In other genera *e.g. Athrotaxis* (Elliot, 1951; Bernan and Doyle, 1956); *Cunnighamia* (Miyake, 1910), *Metasequoia* (Sterling, 1949); *Taxodium* (Coker, 1903; Vasil and Sahni, 1964) and others the pollen is shed at two-celled stage. later during the pollen tube growth, the generative cell divides, producing a **stalk cell** and a **body cell**. The stalk cell loses its cytoplasm and moves ahead of the enlarging body cell. Vasil and Sahni (1964) report that in *Taxodium mucronatum* the generative cell moves into the pollen tube and divides there into stalk cell and body cell. The tube nucleus is the first to move into the pollen tube.

Mostly the pollen tube penetrates directly into the nucellus, but in *Athrotaxis selaginoides* and *Sequoia sempervirens* the pollen tube grows laterally, between the integument and the nucellus, for a considerable distance and then penetrates thd nucellus and reaches the megaspore membrane. The pollen tube grows forward by disorganising the nucellar cells and reaches the archegonial chamber. In *Athrotaxis* the pollen tubes reach the megaspore membrane when the latter is in a free nuclear state. In *Sequoia* it reaches before archegonial development. The body cell divides, shortly before fertilisation, into two equal male gametes. In *Taxodiaceae* a single pollen tube may fertilise two archegonia as the latter lie in close groups. In *Sciadopitys* (Tahara, 1937, 1940) the archegonia are quite separate from each other and the male gametes are unequal. One pollen tube fertilises only one archegonium. In this genus the male gametophyte overwinters during its development.

Megasporangium and Megasporogenesis

The megasporangiate strobili are solitary terminal and consist of a central axis with spirally arranged bract scales that bear axillary seed scale complexes or ovuliferous scales. In most genera the axillary fertile complex consists of a single ovuliferous scale or cone scale bearing from 2–7 ovules but variable in size in relation to the bract scale and variably free from it. In *Cunninghamia* the female cone axis grows beyond into a distal vegetative extension. In this case the bract scale is larger than the ovuliferous scale. The distal edges of the ovuliferous scales are fimbriated. In *Cryptomeria japonica* the mature ovuliferous scale is fan-shaped with 4 or 5 or many free tips. These free tips are seen extending upon the adaxial side of the subtending bract scale. During early stages in development the axillary seed complex or fertile complex is seen as 4 or 5 or more entirely separate morphological entities each with an ovule at their bases. They fuse with each other during development. The bract and the ovuliferous scales are fused except at the tip. The ovuliferous scales bear 3 or 4 ovules with their micropyles directed away from the cone axis. In *Taxodium* the cone scale or the ovuliferous scale does not show any sign of multipartedness at maturity but at any early stage of development there are two separate morphological entities each with an ovule at its base. Perhaps they become fused during their further development.

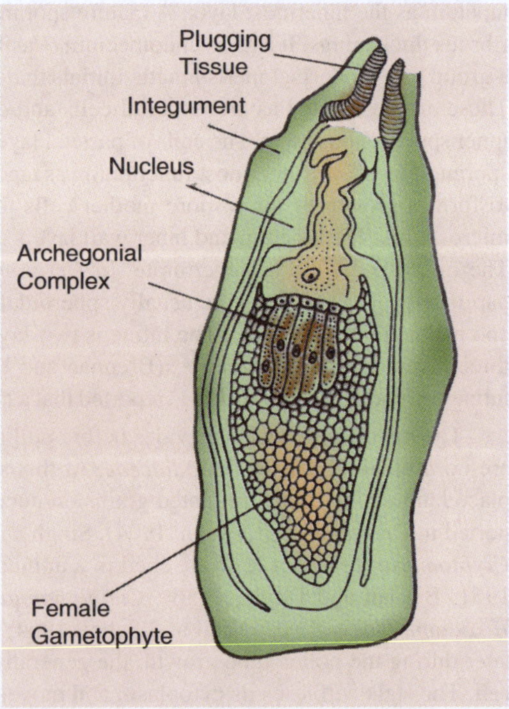

Fig. 12.15. L.S. ovule of *Cryptomeria japonica* (After Singh and Chatterjee)

In *Cryptomeria japonica* (Singh and Chatterjee, 1963) of the youngest ovule appears as a "small hump". The integument grows around the nucellus. In the beginning the growth of the integument is unilateral but later its growth is uniform. At this stage one or two hypodermal archesporial cells appear in the nucellus and divide to form outer parietal cells and inner sporogenous cells. The parietal cells later merge with nucellar tissue, whereas the sporogenous cells divide to form a group of such cells, some of which degenerate. In a mature ovule (Fig. 12.15) the integument is free from the nucellus right upto its base. The micropyle is wide open. The cells in the lower part of the nucellus are distinct by their small size as compared to the large cells in he upper part. Singh and Chatterjee (1953) report the appearance of a construction in the upper large celled portion of the nucellus. They also report the appearance of tannin in the nucellar cells below the constriction. Out of the group of sporogenous cells only one cell enlarges and acts as the **megaspore mother cell**. Tahara (1940) reported a single megaspore mother cell in *Sciadopitys verticillata*. Same is the case in *Taxodium mucronatum* (Vasil and Sahni, 1964) and *Athrotaxis selaginoides* (Brennan and Doyle, 1956). Lobby and Doyle (1942) reported many in *Sequoia* and *Sequoiadendron*.

Usually the megaspore mother cell undergoes reduction division to form a row of three cells. Meiosis-I forms two haploid cells. Of these the upper does not divide and the lower undergoes Meiosis-II resulting in the formation of a linear tetrad of 3 haploid cells. Sometimes the upper nucleus divides but there is no wall formation and is therefore binucleate (*Cryptomeria*). In *Sciadopitys verticillata* out of the three cells the central is binucleate. T-shaped tetrads have been reported in *Athrotaxis* (Brennan and Doyle, 1956). Lobby and Doyle (1942) reported several types of tetrads in *Sequoia sempervirens*. Usually the lowest megaspore in the tetrad functions and enlarges to develop into the female gametophyte. In *Sequoia*, however, all the megaspores are functional and a number of young megagametophytes can be seen in the nucellus. This genus has several megaspore mother cells.

Female Gametophyte

The functional megaspore enlarges in size and its nucleus undergoes extensive free nuclear division (upto 600 in *Sequoiadendron*) followed by centripetal wall formation. Wall formation is by the formation of alveoli in *Cryptomeria japonica* (Kurihara, 1936). A distinct archegonial chamber is formed after the development of archegonial complexes.

Most of the genera develop archegonial complexes at the micropylar end and all the archegonia in a complex are surrounded by a Jacket. An archegonial complex may have 12–20 (*Taxodium*), 15–37 (*Cryptomeria*), or 4–9 (*Taiwania*) archegonia. Singh and Chatterjee (1963) reported rare occurrence of chalazal and lateral archegonial complexes in *Cryptomeria* (Fig. 12.4). Chalazal complexes have also been reported in *Athrotaxis* by Brennan and Doyle (1956). In *Sciadopitys* the archegonia occur singly (Tahara, 1940). Each archegonium has 4-neck cells in *Cryptomeria*, and *Sequoiadendron*, whereas there are only two neck cells in Athrotaxis and Sequoia. The later two genera also do not produce ventral canal nucleus regularly. Singh and Chatterjee (1963) report a definite ventral canal nucleus in *Cryptomeria japonica*. The ventral canal nucleus and the egg nucleus are not separated by a wall. The ventral canal nucleus is epheimeral and degenerates before fertilisation. *Toxodium, Cunninghamia and Sciadopitys* also have a definite but ephemeral ventral canal nucleus. The egg nucleus, in *Cryptomeria*, lies in the upper part of the archegonium (Singh and Chatterjee, 1963).

Fertilisation

Singh and Chatterjee (1963) describe fertilization in *Cryptomeria japonica* as, "The neck of the archegonium begins to degenerate just before fertilisation. The archegonium has dense cytoplasm and a prominent nucleus above the vacuole, which becomes smaller. One or both the male nuclei may enter and fertilisation takes place in the centre of the archegonium. A depression is formed in the upper part of the egg nucleus. The cytoplasm around the fusing nuclei is dense and shows starch grains. Vacuoles now appear in the upper part of the zygote while the vacuole in the lower part starts disappearing." Buchholz (1939) studied fertilisation in *Sequoia* and Brennan and Doyle (1956) studied it in *Athrotaxis*.

Embryogeny

Four types of embryogeny have been reported in the family *Taxodiaceas*. These types are :

1. Usual taxodian type found in *Taxodium, Taiwania, Cryptomeria, Cunninghamia, Glyptost-*

robus and *Sequoiadendron*.

2. *Sequoia sempervirens* type (Fig. 12.16, A–E).
3. *Sciadopitys* type.
4. *Athrotaxis* type.

In the first type the zygote nucleus undergoes 3 divisions to form eight nuclei that organise into two tiers of four cells each. The upper of these two divides again so that three tiers of 4 cells are formed. These are the upper open tier of cells, middle tier of pseudosuspensor cells and a lower embryonal tier. In *Taiwania* even the upper tier is completely enclosed by walls. Primary suspensor has been reported in *Sequoiadendron* (Buchholz, 1940; Sughiara, 1941). In *Cryptomeria* there are two or three cells in the embryonal tier whereas there are 5 or 6 cells in the open tier (Singh and Chatterjee, 1963). The cells of the upper tier soon degenerate. The prosuspensor cells elongate and push the embryonal cells out of the archegonium. The embryo cells also start dividing. The suspensor cells then separate from each other and 2 or 3 of them bear developing embryo cells at their tips. This is the cleavage polyembryony. In *Cryptomeria* the suspensor cells without embryonal cells may divide at the tip to cut off a small embryonal cell or they may undergo transverse septation (Singh and Chatterjee, 1963). The embryo cells may or may not divide transversely into an upper *primary suspensor cell* and a lower embryonal cell.

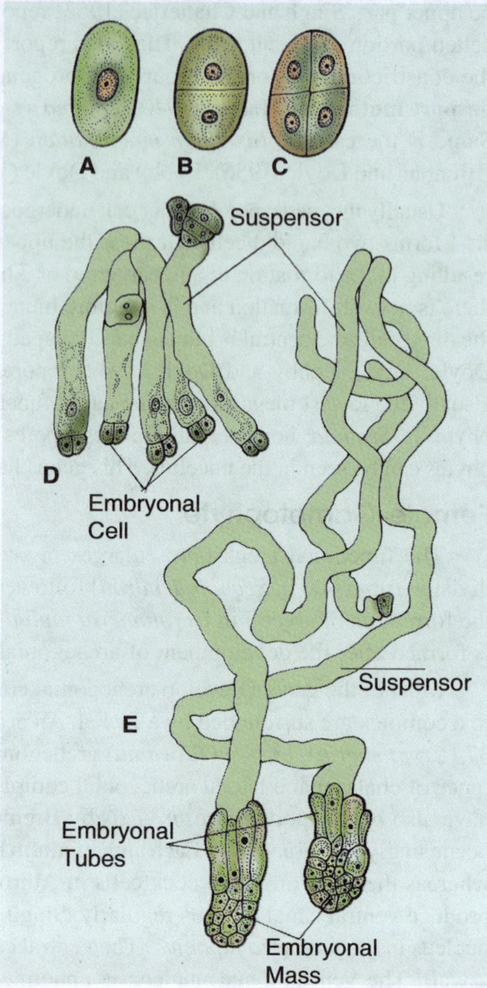

Fig. 12.16. (A–E). Stages in the development of embryo in *Sequoia sempervirens*. (*After Buchholz*)

In *Sequoia sempervirens* (Fig. 12.16) the first division of the zygote nucleus is followed by wall formation thus forming two cells (Buchholz, 1939) which divide to form variously arranged 4 cells. These cells divide to form a primary suspensor cell and and embryonal cell. There is thus no free nuclear stage and no tier system in the proembryo stage.

In *Sciadopitys* 32 free nuclei are formed (Tahara 1937, 1940). Rarely 50–60 free nuclei are formed. All these nuclei arrange themselvs in 3 or 4 tiers with 12–20 cells in the embryonal tier, 7–13 in prosuspensor tier, 2–9 cells in the rosette tier, and 20–30 cells in the upper tier.

In *Athrotaxis selaginoides* (Brennan and Doyle 1956) the zygote forms 4 free nuclei that arrange themselves in 3 tiers. Wall formation takes place and four cells are formed. There is one cell in embryonal, two in prosuspensor and one in upper tier (*i.e.*, 1 : 2 : 1). All the four cells divide vertically resulting in four cells (*i.e.*, 2 : 4 : 1) arranged in the three tiers. There is no cleavage polyembryony in *Athrotaxis* (Elliot, 1951).

The basic haploid chromosome number of *Sciadopitys* is 10 and for the rest of the family it is eleven 11.

FAMILY : CUPRESSACEAE

It is the largest family of the order and includes *twenty genera and 148 species*. Half of them are restricted to the northern hemisphere (*Thuja, Biota, Cupressus, Juniperus*, etc.) and the other half to the southern hemisphere (*Calitris, Libocedrus, Popuacedrus* etc.). They are characterised by the following features :

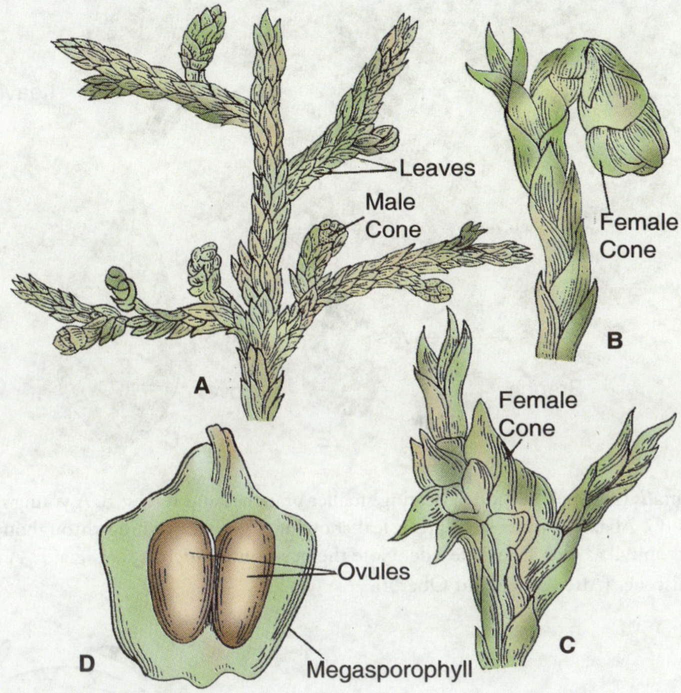

Fig. 12.17 (A–D). *Biota orientalis. A.* Branch bearing male cones; also note the closely adepressed leaves; B–C. Female cones on curved branches; D. Lateral view of megasporophyll bearing ovules. (After Singh and Oberoi).

1. The sporophytes are small or large trees, most of which are cultivated as ornamentals and some as commerical sources of wood.
2. The leaves are small and may be scale-like or linear.
3. The arrangement of leaves is either opposite or whorled except in *Juniperus* and *Actinostrobus*, where the leaves are spirally arranged.
4. There is no shoot dimorphism.
5. The sporophytes may be monoecious or dioecious (*Biota orientalis*).
6. The megasporangiate strobili (Fig. 12.17. B–C) are small and may be dry (*Thuja, Biota Cupressus*) or fleshy (*Juniperus*) at maturity.
7. The ovuliferous scales may overlap (*e.g. Thuja*) or may be valvate (*Cupressus*). Lee (1953) included the former under a sub-family *Callitroideae* and the latter under sub-family *cupressoideae*. Studies on wood anatomy (Boutelje, 1955) also support such a subdivision. The ovuliferous scales are slightly larger than the bract scales.
8. There is a high degree of fusion between the bract scale and the cone scale. The number of ovules varies from 3–20, rarely one or two.

Fig. 12.18. *Biota orientalis* (A–E). A. A branch bearing smaller branches and leaves; B. A young male cone covered by leaves; C. Abaxial surface of a green leaf showing a groove running throughout its length; D. A microsporophyll as seen from one side. Note the position of micro-sporangia; E. A male cone as seen after dehiscene. (After Singh and Oberoi).

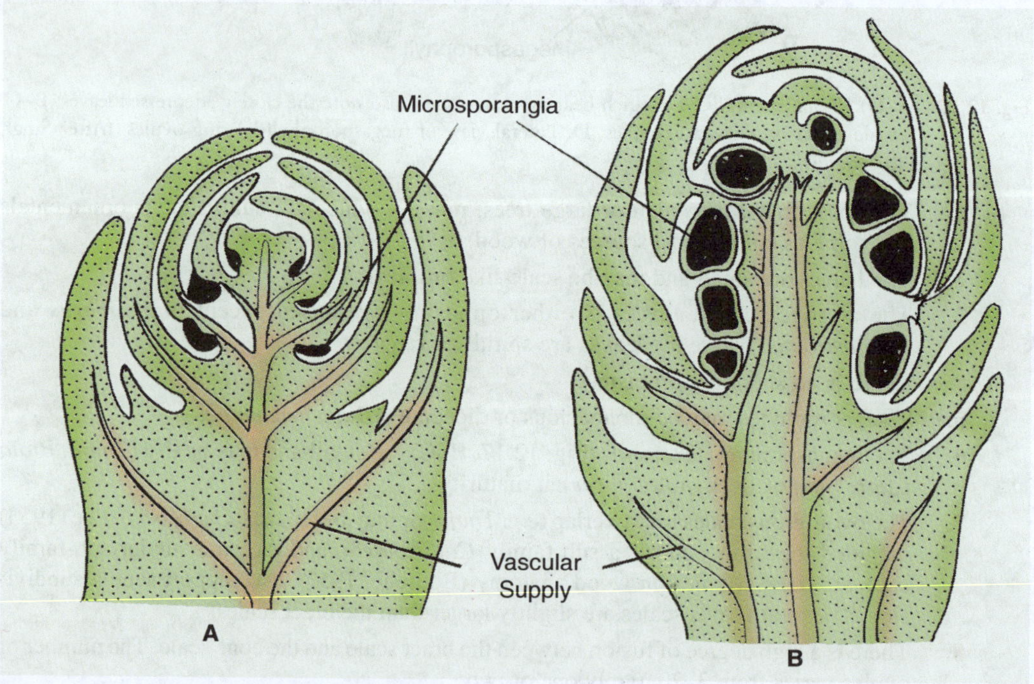

Fig. 12.19. **(A–B).** *Biota orientalis.* A. L.S. of a young male cone; B. L.S. of old male cone showing microsporangia. (After Singh and Oberoi).

9. The ovules are mostly erect. In the section *Microbiota* of the genus *Juniperus* the female cone has as single cone scale with a single ovule. In section *Oxycedrus* the ovules appear to alternate with the cone scales because they are borne laterally. In the section *Sabina* the ovules are borne normall.

10. The microsporangiate strobili are also small and consist of central axis with microsporangiophores arranged in whorls (Fig. 12.18; 12.19, A–B). The microsporangia are nearly spherical and develop from a group of 2 or 3 hypodermal cells. The young sporangium consists of an outer wall layer, a middle layer and a tapetum enclosing a group of mother cells that divide meiotically to form several tetrads of microspores.

11. Each microsporangiophore bears 2–6 sperical microsporangia.

12. The pollen grains are wingless (Fig. 12.21).

13. There is a well developed pollination drop mechanism.

14. The male gametophytes lack prothallial cells (Fig. 12.21).

15. The pollen grains may be released in one-celled stage or (*Juniperus, Actinostrobus, Callitris, Fitzroya, Tetraclinis, Cupressus* and *Widdringtonia*) or in a two-celled stage (*Biota, Thuja, Libocedrus, Thujopsis*, etc.). The two-celled stage consists of a generative cell and a tube cell.

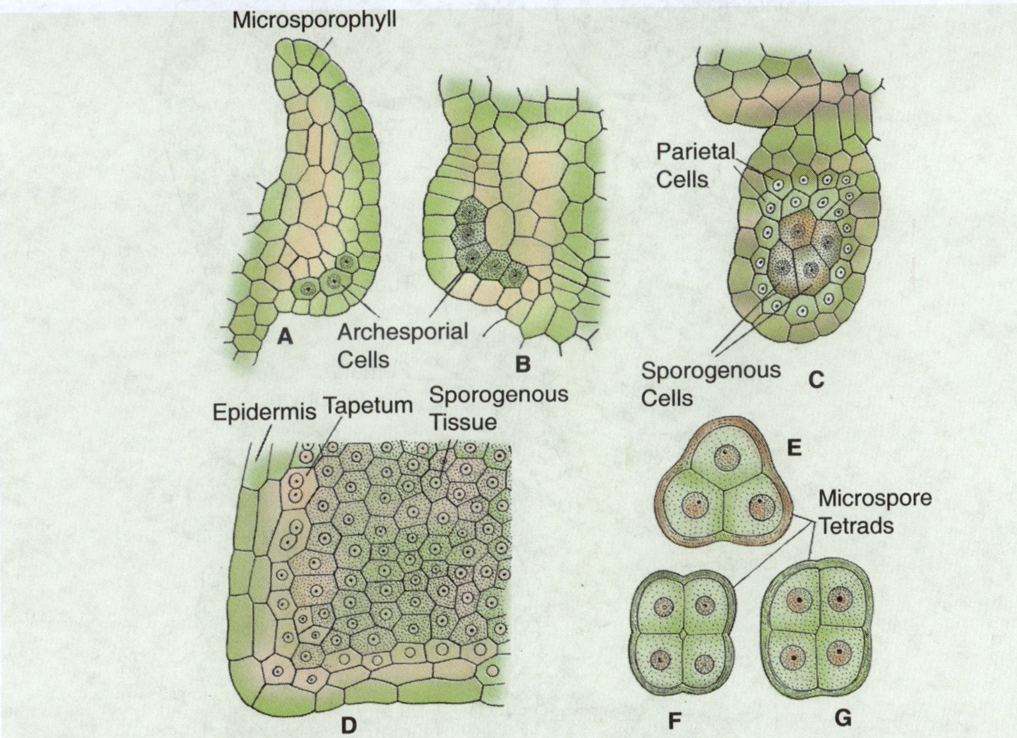

Fig. 12.20 (A–G). *Biota orientalis*. A–B. L.S. of microsporophylls showing a row of hypodermal archesporial cells that will develop into a microsporangium; C. L.S. of microsporophyll showing sporogenous cells and parietal cells; D. L.S. young microsporangium showing wall layers, tapetum and spore mother cells; E–G. Microspore tetrads. (After Singh and Oberoi).

16. The male gametes are equal in size. In *Cupressus sempervirens* (Mehra and Malhotra 1947) the body cell divides into 4–11 male gametes. In *Callitris* both the male gametes are functional (Baird, 1953). A single pollen tube may branch and enter two archegonia

or it may remain unbranched and enter only one archegonium. In *Widdringtonia cupressoides* (Moseley, 1943) and *Callitris* (Baird, 1953) the pollen tube reaches the female gametophyte when the latter is in free nuclear stage. In *Cupressus cupressiformis* Lobby and Doyle (1940) reported the pollen tube to form a groove in the megaspore membrane. The pollen grains that are dispersed by wind are caught in the pollination drop, which then starts diminishing in size and within 10–20 minutes contracts into the micropyle (*Callitris*); Baird, (1953). Strassberger (1892) observed that in *Juniperus* the pollen tube sends several prolongations to several archegonia of the group. Each prolongation has a definite pit at its top. In majority of *Cupressaceae*, pollination, fertilisation and seed development occur within the same season.

17. The mature sporangia in *Biota orientalis* (Singh and Oberoi, 1962) have only the outermost wall layer with cutinised outer walls that also develop annular thickenings (Fig. 12.20). The spore tetrads may be isobilateral or tetrahedral.

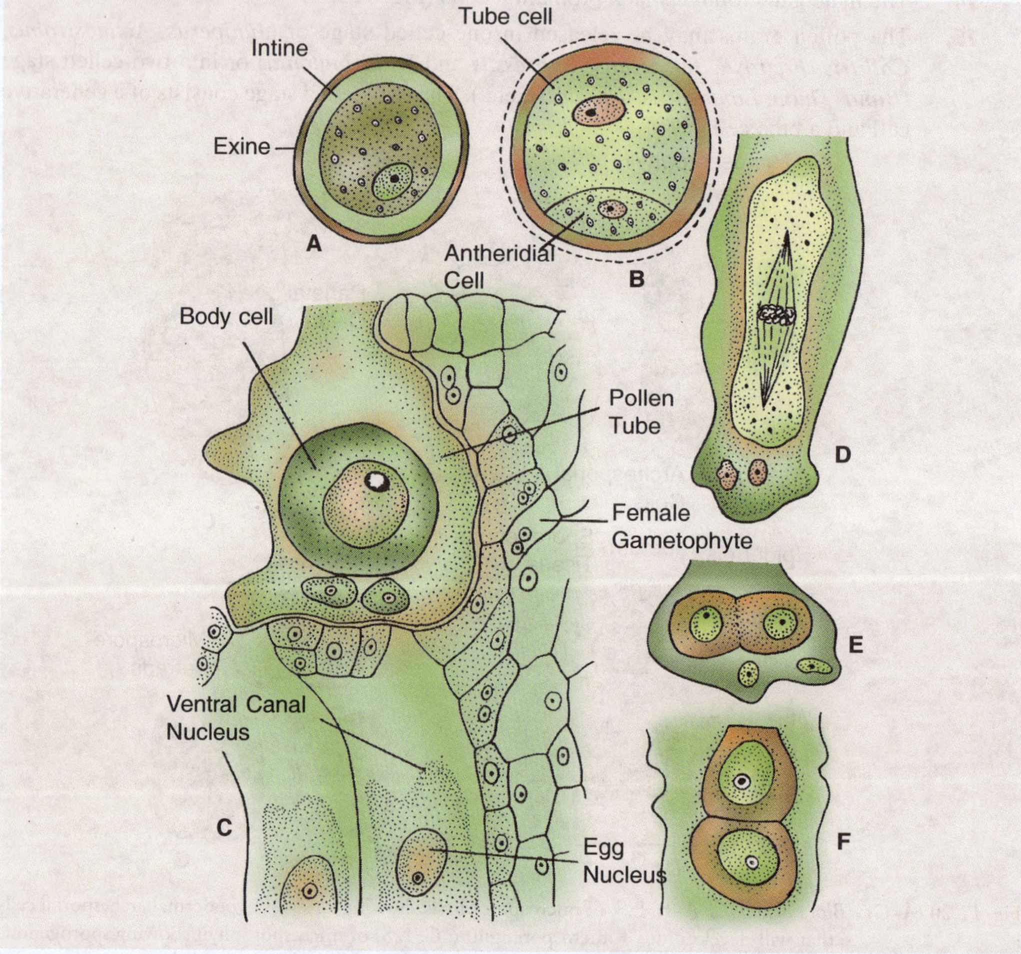

Fig. 12.21 (A–F). *Biota orientalis.* A. Uninucleate microspore; B. Microspore at the time of dehiscence of microsporangium; C. Enlarged portion of pollen tube containing a body cell, a stalk nucleus and a tube nucleus; the archegonia with neck and a part of venter containing egg nucleus are also seen; D. The body cell has divided into two equal male cells; E. Pollen tube with two male cells. (All after Singh and Oberoi).

18. The integument is free from the nucellus right up to its base (Fig. 12.22). In young ovules in *Biota orientalis* (Singh and Oberoi, 1932) the integument is 3–4 cells thick at the base and only one layer thick at the micropylar end. After pollination the cells of the integument lining the micropyle, elongate radially and close the canal. The nucellus shows differentiation of archesporial cells quite deep in its tissue almost near its chalazal end. The number varies from two (*Biota*) to six or ten (*Cupressus sempervirens* and *C. funiberus*). The archesporial cells divide repeatedly to form a roup of sporogenous cells out of which one distinguishes itself (Fig. 12.23. A, B) as a **megaspore mother cell**. (Singh and Oberoi, 1962). In *Biota* this mother cell divides meiotically to form a linear tetrad of three (Fig. 12.23, A) cells (uppermost remains undivided). In *Thuja*, Martin (1950) reported a tetrahedral tetrad of megaspores. Mehra and Mallotra (1947) reported tetrahedral, L-shaped, T-shaped and linear tetrads in *Cupressus sempervirens*.

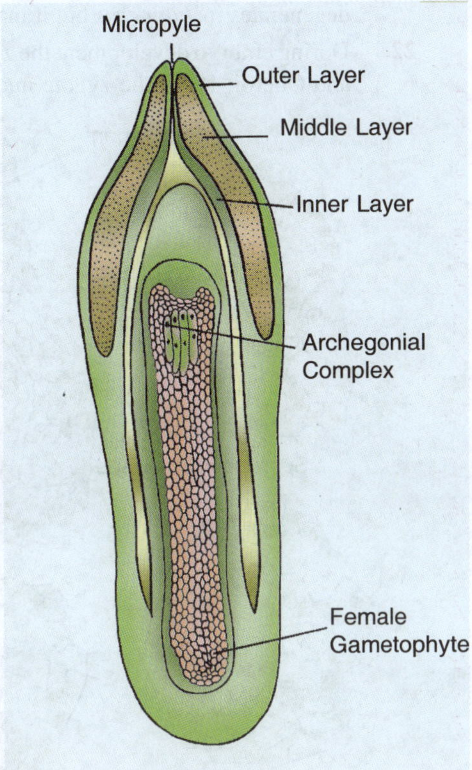

Fig. 12.22. L.S. Mature ovule of *Biota orientalis* (After Singh and Oberoi)

19. The functional megaspore is surrounded by a distinct spongy tissue (Fig. 12.23, D–F) in *Biota* and *Cupressus* (Konar and banerjee, 1963). Baird (1953) did not observe it in *Callitris*.

20. The megaspore enlarges in size and its nucleus undergoes to form 256 free nuclei in *Widdringtonia cupressoides* (Moseley, 1943), 2,000 in *Cupressus funiberus* (Konar and Banerjee, 1963); 2048 in *Thuja* (Martin, 1950); 4,000 (Fig. 12.23, G–H) in *Biota orientalis* (Singh and Oberoi, 1962); and 6,000 in *Cupressus sempervirens* (Mehra and Malhotra, 1947). The nuclei are arranged in a peripheral cytoplasm with a big central vacuole. Wall formation is centripetal. The enlarging gametophyte almost completely consumes the nucellus.

21. The archegonia occur in groups (Fig. 12.24) and form a micropylar, lateral or even chalazal complex. Singh and Oberoi (1962) reported a nonfunctional chalazal complex in *Biota*. This is in addition to the micropylar complex of 15–28 archegonia. They also observed one or two supernumerary archegonia below the micropylar archegonial complex. Mathews (1939) reported a complex of 4–10 micropylar archegonia in *Juniperus virginiana*. *Widdringtonia cupressoides* has 25–100 archegonia in a complex (Moseley, 1943) that is lateral in position and quite deep seated. In *Cupressus funiberus* there are 14–17 archegonia in a micropylar complex. In *Biota orientalis* the archegonial neck has a single tier of 4 cells (Fig. 12.24). In *Thujopsis* there are 8 cells in a single tier. There are cases where there are two to four cells in one or two tiers. In *Biota, Callitris* and *Widdringtonia* the neck cells degenerate before fertilisation. The central cell divides into a ventral canal nucleus and an egg before fertilisation. In *Widdringtonia* the nucleus soon degenerates in *Juniperus* but it may persist in *Callitris*.

22. During embryo development the zygote shows some variations. In Biota orientalis (Singh and Oberoi, 1962) the zygote nucleus divides into eight free nuclei (Fig. 12.26, A, B).

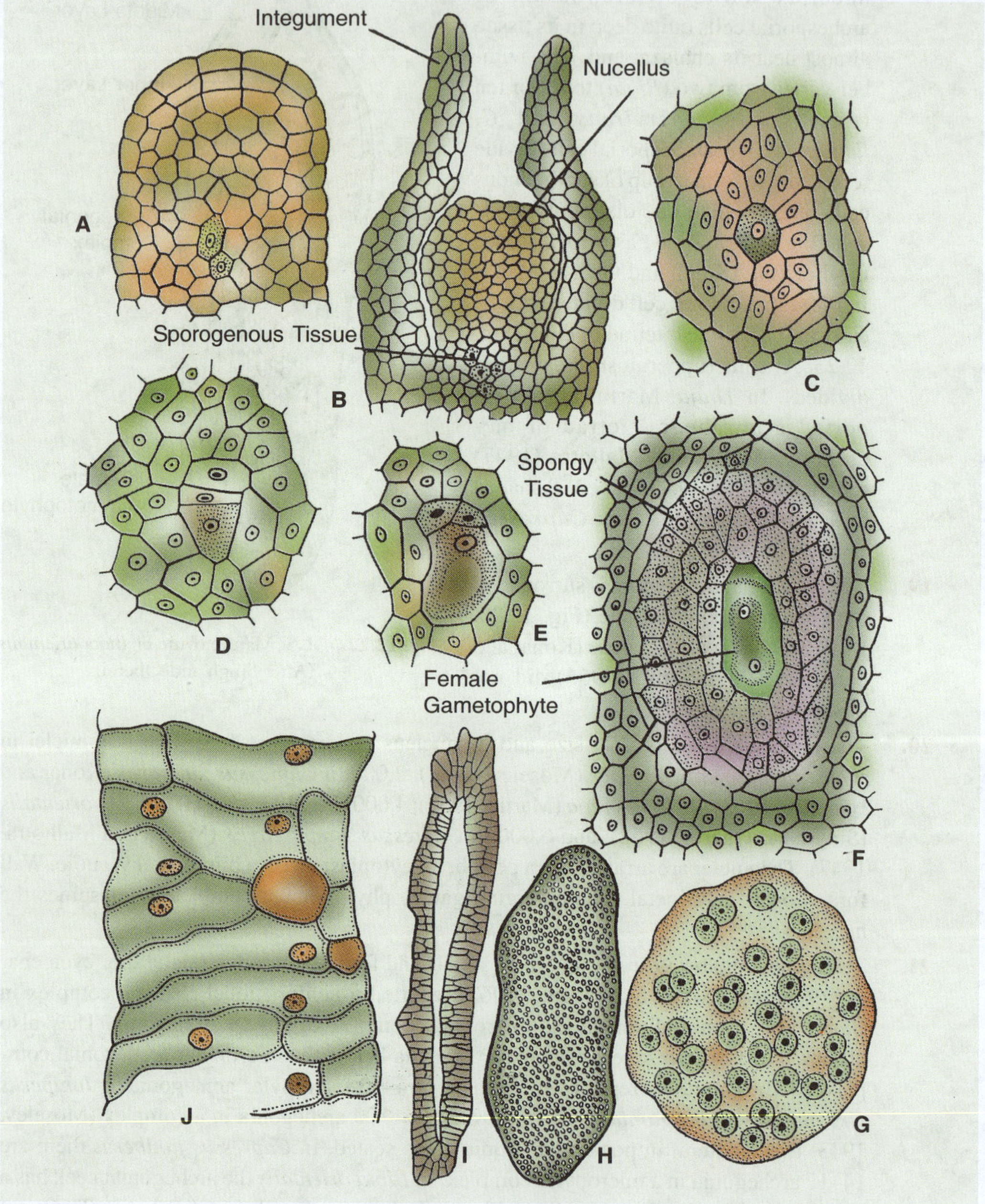

Fig. 12.23. Stages in the development of female gametophyte in *Biota orientalis* (After Singh & Oberoi).

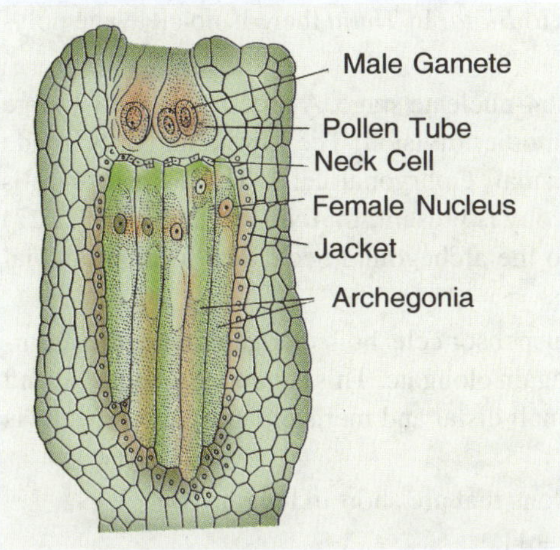

Fig. 12.24. *L.S.* ovule of *Biota* showing female gametophyte and archegonial complex (After Singh & Oberoi).

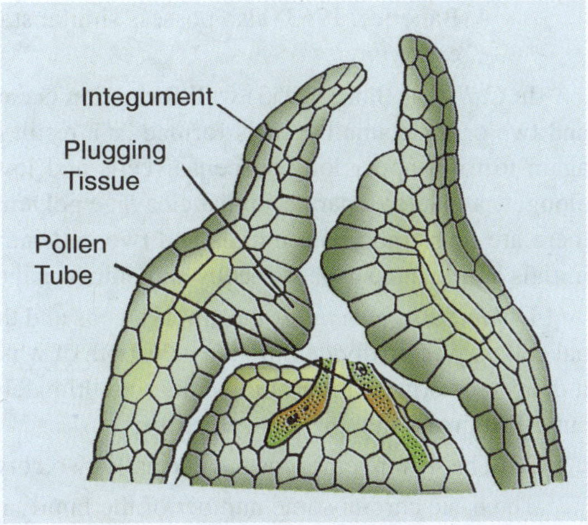

Fig. 12.25. *Biota orientalis.* L.S. of the micropylar part of ovule. Note the plugging tissue (After Singh & Oberoi).

Wall formation results in a lower tier of 4 embryonal cells and an upper tier of 4 cells that are open towards (Fig. 12.26, C–E) the archegonial neck. The cells of the upper tier divide again into a middle prosuspensor tier and an upper open tier. So three tiers of four cells each are organised. The number of cells may vary in each tier. Tetrahedral, isobilateral or even T-shaped arrangement of the four primary embryonal cells has also been recorded in this species. The prosuspensor cells elongate and push the embryonal cells out of the archegoniate complex. Later the primary embryonal cells divide transversely to cut off an upper tier of suspensor cells and a lower tier of embryonal cells. Later suspensor cells start elongating (after the prosuspensor cells stop elongation) and cleavage takes place. So cleavage polyembryony is there in *Biota*. *Thuja, Chamaecyparis* (Sughiara, 1938), *Thujopsis* (Sughiara, 1939), and *Cupressus* (Konar and

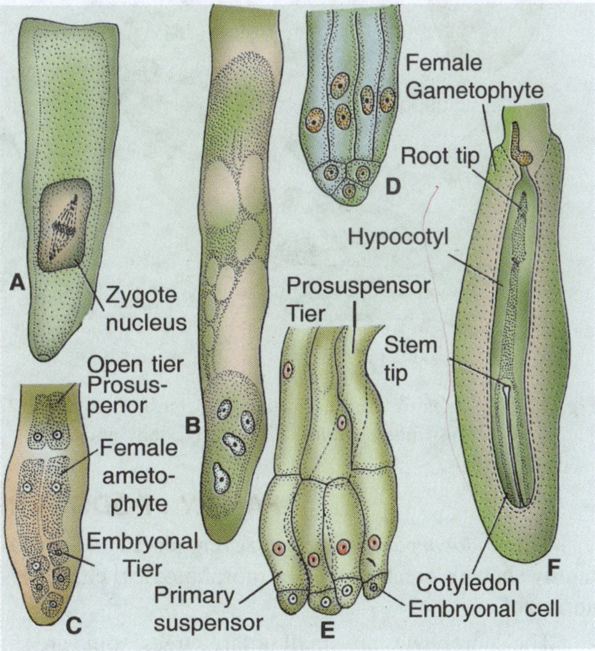

Fig. 12.26. **(A–F).** *Biota orientalis.* Stages in the development of embryo. A. Division of zygote nucleus. It shows intranuclear division; B. Four nucleate proembryo; C. Pro-embryo with three tiers; D. later stage with divided cells of embryo tier; E. It shows prosuspensor, primary suspensor and embryonal cells; F. L.S. female gametophyte showing mature embryo. (All after Singh & Oberoi).

Banerjee, 1963) also possess similar stages to *Biota*. In *Thuja* there is no cleavage polyembryony.

In *Callitris* (Baird, 1953) wall formation occurs at 4-nucleate stage. A proembryo with 4 large and two or three small cells is formed as a result of another division. The lower four cells divide again to form upper long suspensor cells and lower small embryonal cells. The suspensor cells elongate and later separate so that cleavage polyembryony is present. In *Actinostrobus* (Fig. 12.27) there are four embryonic initials and two cells next to the archegonial neck. The four embryonal initials divide into suspensor cells and embryonal cells.

In *Juniperus communis* the embryo cells and the suspensor cells both elongate. Later the terminal cell may again divide into two cells both of which again elongate. This sequence is repeated and it delays the organisation of the embryo. Ultimately small distal and meristematic group of cells is cut off. It grows into the embryo proper.

23. The mature embryos have usually two cotyledons that are short in length.

The basic chromosome number of the family is $n = 11$

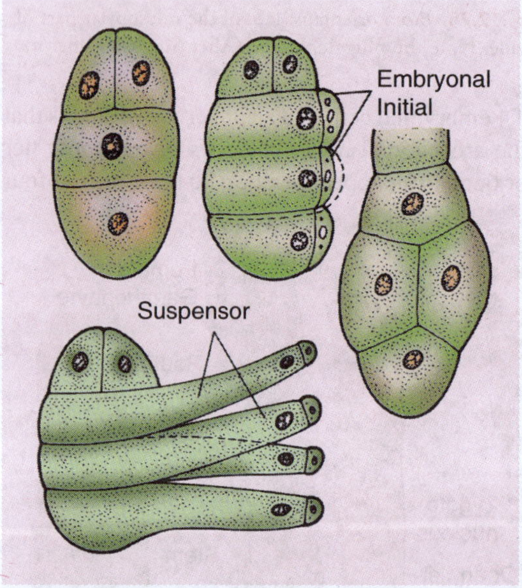

Fig. 12.27. Stages in the development of embryo in *Actinostrobus pyramidalis.* (After Lobby and Doyle)

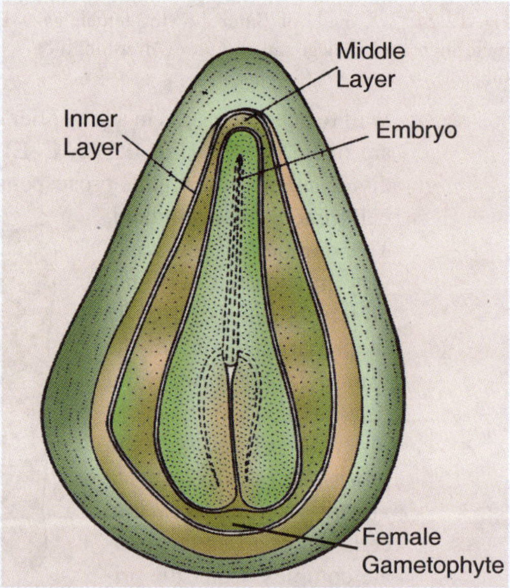

Fig. 12.28. L.S. of Seed of *Biota orientalis.* (After Singh and Oberoi)

FAMILY : PODOCARPACEAE

The *Podocarpaceae* include seven genera and approximately 150 species (see table 12.1). The family shows a greater range of morphological characters and certain interesting developments are noticeable.

The sporophytes are small or large trees, some are shrubs. Some species of *Dacridium* are only a few inches high. Some species of *Prodocarpus* grow as tall as 60 metres. *Podocarpus ustus* (Gray, 1953) is reported to grow as a partial parasite on the roots of *Dacridium taxoides*. The leaves are spirally arranged except in some species of *Podocarpus* they are opposite or sub-opposite. The leaves may be scale-like, needle-like, linear or broadly lanceolate. There is no shoot dimorphism except in *Phyllocladus* where the dwarf shoot is a flattened fusion product called the **cladode**. It is the main photosynthetic organ. In *Microcachyrus* too the leaves are opposite and decusate.

The microsporangiate cones are short or long (*Podocarpus macrophylla*) and bear spirally arranged microsporophylls (Fig. 12.29) that bear two microsporangia each. The male cones are usually axillary in position and in some cases new strobili arise from the bases of older ones. The trees may be monoecious or dioecius.

The pollen grains are winged in all genera except *Saxegotheca* that has wingless pollen. In *Pherosphaera* and *Microcachyrus* the pollen grains have three wings, whereas in all other genera there are two wings.

The development of the microsporangium is eusporangiate like other conifers. The more or less elliptical microsporangia dehisce longitudinally and when young possess a 5–7 layered thick wall with the innermost functioning as the **tapetum**. The tapetum is of glandular type. As the microspores differentiate and mature, all the wall layers and tapetum disorganise, except the outermost layer that acts as endothecium (Doyle and Doyle, 1953).

The megasporangiate strobili generally appear unstrobiloid in appearance and may have one to four or sometime nine bract scales with their axillary complexes. In *P. macrophylla* the female cone is a uni- or bi-ovulate structure borne terminally on a flattened short segment of the axis that bears one or two reduced leaves. This short segment is a lateral shoot. The mature one may have one or two ovules. The ovuliferous scale is folded around the ovule and enclosed it completely or partially and may be fused with it completely or partially. It forms an extra envelope designated as **epimatium**. It may also be fused with the bract scale in some species. In *P. macrophylla* and *Podocarpus spicatus* the epimatium is fused with the integument throughout its length, whereas in *P. ferruginens* the epimatium is free from the integument except at its chalazal end. The female cones in *P. spicatus* and *Dacrydium franklinii* look like loose or lax spikes. The bracts persist as short segments at the base of the ovule. In *Dacrydium araucarioides* a female cone has three elongated bracts, each subtending a single cone scale (ovuliferous scale) that partly envelops the ovule. In most of the species of *Dacrydium* there is a single terminal ovule in a female cone. It is surrounded at the base by a cup-shaped structure formed by the sterile cone scales. In *Phyllocladus* the seed is enveloped by a true aril that is free from the integument and encloses it only partially. In this case there is no epimatium. Epimatium is also absent in *Microstrobus*.

Fig. 12.29. (A-D) : *Podocarpus*, A. A shoot of *Podocarpus spicatus* with male cones; B. A male branch bearing male cones in ovules of bracts. It is a compound branch; C. A single male cone; D. L.S. ovule of *Podocarpus* showing megaspore mother cell (A–C after Hooker; Hooker after Siunot)

The ovules may be erect or anatropous (inverted) and a single integument projects a litele above the level of the nucellus and forms a wide micropylar opening in the young ovule. The epimatium is a little shorter than the integument in *Podocarpus gracilior* (Konar and Oberoi, 1969). As the ovule grows in age the part of the integument lying above the nucellus elongates markedly to form a long and narrow micropylar canal. After the pollination has taken plae the cells of the upper part of the integument lining the canal elongate inwardly and thus close the micropyle. The nucellus encloses the megagametophyte which contains endosperm and archegonial complex. The number and shape of archegonia in a complex is variable.

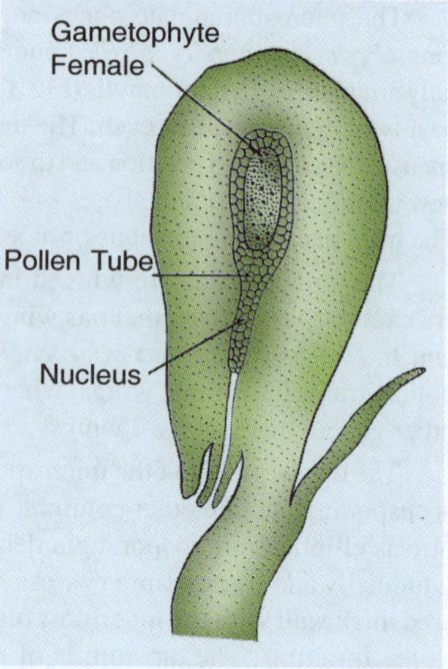

Fig. 12.30. L.S. ovule of Podocarpus gracilior (After Konar and Oberoi).

A megaspore mother cell differentiates quite early in the ontogeny of the ovule. It lies at a depth of 4 or 5 cell layers in the nucellus in *P. gracilior* (Konar and Oberoi, 1969). It originates from a hypodermal archesporial cell that cuts off an outer parietal cell and an inner megaspore mother cell. The upper cells divide by periclinal wlls and push the mother cell deep into the nucellus. Some cells around the megaspore mother cell differentiate into an inconspicuous **tapetum**.

Looby and Doyle (1939, 1944) studied the development of megaspore and female gametophyte in *Saxegothaea* and *Podocarpus andinus*; Boyle and Doyle (1953) studied it in *P. nivalis*; and Quinn (1964) in *P. Lacrydioides*. konar and Oberoi (1969) studied it in *P. gracilior*. The megaspore mother cell usually divides meiotically into a linear row of three megaspores, the upper cell remains undivided. In some genera a linear tetrad of four megaspores in formed (*Microcachrys, Dacrydium, Podocarpus falcatus*, and *P. gracilior*).

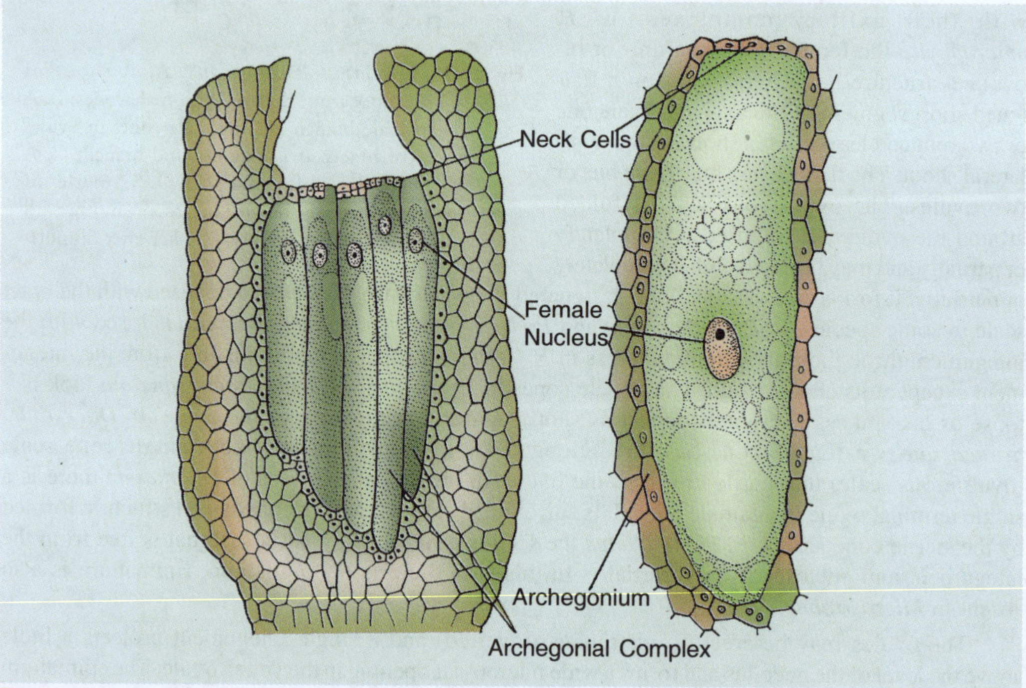

Fig. 12.31 (A–B): *Podocarpus gracilior.* A. L.S. of micropylar region of female gametophyte showing archegonial complex; B. A single archegonium (After Konar and Oberoi).

The megaspore enlarges and undergoes free nuclear division to form several hundred free nuclei. In *Saxagothaea* and *Podocarpus dacrydioides* a tent pole like protuberance arises from the micropylar end of the female gametophyte. The female gametophyte usually overwinters in the free nuclear stage. Wall formation is centripetal and takes place by alveoli.

The archegonia occur in groups (Fig. 12.31) of 20 to 25 (*Podocarpus nivalis*; *P. gracilior* and *P. falcatus*). In *P. andinus* the archegonia are long and pointed at their lower ends. In *P. gracilior* and *P. usumbarensis* the archegonia are oval with rounded base. In *P. gracilior* and *P. falcatus* the shape and position of the archegonia depends upon the early contact of the pollen tube with the gametophyte. The archegonia have a short neck of 16 cells arranged in a single tier in *Podocarpus andinus*, and of 4–6 cells arranged in a single tier in *P. gracilior*. In *P. dacrydioides* two-tiered neck is produced by the division of the cells in a single tier. Ventral canal nucleus is ephemeral in most cases.

The development of male gametophyte is almost similar to that of *Araucariaceae* except that in this case the number of male prothallus cells is small (1–8 in *Podocarpus*; 3–6 in *Dacrydium*, 2–4 in *Microcachrys*; 1–3 in *Phyllocladus*). In *Pherosphaera* there are no prothallial cells. Like the *Araucariaceae* the generative cell divides anticlinally and walls of all cells except the **spermatogenous cell** or the **body cell** break down releasing their nuclei into the pollen tube. The spermatogenous cell divides a little before fertilisation into two unequal or sometimes equal male gametes (*Saxegothaea, Phyllocladus* and *Microcachrys*). Fertilisation and pollination usually occur in the same season except in some species of *Podocarpus* and *Lacrydium*. On reaching the female gametophyte the pollen tube spreads like a disc. The body cell or the spermatogenous cell divides usually into three unequal male gametes. In *P. gracilior* there are two male gametes (Konar and Oberoi, 1969).

Pollination is anemophilous and in most species with winged pollen grains the mechanisms is similar to *Pinaceae*. In *Saxegothaea* the wingless pollen grains fall on the cone scale. The nucellus grows rapidly and bulges out of the micropyle. The pollen tubes grow and penetrate into it. In this genus there is no pollination drop mechanism. The pollen tube may branch extensively (*P. gracilior*; Konar and Oberoi, 1969).

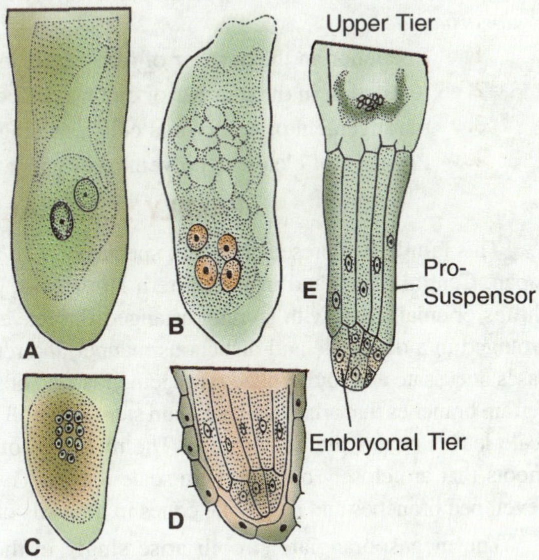

Fig. 12.32. (A–E) *Podocarpus gracilior.* Embryology. A-C, 2-16 nucleate pro-embryo; D. Proembryo with embryonal and suspensor tiers; E. Three-tiered embryo (After Konar and Oberoi)

In *Podocarpus gracilior* the pollen tube generally enters the archegonium laterally. The neck cells do not separate but undergo degeneration. The pollen tube exerts pressure on the archegonial wall and penetrates through it to liberate its contents into the egg cytoplasm. The functional male nucleus and its cytoplasm are the only structures that enter the egg but in rare cases some other contents may also be seen in the archegonium. The male and the female nuclei fuse together in the upper part of the archegonium. During their fusion the nucleoli disappear and the nuclear membrane disappears at the point of contact leading to complete fusion.

Embryo development has been studied in detail in the genus *Podocarpus* (Fig. 12.32). Quinn (1964, 1965, 1966) studied the embryogeny in *Dacrydium*. The embryogeny in other genera is unknown or inadequately known. In *Podocarpus gracilior* (Konar and Oberoi, 1969) the zygote undergoes five free nuclear divisions to form 32 free nuclei (the same is the case in *P. andinus, P. usambarensis, P. nagi*, and *P. amarus*). In some species of *Podocarpus* only 16 free nuclei are formed. In *P. gracilior* some of the free nuclei move into the upper part of the archegonium and are called the **relict nuclei**. They degenerate or persist as such. Wall formation results in the formation of two tiers. The lower tier consists of 9–12 embryonal cells whereas the upper open tier contains 15–22 cells. The embryonal cells are arranged in a single tier. The cells in the upper open tier divide transversely followed by a wall formation. This leads to the formation of an open upper tier and a middle suspensor tier. The embryonal cells undergo internal nuclear division not followed by wall formation so that each embryo cell is binucleate. This condition is also prevalent in *Saxegothaea*, all *Podocarpus* species, *Phyllocladus* and *Dacrydium*. In *Pherosphaera* (Elliot, 1950) binucleate cells were observed but later wall formation takes place to produce a bi-celled stage. The suspensor cells elongate and push the embryonal cells into the gametophyte. Later **cap-like** structures appear at the tips of embryonal cells. The binucleate embryoal cells undergo another nuclear division and wall formation takes place to distinguish tetrads of cells. This is called an **embryo tetrad stage**. There is simple polyembryon in *P. andinus, P. spicatus, P. ferruginens, P. falcatus, Saxegothaea* and *Phyllocladus alpinus*. Cleavage polyembryony is known in *Podocarpus gracilior* and some other species.

The haploid chromosome numbers are variable among the family : *Podocarpus* $n = 10, 11, 12, 13, 17, 18$, and 19; *Dacrydium* $n = 9, 10, 11, 12$ and 15; *Phyllocladus* $n = 9$; *Saxegothaea* $n = 12$.

Konar and Oberoi (1969) outline the following evolutionary trends in the embryology of *Podocarpaceae* :

1. Reduction in the number of free nuclei to 16 from 32.
2. Reduction in the number of embryonal cells.
3. Arrangement of embryonal cells in a single tier.
4. Presence of cleavage polyembryony in some species.

FAMILY : CEPHALOTAXACEAE

This family is represented by six species of *Cephalotaxus* that grow in the subtropical forests of Japan, Central China and in the Eastern Himalayas. The plants are strictly dioecious and are either shrubs or small trees with spirally arranged flat leaves. In *C. drupacea* Sieb. et Zucc. the leaves are arranged in a decussate and distichous manner and remain on the tree for at least three years. In rare cases decussate and four ranked arrangementis also observed (Singh 1961). Singh (1961) also reported certain branches that arise from the main stem near soil level to bear spirally arranged leaves. Numerous scale leaves covers the winter buds. The male tree of this species bear short and unbranched lateral shoots that are closely covered with scale leaves. They arise in the axils of foliage leaves of newly developed branches and bear male cones in the axils of upper scale leaves.

The megasporangiate strobili arise singly in the axils of lowermost two to four leaves on a young branch of the female tree. They are initiated in the middle of October in *C. drupacea*. Each megasporangiate strobilus has a short stalk and is small in size. They appear green when young but become reddish at maturity. It has a central axis that bears 5 to 7 pairs of opposite and decussate bracts. The central axis is fleshy in nature. Each bract bears two ovules in its axil. The lowermost pair of bracts is sterile. A small ridge-like outgrowth which represents a secondary axis is present between the two ovules of each bract. Florin (1938–45) regards the two ovules and the outgrowth between them as a "seed scale complex". The ovules are green when young and enlarge in the following spring and turn red in colour. Usually one or two ovules ripen in one cone.

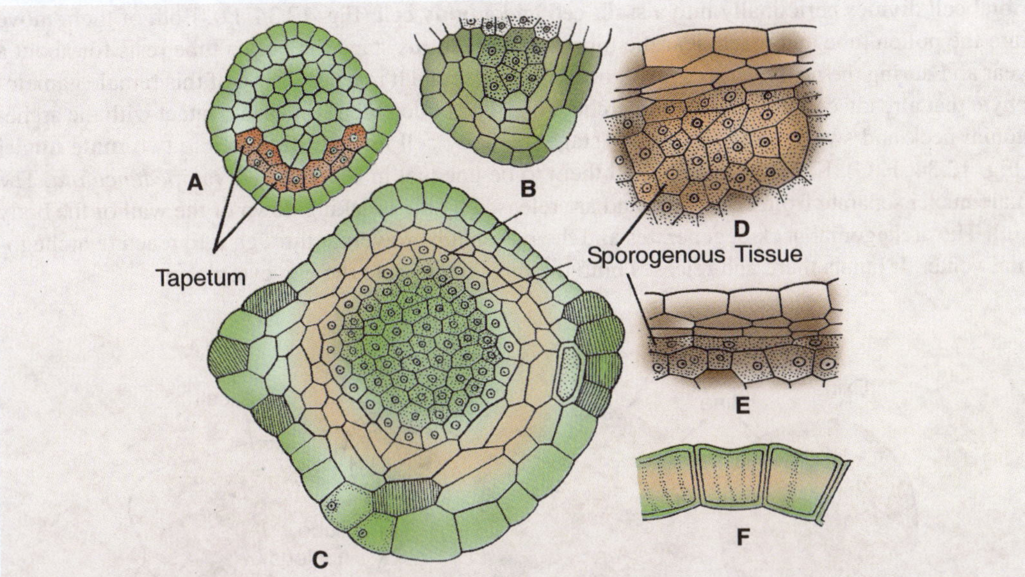

Fig. 12.33. (A-F). *Cephalotaxus drupacea.* Various stages in the development of Microsporangium (After Singh).

The male cones arise in compact clusters of 6–8 cones in the axils of upper scale leaves of the short lateral shoot of the male tree. The microsporangiate cones are sessile and have short central axis that bears 15–20 spirally arranged microsporophylls. Each microsporophyll bears three or four abaxial microsporangia. A sporophyll consists of a short stalk and a distal sterile flattened region which points outwards and upwards. The microsporangia are pointed towards the cone axis. The male cones start developing in August and bear microsporangia in early September.

The microsporangia arise from two to four (Fig. 12.33) groups of *archesporial cells* that are distinguished on the abaxial side of the microsporophyll near its base. They develop into 2 or 4 microsporangia. The archesporial cells divide periclinally into an outer **primary parietal layer** and an inner **primary sporogenous layer** (Fig. 12.33). The former divide anticlinally as well as periclinally to form a 4-layered thick wall. The primary sporogenous cells divide repeatedly to form a group of sporogenous cells. The inner wall layer differentiates into **tapetum** that is of glandular type and whose cells become binucleate at the mother cell stage. The sporogenous cells develop into microspore mother cells which undergo meiosis to form tetrads of microspores. During these changes the tapetum and inner wall layers are absorbed and only outermost wall layer persists. Its cells become highly cutinised and develop fibrous thickenings. Some cells of the outer wall layer of the epidermis facing the sporophyll stalk remain unthickened and help in dehiscence. Mehra and Khoshoo (1956), Sughiara (1940) and Singh (1961) reported 12 bivalents at the diakenesis stage of meiosis—I.

The microspores are small and spherical (Fig. 12.34, A) structures with a centrally located nucleus (Singh, 1961). It has a thick intine and a comparatively thin exine. The nucleus divides to form a small lenticualr **antheridial cell** and a large tube cell. There are no prothallical cells and the grain is wingless. The pollen grains are released at this stage.

The pollen grains are carried by wind and are caught in a pollination drop which dries and sucks them in and deposits them on the top of the nucellus. A short and broad pollen tube develops and starts penetrating the nucellus. The tube nucleus migrates into the pollen tube whereas the antheridial cell divides periclinally into a **stalk cell** and a **body cell** (Fig. 12.34, D). Both of them move into the pollen tube and reach near the tube nucleus. At this stage the pollen tube rests for about a year and during the next year its growth becomes faster and it reaches the tip of this female gametophyte that already contains the mature archegonia. The pollen tube comes in contact with the archegonial neck and swells at its tip. At this stage the body cell nucleus divides into two **male nuclei** (Fig. 12.34, F, G). Kaur (1958) reported them to be unequal in *C. drupacea* var. *pedunculata*. The male nuclei separate from each other and are released by the breaking down of the wall of the body cell. The archegonial neck degenerates and the pollen tube traverses through it to reach te archegonial venter. It bursts there and releases both the male nuclei into the egg cytoplasm.

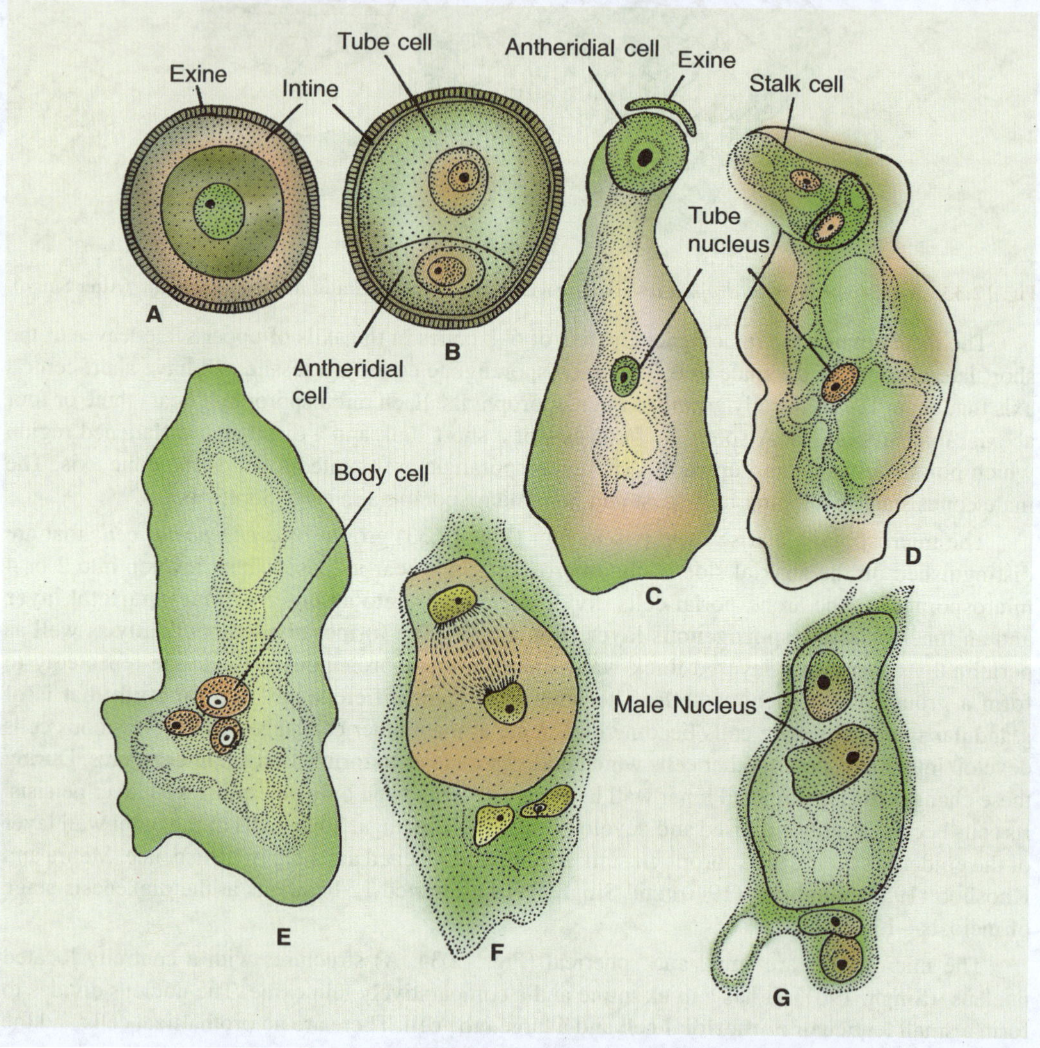

Fig. 12.34 (A–G). *Cephalotaxus drupacea.* Male gametophyte. A. Mature pollen grain; B. Two-celled stage; C. Pollen tube with tube nucleus and antheridial cell; D. The antheridial cell has divided into stalk cell and body cell; E. Pollen tube with body cell, stalk and tube nuclei; F. Body cell dividing; G. Two male nuclei have been formed (After Singh).

Singh (1961) has described the development of ovule and female gametophyte in *C. drupacea*. A bud arises in the axil of a bract scale and grows as a shoot apex. It grows into a short outgrowth or the secondary axis. The two ovules arise as lateral protuberances from this secondary axis. The youngest ovule observed showed a well developed integument bordering a tissue containing 3–6 hypodermal archesporial cells that later divide periclinally into outer **primary parietal** cells and inner **sporogenous cells**. The former divide and redivide to form the massive nucellus that pushes the sporogenous cells down towards the chalazal end. Singh (1961) reported the division of nucellar cells around sporogenous cells to form a distinct tissue called the **pavement tissue**. It gets crushed during free nuclear stage of female gametophyte. A little before pollination a rudimentary pollen chamber develops by the disorganisation of nucellar cells. After pollination the micropyle is partly closed by the radial elongation of a band of cells formed by the division of integument cells lining the pollen chamber (same as in *Podocarpus*). One of the sporogenous cells enlarges and functions as *megaspore mother cell*. It undergoes meiosis to form a linear tetrad of megaspores, out of which the lowermost is functional and enlarges and undergoes free nuclear division to form 4,096 nuclei (Favre-Duchartre, 1957). Wall formation is centripetal and leads to the formation of a well developed oblong gametophyte whose cells are filled with starch grains and oil globules.

The archegonial neck is usually two-celled but sometimes it may become 4-celled or may show two tiers of cells. There is no neck canal. The venter is elongated and has a pointed chalazal end. Its cytoplasm becomes vacuolated (foam stage) and contains many irregular and darkly staining bodies. The vacuoles disappear later and the cytoplasm contains protein bodies. The central nucleus now divides into a **ventral canal** nucleus and an egg nucleus. No wall is laid down between them. The archegonia are separated by 2 to 4 wall-layers of the gametophyte and do not form groups. The ventral canal nucleus disappears before fertilisation and egg nucleus moves to the archegonial base and grows big. The archegonial necks are sunk in pits that grow around them.

The male and female nuclei fuse together, their membranes dissolve at the point of contact and their chromatins approach each other and later merge.

The zygote nucleus divides into 16 free (Fig. 12.35, C) nuclei (the first division of the zygote nucleus is internuclear), that accroding to Favre-Duchartre (1957) arrange themselves into 4 tiers of cells (Fig. 12.35, E–G), namely, (*i*) rosette tier; (*ii*) suspensor tier; (*iii*) embroyonal tier; and (*iv*) cap cells. Singh (1961), however, observed that cap cells do not form any tier but are the lowermost cells of the embryonal tier. He did not observe even the rosette tier in *C. drupacea*, so that there are only two tiers of cells (*i*) the upper tier, and (*ii*) the embryonal tier. later, the upper tier divides into an open tier and the prosuspensor tier. The former degenerates after some time whereas the latter elongates pushing the embryonal tier out of the archegonial venter. There is a cavity in the centre of the female gematophyte. It develops by the degenration of cells and progresses towards the chalazal end. The embryonal cells are pushed into this cavity. Later the cap cells degenerate and their place is taken by enlarging embryonal cells. The embryonal cells may divide once to form secondary suspensor cells above them. They grow further and push the embryonal mass still further. The embryonal cells divide to form a massive embryonal tissue. The prosuspensor cells degenerate after the formation of secondary suspensor cells. There is no cleavage polyembryony. Singh (1961), however, reported three such cases in *C. drupacea*. Prosuspensor embryos have been observed by Singh (1961) in this species (Fig. 12.35, F).

The basic haploid chromosome number is $n = 12$.

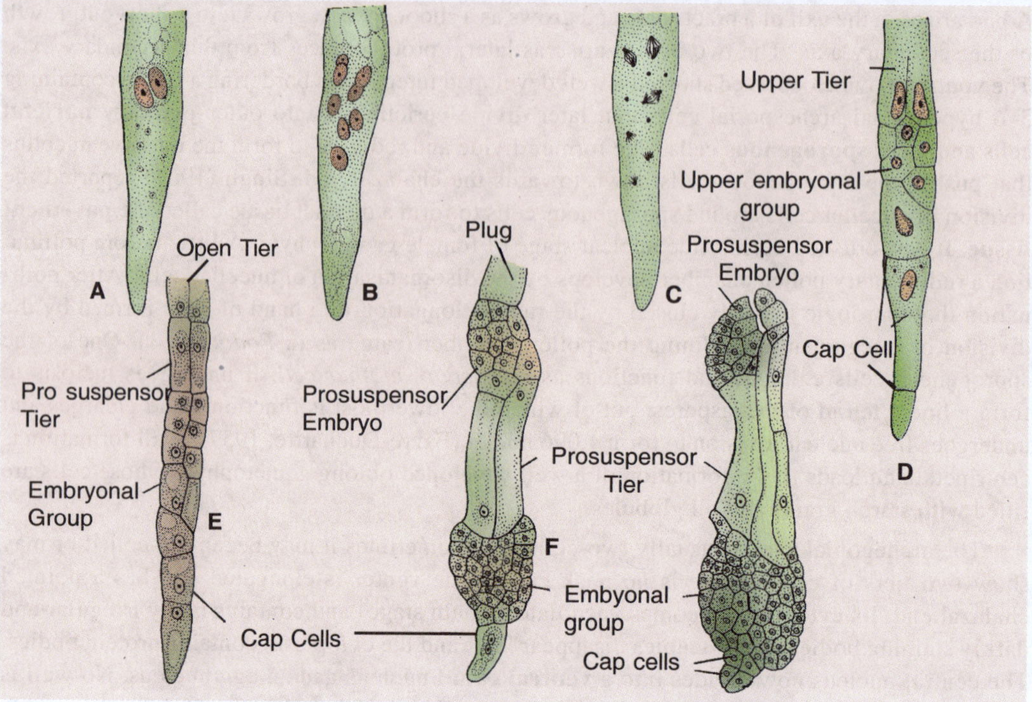

Fig. 12.35 A–G. Stages in the development of embryo in *Cepalotaxus drupacea* (After Singh). A Two nucleate stage; B. Eight nucleate stage; C. The nuclei undergoing further free nuclear division; D. Wall formation has taken place in the proembryo; E. Proembryo showing open tier, prosuspensor tier, and embryonal tier; F. It shows prosuspensor embryo, prosuspensor tier and embryo tier; G. Note the cap cells. (After Singh).

FAMILY : ARAUCARIACEAE

The family is represented now by two genera, *Araucaria* with 16 species and *Agathis* with 20 species (see table 12.1). The former is dioecious whereas the latter is monoecious. The leaves may be linear or broad and are spirally arranged. All species are evergreen and do not show shoot dimorphism. The megasporangiate strobili are almost in *Agathis* (Fig. 12.36, A). The bract scale and ovuliferous scale are intimately fused in *Agathis*. In *Araucaria* the tip of the ovuliferous scale is free (Fig. 12.36, B) and forms the ligule. The ovules have only one archegonium. The integument of the ovule is free from the nucellus right up to its base. The ovules are reflexed and are usually orthotropus and crassinucellate. The male cones may be more than 30 cm in length (*Araucaria rulei*) and consist of a central axis with numerous microsporophylls arranged in a spiral manner. Each microsporophyll bears (5–15 in *Agathis* and 8–15 in *Araucaria*) microsporangia on their abaxial surface. Each microsporophyll has a proximal stalk like portion and a distal sterile broad portion that is directed outwards as well as upwards. The sporangia

Plate V : *Araucaria*

are elongated. Chamberlain reported the pollen output of each cone to be 10,000,000. The pollen grains are wingless. There is no pollination drop mechanism and the pollen grains fall on the ovuliferous scale. They send out pollen tubes that grow towards the micropyle and meet the nucellus whose beak projects through the micropyle. The pollen tube branches extensively on the ovuliferous scale. In *Araucaria angustifolia* pollination takes place much before the ovules start developing whereas in other species the ovule is fully developed and may be with gametophytes in free nuclear stages.

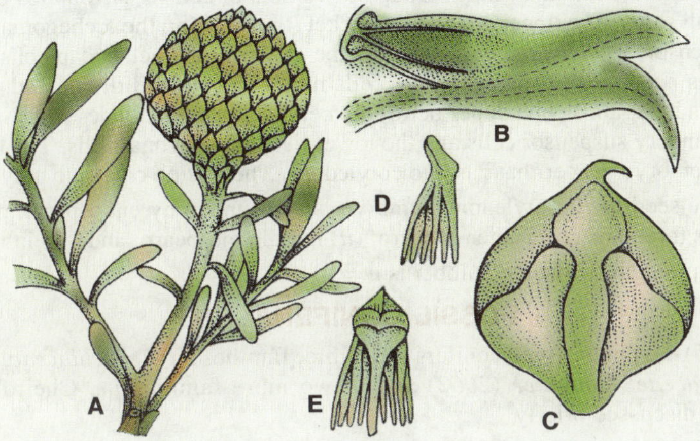

Fig. 12.36. (A-E) Araucariaceae (A–E). A. Shoot of *Agathis australis* bearing a female cone; B. Cone scale of *Araucaria bidwillii* as seen in L.S; C. L.S. scale of B; D–E. microsporophylls of *Araucaria angustifolia*. (A. After Poole and Adams; B. After Hirmer; C. After Pilger and D–E. after Eichler).

The *Araucariaceae* are unique in producing (Fig. 12.37, A, E) many prothallial cells. At first the microspore nucleus divides periclinally into two **prothallial cells** which divide further to produce many prothallial cells. The **antheridial** cell divides into a tube cell and **generative cell**. The latter divides anticlinally into a **stalk cell** and a **body cell**. The cell walls of the prothallial cells, tube cell and stalk cell break down and release their nuclei into the pollen tube. The body cell divides to produce two equal (Ghose, 1924) male gametes. There is only one archegonium and one pollen tube fertilizes only one archegonium. Eames (1913) stated that the male gametes are unequal.

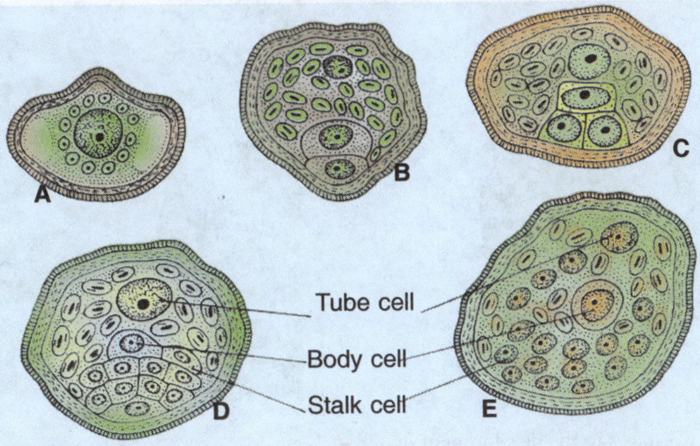

Fig. 12.37 (A–E). *Araucaria conninghamii*. Development of male gametophyte. A. pollen grain; B. Two prothallial cells have been cut off; C. The first prothallial cell has divided anticlinally; D. Prothallial cells have divided further, tube cell, body cell and stalk cell have also been formed; E. The body cell has increased in size, the nuclei of other cells have become free. (After Chamberlain).

Out of the three of four megaspore mother cells that differentiate from the archesporial tissue, only one is functional and undergoes meiosis to form a linear row of 3 megaspores. The chalazal one is functional (Rastogi, 1964) and undergoes free nuclear division followed by wall formation. In *Agathis brownii* Rastogi (1964) observed 4 to 6 archegonia in the female gametophyte. The archegonia have a 2 or more celled neck, a ventral canal nucleus and an egg cell.

The zygote nucleus undergoes free nuclear divisions to form thirty-two (Araucaria) or sixty-four (Agathis) free nuclei that are grouped together in the centre of the egg cell. Walls are laid down in a manner (Fig. 12.37, A) that a central group of two tiers of cells is surrounded by an outer layer or the jacket on all sides. The upper cells of the jacket (those facing the archegonial neck) elongate to form the suspensor, whereas the cells towards the chalazal end act as cap cells (Fig. 12.37, B) whose function is not known. The suspensor cells may elongate and divide and grow out of the archegonial necks in Agathis. The upper tier of the central two tiers divides periclinally to form an upper tier of secondary suspensor cells and the lower tier of embryonal cells. The latter divide and develop into the embryo proper that has two cotyledons. There is no cleavage polyembryony.

In the mature seed the 2 cotyledoned embryo is surrounded by endosperm, nucellus and the seed coat that has three layers (*i*) outer exocarp; (*ii*) middle mesocarp; and (*iii*) inner endocarp.

The basic haploid chromosome number is $n = 13$.

FOSSIL CONIFERALES

Sporne (1965) has put the fossil conifers under three families : (*i*) *Lebachiaceae*, (*ii*) *Voltziaceae*, and (*iii*) *Palissyaceae*. *Gangulee* (*2002*) added two more families *i.e.* Cheirolepidoceae and Protopinaceae of discussed briefly.

FAMILY : LEBACHIACEAE

These are the earliest conifers whose remains were found in the early Permian and the Triassic age and are regarded as the "transition conifers" by several authors (Florin). They illustrate an almost complete and a gradate series of forms intermediate between the cordaitales and the coniferales only so far as the structure of megasporangiate strobili is concerned. The lower Permian forms also show microsporangiate fructifications that are quite comparable to the modern conifers. Lemoigne (1967)and Lemoigne and Schmidt (1968) also suggest a similarity between the wood of these fossil forms with the modern *Araucariaceae*, and their leaves, which were small scale-like and bifid at their tips.

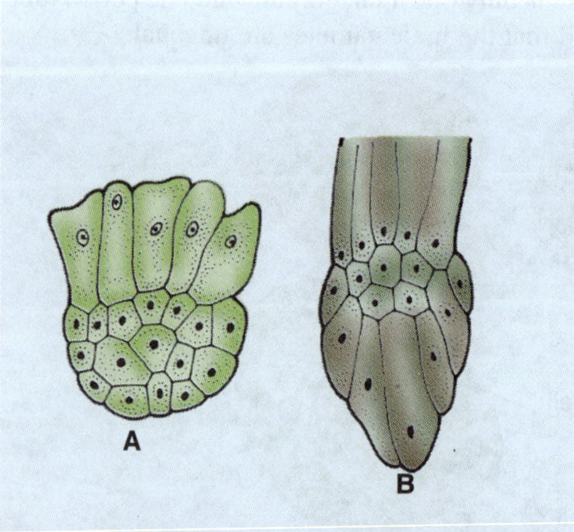

Fig. 12.38 (A–B). Two stages in the development of embryo in *Araucaria augustifolia*. In B, the proembryo has developed cap cells. (After Johansen)

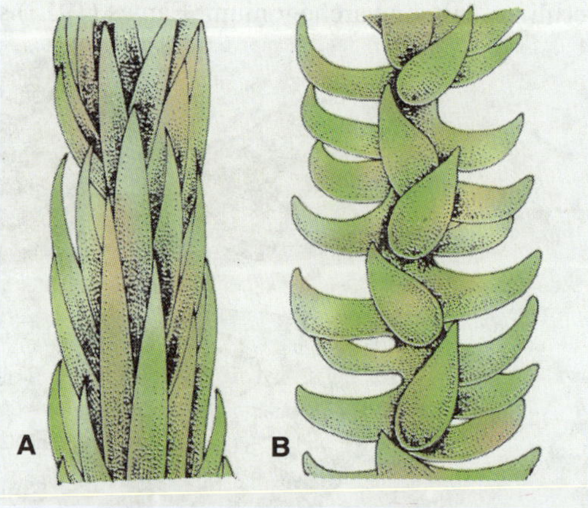

Fig. 12.39 Foliage of *Lebachia piniformii* (A) and *Ernestiodendron filiciforms* (B). (After Florin).

The leaf bearing twigs were encountered in the Permian have been described under the genus *Walchia*. Florin (1949–1951) has made extensive morphological and anatomical studies on these species (more than 25) and has transferred 14 species of *Walchia* to two other genera whose reproductive structures are also known. Thirteen species are included in the genus *Lebachia* and one in *Ernestiodendron*. He is of the opinion that the remaining eleven species will also be attributable to *Lebachia*.

The leaves of *Lebachia* (Fig. 12.39, A) are more closely applied to the stem whereas those of *Ernestiodendron* (Fig. 12.39, B) extend out almost at right angles to the main axis. In their general appearance these plants resemble the modern *Araucaria* excelsa. In *Carpentieria frondosa* the leaves are long and deeply forked. In *Buriadia* (from India and Brazil) the leaves were deltoid and had dichotomous venation.

Fig. 12.40. (A-H) Fossil conifers. Seed bearing shoots. A. seed bearing shoot of *Lebachia goeppertiana*; B. Seed bearing shoot *Walchia germanica* (*Ernestiodendron*); C. Seed bearing shoot of *Ernestiodendron filiciforme*; D. Seed bearing shoot of *Walchiostrobus sp*; E. Seed bearing complex of *Pseudovoltzia liebeana*; F. *Walchiostrobus* sp. bearing seeds; G–H. Upper and under side of seeds of *Ullmannia bronnii*. (After Florin).

The compound megasporangiate strobilus of *Cordaitanthus* can be compared to that of *Lebachia* if it is longitudinally condensed. In *Lebachia piniformis* (Schlotheim) Florin, the megasporangiate strobilus is 6 to 7 cm long and consists of a central axis with several spirally arranged and closely imbricated bracts with bifid tips. Each bract bears a dwarf shoot in its axil. This shoot consists of a short axis with spirally arranged scales and a short stalk bearing a terminal ovule. So in *Lebachia* the megasporangiate strobilus is compact and bears spirally arranged appendages as compared to the loose *Cordaitanthus* strobilus with appendages arranged in two rows. The integument of the ovule was two parted with its tissue continuous with that of the stalk below. In *L. goeppertiana* (Fig. 12.40, A) Florin, a seed cone discovered from the Lower Permiana the strobilus has central axis with numerous spirally arranged and bifid bracts each bearing a dwarf shoot in its axil. These fertile dwarf shoots bear numerous spirally arranged sterile appendages. The one fertile appendage bears an orthotropous ovule on the inner side and not terminally as in *L. piniformis*. This example illustrates a shift of the ovule from its terminal position downwards and with a distinct bilateral symmetry. The other seed bearing genera of the family show variations in the number of sterile and fertile scales on the dwarf shoot and number and position of te ovules.

In another Permian fossil named *Ernestiodendron filiciforme* (Schlothein) Florin, the compact megasporangiate cones were about 2 cms long and were borne terminally on penultimate leaf shoots. The cones had a central axis bearing spirally arranged and forked bracts. Each bract bore an axillary fertile dwarf shoot that in turn bore several fertile appendages each with a terminal ovule. The ovules were erect (Fig. 12.40, C.).

In *Walchia germanica* Florin (Fig. 12.40, B) the ovules were inverted otherwise the cone was exactly similar to *E. filiciforme*.

In *Walchiostrobus* (Fig. 12.40, D, F) the bracts bear expanded and fan-shaped fertile shoots that bore 20 or more sterile scales and three to seven fertile or ovule bearing stalks that bore inverted or erect ovules. In this case there is flattening of the fertile dwarf shoot.

The integuments of the ovules were free from the nucellus. Most of these trees were monoecious.

The microsporangiate strobili were simple and compact structures that had a central axis which bore spirally arranged microsporophylls that had unforked sterile apices and bore two microsporangia. The pollen grains had two wing-like air sacs that partially surrounded the body of the microspore. The male cone can be compared only to the fertile dwarf shoot of the female cone and not to the whole female cone. These cannot be linked with the compound male strobili of the cordaitales and much more material suggesting a link is needed to be discovered before finally accepting Florin's view of the phylogeny of the conifers.

The *Lebachiaceae* became almost extinct by the Upper Permian and were replaced by the *Voltziaceae*.

FAMILY : VOLTZIACEAE

Their megasporangiate strobili show the same organisation as those of the *Lebachiaceae* but their fertile dwarf shoots, that arise in the axils of bracts, show a greater amount of flattening and consequent dorsiventrality.

In *Pseudovoltzia liebiana* (Fig. 12.40, E) the entire bracts bear fertile dwarf shoots in their axils. Each dwarf shoot has a rudimentary axis that bore five sterile scales and three fertile scales. The sterile scales looked flattened and three out of them were large as compared to other two. The fertile ones bore each a reflexed terminal ovule. The bases of sterile scales were decurrent.

In *Ullmannia bronii* Goeppert (Fig. 12.40) that was discovered from Upper Permian the fertile dwarf shoot had a rudimentary axis that bore a single flat and dorsiventral and spathulate sterile scale and a single fertile scale with a terminal inverted ovule. It is an extreme case of reduction. In *Voltziopsis africana* there were five sterile scales and five fertile scales bearing erect terminal ovules.

it was discovered from the Triassic of Mesozoic. In *Voltzia* that was discovered from Triassic rocks the fertile scales were fused with the upper surfaces of sterile ones. In *Schizolepis* the three sterile scales were partially fused and the three fertile ovule bearing appendages were fused to their upper surfaces. In the megasporangiate strobilus of *Hirmeriella rhatoliassica* horhammer, there was a single sterile scale on the rudimentary fertile dwarf shoot. Two inverted ovules were borne on its upper surface. In *Glyptolepis* the fertile axillary complex was composed of two ovule bearing stalks and a fan-shaped unit of the fused scales. A little further loss of integrity of parts in a type such as this would produce a cone scale of the pinaceous type in which the ovules are lateral, adaxial and inverted.

The axillary ovuliferous short shoot in the transition conifers has undergone the following specialisation trends during their course of evolution from the early Permian to the Triassic :

1. Reduction and loss of the axis of the dwarf fertile shoot.
2. A gradual flattening in the plane tangential to the cone axis.
3. The sterile bracts show gradual fusion.
4. The fusion of the fertile ovule bearing appendages with the sterile parts.
5. Reduction in the number of sterile parts to one.
6. Reduction in the number of ovules to two.
7. Loss of ovule stalks so that they are borne directly on the sterile scale.

FAMILY : PALISSYACEAE

It is represented by only two genera : *Palissya* and *Stachyotaxus*. Both were present in the Mesozoic.

In *Stachyotaxus* that was discovered from the late Triassic of Europe and Greenland the megasporangiate strobilus consisted of a central axis that bore spirally arranged entire bracts to which the axillary dwarf or fertile shoot was completely fused and bore two arillate ovules. The ovules were erect. In *Palissya sphenolepis* that was discovered from the Jurassic there were several (about 10) ovules one each fertile shoot. In both the cases there were no sterile scales on the fused fertile shoot. The ovules were arranged in pairs. *Stachyotaxus* elegans has been compared with the arrangement found in the modern *Cephalotaxaceae* that also bear paired ovules on the fertile shoot that has no sterile part. The only difference is that in *Stachyotaxus* the fertile shoot is completely fused with the subtending bract. *Ernestiodendron* of *Lebachiaceae* also lacks sterile appendages on the fertile shoots and is therefore more akin to the modern *Cephalotaxaceae* than other members.

The *Lebachiaceae* and *Voltziaceae* clearly indicate the possible steps of evolution that lead to the development of modern conifer megasporangiate cones especially those of Taxodiaceae (it retains the sterile components of the cone scale complex), Pinaceae, Podocarpaceae and Cuprenaceae. Cephalotaxaceae appear to have affinities with Palissyaceae. Arnold (1961) states, "In summary, the fossil evidence seems to indicate quite clearly that the seed cones of the modern conifers are a much reduced inflorescence in which the ovuliferous scale and seeds borne on its surface are the remnants of a radially symmetrical dwarf shoot. As this evolutionary trend progressed, numerous side line branched out as is evident from the fact that all the examples cited do not fit a perfect sequence."

The possible steps of specialisation heave been listed above.

The families Voltziaceae and Lebachiaceae have been placed under a separate order *Voltziales* because these represent a transition from cordaitales to coniferales coexisting with the cordaitales in late palaeozoic and persisting into the mesozoic 1, these are often called *Transition conifers*. The voltziales include genera that exhibit the reduction series and change in the symmetry of a Cordain thus like fertile dwarf shoot and lead to the ovuliferous scale of living conifers. Florin

(1945, 1951) who made an exptensive study of voltziales, compared the fertile dwarf shoot of Cordainthus with the ovuliferous scale of modern conifers and cited several intermediate stages within the Voltziales.

ECONOMIC IMPORTANCE

The conifers are of immense economic importance. This aspect of this great gymnospermous group will be discussed under the following headings :

Source of Wood

Majority of the conifers possess a light coloured, straight grained light weight wood and is very useful where strength and durability are not essential. They take a fine finish with sharp tools, have good nail taking properties, and take polish and paints satisfactorily. This wood is, therefore, used extensively for cabinet and furniture making, interior decoration, and joinery. The scented wood of some species is extremely valuable. The heart wood of *Cedrus deodara* is strongly scented and resistant to insect and fungus attacks and was extensively used for making railway sleepers and carriages. In North India it is commonly used for making window panes, doors, furniture, electric poles, for flooring in the houses, for building models, boats and storage vats and also for battery separators.

The wood of *Pseudotsuga taxifolia* yields veneer and plywood and gives a clean timber of large dimensions that is used for constructing laminated arches, tanks for breweries and heavy construction work. The wood *Abies pindrow* and *A. spectabilis* is used along with *Picea smithiana* for box making, packing cases and for light camp furniture and slanting roofs in hill houses.

The wood of various species of *Larix* and *Agathis* is hard, heavy and is used for railway sleepers, posts, stakes, boat making and transmission poles.

The wood of *Pinus palustris* is used for ship making, spars, masts, heavy construction work, lorry and railway wagon flooring. The wood of other species such as *P. mugo, P. lambertiana, P. sylvestri, P. monticola, P. resinosa, P. nigra, P. pinaster P. halepensis, P. strobus, P. roxburghii, P. wallichiana*, etc., is used for making cheap furniture, packing cases, agricultural implements, fencing, poles crates, doors, frames, general mill work, etc.

The dark coloured, greasy, dense and durable timber of *Taxodium distichum* is used at places where resistance to decay is required. It is extensively used for making wooden poles, beams, doors, sleepers, blinds, general mill work and especially for chemical tanks and vats because it imparts no colour or taste to the products. It is also praised as fine wood for building cooling towers, green houses and roof planks of dye houses.

The wood of *Callitris verrucosa* of Queensland is olive brown in colour and is used to make decorative furniture. The wood of *Chamaecyparis piscifera* and *C. obtusa* is used in Japan for all purposes where durability and strength are required. The wood is also used for ship building, canoe building, organ pipe making and for match stick making. The wood of *Thuja plicata* is known for possessing certain antibiotics and is resistant to weathers. It is widely employed commercially. The wood of *Juniperus virginiana* is used for making high quality pencils. The strong aroma of the wood makes it excellent for blanket and linen chests, cigar boxes, etc.

Agathis australis (200 ft high and 30 ft diam.) is the largest timber producing tree of the world and especially of Australia. Its hard, strong, fine and even textured wood has lustrous and silky surface and is extensively used for all purposes. The wood of various species of *Podocarpus* are used to obtain plywood. The wood of *Phyllocladus* is also used for mast making, for making sleepers, mine timbers and small vessels. The wood of various species of *Dacrydium* is used for general constructional work. The wood of *Taxus baccata* has a decorative value and has very good mechanical properties.

Source of Resins

The coniferous are the main source of various types of resins that are plant exudates secreted in specialised ducts. They are insoluble in water but readily dissolve in some organic solvents. They find their extensive use in the manufacture of varnishes, paints, lacquers, medicines and in paper sizing. The resins can be obtained from the fossilised material as well as from the living trees by tapping. The various types of resins and their sources are given below.

Rosin

It is also called colophony and is obtained as a residue after the distillation of pine oleoresin or turpentine. *Pinus roxburghii, P. wallichiana, P. insularis* and *P. merkussi* are the chief sources of turpentine in India. The oleo-resin is obtained by tapping the tree and it is distilled to get rosin and teurpentine. The rosin is used in paper sizing, varnish making, enamels, and in the preparation of plasters and ointments. The inferior grades of rosin are used in the manufacture of yellow laundry soap, oil and grease, sealing wax, oil cloth, adhesives, plastics, printing inks, paint driers, insulators, insecticides, disinfectants, shoe polish, furniture and in large number of other commercial articles.

Copal

It is obtained from living or extinct forms of *Agathis australis*. It is tapped from living forms as 'green gum' or 'Candle gum'. It is also obtained by digging suitable sites which yield huge masses weighing up to 100 lbs. These masses were secreted by trees that became fossilised and preserved underground. This type of resin is used for varnish making. It yields a superb type of varnish. *Agathis alba* also yields copal resin. It is called East Indian or Manila Copal. Fossil resin of this type occurs in Indonesia. It is also obtained by tapping living trees. This resin is used in spirit varnishes and in making linoleum, preparation of plastics, polishes and all other articles mentioned for rosin.

Sandarc

It is obtained from various Australian species of *Callitris* and from African *Tetraclinis articulata*. The latter is tapped from living trees and is obtained as pale yellow or orange exudate and contains about 1% volatile oil. The Australian types exude naturally from various species of *Callitris*. It is used as metal varnish giving goods lustre when applied in thin coats; as paper and leather varnish; and as varnish for indoor labels. In present-day pharmaceuticals it is used as pill varnish. In alcoholic solutions on cotton wool it has also been used as temporary filling for teeth. As a medicine it is used by the Arabs for curing diarrhoea.

Canada Balsam

It is obtained for *Abies balsamea*. It is produced by schizogenous cavities located in the bark of this North American species. This resin does not crystallise or granulate on drying and has a high refractive index as that of glass. It is used as a mounting medium for microscopic objects and as a cement for lenses in optical work.

Venice Turpentine

It is obtained from *Larix decidna* and is used for making special types of varnishes, and in histology and veterinary medicines. *Jura turpentine* is obtained from *Picea abies* and is used for making specialised paints and varnishes.

Amber

It is a fossil resin obtained from the fossil member called *Pinites succinifera*. It is obtained from the blue earth near the eastern shore of the Baltic, Madagascar, Sicily and Burma. It can only be obtained by mining and is extremely brittle and hard and has an aromatic odour. It varies in colour from brown to black. The ancients used it as jewellery and considered it to possess certain magical properties. Now-a-days it is used in X-ray therapy. Modern hospitals and laboratories keep blood in amber containers because amber does not permit the blood to coagulate. It is also used to manufacture cigarette holders, mouth pieces of pipes, beads and small ornaments.

Essential Oils

Oil of Hemlock is oblained form *Tsuga canadensis* and *Picea glauca* and is used as deodorant and as a constituent of room sprays. The toilets, shaving soaps, bath preparations, room sprays and deodorants are also scented by Siberian fir needle oil obtained from *Abies sibirica*. Soaps are also scented by Japanese pine needle oil obtained from *Abies sachalinensis*. Oil obtaind from *Cedrus atlantica* and *C. deodara* is also used in perfumery and in medicines to cure bronchitis, tuberculosis, skin diseases and gonorrhoea.

The species of *Pinus* listed before under rosin yield turpentine oil that is put to many uses, *e.g.*, as a solvent, paints and varnish thinner, a chemical raw material and as a pharmaceutical. It is also used as an important constituent of stains, enamels, inks, lubricants, stain removers, wood fillers, liquid and paste polishes, waxes, crayons and insecticides. Pine oil is also extracted from the old pine stumps during the steam and solvent process extraction of wood rosin. This oil is used as a frothing agent in the separation of metals by flotation. It is also used in dyeing cotton, wool, silk and rayon textiles. The waste wood and sawdust of *Dacyridium franklinii* is used as a commercial source of Huon-pine wood oil which has high germicidal properties. The wood oil of *Cryptomeria japonica* is used as a flavouring agent in Japan. In Algeria and Southern France oil is extracted from *Cupressus sempervirens var. stricta*. It is used as a perfumery material. The oil obtained from *Juniperus virginiana* is called Red cedar wood oil and is used for clearing microscopic sections and for oil immersion lenses.

Fatty Oil Production

Some species of *Torreya* (*T. nucifera*), *Cephalotaxus* (*C. drupacea*) and *Pinus* (*P. cembra*) yield fatty oil that is obtained from the seeds. The oil is used for food and also for paint.

Paper and Board

Paper is made from the pulp of wood of some Indian conifers, *e.g.*, *Picea smithiana, Abies pindrow, Pinus roxburghii* and *Cryptomeria japonica*. Various species of Pinus are a source of newsprint almost all over the world. Higher grades of paper are made from the wood pulp of *Abies balsamaea Picea* and *Tsuga*.

Tannin

The barks of *Sequoia sempervirens, Phyllocladus, Araucaria, Picea* and *Dacyridium cupressinum* contains good quantity of tannin but it has not yet been commercially exploited.

Food

The seeds of many conifers are edible (*P. gerardiana, P. monophylla, P. cembroides, P. edulis, P. roburghii*, etc).

Decoration

Many plants such as : *Thuja, Cupressus sempervirens, C. funiberus, Araucaria, Agathis alba* and *Cryptomeria japonica* are grown as garden ornamentals. The wood of many species is used in the manufacture of decorative furniture, toys and other decorative articles.

Plate VI : Genera of Family Pinaceas

I II III

Abies albs

I II III

Abies pinsapa

II III

Picea abies

I II III

Pseudotuga menziesii

Plate VII : Genera of Family Pinaceae

I II III

Larix decidua

I II

Cedrus deodara

I II III

Pinus halepensis

I II III

Pinus pinea

Plate VIII : Vareous Species of Pinus

I

II

III

Pinus pinaster

I

II

III

Pinus nigra

I

II

III

Pinus sylvestris

I

II

Pinus radieta

Plate IX : Genera of Podocarpaceae

I II

Cupressus sempervireus

I II III

Cupressus macrocarpa

I II III

Chamaecyparis Cawsoniana

Plate X : Genera of Cupressaceae and Taxodiaceae

I II III

Thuja Sp

I II III

Juniperus Communis

I II III

Juniperus Communis

Taxodiaceae

I II III

Sequoia sempervirens

Plate XI : Genera of Taxodiaceae

I

II

III

Sequoidendron giganteum

I

II

III

Sequoia sempervirens

I

II

III

Sequoia sempervirens

ARAUCARIACEAE

1 2 3

Araucaria araucana, monkey puzzle tree. These images appear to represent female and male individuals of the dioecious species commonly called monkey puzzle tree. 1 - Female, note large round female cones at the tips of branches, 2,3 - male, note elongated male cones at the tips of branches.

Araucaria columnaris, Cook pine.

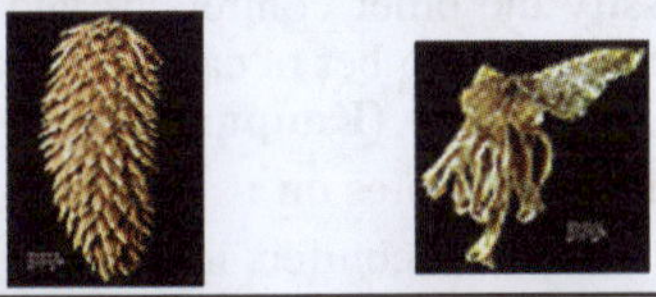

Araucaria columnaris, Cook pine. Male (microsporangiate) strobilus (left) and microsporophyll (right) with several elongated, pendulous microsporangia.

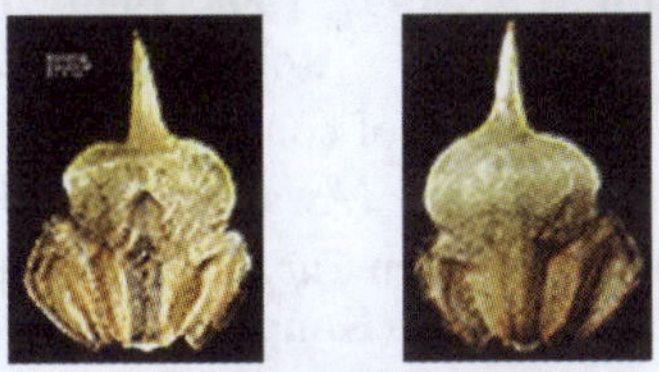

Araucaria columnaris, Cook pine. Imature female (megasporangiate) cone scale/bract units. Left: adaxial (upper) view with small (darker) medial scale visible (mostly fused to the much larger subtending bract). Right: abaxial (lower) view of bract.

Araucaria columnaris, Cook pine. Note the large woody cone and the ovuliferous scale with a single medial seed that is embedded in the scale. Also shown is a seed that has been removed from its scale. In this family the ovuliferous cones disintegrate while still attached to the tree; each cone scale with its seed and fused bract falls from the axis of the cone as a dispersal unit.

Agathis robusta, kauri. 1 - Note the size of the tree, the massive trunk, and the upwardly spreading nature of the crown. 2 - The leaves on this species are quite broad and not typical of conifers in general. The spherical megasporangiate cone contains many 1-seeded ovuliferous scales that fall apart at maturity. 3 - Microsporangiate cone (immature). 4 - Megasporangiate cone (immature).

Plate XII : Genera of Araucariaceae

QUESTIONS

ESSAY TYPE

1. Write detailed account of Economic importance of conifers.
2. Give a brief illustrated account of embryogemy in conifers.
(Kanpur M.Sc, 2001)
3. Give an account of distribution of coniferales in India.
(Kanpur M.Sc, 1999; Rohilkhand M.Sc, 2003; Poorvanchal M.Sc, 1998, 2002)
4. Describe characteristic features of gametophytes of coniferales.
(Kanpur M.Sc, 1997)
5. **Write explanatory notes on :**
(*i*) Seed scale complex in conifers. **(Kanpur M.Sc, 1997)**
(*ii*) Voltziaceae. **(Kanpur M.Sc, 1997)**
(*iii*) Lebachiaceae
(*iv*) Fossil conifers.
6. Describe the salient features of coniferales. Compare the male or female cones of *Pinus, Abies* and *Cedrus*.
(Kanpur M.Sc, 1996, 1994)
7. Classify the conifers giving salient features and examples of different groups. How do these resemble with pteridophytes and Angiosperm. **(Kanpur M.Sc, 1993)**
8. Briefly describe the distribution of conifers in India and mention economic importance of any five of them.
(Kanpur M.Sc, 1991)
9. Write an illustrated essay on the utility of xylotomy in identification of Indian conifers. Give examples.
(Allahabad M.Sc, 2001)
10. Give a comparative account of the male gametophytes of conifers what light does this study alone throw on the phylogenetic position of its different families.
(Gauhati M.Sc, 2002)
11. Give a brief account of the fossil conifers.
12. Describe the distinguishing features of the following families.
(*a*) Taxodiaceae
(*b*) Cupressaceae
(*c*) Podocarpaceae
13. Give a detailed account of *Cephalotaxus*.
14. Describe the genra *Araucaria* and compare it with *Cryptomeria*.

SHORT ANSWER TYPE

15. Briefly discuss the affinities of coniferales.
(Kanpur M.Sc, 1996)
16. Briefly discuss the seed scale complex of conifers.
(Kanpur M.Sc, 2004; Rohilkhand M.Sc, 2000)
17. How are conifers distributed in India ?
(Kanpur M.Sc, 2004)
18. Classify the order coniferales and name at least one member of each family.
(Kanpur M.Sc I, 1999)
19. **Write short notes on :**
(*a*) Distribution of conifers in India.
(Awadh M.Sc, 1997, 1998, 1999, 2001; Rohilkhand, 1991)
(*b*) Ovuliferous scale of conifers.
(Awadh M.Sc, 2000)
(*c*) Distinction between *Cupressus* and *Thuja*.
(Kanpur M.Sc, 1998)
(*d*) *Cryptomeria*. **(Kanpur M.Sc, 1998)**
(*e*) *Podocarpus*. **(Kanpur M.Sc, 1998; Rohilkhand M.Sc, 2000)**
(*f*) Economic importance of conifers.
(Kanpur M.Sc, 1997, 1995, 1993; Allahabad, 2001)
(*g*) Distinction between *Taxus* and *Cephalotaxus*. **(Kanpur M.Sc, 1997)**
(*h*) *Araucaria*. **(Kanpur M.Sc, 1997)**
(*i*) Differentiate between Pollen grains of *Pinus* and *Podocarpus*.
(Kanpur M.Sc, 1995)
(*j*) Female cone of *Podocarpus*.
(Rohilkhand M.Sc, 1991)
(*k*) Female prothallus of Taxodiaceae and Podocarpaceae.
(Rohilkhand M.Sc, 1993)
(*l*) Salient features of Podocarpaceae.
(Rohilkhand M.Sc, 1996)
(*m*) Differentiate between *Araucaria* and *Podocarpus* woods.
(Rohilkhand M.Sc, 1999)

(*n*) General features of taxodiaceae. **(Rohilkhand M.Sc, 2002)**

(*o*) Male gametophyte of *Araucaria*. **(Rohilkhand M.Sc, 2002)**

(*p*) Compare female cones of *cryptomaria* and *cupressus*. **(Allahabad, 2001)**

(*q*) Embryo development in *Taxodium*. **(Allahabad M.Sc, 2001)**

(*r*) *Podocarpus*.

20. Describe pollination and fertilization in *Araucaria*.

21. Draw labelled diagrams of the following :

(*a*) L.S. Ovule of *Cryptomeria*.

(*b*) Pollen grain of *Cephalotaxus*.

(*c*) Development of male gametophyte in *Araucaria*

(*d*) Shoot of *Agathis*.

(*e*) L.S. mature ovule of *Biota*.

(*f*) Leaves of *Pinus, Cedrus* and *Picead*.

(*g*) T.S. leaves of *Sciadopitys verticillata*.

(*h*) T.S. leaf of *Cryptomeria*.

22. Assign the following to their systematic position and point out features of special interest

(*i*) *Voltzia*. **(Rohilkhand M.Sc, 1994)**

(*ii*) *Cephalotaxus*.**(Rohilkhand M.Sc, 1994)**

(*iii*) *Cryptomeria*. **(Rohilkhand M.Sc, 1994)**

(*iv*) *Araucaria*.

(*v*) *Podocarpus*.

(*vi*) *Thuja*.

(*vii*) *Biota*.

23. Write a brief explanatory note on megagametophytes of coniferales.

24. Write the importance of fossil conifers.

OBJECTIVE QUESTIONS

25. Select the correct answer :

(*i*) The tallest conifers is

(*a*) *Pinus*
(*b*) *Sequoia gigantea*
(*c*) *Cedrus deodara*
(*d*) *Thuja*

(*ii*) In addition to *Pinus* the other conifer that has winged pollen grains is

(*a*) *Thuja* (*b*) *Cedrus*
(*c*) *Abies* (*d*) *Araucaria*

(*iii*) Which of the following is a fossil conifer?

(*a*) *Pinus* (*b*) *Cedrus*
(*c*) *Thuja* (*d*) *Voltzia*

(*iv*) *Cryptomeria is a* member of the family

(*a*) Pinaceae
(*b*) Cupressaceae
(*c*) Araucariaceae
(*d*) Podocarpaceae

(*v*) Two fused leaves are present in

(*a*) *Pinus roxburghii*
(*b*) *Cryptomeria Japonica*
(*c*) *Sciadopitys verticillata*
(*d*) *Cephalotaxus*

(*vi*) Which of the following is the largest family in coniferales ?

(*a*) Taxodiaceae
(*b*) Cupressaceae
(*c*) Araucariaceae
(*d*) Podocarpaceae

(*vii*) Red wood tree is

(*a*) *Cedrus* (*b*) *Pinus*
(*c*) *Sequodia* (*d*) *Metasequoia*

(*viii*) Resin canal in the conifers is an example of

(*a*) intercellular space
(*b*) Schizogenous cavity
(*c*) Lysigenous cavity
(*d*) big vacuole

(*ix*) Which of the following is resistant to weather ?

(*a*) *Thuja plicata*
(*b*) *Pinus pallustris*
(*c*) *Agathis australis*
(*d*) *Cedrus deodara*

(*x*) The largest timber producing tree of the world is :

(*a*) *Pinus Syevestris*
(*b*) *Cryptomarice Japonica*
(*c*) *Agathis australis*
(*d*) *Pinus roxburghii*

(*xi*) Which of the following conifer grows aove 3000 metres ?

(*a*) *Pinus roxburghii*
(*b*) *Larix griffithiana*
(*c*) *Thuja orientalis*
(*d*) *Cedrus deodara*

CHAPTER 13

Coniferales (contd.) Life History of Pinus

- Pinus-External Features
- Pinus-Anatomy
- Reproduction in Pinus
- Pollination in Pinus
- Embryogeny in Pinus
- Alternation of Generations

Pinus is the most important genus of the family *Pinaceae* and is represented by ninety to one hundred species in the world. They usually grow on the slopes of hills and form dense and extensive forests of evergreen trees in the North Temperate regions of world and sometimes extend up to the sub-arctic regions and in the tropical hills. The species of pine are of great economic value and are the main sources of resins of various types and of various types of woods that are employed as timber. Seeds of some species are a source of food. In India, the genus is represented by six naturally growing species. Their distribution and general vegetative characters are given below:

PINUS Sp.

1. *Pinus roxburghii* Sarg. (= *P. longifolia Roxb.*) : It is also known as Chir pine and grows in the outer Himalayas from Indus to Bhutan at 1,500 to 7,500 feet above sea level. It is abundant on the hill slopes in both Western and Eastern Himalayas. It is a tall tree with stem displaying the excurrent habit and reaches a height of 50 – 60 metres and varies in diameter from 1.5 to 3 metres. The stem has two types of branches and the green leaves as the needles are confined only to the dwarf shoots that bear three needles. The scale leaves occur on the main stem, long shoots and also the dwarf shoots are brown in colour. The seeds are edible and possess wings that are as long as the seed. The megasporophylls have a distinct hooked beak at the tip of umbo. It is also cultivated in the plains as an ornamental.

2. *Pinus wallichiana* A.B. Jacks (= *P. excelsa*) : It is also known as the blue pine or kail. It forms pure or mixed forests on hill slopes at elevations varying between 6,000 to 11,000 feet above sea level. It sometimes grows with *P. roxburghii* at 5,000 feet. It extends from Garhwal through Jaunsar, the Simla Hills, Ram Pur Bushair to Kulu, Chamba and Kashmir. In the Eastern Himalayas, it grows in NEFA. It is a beautiful tree with a diameter of 8 – 10 feet and a height of 100 – 150 feet. The stem has an excurrent habit with thin, greyish brown and fissured bark in old trees and smooth in young trees and branches. The stem bears whorled branches. The dwarf shoots bear groups of 5 needles on each dwarf shoot. The needles are slightly bent. The scales are lanceolate in younger portions and triangular in older portions. The winter buds are long and cylindrical. The dwarf shoots are about a millimeter in length and arise in the axil of a scale leaf. Each dwarf shoot is covered with spirally arranged 10 – 12 cataphylls. The first formed or outer cataphylls that are also called prophylls are two in number and occur opposite to each other. They are smaller than the inner ones and possess a distinct midrib. The male cones arise in axils of lower scale leaves on a long shoot and are longer in size than those of *P. roxburghii*. The needles in this case are minutely toothed and ridged. Stomata are arranged between the ridges in distinct longitudinal rows.

3. *P. gerardiana* Wall. ex. Lamb. is the chilgoza pine whose seeds are sold in the market as chilgoza or Neoza. It grows in N.W. Himalayas at 5,000 – 12,000 feet above sea level. Dense forests of this species occur in Afghanistan above 10,000 feet and also in Kashmir. The dwarf shoots bear three needles.

Plate I : A tree of *Pinus roxburghii.*

4. *P. insularis* Endl. (= *P. Khasya* Royle) is also known as khasi pine and grows only in Khasia and Chittagong hills in Eastern Himalayas. The tree is totally absent in the Western Himalayas. It also extends into Burma. The trees range in height from 75 feet to 100 feet and are evergreen with excurrent habit.

5. *Pinus merkusii* Jungh. can be distinguished from other Indian species in possessing two needles per dwarf shoot. It grows in Burma but extends into India in the Eastern parts of Himalayas.

6. *P. armandi* Franchet has five needles per dwarf shoot. The needles are sharply bent. It is a Chinese species that extends into NEFA region of Assam.

Key to the above species of *Pinus*.

Leaves in fascicles of five

Leaves 4" to 8" long, cones 6 – 12" long : *P. wallichiana*

Leaves 4" to 6" long, sharply bent, cones 4 to 7" long __________ : *P. armandi*

Leaves in fascicles of three

Leaves stiff, less than 7" long, cons 5" to 9" long________________: *P. gerardiana*

Leaves soft more than 7" long

Leaves 7 – 10" long, rarely fascicle of 2 : cones 2 – 3" long, wings four times as long as seed; umbo of scales with prominent beak : *P. insularis*

Leaves 8 – 14" long; cones 4 – 8 long wings slightly longer than seed, umbo of scales with prominent hooked beak .... *P. roxburghi*

Leaves in fascicles of two ___________ *P. merkusii*

The other common species of *Pinus* that grow elsewhere include : *P. palustris* (Virginia to Florida); *P. caribaea* (Carolina, Florida and Lousiana); *P. banksiana* (Canada); *P. contorta* (N. America); *P. ponderosa* (Texas); *P. monticola* (British Columbia); *P. strobus* (N. America); *P. sylvestris*, *P. nigra* and *P. pinaster* (Northern Europe); *P. halepensis* (Mediterranean); *P. cembra* (North Russia); *P. pumila* and *P. koraiensis* (Manchuria); and North American *P. monophylla*, *P. cembroides* and *P. edulis, P. albicaulis, P. lambertiana*, *P. sabiniana*, *P. coulteri*, *P. torreyana* and *P. Jeffreyi* are other common species that have economic uses.

Plate II : An old tree of Pinus wallichiana.

A general account of the genus is presented below. The main events in the life history pertain mainly to those of *P. roxburghii* and *P. wallichiana* that have now been thoroughly worked out by Sethi (1929), Konar (1960) and Konar and Ramchandani (1958). The account given here has been adopted from these works and those of famous authors like Chamberlain (1935), Haupt (1941), Holloway (1937), Florin (1954) and others.

EXTERNAL FEATURES

The Pine trees vary in height from 70 feet to 200 feet and reach a diameter of about 12 feet or even more. *Pinus aristata* from California (Billings, 1957) is about 4,000 years old. It grows on the slopes where rain-water is easily drained off. As in the pteridophytes, the *pinus* tree is the sporophyte. It is differentiated into roots, stem and leaves. Typically the stem displays the excurrent habit. It is tall, stately and straight and is covered with rugged scaly bark. The main stem is branched. The branches are of two types :

Long branches and the short dwarf ones

Long branches. These are the ordinary or normal branches. They continue indefinitely in active growth by means of an apical bud. They are often called shoots of unlimited growth. They are covered with brown bud scales. These branches arise as buds in the axils of scale leaves at the end of each year's extension growth. The following year these buds develop as an apparent whorl of normal branches. They occur at regular intervals on the main trunk and are horizontal. They become gradually shorter towards the apex. Hence the pine tree presents a graceful tapering or pyramidal appearance. The older parts of the long shoots are covered with scars left by the fallen dwarf shoots and the subtending scales while the younger parts bear brown scale leaves with dwarf shoots in their axils. At regular intervals, it also bears compact rings of bud scales or scars. The portion between these successive rings represents each year's growth.

Short branches. In addition to the main branches or long shoots which arise directly from the trunk, there are very numerous branches of limited growth borne on the ordinary branches in the axils of scale leaves (Fig. 13.1). They are often called the dwarf shoots or shoots of limited growth. The dwarf shoot consists of a short axis terminating in a cluster of three green needles (Fig. 13.1, A).

Fig. 13.1. Pinus sp. A tree; B. A long shoot bearing dwarf shoots or spurs and a cluster of male cones. Each dwarf shoot bears three needles.

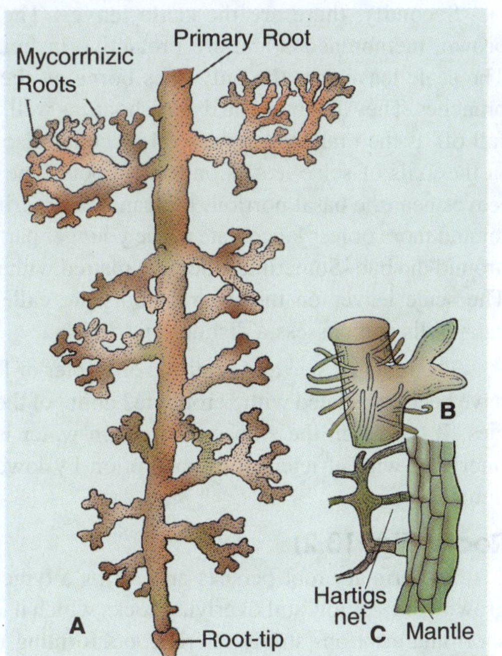

Fig. 13.2. (A–C) *Pinus* sp. mycorrhiza A. The primary or tap root of *Pinus roaxburghii* bearing mycorrhizic roots. B. ectotrophic mycorrhiza; C.T.S of B.

It has an ephemeral terminal bud that stops growing and becomes inactive. It is covered with scales. The dwarf shoot is up to 1/2″ in length and is covered with 10 – 12 scale leaves or *Cataphylls* in *P. wallichiana*. The two outermost cataphylls are called *prophylls* and are opposite to each other as compared to the other spirally arranged ones. The prophylls are smaller than the cataphylls. The innermost cataphylls are the largest.

Leaves (Fig. 13.1C)

The leaves are also of two kinds. The first kind of leaves are the somewhat unusual type being long, narrow (acicular), tough and green and are frequently known as "**Pinus** or **pine needles**." Their surface is smooth. They are not borne on ordinary branches but appear only on dwarf shoots in clusters of two (*P. merkusii*), three (*P. roxburghii*) or five (*P. wallichiana*) according to the species. In *P. monophylla,* only one leaf is produced on the dwarf shoot. The dwarf shoots bearing foliage leaves are known as "Spurs". The spur of *P. roxburghii* with three needles is called **trifoliar** spur. Those of *P. monophylla* are **monofoliar**, of *P. merkusii* are **bifoliar** and of *P. wallichiana* are **penta foliar**. A thin, dry membranous sheath surrounds the base of each cluster of needles. The needles are persistent. they fall only when the spur is shed as a whole. Hence the pine tree is evergreen. The green leaves in *P. wallichiana* are minutely toothed and ridged. The ridges are epidermal projections between which the stomata are arranged in distinct longitudinal rows. The needles are straight on young shoots but spread outwards or drop down in older shoots.

Fig. 13.1C. *P. wallichiana* needle

Secondly, there are the **scale leaves**. They are brown, membraneous and are protective in function. The scale leaves are the only ones borne on the long branches. They are found on dwarf shoots as well. They fall off as the branches mature. Male cones also arise in the axils of sclae leaves on long shoots. The scale leaves near the basal portions of branches are triangular and more or less lanceolate in the younger parts and around the bud. Sometimes they are matted with resin. The scale leaves on the dwarf shoots are called the cataphylls and possess a distinct midrib.

Plate III : Pinus needle borne in

The reduced size and sclerotic character of leaves have been associated with xerophyte habitat of the species. It grows on the slopes where rain water is easily drained off. Extreme cold in winter also interferes with efficient root absorption by lowering temperature and thus causing physiological draught.

Roots (Fig. 13.2)

The primary root persists and forms a typical elongated straight tap root. In case the plant is growing on shallow soil overlying rocks which it is impossible for the tap root to penetrate, it remains short and develops strong lateral roots forming massive root system spread out over an extensive area. Tip on the root has a protective root cap. The root hairs are not well developed. They are developed on the young roots but soon disappear. As the root system becomes fully developed, the new young branches remain short and undergo frequent forking so that clusters of small rootlets are formed. Some of these branch dichotomously, form coralloid masses, and hence ectotrophic mycorrhizae grow. These are the **mycorrhizal roots**. Under the microscope, they are seen to be covered with a close weft of colourless hyphae. Some of these penetrate the outer tissues of young roots. According to Maheshwari and Konar (1971) members belonging to Boletaceac and Agaricaceae constitute the majority of fungi forming ectotropic mycorrhizae. Usually these fungi occur as loose hyphal strands in soil and infect the young root directly. An infected seedling can act as inoculum for the spread of infection. Strong sunlight, adequate soil moisture, aerated and acid soils and low level of soil fertility increase the infection.

The effect of mycorrhizal infection on growth of seedlings and their phosphorus uptake was studied and it was concluded that even a low intensity of infection of *P. sylvestris* seedlings by *Laccaria laccata* greatly stimulates growth of the seedling, possibly due to increase in growth substances produced than its capacity to stimulate phosphorus uptake.

The exact relationship between the hyphae and the root system is still a debatable question. Some physiologists consider the fungus to be merely a parasite. Others hold that both the partners are mutually beneficial to each other. The fungus receives food from the host roots. In return, it helps in the performance of certain physiological functions of the host root. The presence of fungus increases the water absorption by roots and thus takes up the function of root hair. According to some botanists, the presence of fungus makes avilable to the host some of the nitrogen in the organic compounds of soil.

The mycorrhizic roots differ from the normal roots in the following respects:

1. They are short and thick.
2. They lack root hair.
3. They lack root caps.
4. They are more extensively branched or occasionally unbranched.
5. They are covered with fungal hyphae.

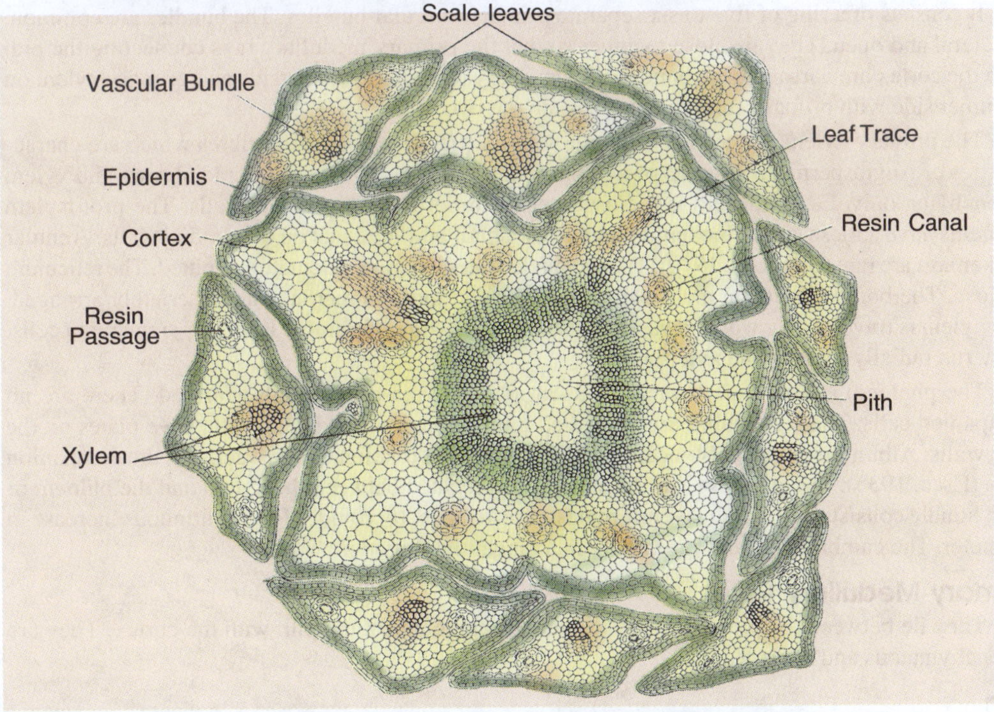

Fig. 13.3. T.S. young stem of *Pinus roxburghii.*

ANATOMY

Primary Structure of Stem

Anatomically the stem shows a considerable advance over the cycadales. Like the stem of dicotyledons, it has a concentric arrangement of primary permanent tissues (Fig. 13.3). It is composed of a small pith, a thick vascular cylinder made up of a ring of separate collateral and open vascular bundles, a thin cortex and an epidermis.

Epidermis

It covers the outer surface of stem and is single layered and irregular in outline. The epidermis cells are highly thickened with walls strongly cutinized.

Cortex

The cortical cells beneath the epidermis are strongly thickened to form the hypodermis. The inner cortex consists of thin-walled parenchymatous cells. It is traversed by large resin ducts or canals (Fig. 13.3) which run both radially and longitudinally through the stem. Each duct or canal is lined by a layer of thin-walled parenchymatous glandular secretary cells constituting the **epithelium**. The epithelial cells usually possess abundant cytoplasm and a conspicuous nucleus. External to the epithelial layer is the **sclerotic sheath** consisting of thick-walled cells. The epithelial cells secrete resin into the canal. This resin flows out when an injury breaks the tissues open. Little is known of the functions of this resin to the plant itself except that on exposure to air, the liquid resin forms a solid covering on the wound until fresh bark is formed. It serves as a valuable protection against the invasion of wounded tissues by Fungi or Bacteria. Resin is the chief source of natural **turpentine**. The latter is a valuable antiseptic reagent. Neither endodermis nor pericycle is recognizable.

Primary Vascular Tissue (Fig. 13.3)

It consists of a ring of five to six separate primary vascular bundles. The bundles are common, collateral and open. They lie close together so that the primary medullary rays connecting the pith with the cortex are narrow. Each bundle has a primary pholem on the outside and primary xylem on the inner side with primary cambium in between the two.

The primary xylem contains neither true vessels (tracheae) nor wood fibres which are characteristics of Angiosperms. It consists of tracheids that are arranged in uniform radial rows and xylem parenchyma only. Each tracheid is a closed cell with bordered pits on its walls. The protoxylem elements have a loose spiral thickening with a few small bordered pits on their radial walls. Annular thickenings are rare (Bierhorst, 1970). The metaxylem elements are reticulate and pitted. The reticulum is close. The bordered pits on their walls are larger and more numerous but uniseriately arranged. The xylem is traversed by **wood rays**, each usually consisting of a single layer of parenchyma cells. They run radially through the xylem.

The pholoem contains sieve tubes. Phloem parenchyma cells are also formed. There are no companion cells. The sieve tubes are elongated more or less pointed cells with sieve plates on the side walls. Albuminous cells also occur and have the same relation to sieve cells on the companion cells (Esau, 1935). The **primary cambium** which is located between the xylem and the phloem of each bundle consists of a single layer of **meristematic cells**. It provides for a continuous increase in diameter. The cambium divides repeatedly in a tangential direction.

Primary Medullary Rays

They lie between the primary vascular bundles and connect the pith with the cortex. They are parenchymatous and narrow.

Pith

Within the ring of vascular bundles is the parenchymatous pith. It forms the core of the upright woody stem of *Pinus*.

The *Pinus* stem and branches (long shoots) grow apically by means of an apical meristem that shows no clear distinction between the **tunica** and **corpus** but there is an initial group of apical cells that divide anticlinally and periclinally to give rise to lateral derivatives and **central mother cell zone** below. In the centre of this zone, there is a gradual differentiation of cells with meristematic properties that give rise to the **peripheral zone** from which arise in part leaf primordia, cortex, procambium, and epidermis. The central mother cell also gives rise basally to the rib meristem which produces the pith. As the stem tissues differentiate, from the base of the apex, provascular strands arise from the peripheral zone and vascular tissues differentiate as collateral bundles in a eustele with endarch primary xylem.

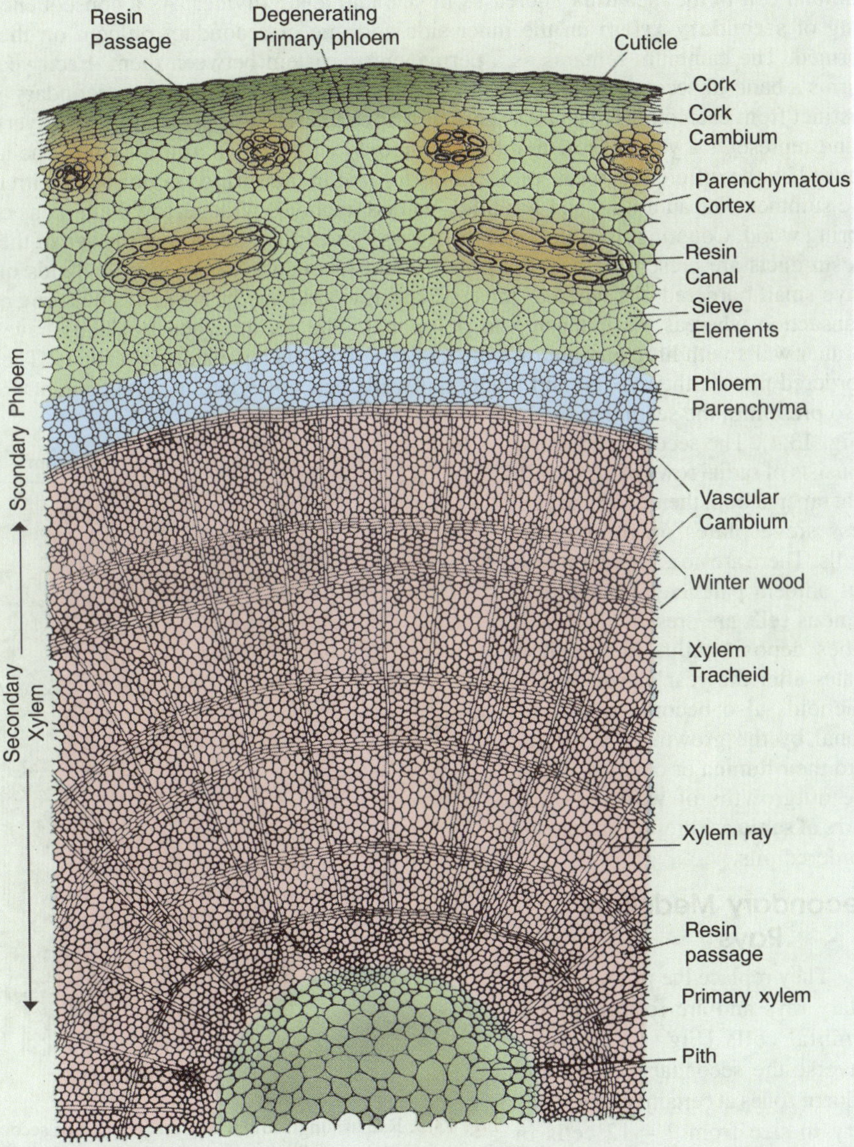

Fig. 13.4. T.S. Portion of old stem of *Pinus roxburghii* showing details of secondary structure.

Secondary Structure (Fig. 13.4)

The secondary tissues are formed in the same way as in the dicotyledons. The primary cambium of the stele extends tangentially. The edges of the bundle cambia come closer and eventually make contact. In this way, a complete or closer ring of cambium is formed. The cambial cells divide repeatedly in a tangential direction. Two cells are formed at each division. Of these, one remains **cambial**. The other, which is called the **tissue mother cell**, may divide once or twice. The cells thus formed become differentiated into xylem or phloem according to their position on the inside or outside of the cambial cell. Such growth is known as **secondary growth**. The new xylem cells which are added to the outside of the primary xylem constitute the **secondary xylem** and the new phloem cells which are added to the inside of the primary phloem form **secondary phloem**. The cambial cell in the meantime increases in size and again divides. As a consequence, a continuous ring of **secondary xylem** on the inner side and one of secondary phloem on the outer side are formed. The cambium remains as a permanent meristem between them. Each year the cambium forms a band of secondary xylem and secondary phloem. The bands of secondary xylem are quite distinct from each other. In a cross section, it appears as a series of concentric layers. Each layer or band represents a year's growth and is commonly called as an **annual ring**. The annual rings are caused by the difference in size of tracheids formed in the spring as compared with those formed in the summer. The autumn wood has thick-walled tracheids with narrow Pamina as compared to the spring wood. Counting the number of these rings gives an indication of the age of the tree or branch. Resin ducts are present both in the primary and secondary wood. The tracheids of autumn wood have small bordered pits whereas those of springwood have larger pits. The former are squarish in transaction whereas the latter appear polygonal. The spring wood tracheids have comparatively thinner walls with little amount of lignifications. The pits have a distinct **torus**. The tracheids have bordered pits on their radial and tangential walls and uniseriate in arrangement. Resin canals are also present in the secondary wood (Fig. 13.4). The secondary phloem consists of radial rows of sieve cells that taper at both their ends and possess sieve plates on their radial walls. There are no companion cells but phloem parenchyma and albuminous cells are present. The sieve tubes deposit callus in the sieve plates after the year's growth. The tracheids also become non-functional by the growth of **tylosoids** into their lumina or cavities. These are outgrowths of wall material. **Bars of sanio** are present around the bordered pits.

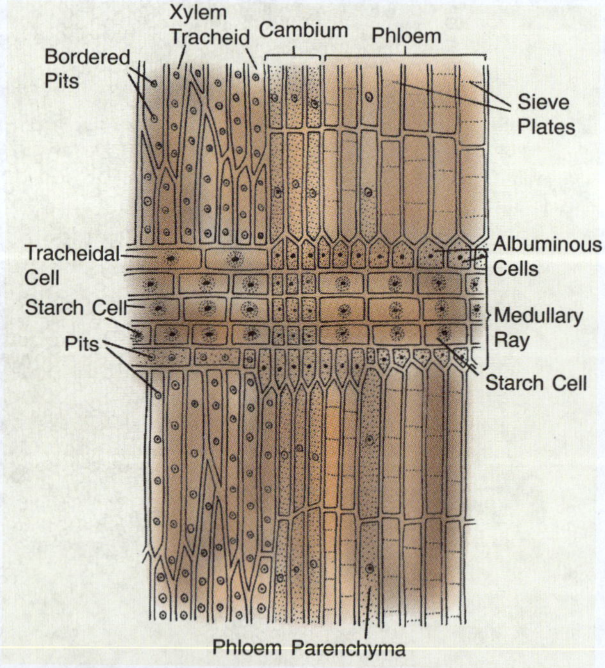

Fig. 13.5. Radial longitudinal section (R.L.S.) of secondary wood and phloem of *Pinus roxburghii* showing the structure of secondary medullary ray.

Secondary Medullary Rays

They replace the primary medullary rays and are formed by the cambial cells (Fig. 13.5). They traverse the secondary xylem and phloem zones at certain places. They vary in size from 2 – 12 cells in height and are only one cell broad. (Figs. 13.5 and 13.6). New medullary rays are formed each year, alternating with those previously laid down. Structurally, the secondary medullary rays are much more complex than those of the dicotyledons. The rays in the secondary wood consist of thick-walled rectangular parenchymatous cells contaiing cytoplasm, a nucleus and starch grains. They possess simple pits. At the upper and lower ends of these rays are elongated, **horizontally** situated one or two rows of short tracheidal cells. They are elongated horizontally and are called the **marginal ray tracheids**. The marginal ray tracheids have bordered pits on their lateral and end walls and thus allow radial diffusion of watery fluids. In the secondary phloem and cambium regions, the ray tracheids are absent. Their place is taken up by large thin-walled cells which are elongated vertically. They are commonly called the **albuminous cells**. They contain protein. In a radial longitudinal section of the wood, the medullary ray is cut longitudinally. It is seen crossing the tracheids in a horizontal direction. The bordered pits are seen in surface view on the radial walls of the tracheids. A longitudinal tangential section of the wood cuts across the medullary ray. The bordered pits are seen in section on the radial walls of the tracheids which are cut through in this case.

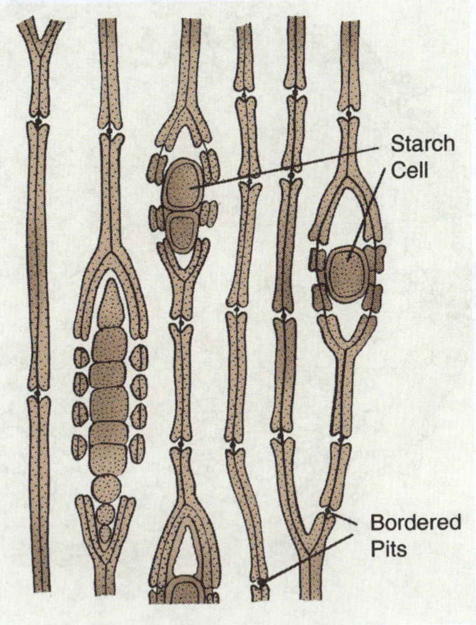

Fig. 13.6. Tangential longitudinal section (T.L.S.) of secondary wood showing the medullary rays of *P. roxburghii.*

Periderm or Cork (Fig. 13.4)

The formation of the secondary vascular tissues stretches the outermost tissues of the stem which eventually rupture and are peeled off as thin strips of scaly bark. Before this takes place a special protective tissue the **periderm** or **cork** is formed. It is formed by a layer of cortical cells near the outside. These cells become meristematic and constitute the **cork cambium** or **phellogen**. It is secondary meristem. The cells of the cork cambium divide by two parallel walls. The inner cells form the additional cortex whereas the outer constitute the **periderm** or **cork**. The latter tissue serves to protect the delicate tissues lying within it when the epidermis is peeled off. It is impervious to water and thus helps to prevent excessive evaporation. It may be mentioned here that cambial activity, xylem and phloem differentiation and formation of cork cambium are all hormone controlled.

The main chemical components of the walls of tracheids in gymnosperms including *Pinus* are : (*i*) cellulose, (*ii*) the non-cellulosic polysaccharides; and (*iii*) lignin. The cellulose makes up the framework of tracheidal wall. The polymeric strands combine to form micelles which, in combination, form the microfibrils. The microfibrils, in turn, form a crystalline lattice that is embedded in an amorphous matrix consisting of noncellulosic polysaccharides, later in combination with the lignin. Lignification usually is initiated within the primary wall at the corners, then spreads intercellularly and only later into the secondary wall.

Structure of the Root (Fig. 13.7)

It consists of piliferous layer, cortex and stele. The arrangement of tissues is in general similar to those of the root of dicotyledons.

Piliferous Layer

It is the outermost layer. The cells of this layer fit tightly against their neighbours.

Cortex

Within the piliferous layer is the cortex. It is a wide zone of parenchymatous cells and thus forms most of the bulk of the root. It is bounded on the inside by the endodermis.

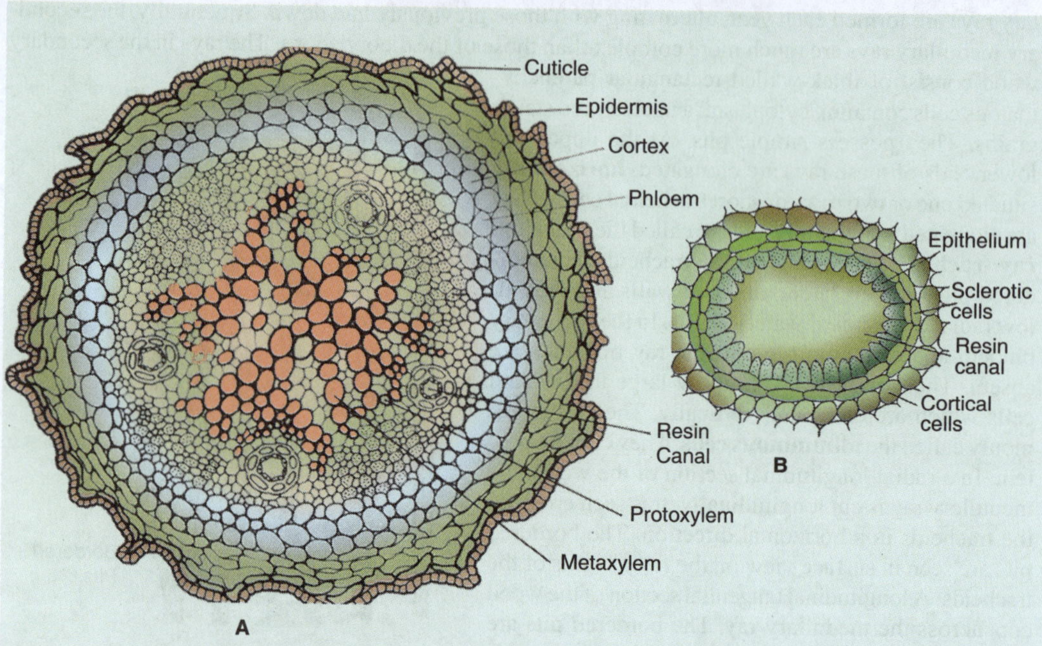

Fig. 13.7. T.S. Primary root of Pinus roxburghii; B.T.S. of a schizogenous resin canal.

Endodermis

It is single layered. The cells are of brownish orange colour. They are suberized and usually impregnated with tannin.

Pericycle

It delimits the stele and is many layered. The pericycle cells contain starch and are usually impregnated with tannin.

Stele

It is **diarch** occasionally **triarch** or tetrarch. The higher number is up to six. An equal number of pholoem bundles alternate with the xylem bundles. The protoxylem groups lie at the ends of metaxylem which occupies the centre of the root. As the development of the metaxylem is centripetal, in young roots, the central cells may be found unlignified constituting the **pith**. Each protoxylem group is forked so that the xylem bundle appears Y-shaped. A resin passage lies between the arrows of the Y.

A cross section through the mycorrhizal root shows that the hyphae run mainly between the cortical cells. They lie thickly in the intercellular spaces. No fungus is present within the endodermis. Most of the hyphae forming a complete layer over the surface of the root are cut transversely. They present the appearance of an outer pseudo-parechymatous tissue.

Secondary Growth (Fig. 13.8)

It takes place by the development of cambial *strips* (two in diarch roots, three in triarch and four in tetrarch) in the parenchymatous cells separating the phloem and the metaxylem. These cambial arcs later become continuous by the pericycle cells external to protoxylem resuming meristematic activity. Secondary xylem is formed towards the centre and secondary phloem towards the outside.

Ultimately the cambium assumes a circular outline and in mature roo, shows a radial arrangement of vascular elements as in the stem. The annual rings are also distinct in the root but are narrower as compared to the stem. The tracheids are longer and thick-walled as compared to the stem. They possess bordered pits like those of the stem.

The phellogen or the **Cork Cambium** arises deep down in the pericycle. It forms periderm or cork towards the outside. As a result, a thick layer of cork is formed separating the stele from the cortex. The latter eventually dies and disappears as **bark**.

Anatomy of the Leaf (Figs. 13.9, 13.10)

The leaf shows a complex and unusual structure very much different from that of the leaves of the cryptogams as well as the typical leaf of the Angiosperms. Its anatomy suggests that it is adapted to endure severe environmental conditions. It is not flat but is about as thick as broad. In a cross section, it is shaped like the tri-sector of a circle with the curved surface facing outwards and the vertex inwards. It reveals the following structure from without inwards.

Epidermis

Forming the outer boundary of the leaf is a single layered epidermis. The epidermal cells are thick-walled and heavily cutinized so that the epidermis is covered on the outside by a thick cuticle. The deeply sunken stomata are present on all sides of the leaf. They are developed in longitudinal rows. Their **guard cells** are situated well below the level of the epidermis so that there is a sunken **pit** leading to the stomatal aperture.

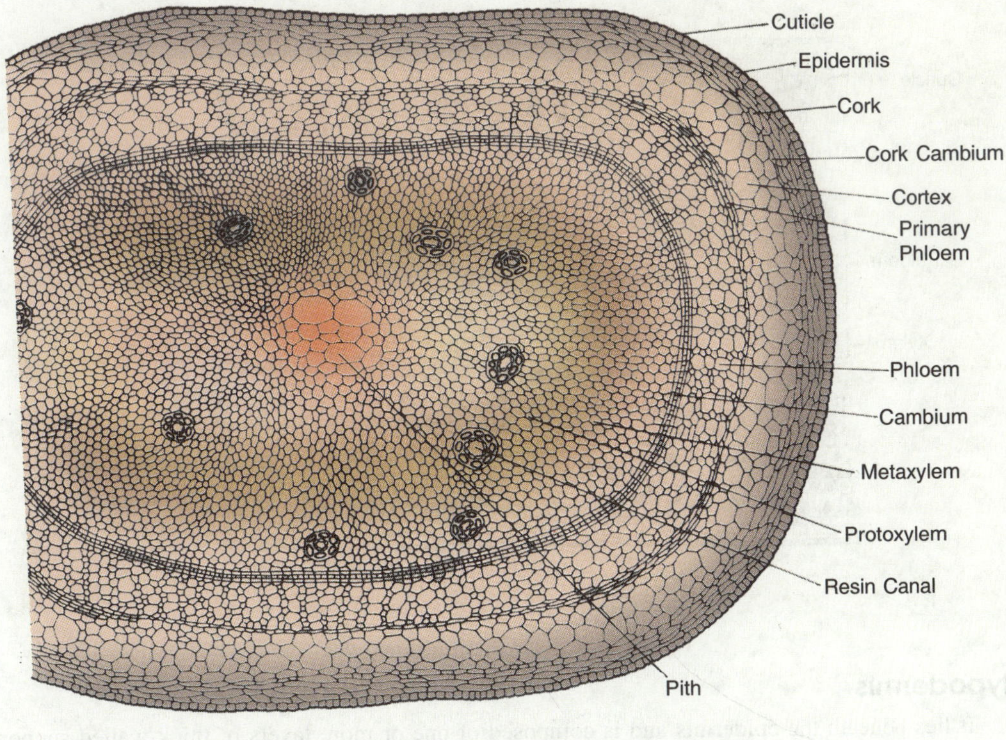

Fig. 13.8. T.S. old root of *Pinus roxburghii* showing secondary structures.

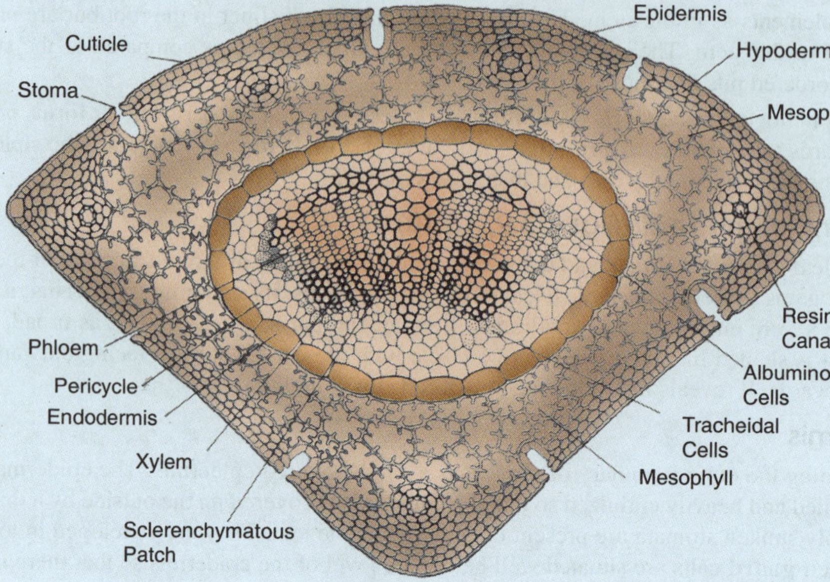

Fig. 13.9. T.S. needle of *P. roxburghii* showing detailed internal structure.

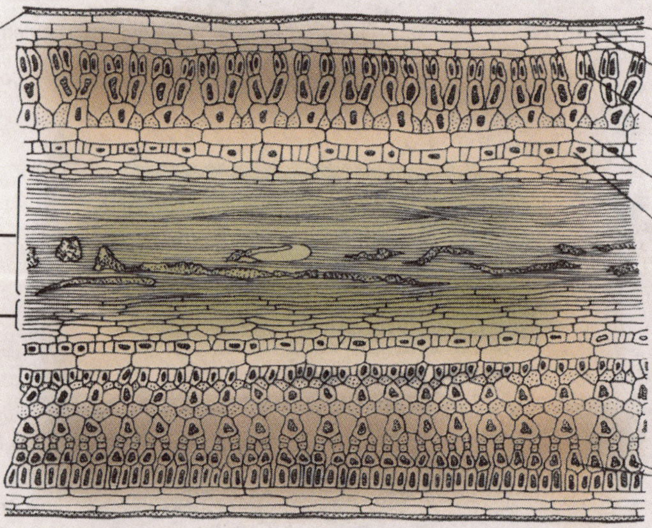

Fig. 13.10. L.S. Needle of *P. roxburghii.*

Hypodermis

It lies beneath the epidermis and is composed of one or more layers of thick-walled support cells or sclerenchyma. The hypodermis is frequently interrupted by air spaces beneath the stomata. It is strongly developed at the angles.

Mesophyll

Within the hypodermis is the parenchymatous **mesophyll**. It is compact and shows no differentiation into palisade and spongy tissues. It consists of thin-walled cells which contain numerous chloroplasts and abundant starch. The mesophyll thus functions as the chlorophyll bearing tissue which manufactures food for the plants. Its cells are peculiar in that their walls have numerous small infoldings which project into the cavities of the cells. These serve to increase the absorptive, aerating and excreting surface of the protoplasm in each cell and thus compensate for reduced leafy surface. The mesophyll contains a number of **resin ducts** immediately under the hypodermis. They are similar in structure to those of the stem.

Endodermis

The central tissue of the leaf is enclosed by a conspicuous single layered endodermis which delimits the mesophyll. The endodermal cells are large and oval and possess casparian strips.

Pericycle

Within the endodermis is the many-layered pericycle. It consists of a number of types of cells. Majority of the cells constituting the pericycle are ordinary parenchymatous cells, containing starch. Embedded in them are two other types of cells. Firstly, there are the other parenchymatous cells rich in proteins. They are called the **albuminous cells**. They abut upon the phloem of the vascular bundles. The albuminous cells serve to pass on the elaborated compounds from the mesophyll to the phloem. The second type of cells resemble the tracheids. They are known as the **tracheidal cells**, and are elongated radially. The tracheidal cells are contiguous to the xylem of the bundle and thus serve to carry watery fluids from the xylem to the mesopyll. These special kinds of cells collectively constitute the **transfusion tissue**. Probably the transfusion tissue makes up for the poor development of the **vascular tissue**. Besides the transfusion tissue, the pericycle includes schlerenchymatous fibres near and between the bundles.

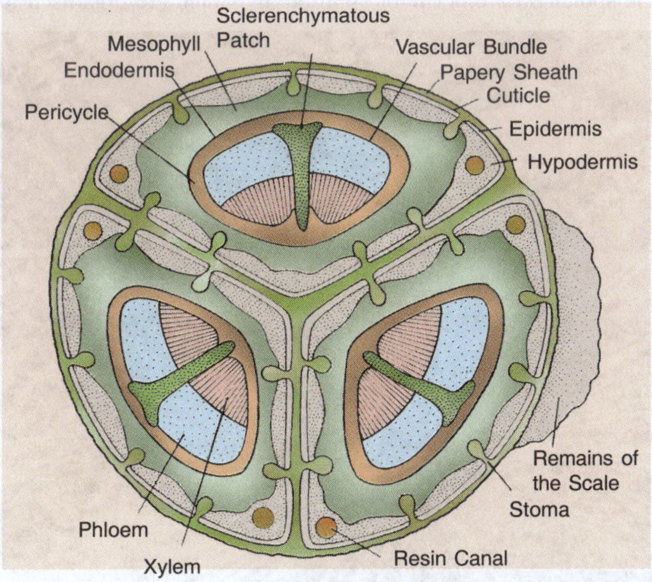

Fig. 13.11. T.S. tip of dwarf shoot showing three needles in transection

Vascular Bundles

Embedded in the pericycle are the two vascular bundles. They are separated by a band of sclerenchymatous fibres. The two bundles run closely parallel from the base to the apex of the leaf and are unbranched. In some species, there is only a single bundle. Each bundle contains xylem and phloem. The phloem faces the curved lower or outer side. The xylem lies above the phloem facing inwards, i.e., towards the pointed end. The bundle is thus collateral. Acicular form of the leaves, which reduces transpiring leaf surface, thick-walled and heavily cutinized epidermis, sclerenchymatous hypodermis which renders the leaf tough in texture, sunken stomata and poor development of vascular tissues are all xerophytic characters which suggest that the pine leaf is not apt to lose water rapidly by transpiration. These xerophytic leaves are associated with the evergreen habit of the pine tree because, they by conservation of water supplies, enable the plant to retain its leaves through winter without injury.

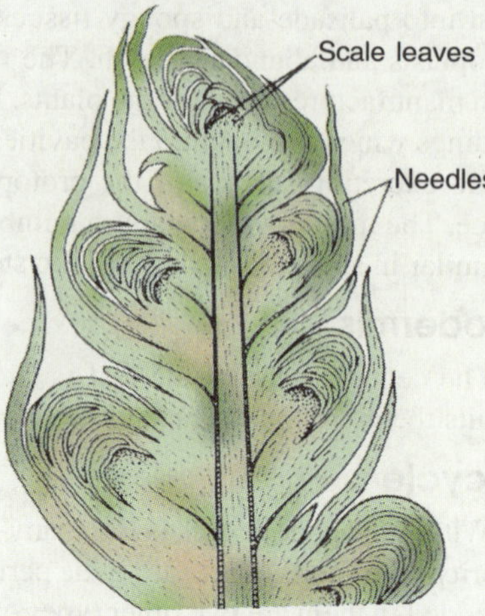

Fig. 13.12. V.S. of tip of long shoot. Note the developing dwarf shoots.

REPRODUCTION

Vegetative reprodcution is unknown in the pines. They reproduce by means of seeds. Reproduction by seeds involves both reproduction by spores and reproduction by gametes. It also comprises the origin, development and growth to maturity of the gametophytic generation.

Reproduction by Spores

All seed plants are **heterosporous** which means that the reproductive organs called the **spores** are of two kinds – **microspores** (pollen grains) and **megaspores** (Embryo sacs). They are produced in specialized structures called microsporangia (pollen sacs) and megasporangia (ovules minus their coats) respectively. The sporangia are borne or sporophylls which are organised to form **cones** or

Plate IV : Portion of tree of *Pinus roxburghii* bearing male cones.

strobili. The latter are of two kinds, one containing only microsporangia and the other megasporangia. It is customary to call the former as **male cones** or **microsporangiate strobili** and the latter **female cones** or **megasporangiate strobili**, even though they are sporophytic in origin and are not in any way sexual in character. It is more appropriate to call the one as microsporangiate or staminate and the other as megasporangiate or carpellate cones or ovulate cones. *Pinus* is monoecious as it bears both types of cones on the same tree on separate branches.

Plate V : Portion of tree of *Pinus roxburghii* bearing male cones.

The Microsporangiate or Staminate or Male Cones (Fig. 13.13, 13.19)

They occur laterally in clusters, each in the axil of a scale leaf at the base of a terminal vegetative bud which they surround. They are concealed amidst last year's foliage. As the vegetative bud grows

Plate VI : Twig of *Pinus roxburghii* bearing a female cone.

and expands it develops short or dwarf shoot in the axils of the upper scales. It means that the male cones replace the dwarf shoots at the base of the developing bud or shoot of the current year. They can be seen on the pine tree in the beginning or middle of March in the hills and in January or

Fig. 13.13. A cluster of very young male cone of *Pinus roxburghii.*

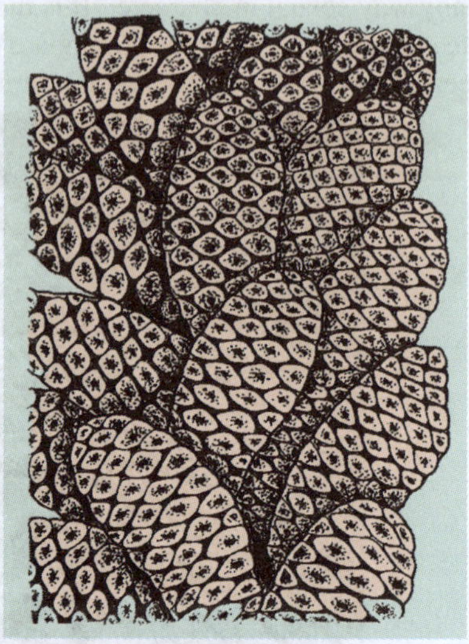

Fig. 13.14. A cluster of male cones of *pinus roxburghii.*

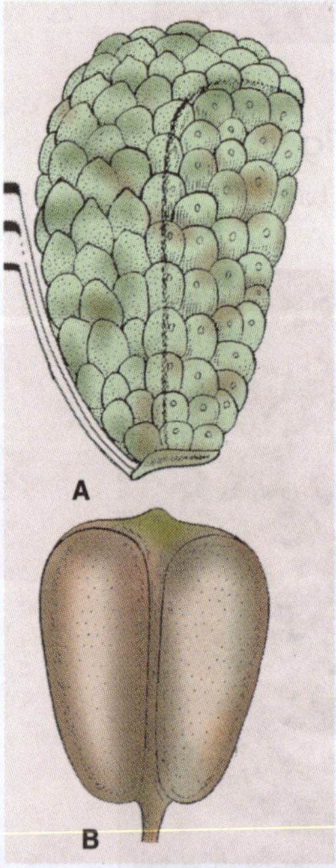

Fig. 13.15. (A – B). *Pinus roxburghii.* A. Fully developed male cone. B. A microsporophyll bearing two undehisced microsporangia.

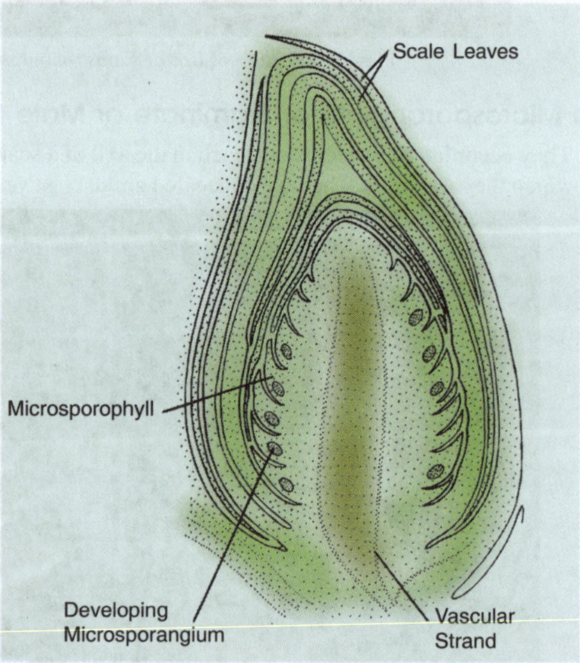

Fig. 13.16. V.S. of a very young male cone of *Pinus roxburghii.* Note the developing sporangia.

February in the plains. In *P. wallichiana,* they are initiated towards the end of October or beginning of November. The shedding of the pollen grains starts towards the end of April and continues till the beginning of June, depending upon the altitude. The lower the altitude, the earlier the cones shed their pollen. The male cones occur in clusters of 15 – 35 in *P. wallichiana* and up to 50 in *P. roxburghii*.

The male cone is shortly stalked and consists of an elongated central axis, bearing a number of small spirally arranged and closely fitting scale-like **microsporophylls**. The latter are equivalent to the stamens of Angiosperms. Each microsporophyll takes its origin from the central axis, runs out horizontally to end in a sterile flattened head, the tip of which is turned up so as to fit over the micorosprophyll above it. It bears two **microsporangia** or **pollen sacs** as swellings on the lower surface of its horizontal position. The microsporophyll is attached to the axis by a short stalk. The sporangia are filled with numerous winged **microspores**. A number of scales at the base of the male cones remain small and sterile, usually 8 – 10. The young cones are green or reddish purple, but become yellow or dark brown at the time of shedding.

Development of Microsporangia (Figs. 13.20, 13.21)

The microsporangium arises as a group of cells on the lower surface of the microsporophyll. The hypodermal cells divide and form the outer **parietal cells** and inner **archesporial cells**. The former divide anticlinally and periclinally to form a 4 layered wall of the sporangium. The archesporial cells divide to form sporogenous cells which ultimately differentiate into **microspore mother cells** that are polygonal in outline and contain prominent nuclei and vacuolated cytoplasm. The innermost wall layer differentiates into **tapetum** whose cells have prominent nuclei and dense cytoplasmic contents. At the tetrad stage the tapetal cells become binucleate. Soon they break down and form a continuous layer of cytoplasm around the spore tetrads. It is consumed by the time the mi-

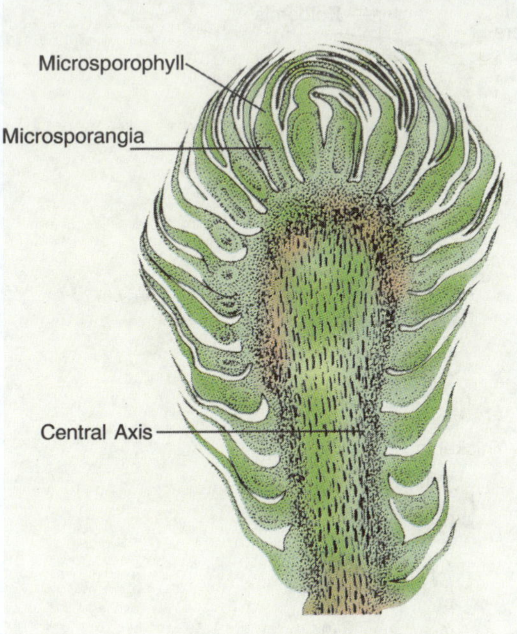

Fig. 13.17. L.S. young male cone with fully developed microsporangia.

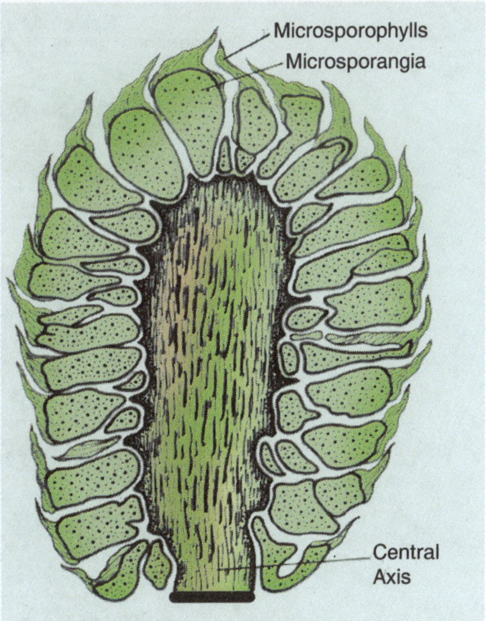

Fig. 13.18. L.S. mature male cone of *P. roxburghii.*

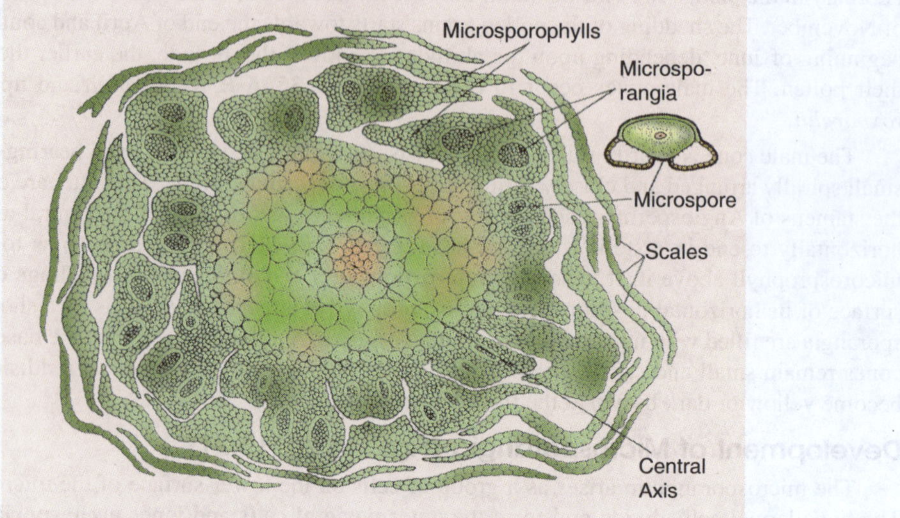

Fig. 13.19. T.S. mature male cone of *P. roxburghii*. Also note the pollen grain or microspore.

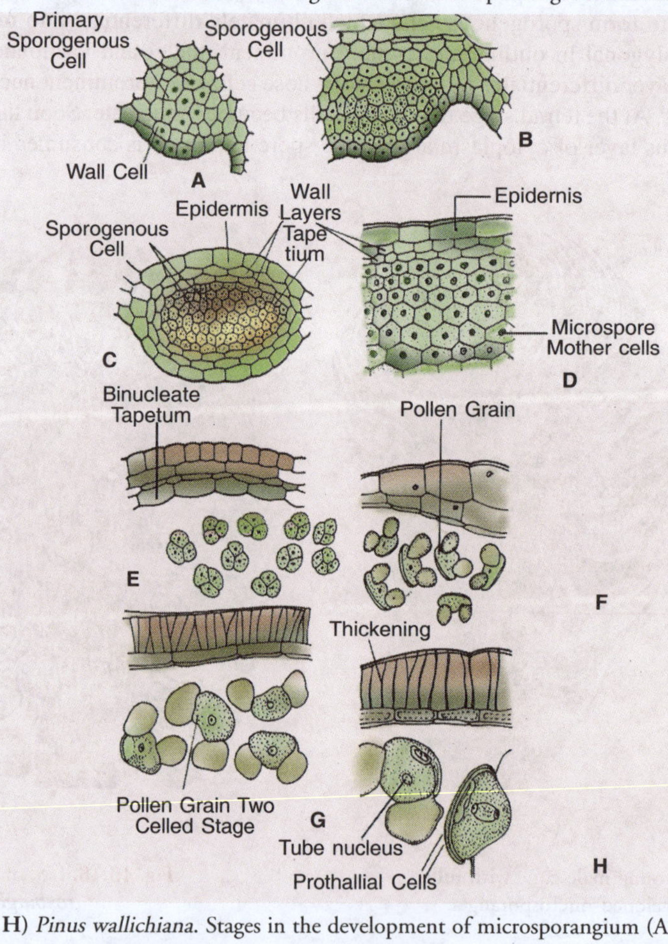

Fig. 13.20. (A – H) *Pinus wallichiana*. Stages in the development of microsporangium (After Konar)

crospores have divided to form the prothallial cells. The outer wall layer is called the epidermis. In a mature microsporangium, its cells become radially elongated and its tangential walls develop bands of thickening (Fig. 13.20, G, H). The cells of the second layer remain thin-walled. The cells of the third layer become tangentially elongated and the first to degenerate. The second layer degenerates when the microspores are mature. The microspore mother cells undergo meiosis to form tetrahedral tetrads of microspores. Sometimes isobilateral tetrads are also formed. Later the spores separate from the tetrads. They are haploid and two winged. The mature microsporangium is an elongated sac-like structure surrounded by its outermost wall or the epidermis and enclosing a large number of microspores or the pollen grains. In *P. roxburghii*, according to Konar (1960) the microsporangium dehisces by a longitudinal slit.

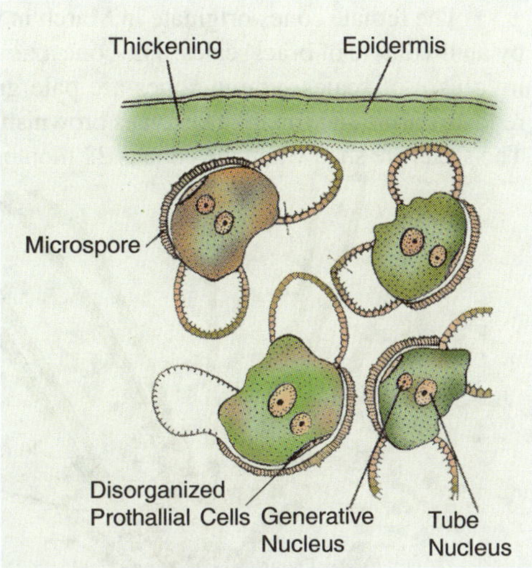

Fig. 13.21. *Pinus wallichiana.* part of mature microsporangium.

Microspore (Fig. 13.20 G, H and Fig. 13.21)

It is unicellular. The spore wall becomes much thickened and differentiated into a three-layered coat. The outer exine, the middle **exointine** and the innermost **intine**. The exine is cuticularized. It covers the spore only on one side. The rest of the spore is covered by the exo-intine. The latter separates from the intine on the side where spores have been in contact with other spores in the tetrad and becomes inflated into two baloon-like expansions or wings. These are not cells but air sacs or bladders. They render the grains extremely light for their size and thus facilitate dispersal by wind. Before liberation, the microspores begin to germinate and form extremely reduced male gametophyte. Their germination is thus **precocious**. The pollen grains are **heteropolar** and radiosymmetric and consist of a body or '**corpus**' and the air sacs or '**sacci**'. It has a single aperture or **tenuitas** at its distal end. The wings or sacci are separated from the body or the corpus by the saccate **nexine** or the inner layer of exine. The nucleus lies towards the distal end.

The Megasporangiate or Ovulate Cones (Figs. 13.22 – 13.28)

They are fewer in number and arise single or in a small cluster or two to four, each as a bud in the axial of a scale leaf towards the end of the new shoots of unlimited growth which do not bear the male cones. They occur below the terminal bud each terminating a short lateral, axillary branch thus replacing in position a newly formed long shoot. The female cones in position, therefore, correspond with the buds which grow out the following year into long shoots. When young, it is a small, erect, reddish to pinkish structure, which stands erect on a short stalk covered with scales (Figs. 13.22 – 13.26). Like the male cone it consists of a central axis (Figs. 13.27, 13.28) which bears paired scales in a close spiral. The lower scale of the pair is small. It is spoken of as the **bract scale**. It is leathery and directly attached to the cone axis. The upper scale of the pair is larger, thicker and stouter. It is the **ovuliferous scale**. It develops from the upper surface of the bract scale and bears two ovules side by side on its upper side as two small white swellings. The bract scale is larger than the ovuliferous scale before pollination but later it grows bigger than the bract scale. The ovuliferous scale is woody and wedgeshaped with its broader sterile end, the **apophysis** directed outwards. The bract scales are thus concealed from view. The surface of the cone is marked by rhomboidal areas each with a small central conical point – the **umbo**. The rhomboidal areas are the outlines of the broad sterile apices or apophysis of the ovuliferous scales. There are about 80 – 90 megasporophylls in a single female cone.

The female cones originate in March in the plains and in February in the hills and are protected by an involucre of bract scales. The cone axis elongates in April and the cone thus protrudes out of its involucre of scales. young cones are pale green but near pollination time, the colour changes to reddish purple and finally it becomes brownish. The cones close and become compact after pollination. The seeds are shed when the cone is 22 months old.

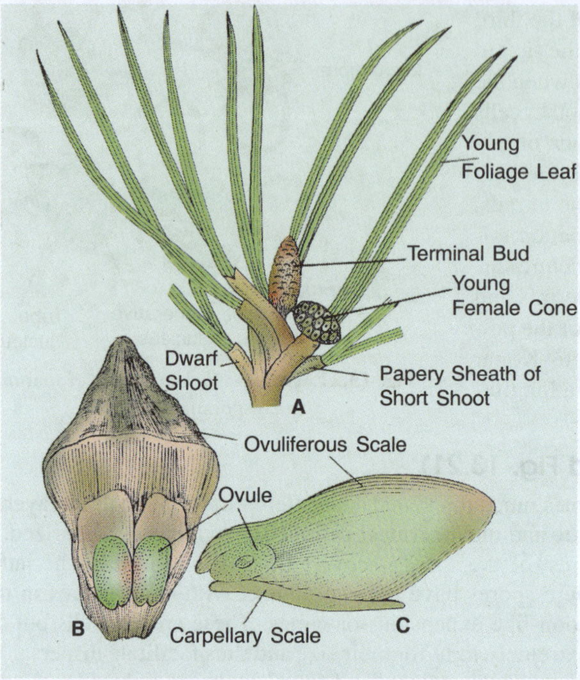

Plate VII: *P. Sylvestris* male cone

Fig. 13.22. *Pinus sylvestris.* A. Showing long shoot bearing dwarf shoots with two needles. Also note the apical bud and a young female cone. B. Megasporophyll bearing two seeds. C. A section through megasporophyll showing ovule.

Fig. 13.23. A very young female cone of *Pinus roxburghii.*

Fig. 13.24. A six month old female cone of *Pinus roxburghii* enlarged.

Development of the cones

Development of the cones and the scales have been studied by Harrison and Slee (1992). The development of the cones start with the transformation of an indeterminate axillary apex into a determinate reproductive apex. A new long shoot terminal bud is formed during February at the tip of rapidly elongating shoot. The axillary apices that emerge on the side of the aptical meristem can later differentiate as follows :

(*i*) *Vegetative dwarf shoot buds.* which become anatomically differentiated into the initiation of three of five needle primordia.

(*ii*) *Reproductive Pollen cone Buds.* in which procambium tissue gradually develops into the microsporophylls. The appearance of microsporophyll initials in the peripheral zone of the axis was designated as stage I (mid march). The size of the apex diminishes and disappears with the development of each sporophyll. At stage II, microsporophyll initiation is complete and developed pollen mother cells are visible.

(*iii*) *Seed cone buds.* become anatomically differentiated with the initiation of bracts, formed directly from conspicuous cell pockets in the peripheral zone of seed cone bud apex (Stage I, Late March – April). Ovuliferous scale primordia are then initiated from hypodermal cells on the adaxial base of bracts. (Stage II). A fused bract, ovuliferous scale complex is formed and the scale become soon displaced from bud-cone axis (Stage III).

Thus long shoot terminal buds may differentiate into vegetative of dwarf shoot buds, male cones or female cones.

Morphology and Nature of the Bract and ovuliferous scales

The nature of the paired scales has long been a debated question and a number of theories have been put forward from time to time to explain the nature of the ovuiferous scales. Robert Brown (1827) was the first who considered ovuliferous scale as an open foliar carpel bearing naked ovules. According to him, this carpel arises in the axil of the bract. This theory was not supported by anyone on the ground that the carpel is not supposed to arise in the axil of another bract (leaf). Schleiden (1829), then opined that the ovuliferous scale is not a leafy carpel but a flattened 'axis' forming the placentum.

A. Brown (1842) throught that the ovuliferous scale represented the first two leaves of an axillary shoot which had become fused. Dickson (1860) supported by Goebel, concluded that the bract was the megasporophyll or carpel bearing at its axil a shoot which was the ovuliferous scale. Later Sachs (1868) who was soon supported by Eichler, concluded that the bract was a carpel and the ovuliferous scale was a ligular outgrowth on its surface. He compared this condition with that found in *Isoetes* or *Selaginella*. On the basis of anatomical study Van Tieghem (1869) concluded that the ovuliforous scale was the first and the only leaf of a suppressed shoot in the axil of the bract. A decade later Celakovsky (1879) thought that the ovuliferous scale represented the fused outer integuments of the ovules while Bessey (1892) supported by Kubart considered it to be equivalent to an axil or enlargement of the chalaza of the two ovules.

Delpino looked upon it as the two lateral lobes of the bract scale which have been turned inwards and fused together. Hirmer opined that the bract and the ovuliferous scales are the two branches of a sporophyll which has forked vertically. It is held by many botanists that since the ovuliferous scale is axillary to the bract scale in postion, it is therefore a branch. According to their view, the bract scale corresponds to the scale leaf and the ovuliferous scale to the dwarf shoot, However, this view is not supported by the vascular supply to the ovuliferous scale.

In the twenteth century, Chamberain (1934) presented his views on the nature of the ovuliferous scale. He concluded that the bract scale is a megasporophyll bearing in its axil the ovuliferous scale which is a modified shoot with or without leaves and which bears the megasporangia or ovules.

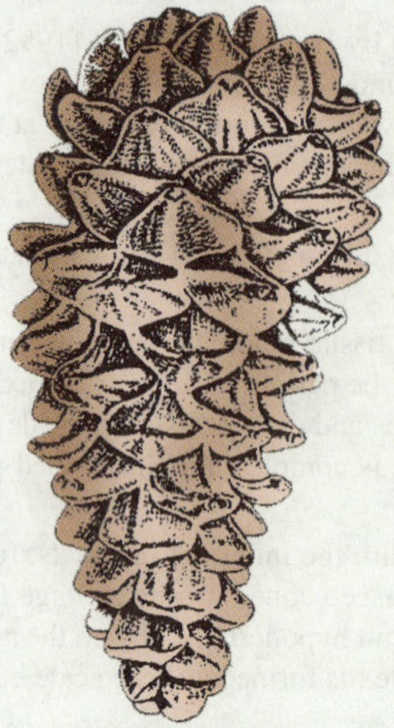

Fig. 13.25. A three year old female cone of *Pinus wallichiana*.

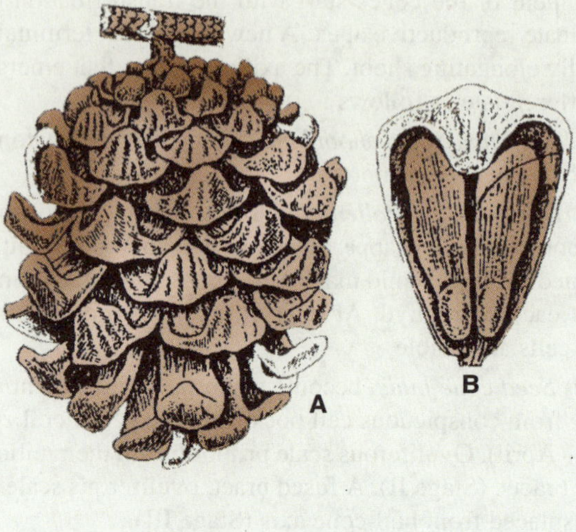

Fig. 13.26. (A—B). *Pinus roxburghii.* A. Four year old female cone of *Pinus roxburghii.* B. An ovuliferous scale.

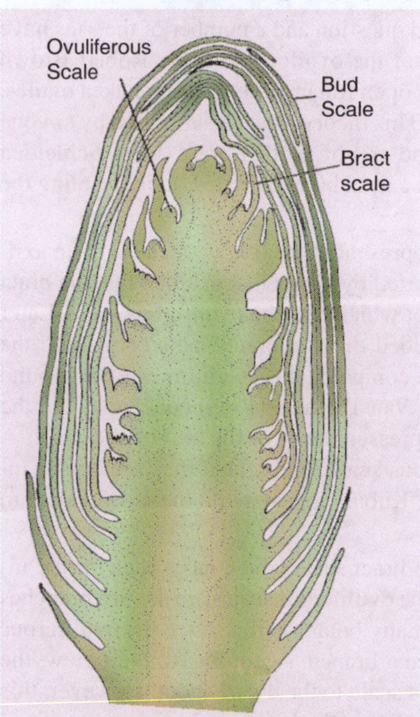

Fig. 13.27. L.S. of a young female cone of *Pinus wallichiana*. Note the well developed bract scales and developing ovuliferous scales.

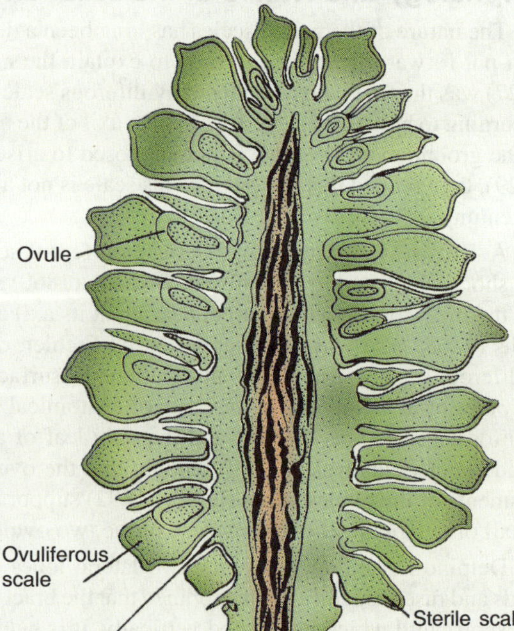

Fig. 13.28. V.S. mature female cone of *Pinus roxburghii.*

For the first time in order to explain the nature of ovuliferous scale, Florin (1951) took the help of a comparative study of fossil conifers known today, beginning from *Lebachia* where the ovules are borne on radial dwarf shoots, ultimately leading to *Ulmannia* and *Voltzia* where the radial shoot become flattened and reduced into a sterile bract partially fused with a flattened fertile ovuliferous bract. He (1951, 1954) convincingly put forward his compound theory on the cone. According to him the inflorescence in *Pinus* bears the bracts (bract scales) on the axils of which are the very much reduced dwarf shoots or flowers. The ovuliferous scale represents a compound structure formed by what were originally two megasporophylls and a few sterile scales. It may, therefore be concluded that the female strobilus is an inflorescence and ovuliferous scale is a compound structure formed by degeneration of two megasporophylls and a few sterile scales. But the debate remains about the the identity of the megasporophyls. Perhaps and most probably, these have fused in the Ovuliferous scale (Florin) or the bract scale represents the megasporophylls (Chamberlain).

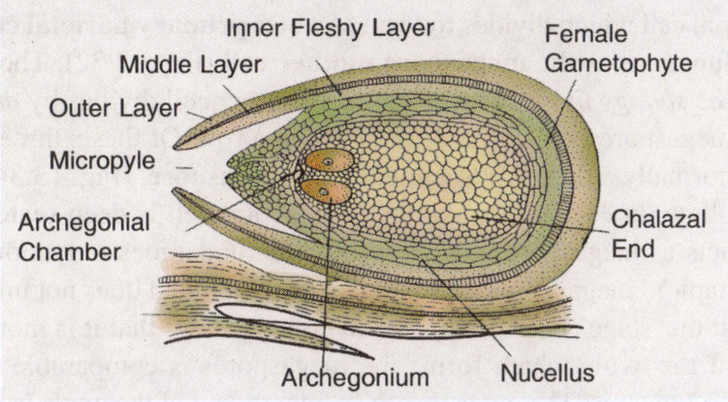

Fig. 13.29. L.S. of a mature ovule of *Pinus roxburghii*. The archegonia have developed.

The ovulate cones takes three years to mature. The mature cones are hard, woody and very large in size.

Development of the Ovule (Figs. 13.29 - 13.30)

On the upper surface near the base of the ovuliferous scale, the ovule arises as a small white rounded swelling consisting of a group of parenchymatous cells, the **nucellus**. The nucellus increases in size and soon a ring-like outgrowth of cells arises from its base and grows up around it forming a covering layer called the **integument**. The latter encloses the nucellus completely leaving only a narrow passage, the **micropyle** at the top. The micropyle faces the base of the scale and is directed obliquely downwards towards the axis of the cone.

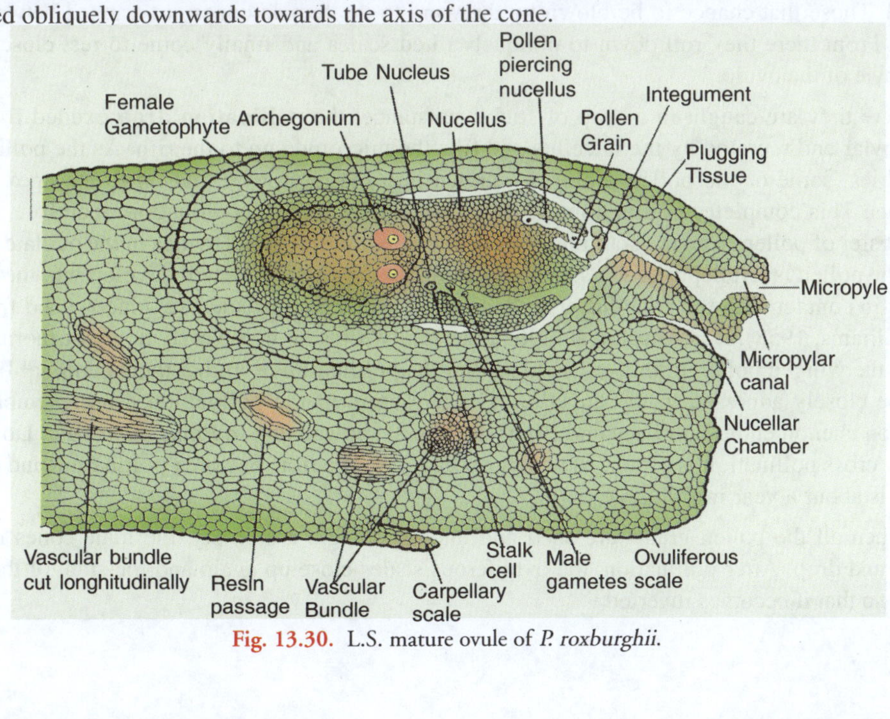

Fig. 13.30. L.S. mature ovule of *P. roxburghii*.

The micropyle leads to the top of the nucellus. The integument is free from it near its micropylar end. About the time, pollination takes place, a hypodermal cell with denser contents appears near the apex of the nucellus in *P. roxburghii* (Konar, 1960). It enlarges to form a single large archesporial cell which divides to form an outer **primary parietal cell** and the inner **sporogenous cell** which functions as the **megaspore mother cell** (Fig. 13.32). The parietal cell divides repeatedly to form the *spongy tissue*. The megaspore mother cell divides by *meiosis* into a tetrad of four small haploid megaspores, (Fig. 13.33) arranged in a row. Of these, three degenerate and only one, the innermost normally enlarges into a functional **megaspore**, (Fig. 13.34) which develops into a female gametophyte. In *P. wallichiana,* the archesporal cell is deep seated (Konar and Ramchandani, 1958) and acts as megaspore mother cell, which undergoes reduction division to form a linear row of three haploid megaspores. In this case the upper diad does not undergo meiosis II. Pollination takes place at this stage. The structure of the ovule shows that it is more than a megasporangium. The nucellus of the ovule which forms the megaspores is comparable to the megasporangium of the vascular cryptogams. The outer sheath or integument of the ovule is an extra structure which is unrepresented in the megasporangium of lower vascular plants (Pteridophyta).

Pollination

It is effected by the wind and is, therefore, **anemophilous**. It takes place towards the end of May or during the month of June in the hills. In plains or at lower altitudes it takes place towards the end of February or mid March. At this period the ovulate cone is young. It is about the time of meiosis which means that the megaspores have been established (Fig. 13.33). The development of the embryo sac is subsequent to this event. The ovuliferous scales which prior to pollination fit tightly over one another, open out providing ready access for pollen grains to the ovules. The microsporangium or pollen sac opens by a medium longitudinal slit along its under side. On slight shaking the tree, which is producing pollen, will emit clouds of yellow dust which rise into the air and colour the whole landscape in the pine forests. The yellow dust is formed by countless numbers of pollen grains rather the immature tiny male gametophytes. They are light in weight and buoyant due to the wings. Upon liberation they float upon the air and settle down only slowly. Much of the pollen is wasted. Those that chance to be blown by the wind to the female cones are trapped between the scales. From there they roll down to obliquely tilted scales and finally come to rest close to the micropyle of the ovule.

Here they are caught in a drop of sticky substance, the **pollination drop** exuded from the micropylar end secreted by the nucellus and fills the micropyle up to the rim. As the pollination drop dries, some of the pollen grains are drawn into the micropyle and on to the apex of the nucellus. This completes pollination. In *Pinus* and other conifers, pollination, therefore, means the transfer of pollen from the staminate cone (microsporangia) to the ovules in the ovulate cones. Here the pollen is left deposited in such a position that the part which will develop into the pollen tube is in contact with the nucellus. The pollination drop contains sucrose, glucose and fructose (McWilliams, 1958). It is secreted at night and usually dries up in the early hours of morning. In *Pinus,* the winged pollen grains are so placed in the pollination drop that their germinal furrows become closely adpresed to the nucellus. The retraction of the pollination drop resembles the guttation phenomenon. Squillace and Bingham (1958) showed that in *Pinus monticola*. Dougl. *ex* Lamb., cross pollination is preferred to self pollination. The time between pollination and fertilization is about a year in *Pinus*.

When all the pollen grains are shed and microsporangia are empty, the male cones rapidly wither and drop. After pollination, the ovuliferous scales close up again and the stalk of the cone curves so that it becomes inverted.

The microsporangiate cones occur on the lower branches and the megasporangiate generally on the upper branches. Thus, position of the cones on the pine tree ensures cross pollination because the pollen does not blow vertically upward.

Reproduction by Gametes

(*a*) Male Gametophyte (Fig. 13.31)

The microspore or the pollen grain is the pioneer structure of the male gametophyte. It begins to germinate even before liberation and develops into a very much reduced male gametophyte. The microspore nucleus divides by a periclinal wall and forms a very small **prothallus cell** on the side of the grain farthest from the wings, (also called **capula**) and a large **central cell** (Fig. 13.31, B). The central cell again cuts off a **second protbhallial cell** on the top of the first (Fig. 13.31, C). The two, prothallial cells represent the last vestige of the vegetative tissue of the gametophyte. They exactly correspond to the gametophyte of a fern but differ in being smaller and parasitic upon the parent generation. The prothallial cells soon degenerate (Fig. 13.31, D) and disappear. The nucleus of the large central cell called the **antheridial cell** divides again and a large rounded cell is formed towards the capula or the degenerating prothallial cells. It is the **generative cell** (Fig. 13.31, D). The other is called the **tube cell** and is towards the **tenuiftas** or the **germ furrow**. In order to see these details mount the mature pollen grains in iron aceto-cermin to which a drop of strong hydrochloric acid (HCl) has been added. The grains should lie with the wings facing downwards. At this stage when the male gametophyte consists of four cells, two small vegetative prothallial cells, a generative cell and a tube cell, the pollen grain is shed.

Further development takes place, when it is lodged in the ovule and has made contact with the tip of the nucellus. The exo-intine ruptures between the wings, the tube cell protrudes and grows out to form a delicate **pollen tube** (Fig. 13.31, D, E, F). The tube nucleus moves down to its advancing tip whereas the generative cell remains in the grain end. The pollen tube grows down into the nucellar tissue upon which it is now dependent for its nourishment and protection. The pollen tube rests for about a year in this condition. The ovuliferous scales, during this period, have thickened and closed the cone completely. The ovule is not yet ready for fertilization. Hence further growth of the male or microgametophyte is arrested. It rests throughout the late summer and following winter resuming activity in the following April. (second year). The pollen tube becomes active again. It penetrates the nucellar tissue. It is an agent of fertilization as is in the angiosperms and does not serve as a nutritive haustorium as is the case in the cycads. The **generative cell** divides to give rise to a barren **stalk cell** (or **sterile cell**) and a fertile **body cell** (or **spermatogenous cell**). The former develops no further (Fig. 13.31, E). The body cell along with protoplasmic contents of the tube cell and stalk cell, passes down the pollen tube. On its way, the body cell divides into two unequal cells with scanty protoplasm and large nuclei. These are the **male cells** or the **gametes**. The gametes are formed only a week before fertilization. Konar (1960) observed that in *P. roxburghii,* the male gametes are equal in size in the beginning but become unequal due to enlargement of one of them.

Many important changes take place in the female cone during the protracted interval of about thirteen months between pollination and the actual act of fertilization. After pollination, the stalk of the cone curves so that the cone becomes inverted. It increases considerably in size and turns green. The cone axis elongates and thickens. The **ovuliferous scales** thicken and become enlarged. Their broad sterile ends fit tightly over one another so as to form rhomboidal areas on the outer surface of the cone. The **bract scales** remain small and concealed. The development of the ovules within continues. The cones are about 4 – 5 cm long by the end of the first year. The growth is continued rapidly in the second year.

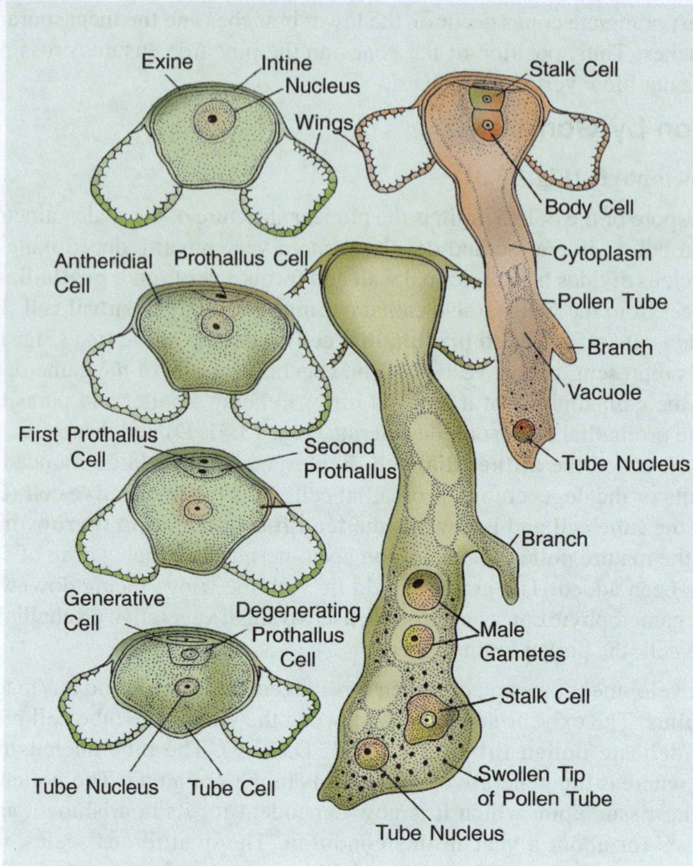

Fig. 13.31. Germination of the microspore or pollen grain of *P. roxburghii*.

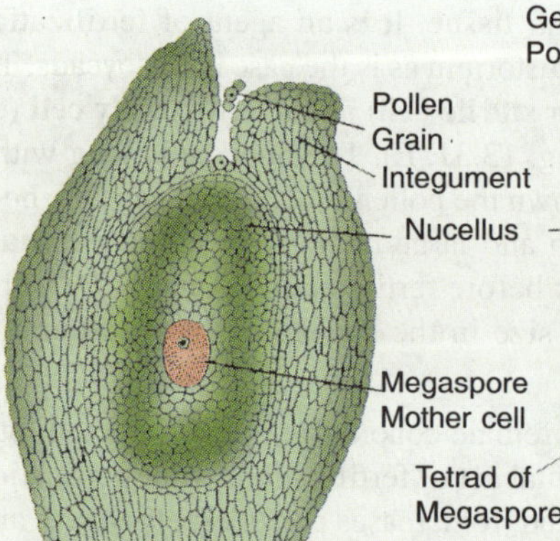

Fig. 13.32. V.S. young ovule of *P. roxburghii* showing a megaspore mother cell in the nucellus.

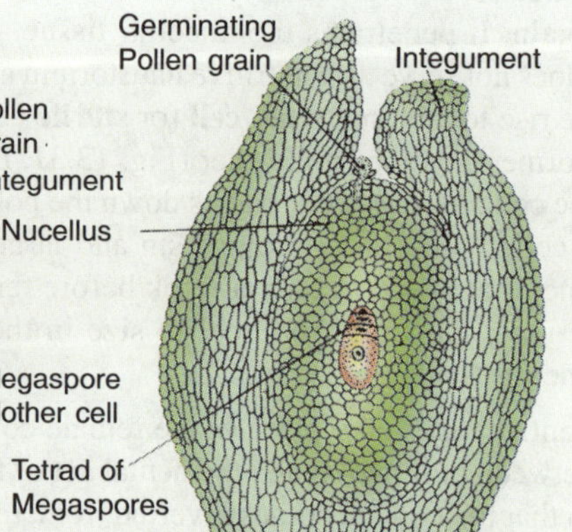

Fig. 13.33. V.S. young ovule of P. *roxburghii* showing a tetrad of megaspores (haploid) in the nucellus.

(*b*) Female Gemetophyte (Fig. 13.32 - 13.36)

The haploid **megaspore** is the first cell of the female gametophyte. It is never shed and is retained permanently within the nucellus. By the end of first year i.e., after pollination, it starts to

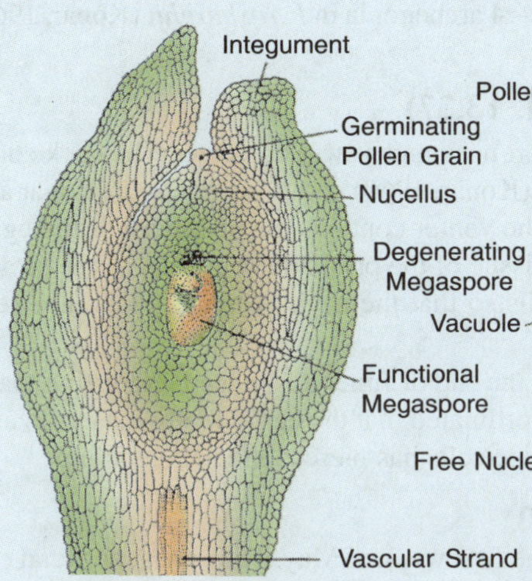

Fig. 13.34. V.S. young ovule of *P. roxburghii* showing enlargement of functional megaspore and degeneration of other three.

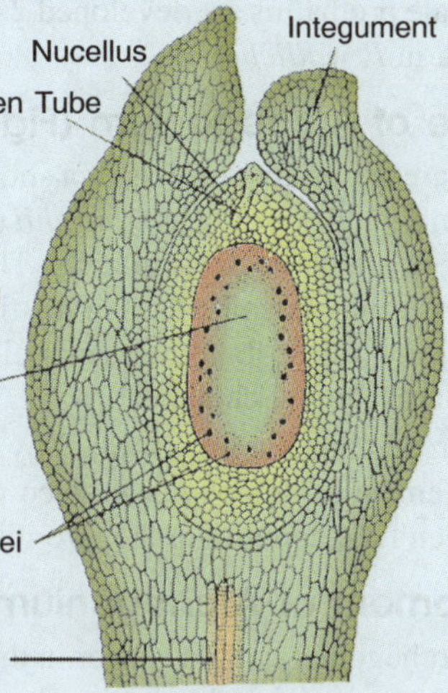

Fig. 13.35. V.S. ovule of *P. roxburghii* with female gametophyte in free nuclear stage.

germinate. The germination of the megaspore and the growth of the female prothallus is very slow. There is an interval of about 13 months between the origin of megaspore and the development of the mature female gametophyte. In *P. wallichiana* and *P. roxburghi* the megaspore forms a few free nuclei before entering upon a period of rest. The rest of development takes place later. The megaspore increases in size. Its nucleus undergoes a series of divisions approximately eleven times in succession. This results in the production of about two thousand five hundred nuclei. These daughter nuclei, according to the old conception, become arranged in a peripheral layer surrounding a large central vacuole. The walls are then laid between these nuclei. They first appear at the periphery of the megaspore. Later the gametophyte is entirely filled up with free nuclei. There is also a gradual accumulation of reserve food materials in the gametophyte. The embryo sac is surrounded by a few layers of spongy tissue that are absorbed completely later. The walls are then laid between them. Wall formation starts by the second week of May. Starting at the micropylar end and extending to the base. Wall formation is effected by a number of alveoli which grow centripetally. Nuclear divisions followed by cross walls result in an increase in the number of cells. The megaspore

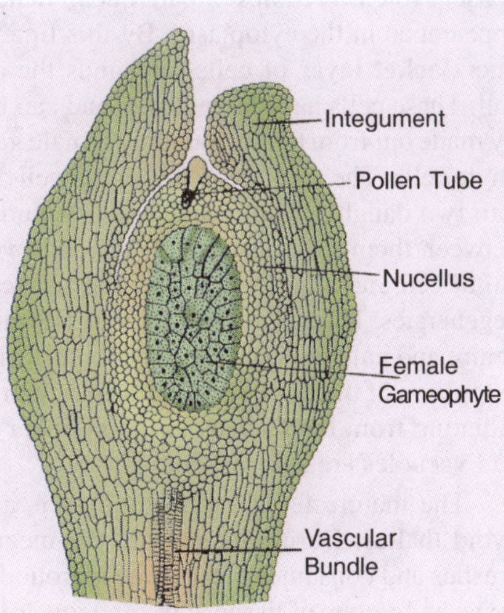

Fig. 13.36. V.S. ovule of *P. roxburghii* with young cellular female gametophyte.

is now filled with a solid mass of thin-walled parenchymatous tissue. It is evidently the female prothallus but sometimes erroneously described as the endosperm. It has neither chlorophyll nor rhizoids. The megaspore containing the female prothallus is enclosed in the nucellus (diploid tissue). Neither of them is detached from the parent plant. Its growth is, therefore, parasitic. From the superficial cells of the female prothallus are developed 2 – 4 archegonia in *P. roxburghii* (Konar, 1960) and one to two archegonia in *P. wallichiana*.

Structure of Archegonium (Fig. 13.37)

Each archegonium consists of a short neck and a swollen venter. The neck consists of four cells arranged in a single tier in *P. roxburghii* (Konar, 1960) and *P. wallichiana* (Konar and Ramchandani, 1956). There are no neck canal cells. The venter contains the **oosphere** or the egg and the **ventral canal cells**. Venter is embedded in the tissue of the prothallus but cells of the prothallus adjacent to the neck, grow faster than the neck cells so that the neck eventually lies in a depression, usually called the archegonial chamber.

The archegonia are fully formed by the end of June. It is indeed interesting that the formation of male and female gametophytes is so co-ordinated that the egg is ready for fertilization before the tip of the pollen tube containing the two male cells has pierced the archegonium.

Development of Archegonium

The archegonial initials appear in the first week of May. They are superficial cells larger in size, with prominent nuclei and vacuolated contents. The initial divides by a periclinal wall into an upper **primary neck cell** and a lower **central cell**. The former divides by two verical walls into a tier of four cells that constitute the neck of the archegonium. The central cell undergoes rapid enlargement and its cytoplasm shows conspicuous vacuoles, which later disappear and the cytoplasm becomes dense. Proteid vacuoles or para nuclei make their appearance in the cytoplasm. By this time a distinct **Jacket layer** of cells surrounds the central cell. These cells have large nuclei and can be easily made out from the surrounding female gametophyte cells. The nucleus of the central cell divides into two daughter nuclei and a wall is laid down between them to form a small lenticular **ventral canal cell** and a large **egg cell**. The former soon degenerates. The egg nucleus migrates towards the centre and enlarges considerably. At maturity, the cytoplasm of the egg becomes fibrillar with fibrils radiating from the nucleus to the periphery. Proteid vacuoles are also present.

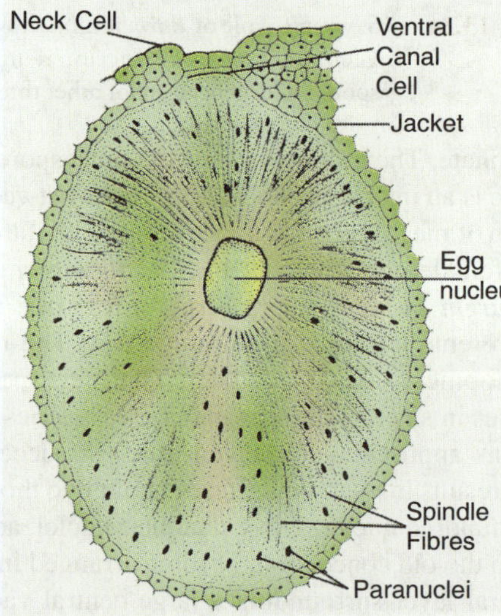

Fig. 13.37. A mature archegonium of *Pinus wallichiana* (After Konar).

The mature female gametophyte is a small ovoid thallus. During its growth, the megaspore crushes and consumes most of the surrounding nucellar tissue reducing it eventually to a thin membrane and a cone of tissue at the end towards the micropyle. It is the **nucellar beak**. Its cells contain abundant starch.

The megaspore with its contained female prothallus corresponds to the embryo sac of the angiosperms. The latter is also a gametophyte. The female gametophyte of the *Pinus* differs from that of an angiosperm in certain respects. It is a more massive structure consisting of a large number of cells. Besides the reproductive structures, *i.e.*, the archegonia are far more complex.

The archegonia of *Pinus* are essentially like those of the ferns differing in, (*i*) both the venter and the neck are embedded in the tissue of the prothallus, (*ii*) the larger size of the venter and the enclosed oosphere and (*iii*) a smaller and simpler neck without neck canal cells. The fern prothallus is a large green independent structure which is anchored to the substratum by means of rhizoids. It bears both megagametes and microgametes. The *Pinus* female prothallus is small ovoid prothallus which is neither green nor has any rhizoids. It is parasitic upon the parent plant and bears only megagametes.

Fertilization

The pollen tube, whose growth is arrested in the winter of the first year after it has penetrated a short distance into the nucellus, resumes growth in the month of April when the female cone is running into its second year. The generative cell divides into a **stalk** and **body** cell. The body cell then divides into two male cells, or microgametes but no wall is formed separating the cytoplasm of one from the other. The male cells are non-motile. In this respect, these differ from the motile gametes of the Cycads. They float within the pollen tube. Meanwhile the pollen tube has passed entirely through the nucellus and reached the female prothallus. As in the Cycads, it has to pierce a relatively short distance from the tip of the nucellus to the female prothallus whereas in the angiosperms, it must penetrate the tissue of the style and the cavity of the ovary to reach the ovule. From there it passes into the micropyle and penetrates the nucellus. Finally the tip of the pollen tube enters the neck of the archegonium after rupturing it and swells at its tip. It is evident that the pollen tube in *Pinus* is an agent of fertilization whereas in the Cycads it does not enter the archegonium. The microgametes in Cycads are discharged in the liquid in the archegonial chamber over the archegonia and enter their necks to effect fertilization. In the *Pinus* the two non-motile cells and the other contents of the pollen tube are discharged into the venter, where all but one male cell disintegrate. The pollen tube ruptures in the venter. The nucleus of the surviving male cell unites with that of the egg to accomplish fertilization which process is usually completed in the last week of June. The fertilized egg secretes a wall round it and becomes **oospore** or **zygote**. This act of fertilization restores the double chromosome number in the nucleus. The zygote, therefore, is the first cell of the sporophyte. More than one archegonium may be fertilized in a single ovule but only one comes to maturity, others perish.

McWilliams (1958) gave a vivid account of fertilization in *Pinus*. Shortly before the contents of the pollen tube are released into the venter, a receptive vacuole appears in the upper part of the egg cytoplasm. The functional male nucleus approaches the female nucleus and they come in contact or sink into each other. Their membranes at the point of contact dissolve and the chromatin material in each nucleus condenses to form a prophase-like configuration and later the chromatin transforms into distinct chromosomes.

During the winter of second year, the cones change their colour from green to brown and the following spring, *i.e.*, during its third year the woody cones reach maturity. They become dry. Their ovuliferous scales separate from one another. The seeds are then shaken out.

EMBRYOGENY (Figs. 13.38, 13.39)

The zygote immediately germinates. It undergoes a series of divisions still enclosed within the ovule, to form a relatively elaborate **Pro-embryo**. Before divisions, the nucleus of the zygote is in central position. It divides twice to form four nuclei, which migrate to the chalazal end of the archegonium. They divide again. The resultant eight nuclei arrange themselves in two tiers of four each. Walls appear between the four basal nuclei. The upper four nuclei take no further part in the development of the embryo. Each of the four small cells of the basal tier divides once by a transverse wall to form two tiers of four cells each. The cells of the lower tier divide again. In this way, three tiers of four

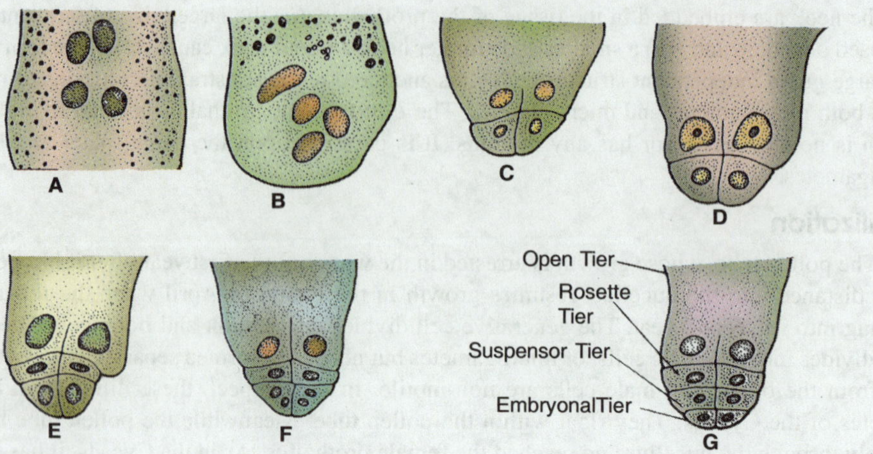

Fig. 13.38. (A-G). Early stages in the development of embryo in *Pinus wallichiana* (After Konar)

cells each are formed. The fourth uppermost tier is formed by the four free nuclei which as stated above take no share in the development of the embryo and are separated by imperfect walls. They form an open upper tier without walls on one side. This symmetrical group of sixteen cells is called the **pro-embryo**. The four lowermost cells farthest from the micropylar end constitute the **embryo tier**. Each of these may develop into an embryo. The cells next above the embryonal cells constitute **suspensor tier**. The third tier from below is called the **rosette tier**. The suspensor cells elongate considerably and push the embryonal cell out of the archegonium and deep into the tissue of the female prothallus, at the expense of which they grow. The four suspensor cells undergo cleavage on fourth elongation and separate from each other. Each suspensor cell has a rapidly dividing embryonal cell at its tip. As division continues in the embryonal cell, it passes through the quadrant and then an octant stage. Each of the octants is a **potential embryo**. It is formed from one embryonal cell only. There are thus four potential embryos formed from one fertilized egg. This curious feature of **embryogeny** is termed as **cleavage polyembryony** in contra-distinction to the similar condition which may be the result of fertilization of more than one archegonium. This is **simple polyembryony**. In the course of further

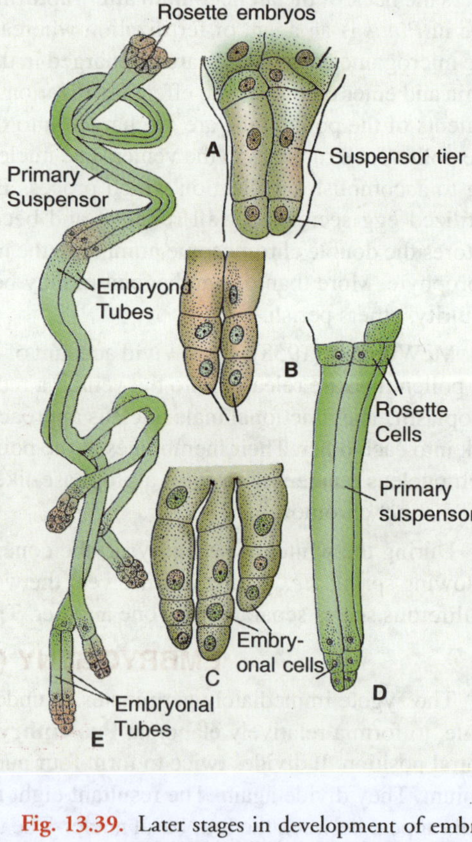

Fig. 13.39. Later stages in development of embryo in *Pinus wallichiana*. (After Konar)

development of the potential embryos, one takes the lead and grows more rapidly than the others. The actively growing potential embryo by cell division, cell growth and differentiation becomes the embryo of the seed, others die off, while in this embryo development is taking place. The female prothallus grows, enlarges and inroads in the nucellus. Its cells become packed with food material which is used by the growing embryo and also serves a reserve in the seed. Sometimes the cells of the rosette tier also develop into embryos and this is called **rosette polyembryony**.

Formation of Seed (Figs. 13.40 - 13.43)

It is rather complicated structure and is formed when the cone is running into its third year. It is formed from the mature ovule (containing the female prothallus and embryo) which with its contents falls from the tree as a seed.

The growing embryo feeds on the central tissue of the female prothallus. The space so formed is filled with the folds of the massive suspensor. As the embryo matures, the suspensor eventually dries to form a coiled thread towards the root end of the embryo. The embryo at this stage becomes dormant. The outer unused layers of the female prothallus tissue persist in the mature seed and function as a nutritive tissue surrounding the embryo. It is the so called **endosperm**. It is different in origin from the endosperm of angiospermic seeds but has the same function. It will be recalled that as the female prothallus enlarges and becomes laden with food materials, it invaded the surrounding nucellar tissue which is invariably crushed except for a thin cap-like structure at the micropylar end of the endosperm. It is called the **nucellar cap**. The integument, which at an early stage of development, consists of a median stony layer with inner and outer layers of soft cells, dries up. The outer layer of the soft cells disappears, the middle stony layer hardens to form the **testa** and the inner soft layer forms a thin brown membranous covering called the tegmen. The seed now consists of the hard testa, the papery tegmen, endosperm and the embryo with a nucellar cap at the micropylar end. As the seed matures, a thin layer of tissue from the upper surface of the ovuliferous scale separates as a membranous wing which is attached to the testa and aids in seed dispersal. The pine seed is now ready to be liberated. At this stage, the female cone is in its third year. It is large, dry, brown and woody. Under dry atmospheric conditions, the ovuliferous scales spring apart and the winged seeds, as they

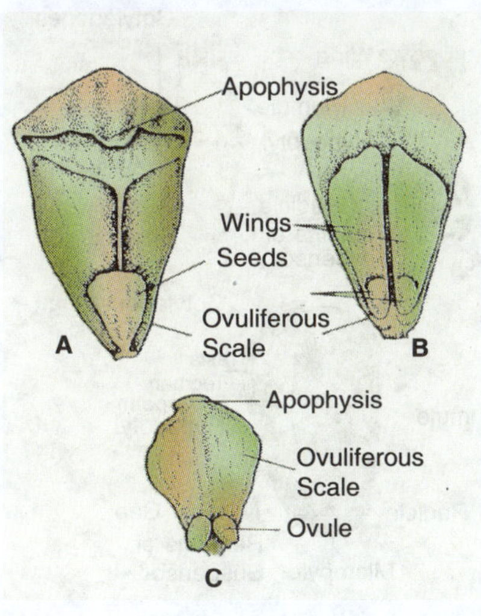

Fig. 13.40. (A – C). Mature megasporophyll of *P. roxburghii* as seen from ventral side. B. Megasporophyll bearing two seeds. C. Megasporophyll bearing two ovules.

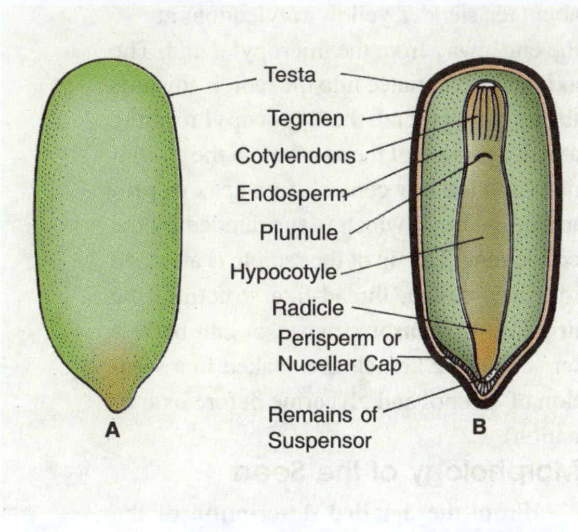

Fig. 13.41. (A – B). A. Seed of *P. gerardian*. B. L.S. of seed.

escape from the cones, are blown away by the wind. The opened out ovuliferous scale is a large, thick woody structure with two seed pits at its base and well marked wing impressions on its upper surface in continuation with the seed cavities.

The seeds of *Pinus gerardiana*, which are commonly known as chilgoza, are fairly large and used as human food.

Structure of the Seed

There is the outer hard shell or seed coat called the **testa**. It protects the inner delicate parts of the seed from drying and mechanical injury. Within the testa is the brown papery inner seed coat called the **tegmen**. The latter surrounds a white fleshy kernel which on analogy with angiospermic seeds is often called the **endosperm**. Towards the pointed end of the endosperm is the nucellar cap which represents the remains of the nucellus. In the centre of the endosperm is a distinct central cavity in which lies the embryo. The embryo is straight and consists of a short axis bearing a ring of about ten slender, yellow cotylendons at the end, away from the micropylar end. The axis is differentiated into the radicle towards the micropylar end, the **hypocopyl** forming a major portion of the axis below the cotyledons and the tiny conical stem apex or **plumule** (epicotyl) which is surrounded by the cotyledons. The tip of the radicle is attached to a dry, coiled, thread-like structure, the dried up **suspensor** (suspensor can be better seen if the kernels are soaked in a solution of alcohol and glycerine before examination).

Morphology of the Seed

From the detailed description of the structure of the seed, it is evident that the seed of *Pinus* or other gymnosperms is not a single structure. It is made up of tissues

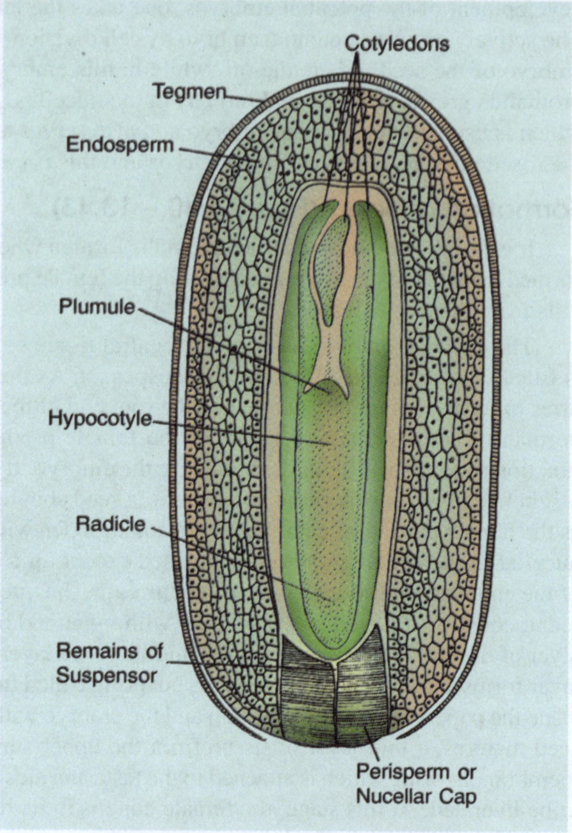

Fig. 13.42. L.S. seed of *Pinus roxburghii.*

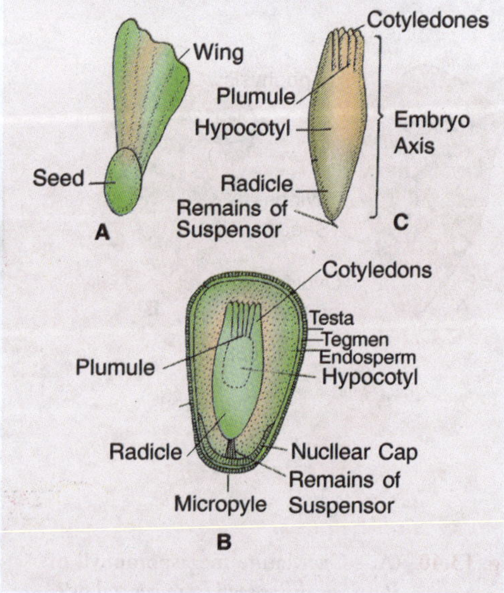

Fig. 13.43. (A – C). A. Winged seed of *P. roxburghii.* B. L.S. seed. C. The embryo after taking it out from the seed.

which do not belong to the same generation. It represents three generations as follows :

The testa or seed coat which is the matured integument and the nucellar cap, which represents the remains of the nucellus (megasporangium) are parts of the old sporophyte (pine tree); the food storage tissue of the female prothallus which is often called the endosperm and surrounds the embryo represents the next generation, the **female gametophyte**. The third generation is an immediate off-spring of the pine tree sporophyte which may develop into a new tree. This is the embryo and is for the time being dependent for its nutrition (parasitic) upon its parent, the female gametophyte represented by endosperm, which in its turn, is protected by the seed coats and is dependent on the old sporophyte. The pine seed is, therefore, a representative of three generations.

Origin of the Seed

The formation of this important reproductive organ, **the seed**, in the spermatophytes, is the result of three conditions : heterospory and its further specialization, retention and germination of the megaspore within the nucellus of ovules, and formation of the dormant embryo.

(*i*) Heterospory. It ocurs in a number of lower vascular plants such as *Selaginella, Marsilea*, etc., in which the two kinds of spores, the microspores and the megaspores are produced in separate sporangia. On germination, they give rise to reduced and endosporic male and female gametophytes respectively. In the spermatophytes, the male gametophyte is further reduced but it is the female gemetophyte which has undergone remarkable changes.

(*ii*) Retention of Megaspore. In the spermatophytes, the megaspore is not at all discharged from the parent plant, because the nucellus which encloses it never opens. The megaspore thus germinates *in situ*, to give rise to the female prothallus representing the female gametophyte. The latter in the gymnosperms produces female gametes (eggs) in archegonia. The female prothallus and the enclosing nucellus are surrounded by a protective covering the integument. This whole structure is called the ovule. This ovule is the potential seed. The ovule remains attached to the parent plant till it is matured. The maturity of the ovule is dependent upon the transfer of the microspores, containing little microgametophytes within them through the air, in a position near the female gametophyte which is buried in the nucellar tissue and eventually the carriage of the microgametes through nucellar tissue to the egg in the archegonium. A new structure, the **pollen tube**, develops from the male gametophyte. It helps in accomplishing fertilization by carrying the male gamete to the egg in the archegonium.

(*iii*) Formation of dormant embryo. The fertilized egg immediately develops into the embryo which, as it grows, feeds on the surrounding tissue of female gametophyte. The rest of the tissue of the female gametophyte whose cells are gorged with food, functions as the endosperm. The integument hardens to form the seed coat. The ovule is now mature, and ready to fall from the tree as a seed. As pointed out above, the seed is a reproductive structure containing a dormant embryo—the future sporophyte, well supplied with food reserve in the form of endosperm and heavily protected against drying and mechanical injury by the hard seed coat or testa. The formation of seed thus involves the nutrition of embryo sporophyte, through the much reduced female gametophyte, by the parent sporophyte of the previous generation. The gametophyte, therefore, merely functions to transfer food. It is not a supplier of food.

Biological Importance of Seed

The seed is a special structure linked with the adaptation of land habits. As a matter of fact perfect adaptation to life on land and subsequent successful distribution on the surface of the globe has much to do with the formation of seed in the spermatophytes. The bryophytes and pteridophytes suffer from two great handicaps in their successful adaptation to life on land. These are the presence of the exernal water at the time of fertilization as their sperms are motile and the forced germination of their embryos in external conditions whether they are favourable or not. In the spermatophytes the seed overcomes both these handicaps. The first handicap, *i.e.*, presence of external water at the time of fertilization is eliminated by the formation of non-motile male gametes in the male gametophyte. The latter in a state of partial development in the pollen grain is transported through the air to the micropyle of the ovule and from there on to the top of the nucellus. Here the pollen grain puts out a pollen tube which carries the male gametes, and traverses through the nucellus to reach the archegonia on the female gametophyte. This makes fertilization more certain. The second difficulty has been overcome by the formation of a dormant embryo, well supplied with food materials and protected by a heavy seed coat which enables it to live in a state of dormancy until the external conditions become favourable for its growth.

Advantages of Seed

The seed has many advantages over similar reproductive structures such as the spores, the zygotes and the like. Its heavy often hard and impervious seed coat protects the innermost delicate parts from drying and mechanical injury. Owing to a very low percentage of water within it, the seed is able to withstand a long period of low temperature. The reserve food within the seed enables the embryo to remain dormant but alive until the external conditions are favourable for growth. The food stored within the seed supplies a requisite amount of energy at the time of germination. The differentiation of the dormant embryo in advance of dormancy into the radicle, the plumule and the cotyledons hastens the establishment of the primary root and the primary leaf at the time of germination. The seeds of pine and many angiosperms develop structures like wings, plumes or hair etc., which help in their dissemination away from the parent plant to more suitable localities. This leads to the distribution of the species over a wide area.

Germination of the Seed (Fig. 13.44)

The seed may germinate at once if it falls on a suitable soil where it can get moisture. The resting period if any, is not long. The hypocotyl elongates absorbing food from the endosperm and the testa splits open. The micropylar end of the hypocotyl emerges first through the crack. Its tip-radicle grows downwards and penetrates the soil to become the primary root. This is followed by the elongation of the rest of the hypocotyl which may at first be arched as it lifts the seed out of the soil because its one end (radicle) is anchored in the soil and the other (epicotyl) is put in the seed. The arched hypocotyl then straightens out pushing the cracked testa, remains of nucellus if any and what is left of the endosperm through the ground ahead of the cotyledons. As the germination proceeds cotyledons become green, their tips absorbing what is left of the endosperm. Finally, a thin film of the latter is left which dries and breaks away as the cotyledons unfold and eventually drop off and decay along with the testa. The tiny plumule is very slow to show any signs of growth and practically no difference is seen in it until the cotyledons are

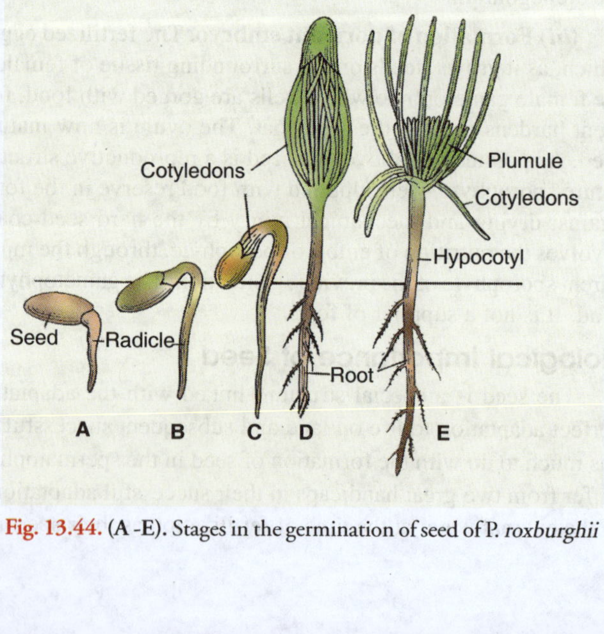

Fig. 13.44. (A -E). Stages in the germination of seed of P. *roxburghii*

unfolded. The green unfolded cotyledons begin to manufacture food. The embryo seedling is now an independent plant. The cotyledons eventually wither and fall off. The plumule grows into a shoot of unlimited growth. It bears only green needle-like leaves in a spiral manner. When the seedling is about 3 to 4 inches in height, first foliar spurs appear in the axils of the green needle-like leaves. As the seedling grows, the green needles become progressively smaller in size until they gradually merge into the scale leaves characteristic of the pine tree. After a long period of growth, the seedling gradually develops into a mature sporophyte consisting of a stem, branches, leaves and cones. The life cycle of *Pinus* is graphically represented in Fig. 13.45.

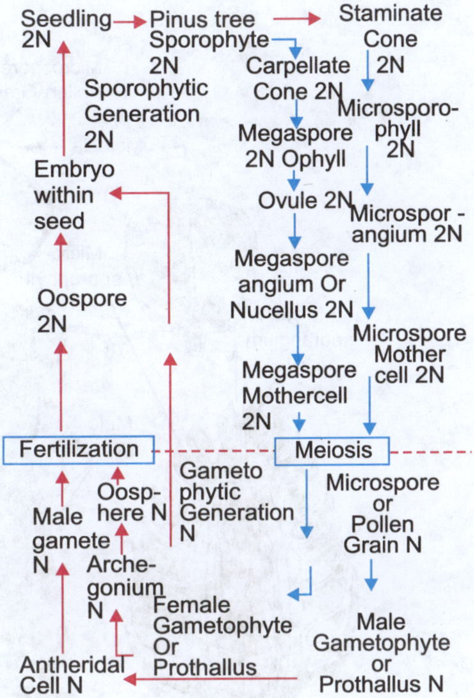

Fig. 13.45. Graphic life cycle of *Pinus*.

Duration of the reproduction

In Northwest Himalayan species of *Pinus* growing at an altitude between 500 – 2500 meters especially *P. roxburghii* the cones are initiated in september. Pollination takes palce at the four celled stage of pollen grains in March next (Ist year) the germination of the pollen takes place immediately on the nucellus. Then follows a rest period of about ten months. Growth is resumed in March (IInd year) and fertilisation occurs in April (IInd year).

The female cones are initiated in February. Soon after pollination in March, the cone undergoes a period of rest for about ten months. During January (IInd year) the female gametophyte is at free nuclear stage. Growth of the female gametophyte is resumed and fertilisation occurs in April (IInd year). The embryo is formed and matures by December (IInd year). After fertilisation, there is a gap of about 4 – 5 months and the cones open and shed their seeds in April – May (IIIrd year). Usually it takes three years for the matration and shedding of seeds. Undehisced cones may also contain mature seeds.

Plate VIII : A young plant of *Pinus roxburghii.*

Alternation of Generations

The life cycle of *Pinus* displays the same alternation of generations as we have seen in other types lower in organisation than the gymnosperms. The sporophyte phase is of long duration and is represented by the huge pine tree which bears the staminate and ovulate cones. The staminate

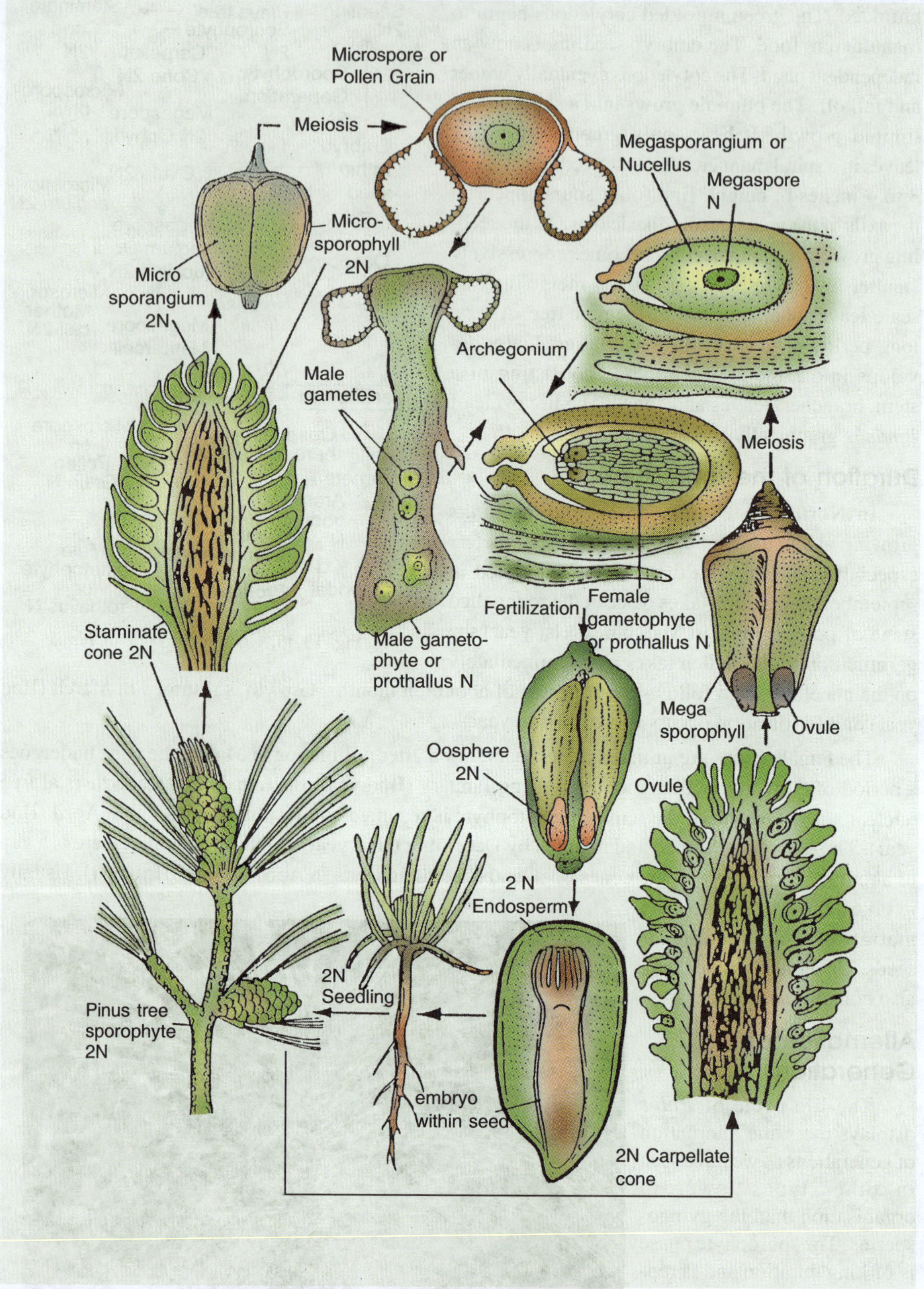

Fig. 13.46. Diagrammatic life cycle of *Pinus roxburghii*.

cone produces pollen grains in pollen sacs on the microsporophylls. The female cone produces megaspores one in each ovule borne on the upper surface of the ovuliferous scale. Since meiotic divisions occur at the time of the differentiation of the pollen grains and megaspores, the latter sturctures represent the first cells of the second generation—the gametophytes. Both the gametophytes (male and female) in *Pinus* are further reduced and are parasitic upon the sporophyte. They are retained within the respective spores. The vegetative or sterile phase of the male gametophyte is represented by the prothallus cells. The female gametophyte, no doubt, consists of an extensive tissue forming the female prothallus but its chief function is to provide nutrtion to the embryo. It obtains the food for this purpose from the sporophyte plant. It has neither chlorophyll nor rhizoids, hence it is unable to carry on independent assimilation. The most remarkable feature is that it is retained within the sporophyte tissue—the nucellus and not discharged. The little microgametophyte is carried by the wind to a position in the vicinity of the ovule from where, it is shifted on to the apex of the nucellus. It then puts out a pollen tube which carries the male gametes on to the archegonia developed at the micropylar end of the female prothallus. The male gametes unite with the egg in the archegonium to effect fertilization. The fertilized egg is a diploid structure. It is the first cell of the furture sporophyte. It develops into the embryo which is the new sporophyte.

QUESTIONS

ESSAY TYPE :

1. Describe the structure of seed and development of embryo in *Pinus*.
(Lucknow, 1993; Punjab, 2003)
2. Compare the internal structure of young and old stem of *Pinus*.
(Lucknow, 1994)
3. Write an illustrated account of the female reproductive structures of *Pinus*.
(Lucknow, 1995; Vinoba Bhave, 1997)
4. (*a*) By means of labelled diagrams only, show the development of embryo in *Pinus*.
(Lucknow, 1996; Meerut, 1997; Rohilkhand, 2004)
(*b*) With neat and labelled diagrams give a detailed account of stem anatomy and reproduction in *Pinus*.
(Rohilkhand, 2001)
5. Describe the structure of female cone of *Pinus* and comment upon the morphological nature of ovuliferous scale.
(Lucknow, 1999; Agra, 1995; Rohilkhand, 1997; Bhagalpur, 1990; Agra, 2000)
6. Give an illustrated account of the internal structure of *Pinus* needle and point out its xerophytic characters.
(Kanpur, 1996; Agra, 1994; Rohilkhand, 2002)
7. Explain the life cycle of *Pinus* with the help of labelled diagrams only.
(Kanpur, 1997; Allahabad, 1998; Rohilkhand, 1999)
8. Give the fertilisation in *Pinus*. Describe the development of embryo in *Pinus*.
(Kanpur, 1998; Poorvanchal, 1999)
9. Describe the development of male gametophyte of *Pinus*.
(*Kanpur*, 1999; *Awadh*, 1993; *Poorvanchal*, 2000; *Bhagalpur*, 1994; *Himachal Pradesh*, 2000)
10. Explain pollination and fertilisation in *Pinus*. Describe the development of embryo.
(*Kanpur*, 2000; *U.P. College*, 1996).
11. Describe the post fertilisation/Pollination changes in the ovule of *Pinus* till seed formation.
(Kanpur, 2001, 2003; Awadh, 1996; Poorvanchal, 2001; Agra, 2002)
12. Describe the development of female gametophyte of *Pinus* with labelled diagrams.
(Kanpur, 2004; Magadh, 1992)
13. Give an account of embryo in *Pinus* and compare it with that of *Cycas*.
(Allahabad, 1992)
14. With the help of suitable diagrams describe the structure of wood of *Pinus* as is seen in transverse, radial and tangential longitudinal sections. **(Allahabad, 1994, 1999; Meerut, 1997; Gorakhpur, 1999)**

15. With the help of suitable diagrams, describe the anatomy of young and old stem of *Pinus*.
(Allahabad, 1998; U.P. College, 1998; Gorakhpur, 1996)
16. Write detailed account of Economic importance of conifers.
17. Discuss the structure of male cone of *Pinus* with suitable diagrams.
(U.P. College, 1997; Vinoba Bhave, 1997)
18. Describe the structure of ovule, mode of pollination and fertilisation in *Pinus*.
(U.P. College, 1999)
19. Give a comparative account of the male gametophyte of *Pinus* and *Ephedra*.
(Agra, 1991; 1994)
20. Describe secondary growth with diagrams in *Pinus* **(Agra, 1992)**
21. Give a comparative account of the post fertilisation developments leading to the formation of seeds in *Pinus* and *Cycas*.
(Gorakhpur, 1997)
22. Define alternation of generations. Explain it with the help of labelled diagrams only in the life of *Pinus*. **(Rohilkhand, 1996)**
23. Describe the structure of seed cone of *Pinus* and discuss its evolutionary advancement over the fossil conifers.
(Rohilkhand, M.Sc. 1997)
24. What is Polyembryony ? Explain with the help of embryo development in *Pinus*.
(Garhwal, 2004)
25. Compare the ovules of *Pinus and Gnetum*.
(Vinoba Bhave, 1999)
26. Describe the female gametophyte of *Pinus*, upto the formation of seed.
(Punjab 1998, 2001)
27. (*a*) Giving the systematic position of *Pinus* state its occurence and external morphology.
(*b*) Draw a diagram of T.S. *Pinus* stem.
(Punjab, 1992)
28. (*a*) Describe structure of ovule of *Pinus* before fertilisation.
(*b*) Describe male cone of *Pinus*.
(Punjab, 2002)

SHORT ANSWER TYPE :

29. Draw labelled diagrams of the following :
(*i*) T.S. Dwarf shoot of *Pinus*.
(Lucknow, 1992, 1994, 1997, 2003; Allahabad, 2003; Kumaon, 2000; Awadh, 1998, 2003; Gorakhpur, 1990, 1992, 1996, 1999 Rohilkhand, 1993)
(*ii*) Medullary rays in *Pinus* stem.
(Lucknow, 1992)
(*iii*) T.S. *Pinus* needle.
(Lucknow, 1993, 1996, 1998, 2001; Kanpur, 1995, 1997, 1998; Allahabad, 1994, 1997, 1998, 2000, 2002, 2004; Kumaon, 2003; Awadh, 1994; U.P. College, 1997; Poorvanchal, 1996, 2001; Agra, 1991, 1996; Meerut, 1999; Gorakhpur, 1990, 1993, 1995, 1995; Rohilkhand, 1997; Kanpur M.Sc, 2001; Raipur,1996; Kerala, 1999; Bhagalpur, 1993)
(*iv*) R.L.S. of *Pinus* stem.
(Lucknow, 1997; Kumaon, 2002; Poorvanchal, 1996, 1999; Agra, 1991, 1995)
(*v*) L.S. female cone of *Pinus*.
(Lucknow, 2002; Kanpur, 2000; Allahabad, 1993, 2004; Awadh, 1997, 2000; Poorvanchal, 2003; Gorakhpur, 1994, 1998, 2003; Rohilkhand, 2000; Garhwal, 2003)
(*vi*) T.S. of *Pinus* young stem.
(Kanpur, 1996, 1999; Allahabad, 1996; Meerut, 2000)
(*vii*) Needle of *Pinus*. **(Allahabad, 1992)**
(*viii*) T.S. *Pinus* root having secondary growth. **(Allahabad, 1993, 2000)**
(*ix*) L. S. male cone of *Pinus*
(Allahabad, 1994, 1998; Awadh, 1992, 1995, 1996; Poorvanchal, 2002; Agra, 1995)
(*x*) Dry and wet pollen grains.
(Allahabad, 1995)
(*xi*) Pollen grains of *Pinus*.
(Allahabad, 1997; Awadh, 1995)
(*xii*) T.L.S. and R.L.S. Wood (stem) of *Pinus*.

(Allahabad, 1997, 2002; Kumaon, 2004; Awadh, 1995; Agra, 1993; Rohilkhand, 1995; Garhwal M Sc, 2004; Kerala, 1999; Poorvanchal, 1996)

(*xiii*) T.S. young root of *Pinus*.
(Allahabad, 2003; Awadh, 1996; 1998; Punjab, 1993).

(*xiv*) L.S. *Pinus* ovule.
(Poorvanchal, 2001; Raipur, 1996; Punjab, 1993; Bhagalpur, 1994; Himachal Pradesh, 2000)

(*xv*) L.S. of II year of female cone of *Pinus*.
(Kumaon, 2001; Agra, 1994)

(*xvi*) Winged Pollen.**(Poorvanchal, 1996)**

(*xvii*) Seed of *Pinus*.
(Gorakhpur, 1990; Punjab, 1993)

(*xviii*) Dwarf shoot of *Pinus*.
(Gorakhpur, 1998)

(*xix*) Ovuliferous scale of *Pinus*.
(Gorakhpur, 2003)

(*xx*) T.S. of *Pinus* leaf.
(Rohilkhand, 2003; Meerut, 1998)

(*xxi*) Cones of *Pinus*.
(Mahatma Gandhi, 1998)

(*xxii*) Life cycle of *Pinus*.
(Gurunanak Deo, 1992, Calicut, 2001; Kerala, 2001)

30. Write short notes on :

(*i*) Morphological nature of female cone of *Pinus*.
(Lucknow, 1992; Allahabad, 1997, 2000; U.P. College, 1997; Gauhati, 2003; Utkal, 2003)

(*ii*) Pycnoxylic wood.
(Lucknow, 1994; U.P. College, 1999)

(*iii*) Polyembryony in *Pinus*.
(Lucknow, 1995; Kanpur, 1996, 2002; Allahabad, 1992, 1993, 1996, 1997; Kumaon 1996; 1997; Awadh, 2000; U.P. College, 2000; Poorvanchal, 2000; Gorakhpur, 1991, 1993, 1994; Rohilkhand, 2002; Kanpur M.Sc, 1995; Meerut, 2001; Mahatma Gandhi, 1998; Vir Kumar Singh, 1993)

(*iv*) Economic importance of *Pinus*.
(Lucknow, 1996, 1997; Kanpur, 2003; Allahabad, 1993, 1997, 1998, 2004; Awadh, 1999; Poorvanchal, 1998; Gorakhpur, 1992; Punjab, 2003)

(*v*) T.S. Dwarf shoot of *Pinus*.
(Lucknow, 1997; Allahabad, 1992, 1996, 2001; Kumaon, 1995; Awadh, 2000; Meerut, 2000)

(*vi*) Internal structure of the needle of *Pinus*.
(Lucknow, 1999, 2002; Bangalore, 2001, 2002; MG University, 2004)

(*vii*) Morphological nature of ovuliferous scale in *Pinus*.
(Lucknow, 2000, 2003, 2004; Kanpur, 1995, 2002; Agra, 1992; Mahatma Gandhi, 1998; Himachal Pradesh, 2000)

(*viii*) Male gametoplyte of *Pinus*.
(Lucknow, 2001; Mahatma Gandhi, 1998)

(*ix*) Male cone of *Pinus*.
(Kanpur, 1995; Kumaon, 1999; Poorvanchal, 1997; Gorakhpur, 1995; Rohilkhand, 2002; Kerala, 1999)

(*x*) Medullary rays of *Pinus*.
(Kanpur, 1997)

(*xi*) L.S. female cone of *Pinus*.
(Kanpur, 1999)

(*xii*) Embryogeny in *Pinus*.
(Vinoba Bhave, 1998; Kerala, 1998)

(*xiii*) Ovuliferous scale.
(Kumaon, 1996; Poorvanchal, 1999)

(*xiv*) *Pinus* seed.
(Kumaon, 1998; Awadh, 2000; U.P. College, 2000; Poorvanchal, 1997; Gurunanak Deo, 1997, 1998)

(*xv*) Pollen grain of *Pinus*.
(Awadh, 1999; Rohilkhand, 1998; Mahatma Gandhi, 1998; Bhagalpur, 1993)

(*xvi*) Cleavage polyembryony.
(U.P. College, 1997, Rohilkhand, 1996)

(*vii*) Pollen grains of *Pinus* and *Ephedra*.

(U.P. College, 1998; Garhwal M.Sc, 2004)

(*xviii*) *Pinus* needle leaf. **(U.P. College, 2000; Gorakhpur, 1990; Vinoba Bhawe, 1999)**

(*xix*) Anatomy of *Pinus* root. **(Gorakhpur, 1993, 1996)**

(*xx*) Female cone of *Pinus*. **(Gorakhpur, 1996; Rohilkhand, 1997; Agra, 2002)**

(*xxi*) Development of embryo in *Pinus*. **(Rohilkhand, 1999; Mahatma Gandhi, 1998)**

(*xxii*) Geographical distribution of *Pinus* in India. **(Rohilkhand, 2004; Calicut, 2003)**

(*xxiii*) Dwarf shoot of *Pinus*. **(Kanpur M.Sc, 1995, 1992; Mahatma Gandhi, 1998; Kerala, 2000; Virkumar Singh, 1991; Agra, 2000; Poorvanchal M.Sc, 1996)**

(*xxiv*) Seed scale complex in *Pinus*. **(Kanpur M.Sc, 2000)**

(*xxv*) L.S. of *Pinus* stem. **(Gauhati, 2003; Utkal, 2003)**

(*xxvi*) Development of female gametophyte in *Pinus*. **(Devi Ahilya, 2001)**

(*xxvii*) Male cone of *Pinus* and pollen grains. **(Bhagalpur, 1995)**

(*xxviii*) Vascular bundle of *Pinus* needle. **(Calicut, 1999)**

(*xxix*) Cone of *Pinus*. **(Utkal, 2003)**

(*xxx*) RLS wood of *Pinus*. **(Agra, 2000)**

31. Compare between:

(*a*) Fertilisation in *Cycas* and *Pinus*. **(Meerut, 1997)**

(*b*) Megasporophylt of *Pinus* and *Cycas*. **(Rohilkhand, 1992)**

(*c*) Foliage leaves of *Cycas* and *Pinus*. **(Rohilkhand, 1993, 1994)**

32. What are the functions of wings in the pollen grains of *Pinus* ? **(Gorakhpur, 1997)**

33. What are resin canals ? **(Gorakhpur, 1994, 1998)**

34. What is the significance of unfoldings of mesophyll in *Pinus* leaf ? **(Gorakhpur, 2003)**

35. Differentiate between:

(*a*) Archegonium of *Cycas* and *Pinus*. **(Rohilkhand, 1992)**

(*b*) Prepollination changes in the male gemetophyte of *Cycas* and *Pinus*. **(Rohilkhand, 1992)**

(*c*) T.L.S. and RLS of *Pinus* Stem. **(Rohilkhand, 1998)**

36. What is chilgoza and from which plant it is obtained. **(Kanpur M.Sc, 1996)**

37. Write names of *Pinus species* found in Uttaranchal. **(Garhwal, 2004)**

38. Name three species of *Pinus*. **(Punjab, 1998)**

39. Comment on Microspores of *Pinus*. **(Kerala, 1998)**

40. Give the xerophytic characters of *Pinus* needle. **(Punjab, 2003)**

41. Write the nature of endosperm, embryo and Nucellus in *Pinus*. **(Punjab, 2002)**

42. What are mycorrhiza ? In which gymnosperm are they present. **(Calicut, 2003)**

43. What are the three different functions which are attributed to the sacci in pollen grains of *Pinus*. **(Lucknow, 1996; Allahabad, 1991, 1996; Awadh, 2003)**

44. Write in not more thirty words :

(*a*) Structure of pollen grains in *Pinus*. **(Lucknow, 1999)**

(*b*) Resin canals. **(Lucknow, 2002, 2003; Kanpur, 2000)**

(*c*) *Pyenoxylic wood*. **(Lucknow, 2004)**

45. Name the plant which yields chilgoza. **(Lucknow, 2001; Allahabad, 2004)**

46. In which gymnosperm polyenbryony is present. **(Lucknow, 2001; Kanpur, 1995, 1997)**

47. Which gymnosperm has winged pollen grains ? **(Kanpur, 1995, 1998, 2000; Allahabad, 2003; Awadh, 1994, 2003; U.P. College, 1998; Poorvanchal, 2002; Gorakhpur, 1994)**

48. Name a gymnosperm which has ovuliferous scale in female cone. **(Kanpur, 1995; U.P. College, 1997)**

49. Name a gymnosperm which has dwarf shoots. **(Kanpur, 1996)**

50. Name two species of *Pinus* which yield turpentine oil. **(Kanpur, 1996)**

51. Name the species of *Pinus* whose seeds are dible. **(Kanpur, 1997)**

52. Useful wood is obtained from which gymnosperm. **(Kanpur, 1998)**

53. Give the classification of *Pinus*. **(Kanpur, 1998, 2000)**

54. What is special with the mesophyll of *Pinus* needle .**(Kanpur, 1999)**

55. Describe the structure of dwarf shoot in *Pinus*. **(Kanpur, 2001)**

56. Describe the Indian species of *Pinus*. **(Kanpur, 2004)**

57. Why it is difficult to extinguish fire in Pine forest ? **(Allahabad, 1991; Gorakhpur, 1999)**

58. Where does *Pinus gerardiana* grow ? **(Allahabad, 1991)**

59. By labelled diagrams only differentiate between:
    (*a*) Haploxylon and diploxylon *Pinus*. **(Allahabad, 1991, 1995)**
    (*b*) Pollen grains of *Cycas* and *Pinus*. **(Allahabad, 2004)**
    (*c*) Female cones of *cycas* and *Pinus*. **(Meerut, 1997)**

60. In which plant resin canals are present ? **(Poorvanchal, 2003; Gorakhpur, 1995)**

61. With the help of diagrams only:
    (*i*) Differentiate between T.S. Dwarf shoot and needle of *Pinus*. **(Allahabad, 1991)**
    (*ii*) Describe R.L.S. of *Pinus* stem. **(Allahabad, 1991)**
    (*iii*) Compare R.L.S. and T.L.S. of *Pinus* stem. **(Kumaon, 2000)**
    (*iv*) Describe T.S. of *Pinus* leaf. **(Poorvanchal, 2002)**
    (*v*) Differentiate between the roots of *Cycas* and *Pinus*. **(Gorakhpur, 1992)**

62. Write the botanical name of chilgoza pine. **(Allahabad, 1995, 2000; Awadh, 1995; Poorvanchal, 2001; Agra, 1996)**

63. Mention two economic importance of *Pinus*. **(Allahabad, 1996)**

64. (*a*) The foliage spurs of *Pinus* are deciduous, yet the tree appears evergreen. Explain
    (*b*) What is meant by the term 'Harmomegathy' ? How do you explain the term in referene to the pollen grains of *Pinus* ?
    (*c*) The foliage spur of *Pinus* is not a leaf. Explain.
    (*d*) In the Himalayan pine forests, one sees a sulphur like dust underneath the trees in the months of April and May. What is it due to ?
    (Allahabad, 1999; Gorakhpur, 1997, 1999)

65. Mention the name of long lived *Pinus* species. **(Allahabad, 2001)**

66. Mention the name of species of *Pinus* which is found at higher altitudes of three thousand feet in western Himalayas. **(Allahabad, 2001)**

67. Name a plant which encloses a resin duct in its vascular rays. **(Allahabad, 2002)**

68. In which gymnosperm polycotyledonous embryo is present ? **(Allahabad, 2002)**

69. In which species of *Pinus*, dwarf shoots have five nedles. **(Allahabad, 2003)**

70. Why does the wood of *Pinus* burn easily than the *Cycas* wood ? **(Allahabad, 2003)**

71. Which species of *Pine* is called chilgozapine ? **(U.P. College, 1999)**

72. Write the significance of ovuliferous scale in *Pinus*. **(U.P. College, 2000)**

73. In how many years the female cone of *Pinus* matures ? **(Poorvanchal, 1996)**

74. What is chilgoza and from which species it is obtained ? **(Poorvanchal, 1997)**

75. Describe spur branches of *Pinus*. **(Poorvanchal, 1998)**

76. (*a*) What is thin dry brown, membranous structure towards the micropylar end in *Pinus* seed ? Give its name.

(*b*) Name the plant which shows sulphur grain ?

(Poorvanchal, 2000; Gorakhpur, 1991, 2003)

77. From which gymnosperm commercial timber is obtained ? **(Meerut, 1997)**

78. What is the age of female cone of *Pinus* ? When the seeds are shed ? **(Gorakhpur, 1990, 1991, 1996)**

79. Draw a diagram to show bordered pits as seen in R.L.S. of *Pinus* wood. **(Gorakhpur, 1993)**

80. What are the two types of shoots found in *Pinus* ? **(Gorakhpur, 1994)**

81. Explain the xeromorphic characters of *Pinus* needle. Draw a neat labelled diagram of T.S. of *Pinus* needle. **(Calicut, 2003)**

OBJECTIVE TYPE

82. Fill in the blanks :

(*a*) Winged pollens are found in ________________ .

(*b*) The development of several embryos from a single zygote in *Pinus* is called _____ .

(*c*) The edible seeds are produced in *Pinus* ____________________ .

(*d*) Turpentine and resin is obtained from ____________________ .

(*e*) Chilgoza fruits are obtained from ____________________ .

(*f*) Ovuliferous scale is found in ______________________ .

(*g*) Winged pollen is characteristics of ____________________ .

(*h*) If there are five needles in a spur of *Pinus*, then it will be ________ species of *Pinus*.

(*i*) Production of more than one embryo in a seed is called ____________ .

(*j*) Edible part of seed of *Pinus* is __________________ .

(*k*) Seeds of *Pinus gerardiana* are used as dry fruits called _____________ .

83. Select the Correct Answer :

(*i*) Polycotyledonay condition is found in

(*a*) *Cycas* (*b*) *Pinus*
(*c*) *Ephedra* (*d*) All of the above.

(*ii*) The *Pinus* wood (secondary xylem) is

(*a*) Monoxylic (*b*) Pycnoxylic
(*c*) Polyxylic (*d*) None.

(*iii*) In which winged pollen grains are found ?

(*a*) *Pinus* (*b*) *Cycas*
(*c*) *Equisetum* (*d*) *Ephedra.*

(*iv*) What are winged structures in *Pinus* ?

(*a*) Pollen grains (*b*) Seeds
(*c*) Both (*d*) None of the above.

(*v*) Ovuliferous scale of *Pinus* is a part of

(*a*) Megasporophyll
(*b*) Microsporophyll
(*c*) Ovule (*d*) Dwarf shoot.

(*vi*) The type of Polyembryony in *Pinus* is

(*a*) Cleavage (*b*) Synergid
(*c*) Nucellar (*d*) None of the above.

(*vii*) Which one of the following has resin canals in the stem and leaves ?

(*a*) *Cycas circinalis*
(*b*) *Cycas revoluta*
(*c*) *Pinus*
(*d*) *Ephedra.*

(*viii*) The ovule of Pinus is naked due to the

(*a*) absence of integuments
(*b*) absence of nucellus
(*c*) absence of ovary
(*d*) absence of pereanth.

(*ix*) *Pinus* leaves are of two types

(*a*) Prophylls and cataphylls
(*b*) sheath and cataphylls
(*c*) Spur and Sheath
(*d*) None of the above.

(*x*) Chilgoza is found in

(*a*) Seed of *Pinus gerardiana*
(*b*) Seed of *Thuja*
(*c*) Seeds of *Cycas revoluta*
(*d*) seeds of *Pinus wallichiana.*

(*xi*) Resin and Turpentine is obtained from

(*a*) *Cedrus* (*b*) *Cycas*

(*c*) *Gnetum* (*d*) *Pinus*.

(*xii*) In which of the followign, branches as well as leaves are dimorphic ?

(*a*) *Ephedra* (*b*) *Gnetum*

(*c*) *cycas* (*d*) *Pinus*.

(*xiii*) Botanical name of chilgoza pine is

(*a*) *Pinus monophylla*

(*b*) *Pinus roxburghii*

(*c*) *Pinus insularis*

(*d*) *Pinus gerardiana*.

(*xiv*) Sex chromosomes are found in the plants of

(*a*) Cycas (*b*) *Pinus*

(*c*) *Equisetum* (*d*) *Marsilea*.

(*xv*) Which does not belong to sporophyte generation in *Pinus* ?

(*a*) Dwarf shoot (*b*) Long shoot

(*c*) Endosperm (*d*) Root.

(*xvi*) In *Pinus* seeds there are

(*a*) two cotyledons

(*b*) three cotyledones

(*c*) many cotyledons

(*d*) no cotyledons.

(*xvii*) Chilgoza of commerce is the

(*a*) fruit of *Cycas revoluta*

(*b*) seeds of *Pinus gerardiana*

(*c*) seeds of grasses

(*d*) fruit of *Thuja*.

(*xviii*) Pollen grains are discharged in the form of sulphur powder in

(*a*) *Cycas* (*b*) *Pteris*

(*c*) *Pinus* (*d*) *Ephedra*.

C H A P T E R 14

Coniferopsida–Taxales

- Distribution, Habit and Habitat
- Anatomy
- Reproduction
- Embryogeny
- Economic Importance

The order includes a single family *Taxaceae* that has been removed from the order Coniferales (Florin 1954, 55) and placed under a separate order due to certain notable differences in the organisation and structure of their reproductive organs. The family is characterised by the following features :

1. The sporophytes are either evergreen small trees or shrubs that branch extensively.
2. The leaves are simple, solitary, flat apparently distichous and spirally arranged.
3. The secondary wood is pycnoxylic and is formed by the activity of a single cambium. It has no parenchyma.

TORREYA

4. The tracheids have abundant tertiary spirals and uniseriate pits. The wood is elastic due to the tertiary spirals.
5. The shoot apex shows no distinction into tunica and corpus.
6. The wood rays are homogenous and are composed of heavily thickened walls.
7. The trees are mostly unisexual but sometimes monoecious trees also occur.
8. The female strobilus is represented only by a single ovule that is borne terminally on a dwarf shoot that arises in the axil of a leaf on the main branch. This dwarf shoot bears scale leaves arranged in an opposite and a decussate manner. The ovule bears 10 or even up to 25 archegonia each consisting of 2–4 celled neck (in some genera the neck may have 8 cells in a single tier) and venter with egg. There is no ventral canal cell in *Taxus*, but it is present in some species of *Torreya*.
9. There is a distinct aril at the base of the ovule.
10. The integument is fused to the nucellus, except at the apex and becomes stony. It is traversed by two phloem strands that have no xylem. The xylem only extends upto the chalazal end.
11. The aril is also innervated by two rudimentary vascular strands.
12. The male strobilus consists of a small central axis terminating a dwarf shoot. It bears a number of peltate sporangiophores. Each sporangiophore bears 6–8 pendulus sporangia that bear wingless pollen and dehisce by means of a longitudinal slit. In *Taxus* there are 6–14 microsporophylls and in *Nothotaxis* there are 10 microsporophylls or sporangiophores; numerous in *Torreya*; 1–5 in *Austrotaxus*.
13. The male gametophytes lack prothallial cells.
14. Pollination involves pollination drop mechanism. The pollen tubes form a characteristic sac-like expansion that covers the entire apex of the female gametophyte. The male gametes are unequal.
15. The megaspore mother cell lies in the centre of the sporogenous cells. It divides by meiosis to form a linear tetrad of 4 megaspores. All of them or the chalazal one becomes functional and undergoes free nuclear divisions to form 128 (*Taxus baccata*) or 256 (*T. cuspidata*) free nuclei. Later wall formation takes place. *Tent pole* has also been observed at the apex of the gametophyte.
16. The embryos have two cotyledons.
17. Cleavage polyembryony is not of regular occurrence. It has been reported in *Torreya nucifera* (Tahara, 1940). Some species also show rosette polyembryony. In *Taxus* 16 free nuclei are formed; in *Austrotaxus* 8–16 free nuclei are formed and in *Torreya* 4–8 free nuclei are formed before wall formation starts.
18. The seeds are endospermic. The endosperm is homogeneous.

The family is represented by five genera with about 15 to 18 species : *Austrotaxus, Amentotaxus, Nothotaxus, Taxus* and *Torreya.* Florin (1948) raised *Taxus chienii* to the rank of a new genus which he named as *Nothotaxus*. Chuang and Hu (1963) made a detailed observation on the cytology and taxonomy of *Amentotaxus argotaenia* Pilger and suggested that this plant differs from *Taxus* and *Torreya* in its chromosome morphology and chromosome number and also in its chemical constituents. They gave a strong support to the view that this genus should be put under a separate family Amentotaxaceae. Earlier Lee (1952) created this as a separate family. Florin (1952) and Koidzumi (1942) do not subscribe to this view.

According to Gangulee (2002) the family has been divided into two tribes as follows :

Tribe Torreyeae with two genera

Genus : (*i*) *Torreya* has five species, known to grow in china, Japan, California and Florida; also known as fossil since upper Jurassic.

Genus : (*ii*) *Amentotaxus* with a single species, from China, also known as tertiary fossil.

Tribe Taxeae.

Genus (*iii*) *Austrotaxus* with a single species, from Caledonia .

(*iv*) *Nothotaxus* with a single species, from China.

(*v*) *Taxus* with 9 species; grows in England, Continental Europe, Asia Minor Himàlaya, China, Japan, Phillipines. North America.; Mexico etc.; also knows as fossil since Jurassic.

Fossil genera assigned to the family are :

Palaeotaxus from upper Triassic; *Taxoxylon* (Wood) from cretaceous; *Taxites* (twigs) from Jurassic to recent; *Taxites* from Jurassic Indian upper Gondwana in the Rajmahal series.

A detailed account of the life history of *Taxus* is given in this text.

TAXUS Linn.

Distribution, Habit and Habitat

The genus is represented by nine species (Dallimore and Jackson, 1948) that are distributed over Europe, North America, Asia and extend up to Malaysia. *Taxus baccata* L., is a small tree that grows in the Himalayas in India. It grows in moist and shady places above 1,800 metres all along the Himalayas, Khasi-Jaintia hills and Naga Hills and may grow up to a height of twenty metres and in diameter up to 7 metres. In India the usual height attained by this species varies between 9–15 metres. Raizada and Sahni (1960) recognised *T. wallichiana* that also grows in Himalayas as a separate species. It is also a small tree reaching in height upto 12 metres or so. *Taxus cuspidata* Siebold and Zucc.; *T. chinesis* Rehder, *T. brevifolia* Nutall; *T. Floridiana* Chapman; *T. canadensis* Marshall, *T. globosa* Schl.; *T. media* Rehder; and *T. hunnewelliana* Rehder are other species of *Taxus*. *T. wallichiana* Zucc. is the tenth species.

(a) (b) (c)

Plate I : *T. baccata*-(a) A tree; (b) Branches with ♂ strobili (c) Trunk

The stem is monopodial with short and thick stem that gives off horizontal and spreading branches that are all long shoots. Only reproductive shoots that bear male and female structures are dwarf. There is no distinction between vegetative shoots into dwarf and long shoots. All grow indefinitely and form a dense canopy, thus making it a shade giving tree. The scale leaves are restricted only to the reproductive dwarf shoots. The vegetative branches bear only the green leaves.

The green leaves are linear, 2–3 cm long, dorsiventral, narrow and spirally arranged and persistent (Figs. 14.1, 14.2). They have an entire margin, a pointed apex and a shining dark green upper surface. The ventral side of the leaf is pale green or reddish brown in colour. The leaves have a single vein and are almost sessile or sub-sessile and often assymetrical and very poisonous. The scale leaves are arranged in an oppposite and a decussate manner on the reproductive shoots, that are devoid of foliage leaves (Figs. 14.1, B and 14.2, B). The plant of *Taxus canadensis* is a low and straggling shrub with green or reddish brown branches; whereas in the 50 feet high *T. cuspidata* the spreading branches are upright. In *T. brevifolia* (45–50 ft. high) the branches are drooping with mature branchlets yellowish green or light brown. The stem is covered with a scaly and reddish brown bark. The winter buds in all species are covered with imbricate scales, that are abtuse in *T. baccata* and distinctly keeled in *T. cuspidata, T. canadensis* and *T. brevifolia*. The leaf base is flat and decurrent with the branch. They remain on the stem when the rest of the leaf falls. They look like ridges on the stem. The leaf base shows a slight twist. The tip of the leaf accumulates silica and is sharply pointed.

The root is a deep feeder and much branched tap root. Root cap and root hair are distinct in younger regions. Mycorrhizic roots as those of *Pinus* are absent.

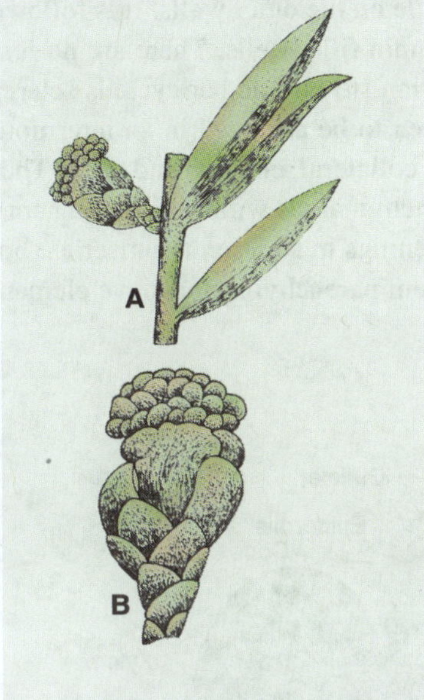

Fig. 14.1. Branch of *Taxus baccata* bearing male *strobili*. B. A male strobilus.

Fig. 14.2. A. Branch of *Taxus baccata* bearing two seeds. B. A mature seed.

(A)

A branch of *Taxus baccata* bearing male strobili

(B)

A male strobilus

(C)

A branch of *Taxus baccata* bearing seeds

(D)

Mature seeds

Plate II : ***Taxus baccata***

ANATOMY

Stem (Fig. 14.3)

It shows an irregualr circular outline due to the presence of persistent leaf bases. The epidermis has a single layer of parenchymatous cells with thick cuticle on the outer walls. It is followed by the cortex that is mainly parenchymatous and has some tannin filled cells. There are no resin ducts. Its last layer is the endodermis whose walls lack casparian strips. The pericycle is sclerenchymatous. In a very young stem the vascular bundles are seen to be arranged in an interrupted ring. The medullary rays are narrow. The bundles are conjoint, collateral, endarch and open. There is a distinct pith in the centre. It is, however, narrow and parenchymatous with cells having small intercellular spaces. The xylem has tracheids with spiral thickenings in addition to uniseriate bordered pits. The phloem is devoid of companion cells but phloem parenchyma and sieve elements are present.

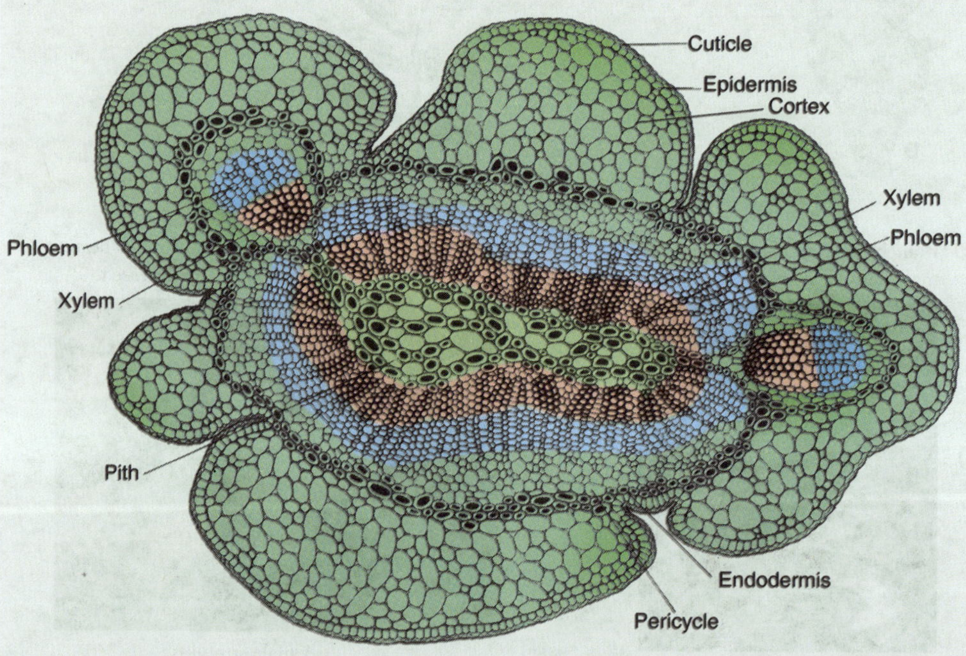

Fig. 14.3. T.S. of young stem of *Taxus baccata.*

The secondary growth (Figs. 14.4 and 14.5) is effected by a single persistent cambium that cuts off secondary xylem towards the pith and secondary phloem towards the cortex. The secondary wood is devoid of wood parenchyma and resin canals. Its tracheids are characteristic in showing tertiary spirals in addition to the uniseriate pits. The medullary rays are also uniseriate and homogeneous. The wood is dense, strong, heavy and elastic. Its elasticity is due to the presence of tertiary spirals in the tracheids and is used in the manufacture of bows. The cells in the medullary rays have heavily thickened walls. The annual rings are clear. The oldest tree of *Taxus baccata* is 3,000 years of age and is still growing in Scotland.

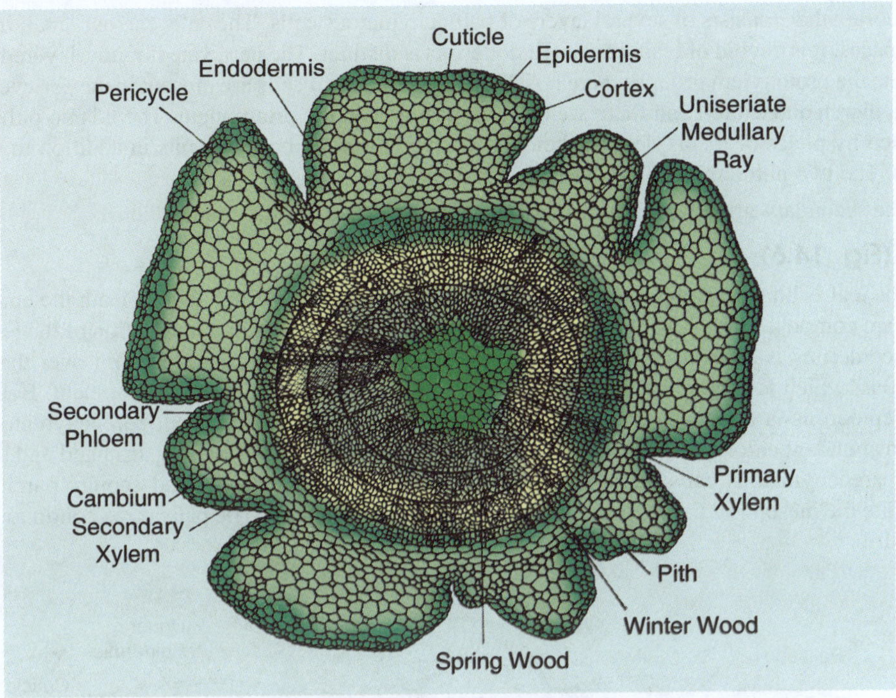

Fig. 14.4. T.S. of old stem of *Taxus baccata* showing secondary structure.

Root

The root has an outer layer known as *epiblema* or *piliferous* layer in younger regions because here its cells possess long an unicellular root hair. A thin cuticle is also found. Next to the epidermis

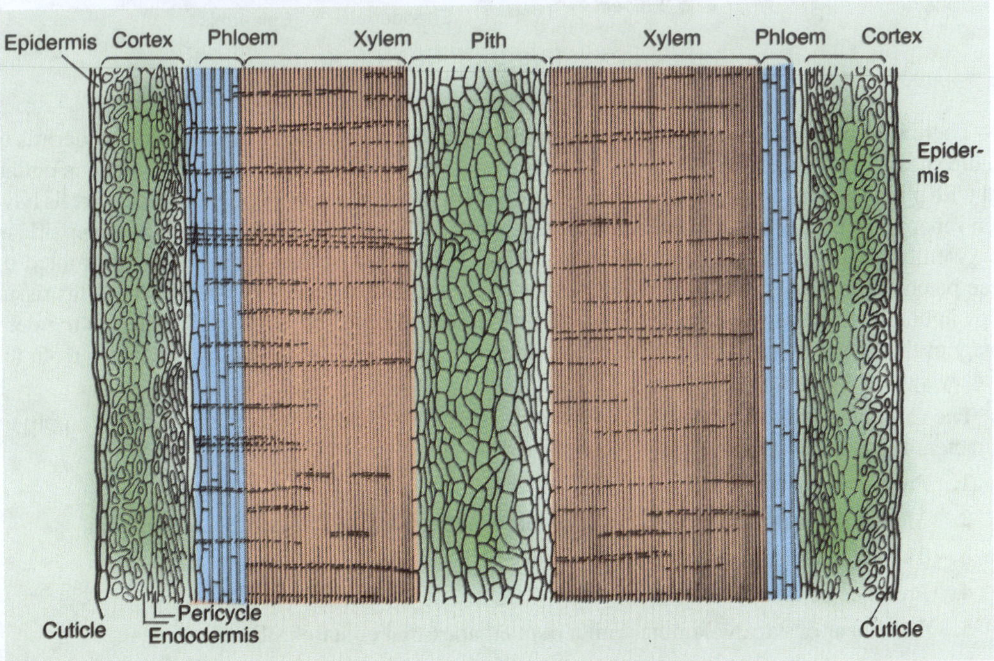

Fig. 14.5. L.S. of old stem of *Taxus baccata*.

is the cortex that consists of several layers of parenchymatous cells. The cells enclose small intercellular spaces. It is devoid of resin canals. Endodermis is distinct. The pericycle is many-layered except opposite the protoxylem groups where it is 2 or 3 layered. Resin cells are present in the pericycle. The stele is diarch and exarch and there are no resin passages in the protoxylem. There is no pith as it is occupied by plates of metaxylem tracheids that have uniseriate bordered pits in addition to tertiary spirals. The two phloem bundles alternate with the xylem.

The secondary growth in the root is exactly similar to that of the root in *Pinus*.

Leaf (Fig. 14.6)

The leaf is bifacial and possesses the upper and the lower epidermal layers. Both the epidermal layers are composed of parenchymatous cells that appear rectangular in a transection of the leaf. The upper epidermis is covered with a thick cuticle. The cuticle is comparatively thin over the lower epidermis, which is perforated by sunken stomata that are haplocheilic in development. Below the upper epidermis is a two-layered palisade tissue whose cells are elongated, parenchymatous and contain abundant chloroplasts. The spongy parenchyma is made up of loosely arranged and slightly lobed parenchymatous cells that also contain chloroplasts. The palisade and spongy parenchyma constitute the mesophyll tissue. It is the main seat of photosynthesis. The palisade is continuous over the midrib.

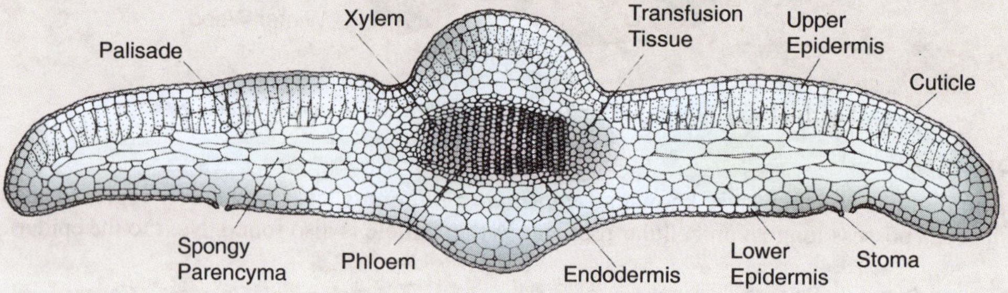

Fig. 14.6. V.S. of leaf of *Taxus baccata*.

There is a single vascular bundle in the midrib region. It is enclosed by a distinct endodermis or bundle sheath that is made up of barrel-shaped parenchymatous cells. The vascular bundle is collateral with phloem towards the lower epidermis and xylem facing the upper epidermis. Resin cells have been reported in the spongy parenchyma cells. A patch of lignified cells is present on either side of the xylem. They represent the transfusion tissue. A single layer of thin-walled cells interrupted by these patches constitutes the pericycle which also forms a part of the transfusion tissue. This tissue helps in conduction. transfusion cells also occur external to the endodermis and make up the deficiency of the lateral veins. A single leaf trace enters the leaf. It is also collateral and arises from the primary xylem of the stem.

The vasculature of the cotyledons is, however, mesarch. The leaf shows the following xerophytic characters :

1. Presence of a thick cuticle.
2. Mesophyll is differentiated into a two-layered palisade and a spongy parenchyma.
3. The stomata are sunken.
4. Presence of transfusion tissue.
5. The linear or narrow lamina with a pointed apex that contains silica.

REPRODUCTION

Propagation by vegetative means is entirely lacking in *Taxus*. The male and female plants are separate *i.e.*, *Taxus* is a *dioecious* tree. There is no distinction in the vegetative organisation of the male and the female plants. They are exactly similar and can only be distinguished from each other when they bear microsporangiate and megasporangiate structures that are popularly known as male and female flowers. February and March are the usual months in which the trees bear reproductive structures and are sexually distinct. The male reproductive organs are organised into cones or strobili whereas the female reproductive organs are not. The term female flowers is often used to designate them.

Male or Microsporangiate Strobilus (Figs. 14.1 A, B and 14.7 to 14.10)

The male strobilus terminates a lateral short shoot that arises in the axil of a foliage leaf of a vegetative branch. This short shoot bears about ten decussate sterile leaves and is produced into a short cone axis which bears 6–14 symmetrical and perisporangiate microsporophylls or sporangiophores (Figs. 14.8, 14.9, 14.10), each bearing 6–8 sporangia. The cone axis has a broad apex and is consumed in the formation of a terminal microsporophyll. The microsporophylls appear to be spirally arranged and are coloured yellowish green when young. Mature cones are pale yellow in colour. In a young cone the sporophylls are compactly arranged, their peltate heads fitting lightly. At maturity the cones become loose and the sporangia undergo dehiscence. In a young cone the sterile scales cover and protect the microphylls. Later due to active growth of the cone axis the sporophylls are exposed and the cone rises above the sterile scales. The sterile scales are smaller towards the base of the dwarf shoot and become larger towards the apex.

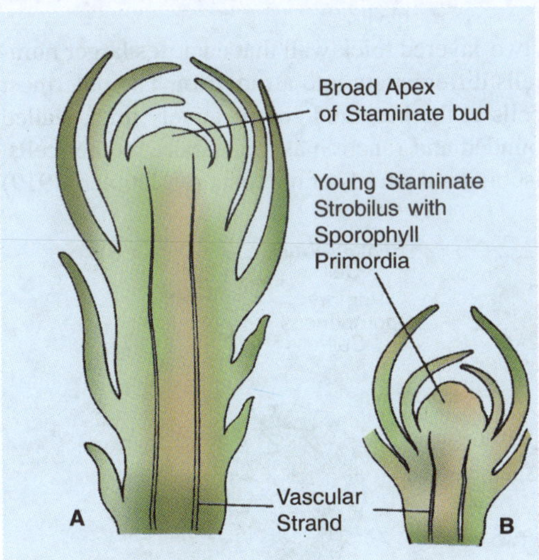

Fig. 14.7. (A–B) L.S. of young staminate buds of *Taxus baccata*.

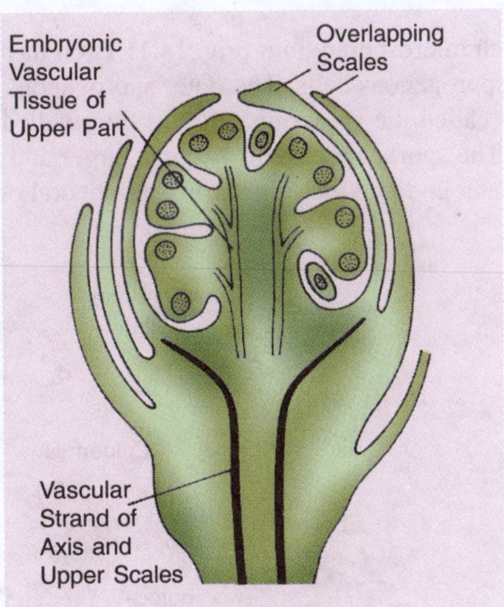

Fig. 14.8. L.S. of young male strobilus of *Taxus baccata*. The sporangia at this stage are young and contain sporogenous tissue.

A simple microsporophyll (Figs. 14.9, 14.10) looks like the sporangiophore of *Equisetum*. It consists of a short stalk attached to the centre of a polygonal and disc-like head i.e., why it is called peltate. Six to eight microsporangia arise from the periphery of the disc on its under side. The sporangia are thus pendent, elongated and laterally concrescent. They are also coherent with the sporophyll stalk. The sporangia are sessile and contain numerous yellow microspores or pollen grains.

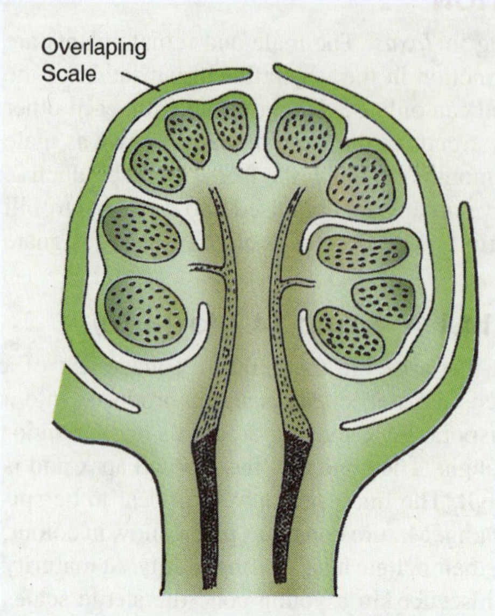

Fig. 14.9. L.S. Older strobilus of *Taxus baccata*. The sporangia contain microspores.

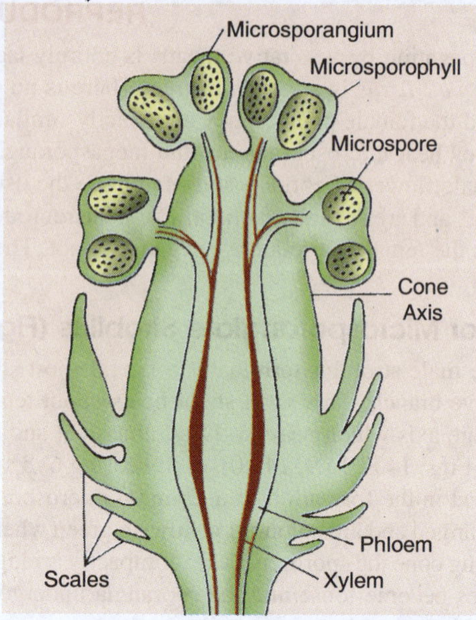

Fig. 14.10. Median L.S. of mature strobilus of *Taxus baccata* just before dehiscence of sporangia. It shows only four microsporo phylls.

Each microsporangium (Fig. 14.11 F, G) has a two-layered thick wall that encloses larger number of sporogenous cells. The outer sporogenous cells differentiate into a **tapetum**. The outermost layer is called the epidermis and has thin-walled cells at the base and comparatively thick-walled above. The sporogenous cells become larger and rounded and function as microspore mother cells, which later undergo meiosis to form tetrahedral or isobilateral tetrads of microspores. Dupler (1919)

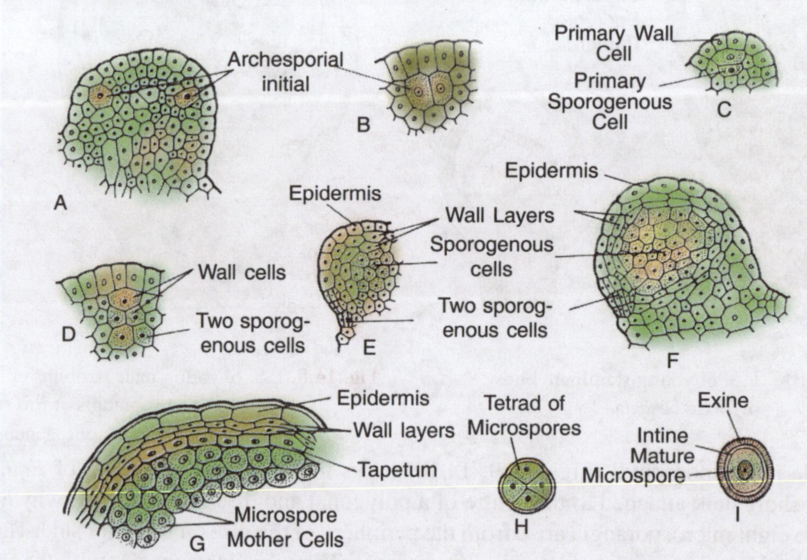

Fig. 14.11 (A–I). *Taxus canadensis.* A–G. Various stages in the development of microsporangium. H. Tetrad of microspores. I. A microspore (After Dupler).

studied the development of sporangium in *T. canadensis* and reported it to develop from a single or rarely two hypodermal **archesporial** cells. The development is eusporangiate (Fig. 14.11) like other conifers. He reported that the mother cells are fashioned in October and the microspores are fully formed after a fortnight or so.

The microsporangium dehisces along a longitudinal slit that appears towards the sporophyll stalk. The dehiscence starts from the base upwards and the ruptuerd edges diverge from each other and open the microsporangium. The microsporangia separate from the sporophyll stalk and along with the opened sporangia the sporophyll appears like a mature mushroom. The microspores are released and are carried by wind. The hygroscopic role of the thickened epidermal cells also helps in sporangial dehiscence.

Each cone is supplied by a single vascular strand that is collateral and endarch as it enters the cone. During its course upwards it gives off branches to the sterile leaves and to the sporophylls. Near the cone apex the vascular bundle becomes concentric and has a mesarch xylem.

When young the strobili are erect in the axils of the leaves, but as they develop they become oriented in such a way as to hang pendent on the lower side of the shoot, the fertile portion of the strobilus being directed downwards.

The male cone of *Taxus* resembles that of *Pinus* and other conifers in possessing an elongated axis, sterile scales and spiral arrangement of microsporophylls. It differs mainly in the peltate nature of the sporophylls and pendent sporangia that are laterally concrescent. The male cone differs from an angiospermic flower in possessing, (*i*) an elongated thalamus (in some angiosperms *e.g.*, *Gynandropsis, Cleome*, the thalamus becomes elongated); (*ii*) the microsporophylls are peltate; (*iii*) the number of microsporangia varies from 6–8 as compared to 2–4 in angiosperms; (*iv*) there is no perianth; and (*v*) the flowers are always unisexual whereas in angiosperms hermaphrodite flowers are extremely common. The mesarch character of the terminal portion of the scale bundles and the appearance of centripetal xylem in the terminal portion of the sporophyll bundle, where the bundle may be mesarch, exarch or concentric, are the significant features of male strobilus, besides their peltate nature.

Female Flowers (Fig.14.12, A–H)

The female reproductive organs are not organised into strobili or cones and hence they differ markedly from the other conifers. A vegetative shoot bears a bud in the axil of one of its foliage leaves (Fig. 14.12, A). This bud is protected by a number of imbricate scales. It is known as the

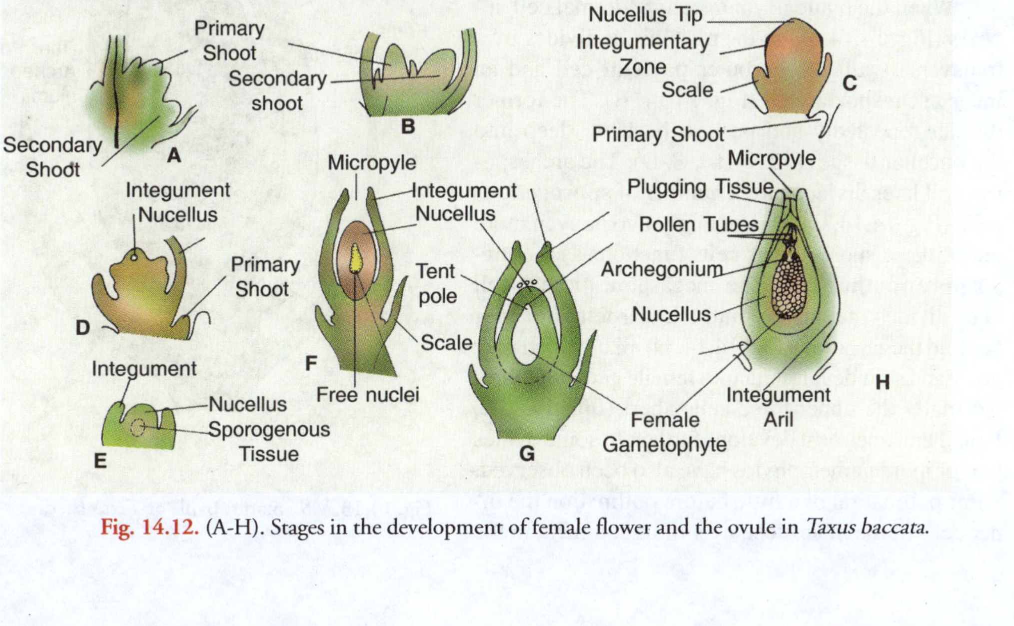

Fig. 14.12. (A-H). Stages in the development of female flower and the ovule in *Taxus baccata*.

primary axis (Fig. 14.12, A, B). The growth of this bud is arrested *i.e.*, it does not grow further and remains short and abbreviated. The uppermost scale of this bud bears an axillary bud which is covered with three pairs of scale leaves that are opposite and decussate in arrangement. This second bud grows into a **secondary axis** (Fig. 14.12, B, C) that pushes aside the bud of the primary axis and becomes continuous with its short axis. The apex of the secondary axis is consumed in the formation of an apical ovule that is **orthotropous** (Fig. 14.12, C–H). Sometimes the apical bud of the primary axis bears anoter bud in the axis of its penutimate bract that devlop into another axis resumes growth (usually after the first formed ovule has matured into a seed) and produces another secondary axis that bears a terminal ovule in the next year. Sometimes the secondary axis bearing another ovule. Like this two ovules are sometimes seen to arise on the same secondary axis.

Structure of Ovule (Fig. 14.13)

The ovule is orthotropus and sessile. It consists of a massive nucellus (crassinucellate) enveloped by a single thick integument (unitegmic) that is free from the nucellus right upto its base. The integument overgrows the nucellus and forms a distinct and long micropylar canal. The integument is distinguished into an **outer fleshy, middle stony** and **inner fleshy layer**. In mature ovules the integument may become fused with the nucellus near its lower region. A ring like outgrowth arises from the base of the integument and slowly grows around it and covers the entire ovule after the embryo has developed. This extra envelope is called the '**aril**' or the '**cupule**'. the aril arises from the periphery of a **plate meristem** situated entirely across the axis (Bierhorst, 1971). The aril assumes a red colour and becomes juicy in the mature seed. The stony layer of the integument is supplied by two vascular strands that consist of phloem alone, they do not extend beyond the chalazal end. Two minute and rudimentary vascular strands also supply the aril. The nucellus has no pollen chamber and no nucellar beak.

The mature ovule (Fig. 14.13) which is ready for fertilisation consists of a well developed female gametophyte whose apex is produced into a flask-shaped structure also called the '**tent pole**'. (Favre Duchartre, 1958). It appears during early stages of female gametophyte and persists upto the cellular stage. It disappears later. There is no spongy zone or tapetum around the embryo sac. The female gametophyte consists of 3–12 archegonia in *Taxus baccata* and *T. cuspidata* (Favre Duchartre, 1958). Sometimes the number goes upto 25 in *T. cuspidata*.

When the ovule is young a hypodermal cell appears (Fig. 14.14, A) in the nucellus, it divides by a transverse wall into an outer **parietal cell** and an inner **archesporial cell** (Fig. 14.14, B). The former divides repeatedly and pushes the latter deep into the nucellar tissue (Fig. 14.14, C, D). The archesporial cell later divides into a number of **sporogenous cells** (Fig. 14.14, C). One or rarely two or even more out of these sporogenous cells functions as the **megaspore mother cell**. The megaspore mother cell later divides meiotically into a linear tetrad of four haploid megaspores, (Fig. 14.14, D) all of which may grow and start developing into female gametophytes. Normally the upper three cells abort (Fig. 14.4, E, F) and the innermost develops further. In some ovules two or more gametophytes have also been observed. After pollination or a little before pollination the inner cells of the innermost layer of integument, in the

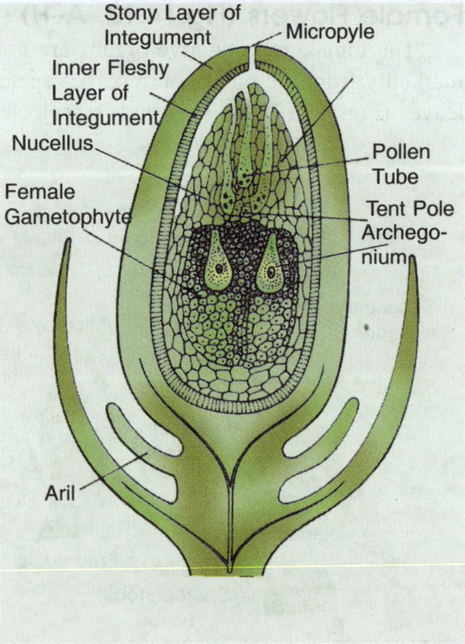

Fig. 14.13. V.S., mature ovule of *Taxus baccata*.

micropylar region grow out in a radial direction. They grow, rapidly after pollination has taken place and close the micropyle. This is called the **plugging tissue** (Dupler, 1920).

Morphological Nature of the Aril

The morphological nature of the aril has been a debatable question. The following explanations have been put forth by various workers :

1. Richard (1826) regarded the aril as an accessory structure formed from the strobilar axis and considered it to be equivalent to the collar in the ovule of *Ginkgo*.
2. Blume (1847) regarded it as a carpel.
3. Baillon (1860) regarded it as an expansion of the axis surrunding the ovary.
4. Parlatore (1864) regarded it as an equivalent of the ovuliferous scale of other forms.
5. Celakovsky (1890) and Worsdell (1899) claimed the ovuliferous scale of conifers to be the morphological equivalent of the "epimatium" of the podocarps, of the outer fleshy layer of the ovule of *Torreya* and *Cephalotaxus*, and the aril of *Taxus*.

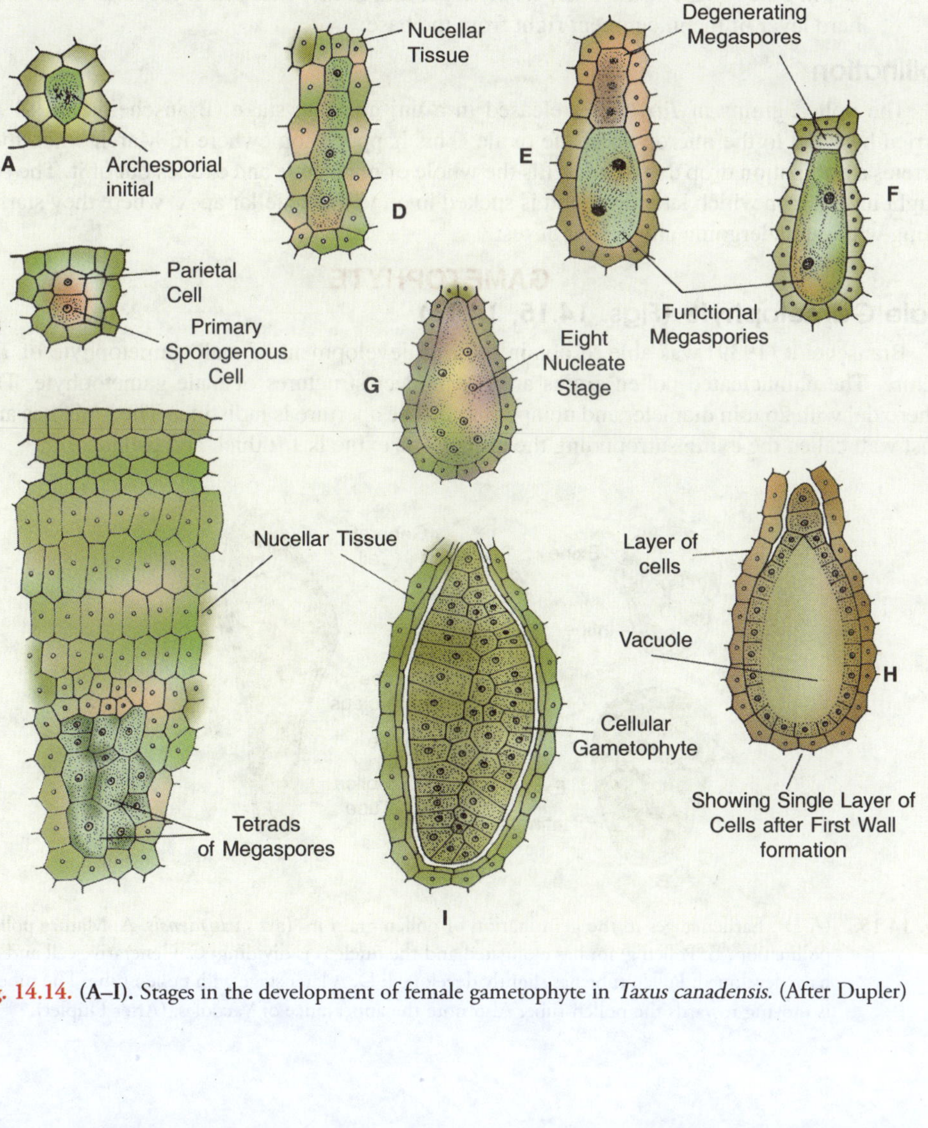

Fig. 14.14. (A–I). Stages in the development of female gametophyte in *Taxus canadensis.* (After Dupler)

6. Strasburger (1872) agreed with Baillon in regarding the aril as an outgrowth of the axis, discoid in nature.
7. Bertrand (1878) and Schumann (1902) regarded the aril to be a special structure.
8. Coulter and land (1905) have described the outer fleshy layer of the ovule of *Torreya taxifolia* as the outer integument.
9. Jager regarded the aril to be a second or the outer integument, because aril originates in the same way as the integument.
10. Sinnot (1913) pointed to the aril of *Taxus* as equivalent to the epimatium of podocarpaceous ovules.
11. Coulter and Chamberlain (1917) stated, it is a natural thing to see in these three layers characteristics of the testa in cycads, *Ginkgo*, and the older gymnosperms; and to conclude that the two integuments have arisen from a single one by delaying the development of the region that becomes the outer fleshy layer. These facts and the inference seem to hold good also in the case of *Taxus*, the only difference being that the outer fleshy layer (aril in this case) remains distinct from the inner one. Dupler (1920) also subscribed to this view that the aril in *Taxus* ovule is the outer fleshy layer and in this case this layer (aril) is free from the hard layer of the integument right from the base.

Pollination

The pollen grains in *Taxus* are released in a uni-nucleate stage (Branscheidt, 1939) and are carried by wind to the micropyle of the ovule. This happens somewhere in March when the ovule secretes a pollination drop that almost fills the whole of micropyle and exudes out of it. The pollen is caught in this drop which later dries and is sucked in on to the nucellar apex, where they start germinating without undergoing any period of rest.

GAMETOPHYTE

Male Gametophyte (Figs. 14.15, 14.16)

Branscheidt (1939) was able to obtain the full development of male gametophyte of *Taxus* in culture. The uninucleated pollen grains are the pioneer structures of male gametophyte. They are spheroidal with 26 μ in diameter and uniaperturate. The aperture is indistinct. The grain has an outermost wall called the **exine** surrounding the intine. The exine is 1 μ thick and psilate.

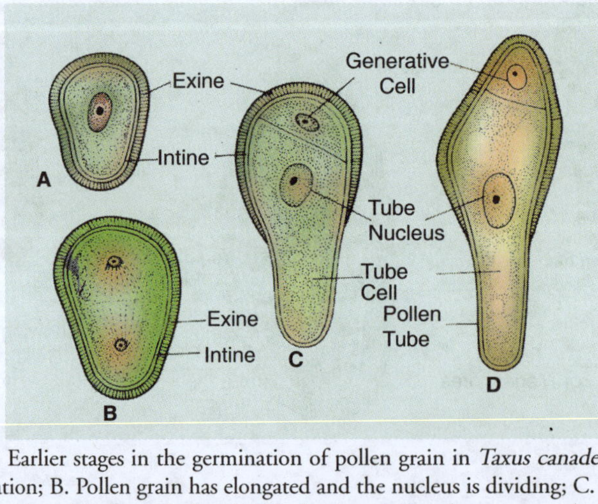

Fig. 14.15. (A-D). Earlier stages in the germination of pollen grain in *Taxus canadensis.* A. Mature pollen after pollination; B. Pollen grain has elongated and the nucleus is dividing; C. Generative cell and tube cell have developed. Pollen tube has slightly developed; D. A later stage with pollen tube. The tube nucleus is moving towards the pollen tube. Also note the appearance of Vacuoles. (After Dupler).

The pollen is released in uninucleate condition (Fig. 14.15, A) and starts germinating on the nucellus. Its nucleus divides into a **tube cell** and a **generative cell** (Fig. 14.15, B). No prothallial cells are formed. The intine protrudes out through the germ pore and grows into a short pollen tube (Fig. 14.15, C) that penetrates the nucellus. The tube nucleus moves into the pollen tube and lies at its tip. The generative cell remains in the grain end and divides into a **stalk cell** and a **body cell** (Fig. 14.16, A). Sterling (1948) noticed that in T. *cuspidata* the body cell and stalk nucleus (liberated by the rupture of the wall of stalk cell) move into the pollen tube, the stalk nucleus still sticking to the body cell. The tube and the stalk nucleus are seen ahead of the body cell. Later the body cell divides into two unequal male gametes. Favre Duchartre (1960) did not observe any wall around the two male gametes and regarded them as male nuclei and not male cells. The pollen tube penetrates down the nucellus and ultimately reaches the archegonial neck. At this time its tip contains the tube nucleus, the stalk nucleus and the two male nuclei or gametes (Fig. 14.16, B, C).

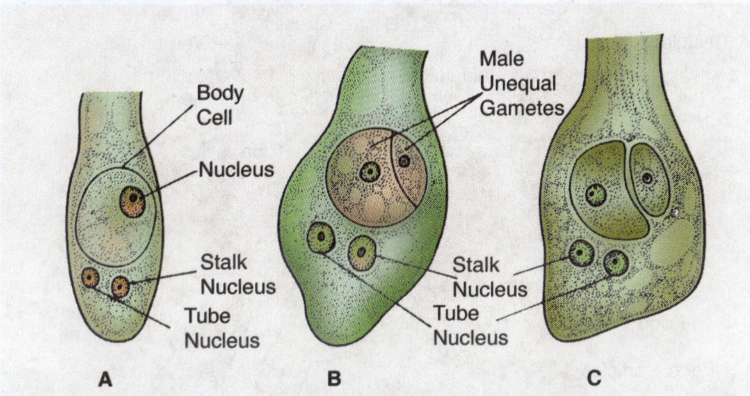

Fig. 14.16. *Taxus canadensis.* Later stages in the development of male gametophyte. A. Body cell, stalk nucleus and tube nucleus have migrated to the tip of pollen tube; B–C. Division of body cell into two male gametes. (After Dupler).

Female Gametophyte (Fig. 14.14)

The functional megaspore that outgrows others, enlarges considerably and its nucleus undergoes repeated free nuclear divisions to form 128 free nuclei in *T. baccata* (Favre Duchartre, 1958) and 256 in *T. cuspidata*. Tapetum or the spongy tissue does not develop. The **tent pole** makes its appearance on a flask-shaped cellular structure from the apex of the gametophyte and persists till it becomes cellular. The free nuclei lie in a parietal mass of cytoplasm because a big central vacuole develops. Later the central vacuole gradually disappears and there is a centripetal wall formation by the appearance of alveoli. This makes the whole tissue cellular. It is the female gametophyte enveloped within the megaspore membrane. Its cells are filled with reserve food materials. During its development the female gametophyte encroaches upon the surrounding nucellar tissue and consumes much of it so that it forms only a thin layer. Some cells in the female gametophyte remain elongated and form a longitudinal strand. It may be regarded as the conducting strand. Saxton (1936) described that in some cases the female gametophyte, while still in free nuclear stage sends out a tube (Fig. 14.20, C) to meet the tip of pollen tube. He compared this outgrowth to the prothallial tubes in *Welwitschia*.

Archegonia (Fig. 14.17)

As stated before the number of archegonia varies between 3–12. Each archegonium develops from a superficial initial (Fig. 14.17, A) which divides by a transverse wall into an upper **primary neck cell** and a lower **central cell** (Fig. 14.17, B). The former divides once or twice by intersecting walls to form an archegonial neck of two or four cells arranged in a single tier (Fig. 14.17, C, D). The central cell functions directly as the egg (Saxton, 1936; Sterling, 1948; Favre Duchartre,

1958) *i.e.*, there is no ventral canal cells or nucleus. The central cell enlarges. Its nucleus functions as the female nucleus (Fig. 14.17, D). As this nucleus begins to move down, the cytoplasm in that part of the archegonium stains more lightly and shows conspicuous vacuoles. Meanwhile the large vacuole in the lower part of the egg cell is invaded by cytoplasmic strands which later form a foam-like mass. Thus the entire egg becomes uniformly foam-like with the vacuoles appearing as large or small globules. Later a dense condensation of the cytoplasm appears below thc cgg nuclcus cxtending in some cases to the base of the archegonium.

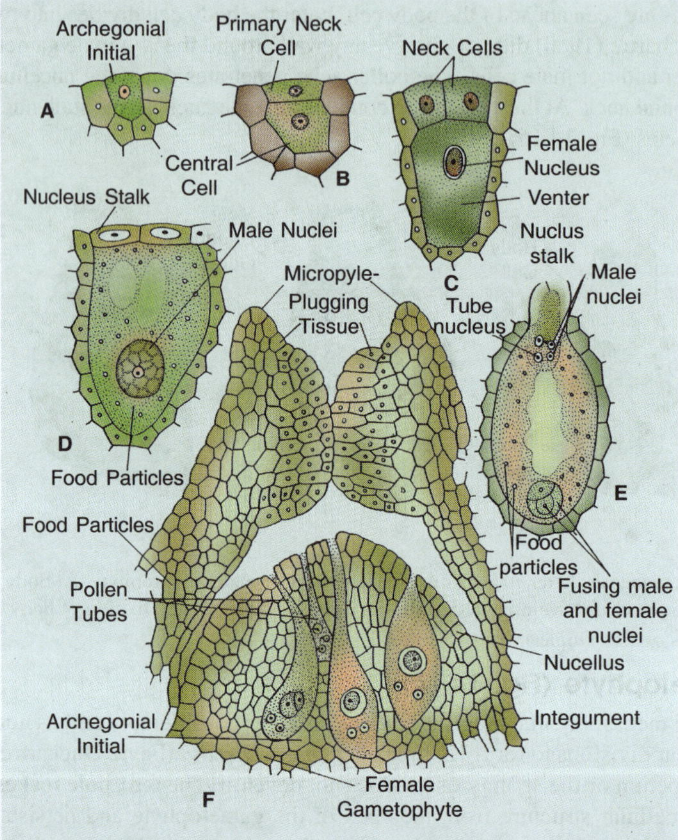

Fig. 14.17. (A–E). stages in the development of archegonium in *Taxus baccata* (After Dupler); F. V.S. part of mature ovule showing plugging tissue and germinated pollen tubes; E. It shows fertilization.

Fertilization (Fig. 14.17, E)

The pollen tube penetrates through the neck of archegonium and bursts releasing its contents into the archegonial ventre. It has been described by Dupler (1917) and Saxton (1936) that in *T. baccata* and *T. canadensis* downwardly moving pollen tube first meets or comes in contact with the tent pole or the prothallial tube. (Fig. 14.19, A). This happens when the female gametophyte is still in free nuclear stage. Later the prothallial tube on the tent pole disorganises and pollen tube comes in contact with the archegonial neck. In *Welwitschia* also prothallial tubes develop. The larger male gamete moves down the egg cytoplasm, leaving its cytoplasmic envelop at the apical part of the archegonium. It reaches the female nucleus and fuses with it to effect fertilization or syngamy. The smaller male gamete, stalk nucleus and the tube nucleus, all degenerate in the apcial part of the archegonium. Fertilization is usually effected two months after pollination, somewhere in early summer. Saxton (1936, p. 521, Fig. 5) has described an abrormal case of pollen tube traversing along the side of the female gametophyte and reaching its chalazal end (Fig. 14.19. B). It however, does not penetrate the female gametophyte.

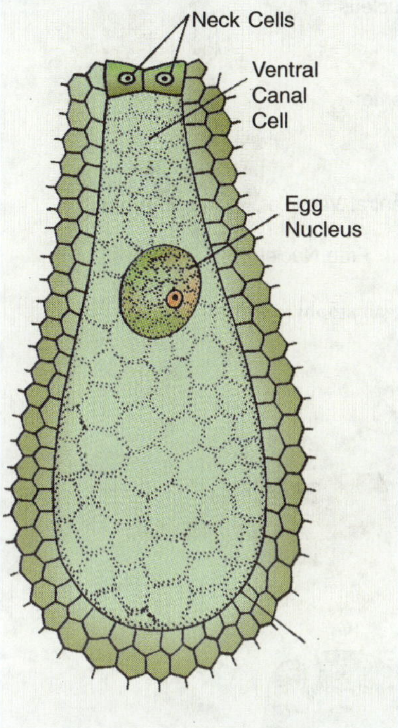

Fig. 14.18. A mature archegonium of *Taxus baccata.*

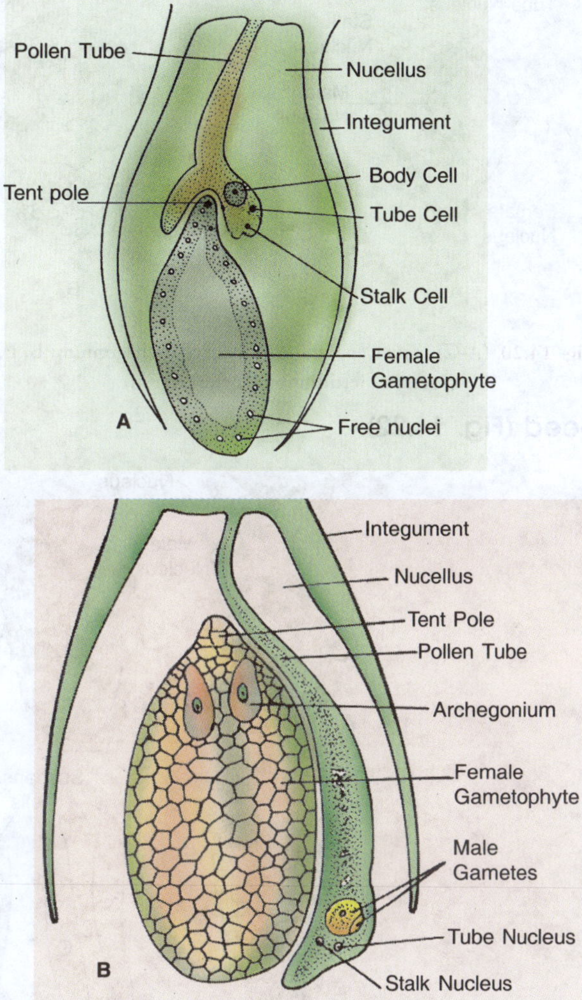

Fig. 14.19. *(A-B) Taxus baccata* A. usual method of pollen tube approach in *T. canadensis*; B. Abnormal condition of pollen tube approach in *T. baccata.* The pollen tube pierces through the chalazal end. (After Saxton)

EMBRYOGENY (Fig. 14.21)

In *Taxus cuspidata* the zygote nucleus divides by 4 divisions (Fig. 14.21, A–D) into sixteen free nuclei (Sterling, 1948). Now wall formation starts and the proembryo comprises the upper tier of 9–13 cells and a lower fully walled embryonic tier consisting of 3–7 cells (Fig. 14.21, E, F). Another division in the upper tier results in the formation of three tiers viz., the lower **embryonal tier**, the middle **suspensor tier** and the upper **open tier**. Sometimes wall formations take place at the 32-nucleate stage (Sterling, 1948). After this the suspensors elongate and push the embryonal tier out of the archegonium, deep into the tissue of the food laden female prothallus. Sterling (1948) has reported the cleavage of the embryo system in a significant percentage of embryos in *T. cuspidata*, but it is not a phenomenon of regular occurrence. Sterling (1948) described later stages in embryo development in *T. cuspidata*. The embryo in this case is dicotyledonous (Fig. 14.21, L). Simple polyembryony is there besides cleavage polyembryony that is not a rare phenomenon. The mature embryo is straight and has radical towards the micropylar end, a hypocotyle comes next and then the plumule concealed in-between the two cotyledons. Only one embryo matures in one ovule. The embryo consumes the entire female prothallus so that the seed in non-endospermic.

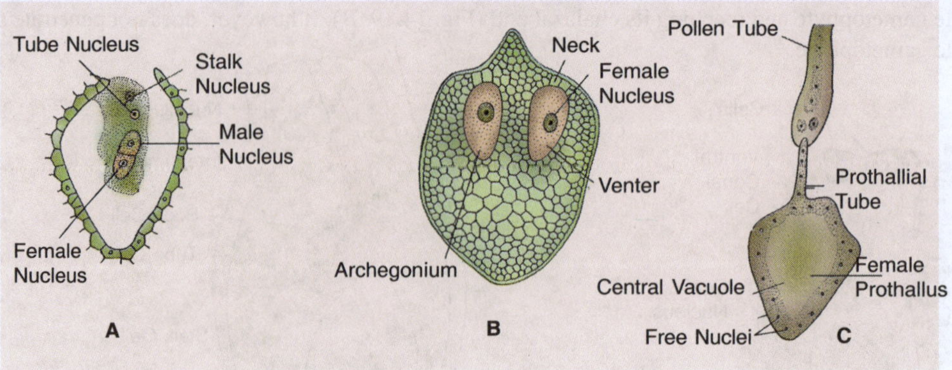

Fig. 14.20. (A-C). *Taxus baccata.* A. Fertilized archegonium; B. Female gametophyte with two archegonia; C. Process of fertilization. (After Saxton)

Seed (Fig. 14.22)

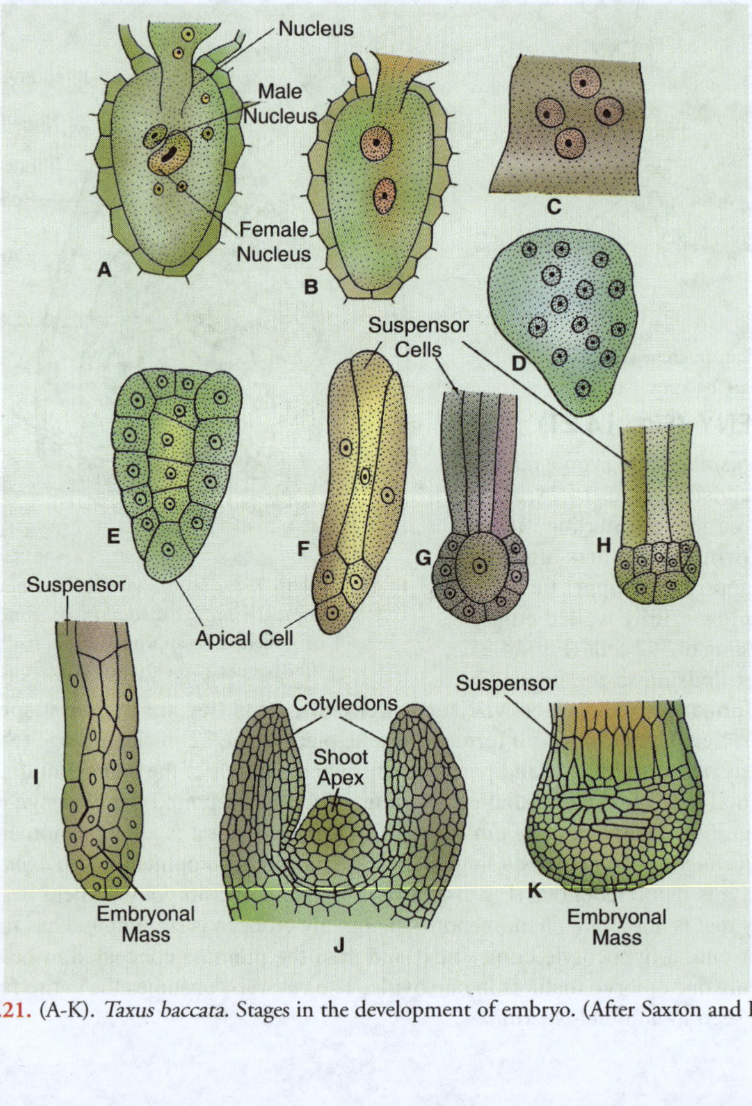

Fig. 14.21. (A-K). *Taxus baccata.* Stages in the development of embryo. (After Saxton and Dupler)

During seed formation the outer integument becomes brown and papery whereas the stony layer develops into testa. The inner integument also shrivels up. The nucellus too is not represented by any structure. It is completely consumed by the female gametophyte. The young embryo is surrounded by the endosperms, but as the embryo matures it consumes the whole of endosperm and thus becomes non-endospermic. The aril also grows up, becomes fleshy and edible and assumes a scarlet red colour. It is edible and is eaten by the birds that also help in the dispersal of the seeds. The seed proper is not edible. The colour of the seed proper is dark olive-green. It represents the usual three generations.

1. The testa represents the parent sporophytic generations.
2. The endosperm when present represents the gametophytic generation.
3. The young embryo is the future sporophytic generation.

The seeds usually mature quite late in the summer and germinate to give rise to a young seedling that has cotyledonary leaves larger than the normal foliage leaves. The plumule grows into the shoot that bears spirally arranged foliage leaves. The cotyledonary leaves fall down as soon as the seedling starts producing its normal foliage leaves. The germination is epigeal.

ECONOMIC IMPORTANCE

Medicinal Importance

The leaves of *T. baccata* are used in asthma, bronchitis, hiccough, epilepsy and for indigestion. The leaves are also used as fodder inspite of the presence of alkaloid. The seeds are used as sedative. The whole plant is poisonous and is also used as a fish poison. The alkaloid **taxine** is the toxin

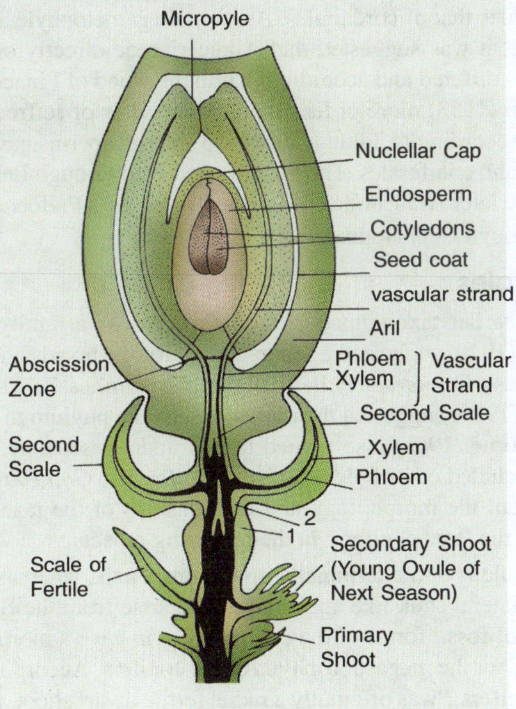

Fig. 14.22. L.S. seed of *Taxus* while attached to the shoot. Note the vascular supply. 1, 2 and 3 are the traces to first, second and third pairs of scales. (After Dupler)

principle contained in the leaves, shoots and seeds. Traces of ephidrine are also present in the leaves. Taxine is a vigorously active heart poison. The aril is eaten by birds.

Ornamental Uses

The plants are usually cultivated as ornamentals. It is also used as a hedge plant. Articles made from its wood are also used as decorative pieces.

Industrial Importance

The timber is oily and has irregular growth rings. It is the heaviest of the soft woods and has very good mechanical properties. It is durable. It is also elastic and in the past it was used to make bows. These days it is used for making decorative veneers, flowers, turnery, and posts. It is also uses for poles, ploughs, cartaxles etc.

DISCUSSION

(A) Origin of Conitferales and Taxales

It has now been accepted by most scientists that the coniferales (including Taxales) originated from cordaitales. In 1951 Florin has shown by comparison of the fossil forms of cordaitales with those of coniferales that the female strobilus of Pinaceae has evolved from cordaitales through Lebachiaceae and Voltziaceal of coniferales. It has also been taken for granded that the needle leaves of *Pinaceae* might have evolved through reduction of the broader leaves of cordaitales.

The two primitive families of coniferales are Pinaceae and Araucariaceaee. The secondary wood of Araucariaceae reseubles that of cordaitales. As its male gametophyle as well as embryo are more primitive than Pinaceae, it was suggested that Araucariaceae directly originated from cordaitales. However, Jeffrey (1917) differed and according to him the wood of Pinaceae is more directly related to cordaitales, Takhtajan (1953) more or less supports the view of Jeffrey but has also incorporated the findings of Florin. According to Takhtayans (1953) pteridosperms gave rise to cordaitales which gave rise to ginkgoales and coniferales. The Pinaceae evolved through Lebachiaceae and Voltziaceal simultaneously or a little later Araucariacal originated along with Podocorpaceal. This branch terminated into Taxaceae through Cephalotaxaceae (Fig. 14.23).

(B) Position of Taxales.

Pilger (1926) and earlier taxonomists treated *Taxaceae* as a family of the order Coniferales. Pulle (1937) and Sahni (1920) suggested a separate rank for the taxads and treated them as a distinct order Taxales of the class *Coniferae*. Pulle included the families *Taxaceae* and *Cephalotaxaceae* under this order. Sahni (1920) suggested the rank of a separate phylum to the taxads equal in rank to the coniferae. Later Sporne (1965) also treated taxads under a separate order *Taxales* of the class *Coniferopsida* which included other orders such as *Coniferales, Ginkgoales* and *Cordaitales*. Florin (1948) discussed in detail the morphology and relationships of the taxads and concluded that the taxads stand apart from the 'true conifers' in the following repects :

1. In taxads the ovule is borne terminally on the floral axis, whereas in the conifers the ovules arise terminally on the lateral stalk like appendages that arise from the floral axis (cone axis). As a result of detailed study of fossil forms of the conifers, Florin gave a morphological interpretation of the "seed scale complex" or the megasporophylls of the conifers. According to Florin the "seed scale complex" of the true conifers, "was originally a radial fertile dwarf shoot, placed in the axil of a bract on the axis of an inflorescence; the axis of the flower carried several spirally arranged sterile scales and intermingled with them, macrosporophylls (seed scales), each with a single terminal and erect ovule. The different subsequent types apparently have arisen from this basic complex by various transformations and reductions." So taxads lack cone organisation and thus differ markedly from the conifers that possess distinct cones or strobili.

2. In conifers the female flowers arise in the axils of bract scales that are borne on the inflorescence or cone axis; whereas in taxads the female flowers arise singly in the axils of bracts on the vegetative shoots. There is no cone like organisation in the taxads.

3. Unlike the conifers the taxads bear erect ovules that are enclosed by a specialised outgrowth called the *aril*.

4. The male flowers are radially symmetrical in the taxads and have perisporangiate microsporophylls as compared to the dorsiventral and hyposporangiate ones in the conifers.

5. The taxads have wingless pollen grains whereas majority of conifers have winged pollen grains.

The vasculature of the female flower and the ovule in the taxads suggests that the seemingly single integument of the ovule is actually of multiple origin *i.e.*, it has arisen from a ring of sterile appendages that are all uni-nerved at the base, but became laterally fused into a single integument by

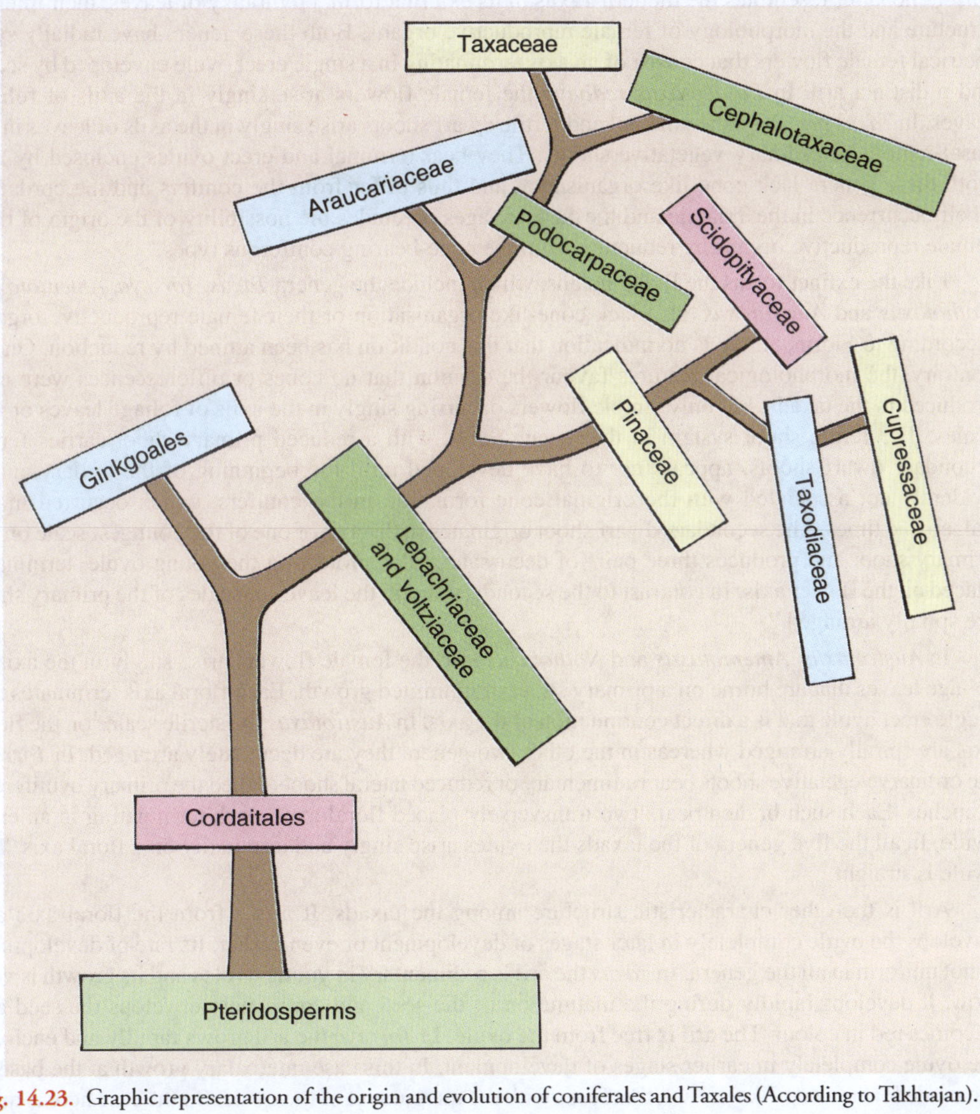

Fig. 14.23. Graphic representation of the origin and evolution of coniferales and Taxales (According to Takhtajan).

the growth of the underlying meristematic tissue. Florin has suggested that vasculature of the ovule in *Austrotaxus* suggests that each of these sterile appendages is equivalent to the entire megasporophyll of the cordaitalean and coniferous strobilus the integuments in whose ovules are formed by two opposite branches of the megasporophyll.

Due to the differences listed above Florin suggested the exclusion of taxads from the conifers and to include them in a separate class of the gymnosperms that may be called *Taxineae*.

Florin arrived at these conclusions after carefully studying the morphology of the reproductive organs in the extinct as well as extant taxads and conifers. Almost all the fossil conifers discovered from the Upper Paleozoic and the Lower and Middle Mesozoic show cone-like organisation of their female reproductive organs. The same is the case with the *Cordaitales*. The fossile taxads are represented by two genera discovered from the Triassic and jurassic horizons of Sweden and Yorkshire. These are *Palaeotaxus rediviva* Nath., from South Sweden and *Taxus jurassica* Florin from Yorkshire. The latter resembles the modern Taxus in its external form, phyllotaxy of leaves, their internal structure and the morphology of female reproductive organs. Both these genera have radially symmetrical female flowers that consist of an axis terminating in a single erect ovule enveloped by scales and a distinct aril. In *Palaeotaxus rediviva* the female flowers arise singly in the axils of foliage leaves. In *Taxus jurassica* the uniaxial and fertile dwarf shoots arise singly in the axils of leaves in the basal regions of ordinary vegetative shoots. They bear terminal and erect ovules enclosed by aril. Both these genera lack cone-like organisation and thus differ from the conifers and the cordaites. Their occurrence in the Triassic and the Jurassic ages precludes the possibility of the origin of their female reproductive organs by reduction from the cone-bearing coniferous types.

Like the extinct forms the living taxads, which include the genera *Taxus, Torreya, Amentotaxus, Nothotaxus* and *Austrotaxus*, also lack cone-like organisation of their female reproductive organs. According to Florin, "there is no indication that this condition has been attined by reduction. On the contrary, the morphological features favour the opinion that no cones or inflorescences were ever produced in the taxads, but only simple flowers occurring singly in the axils of foliage leaves or leaf scales. The fertile shoot system in the recent *Taxus*, with a reduced primary shoot carries fertile secondary dwarf shoots, appears not to have developed until the beginning of the tertiary and is evidently not associated with the original cone formation in the conifers, which occurred in the Palaeozoic times. The secondary dwarf shoot originates in the axil of one of the youngest scale on the primary shoot and produces three pairs of decussate scales protecting the young ovule, terminally placed on the flower axis. In contrast to the secondary shoot, the leaves or scales of the primary shoot are spirally arranged."

In *Austrotaxus, Amentotaxus* and *Nothotaxus* also the female flowers arise singly in the axis of foliage leaves that are borne on a primary shoot of unlimited growth. Each floral axis terminates in a single erect ovule that is a direct continuation of the axis. In *Austrotaxus* the sterile scales on the floral axis are spirally arranged whereas in the other two genera they are decussately arranged. In *Torreya* the ordinary vegetative shoots bear rudimentary or reduced lateral shoots called the primary ovuliferous branches. Each such branch bears two transversely placed floral axes, each terminating in an erect ovule. In all the five genera of the taxads the ovules arise singly and terminally on a floral axis. The ovule is straight.

Aril is the other characteristic structure among the taxads. It arises from the floral axis and envelops the ovule completely in later stages of development or even earlier. Its rate of development is not uniform in all the genera. In *Taxus* the aril is rudimentary in young ovules and its growth is very slow. It develops rapidly during the maturation of the seed and completely envelops the seed and becomes red in colour. The aril is free from the ovule. In *Torreya* the aril grows rapidly and encloses the ovule completely in earlier stages of development. In this case intercalary growth at the base of the ovule forms a cup-like growth that fuses with the integument of the ovule. So the case here is quite different from *Taxus*. In this case the upper free part that encloses the ovule is the aril proper and the lower fused portion is the cup-like extension of the floral axis. In *Amentotaxus* this basal part too is free from the testa. In *Austrotaxus* and *Nothotaxus* the condition is similar to that of *Taxus*. The condition in these three genera is regarded as primitive.

The vasculature of the ovule among the taxads presents certain interesting features. At the top of the floral axis in the female flower there is a ring of collateral bundles that number four in *Taxus, Nothotaxus, Torreya* and *Austrotaxus* and seven to eight in *Amentotaxus*. In *Amentotaxus* there is no fusion of the bundles and all of them enter separately into the base of the ovule; whereas in other four genera the bundles fuse in pairs so that only two double bundles enter the base of the ovule. There is no vascular supply to the aril. In *Taxus, Austrotaxus* and *Nothotaxus* these two fused bundles enter the integuments directly and give off horizontal branches at the base of the female gametophyte, whereas the main strands continue their course upwards into the integument and reach the micropylar end. In *Austrotaxus* these bundles ramify in the integument, whereas in *Taxus* and *Nothotaxus* they do not give off ramifications. This is indicative of a slight trend towards reduction in thevasculature of the integuments. It is further evident in *Torreya* where the vascular strands do not reach the micropylar end, the peak of reduction is illustrated in *Amentotaxus* where no bundle enters the integument.

The microsporangiate organs or the so-called male flowers of taxads present certain variations among themselves and further help to elucidate the differences between the taxads and the conifers. The radial symmetry of the microsporophylls and their occurrence amidst sterile scales highlight these differences. A brief description of taxaceous male flowers is given below to elucidate the differences :

The male flowers in *Taxus, Nothotaxus* and *Torreya* arise singly in the axils of foliage leaves. In *Austrotaxus* and *Amentotaxus* the male flowers are arranged in spike-like inflorescences. Each flower arises in the axil of a bract borne on the inflorescence axis.

In *Nothotaxus* the male flower consists of an axis that bears eight opposite end decussate sterile scales at its base. The upper part of the axis bears distinct whorls of microsporophylls. The upper whorl may have one or two microsporophylls, the next lower may have two, and others below may have three and four microsporophylls in each whorl so that there are in all about ten microsporophylls arranged in 3 or 4 whorls. Above the lowermost whorl there are two sterile scales that may form a single whorl or they may occur mixed with the microsporophylls. The microsporophylls are radially symmetrical and perisporangiate. Each microsporophyll bears 4 to 6 united microsporangia at their tips. This mingling of sterile scales and microsporophylls, in the fertile part of the floral axis, has been compared to the male strobili of *Cordaianthus penjoni* by Florin (1948).

In *Taxus* there is no mingling of sterile scales and microsporophylls. In this case the male axis bears 8–10 decussate sterile scales only at its base. The upper fertile part of the axis bears only six to fourteen microsporophylls, that are radially symmetrial and perisporangiate. Each microsporophyll bears six to eight fused microsporangia.

The floral axis in *Torreya* bears opposite and decussate sterile scales at its base. The upper fertile region bears numerous stalk-like microsporophylls that are arranged in distinct whorls of four each. Each microsporophyll is flattened in a dorsiventral manner at its tip. This flattened and scale-like upper part of the sporophyll is indented adaxially and bears four, free and pendent microsporangia on its abaxial side. Coulter and Land (1905), however, report that there are three adaxial and four abaxial microsporangia. The former (adaxial ones) do not develop to maturity and are ultimately replaced by a resin cavity.

In *Austrotaxus* the male flowers arise singly in the axils of bracts that are borne on an inflorescence axis. Each flower has an extremely short axis that bears one to three microsporophylls that are radially symmetrical and each bears three united sporangia at their apices.

In *Amentotaxus* the male flowers appear hemispherical and airse singly in the axis of bracts that are borne on an inflorescence axis (same as in *Austrotaxus*). The inflorescence axis may be undivided or it may bifurcate once. The inflorescence axes arise in group of two to four at the end of vegetative shoots. Each inflorescence axis arise in group of two to four at the end of vegetative shoots. Each inflorescence axis bears two or four rows of distinctly keeled and stiff sterile scales. Each mele flower has reduced axis that bears microsporophylls with flattened and dorsiventral tips as in *Torreya*. The dorsiventral tip carries three to five pendent sporangia.

Florin regards radially symmetrical and perisporangiate microsporophylls of *Taxus* and *Nothotaxus* as more primitive than the dorsiventral and hyposporangiate ones in *Torreya* and *Amentotaxus*. According to Florin, "the perisporangiate microsporophyll may in its turn be derived morphologically from an indifferentiated, dichotomising branch system, a sporangial truss, with terminal and erect sporangia. Such male sporangial trusses were, therefore, probably characteristic of the oldest representatives of the taxad group or their fore-runners."

There is no fossil record of the microsporangiate organs of the taxads from the Upper Palaeozoic or Lower mesozoic strata and hence it is difficult to draw any far-reaching phylogenetic conclusion. So far as the conifers and cordaites are concerned much fossil record is known and the male reproductive organs in these two groups present a wide gap. In the cordaites the male flowers are united into inflorescences whereas in the Upper Carboniferous and Lower Permian conifers the male flowers occur singly and terminally on vegetative lateral shoots of the last order. In the cordaites the microsporophylls are mixed with sterile scales and are not peltate but bear terminal clusters of 4–6 erect microsporangia (aerosporangiate). In the oldest known conifers like *Lebachia* the axis of the male flower bears only the microsporophylls and no sterile scales. The microsporophylls are bisporangiate, hyposporangiate and hypo-peltate.

To sum up in the words of Florin, "a contrary to the conditions in the taxads, the true conifers, including all the hitherto known, fossil types, have dorsiventral and hyposporangiate microsporophylls and the flower axis always carries exclusively such sporophylls in the fertile region. The microsporophylls of the conifers have probably also evolved from radial, repeatedly dichotomising sporangial trusses with terminal, erect sporangia. But as regards the phylogeny of the male flower the conifers seem to differ from the taxads in the absence (or possibly much more rapid passing) of that interesting intermediate stage between the acrosporangiate and the hyposporangiate forms of the microsporoyhylls retained in *Taxus* and *Nothotaxus*."

In the present text the authors have followed Sporne (1965) in raising *Taxaceae* to the rank of a separate order *Taxales* equal in rank to the orders *Ginkgoales, Cordaitales* and *Coniferales*, all belonging to a class Coniferospsida. Such a treatment is justified in view of the important differences in the reproductive organs of *Coniferales* and the taxads. There are, however, many resemblances between the taxads and the true conifers in the vegetative organs, their external morphology and histology.

QUESTIONS

ESSAY TYPE :

1. Describe the distinguishing features of Taxales. Discuss the position and relationships of this order among the gymnosperms. **(Rohilkhand M.Sc, 1990)**
2. Discuss salient features of Taxales and compare this group with the coniferales. **(Rohilkhand M.Sc, 1996)**
3. Give a comparative account of the male gametophytes of *Pinus, Taxus* and *Gnetum*. **(Vinoba Bhave, 1998; VirKumar Singh, 1994)**

4. Give a comparative account of wood structures of *Pinus, Taxus and Gnetum*. **(VirKumar Singh, 1991, 1996)**

5. Describe the structure and reproduction in *Taxus*. **(VirKumar Singh, 1995)**

6. Give a comparative account of leaf of *Pinus, Taxus* and *Gnetum*. **(VirKumar Singh, 1995)**

7. Describe the salient features of leaf and stem of *Taxus*. In what characters its leaves are different from those of *Pinus*. **(VirKumar Singh, 1996)**

8. Give a comparative account of the development of male gametophyte in *Pinus* and *Taxus*. **(VirKumar Singh, 1998)**

9. Describe with diagrams the life cycle of *Taxus* and point out why Taxales has been kept apart from conifers. **(VirKumar Singh, 1998)**

10. Discuss the affinities of Taxales.

11. Describe the distribution, habit and habitat and external characters of *Taxus*.

12. Describe in detail the internal structure of stem, leaf and root of *Taxus*.

13. Describe the pollination, fertilisation and embryogeny of *Taxus*.

14. Describe the male and female strobili of *Taxus*.

15. Describe the structure and development of male gametophyte of *Taxus*.

16. Give an account of post-fertilisation charges in *Taxus*.

SHORT ANSWER TYPE :

17. **Write short notes on**
   (*i*) *Taxus baccata*. **(Rohilkhand MSc, 1995; 2000; Kanpur M.Sc, 1998; Garhwal M.Sc, 2004)**
   (*ii*) Distinguish between *Taxus* and *Cephalotaxus*. **(Rohilkhand M.Sc, 1999)**
   (*iii*) Male strobilus of *Taxus*. **(Rohilkhand M.Sc, 2002)**
   (*iv*) Differentiate between ovules of *Taxus* and *Ginkgo*. **(Kanpur M.Sc, 1992, 1994)**
   (*v*) Distinction between *Taxus* and *Cephalotaxus*. **(Kanpur M.Sc, 1997)**
   (*vi*) Wood Anatomy of *Taxus*. **(Vinoba Bhave, 1998)**
   (*vii*) Secondery growth in *Taxus*. **(Vinoba Bhave, 1999)**
   (*viii*) Female cone of *Taxus*. **(VirKumar Singh, 1994)**
   (*ix*) Ovule of *Taxus*. **(VirKumar Singh, 1996)**
   (*x*) Male flower of *Taxus*. **(VirKumar Singh, 1998)**
   (*xi*) *Taxus* leaf. **(Magadh, 1990)**

18. **Draw well labelled diagrams of :**
   (*a*) L.S. Ovule of *Taxus*
   (*b*) T.S. Stem of *Taxus*
   (*c*) L.S. Seed of *Taxus*
   (*d*) T.S. Old stem of *Taxus*
   (*e*) Mature microspore in *Taxus*
   (*f*) *Taxus* leaf

19. Draw a well labelled diagram of T.S. Taxus leaf. **(Magadh, 1991)**

20. Write external features and economic importance of *Taxus*. **(Himachal Pradesh, 2000)**

21. Why has Taxaceae been Separated from coniferales ?

22. Discuss in brief the position of *Taxus* in coniferales.

23. Name the fossil genera of Taxaceae.

24. Give the name of the alkoloid derived from *Taxus*.

25. Describe wood of *Taxus*.

26. "The family Taxaceae constitutes an independent Taxon, Taxopsida, ranking with the coniferae". Discuss this statement.

OBJECTIVE TYPE :

27. Select the correct Answer:

(*i*) *Taxus baccata* yields an alkaloid known by the name

(*a*) Taxol (*b*) Taxine

(*c*) Ephedrine (*d*) Toxin

(*ii*) The ovule in *Taxus* is

(*a*) orthrotropus and sessile

(*b*) orthrotropous and stalked

(*c*) anatropous and stalked

(*d*) anatropous and sessile.

(*iii*) An aril is present in the ovule of

(*a*) *Cycas* (*b*) *Taxus*

(*c*) *Ephedra* (*d*) *Selaginella*.

(*iv*) Wingless pollen grains are characteristic of

(*a*) *Taxus baccata*

(*b*) *Pinus triphylea*

(*c*) *Pinus wallichiana*

(*d*) *None of the above*.

(*v*) The mature archegonium in *Taxus* is

(*a*) one celled

(*b*) two celled

(*c*) three celled and a nucleus

(*d*) four celled.

CHAPTER 15

Family Ephedraceae – Ephedra

EPHEDRA SP.

- Family Ephedraceae :
- Anatomy and Reproduction in Ephedra
- Embryology
- Economic Importance of Ephedra

The class Gnetopsida as considered in this text includes three families : the *Ephedraceae, Gnetaceae* and *Welwitschiaceae*; each represened by a single genus, *viz., Ephedra, Gnetum* and *Welwitschia*, respectively. All the three families have been included under the order Gnetales. In doing so the authors have followed Sporne (1965). The fossil history of this order is almost unknown except for the fossil remains of pollen grains of *Ephedra* from the Permian and the Eocene (Delevoryas, 1962). Previously these three genera were included under the order Gnetales and no separate families were recognised. They were all placed under a single family *Gnetaceae*.

This was due to certain noteworthy features that are common to all the three genera :

1. The secondary wood contains vessels.
2. The flowers are arranged in compound strobili and are unisexual and usually dioecious.
3. The male flowers are surrounded by perianth and the structure of microsporangial wall is similar in all the three genera.
4. The ovules are also surrounded by several envelopes that are variously interpreted as extra integuments or as perianth.
5. The micropylar tube is extremely elongated and froms a bristle-like tube.
6. The ovules are orthotropous.
7. The embryo has two cotyledons and there is a unicellular primary suspensor in all the three genra.

There are many points in which these three genera differ from each other, the most outstanding being :

1. The stomatal development in *Ephedra* is haplocheilic like those of Cycadales, Cordaitales, Coniferales and Ginkgoales, whereas those of *Welwitschia* and *Gnetum* are syndetocheilic like those of the Angiosperms and the Cycadeoideales (Florin 1931, 1933).
2. The ovule of *Ephedra* is appendicular (Eames, 1952) and not cauline as in *Gnetum* and *Welwitschia*.
3. The development in the female gametophyte of these genera shows many differences.

In addition to these there are many morphological differences of far-reaching consequence and hence Eames (1952) recommended that the Gnetales should be broken into three separate orders : Ephedrales, Gnetales and Welwitschiales, each having a single family with a single genus. In this text, the three families are retained under a single order Gnetales. The order Gnetales is characterised by the following features :

1. The sporophytes may be shrubs, trees or even woody climbers. In *Welwitschia* the stem is turnip-like and is partly subterranean.
2. The leaves may be spirally arranged, opposite or even whorled.
3. The leaves are simple and may be scale-like, strap-shaped or even ovate or elliptic. In *Welwitschia* they become long and woody.
4. The secondary wood has vessels. These vessels unlike those of angiosperms are derived from tracheids with pitted thickenings. There are no resin canals.
5. The flowers are unisexual.
6. The trees are dioecious except in some species of *Gnetum*.
7. The reproductive organs are organised into compound strobili or inflorescences.
8. The male flowers are surrounded by sterile perianth scales.
9. The ovule is also surrounded by extra envelopes.
10. The inner integument of the ovule is produced into a long and tube-like micropyle. The gametophyte in *Gnetum* and *Welwitschia* is tetrasporic.
11. The sporangiophores or antherophores in the male flowers have one to eight synangia.
12. The primary suspensor is unicellular.
13. Fertilisation is effected by means of a pollen tube. The pollen grains are devoid of wings.
14. The embryo has two cotyledons.

FAMILY EPHEDRACEAE

Habit

The family is represented by a single genus called *Ephedra*. It is represented by forty species that are widely distributed throughout the globe and are mostly xerophytic shrubs or woody climbers and lianes. One species grows into a small tree *e.g.*, *E. triandra* Tul. *E. gerardiana* is a perennial herb. *E. foliata* is a scrambling shrub that may attain a height of up to 6 metres. It climbs up the neighbouring trees or even walls. In the absence of a support it spreads along the ground. Other Indian species, *e.g.*, *E. intermedia, E. saxatilis, E. regeliana* are shrubs.

Distribution : *Ephedra* is widespread and sporadic in distribution in arid regions of tropics and sub-tropics of northern and southern hemispheres. About eighteen species occur in the old world (France, Canary islands, around the Mediterranean east to Persia, India and China), and about 22 in the New World with 10–12 in North America and 7–9 in South America (Bolivia to Patagonia). The North American species are most abundant in California, Arizona, and New Mexico, with ranges extending north and south into Mexico, and 3–4 species occurring as far east as Texas.

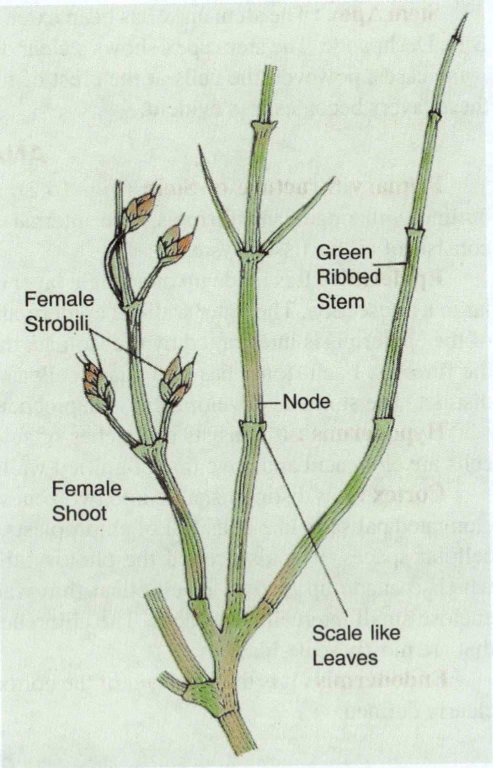

Fig. 15.1. *Ephedra foliata.* A branch bearing leaves and young female strobili.

In India the genus is represented by *E. foliata* Boiss et kotschy, *E. gerardiana* Wall., *E. intermedia* Schrenk et Mey., *E. nebrodensis* Tineo, *E. regeliana* Florin, and *E. saxatilis* Royle ex Florin. *E. foliata* is a scrambling shrub found in the drier parts of Rajasthan, Panjab and Haryana. *E. gerardiana* grows at Chakrata (Himalayas, in U.P.) in the crevices of rocks as a perennial herb. It also grows in Baluchistan. It grows up to an altitude of 7,000–16,000 feet above sea level and extends from the drier regions of temperate and alpine Himalayas, from Kashmir to Sikkim, Frequently met with in Pangi, Lahaul, Spiti, Chini and Kilba-Kailash ranges of Kanawar, Shali-hills (north of Simla), Kashmir and Ladakh. *E. intermedia* occurs along Pangi hills, Kanawar and to a lesser extent in Kashmir, Kulu and Jaunsar. *E. nebrodensis* grows in Kashmir, Lahaul and Spiti. *E. saxatilis* also grows in the drier ranges of the Himalayas especially in the Lahaul and Spiti valleys.

External Features : The stem is green, hard, ribbed, glabrous and much branched. It is distinctly jointed, slender and has long internodes (Fig. 15.1). The branches are also green and photosynthetic. They remain green for several seasons. The leaves are deciduous, opposite or whorled, more or less connate basally and usually reduced to membranous sheaths. In the case of whorled phyllotaxy each node bears 3 or 4 leaves. Each leaf is inervated by two unbranched veins that run parallel to each other. The scale leaves bear a bud each in their axils. These axialary bud grow into branches. The internodes grow by means of a basal or an intercalary meristem that is present at the base of each node. Sometimes this basal or intercalary meristem becomes non-functional and hard and forms an abscission zone. This usually happens at the close of the growing season and results in the fall of the branches. The plants are usually dioecious and only rarely monoecious. Eames regards the monoecious condition as an abnormal state of development and calls such plants as monstrosities. *E. campylopoda* has pendulous branches andis commonly grown in hanging baskets. In *E. gerardiana* one to several short thick stems arise from the ground and branch profusely above so as to give a characteristic hemispherical look to the plants. The roots are long, branched and deep feeders. They are tap roots that grow pretty deep into the soil. There is no report of a mycorrhiza.

Stem Apex : The stem apex has been extensively studied by Gifford (1943), Seeliger (1954) and B.D. Deshpande. The stem apex shows a clear demarcation between **tunica** and the **corpus** layers. In some cases, however, the cells at the crest of tunica divide periclinally and the distinction between these layers becomes less evident.

ANATOMY

Primary Structure of Stem (Fig. 15.2). A transverse section of the stem reveals a circular outline with ridges and furrows. The internal structure clearly portrays its xerophytic nature and consists of all the tissue systems.

Epidermis : It is made up of a single layer of parenchymatous cells that appear nearly rectangular in a transection. The outer walls of epidermal cells are covered with a thick cuticle. The continuity of the epidermis is interrupted by the stomata that are sunken in circular pits and are locate donly in the furrows. Each stoma has two guard cells and a substomatal cavity. The subsidiary cells are not distinct. The stomatal development is haplocheilic.

Hypodermis : It consists of patches of sclerencymatous cells located below the ridges. These cells are elongated and have thick lignified walls. They give mechanical strength to the stem.

Cortex : It is distinguishable into two zones. The outer chlorenchymatous cortex that consists of elongated palisade-like cells full of chloroplasts. These are loosely arranged and enclose lot of intercellular spaces. It is also called the photosynthetic zone. Next to it is the parenchymatous cortex which is made up of oval or elliptical thin-walled cells that also possess chloroplasts. They also enclose small intercellular spaces. The chlorenchymatous cortex makes up the deficiency of leaves that are mostly scale-like.

Endodermis : It is the last layer of the cortex and is quite distinct in young stem. Pericycle is not clearly defined.

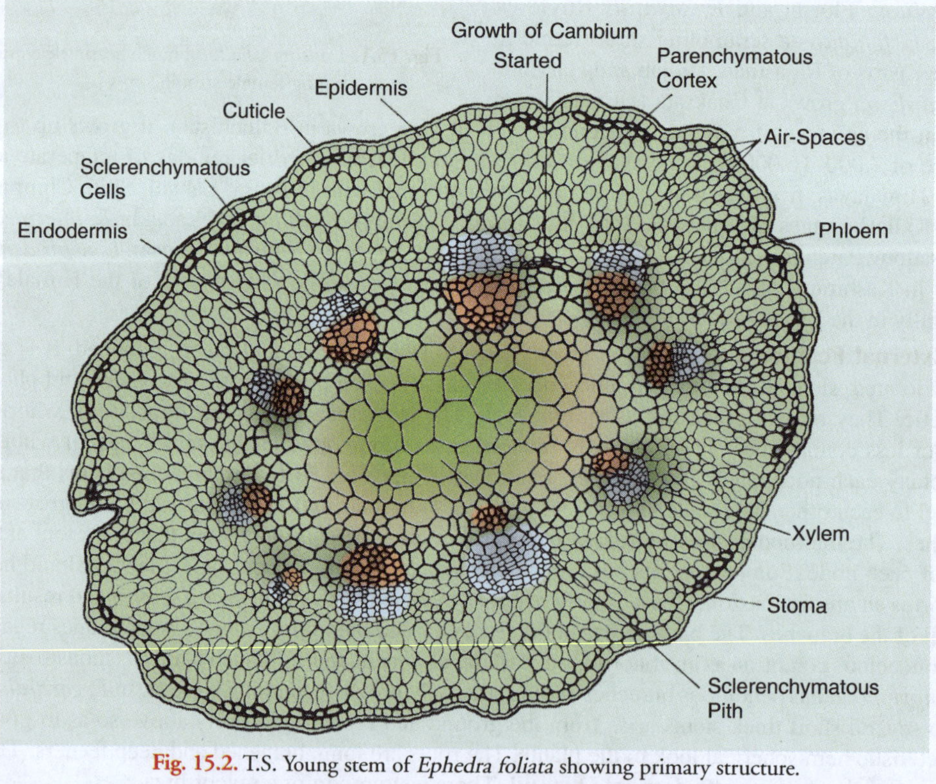

Fig. 15.2. T.S. Young stem of *Ephedra foliata* showing primary structure.

Vascular Region : It consists of an interrupted ring of conjoint, collateral, endarch and open vascualr bundles. The cambium consists of a single layer of meristematic cells. There are 8–12 vascular bundles in the internode. Marsden and Steeves (1955) reported that there are alternate pairs of large and small bundles. Each pair of small bundles supplies a leaf at the upper node. Just above the node one bundle produces as smaller one by tangential division towards the next larger one so that the paired alternation is re-established and the two leaf traces extend the entire length of the next internode. Just below the level of departure of the traces to the leaves there is a complete girdle of primary vascular tissue that inter- connects all boundles of the rings. Branch traces are also two in number and are given off by adjacent large bundles immediately above the point of departure of leaf traces. In some species there is an alternation of three small and two large bundles. The nodes are two-trace and unilacunar type. This is also the case in *Ginkgo* and *Pteridosperms*.

The xylem consists of tracheids, vessels and xylem parenchyma. The vessels originate from the pitted tracheids. Certain groups of pits on long and sloping walls near the ends of tracheids are larger in diameter, loose their pit closing membranes, and make up the foraminate perforation plates characteristic of the vessel members of *Ephedra*. The plates show a uniseriate arangement of pores in early formed elements, but in later formed ones most pores are two ranked (*E. major*) or irregularly grouped (*E. californica*). The vessels have bordered pitted thickening. The tracheids may be annular spiral or have uniseriate or irregularly arranged bordered pits on their radial walls.

The phloem consists of sieve cells, phloem parenchyma and albuminous cells. The albuminous cells are associated with the sieve cells and look like companion cells of the angiosperms. They are however, different in origin and derived from a different cell and not from the same mother cell.

Pith : It occupies the centre of the stem and consists of parenchymatous cells that enclose small intercellular spaces. In older branches the cells become thick-walled in the nodal region. This is the intercalary meristem whose cells become thick-walled and help in abscision of the branches. It is also termed as the peridermal diaphragm. The medullary rays between the vascular bundles are narrow.

Root (Fig. 15.3) : The root has the usual structure and consists of **epiblema, cortex, endodermis, pericycle** and a **diarch** vascular region. There are no resin canals. Mucilage canals are, however, present here and there in the cortex.

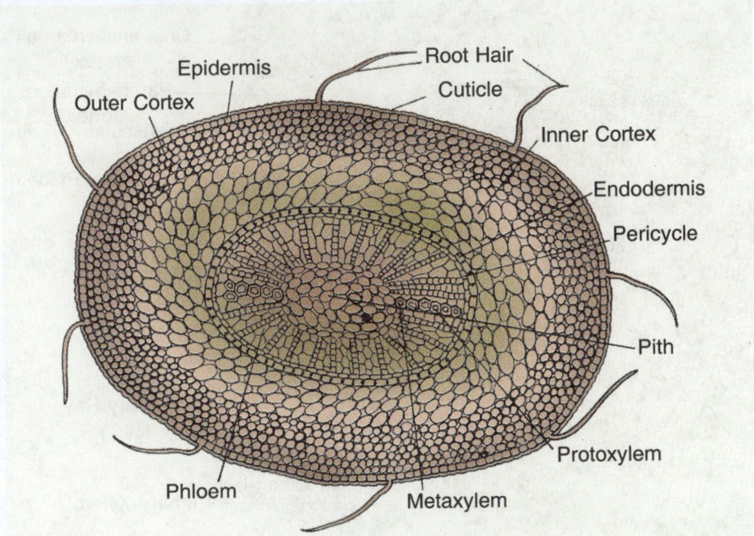

Fig. 15.3. T.S. Young root of *Ephedra foliata.*

Secondary Growth in Stem (Fig. 15.4). It is affected by means of vascular cambium present in between the primary phloem and primary xylem. The cambium forms a complete ring and starts dividing horizontally to cut off secondary phloem towards the outerside and seconday xylem towards the innerside. The amount of secondary xylem produced is more than the secondary phloem. The cambium is persistent and keeps pace with the growing cylinder by the anticlinal division of its cells. It has the usual two types of cells—the **ray initials** and **fusiform initials**—the former giving rise to the phloem and xylem rays, whereas the latter giving rise to the secondary vascular elements. The activity of the cambium varies with the season and hence there are distinct annual rings. The secondary wood is markedly ring porous.

The secondary xylem consists of vessels, tracheids and scanty amount of xylem parenchyma. Xylem fibres are absent. The vessels in *Ephedra* originate from the long tracheid like cells by the loss of closing membranes from a series of circular bordered pits located near the ends of such elongated cells. Such a step leads to the creation of a foraminate perforation plate. Transitional conditions leading to such a state are noticeable in *E. californica* and *E. major*. In the latter case end walls with both perforations and ciruclar bordered pits occur. This illustrates that the perforations originated by the loss of closing membranes of the circular bordered pits. There are clear transitional forms between tracheids and vessel members. These perforation plates may be arranged in a uniseriate manner or biseriae manner or may be even irregular in arrangement. In te angiosperms, on the other hand, the vessels originate from tracheid-like cells with transversely elongated or scalariform bordered pits located at their tapering end walls. These elongated pits loose their closing membranes. Later elaboration results in the development of well defined scalariform perforation plates on the end walls. Elimination of the bars between these scalariform plates leads to the establishment of a simple oval or circular perforation plate. These developmental events were studied by Thompson in 1918 and revealed that the vessel members in Angiosperms and *Ephedra* (along with other gnetales) have a different ontogeny (Marsden and Steeves, 1955) and have a different origin. The superficial resemblance between the vessels of Angiosperms and Gnetales is no more that an example of convergent evolution (Marsden and Steeves 1955; Bailey 1953).

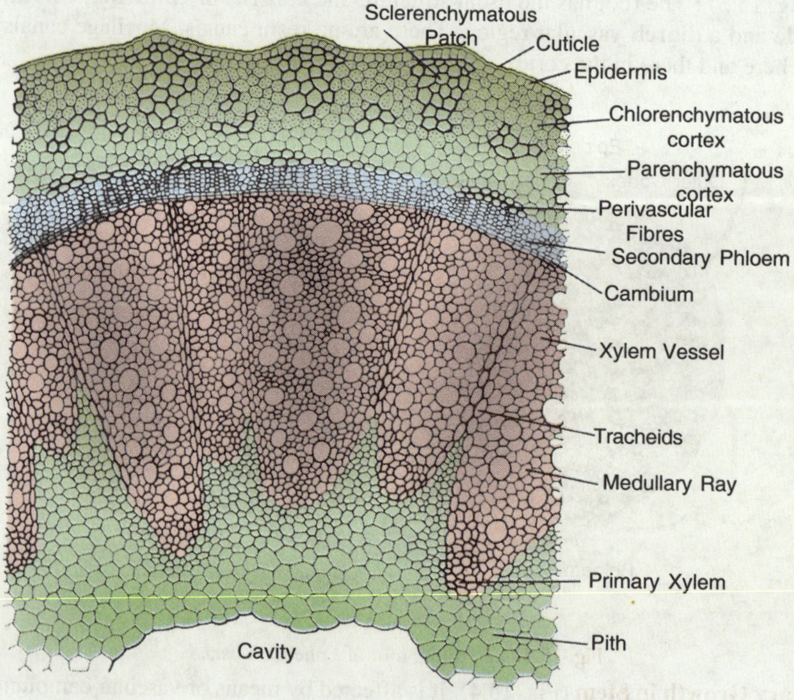

Fig. 15.4. T.S. old stem of *Ephedra* showing secondary growth.

The spring wood abounds in vessels and is ringporous, whereas the autumn wood has only a few vessels and more tracheids. The vessels and tracheids have bordered pits that may have uniseriate arrangement or may beirregularly scattered. Spiral, annular or reticulate tracheids occur in the protoxylem elements of the primary xylem. The nodes especially have reticulate thickenings on the tracheids. Some cells with crystals of calcium oxalate are also present in the secondary xylem. There is no distinction into hard wood and soft wood. Most of the branches fall down after 3 or 4 yeas, due to abscission.

Alosi and Alfieri (1972) studied the secondary phloem in *E. californica* and *E. viridis*. The secondary phloem is a typical of the gymnisperms in general and exhibits some angiospermic characters. The conducting phloem which lies just outside the cambial zone contains parenchymatous rays, sieve cells and axial parenchyma including albuminous cells. The **albuminous cells** in *Ephedra* are found in the axial system and are the product of fusiform initials that also give rise to other vascular elements like sieve cells and phloem parenchyma. These albuminous cells are found along with the sieve cells and may be of the same length as the sieve cell or half its length. These cells cannot be regarded as companion cells because they do not occur in the same radial rows as the sieve cells. Their origin may also be from different cells. They form a conspicuous element of the axial system and exist in radial files or may be scattered in the conducting phloem. They result from the transverse or oblique division of a fusiform phloem mother cell and have one slightly inclined end wall and the other end wall is tapering. They have denser cytoplasmic contents and have connections with small sieve areas of an adjoining sieve cell. They also contain besides the nucleus an ovoid slime body. The sieve cells and sieve elements resemble each other : (*i*) in possessing an ovoid proteid body; (*ii*) dispersing slime; (*iii*) general shape; (*iv*) a similar position arrangement in the tissue.

The sieve cells in *Ephedra* may have blunt or tapering ends and have their walls with no secondary or nacreous walls. Alosi and Alfieri (1972) recognise two types of sieve cell in *Ephedra* : (*i*) those that are about 400 μ long; and (*ii*) those that are about 220 μ long. The sieve areas (formerly called sieve plates by Boodle and Worsdell, 1894) occur on the end walls as well as on the lateral walls. Each sieve area has pores that are generally aggregated into groups. These groups of pores have been termed as sieve fields by Hill (1901) and Everest and Alfieri (1965). These sieve areas do not occur between sieve cells and albuminous cells. The mature sieve elements have several "nacrotic-appearing" nuclei that are generally flattened and are found around the periphery of the parietal cytoplasmic layer. Sometimes these nuclei are rounded and occur in the centre of the cell. Slime body is present in the young sieve element but it is lost in the mature elements which contain a P-protein. The P-protein may appear in the form of a parietal cylinder, thread-like strands, as a slime plug or in the form of amorphose globules.

The radial system of the conducting phloem consists of secondary medullary rays that are mainly parenchymatous and multiseriate and are usually bordered by short fusiform cells. Uniseriate and biseriate rays are also present, but mainly in stems with a year or two of secondary growth. Multiseriate condition results as secondary growth continues for more years. In mature stem only multiseriate rays are found. Thick-walled sclereids are also met with in the rays. These selereids are living and store starch.

The xylem rays are also uniseriate or biseriate in the beginning and become multiseriate later.

That part of the phloem in which the sieve elements have stopped functioning is called the non-conducting phloem. In *Ephedra* this type of phloem forms a small zone external to the functional region. Some cells of this phloem change into cork cambium or phellogen whereas others become sclerenchymatous.

The annual periderm arises within the non-conducting phloem at about the same time that the vascular cambium is reactivated in spring. The cork cambium arises from the living parenchyma cells of the non-functionary phloem and cuts off suberised cells towards the outer side. Fusiform parenchyma cells lying internal to the new periderm develop into fibre sclereids. These cells possess crystal sand in the region of the middle lamella.

The pith becomes narrow in the stems that have undergone secondary growth and may ultimately disappear in very old stems, its place being taken by the secondary xylem.

The secondary growth in the root is normal.

Leaf : All the species of *Ephedra* have reduced and membranous scale leaves. A transverse secion of the leaf (Fig. 15.5) reveals an almost oval outline. There is a thick cuticle that covers the single layered epidermis of parenchymatous and slightly radially elongated cells. A single stoma was observed in the section. It is sunken. Next to the epidermis is the palisade tissue made up of two or more layers of oval cells that are full of chloroplasts. The centre of the leaf is occupied by parenchymatous tissue. The palisade and the central tissue enclose small air spaces. Two vascular bundles are present in the central parenchyma.

REPRODUCTION

All the species of *Ephedra* are dioecious and bear male and female reproductive organs on different plants. These plants do not show any difference in their vegetative organisation and can be recognised only when reproductive structures appear. Monoecious specimens have also been reported (Pearson, 1929). Eames (1952) regards them as 'monstrosities' or cases of abnormal development. Usually the "angiosperm-centred" terminology is used to interpret reproductive structures in *Ephedra* and other Gnetales. The terms inflorescence, flower, perianth, stamen and another are commonly used. This type of terminology creates confusion while describing a plant that has gymnospermous morphology. Eames (1952) has strongly criticised such a practice. In this text, the term flower has been used, but other angiospermic terms have been discarded.

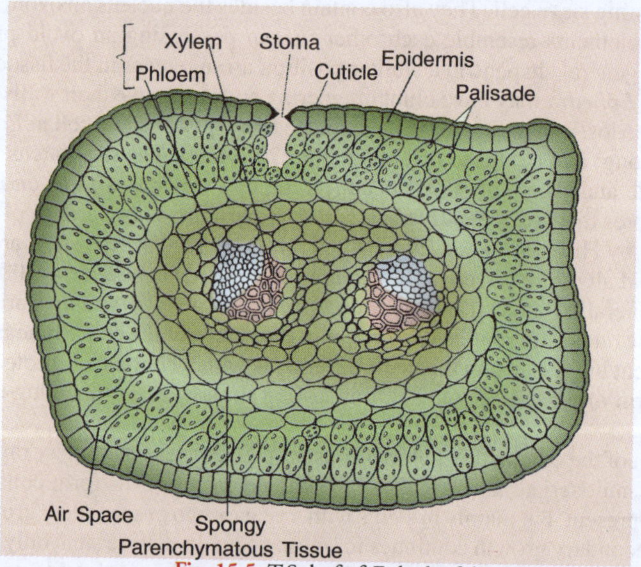

Fig. 15.5. T.S. leaf of *Ephedra foliata*.

Microsporangiate or Male Strobilus (Figs. 15.6–15.10). The microsporangiate strobilus of *Ephedra* is regarded as a compound structure. These strobili arise in clusters from the nodes of branches. Each strobilus arises in the axil of a scale leaf. Their number at the node depends upon the number of scale leaves. Each strobilus consists of a central axis that bears two to eight pairs of decussately arranged simple, broad and cupped bracts. The basal one or two pairs of these bracts are sterile, whereas the upper ones bear each a solitary male shoot or microsporangiate shoot in their axils. This shoot is continued into a short axis or the microsporangiophore and bears at its base a pair of fused bracts. The microsporangiophore bears terminally 8–12 microsporangia that are sessile and dehisce terminally. These fused basal bracts are also called called bracteoles or perianth parts. The male flower is also called a simple strobilus. A compound male strobilus, therefore, consists of many such simple strobili. There are slight variations regarding the number of male flowers or simple strobili arising in the axil of each bract of the compound strobilus. In *E. gerardiana* and *E. foliata* one simple strobilus arises in the axil of a bract, but in this case there is a slight external expression of a fusion of two sporangiophores to form one. The male shoot is single but according to Eames (1952) it bears two fused bracteoles and two sporangiophores that become ontogenetically fused (the signs of fusion only slightly clear externally). In *E. antisyphilitia* the single sporangiophore shows no external signs of fusion. In *E. intermedia* and *E. distachya* two or more sporangiophores arise separately in the axil of the two fused bracteoles. The latter condition is regarded by Eames as primitive and that found in *E. foliata* as intermediate, whereas the one found in *E. antisyphilitica* as advanced.

The vascular supply to the cone consists of two pairs of vascular bundles (Fig. 15.9) at the base of cone axis in *E. gerardiana*. Later the bundles of each pair split into five so that there are 10 vascular bundles in all (Singh and maheshwari, 1962). Further splitting of bundles occurs in two bundles lying laterally to the middle bundle of each side so as to form a total of fourteen bundles. Four of these supply the bract scales subtending the male flowers. Of the rest 19 bundles three enter the sporangiophore. They unite into an arc-shaped vascular bundle at the base of the sporangiophore. The perianth is not supplied any vascular trace. At the tip of the sporangiophore, each sporangium is supplied by its own vein ending, which is a terminus of a dichotomous truss.

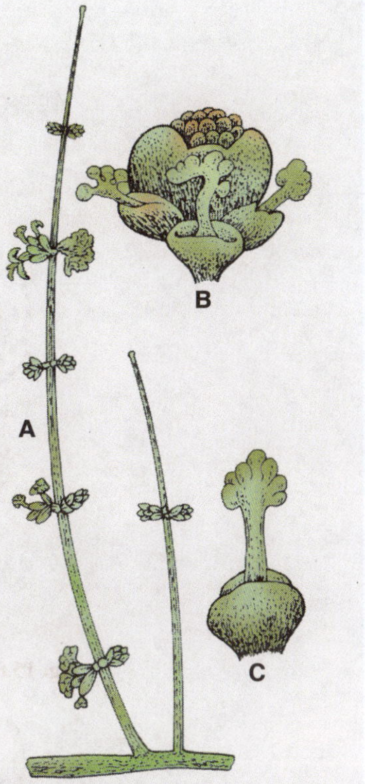

Fig. 15.6. *Ephedra foliata.* A branch bearing male strobili; B. A male strobilus with paired bracts and microsporophylls; C. A microsporophyll bearing microsporangia. Also note the pair of bracts.

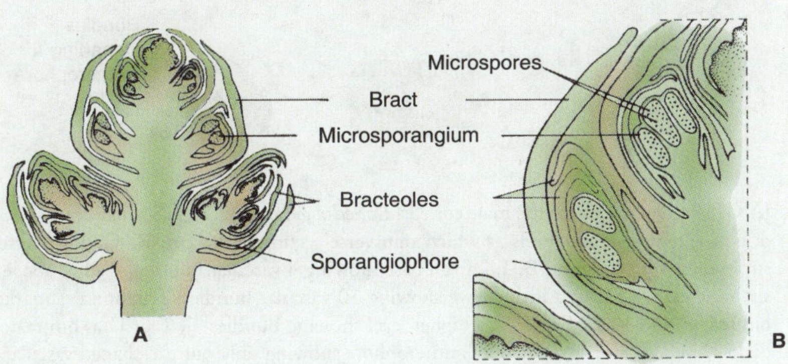

Fig. 15.7. *Ephedra foliata.* A. L.S. trough the apex of male reproductive shoot showing microsporangiate strobili, each arising in the axil of a bract, B. Portion of A magnified to show details.

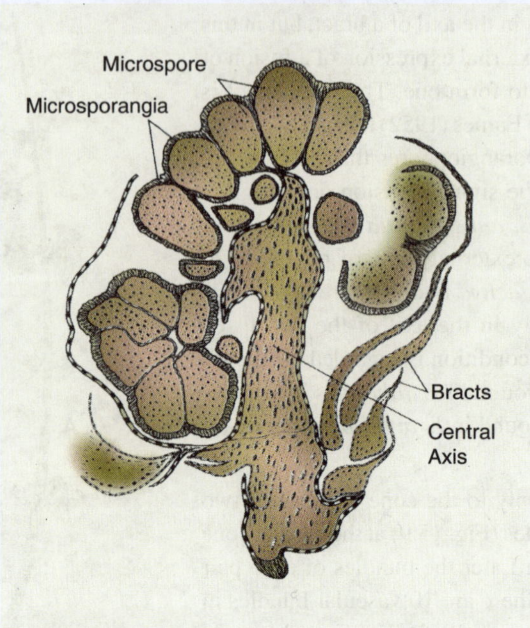

Fig. 15.8. L.S. Male strobilus of *Ephedra foliata*.

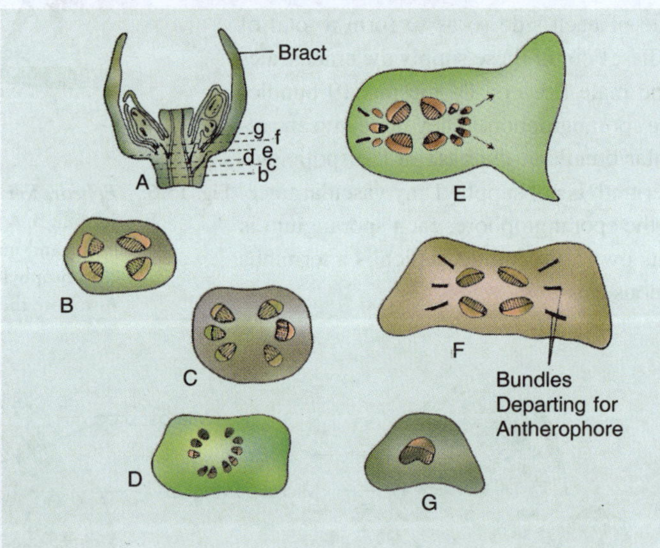

Fig. 15.9. (A-G) Vascular supply to the male cone in *Ephedra gerardiana*. A. L.S. male cone. The small figures b, c, d, e, f, g represent the levels at which transverse sections were cut; B.–G. Transverse sections of male strobilus at various levels; B. T.S. from base showing 4 vascular bundles; C. T.S. above the base showing six bundles. D. T.S. at a little above showing 10 vascular bundles; The bundles marked 1, 2 in C have bifurcated into 1', 2'; E. T.S. at a higher level showing bundles 1', 2' of D as bifurcated; F. T.S. at a still higher level; G. T.S. at the base of antherophore showing only one arc-shaped vascular bundle. (All after Singh and Maheshwari)

Microsporangium (Fig. 15.10, A–K). The sporangiophore arises as small protuberance in the axil of the fertile bract of the male strobilus. The protuberance shows a clear distinction into tunica and corpus in the earlier stages of development but later due to periclinal divisions in the tunica this

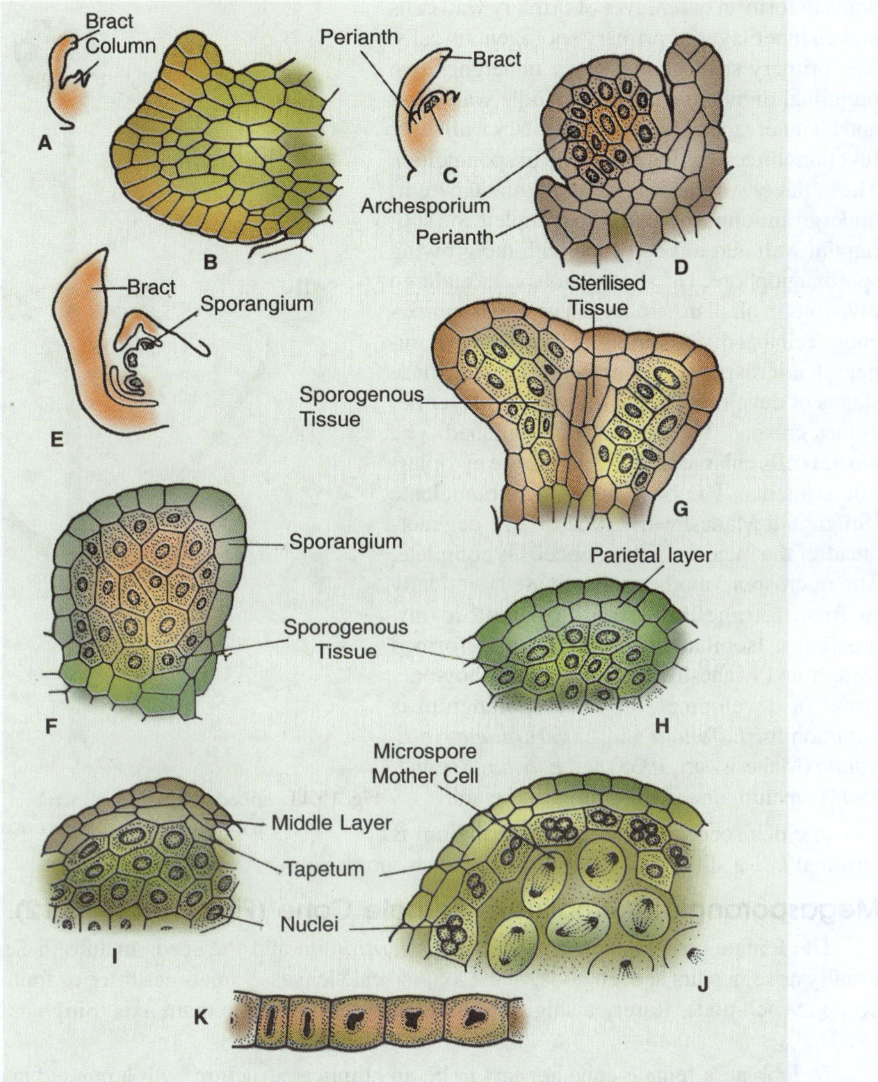

Fig. 15.10. (A-K). *Ephedra gerardiana* A, C and E are the diagrammatic longitudinal sections of male strobili at various stages of development; B. Enlarged view of column as in A; D. L.S. of sporangium showing differentiation of archesporial cells and wall layer. Also note the bracteoles or perianth lobes; F. L.S. of sporangium showing sporogenous cells and wall layer; G. It shows the appearance of sterile cells in the sporogenous tissue dividing the sporangium into two chambers; H. T.S. of portion of sporangium showing two wall layers and sporogenous tissue; I. T.S. showing division of inner wall layer into a tapetum; J. T.S. showing tapetal cells with two to four nuclei. Also note the microspore mother cells; K. Portion of outer wall layer. (All after Singh & Maheshwari)

distinction becomes obscure. The two perianth lobes arise growing for some time. Each lobe represents a sporangium. A hypodermal group of archesporial cells becomes apparent in each lobe and consists of larger cells with distinct nuclei and denser contents. After a little period of growth each archesporial group of cells in each sporangium becomes divided into two groups by the appearance (Fig. 15.10, G) of a band of sterile cells (Singh and maheshwari, 1962). This indicates that a single sporangium becomes two chambered. Now each group of archesporial cells divides by periclinal walls to form an outer layer of primary wall cells and an inner layer of primary sporogenous cells. The primary sporogenous layer undergoes two periclinal divisions to form a middle wall layer and an inner tapetal layer. The primary wall layer functions directly as the outer wall of sporangium. These three wall layers (including the tapetum) undergo anticlinal divisions to complete the sporangial wall and to keep pace with the growing sporoangiophore. The sporogenous cells undergo divisions in all planes to form a group of sporogenous cell that divide further to form a large number of microspore mother cells. During these stages of development the middle wall layer becomes crushed. Its cells become flattened. The tapetal cells enlarge and develop dense cytoplasmic contents. The become bi-or tetranucleate (Singh and Maheshwari, 1962). They degenerate after the meiosis of mother cells is complete. The microspore mother cells divide meiotically to form tetrahedral tetrads of haploid microspores. Isobilateral tetrads are also formed (Singh and Maheshwari, 1962). The above account of development of microsporangium is common for *E. foliata* and *E. gerardiana*. In *E. foliata* (Maheshwari, 1935) and *E. trifurca* (Land, 1904) tapetum arises from the wall layers.

Fig. 15.11. *Ephedra foliata*. A branch bearing ovulate strobili in which the ovules have matured into seeds.

The dehiscence of the microsporangium is terminal *i.e.*, a slit appears at the apex of each sporangium.

Megasporangiate Strobilus or Female Cone (Figs. 15.11, 15.12).

The female cones appear in April in *E. gerardiana* and the seeds mature in September. They usually arise in pairs at each node, in the axil of scale leaves. Sometimes three or four strobili may be borne at each node. Rarely a single strobilus may terminate the main axis (Singh and Maheshwari, 1962)

To look at, a female cone appears to be an elliptical structure with a pointed apex. It seems to retain the same compound strobilar organisation as the male strobilus and consists of a short axis to which are attached three or four pairs decussate bracts. The lower bracts are sterile whereas the upper two bear an ovule in their axils (one bract bears one ovule). Each ovule terminates a short stalk. The lower bracts are sterile whereas the upper two bear an ovule in their axils (one bract bears one ovule). Each ovule terminates a short stalk. The stalk and the terminal ovule are also termed as a *female flower*.

In *E. gerardiana* (Singh and Maheshwari, 1962) each female cone is supplied by four vascular strands (Fig. 15.13, A, B) that enter its base and then bifurcate into eight vascular strands that appear to be arranged in two groups of four each (Fig. 15.13, C). The two median bundles of C bifurcate into small bundles (Fig. 15.13, D). In each group there are two small median strands that fuse with each other (Fig. 15.13, E) and enter the first pair of sterile bracts. The fused small bundles again merge with the four bundles of the cone axis so that once again there are four bundles (Fig. 15.13, F) in the cone axis (four bract traces are also seen). They are arranged in a ring. A little above, these bundles again bifurcate into two groups of four bundles each (Fig. 15.13, G). These are arranged at right angles to the previous bifurcation. In this case again the small median pairs fuse and enter the decussate bracts (Fig. 15.13, G, H). Like this the third and the fourth pairs of bracts are supplied. At the end there are six bundles (Fig. 15.13, I) left in the cone axis because after supplying the last pair of bracts the medial bundles do not merge with larger bundles of the cone axis. These six bundles supply the ovules, i.e., three enter each ovule.

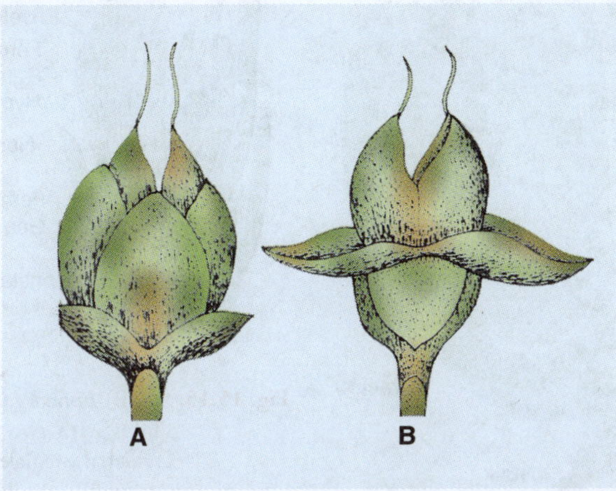

Fig. 15.12. (A-B) *Ephedra foliata*. Views of ovulate cones or strobili showing the bracts and ovules (A) and seeds (B) each with an elongated tube around the micropyle, which appears like a style of an angiospermous pistil.

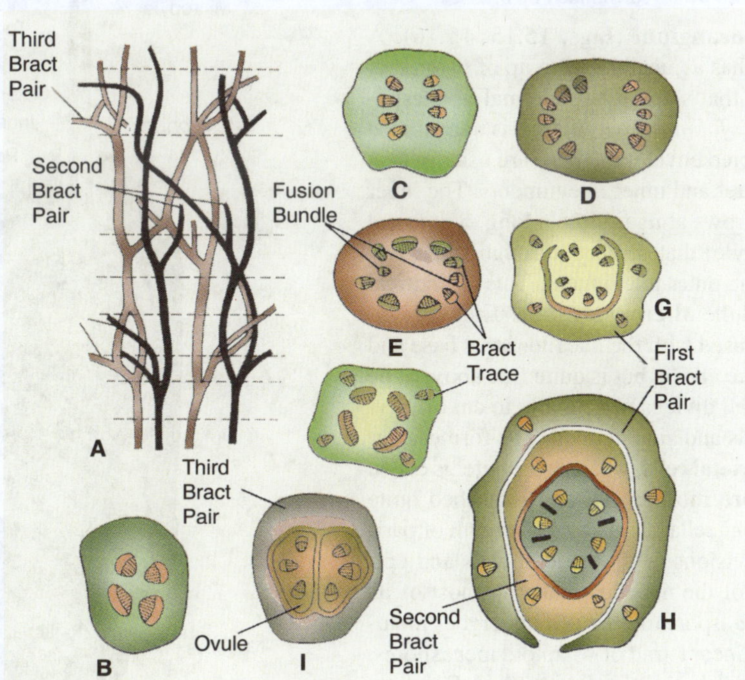

Fig. 15.13. (A-I) *Ephedra gerardiana*. It illustrates vascular supply to the female cone. For explanation see text. (All after Singh & Maheshwari).

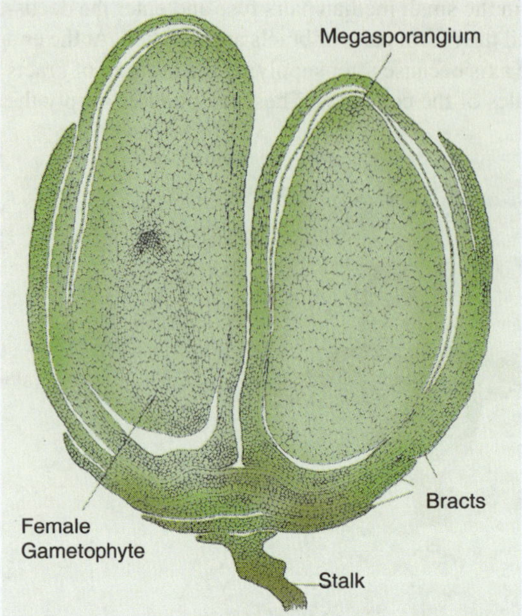

Fig. 15.14. *Ephedra foliata.* L.S. female strobilus showing two ovules surrounded by bracts.

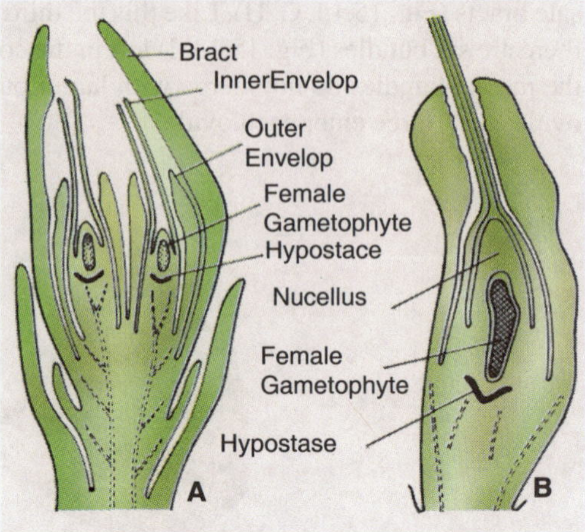

Fig. 15.15. (A-B) *Ephedra foliata* A. L.S. female strobilus with two ovules. B. L.S. ovule in which the female gametophyte has become cellular.

Megasporangium (Figs. 15.15, 15.16). A young ovule has a nucellus made up of parenchymatous cells that show a hypodermal **archesporial cell** in *E. foliata*. The nucleus is surrounded by a two-layered envelope. These are usually designated as outer and inner integuments. The inner integument grows at its tip into a long cylindrical tube (micropyle) that projects through the apical opening of the outer integument. This cylindrical tube is called the **Micropylar tube**. The inner integument is fused with the nucellus at its base and a little distance above, but is quite free above. The arhesporial cell divides periclinally to cut off several outer cells and an inner cell. The former constitute the **parietal cells**, whereas the latter is called the **megaspore mother cell** and is pushed quite deep into the nucellar tissue by the growth of parietal cells. Divisions of the parietal cells and epidermal cells of the nucellus make up most of its bulk. The megaspore mother cell undergoes meiosis to form a linear tetrad of 4 haploid megaspores. 'T'-shaped tetrads have been noticed in *E. foliata* (Maheshwari, 1935). In *E. distachya* wall formation is delayed between the four megaspores so that a false tetrasporic sac is indicated (Lehmann, 1967). Usually the lowest or the chalazal megaspore is

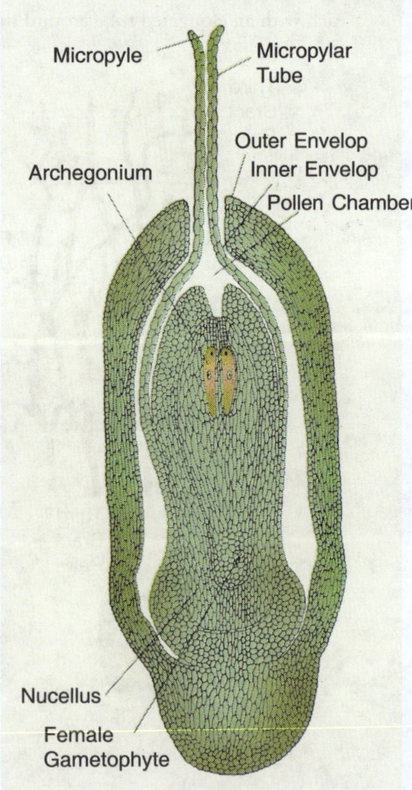

Fig. 15.16. L.S. mature ovule. of *Ephedra* sp. Note the two archegonia.

functional and develops into a cellular female gametophyte that has a broader micropylar end and a narrow and pointed chalazal end (*E. foliata* and *E. gerardiana*). One to three archegonia develop at the micropylar end of the gametophyte. A distinct plate of colourless and thin-walled cells is present at the chalazal end of the ovule (Figs. 15.15, 15.16). This is called the **hypostase** (Singh and Maheshwari, 1962). Below this plate xylem tracheids with spiral and pitted thickenings are present. In *E. intermedia* (Mehra, 1950) the archesporial cell directly functions as the megaspore mother cell.

Singh and Maheshwari (1962) report that in E. gerardiana the nucellar cells in the micropylar region become binucleate and their cytoplasmic contents stain deeply. They also report the development of small papillate outgrowths from the outer integument. These outgrowths are directed towards the inner integument (Fig. 15.17). These outgrowths grow more and become thick-walled as the ovule matures. They close the space between the two integuments and may also help in sealing of the micropyle by pressing in the micropylar tube, after the pollination has been affected.

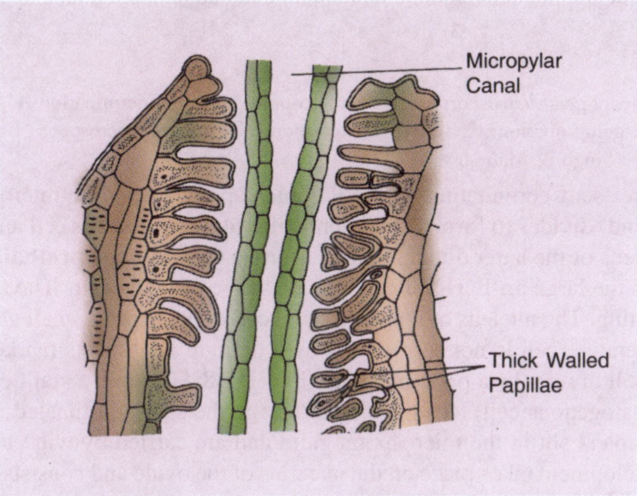

Fig. 15.17. L.S. through micropylar canal of ovule of *E. gerardiana* showing papillate projecting (After Singh and Maheshwari).

The ovules vary in shape and vascularisation. In *E. gerardiana* three bundles enter each ovule (Fig. 15.15, A, B) and after contributing a few tracheids to the nucellar base innervate the outer envelope or integument (Singh and Maheshwari, 1962). The inner integument is supplied by two bundles that end at its base or extend up to teh level of separation from the nucellus. According to Eames (1952) the vascular supply to the outer integument indicates that it represents another pair of fused bracts. He does not regard it to be an integument because it is innervated by an anterior and a posterior vascular system. Bierhorst (1971) is of the view that the term integument can be applied to any complete ovular envelope.

GAMETOPHYTES

The male and the female gametophytes develop from the microspores and the megaspores, respectively. The spores are, therefore, the pioneer structures of the gametophyte generation.

Microspore and the Microgametophyte (Figs. 15.18–15.20). The microspore in *E. foliata* is elongate (Fig. 15.19) with longer diameter 72 μ and shorter diameter 23 μ. The exine is 1.4 μ thick with plicate surface and is provided with parallel ridges (12 in number) along the longer axis of the grain. In *E. gerardiana* there are seven ridges and the microspores measure 57 × 25 μ and in *E. saxatilis* 48 × 28 μ. The pollen grains are wingless and inaperturate. They are usually ovoid in form. The nucleus is first centrally located and later moves to one end. The microspore is yello in colour.

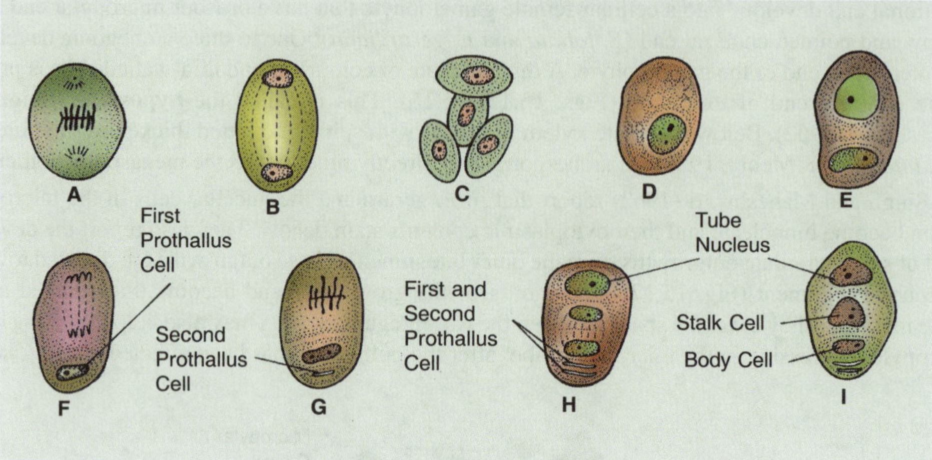

Fig. 15.18. (A-I) *Ephedra gerardiana.* Formation of microspores and their germination A–B. Microspore mother cells undergoing division; C. Tetrad of microspores; D. mature Microspore; E–I. Stages in germination. (After Singh & Maheshwari)

The microspores start germinating within the microsporangium. The microspore nucleus (that has moved to one end) divides to form a small lenticular **male prothallus** cell and a larger cell (Fig. 15.18, E). The nucleus of the latter divides again to form a **second male prothallus cell** (Fig. 15.18, F) marked off from the large **antheridial** cell by a cleavage of cytoplasm. The two male prothallus cells start degenerating. The nucleus of the antheridial cell divides into a small **generative cell** and a **tube cell**. The generative cell is not surrounded by a definite wall but is marked off by a distinct cytoplasmic sheath. It divides by a periclinal wall (Fig. 15.18, G–I) into a **stalk cell** (sterile cell) and a **body cell** (spermatogenous cell). At this 5-celled stage the semi-germinated microspores are released through an apical slit in the microsporangium and are carried by wind to the micropyles of ovules. Further development takes place on the nucellus of the ovule and consists in the development of the pollen tube and formation of two unequal male gametes by the division of the body nucleus.

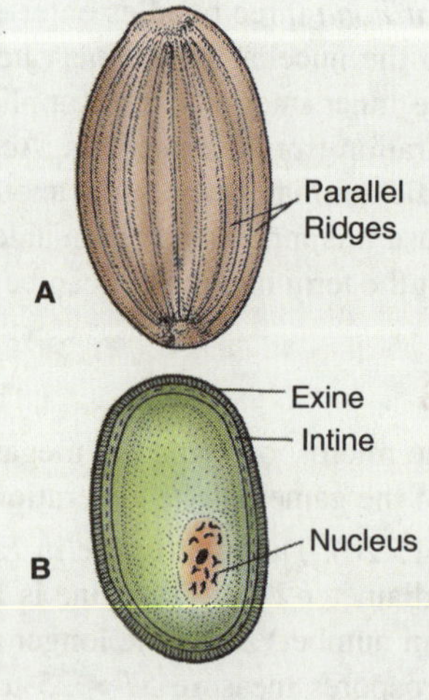

Fig. 15.19. Pollen grains of *E. foliata. Note* the parallel ridges on the exine in A.

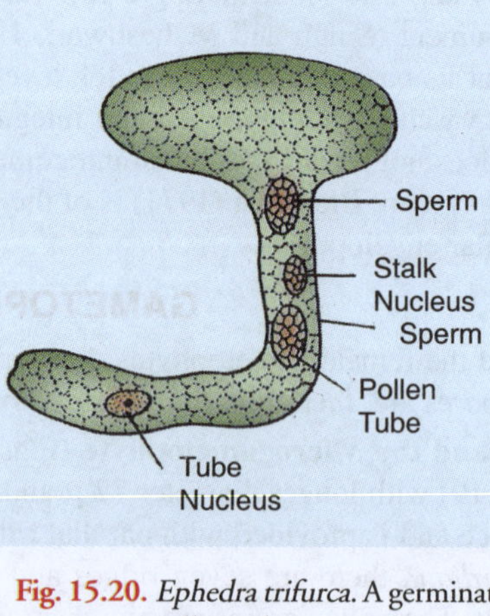

Fig. 15.20. *Ephedra trifurca.* A germinated pollen grain with pollen tube containing two male gametes, tube nucleus and stalk nucleus.

Previous literature regarding the development of male gametophyte in *Ephedra* reveals a number of variations, that have been reported in different species of the genus by different authors. Maheshwari (1935) reported that in *E. foliata* the second prothallial cell is actually a prothallial nucleus because it is not surrounded by a cell wall. Mehra (1938) observed that a wall is present around the second prothallial cell as soon as it is formed but it breaks down as soon as it is formed. Mehra (1938, 1947) observed distinct male prothallus cells in *E. gerardiana* and *E. saxatilis*. Singh and Maheshwari (1962) also confirm this observation by Prof. Mehra in *E. gerardiana*. Similarly there is a controversy as to whether there is a stalk cell or a stalk nucleus. Berridge (1909) reported the presence of a stalk cell in *E. distachya*; but in *E. saxatilis* Prof. Mehra (1947) reported that it is a stalk cell in the beginning but gradually the separating wall disappears and only the nucleus is left behind. In *E. gerardiana* also Prof. Mehra (1938) observed a stalk nucleus, but Singh and Maheshwari

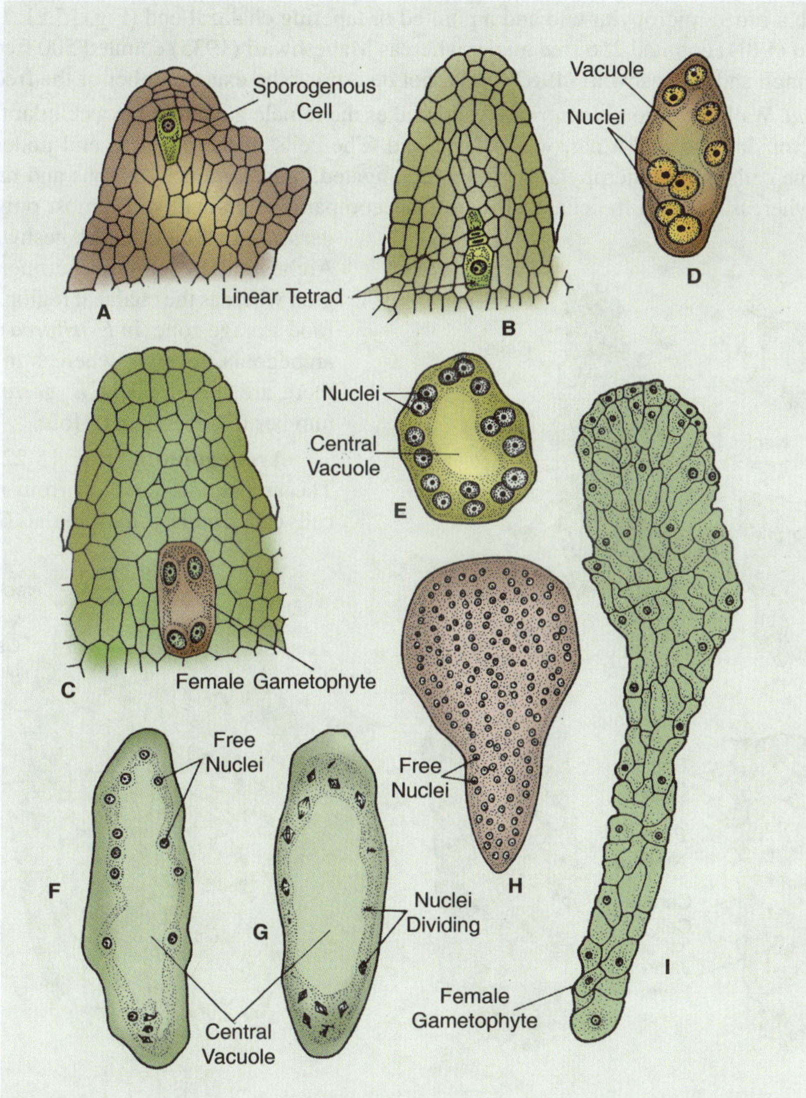

Fig. 15.21. (A-I). *Ephedra gerardiana* . Stages in the development of female gametophyte. For explanation see text. (All after Singh & Maheshwari)

(1962) observed a distinct stalk cell. Prof. Mehra (1947) reported diploid pollen grains in *E. saxatilis, E. intermedia, E. altissima* and *E. foliata*. These are formed by incomplete partitioning of the mother cell after four free nuclei are formed in it. One cell contains two nuclei that fuse to form a diploid one.

Megagametophyte or the Female gametophyte (Fig. 15.21). The chalazal megaspore is usually functional (Fig. 15.21, B). It enlarges and its nucleus divides into two. These nuclei move towards the opposite poles and are separated by a large central vacuole. Later four (Fig. 15.21, C), eight (Fig. 15.21, D) and sixteen free nuclei (Fig. 15.21, E—G) are formed. These are arranged in a peripheral layer around the central vacuole. The free nuclear divisions are not strictly simultaneous. Singh and Maheshwari (1962) reported that in *E. gerardiana* some nuclei are in metaphase and others in anaphase and telophase. Later the central vacuole disappears and the free nuclei are evenly distributed throughout (Fig. 15.21, H). In *E. foliata* and *E. gerardiana* the mature gametophyte is almost obovate with a broad micropylar end and a pointed or tapering chalazal end (Fig. 15.21, H, I). In *E. trifurca* Land (1904) counted 256 free nuclei whereas Maheshwari (1935) counted 500 free nuclei in *E. foliata*. Singh and Maheshwari (1962) could not determine the exact number of the free nuclei in *E. gerardiana*. Wall formation is centripetal and makes the female gametophyte a cellular tissue with a broad micropylar end and a narrow chalazal end. The cells in the chalazal end undergo active divisions. The cells in the micropylar region are elongated, large, parenchymatous and have scanty cytoplasm whereas those in the chalazal region are comparatively small and almost polygonal (*E. gerardiana*, Singh and Maheshwari, 1962). Archegonia develop in the micropylar region whereas the chalazal region acts as the food storage zone. In *E. trifurca* one or two archegonia develop, whereas in *E. foliata* there are three and in *E. gerardiana* the number may be three or four.

Archegonium (Fig. 15.22, 15.23) : The archegonia develop from superficial cells called archegonials initials (Fig. 15.22,

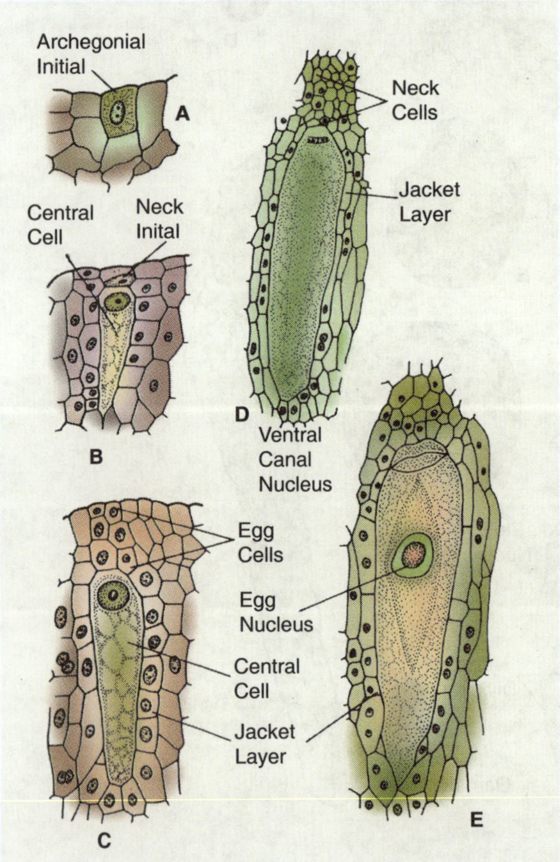

Fig. 15.22. (A-E). *Ephedra gerardiana* Stages in the development of archegonium (After Singh and Maheshwari)

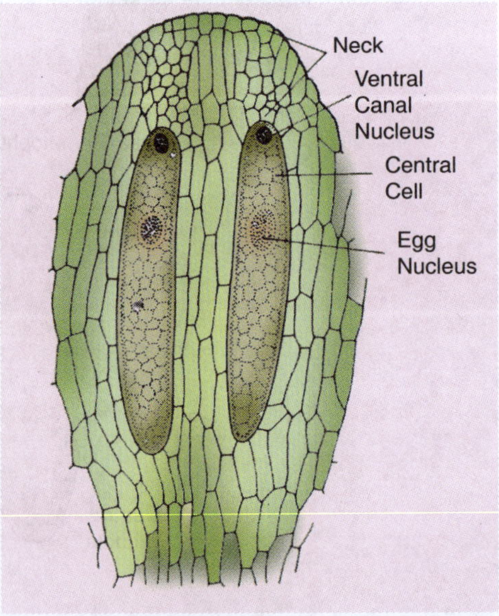

Fig. 15.23. V.S. portion of nucellus showing two archegonia of *E. gerardiana*.

A). Each archegonial initial divides by a transverse wall to form an outer **primary neck cell** or **neck initial** and an inner **central cell** (Fig. 15.22, B). The neck cell undergoes a number of divisions to form a long and massive neck of about thirty to forty cells (Fig. 15.22, C, E). It encloses no neck canal. The neck of archegonium in *Ephedra* is the longest in the gymnosperms. The cells of the neck usually merge with the surrounding gametophytic cells (Fig. 15.22, E) and are hence not very distinct. The central cell enlarges in size and assumes an elliptical appearance with pointed or narrow chalazal end. Its nucleus divides into a **ventral canal nucleus** and an **egg nucleus** (Fig. 15.22, E). The latter moves to the centre of the archegonium. The ventral canal nucleus degenerates. The cells adjacent to the central cell divide transversely to form a distinct jacket layer, which may become two to three layered thick. Singh and Maheshwari (1962) observed that in some cases all the four archegonia in *E. gerardiana* get surrounded by a common jacket. usually every archegonium has its own jacket.

Pollination : The pollen is released in the five-celled stage and is carried by wind to the aperture of the micropylar tube. The pollen lands on a pollination drop and is pulled through the micropyle and into the lysigenously formed pollen chamber and directly on to the female gametophyte. Further germination is resumed after a few hours and consists in the formation of a short pollen tube that grows out through the rupture exine. The body cell divides into two unequal male gametes, the prothale cells degenerate completely. The tube nucleus and the stalk nucleus persist. The male gametes and these nuclei move into the short pollen tube.

Fertilisation : The pollen tube along with its four nuclei (2 male nuclei plus tube and stalk nuclei) grows through the archegonial neck and discharges its contents into the cytoplasm of the egg. Its tip swells and bursts. The larger male nucleus migrates toward the female nucleus. The latter is surrounded by a dense cytoplasmic sheath. The male nucleus moves through this sheath and fuses with the female nucleus to form a fusion nucleus. The fusion nucleus along with its surrounding cytoplasm is termed as **oospore** or the **zygote**.

Khan (1940, 1941) and Mulay (1941) observed the fusion of second male gamete with the ventral canal nucleus in *E. foliata*. The fusion nucleus formed in this way does not develop into embryo.

EMBRYOLOGY (Figs. 15.24, 15.25)

The zygote nucleus divides into eight free nuclei in *E. trifurca* (Land, 1907; and Johansen, 1950) and *E. foliata* (Khan, 1943). These eight nuclei are of unequal size in *E. trifurca*. These unequal are unevenly distributed in the cytoplasm of the archegonium. They do not move to the base of the archegonium. Johansen (1950) observed that in *E. trifurca* three to five of these nuclei become individually enclosed in somewhat irregular walls. These become somewhat globular cells at a later stage. In *E. foliata*, six out of these eight become enclosed within walls and form (Fig. 15.24) six globular cells that are all potential embryos (Khan, 1943). In *Ephedra* this process of polyemryony is unique among the gymnosperms. These proembryos develop further individually and form embryos. Only one of them is able to grow to maturity. Each proembryo grows into a tubular structure called the suspensor tube (Fig. 15.25, A). The nucleus of the proembryo divides into two (Fig. 15.25, B). Both these nuclei move into the tube. The tube elongates considerably and pierces the archegonial venter and enters the gametophyte tissue. A wall separates two daughter nuclei, one of which moves backwards into the micropylar end of the tube whereas the other moves towards the chalazal narrow end of the tube (Fig. 15.25, B). The former disintegrates. The tube grows more, carrying the lower cell deep into the female prothallus

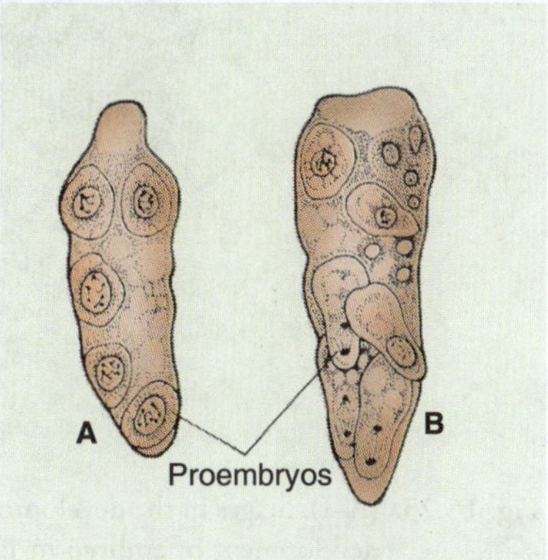

Fig. 15.24. L.S. of egg cells of an archegonium showing five (A) and six (B) proembryos of *E. foliata* (After Khan).

tissue. This lower uninucleate cell is called the **embryonal cell**. It divides into a proximal **suspensor cell** and a distal **embryo cell** (Fig. 15.25, C). The suspensor cell divides longitudinally (Fig. 15.25, D) into two longer cells. The embryo cell divides by two obliquely intersecting longitudinal walls thus cutting off an apical cell (Fig. 15.25, E). Further cell division results in the formation of a massive apical region (Fig. 15.25, F—H). The apical cell ultimately loses its identity. Cells in the proximal part of this apical massive regionelongate to form secondary suspensors (Fig. 15.25, H). At this stage the suspensor tube callapses and loses its identity. The distal cells grow into embryo proper that (Fig. 15.25, I) has two cotyledons, The root is distinguished close to the secondary suspensor cells. The above account refers to *E. trifurca*.

In *E. foliata* the embryonal cell divides transversely into a terminal and a subterminal cells. The terminal cell divides vertically into two cells so that there are three cells in all (including the subterminal cell). These cells divide further to give rise to a massive apical part whose proximal cells develop into secondary suspensors, whereas distal cells grow into embryo proper that has two colyledons. In *E. foliata* about 18–19 embryos were observed by Khan (1943) but only one reaches maturity.

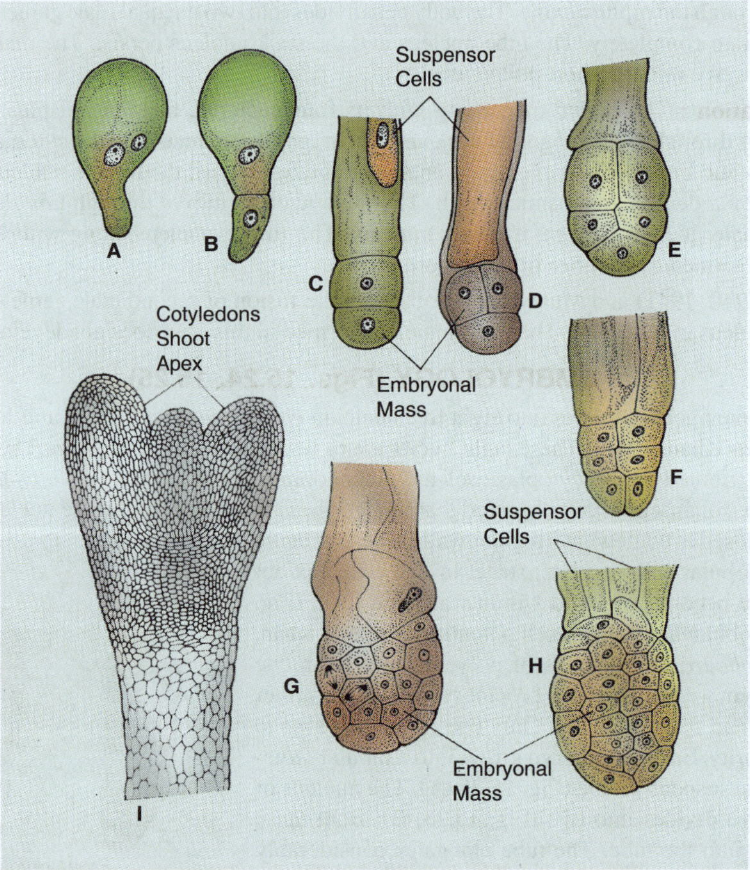

Fig. 15.25. (A–I). Stages in the development of embryo in *Ephedra trifurca* (A–H. after Johansen). I. Later stage of development of embryo in *E. altissima* (After Strasburger).

Seed : The seed is enclosed within an outer fleshy envelop formed by the two subtending bracts of the megasporangiate strobilus. Next to it are the two seed coats developed from the two integuments of the ovule. These enclose the food laden female gametophyte that contains embedded in it the dicotyledonous embryo. The cotyledons are large. The seed is, therefore, protected by three envelopes (Fig. 15.26). The nucellus persists as a nucellar cap at the micropylar end of the seed. The micropylar tube withers.

The germination of the seed (Fig. 5.27) is epigeal and the cotyledons grow large and green and are traversed by two parallel veins. These veins are fused at the base of the cotyledon. The cotyledons persist during the earlier stages of development, but wither away at a later stage. The seed undergoes no resting period and germinates immediately on falling on the ground. In *E. trifurca* the seed may even germinate within the parent strobilus (Land, 1907). In the young seedlings (Voth, 1934) one of the two exarch poles of xylem of the root enters each cotyledon after bifurcating. The two exarch traces of each cotyledon twist around nearly 180 degrees in their longitudinal course to become endarch. Such a twisting within cotyledons or hypocotyl, is characteristic of angiosperms also.

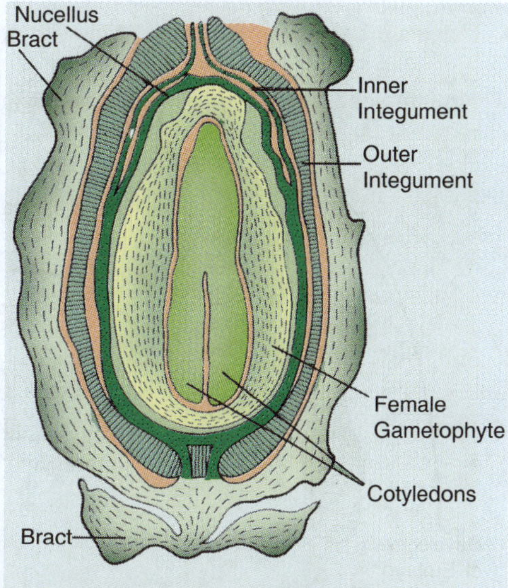

Fig. 15.26. L.S. Seed of *E. foliata.*

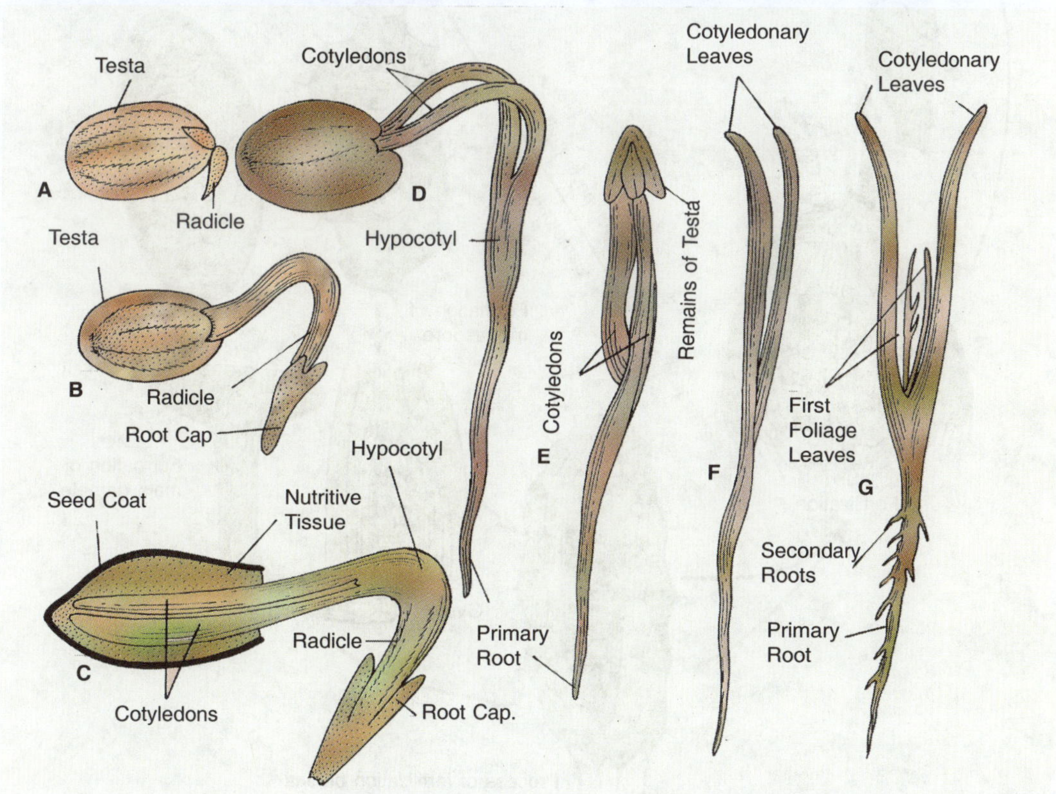

Fig. 15.27. (A-G). Stages in the germination of seed in *E. foliata.*

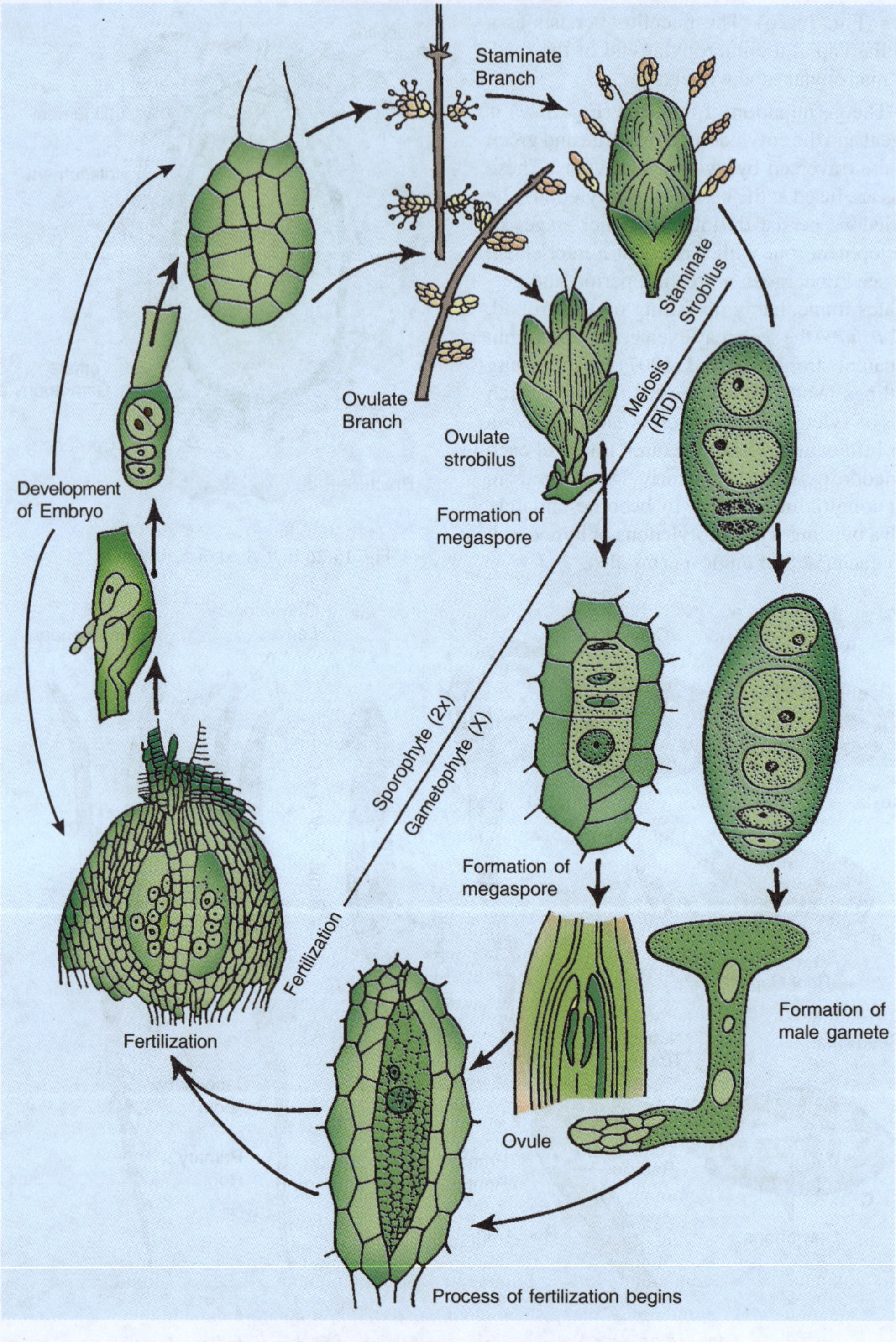

Fig. 15.28. Diagrammatic life cycle of *Ephedra*.

Economic Importance : An alkaloid called ephedrine is obtained from the species of *Ephedra*. It is used in the preparation of medicines that cure cough, bronchitis and asthma. It also cures hay fever. The Indian species that yield ephedrine include *E. gerardiana, E. intermedia*, and *E. nebrodensis*. Tincture of *E. gerardiana* is also used as a cardiac and circulatory stimulant. Decoction of the stems and roots are used to cure rheumatism and syphilis in Russia. The juice of its berries is used to cure respiratory affections. Some species are grown as ornamentals.

Chromosome Number : Mehra (1946 and 1947) established the basic chromosome number for the genus as $n = 7$. He also concluded that polyploidy has been of frequent occurrence in the evolution of the species. Maheshwari (1935) also counted $n = 7$ in *E. foliata*.

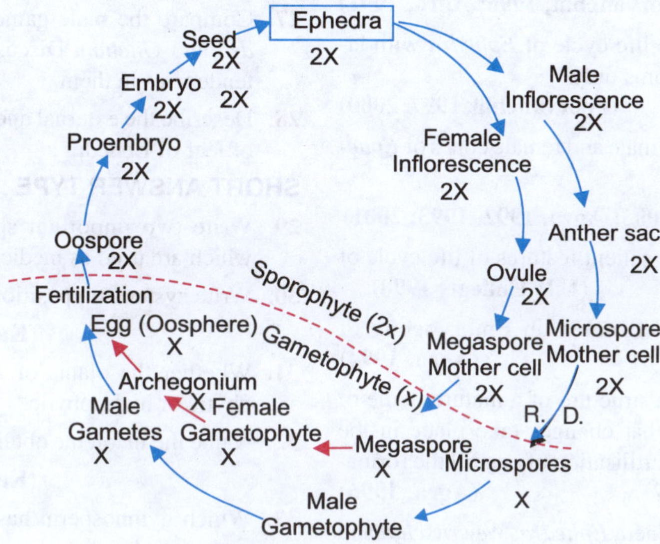

Fig. 15.29. *Ephedra* Graphic life cycle.

QUESTIONS

ESSAY TYPE :

1. Draw labelled diagram of the internal structure of the stem of *Ephedra* and comment upon it. **(Kanpur, 1995; Poorvanchal, 1998)**
2. Explain the life cycle of *Ephedra* with the help of suitable diagrams. **(Kanpur, 1998, 2004; Rohilkhand, 1995, 1998, 2002; Agra, 1996; GuruNanak Deo, 2004)**
3. Give a brief account of economic importance of *Cycas, Pinus* and *Ephedra*. **(Lucknow, 1993)**
4. Give a comparative account of the structure of the ovule of *Cycas* and *Ephedra*. **(Lucknow, 1993)**
5. Describe with the help of suitable sketches, the staminate strobilus of *Ephedra*. **(Lucknow, 1995)**
6. Give an illustrated account of female flower and structure of ovule of *Ephedra*. **(Lucknow, 1997)**
7. Give a comparative account of the seed of *Cycas, Pinus* and *Ephedra*. **(Lucknow, 1999, 2001, 2004)**
8. Describe sexual reproduction in *Ephedra*. **(Rohilkhand, 1992, 2000)**
9. Give a comparative account of the development of female gamenotype of *Cycas, Pinus* and *Ephedra*. **(Rohilkhand, 1994)**

10. Give a comparative account of the post-fertilisation changes in *Cycas, Pinus* and *Ephedra*. **(Kumaou, 2001; Punjab, 1991, 1993)**
11. Describe the mode of pollination and fertilisation in *Ephedra*. **(Awadh, 2003)**
12. Give a comparative account of the male gametophytes of *Cycas, Pinus* and *Ephedra*. **(Poorvanchal, 1996; Agra, 1991)**
13. Illustrate the life cycle of *Ephedra* with labelled diagrams only. **(Poorvanchal, 1997, 2000)**
14. Describe the male and female cones of *Ephedra*. **(Poorvanchal, 2003; Agra, 1992, 1993, 2001)**
15. Describe the salient features of life cycle of *Ephedra*. **(U.P. College, 1998)**
16. Write detailed notes on Embryogeny of *Ephedra*. **(Agra, 1994)**
17. Describe the structure of a mature ovule of *Ephedra*. What changes take place in the ovule after fertilisation of egg till the formation of seed? **(Agra, 1996)**
18. The three genera *Ephedra, Wewritschia* and *Gnetum* are generally placed together in a single order the *Gnetales*." Discuss the reasons for and against this statement. **(Rohilkhand M.Sc, 1989)**
19. Give a comparative account of reproduction in *Gnetum, Ephedra* and *Welwitschia*. **(Rampur M.Sc, 2000; Awadh M.Sc, 1998)**
20. Explain embryo development in *Gnetopside*. Give diagrams. **(Allahabad M.Sc, 2000)**
21. Compare the ovules of *Gnetum* and *Ephedra* and add a note on the affinities of *Ephedra*. **(Mahatma Gandhi, 1998)**
22. Describe the development of microsporangium and development of male gamutophyte in *Ephedra*. **(GuruNanak Deo, 1992)**
23. Describe the structure and development of reproductive structures of *Ephedra*. **(Punjab, 1998)**
24. Give a detailed and illustrated account of the male and female strobili (flowers) of *Ephedra* with special reference to position and structure. **(Punjab, 1992, 2003)**
25. Describe the structure and germination of seed in *Ephedra*. **(Punjab, 2002)**
26. Compare the female gametophyte and embryogeny in *Pinus* and *Ephedra*. **(Calcutta, 2002)**
27. Compare the male gametophytes of *Ephedra* and *Gnetum*. Discuss the evolutionary tendencies in them. **(Calcutta, 2001)**
28. Describe the external and internal structures of leaf of Ephedra :

SHORT ANSWER TYPE :

29. Write two important species of *Ephedra* which are used as medicine.**(*Kanpur*, 1995)**
30. Write systematic position of *Ephedra*. **(Kanpur, 1995, 1996)**
31. Whether the plants of *Ephedra* are Xerophytic or hydrophytic? **(Kanpur, 1996)**
32. Name the medicine obtained from *Ephedra*. **(Kanpur, 1997, 1999)**
33. Which gymnosperm has vessels in its secondary Xylem ? **(Kanpur, 1998; Punjab, 1999)**
34. What is special character in the secondary xylem of old stem of *Ephedra*? **(Kanpur, 1999)**
35. Name the medicine for respiratory disease which is obtained from a gymnosperm. **(Kanpur, 2000)**
36. Describe the structure of male and female strobili of Ephedra. **(Kanpur, 2002)**
37. Describe the structure of mature male gametophyte of *Ephedra*. **(Kanpur, 2003)**
38. Describe the development of male gametophyte in *Ephedra*. **(Kanpur, 2003)**
39. Describe double fertilisation in *Ephedra*. **(Kanpur, 2004)**
40. Write in thirty words :

    (*a*) Economic use of *Ephedra*. **(Lucknow, 1999)**

(*b*) Angiospermic features of *Ephedra*.

(Lucknow, 2001, 2003; Awadh, 1998; Poorvanchal, 1999)

41. Which gymnospermic plant fyelds ephedrine ? What is its economic importance ?

(Agra, 1999; Bundelkhand, 1999)

42. **Differentiate between :**

(*a*) Leaf of *Ephedra* and *Pinus* neede.

(Lucknow, 1996)

(*b*) Xylem of *Ephedra* and *Pinus*.

(Gorakhpur, 1991)

(*c*) *Ephedra* and *Gnetum* Ovules.

(Rohilkhand, 1999)

(*d*) Female reproductive structures of *Gnetum* and *Ephedra*.

(Rohilkhand, 1990)

43. Name a gymnosperm which has vessels.

(Poorvanchal, 1998, Meerut, 1997; Punjab, 1999)

44. Which drug is obtained from *Ephedra* ? Give one medicinal importance of that drug.

(Poorvanchal, 2000)

45. Give a short account of development of Embryo in *Ephedra*. **(Poorvanchal, 2001)**

46. According to Pant's classification, to whichdivision *Ephedra* belongs ?

(U.P. College, 1998)

47. Name the plant from which ephedrine is obtained. **(Meerut, 1997)**

48. Write one character of *Ephedra* in which it resembles Augiosperms.

(Gorakhpur, 1990; Bangalore, 2001)

49. Name the alkaloid obtained from *Ephedra*.

(Punjab, 1991)

50. From which plant a drug ephedrin is extracted.

(GuruNanak Deo, 2004)

51. **Give a comparative account of the following :**

(*a*) Xerophytic characters of leaves of *Cycas* and *Ephedra*. **(Gorakhpur, 1991)**

(*b*) Male gametophyte of *Pinus* and *Ephe dra*. **(Gorakhpur, 1991)**

(*c*) Stem of *Ephedra* and *Gnetum*.

52. **Write short notes on the following :**

(*i*) Male and female flowers of *Ephedra*. **(Kanpur, 1995; 1996; Awadh, 1997)**

(*ii*) Male flower of *Ephedra*.

(Lucknow, 1992; Rohilkhand, 1996; Kumaon, 1995; Poorvanchal, 1996, 1997, 1999, 2003; U.P. College, 1996; Gorakhpur, 1990; Jiwoji, 1994)

(*iii*) Pollination and fertilisation in *Ephedra*.

(Lucknow, 1994; *Rohilkhand* M.Sc, 1998, 2003)

(*iv*) Female gametophyte of *Ephedra*. **(Lucknow, 1995, 2004)**

(*v*) Vessels in gymnosperms.

(Lucknow, 2003)

(*vi*) L.S. female flowers of *Ephedra*.

(Lucknow, 2003)

(*vii*) Secondary growth and vessels in *Ephedra*. **(Rohilkhand, 1995)**

(*viii*) Female strobilus of *Ephedra*. **(Rohilkhand, 1999, 2001; Kumaon, 1998)**

(*ix*) Staminate (males) strobilus of *Ephedra*.

(Rohilkhand, 2003; Kumaon, 2002)

(*x*) Double fertilisation in *Ephedra*. **(Rohilkhand, 2004; Kanpur M.Sc, 2000; 2001)**

(*xi*) Stomata of *Ephedra*.

(Rohilkhand, 2004)

(*xii*) Ovule of *Ephedra*. **(Kumaon, 1996)**

(*xiii*) Internal structure of *Ephedra* stem.

(Kumaon, 1999)

(*xiv*) *Ephedra* sp. is nearer to Angiosperm. **(U.P. College, 1996)**

(*xv*) Significant features of stem of *Ephedra*. **(U.P. College, 1997)**

(*xvi*) Occurrence of *Pinus* and *Ephedra* in India. **(U.P. College, 1998)**

(*xvii*) Seed of *Ephedra*.

(U.P. College, 1999; Agra, 2000)

(*xviii*) Development of male gametophyte of

Ephedra. **(U.P. College, 1999)**

(*xix*) Male gametophyte in *Ephedra*. **(U.P. College, 2000)**

(*xx*) Economic importance of *Ephedra*. **(Meerut, 1998)**

(*xxi*) *Ephedra* and *Welwistschia*. **(Rohilkhand M.Sc, 1990)**

(*xxii*) Seed germinatiori in *Ephedra*. **(Agra, 2001)**

53. Mention special features of

(*a*) Habit of *Ephedra*. **(Mahatma Gandhi, 1997)**

(*b*) Affimities of *Ephedra*. **(Mahatma Gandhi, 1996)**

(*c*) Male flower of Ephedra. **(Mahatma Gandhi, 1996)**

(*d*) Embryo development in *Ephedra*. **(Mahatma Gandhi, 1996)**

(*e*) External features of *Ephedra*. **(Himachal Pradesh, 2000)**

54. Draw labelled diagrams of the following :

(*i*) L.S. Ovule of *Ephedra*.

(Kanpur, 1995, 1996, 2002; Lucknow, 1992; Rohilkhand, 1997; Kumaon, 2001; Awadh, 1994, 1997; Poorvanchal, 1996, 2003; Agra, 1991, 1995; Garhwal, 2004)

(*ii*) T.S. Young stem of *Ephedra*.

(Kanpur, 1997; Lucknow, 1995, 1991, 2001, 2002, 2003; Rohilkhand, 1994, 1998, 1997, 1999, 2002; Meerut, 1998; Agra, 1991; Punjab, 2002; Agra, 2000; Jiwaji, 1995)

(*iii*) T.S. old stem of *Ephedra*. **(Kanpur, 1998; Lucknow, 1997, 1999)**

(*iv*) Male flower of *Ephedra*. **(Kanpur, 2001; Rohilkhand, 1993, Kumaon, 2003; Awadh, 1998)**

(*v*) L.S. female flower of *Ephedra*. **(Lucknow, 1999, 2000, 2002)**

(*vi*) V.S. of leaf of *Ephedra*. **(Lucknow, 2000)**

(*vii*) L.S. male strobilus of *Ephedra*. **(Kumaou, 2000)**

(*viii*) L.S. *Ephedra* seed. **(Awadh, 1992; Poorvanchal, 1997)**

(*ix*) Internal structure of *Ephedra* stem. **(Jiwaji, 1999)**

(*x*) T.S. Root and leaf of *Ephedra*. **(Bundelkhand, 1990)**

OBJECTIVE TYPE :

55. Select the correct answer :

(*i*) The largest seeds are found in

(*a*) *Pinus* (*b*) *Ephedra*
(*c*) *Cycas* (*d*) *Gnetum*.

(*ii*) *Ephedra* polelen grains are shed at which stage ?

(*a*) Binucleate (*b*) Trinucleate
(*c*) Tetra nucleate (*d*) Penta nucleate.

(*iii*) Which of the following plant resembles *Equisetum* ?

(*a*) *Cycas* (*b*) *Pinus*
(*c*) *Gnetum* (*d*) *Ephedra*.

(*iv*) Which plant has green, branched jointed stem ?

(*a*) *Cycas* (*b*) *Pinus*
(*c*) *Ephedra* (*d*) All the three.

(*v*) How many female cones develop on each node of *Ephedra* ?

(*a*) 2–4 (*b*) 5–8
(*c*) 10–15 (*d*) 1–∞

(*vi*) What type of development of microsporangium is found in Ephedra ?

(*a*) Eusporangiate
(*b*) Leptosporangiati
(*c*) Both
(*d*) None

(*vii*) Which of the following is most highly evolved ?

(*a*) *Ephedra* (*b*) *Cycas*
(*c*) *Pinus* (*d*) *Lycopodium*

(*viii*) Ephedrin is obtained from

(*a*) *Equisetum* (*b*) *Ephedra*

(*c*) *Ectocarpus* (*d*) *Isoetes*.

(*ix*) Male and female flowers are found in

(*a*) *Cycas* (*b*) *Ephedra*

(*c*) *Pinus* (*d*) None of the above.

(*x*) Name the gymnosperm in which companion cells are present.

(*a*) *Pinus* (*b*) *Ginkgo*

(*c*) *Ephedra* (*d*) *Pinus*.

(*xi*) In which of the following female cone is not organised ?

(*a*) *Ephedra* (*b*) *Pinus*

(*c*) *Ginkgo* (*d*) *Cycas*

(*xii*) Kaiffer bread is prepared from :

(*a*) *Cycas* (*b*) *Ginkgo*

(*c*) *Ecephalaratos* (*d*) *Gneturs*.

(*xiii*) Which among the following has medicinal importance ?

(*a*) *Cycas* (*b*) *Ephedra*

(*c*) *Cupressus* (*d*) Welwitschia.

56. Fill in the blanks :

(*i*) Ephedrin is obtained from the ____________ of ____________ .

(*ii*) The *Equisetum* and *Ephedra* stem have similarity in having _______and _____.

(*iii*) Female cones are not formed in ______________ .

(*iv*) Secondary xylem of *Ephedra* is characterised by the presence of __________ .

(*v*) *Ephedra* shows some characters in which it ______________________ .

(*vi*) *Ephedra* belongs to order ________________________ .

(*vii*) Leaves are dimorphic in ________________________ .

(*viii*) Stem in *Ephedra* is modified into a ________________________ .

(*ix*) Resin conals are ________________________ in *Ephedra*.

(*x*) In *Ephedra* stomata are ________________________ .

CHAPTER 16

Gnetaceae

- Gnetum : Distribution and Habit
- Structure of the Stem
- Structure of the Root
- Anomalous Secondary Growth in Stem
- Reproduction
- Endosperm Formation
- Economic Importance

GNETUM MONTANUM

Genus : Gnetum

Distribution

The genus includes 30–35 species of woody trees, shrubs or even climbers. These are dioecious tropical species that occur mostly in the rain forests. *Gnetum* is represented in India by 5 species. These include :

1. *Gnetum contractum* Mgf. is confined to southern regions of India e.g., in Kerala, Nilgiri hills and Coonoor in Madras. It is a scandant shrub.
2. *G. gnemon* L. is a tree that has six varieties out of which two occur in India. *G. gnemon*

var. *brunonianum* (Griff.) Mgf. is found in the Naga hills and Golghat areas of Assam; whereas var. *griffithii* (Part.) Mgf. is found in Naga hills and Kungaba areas of Assam. The other varieties include var. *domesticum*, var. *silvestre*, var. *tenerum* and var. *gracile*. They occur in Celebes, Moluccas, Philippines, Borneo, Malaya and Sunda Islands. These are also cultivated as ornamentals.

3. *G. montanum* Mgf. is found in Sikkim, Assam, Burma, Siam and China. It is a robust, woody climber and is characterised by possessing branches that are swollen at the nodes. It is also found in Orissa.
4. *G. latifolium* Bl. has five varieties all of which are woody climbers with dark green leaves. *G. latifolium* var. *macropodum* (Kurz.) Mgf. and *G. latifolium* var. *funiculare* (Bl.) Mgf. occur in the Nicobar and Andaman Islands.
5. *G. ula* Brongn. is also a woody climber found commonly in various parts of India *e.g.* Western Ghats, Coorg, Kerala, Nilgiri hills, Pulni hills, Orissa and Godavari district of Andhra.

According to Pearson (1929) one of the species *G. trinerve* is apparently parasitic in nature.

Characteristics of Indian species of *Gnetum*.

1. *Gnetum contractum* : Leaves black on drying, elliptical or ovate, male inflorescence one branched. Seeds oblong-ovate, 3 – 3.5 × 1.5 cm in size, long, acuminate, longer than stalk, similar to *G. ula* in vegetative condition.
2. *G. nemon*. Leaves yellowgreen to brown when dry, upto 20 × 9 cm in size oblong or elliptic. Seeds 1 – 2.5 cm long, smooth or pappitose, ellipsoid, in lax inflorescence in var. *brunonianum*; in var. *griffithii* almost round in contracted branched inflorescence.
3. *G. montanum*. Leaves black on drying upto 30 × 12 cm in size, ovate to oblong ovate, male and female inflorescence branched. Seeds ellipsoid, 1.5 × 1 cm in size, longer than stalk.
4. *G. latifolium*. Leaves black to ovate on drying, elliptic to broadly ovate; male inflorescence branched once only, female inflorescence branched once or wice, seeds broad or narrowly ellipsoid, 2 – 2.5 × 1.5 cm in size, shorter than stalk.
5. *G. ula*. Leaves black on drying 8 – 15 × 4 – 7.5 cm in size, ovate-elliptic; female inflorescence simple, seeds ellipsoid, rugose, 3 – 4 × 1.5 – 2 cm in size.

Habit

Majority of the Indian species are woody climbers (*G. ula, G. latifolium, G. montanum*), a few are scandant shrubs (*G. contractum*), whereas *G. gnemon* is a medium sized tree. Spruce (1908) reported that *G. trinerve* Spruce of Cinchona forests in Chimboraze is a parasite.

Fig. 16.1. A branch of *Gnetum ula* showing opposite and decussate arrangement of leaves.

The stem is woody and branched. The branches are of two types :

1. Dwarf shoots or branches of limited growth that are divided into nodes and internodes and bear foliage leaves that arise in decussate pairs. The number of leaves on a dwarf shoot may be 9 or 10.
2. Long shoots or shoots of unlimited growth that bear reduced and cup-like pairs of scale leaves. The nodes are swollen and the branches become articulated in many species.

The leaves are also of two kinds : the scale leaves that occur only on the long shoots; and the foliage leaves that are borne in an opposite and decussate manner on the dwarf shoots. The foliage leaves bear axillary buds. Weterkeyn (1959) reported accessory buds, in addition to axillary buds in *G. africanum*. The shape of the lamina varies with species and may be oblong, elliptical or even lanceolate. The leaves are subsessile or shortly petiolate. The lamina is traversed by a reticulate system of veins and has entire margins and acute apex.

In *G. gnemon* there is no distinction into two types of branches. The branches are all of one type and are distinctly jointed.

The *Gnetum* sporophyte resembles very much to a dicotyledonous angiosperm because of its leaf shape and reticulate venation.

ANATOMY

Stem

(Fig. 16.2) : The apex of the stem shows a clear demarcation between **tunica** and the **corpus**, a

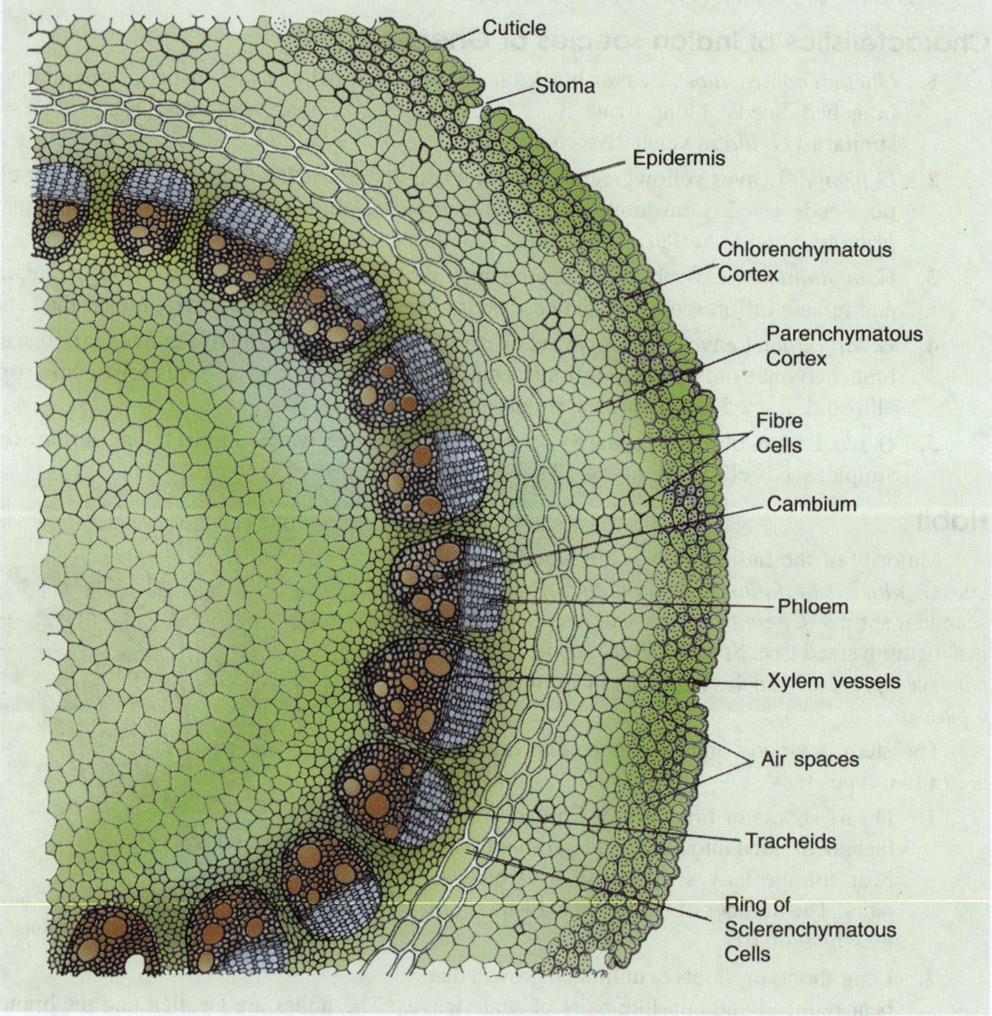

Fig. 16.2. T.S. Young stem of *Gnetum ula*, showing primary structure.

character which it shares with the dicotyledonous angiosperms. The tunica is a single layered outermost covering that extends over the shoot apex and starts from the axil of the first pair of leaf primordia. The corpus consists of **sub-apical initials** lying immediately below the tunica, **central mother cell zone flanking layers** and **pith rib meristem**. The primary permanent tissues are derived from these apical regions of the stem.

Primary Structure of the Stem

(Fig. 16.2). A transverse section of the young stem in *G. gnemon* presents a roughly circular outline and reveals the following tissues from outwards within :

Epidermis

It consists of a single layer of rectangular cells and its continuity is interrupted by sunken stomata. The cells also bear papillate outgrowths. The cuticle is thick. The outer walls of the epidermal cells are also very thick.

Cortex

It consists of three distinct regions :

1. **Outer Chlorenchymatous cortex** that is made up of 5–7 layers of polygonal or rounded cells that contain choroplasts and enclose small intercellar spaces.
2. **Middle Parenchymatous Cortex** that consists of few layers of thin-walled cells. The cells are compactly arranged. This part of the cortex contains scattered groups of fibrous cells. The cells mainly store food materials.
3. **Inner Sclerenchymatous Ring** (Fig. 16.2) that consists of 2 to 5 layers of thick-walled cells that appear polygonal in outline. The cells have branched or unbranched pit canals. This layer is very clear in slightly old stems.

Endodermis

Endodermis is not clearly identifiable in young stems.

Pericycle

It is also not clearly defined.

Vascular Region

It consists of a variable number of almost oval vascular bundles that are arranged in the form of an interrupted ring. The bundles are conjoint, collateral, endarch and open. The xylem consists of tracheids and vessels. Xylem fibres are absent and xylem parenchymais scanty or even absent. The protoxylem consists of vessels with bordered pits occurring in early matured helical elements. The pits are circular in outline. The perforation plates bear fewer pores than those of *Ephedra*. The protoxylem elements that develop in the beginning have their walls thickened in a spiral, annular or reticulate manner. The elements that develop a little later possess, in addition, a row of bordered pits on the oblique walls. The tracheids with reticulate thickenings are abundant in the nodal region. The metaxylem vessels bear still fewer pores, which are in contact with each other. The primary phloem consists of sieve cells and phloem parenchyma.

The vascular bundles are separated by broad parenchymatous regions (Fig. 16.2). These are the medullary rays that are many cells broad and many cells high. The number of vascular bundles is variable. It is 24 or 28 in *G. gnemon* and varies between 8–20 in other species.

Pith

It occupies the centre of the stem. In young stems the pith is extensive and entirely parenchymatous. The cells are polygonal in outline. In older stems the cells in the vicinity of vascular bundles may become thick-walled and develop pits.

Secondary Growth in Stem

Figs. 16.3, 16.4, 16.5). The secondary growth in the tree-like and shrubby species is of the normal type. The cambium forms a complete ring in the usual manner and its cells divide in a

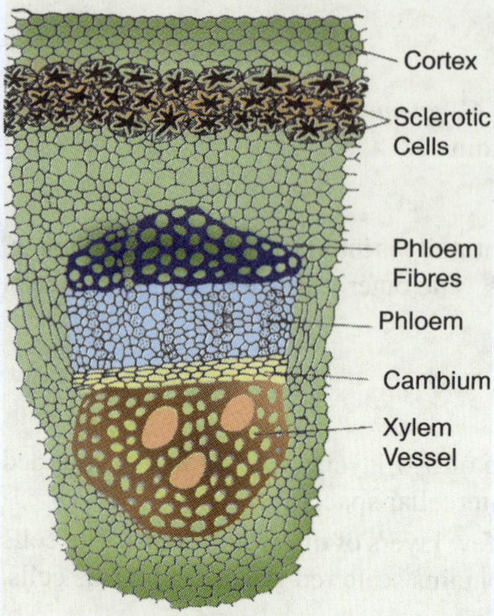

Fig. 16.3. *Gnetum ula* T.S. of portion of slightly old stem showing a part of cortex and a vascular bundle. Note the sclerotic cells in the cortex and phloem fibres. The bundle has three xylem vessels. It has undergone some secondary growth.

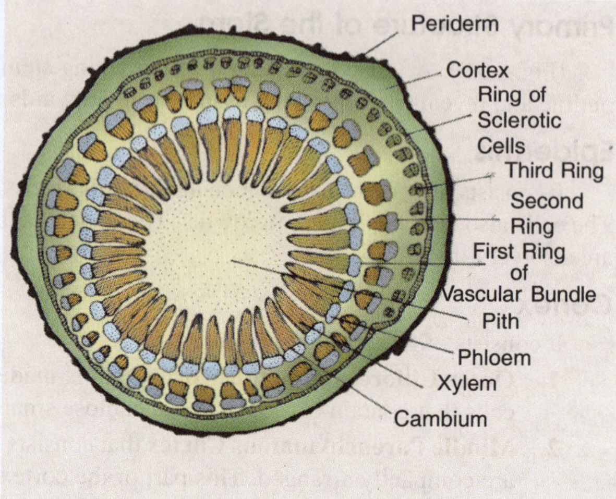

Fig. 16.4. T.S. of old stem of *Gnetum ula.* showing abnormal secondary growth (diagrammatic)

horizontal manner to cut off secondary phloem towards its outer side and secondary xylem towards the inner side. The ray initials of the cambium form secondary medullary rays. The amount of secondary xylem formed is much more than the secondary phloem.

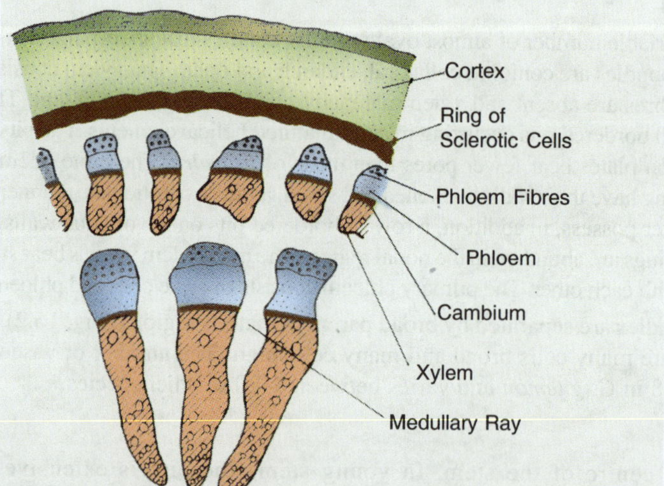

Fig. 16.5. *Gnetum ula.* Portion of stem that has undergone abnormal secondary growth. It shows two rings of vascular bundles. Cortex and a ring of sclerotic cells are also shown (After Maheshwari & Vasil).

The secondary phloem mother cells mature into three types of phloem elements :

1. The sieve cells are arranged in uniform rows and are elongated cells with thin walls and blunt or tapering ends. The sieve areas (sieve plates) occur on the end walls as well as the lateral walls. Each sieve area has pores that are aggregated into groups. These groups of pores have been termed as sieve fields by Hill (1901) and Evert and Alfieri (1965). The young elements contain a slime body which is usually lost in the mature elements. P-proteins are found in the mature sieve elements. The P-protein may be present in the form of a parietal cylinder, thread-like strands, as a slime plug or in the form of amorphous globules.
2. Phloem parenchyma occurs in the form of distinct rows of cells that alternate with the sieve elements. They are smaller in size and are located in the corners of sieve cells. The cells are vertically elongated, thin-walled and devoid of starch content. They look like companion cells, but do not occur in the same radial row. They have no connection with the sieve cells.
3. Albuminous cells may occur in radial files or they may be scattered in the conducting phloem. They are always associated with the sieve cells and may be slightly smaller or half the length of the sieve cells. Their diameter is, however, narrower. These cells contain an ovoid slime body which is absent in the phloem parenchyma cells. The albuminous cells have connections with small sieve areas of adjoining sieve cells. Phloem parenchyma cells have no such sieve area connections. The albuminous cells have denser cytoplasmic contents, an ovoid protean body and dispersing slime.

The two daughter cells that result from the transverse division of a fusiform derivative do not necessarily differentiate into the same cell type. One may develop into a sieve cell, the other into an albuminous cell or phloem parenchyma cell.

In *G. ula* the first formed phloem cells lying towards the periphery get crushed or flattened and appear as arc like dark areas. In this species phloem fibres also appear in the secondary phloem (Fig. 16.30). They are differentiated at the periphery of each bundle and lie close to each other so that they look like a sclerenchymatous patch above each secondary vascular bundle.

The secondary xylem in *Gnetum* consists of three types of elements. These are :

1. **Tracheids** are long cells with thick and lignified walls. The lumen is narrow. They have circular bordered pits that are arranged in a uniseriate manner on the tangential as well as radial walls.
2. **Xylem parenchyma** cells have thin cellulosic walls. They are usually living cells with protoplasmic contents and nuclei. The cells may be rounded or elongated with simple pits.
3. **Vessels** (Figs. 16.6, 16.7) are also present in the secondary wood. These are elongated cells with tapering ends and possess bordered pits. The end walls are perforated. There may be a single (Fig. 16.6, A) row of perforations (*G.*

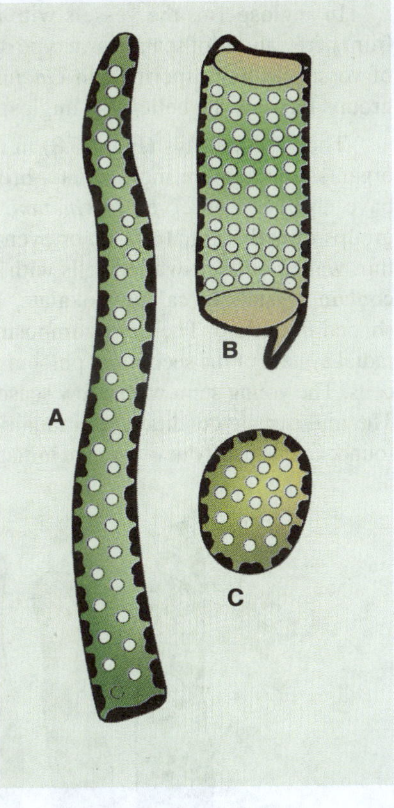

Fig. 16.6. (A-C). *Gnetum ula.* A & C. Thick-walled sclerotic cells from the phloem; B. Entire vessel. (After Maheshwari and Vasil)

africanum, G. gnemon) or two rows of simple perforations (*G. paniculatum*) at the end walls. In *G. ula* there is a single large and simple perforation (Fig. 16.6, B) at the end walls. Such a condition simulates the vessels in angiosperms. It has been noticed that some tracheids also possess a row of closed bordered pits on their end walls. There are also vessels in which the end walls possess some pits that are closed while others are perforated. This indicates transitional stages between tracheids and vessels. It appears that tracheids develop into vessels. The closed bordered pits at their ends develop pores and they become vessels. Moreover further stages have also been observed in which the vessels with many perforations develop into vessels with a single long perforation at their end walls (Fig. 16.7). The end walls may be inclined or rounded. In *Gnetum* the vessels develop in the following manner:

1. The vessel elements increase in size, both length and the breadth.
2. The bordered pits on the end walls increase in size.
3. The border of the pit reduces in extent.
4. The middle lamella and the torus forming a circular perforation.
5. These perforations enlarge and fuse to form a single large perforation (Fig. 16.7 B—D).

In angiosperms the vessels with a single perforation are formed by the disappearance of bars from perforations of scalariform type of pithing. There is thus a difference in the mode of development of vessels in angiosperms and *Gnetum*. This difference in the development of vessels in the two groups falsifies the belief that angiosperms originated from the Gnetales.

The vascular rays (Fig. 16.8) in the xylem and the phloem region are similar in structure and organisation and are multiseriate, broad and of considerable height. Uniseriate and biseriate rays have also been observed in *Gnetum*. In some cases the rays possess scattered sclerenchymatous groups of cells (Fig. 16.8, B) or even some xylem and phloem elements. Usually the rays possess thin-walled or thick-walled cells with simple pits. The cells are radially elongated and some of them contain crystals of calcium oxalate. In a tangential section the rays appear (Fig. 16.8, A) boat-shaped in outline. The cells composing them appear polygonal in outline. The rays constitute the radial system of the secondary phloem and the xylem. The rays are usually bordered by short fusiform cells. The young stem with a few seasons of secondary growth has only uniseriate and biseriate rays. The multiseriate condition is gradually built up until, in the mature stem, only multiseriate rays are found. Occasionally new rays are initiated, but they are quickly converted to the multiseriate condition.

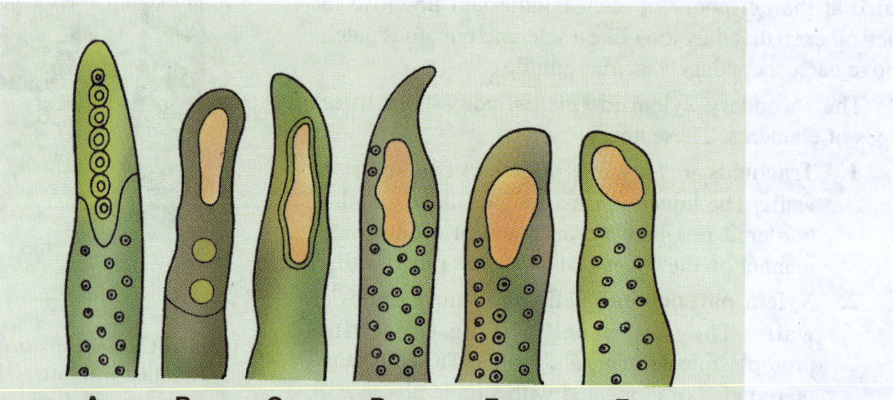

Fig. 16.7. (A-F). *Gnetum africanum*. It illustrates the perforations of the end walls of vessels.

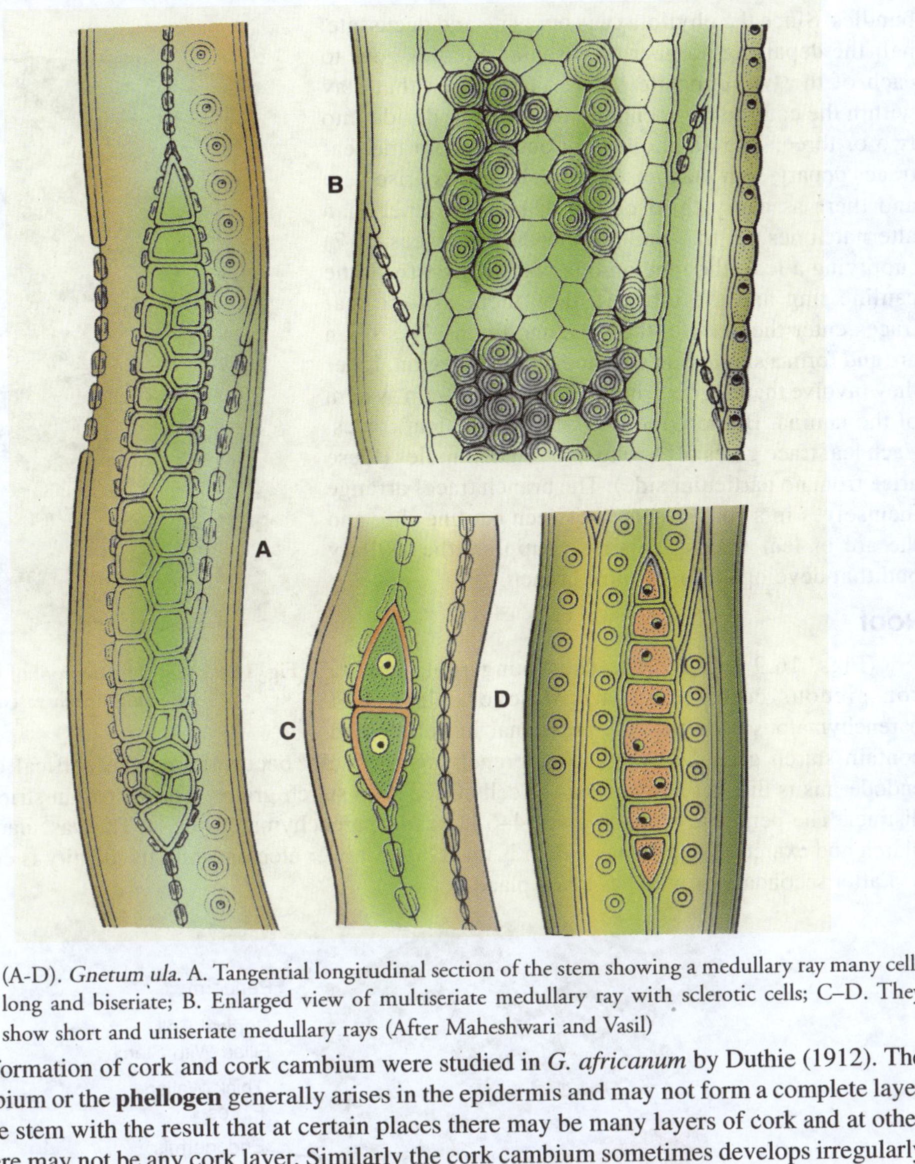

Fig. 16.8. (A-D). *Gnetum ula*. A. Tangential longitudinal section of the stem showing a medullary ray many cells long and biseriate; B. Enlarged view of multiseriate medullary ray with sclerotic cells; C–D. They show short and uniseriate medullary rays (After Maheshwari and Vasil)

The formation of cork and cork cambium were studied in *G. africanum* by Duthie (1912). The cork cambium or the **phellogen** generally arises in the epidermis and may not form a complete layer around the stem with the result that at certain places there may be many layers of cork and at other places there may not be any cork layer. Similarly the cork cambium sometimes develops irregularly at various depths. A strip of phellogen may develop in epidermal layers, another in sub-epidermal and still another quite deep in the cortex. These strips may join laterally and form an irregular layer of cork cambium. Such a condition results in the formation of islands of sclerenchymatous cells and thin-walled cells embedded in the cork layers. Laticiferous cells are present in the cortex and these may also be embedded in the deeper layers of cork. The pith is narrow and may also contain laticiferous elements that develop from files of parenchyma cells by the dissolution of their end walls.

Leaf and Branch Traces

The internode of the stem has 24–28 bundles in *G. gnemon* and 8–20 in other species (Pearson, 1929). Any additional bundle departs as the leaf trace. Each leaf is supplied by 3–7 bundles depending upon the species. In *G. africanum* each leaf is innervated by 4 leaf trace bundles. Since the phyllotaxy is opposite and decussate, half the departing number of vascular bundles goes to each of the two opposite leaves. The traces that stay within the cauline ring remain undivided or divide into two or three. The divisions may occur before the leaf traces depart; so it may appear that the traces arise here and there as every third or fourth bundle rather than alternate ones. Of the total number of leaf traces (3–7) supplying a leaf, the median ones depart first from the cauline ring and the others follow them. As the leaf traces enter the petiole, they arrange themselves in an arc and form a single line within the midrib region. Later they involve themselves within the reticulate vein system of the lamina. Branch traces arise from the leaf traces. Each leaf trace gives off one to four small bundles (these arise from no particular side). The branch traces arrange themselves in a ring situated between cauline ring and the arc of leaf traces. This ring supplies the axillary bud that develops into a future branch.

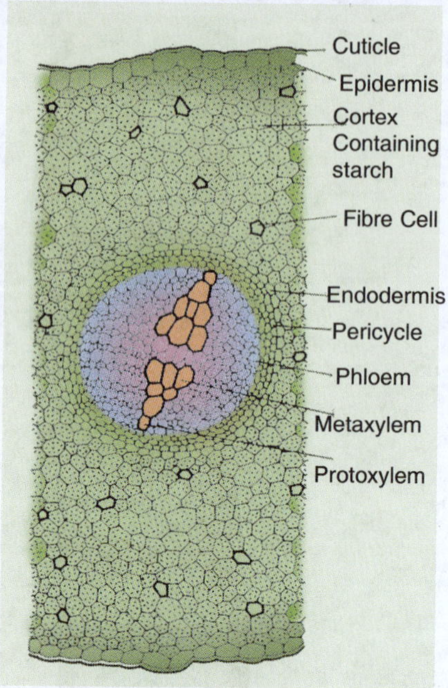

Fig. 16.9. T.S. of young root of *Gnetum sp.* showing diarch vascular region.

Root

(Figs. 16.9 and 16.10). The young roots have a conspicuous cortex made up of several layers of parenchymatous cells that are polygonal in outline and contain starch grains. Groups of sclerenchymatous cells occur among the cortical cells. The endodermis is distinct and consists of cells filled with starch grains. The casparian strips are also distinct. The pericycle is made up of 4–6 layers of parenchymatous cells. The vascular region is diarch and exarch. The primary xylem is made up of fewer elements and its identity is completely lost after secondary growth has taken place.

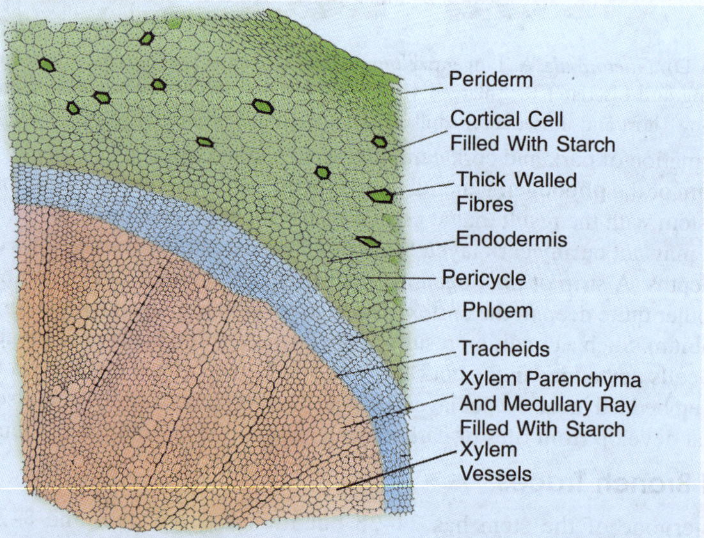

Fig. 16.10. T.S. of portion of old root of *Gnetum sp.* showing secondary growth.

The secondary growth is normal and is effected by a complete cambial arc formed internal to phloem groups and external to the xylem groups. The secondary xylem consists of tracheids, vessels and xylem parenchyma. The tracheids and vessels of the root are larger than those of the stem. The tracheids are elongated cells with tapering ends and possess circular bordered pits on their tangential and radial walls. The pits are arranged in a uniserdite manner. Conspicuous bars of Sanio are present around the pits. In the vessels the pits are arranged in a multiseriate manner and have smaller diameters.

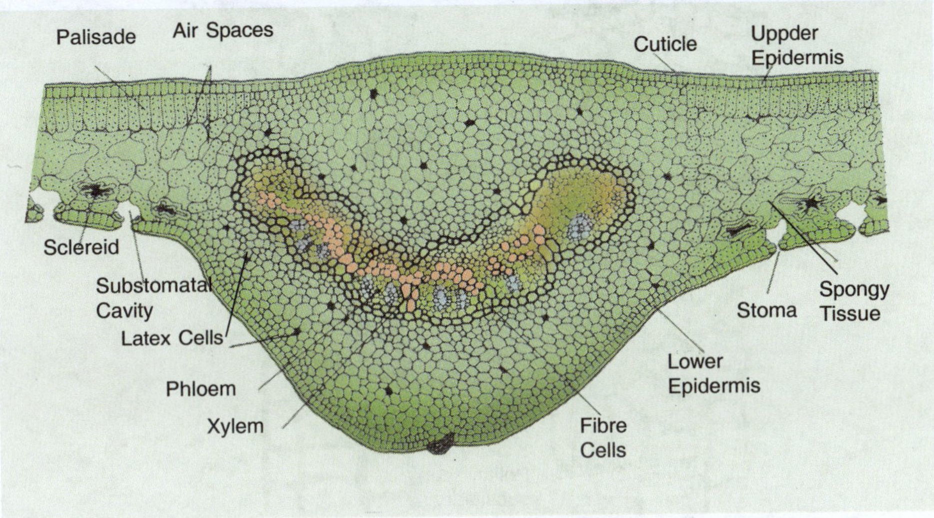

Fig. 16.11. T.S. of leaf of *Gnetum sp.*

Bars of Sanio are absent or inconspicuous in the vessels. The xylem elements in the root contain starch grains at maturity. The parenchyma cells are living and have thin or thick walls with simple pits. The phloem consists of sieve cells and phloem parenchyma cells. The parenchyma cells form distinct rows alternating with those of sieve cells. The secondary medullary rays are multiseriate and are exclusively made up of thin-walled cells. This accounts for the softness of the wood of root. The ray cells are full of starch grains. The root develops periderm that is peripheral in position and develops by the activity of a cork cambium that originates in the outermost cortical layer.

Leaf (Fig. 16.11)

The leaf is bifacial or dorsiventral and consists of distinct upper (dorsal) and lower (ventral) epidermal layers. The epidermal cells have undulate walls. A thick layer of cuticle covers the upper epidermis. Cuticle is comparatively thinner over the lower epidermis. The stomata are restricted to the lower epidermis, but are absent over the midrib and the veins. The stomatal development is of syndetocheilic type, *i.e.*, the same epidermal cells gives rise to the two subsidiary cells and the guard cells (Takeda, 1913; Florin, 1931). The subsidiary cell may divide further so as to form more than two subsidiary cells. Maheshwari and Vasil (1961) studied the development of stomata in *G. gnemon* and *G. ula* and found it to be of haplocheilic type. They also reported stomata in the epidermal cells of the collars, axes of the male and the female cones and on the outer two envelopes of the ovules.

The mesophyll of the leaf is distinguished into a palisade layer and a spongy parenchyma. The palisade is composed of a single layer of elongated palisade cells that are full of Chloroplasts. The spongy parenchyma consists of :

1. Loosely arranged and lobed cells that contain chloroplasts and are thin-walled. They enclose large air spaces.

2. Stellate sclereids with thick and lignified walls are confined to a portion of the spongy parenchyma near the lower epidermis.
3. Scattered fibres with lignified walls occur in groups around the midrib region.
4. Latex tubes are also found scattered around the midrib region.

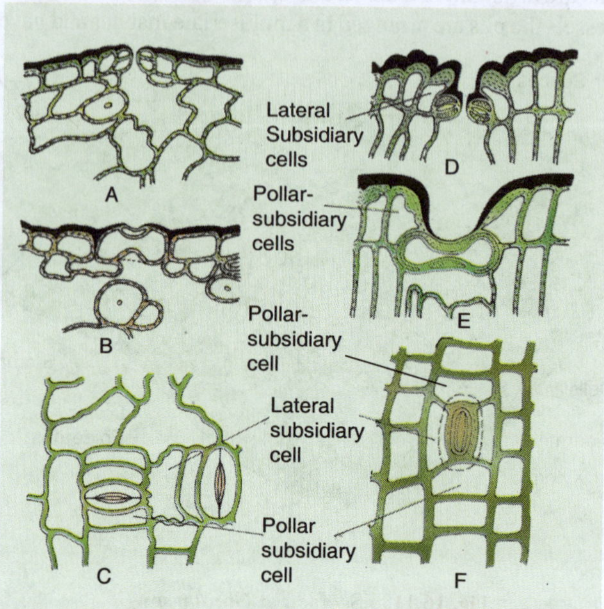

Fig. 16.12. (A-F) It illustrates the structure of stomata in *Welwitschia* (D–F) and *Gnetum gnemon* (A–C) (After Florin).

The spongy parenchyma cells around the midrib are compactly arranged and appear polygonal in shape.

The vascular region in the midrib consists of an arc of vascular bundles. In young leaves the identity of vascular bundles is obscure, whereas in older leaves distinct vascular bundles are observed (Fig. 16.11). The phloem lies towards the lower epidermis whereas the xylem faces upper epidermis. The bundles are conjoint, collateral and endarch. Each bundle has a patch of thick-walled cells capping the phloem in *G. ula*. Patches of thick-walled cells are also present above the protoxylem. They are distinct only in mature leaves. The phloem consists of sieve cells and phloem parenchyma. These elements are arranged in distinct files. Some phloem elements towards the lower side become crushed. They represent the protophloem. The xylem consists of tracheids, vessels and xylem parenchyma.

Anomalous Secondary Growth in Stem (Figs. 16.4 and 16.5).

In the climbing species of *Gnetum* (*G. ula, G. africanum, G. latifolium, G. montonum*) the secondary growth is normal in the beginning, but at a later stage successive rings of cambium distinguish in the deeper layers of cortex. These cambia cut off secondary rings of phloem and xylem. Broad medullary rays develop in the secondary tissue so that distinct, wedge-shaped vascular bundles make their appearance. The first ring of cambium stops functioning and the second ring appears outside it and produces a ring of secondary vascular elements separated by medullary rays, into distinct bundles. Likewise, it stops functioning and the third cambial ring 'appears outside' the second and produces the third ring of vascular bundles. The process continues and successive rings of cambia are produced. They give rise to successive rings of vascular bundles. Sometimes the successive rings are incomplete and result in the formation of eccentric rings.

REPRODUCTION

The sporophyte of *Gnetum* is dioecious and bears male and female reproductive organs on separate plants that may be trees, shrubs or woody climbers. The reproductive organs are organised into well-defined cones or strobili. These strobili are organised into well-defined inflorescences, which are usually panicles, that arise singly or in fascicles in leaf axils (Fig. 16.13). They may also be terminal in position. The cones arise in the the axils of paired and decussate scale leaves. These leaves are fused at their bases to form a boatshaped structure. They are also called **bracts**. The bracts bear auxiliary or accessory buds that may develop into new inflorescence in addition to the strobilus.

Male Strobilus and Male Flowers (Figs. 16.13 to 16.15)

It consists of an elongated axis that is clearly distinguishable into nodes and internodes. The internodes are very short in young cones. The nodes of the axis bear scaly bracts arranged in whorls. These nodal bracts fuse together to form a cup-like structure called the **cupule** or the **collar**. So every node of the axis bears a collar of fused bracts. The number of collars corresponds to the number of nodes on a cone axis and usually this number varies between ten to twenty-five. Each collar bears above it 3 to 6 rings of male flowers, each ring possessing a number of male flowers.

The male flowers in the rings are arranged alternately. Above the rings of male flowers there is one ring of abortive female flowers or ovules. In rare cases in *G. africanum* the ovules in this ring may be fertile so that the cone becomes bisexual. The cones look very compact in a young state, but as the internodes elongate the collars separate from each other. In *G. ula* the apical region of the male cone is bare or naked. In *G. gnemon* two or three collars at the apex of the cone are reduced and sterile. The young male cone is completely enveloped within the basal bracts that are connate at base. Later the male cone grows and emerges out of the fused bracts. It is extremely compact at this stage. The collars also contain sclereids of various shapes and sizes. Laticiferous tubes are also present in the collars and the axis.

Fig. 16.13. *Gnetum latifolium*. a branch bearing opposite and decussate leaves and a cluster of three staminate strobili.

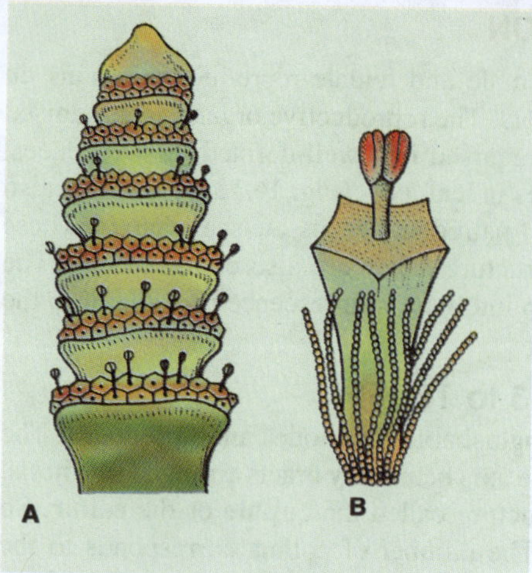

Fig. 16.14. (A-B) *Gnetum ula*. A. portion of male strobilus; B. Single male flower.

Each male flower (Fig. 16.14, B) is enclosed within a perianth that appears sheath-like. It consists of a stalk that bears two anthers at its apex (Fig. 16.15, B). Each anther has a single locule or a microsporangium (Fig. 16.16). At maturity and a little before dehiscence the stalk elongates and projects beyond the perianth. At this stage the anthers and the stalk are visible outside the perianth. The male flowers are interspersed with uniseriate and multicellular hair. The perianth of the male flowers also consists of sclereids that may be very small in size or become profusely branched and star-shaped. Branched or unbranched latex tubes have also been reported in the perianth lobes. These latex tubes contain latex. Some variations have also been reported in the male flowers.

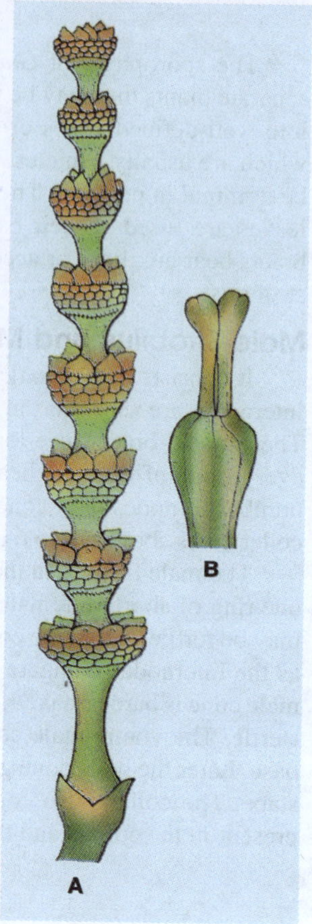

Fig. 16.15. (A-B) *Gnetum gnemon*. A male cone bearing an ovular ring above the male flowers above each collar; B. A male flower (After Vasil).

In *G. gnemonoides* the stalk bears only one anther lobe. In *G. ula*, *G. gnemon* and *G. africanum* the number of anthers on one stalk varies from two to four. The two anthers were unequal in one flower of *G ula* (Maheshwari and Vasil, 1961). In a few cases these authors observed two stalks enveloped by the same perianth, each stalk bearing two anthers. In some cases (in the same species) there were two stalks, within the same perianth, one bearing two anthers and the other bearing one anther. Madhulata (1960) observed fusion between the adjacent flowers in *G. gnemon*.

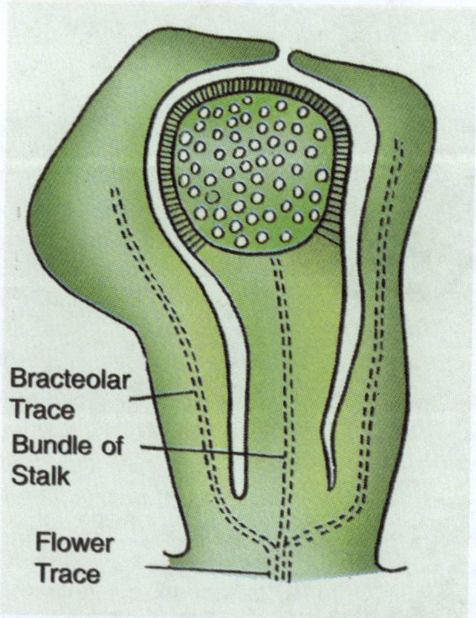

Fig. 16.16. *Gnetum gnemon*. L.S. of male flower showing one another with a single locule containing microspores. Also note the vascular supply. (After Madhulata)

The following abnormalities have been observed in the male cones of various species ;

1. Madhulata (1960) observed two rings of abortive ovules above the rings of male flowers in *G. gnemon*.
2. The same author observed that in some male cones the lower collars do not bear any male flowers and instead bear sterile hair.
3. In some cases (*G. gnemon*) the collars bear male flowers on one side only whereas the other side is devoid of them.
4. In *G. buchholzianum* some of the male cones bear only two collars (pearson, 1912).
5. Madhulata (1960) and Pearson (1914) reported that in *G. gnemon* the male cones may possess reduced number

of male flowers. In such cases these authors also observed that some sterile ovules may become fertile. The number of rings of male flowers may be reduced to one or two or even to a few flowers among the ovules.

6. In some cases the lower collars of male cones bear only one or two fertile ovules whereas the upper collars of the same cone bear normal male flowers.
7. In some cases the ovules in the lower collars are replaced by short male cones that have only few collars and remain dwarf or stunted. The collars in these secondary cones bear one or two rings of male flowers and a ring of sterile ovules above.

Development

A longitudinal section of a very young cone shows that a few cells lying below each collar become meristematic and appear all around below the collar (Fig. 16.17, A). They divide repeatedly and form a small hump-like annular outgrowth below the collar (Fig. 16.17, B). This happens below each collar. This outgrowth becomes prominent and appears to lie between the two collars (one below which it appears and the other lower collar). The upper cells of this annular outgrowth differentiate into initials of the ovules (Fig. 16.17, C). They develop into abortive ovules that form the uppermost ring. The lower cells of the meristematic annular ring differentiate into primordia that develop into rings of male flowers (Fig. 16.17, C, D). Successive rings of male flowers develop from different rings of primordia that develop below the upper ring.

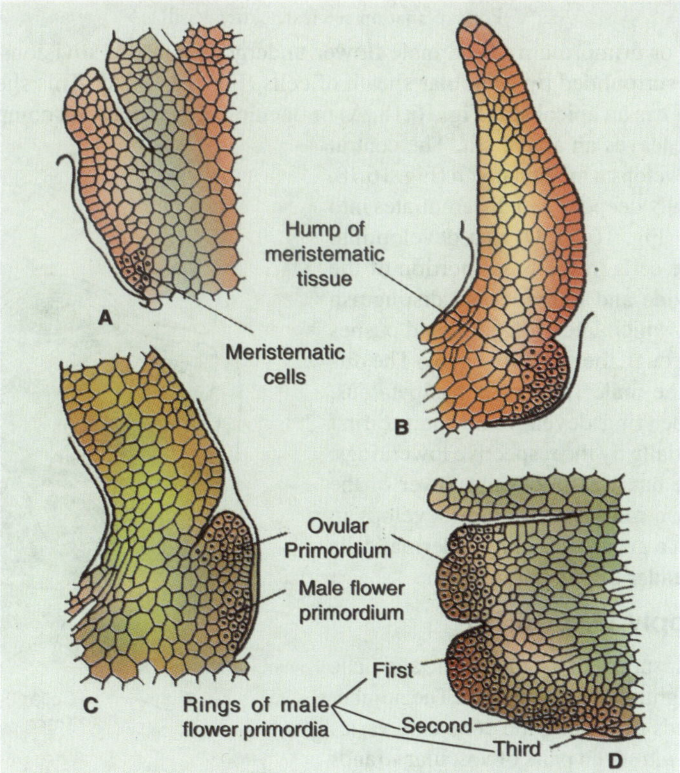

Fig. 16.17. (A-D). *Gnetum ula* Various stages in the development of male flowers. For explanation see text. (After Vasil).

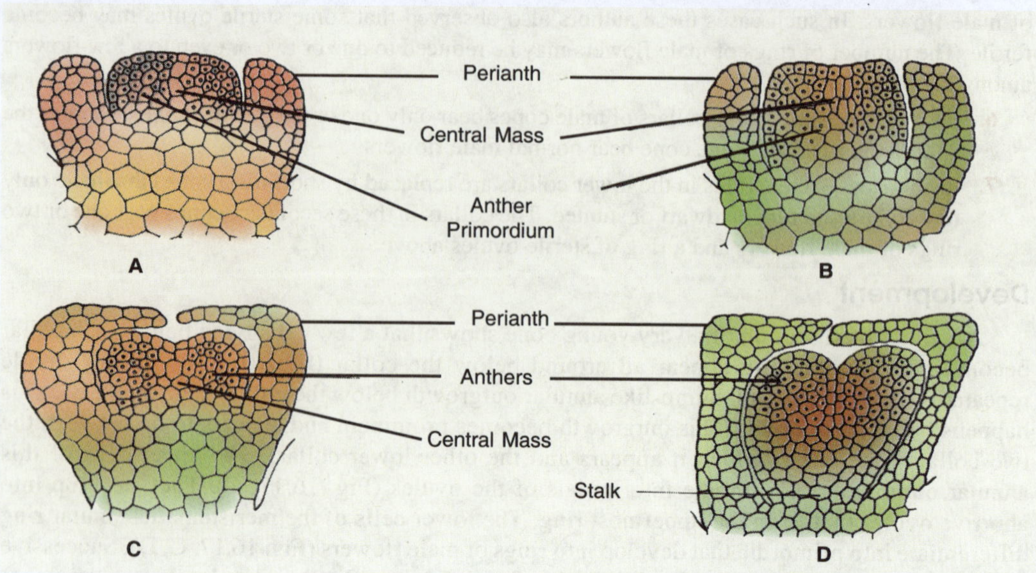

Fig. 16.18. (A-D). *Gnetum ula* Stages in the development of male flower. For explanation see text. (After Vasil)

Each initial or primordium of the male flower undergoes repeated divisions to form a central cushion of cells surrounded by a circular sheath of cells (Fig. 16.18, A). This sheath develops into the perianth, that has an apical slit (Fig. 16.18, A) or opening, *i.e.*, it does not completely enclose the central mass but leaves an apical slit. The central mass, by now, develops a median notch (Fig. 16.18, B, C) that gradually deepens and differentiates into two small lobes (Fig. 16.18, D) that develop into two anthers. The cells in the basal portion of the central mass divide and redivide and distinguish into a short stalk which later elongates and pushes the upper part through the slit (Fig. 16.16) The differentiation of the male flowers is basipetalous, those of the upper ring develop and mature first followed sequentially by the respective lower rings, each anther lobe has a distinct outer layer or the epidermis. Later a microsporangium develops in each anther from a group of two hypodermal cells (for details see under microsporangium).

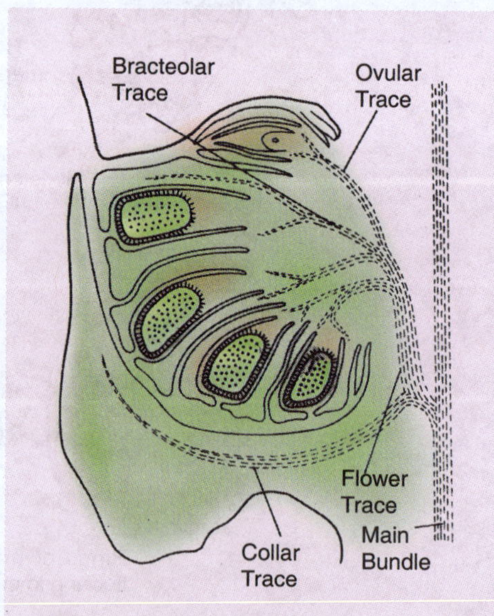

Fig. 16.19. *Genetum ula.* It illustrates vascular supply to the male strobilus. (After Vasil).

Vascular Supply (Fig. 16.19)

The vascular supply to the male cone and male flowers varies slightly with the species. The number of vascular strands traversing the cone axis varies from 8–20. In *G. ula* seven pairs of vascular strands arise from the axial bundles, a little below the origin of paired bracts. These are the **bract traces** and supply the bracts. The bundles of the axis continue their course upwards and during this their number may increase, but near the apex of the axis their number decreases to two.

As the axial bundles traverse the axis of the cone, each one of them gives out a **collar trace** that innervates the collar at the node. Each collar bundle enters the collar and undergoes further divisions. Before entering the collar each collar trace gives off an irregularly oriented or inversely oriented trace. This trace travels towards the groups of male flowers and the abortive female flowers that arise from the internodal region and divides to give off traces to each male flower and abortive female flowers. These are the **flower traces**. Each flower trace reaches the base of the male flower and divides into two pairs of strands at right angles to each other, so that in longitudinal section we can see only two of them. Out of these two pairs, one pair enters the perianth and reaches almost its upper end. The other pair enters the stalk of the male flower and stops at the base of the anther (Fig. 16.19).

Female strobilus (Figs. 16.20, 16.21)

The organisation of the female strobilus is quite like that of the male strobilus except that in this case there is only a single ring of four to ten ovules above each collar. In younger stages of development

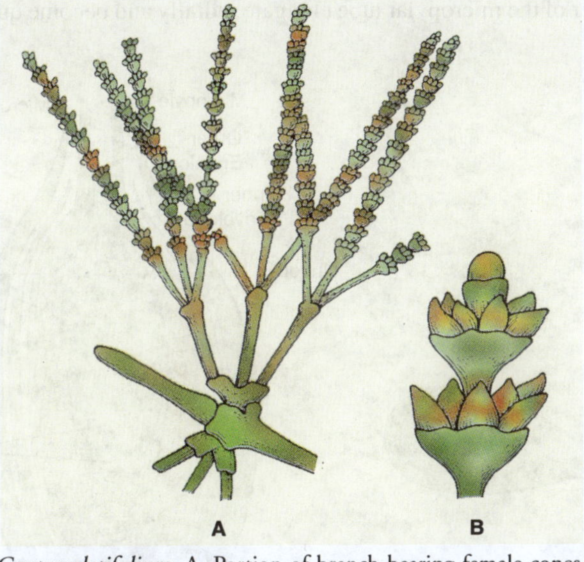

Fig. 16.20. (A-B). *Gnetum latifolium*. A. Portion of branch bearing female cones. B. Portion of female cone enlarged to show ovules.

the female cones cannot be distinguished from the male strobili. It is only when the ovules grow sufficiently and project beyond the collar that the female cone can be clearly marked off. During younger stages of development all the ovules are of the same size. Later some ovules grow bigger while others remain smaller and may fall down. All the ovules never develop into seeds. Only a few of them reach maturity. The ovules may not develop in the uppermost collars. Several uniseriate hair are interspersed among the ovules.

Ovule (Fig. 16.22)

Each ovule consists of a central mass of cells called the nucellus. The nucellus is surrounded by three envelopes. The inner envelope grows beyond the middle one and forms a narrow and cylindrical tube called the micropylar canal or the 'style'. The nucellus contains the female gametophyte. A rudimentary pollen chamber is present at the nucellar apex. There is no nucellar beak in *Gnetum*.

Out of the three envelopes, the outermost is called the perianth, the middle is designated as the outer integument whereas the innermost is called the inner integument. The inner integument is fused with the nucellus in its lower part and is free above it. Stomata are found in the epidermal layer of outer two envelopes only. Sclereids and laticiferous ducts are abundant in the outer two envelopes alone. They are absent in the inner envelope. The outer envelope or the perianth becomes thick and fleshy in the seed. The middle forms the stony layer. In *G. gnemon*, Madhulata (1960) observed the formation of a circular rim from the outer epidermis of the inner integument. This rim develops at a level where the inner integument pierces through the middle envelope (outer integument). This rim sometimes grows massive and covers the tip of the middle envelope. A second rim of this type was observed by Thoday (1921) at a higher level. The micropylar tube may be straight or sometimes bent in a manner to form a lid-like structure above the nucellus. In *G. ula* the cells of the inner layer of the micropylar tube elongate radially and become quite conspicuous by their larger size. After pollination these cells elongate more and interlock with each other and close the micropylar canal. In *G. gnemon* these cells become thick-walled and may develop simple pits (Madhulata, 1960). The ovules are stalked in *G. ula* and may be sessile or sub-sessile in other species.

Plate I : *Gnetum* : female strobilus and fruit.

Fig. 16.21. *Gnetum gnemon*. A female strobilus bearing two seeds.

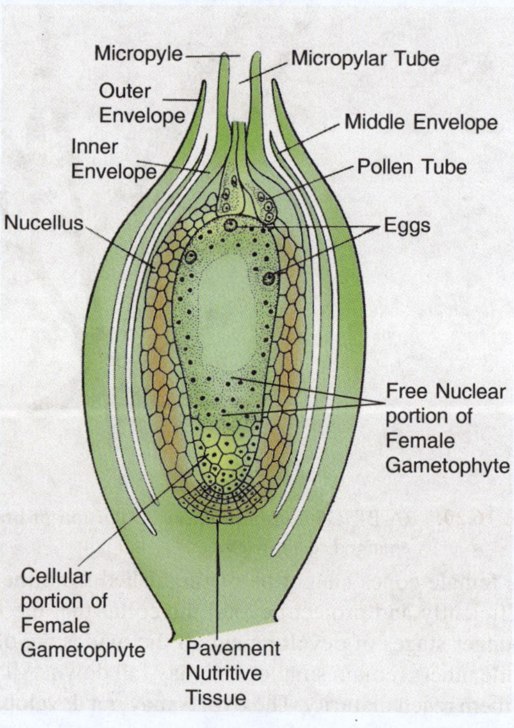

Fig. 16.22. L.S. of ovule of a *Gnetum sp.*

The female gametophyte is tetrasporic in development and as many as 256 to 1,500 free nuclei are formed. Only two to four female gametophytes start active development but ultimately one reaches maturity. It is broader above and tapers towards the chalazal end (Fig. 16.22). The free nuclei are peripherally disposed with a big central vacuole. There are no archegonia in any species of *Gnetum*. One or more nuclei become larger at the micropylar end of the gametophyte and act as egg nuclei. There is no wall formation in upper part of the gametophyte. The lower of chalazal end becomes cellular forming about 12 cells.

Morphology of the Envelopes

The following views have been put forth to explain the nature of the three ovular envelopes :

1. Van Ticgham (1869) considered the outer envelope as an ovary or something analogous to it, whereas the other two were regarded as the two integuments.
2. Strasburger (1872) regarded them to be three layers of an integument that differentiated as a result of splitting up of a single integument. This view is not supported by any developmental or structural evidence.
3. Beccari (1877) regarded the inner two envelopes as outer and inner integuments and the outer envelope as the perianth. Coulter (1908) supported this view and is even now regarded to be more plausible.
4. Lignier and Tison (1912) and Thompson (1916) put forward a different explanation. They regarded the inner envelope as an ovary and the outer two as perianth lobes.
5. Vasil's (1959) observations on *G. ula* also support the 4th view put forward by Lignier and Tison. She observed that in some specimens of *G. ula* the innermost envelope is abortive or partially developed and in its place the middle envelope develops into the micropylar tube or the 'style'. Sometimes the middle as well as the inner envelopes were observed to grow into the 'styles' or the micropylar tubes. She regarded the inner and the middle envelopes as ovaries, because of their capacity to form styles. In some ovules four envelopes were observed. The fourth one was situated between the middle and the inner envelopes and sometimes grew as long as the micropylar tube.

Madhulata (1960) observed twin ovules in *G. gnemon*. These were enveloped by a common outer envelope.

The cells of the nucellus lying below the female gametophyte divide to form a fan-shaped tissue made up of several rows of cells. This was named as **pavement tissue** by Coulter (1908). This tissue is later absorbed. The nucellar cells lying between the female gametophyte and the pollen chamber contain abundant starch. Nucellus is almost completely absorbed by the growing female gametophyte except for a small part in the micropylar region where its cells develop cuticle. This part persists in the seed as a nucellar cap.

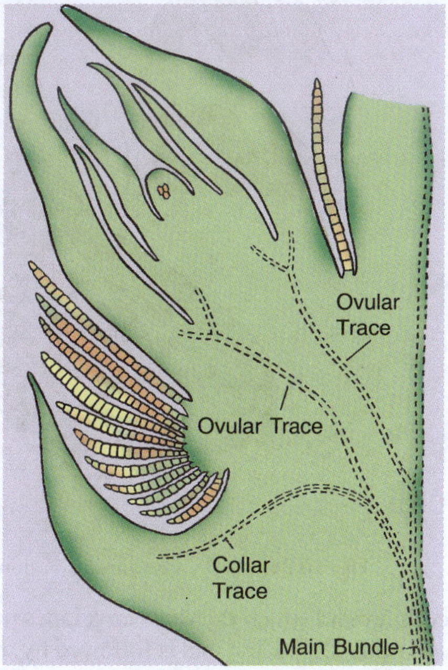

Fig. 16.23. *Gnetum gnemon.* It illustrates vascular supply to the female strobilus (After Madhulata).

Vascular Supply (Fig. 16.23)

There is a great variation in the vasculature of the cone and the ovule in the various species. The vasculature of the cone, bracts and the collar is the same as in male cone. In *Gnetum ula* the collar trace before entering the collar divides into two strands. One of them is inversely oriented whereas the other is normal. The former traverses the upper side of the ovule, whereas the latter enters the lower side of the ovule. Additional bundles also arise directly from the axial bundles (bundles of cone axis) and join the inversely oriented bundle supplying the upper side of the ovule. Madhulata (1960) observed a similar condition in *G. gnemon*. The vascular strands at the base of the ovule are quite distinct. In *G. gnemon* they arrange in the form of a ring of 12–14 vascular bundles (Berridge, 1912) that supply the three ovular envelopes.

Abnormalities

Some female cones with spirally arranged collars have been observed in *G. gnemon*. In some cones both cyclic and spiral arrangements of collars have been observed.

SPORANGIA

The Megasporangium (Fig. 16.24)

The structure of the ovule has already been discussed in detail. Now only developmental stages will be discussed. The ovular primordia arise on the annular meristematic ring that develops below each collar, in the same manner as in the male cone. The ovular primordium divides to form a mass

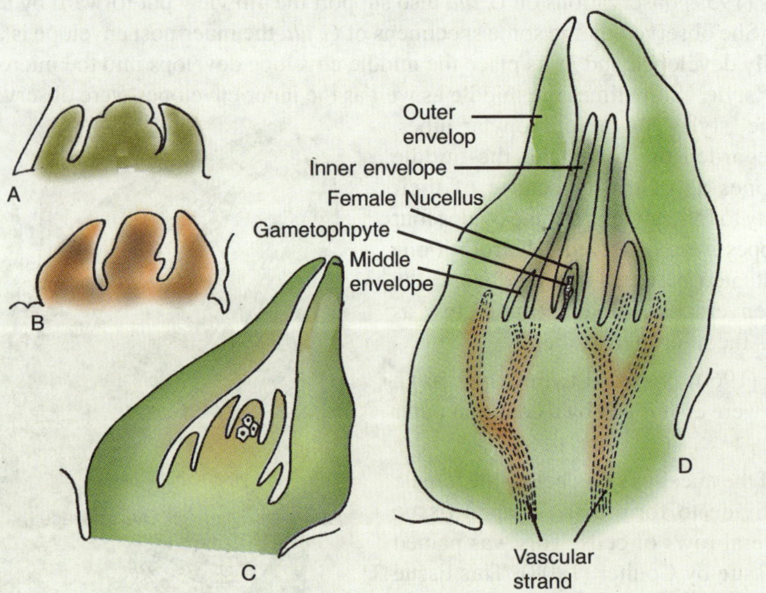

Fig. 16.24. (A-D). *Gnetum ula* Various stages in the developments of ovule. (After Vasil).

of cells around which the three envelopes develop one after the other (Fig. 16.24, A–D). The outer envelope develops first and is followed by the second and the third envelopes. The third or the inner envelope is fused with the nucellus in the basal region, but is free in the upper region and grows into a long micropylar tube or the 'style' (Fig. 16.24, D). At a later stage the cells lining the micropylar tube differentiate into a nutritional tissue and a group of radially elongated cells that block the micropylar tube after the act of pollination. The mass of cells surrounded by the three envelopes develops into the nucellus. A distinct epidermal layer distinguishes this tissue. It is the nucellar epidermis. A group of two to four archesporial cells develop (Fig. 16.25, A) below the epidermis. This happens at the same time when the inner envelope is developing. The archesporial cells divide periclinally to cut off outer primary parietal cells and inner sporogenous cells (Fig. 16.25, B). The nucellar epidermis and the parietal cells undergo further periclinal and anticlinal divisions to form the massive nucellus and push the sporogenous cells deep into its tissue (Fig. 16.25, C). The sporogenous cells may divide once or twice to form 8–16 or rarely 20 sporogenous cells arranged in linear rows (Fig. 16.25, D). They function directly as megaspore mother cells. All of them may undergo meiosis to form tetrasporic embryo sacs (Fig. 16.25 E, F) but ultimately one develops to maturity, the others are seen aborting. Free nuclear divisions produce 256 to 1,500 nuclei that are peripherally arranged around a central vacuole. The embryo sac or the female gametophyte assumes a spatulate form, i.e., broad above and narrow below. At a later stage the nuclei lying towards the chalazal end get enclosed by walls to form distinct cells. The nuclei at the micropylar end remain free. So the gametophyte is partly cellular and partly invested. Some free nuclei act as eggs.

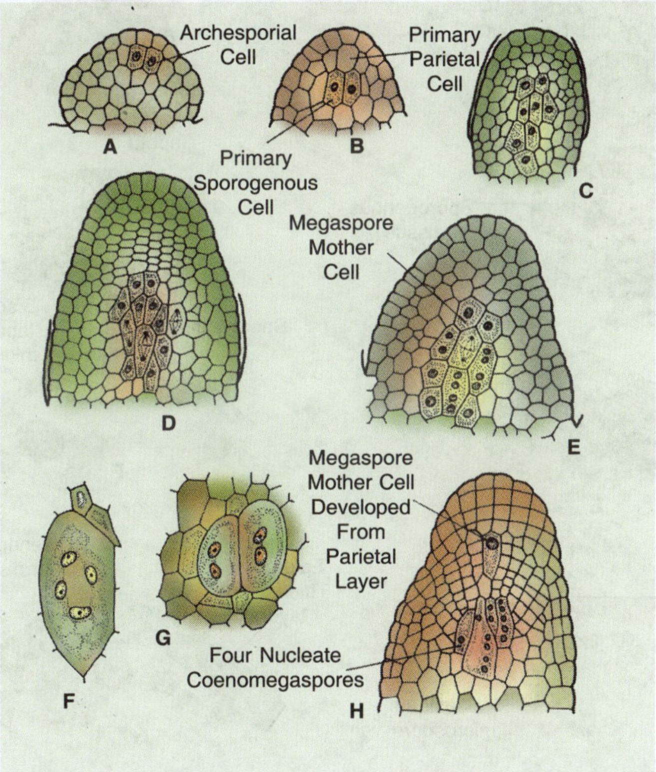

Fig. 16.25. (A–H). *Gnetum ula* Stages in the development of ovule and female gametophyte. For explanation see text. (After Vasil).

A pavement tissue consisting of several rows of cells develops at the base of the young embryo sac but gets dissolved during later stages of embryo sac (female gametophyte) development.

There is no nucellar beak in *Gnetum*, but the apical cells undergo degeneration to form a small pollen chamber. Some cells at the base of the pollen chamber may form papillate outgrowths between which the pollen grains get lodged after pollination.

In *Gnetum* megaspore tetrads are never formed. The mother cells undergoes meiosis without wall formation so that the 4 haploid nuclei lie free within the wall of the mother cell. Such gametophytes are called **tetrasporic**.

The Microsporangium (Fig. 16.26)

Vasil (1959) studied in detail the development of microsporangium of *G. ula*. Below the epidermis of the anther two **archesporial cells** are distinguished (Fig. 16.26, A). They divide and redivide to form many celled **archesporium** (Fig. 16.26, B). The cells in the outermost layer of this archesporium undergo periclinal divisions to form an outer layer of **parietal cells** and an inner layer of **sporogenous cells**. (Fig. 16.26, C). The parietal layer undergoes another periclinal division to form an external **wall layer** (lying beneath the epidermis of the anther lobe) and an internal **tapetal layer** (Fig. 16.26, C). The wall layer may become two layered (Fig. 16.26, E) at certain places. The

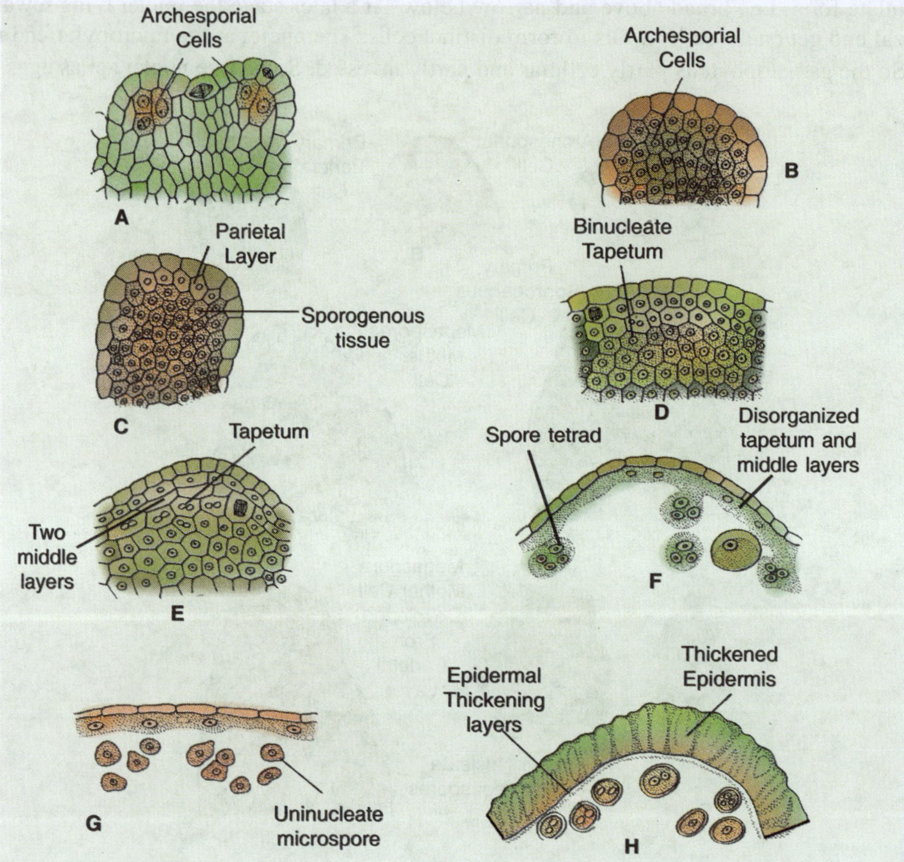

Fig. 16.26. (A-H). *Gnetum*. Various stages leading to the development of microsporangium. (After Madhulata and Vasil).

sporogenous cells undergo a few divisions to form the **microspore mother cells**. They contain dense cytoplasmic contents and distinct nuclei. At this time the tapetal cells become conspicuous and large in size and have dense cytoplasmic contents. They become bi-nucleate (Fig. 16.26, E). These nuclei may fuse to form polyploid nuclei and may have many nucleoli. The wall layers become compressed due to the pressure of the enlarging spore mother cells (Fig. 16.26, F, G). The microspore mother cells undergo meiosis and form isobilateral, decussate or tetrahedral tetrads of haploid **microspores** (Fig. 16.26, F). The microspores enlarge and become free after breaking the original microspore mother cell wall (Fig. 16.26, G.H). The microspores are uni-nucleate.

As the mother cells are undergoing meiosis the wall cells and the tapetal cells start degenerating (Fig. 16.26, F, G). When the microspores have completely developed the tapetum and the wall layers have disorganised. Their remnants can, however, be seen in the cytoplasm. The epidermal cells persist and their outer walls become thick and cutinised. They become radially elongated and fibrous bands of the thickenings arise from the inner tangential walls and run upwards and touch the outer wall. These thickening bands are also present in the endothecium of the angiospermous anthers. Small rounded structures appear on the inner surface of the epidermis at the time of differentiation of microspores and persist till the anthers dehisce.

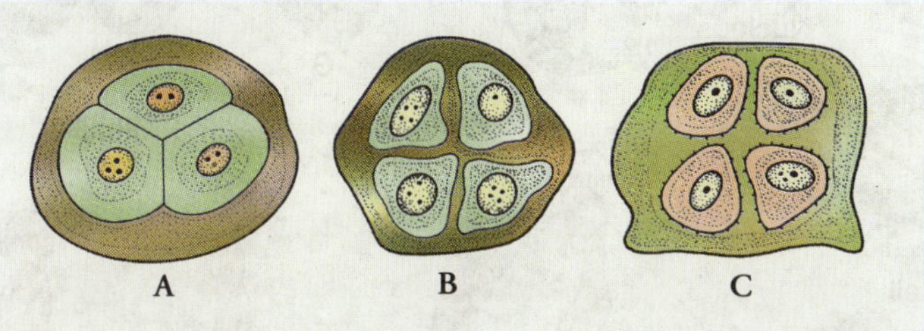

Fig. 16.27. *Gnetum ula*. It shows different types of microspore tetrads. (After Vasil)

The dehiscence of the anther takes place along a double row of small cells that extend from the tip of the anther downwards towards its base.

Some variations have been recorded in the development of microsporangium in *Gnetum*. Thompson (1916) reported that tapetum originates from the sporogenous tissue and there are two wall layers between the epidermis and the tapetum. Karstin (1893) reported two tapetal layers in *G. gnemon*. Fagerlind (1941) reported that tapetum may be several layered thick in some specimens. Two layers of tapetum have also been reported in *G. gnemon* (Madhulata 1960) and *G. africanum* (Weterkeyn, 1959).

GAMETOPHYTE

Male Gametophyte (Figs. 16.28, 16.29)

Microspore or the pollen grain is the pioneer structure of the male gametophyte. It has a roughly spherical shape when young. It is uni-nucleate and is enveloped by a thick and spiny exine and a thin intine. It is wingless and is released in a three nucleate condition. Karstin (1893), however, mentions that the pollen grains are two nucleate at the time of their release from the microsporangium.

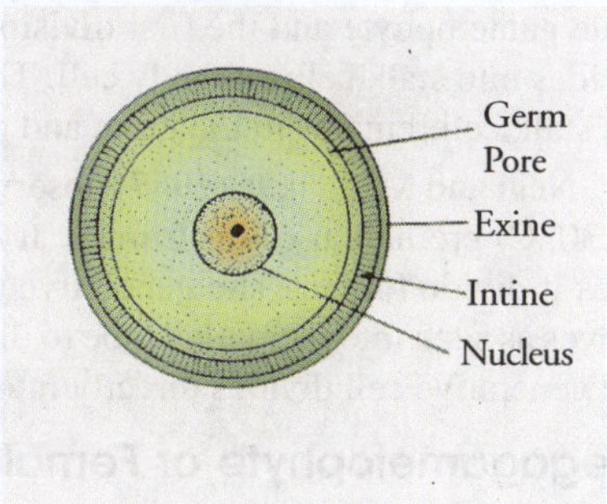

Fig. 16.28. A pollen grain or microspore of *Gnetum ula*.

Pearson (1912, 1914) studied the development (Fig. 16.30, A) in microgametophyte in *G. gnemon* and *G. africanum* and reported that the microspore nucleus divides into two daughter nuclei that lie free in the cytoplasm or an evanescent cell plate is formed between them. One of them divides again so that there are three nuclei. He named these nuclei as prothallial nucleus, generative nucleus and the tube nucleus. At this stage the pollen grains are released and further germination is effected in the pollen chamber. The exine ruptures and the intine grows into a pollen tube. The tube nucleus migrates first into the pollen tube. The generative nucleus now gets surrounded by a distinct cell membrane and forms the generative cell. It also migrates into the pollen tube and later divides into two unequal **male gametes**. The prothallial nucleus never enters the pollen tube and usually disappears before its formation.

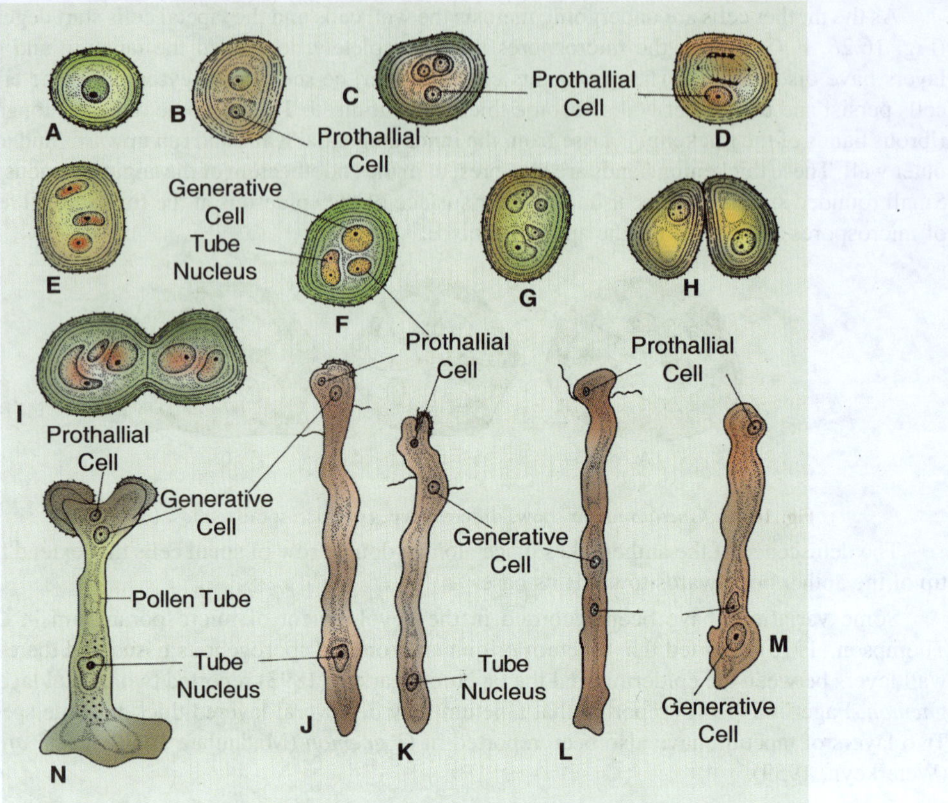

Fig. 16.29. (A-N) *Gnetum ula*. Stages in the development of male gametophyte or the germination of pollen grains. (After Vasil)

Thompson (1916), on the other hand, states (Fig. 16.30, B) that there is no prothallial cell in the male gametophyte and the first division results in a tube nucleus and a generative cell which later divides into stalk cell and body cell. The latter divides into two equal male cells. According to him the stalk cell remains in the grain and only the body cell and tube nucleus enter the pollen tube.

Negi and Madhulata (1957) observed that in both *G. ula* and *G. gnemon* a small lenticular (Fig. 16.30, C) prothallial cell is formed. It later becomes spherical and never enters the pollen tube nor does it divide further. The generative cell is well defined before the formation of pollen tube and moves as such into the pollen tube to divide later into two male gametes. According to these authors the generative cell divides directly into male gametes and does not form stalk cell and body cell.

Megagametophyte or Female Gametophyte (Figs. 16.31, 16.32, 16.33).

It has been stated before that the embryo sac of female gametophyte is terasporic in development and large number (usually 8–16) of megaspore mother cells start developing within the nucellus. Each one of them enlarges and undergoes meiosis to form four haploid nuclei (Fig. 16.31) that are not separated by any wall and thus lie free in the mother cell cytoplasm. So no megaspore tetrads are organised in *Gnetum*. Such a development is called tetrasporic and characteristic of some dicotyledonous angiosperms.

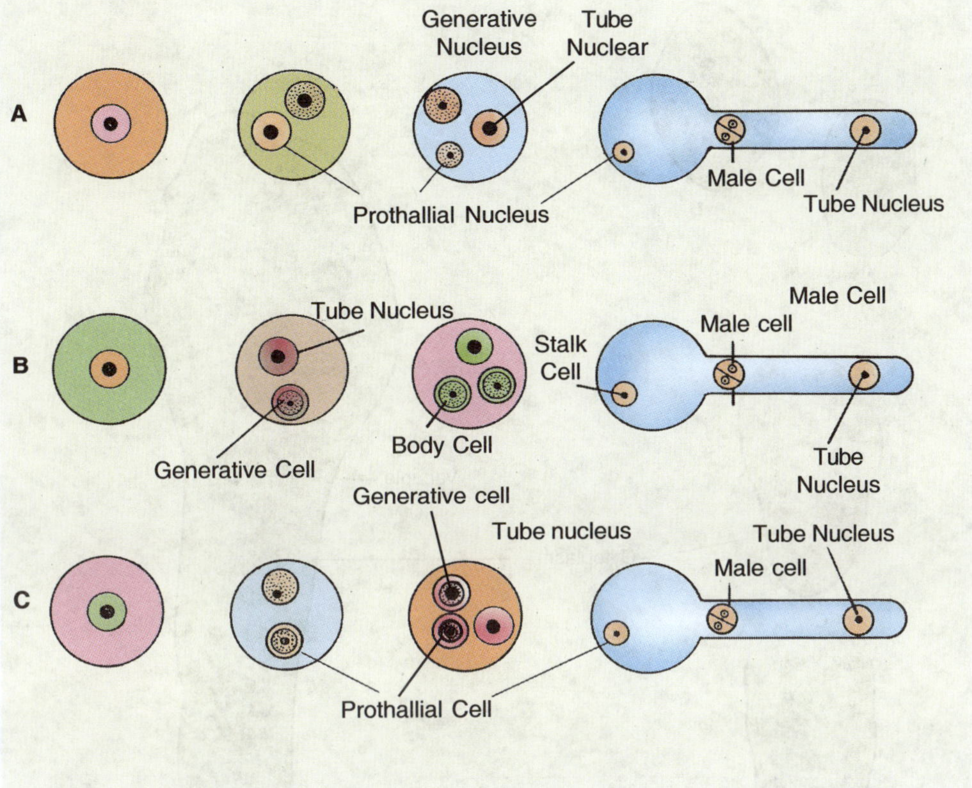

Fig. 16.30 (A–C). The figure is a diagrammatic illustration of the three prevalent views regarding the development of male gametophyte. (After Maheshwari and Vasil)

The free haploid nuclei divide further (Fig. 16.31, A, B) and form 16 free nuclei. At this state most of the developing gametophytes degenerate and only two or three proceed further and undergo free nuclear divisions to form 256 (*G. gnemone*), 512 (*G. africanum*) or 1,500 (*G. ula*) free nuclei (Fig. 16.31, D). The female gametophyte at this stage elongates towards the chalazal end and appears almost spatulate in outline. It has a big central vacuole and the free nuclei are arranged in the peripheral layer of cytoplasm. Further developments usually take place only in one female gametophyte and the others completely degenerate.

The nuclei at the chalazal end get surrounded by cell walls (Fig. 16.33) and form a cellular tissue, whereas those towards micropylar end remain free. So the gametophyte is partly cellular and partly nuclear. There are no archegonia in *Gnetum*. Some nuclei at the micropylar end function as eggs. They become bigger in size and can be distinguished from the other nuclei.

Vasil (1959) noticed an interesting behaviour of the female gametophyte in *G. ula*. She observed that one more groups of cells differentiate at the micropylar end (these are in addition to the lower cellular tissue). These cells have denser cytoplasmic contents. One cell in each group functions as the egg (Fig. 16.32, C, D, E). These changes take place only when the pollen tubes have reached the female gametophyte.

The rest of the gametophyte becomes cellular after fertilization.

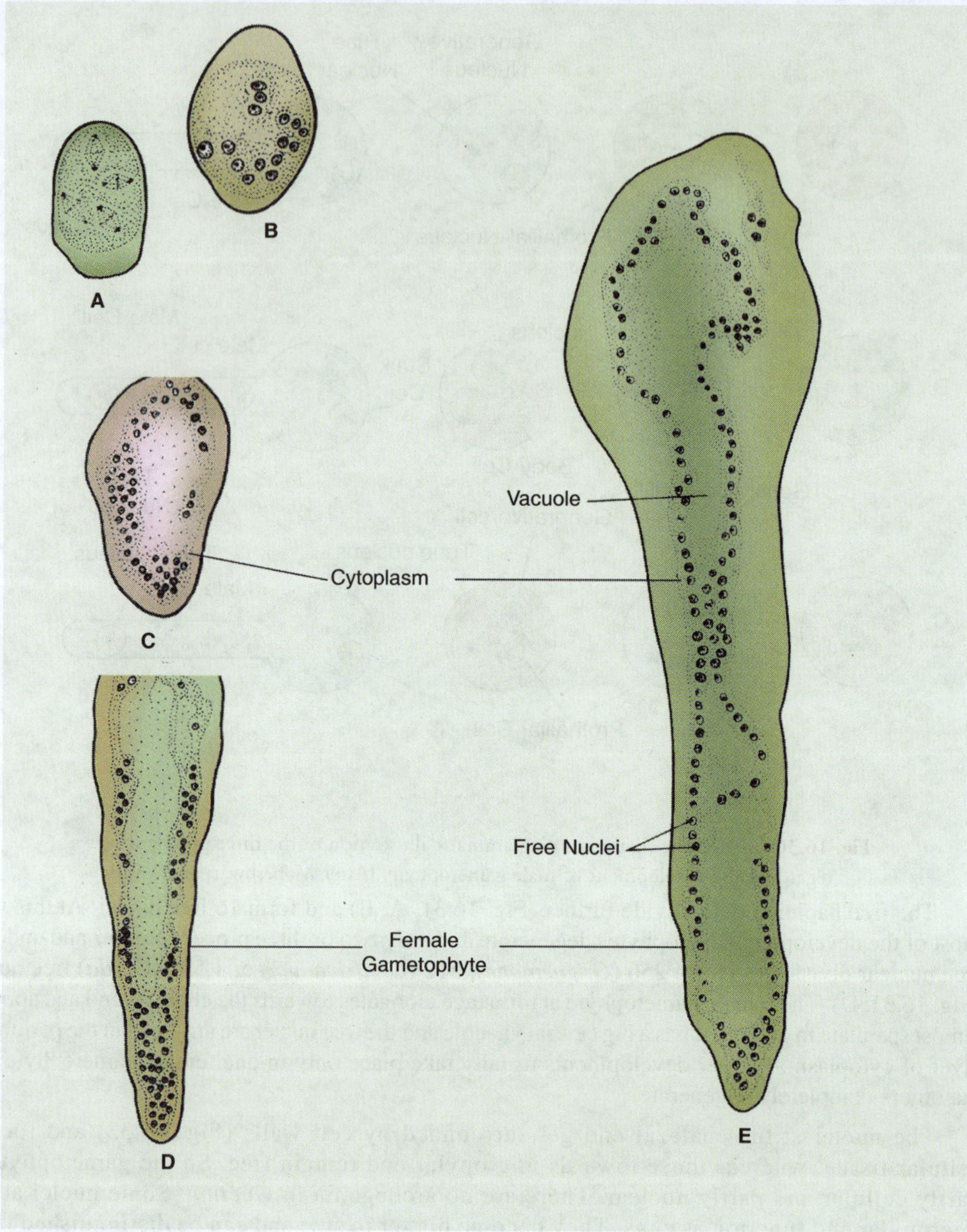

Fig. 16.31. (A—E). Earlier stages in the development of free nuclear female gametophyte in *Gnetum ula* (After Vasil).

Pollination

The pollen grains are carried by wind to the micropylar tube of the ovule. The micropylar tube secretes a drop of sweet fluid at its tip. The pollen grains are caught in this liquid which later evaporates and the pollen grains are withdrawn into the tube and down to the pollen chamber.

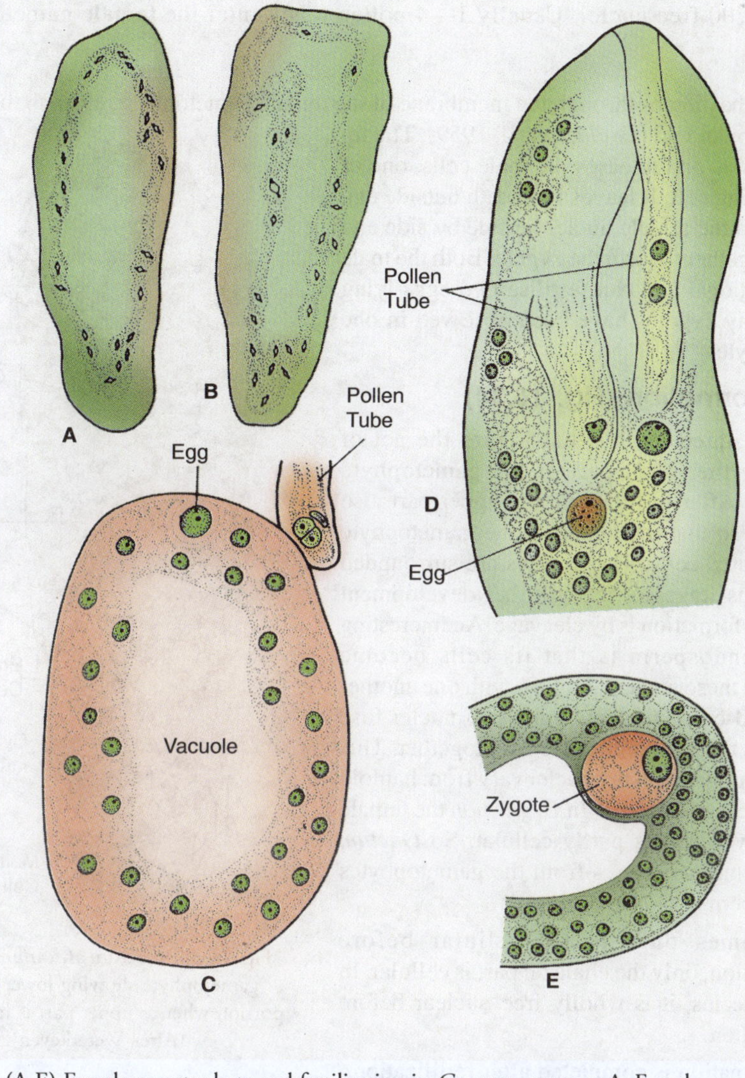

Fig. 16.32. (A-E) Female gametophyte and fertilization in *Gnetum gnemon*. A. Female gametophyte with many free nuclei and a central vacuole (After Madhulata); B. Stage next to A, showing more nuclei. In both A and B there is intra-nuclear division (After Madhulata); C. Mature female gametophyte with an egg and a pollen tube lying adpressed to it (After Thompson); D. Process of fertilization (After Thompson); E. portion of female gametophyte with zygote (After Madhulata).

The exine has no germ pores and is completely discarded at the time of germination. The intine grows into a pollen tube opposite the point where the tube nucleus lies. The pollen tube traverses through the intercellular spaces between the nucellar cells. The cells of the nucellus below the pollen chamber are full of starch. The growing pollen tube scretes a starch dissolving enzyme which dissolves the starch and the tube absorbs it as its nutrition. Tube nucleus and the generative cell move into the pollen tube. The generative cell divides into two unequal male gametes when the pollen tube has travelled more than half of the nucellar tissue. Now the growth of the tube shows slackness and the male cells move to the tip of pollen tube. The tube nucleus lies behind them. In *G. ula* the pollen tube reaches the female gametophyte when the latter has about 700 free nuclei. Usually 1—4 pollen tubes enter the female gametophyte.

Fertilization

The pollen tube pierces through the membrane of the female gametophyte, close to the group of cells at the micropylar end (*G. ula;*) Vasil, 1959). The tip of pollen tube bursts and releases the male cells, one of which enters the egg cell. It leaves its sheath outside egg cell. The male and the female nuclei lie side by side and ultimately fuse together to form the zygote. Both the male gametes are functional and can fertilise two eggs lying side by side. Many zygotes have been observed in one female gametophyte.

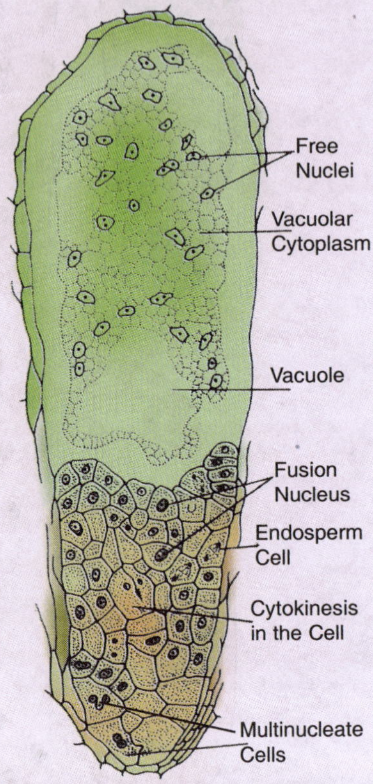

Fig. 16.33. *Gnetum africanum.* Female gametophyte showing lower cellular portion whereas upper part is free nuclear (After Weterkeyen).

Endosperm Formation (Fig. 16.34)

It has been stated earlier that before the act of fertilisation only the lower part of the gametophyte becomes cellular. After fertilization the upper part also becomes cellular so that the whole of the gametophyte consists of nucleated cells. The zygotes are surrounded by cells in various stages of their further development into embryo. Wall formation is by cleavage. An interesting feature of the endosperm is that its cells become multinucleate and these nuclei later fuse with one another to form polyploid cells. In some cells two nuclei fuse whereas in others three or even more fuse together. This is called **gradate ploidy** and the nuclei vary from haploid to 12 **n** in the same gametophyte. In *G. gnemon* the female gametophyte only becomes partly cellular. So *Gnetum* shows the following departures from the gametophytes of other gymnosperms :

1. It becomes only partly cellular before fertilization, only the chalazal part is cellular. In some species, it is wholly free nuclear before fertilization.
2. Cell formation is completed after fertilization.
3. The cells are mulinucleate and the nuclei fuse with each other resulting in gradate ploidy.
4. Free nuclei or as in *G. ula* specialised cells act as eggs. There are no archegonia or female sex organs.
5. The gametophytes are tetrasporic in development.

Gnetum resembles angiosperms in that the endosperm formation is completed only after fertilization. The endosperm in *Gnetum*, however differs from angiosperm in the following respects:

1. The endosperm in some species *e.g.*, *G. ula* becomes partly cellular before sterilization. It illustrates a stage partly resembling gymnosperms and partly angiosperms.
2. In *Gnetum* there is no fusion of nuclei before endosperm formation as occurs in angiosperms where the two polar nuclei fuse with each other to form a secondary nucleus which also fuses with the second male gamete to form the primary endosperm nucleus (double

fertilisation). In *Gnetum* nuclear fusions take place between nuclei within endosperm cells resulting in gradate ploidy, there being no male nucleus taking part in these fusions.

3. In *Gnetum* all the gametophytic nuclei take part in the formation of endosperm, whereas in angiosperms only two polar nuclei take part in endosperm formation.

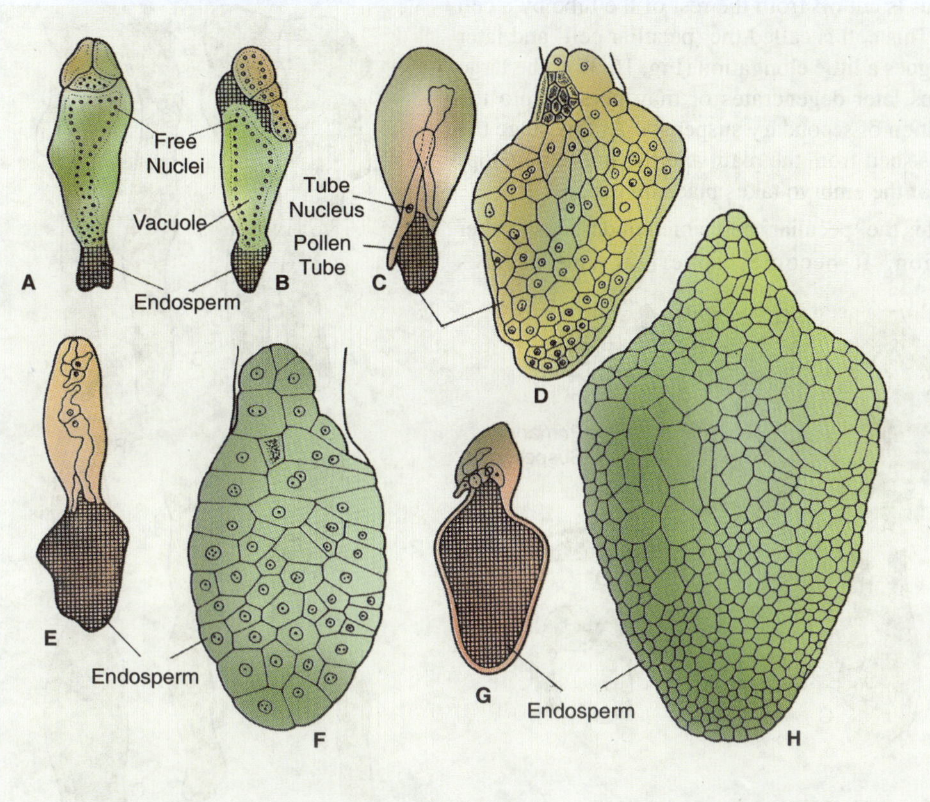

Fig. 16.34. (A-H) Formation of endosperm in *Gnetum gnemon.* A, B, C, E and G. Show development of female gametophyte (L.S) with upper free nuclear and lower cellular parts; D. it is lower cellular part of C where all nuclei have degenerated; F. Endosperm part of E; H. It shows details of G where endosperms develops throughout the female gametophyte. (After Madhulata).

EMBRYOGENY (Figs. 16.35—16.39)

There are many variations in the development of embryo in various species of *Gnetum.* In this text *G. ula* is selected as a type to illustrate the embryogeny of the genus. Embryogeny in this species has been extensively studied by Vasil (1959). The following account is based upon her observations.

The zygote in this species is a large cell with a prominent nucleus and dense cytoplasmic contents. The zygote divides by a wall into two cells. Both these cells elongate to form tube-like structures (Fig. 16.32, B). One of the two tubes thus formed divides into two or three cells, earlier than the other tube (Fig. 16.35). These tubes are designated as **primary suspensor tubes**. The upper part of the zygote becomes bulbous (Fig. 16.35). These tubes elongate considerably and may even branch repeatedly. The tip of each tube contains dense cytoplasm and a single nucleus. Large number of these tubes are present in the endosperm as a result of branching of original tubes. Some of them penetrate through the endosperm, others pierce through the female gametophyte and penetrate the nucellus. Such tubes usually degenerate. Majority of the tubes grow down into the centre of the endosperm and can be easily dissected out. These primary suspensor tubes become variously coiled. It takes six to seven months for the zygote to reach this stage. After this the tubes cease to branch.

The nucleus at the tip of the tube divides into two unequal nuclei (Fig. 16.37, A). The smaller nucleus is cut off from the rest of the tube by a cell wall. This cell is called the 'peculiar cell' and later undergoes a little elongation (Fig. 16.37). The large nucleus later degenerates or may persist upto the formation of secondary suspensor. At this stage the seed is shed from the plant and the future development of the embryo takes place on the soil.

It is the 'peculiar' cell which undergoes further division. It becomes spherical and divides periclinally into cells that lie one above the other (Fig. 16.37, B). Sometimes it may divide anticlinally so that the resulting two cells lie side by side. The second division results in the formation of four cells (Fig. 16.37, C). By this time the upper part of the wall of the tube becomes thick. The wall at the tip remains thin. The four cells that formed may be of equal size or the upper ones may be larger than the lower ones. The third transverse division leads to the formation of an octant (Fig. 16.37, C). Further divisions are irregular and form a mass of cells at the tip of the tube (Fig. 16.37, D). This mass increases in size by further divisions. The cells lying towards the tube increase in size to form the elongated **secondary suspensors** (Fig. 16.38, A).

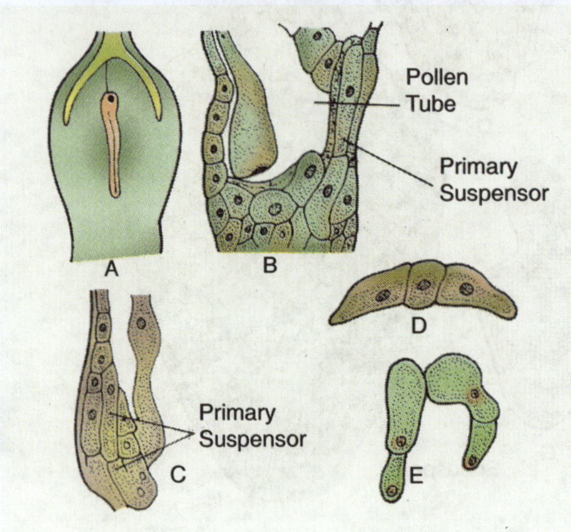

Fig. 16.35. (A-E). *Gnetum ula*. A. L.S. ovule after fertilization; B. The zygote has divided to form two primary suspensor tubes; C. It shows three primary suspensor tubes in one gametophyte; D. Two primary suspensor tubes. One of them has divided into two; E. Both the primary suspensor tubes become two celled. (After Vasil)

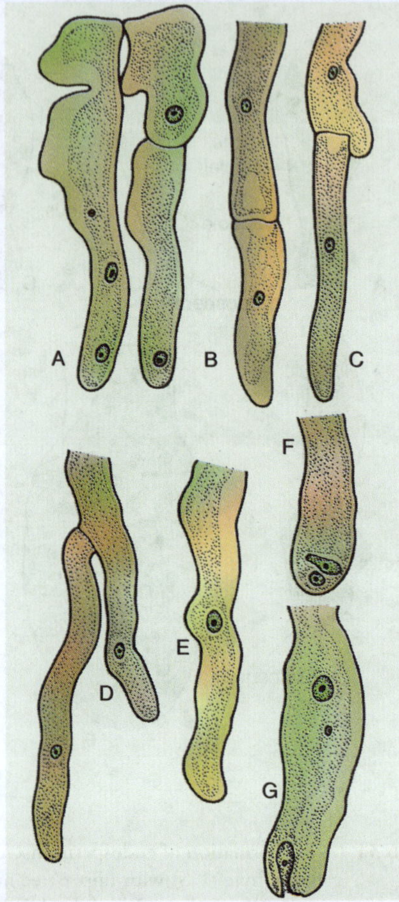

Fig. 16.36. (A-G). *Gnetum ula*. A. The primary suspensor tubes have divided to become two celled; B. two celled primary suspensor tube has elongated; C–D. The primary suspensor tubes divide to form additional primary suspensor tubes; E. Portion of primary suspensor tube; F. Two nuclei at one tip of the primary suspensor tube; G. Formation of a peculiar cell. (After Vasil)

These cells are thin-walled, uni-nucleate and have vacuolate contents. The cells at the tip remain small and have denser cytoplasmic content (Fig. 16.38). These develop into the embryo proper. The secondary suspensor cells elongate and push the embryonal cells deep into the endosperm tissue. The primary suspensor tubes cease to elongate. The secondary suspensor cells grow considerably and divide to form a slightly folded plate of cells (Fig. 16.38, C). Large number of primary suspensor cells develop but only one or two reach this stage. The rest degenerate. The stem tip differentiates at the end of the embryonal mass. It is surrounded by two cotyledons that develop from

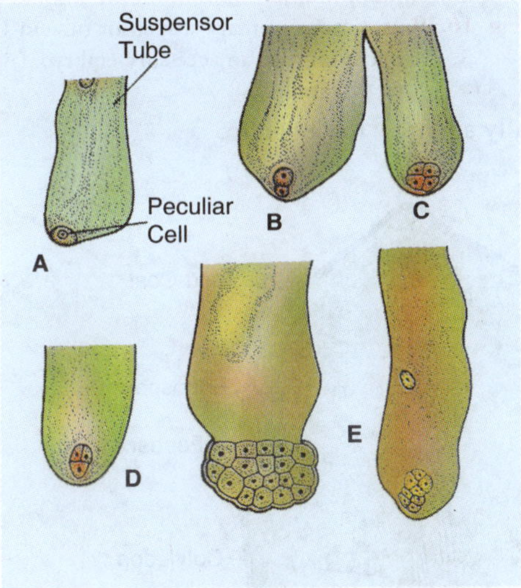

Fig. 16.37. (A-E). *Gnetum ula*. A–E. Stages in embryology illustrating the formation and division of peculiar cells at the tip of primary suspensor tubes. (After Vasil).

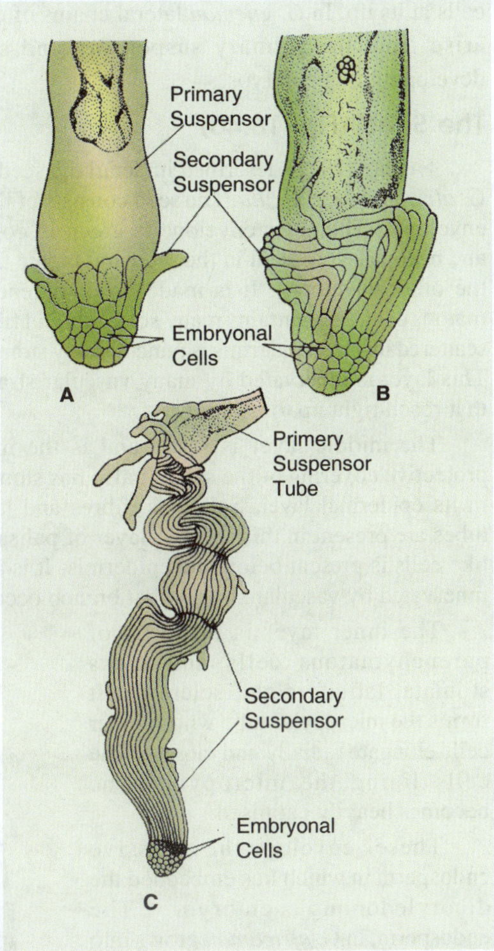

Fig. 16.38. (A-C). Later stages in embryo development in *Gnetum ula*. A–B. Secondary suspensors formed by the division of upper cells formed from the peculiar cell; C. Portion of primary suspensor bearing large numbers of secondary suspensors that have elongated and coiled. Also note the embryonal cells at the tip.

cells surrounding it. The root tip differentiates lower down at the other end so that the cells of the root cap are in touch with the secondary suspensor cells. A small hump appears between the stem tip and the root tip. It later enlarges and becomes distinct and is called the 'feeder'. The feeder has its own epidermis, cortex and vascular tissue. There is a distinct hypocotyl, which in earlier stages of development is longer than the hump but at a later stage the feeder becomes more prominent. The suspensors shrivel up and appear as fine threads attached to the root tip or the radicle.

There is polyembryony in *Gnetum* because large number of zygotes start development and moreover each zygote produces many tubes, the tip of each tube develops into an embryo. Ultimately only one reaches maturity. Additional embryo may also develop from secondary suspensors. Sometimes the primary suspensor tubes (Fig. 16.39) may produce more than one mass of embryonal cells at its tip. In *G. gnemon* lateral chains of cells arise from the primary suspensors and start developing into embryos.

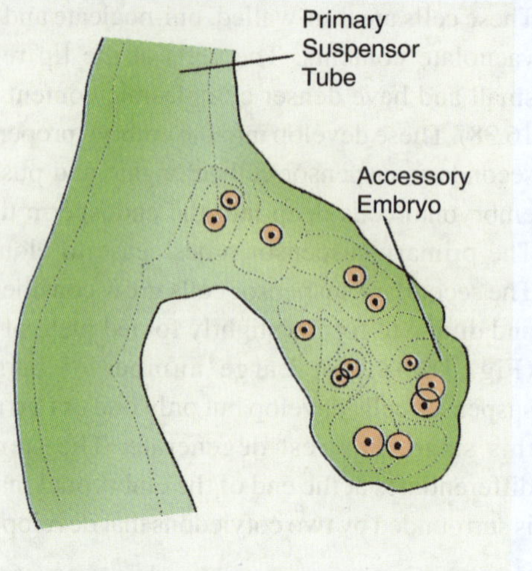

Fig. 16.39. It shows a primary suspensor tube in *G. gnemon*, with an accessory embryo. (After Madhulata)

The Seed (Fig. 16.40)

Thoday (1911) described in detail the seeds of *G. africanum* and *G. ula*. The seed consists of three envelopes. The outer envelope is green in colour and becomes succulent in the seed and is free from the other envelopes. It is made up of parenchymatous cells and contains many sclereids and fibres scattered in it. Its epidermis is punctured by stomata. This layer is innervated by many vascular strands that reach right up to its apex.

The middle layer is stony and is the main protective covering of the seed. It also has stomata in its epidermal layer. Sclereids, fibres and latex tubes are present in this layer. A layer of palisades like cells is present below the epidermis. It is also innervated by vascular strands that branch occa- sionally and may reach its lip.

The inner layer is made up of parenchymatous cells and lacks stomata, laticifers and sclereids. It forms the micropylar tube whose inner cells elongate radially and close it. The cells lining the micropylar tube becomes heavily cutinised.

These envelop the massive endosperm in which lies embedded the dicotyledonous embryo. The endosperm, in *G. africanum* grows into a **tent-pole** like protuberance at its apex. This protuberance touches the nucellar cap.

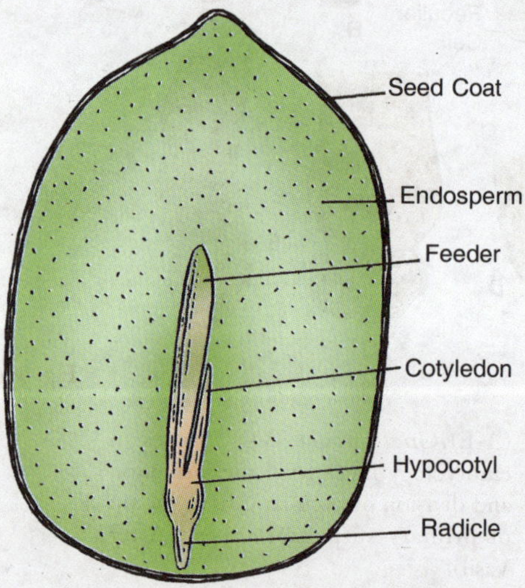

Fig. 16.40. L.S. Seed of *Gnetum* sp.

The germination of the seed is epigeal (Fig. 16.41). The hypocotyl elongates and pushes the cotyledons out of the seed and then carries them above the soil. The feeder remains within the seed. The cotyledons form the first green leaves. Later the plumule grows and produces the first pair of foliage leaves that lie at right angles to the cotyledons. The cotyledons wither at a later stage.

Chromosome Numbers

The haploid number 12 has been reported for *G. gnemon* (Coulter, 1908), *G. ula*, *G. africanum* and *G. buccholzianum* (Pearson, 1912). Fagerlind (1941) reported $n = 22$ in *G. gnemon*. Weterkeyn (1959) reported $n = 20$ to 25 in *G. africanum*.

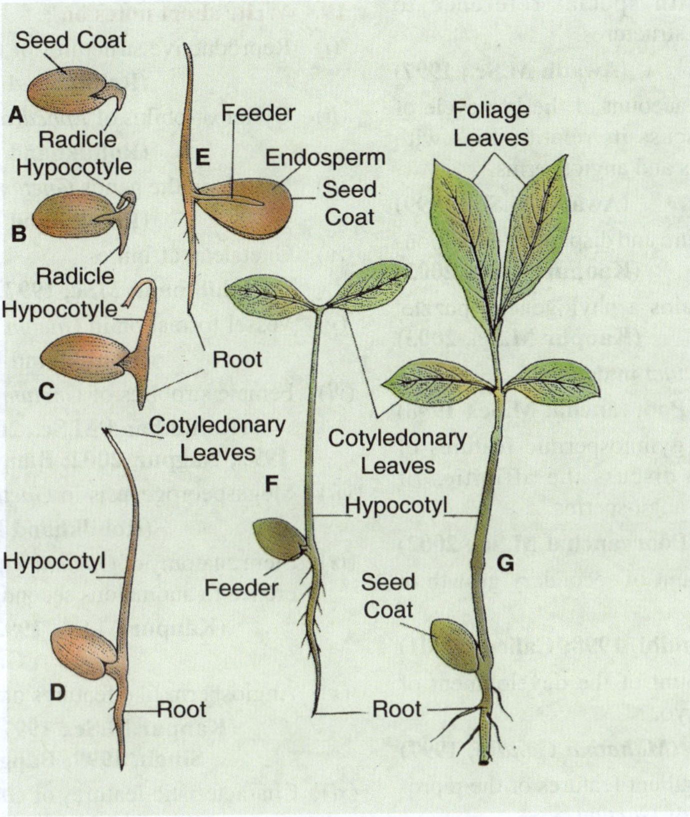

Fig. 16.41. (A-G) Various stages in the germination of seed in *Gnetum* sp.

Economic Importance

The seeds of *G. gnemon, G. ula* and *G. lallipfolium* are roasted and eaten. The young leaves and strobili of *G. gnemon* are cooked as vegetable. The bark of this species, yields fibre of high tensile strength. These are used for making ropes, fishing nets. The fibres are also beaten into pulp that is used for making paper. The kernels of *G. ula* yield an oil used for illumination and also as a massage in rheumatism. It is also used for cooking purposes. The plant of *G. montanum* is used as fish poison.

QUESTIONS

ESSAY TYPE

1. Mention the angiosperm like characters of *Gnetum* and justify its inclusion in gymnosperms. **(Rohilkhand M.Sc., 1994)**
2. Describe the important features of the order Gnetales and discuss its affinities with other seed plants. **(Kanpur M.Sc., 1999)**
3. Discuss the phylogeny and inter-relationships of Gnetales. **(Kanpur M.Sc., 1990, 1994; Mahatma Gandhi, 1997)**
4. Describe the order Gnetales with special reference to its affinities and relationships to the Gymnosperms and Angiosperms. **(Awadh M.Sc., 1997)**

5. Compare the three genera of order Gnetales with special reference to reproductive structures. **(Awadh M.Sc., 1997)**
6. Give a brief account of the life cycle of *Gnetum*. Discuss its relationships with gymnosperms and angiosperms. **(Awadh M.Sc., 1999)**
7. *Gnetum* is a lure and dispair for evolutionists Comment. **(Kanpur M.Sc., 2002)**
8. *Gnetum* remains a phylogenetic puzzle. Discuss. **(Kanpur M.Sc., 2003)**
9. Describe *Gnetum* in detail. **(Poorvanchal M.Sc., 1998)**
10. Mention the gymnospermic features of *Gnetum* and discuss the affinities of *Gnetum* with angiosperms. **(Poorvanchal M.Sc., 2002)**
11. Give an account of secondary growth in *Gnetum*. **(Mahatma Gandhi, 1998; Calicut, 2001)**
12. Give an account of the development of *Gnetum* embryo. **(*Mahatma Gandhi*, 1997)**
13. Mention the salient features of the reproductive parts of *Gnetum*. **(VirKumar Singh, 1997)**
14. Give an account of the development of female gametophyte of *Gnetum*. How does it approach near angiospermic level ? **(VirKumar Singh, 1998)**
15. Write the important anatomical features of *Gnetum* and point out its similarities with angiosperm. **(Bhagalpur, 1991)**
16. Explain the female cone of *Gnetum* and add a note on the ovule structure. **(Bangalore, 2002)**
17. Why *Gnetum* wood is considered advanced gymnospermic wood ? Discuss. **(Osmania, 2002)**
18. Describe the salient features of *Gnetum* and point out its resemblances with angiosperms. **(Utkal, 2003)**

SHORT ANSWER TYPE

19. **Write short notes on :**

(*i*) Reproductive structures of *Gnetum*. **(Rohilkhand M.Sc., 1991)**

(*ii*) Gymnostrobilus of *Ephedra* and *Gnetum*. **(Rohilkhand M.Sc., 1993)**

(*iii*) Vessels in the genus *Gnetum*. **(Rohilkhand M.Sc., 1995)**

(*iv*) Gnetalean affinities. **(Rohilkhand MSc, 1997, 2001, 2002)**

(*v*) Vessel formation in *Gnetum*. **(Rohilkhand M.Sc., 1998)**

(*vi*) Female strobilus of *Gnetum*. **(Rohilkhand M.Sc., 2002; Calicut, 1997; Nagpur, 2002; Bangalore, 2001)**

(*vii*) Megaspeorogenesis in *Gnetum*. **(Rohilkhand M.Sc., 2003)**

(*ix*) Stem anatomy of *Gnetum* with special reference to anomalous secondary growth. **(Kanpur M.Sc., 1993; Mahatma Gandhi, 1996)**

(*x*) Angiosperm like features of *Gnetum*. **(Kanpur M.Sc., 1997; VirKumar Singh, 1999; Bangalore, 2001)**

(*xi*) Characteristic features of *Gnetum*. **(Kanpur M.Sc., 1998)**

(*xii*) Embryogeny in *Gnetum*. **(Poorvanchal M.Sc., 2001)**

(*xiii*) Resemblances of *Gnetum* to angiosperms. **(Mahatma Gandhi, 1998)**

(*xiv*) Secondary growth in *Gnetum*. **(Vinoba Bhave, 1998)**

(*xv*) Male gametophyte of *Gnetum*. **(Vinoba Bhave, 1999)**

(*xvi*) Ovule of *Gnetum*. **(Kerala, 2000)**

(*xvii*) Male cone of *Gnetum*. **(VirKumar Singh, 1995; Calicut, 1999; Garhwal M.Sc., 2004; Nagpur, 2001)**

(*xviii*) Structure of Root in *Gnetum*. **(VirKumar Singh, 1995)**

(*xix*) Flower of *Gnetum*. **(VirKumar Singh, 1998; Bhagalpur, 1991; Osmania, 2002; Utkal, 2003)**

(*xx*) Distribution of *Gnetum* species in India. **(Calcutta, 2002)**

(*xxi*) Seed of *Gnetum*.
(*xxii*) Development of embryo.
(*xxiii*) Economic importance of *Gnetum*.
(*xxiv*) Wood in *Gnetum*.

20. Draw labelled diagrams of the following :

(*a*) L.S. Ovule of *Gnetum*. **(Poorvanchal M.Sc., 1996)**
(*b*) T.S. RLS and TLS of *Gnetum* wood. **(Bhagalpur, 1990)**
(*c*) *Gnetum* Ovule, L.S. and label part.
(*d*) T.S. young stem.
(*e*) T.S. old stem.
(*f*) T.S. Leaf.
(*g*) Stomata in *Gnetum*.
(*h*) Pollen grain in *Gnetum*.
(*i*) L.S. seed of *Gnetum*.

21. Draw a neat labelled diagram of *Gnetum* L.S. Ovule. Describe its parts. **(Calicut, 2001)**

22. Mention special features of :

(*a*) Nutrition of *Gnetum* Embryo sac. **(Mahatma Gandhi, 1998)**
(*b*) Venation in *Ephedra*. **(Mahatma Gandhi, 1996)**
(*c*) Secondary growth.
(*d*) Stem of *Gnetum*.
(*e*) Habit of *Gnetum*.

23. Name any parasitic gymnosperm. **(Poorvanchal M.Sc., 1998, 2003)**

24. Name the author of the book "*Gnetum*". **(Poorvanchal M.Sc., 2003)**

25. Which of the following has longest micropyle among plant kingdom ?

(*a*) *Gnetum*
(*b*) *Ephedra*

26. Name any two gymnosperms in which the secondary wood contains vessels in addition to tracheids. **(Poorvanchal M.Sc., 1995)**

27. Discuss the similarities of the leaves of *Gnetum* to that of angiosperms. **(Calicut, 1998)**

28. What is the peculiarity of secondary thickening in *Gnetum* ? **(Calicut, 1998)**

29. Draw L.S. of a mature ovule of *Gnetum* at the time of fertilisation and label the parts. Briefly describe its structure. **(Calicut, 1999)**

30. Describe the advanced characters of *Gnetum*. **(Mahatma Gandhi, 2004)**

31. Mention the venation of leaves in *Gnetum*.

32. Describe the structure of pollen grain in *Gnetum*.

33. Despite the presence of a carpel like envelope in the female flower of *Gnetum*, the genus is not included in angiosperms. Give reasons.

34. How do the vessels in *Gnetum* differ from those of angiosperms ?

35. Name a gymnosperm which has only two leaves.

36. Name the species of *Gnetum* in which double fertilisation takes place.

OBJECTIVE TYPE

Select the correct answer :

(*i*) *Gnetum* when not in flowering can easily be mistaten for
(*a*) Treefern
(*b*) Dicot plant
(*c*) Monocot plant
(*d*) None of the above.

(*ii*) Vessels are present in
(*a*) *Gnetum* (*b*) *Ginkgo*
(*c*) *Araucaria* (*d*) *Zamia.*

(*iii*) Reticulate Venation is present in
(*a*) *Ginkgo* (*b*) *Gnetum*
(*c*) *Araucaria* (*d*) *Cycas.*

(*iv*) According to Maheshwari and Vasil the development of stomata in *Gnetum*
(*a*) Syndetocheilic (*b*) Haptocheilic
(*c*) Both (*c*) None.

(*v*) The pollen grain at the time of discharge from microsporangium is
(*a*) uninucleate (*b*) binucleate
(*c*) trinucleate (*d*) tetranucleate.

(*vi*) The fused bracts in male flowers of *Gnetum* are comparable to
(*a*) bracts of Angiosperms
(*b*) the perianth of angiosperm

(*c*) leaves of Angiosperms
(*d*) the bracteoles of angiosperm.

(*vii*) *Gnetum trinerve* is
(*a*) tree (*b*) shrub
(*c*) parasite (*d*) climber.

(*viii*) The female gametophyte of *Gnetum* follows the developmental pattern of
(*a*) Monosporic type
(*b*) Bisporic type
(*c*) Tetra sporic type
(*d*) None of the above.

(*ix*) Eccentric rings are the characteristic features of the wood of *Gnetum* because
(*a*) Secondary growth is normal
(*b*) Secondary cambia generally originate from the cortex
(*c*) Activity of cambium is not uniform on the sides
(*d*) Cambium does not function for a long time.

(*x*) In spite of several angiospermic features, *Gnetum* is a gymnosperm because
(*a*) the seed is naked
(*b*) polyembryony is a common feature
(*c*) prothalleal cell is present
(*d*) the stem shows anomalous secondary growth.

(*xi*) The pavement tissue in the ovule of *Gnetum* is developed from
(*a*) Nucellus
(*b*) female gametophyte
(*c*) integument
(*d*) Male gametophyte.

CHAPTER 17

Welwitschiaceae

- Welwistschia : Distribution, Anatomy and Reproduction
- Difference Between
- Gnetum and Ephedra
- Gnetum and Welwitschia
- Gnetum and Cycadeoidales
- Gnetum and Angiosperms
- Affinities of Gnetopsida

WELWITSCHIA SP.

GENUS : WELWITSCHIA

Distribution

Welwitschia bainesii (= *W. mirabilis*) is the only representative of the family that has a restricted distribution. The marvellous plant resides only along a narrow coastal belt in South West Africa. It is an example of an extreme xerophyte that usually grows in sand or on bare rocks or rocky beds of streams that are almost always dry. It has been found growing at localities where annual rainfall does not exceed one inch. In Walvis Bay locality of Welwitschia the rainfall during 10 years was recorded to be $\frac{1}{3}$.

Plate I : *Welwitschcia bainesii*

Habit : The sporophyte (Fig. 17.1) has peculiar habit. It is differentiated into stem, root and leaves. The stem or the so called 'axis' is mostly underground, sometimes projecting a little above ground. It looks like a huge turnip whose outline appears to be elliptical when seen from above. It reaches a maximum diameter of about 1.2 meters and may be completely subterranean or a portion may grow above ground. It has distinct depression at its apex. The portion of the stem around the depression forms the rim. The underground portion of the stem tapers abruptly into a narrow portion that continues into an extremely long tap root that reaches the water table. The plant bears only two leaves that are opposite and decussate and grow continuously from a basal meristem. These leaves appear along the upper rim of stem. The plant produces only three pairs of foliar organs during its life time. The first is a pair of cotyledons (Fig. 17.2, B) that live only for a short while, then appear two large leaves and after these two more leaf primordia appear, a little before the apical meristem ceases to function. All these are opposite and decussate in arrangement. The leaf primordia that appear last of all form the characteristic 'scaly bodies' which have been regarded by some as buds in the axils of cotyledons and by others as cones. Martens and Waterkeyen (1963) studied the stem apex of *Welwitschia* and observed them to be the leaf primordia.

The principal leaves that appear after the cotyledons persist throughout the life of the plant that is estimated to be one thousand years. They stand erect in young plants (Fig. 17.2) but in old plants they reflex and grow along the substratum. (Fig. 17.1). The young leaves are entire (Fig. 17.2, A, B) but as they grow old they become thick and leathery and split along their length. The leaves are parallel veined but anastomosing lateral veins also appear (Rodin, 1953, 58). The split regions soon become corky so that splitting causes no harm to the plant. Old plants possess leaves as long as two meters.

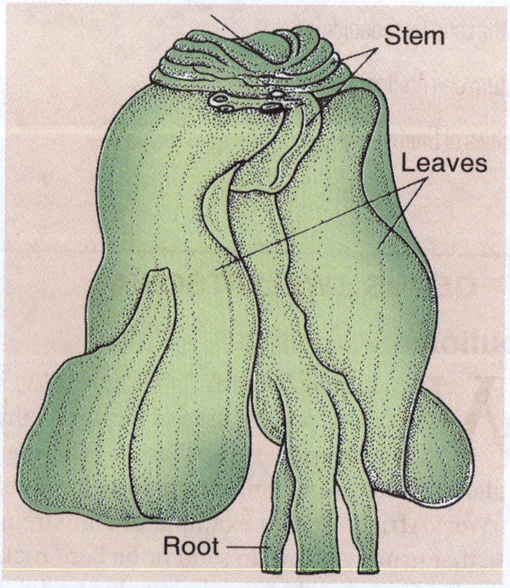

Fig. 17.1. An old plant of *Welwitschcia bainesii* (= *W. mirabilis*).

The stem is covered with a thick bark and may be as thick as two centimeters in the upper region. Portion of the stem above the insertion of leaves is marked by a series of ridges and each of them extends almost half the circumference of the stem. The ridges bear scars of old and fallen inflorescences. Such ridges also appear on the stem below the point of insertion of the leaves but these are not so prominent. The portion of the stem above the leaves is called the **crown** whereas the part below the point of insertion of the leaves is called the **stock**.

At the base of each leaf and well within a slit or the apical groove, there is a meristematic zone that adds to the leaf base as well as to the tissue inside the apical groove. This tissue expands and matures up and over the edges of the groove.

The apical meristem is quite active in the young meristem but shows no definite zenation because of the appearance of periclinal walls at levels. This meristem later becomes functionless and the growth is restricted to the periphery affected by the lateral meristem and causes only increase in the thickness of the stem. Meristem at the base of leaf may also frequently become injured and discontinuous. In such cases the individual leaf segments (formed by splitting) continue to grow independently.

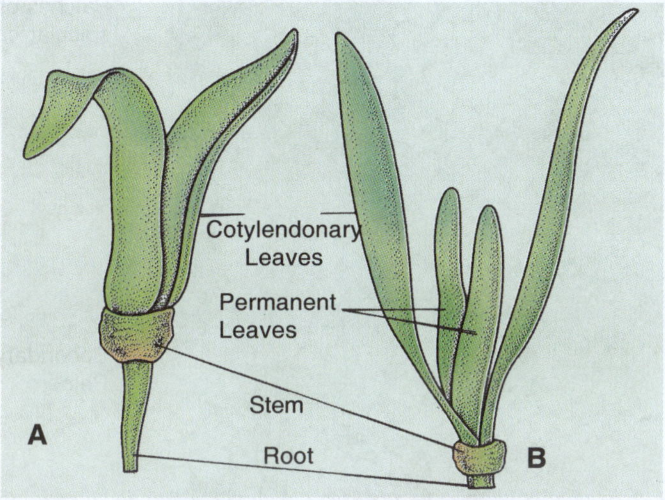

Fig. 17.2 (A-B). *Welwitschia bainesii.* A young plant with two erect leaves; B–a A seedling with two cotyledons and two permanent leaves.

The inflorescence axes usually arise on the series of transverse that mark the crown. They are produced mostly on the inner, but a few on the outer side of the plant crown. The fallen inflorescence axes leave scars on the ridges.

Plate I(A) : Plant of *Welwitschia bainesii*

ANATOMY

Root (Fig. 17.3) : The young root has an extensive cortex with a distinct endodermis. The stele is diarch. There is abundant secondary growth. Cork cambium is formed at a later stage and produces thick layers of cork. Spicular cells are abundant.

Stem : The young stem (Fig. 17.4) has a ring of conjoint collateral, endarch and open vascular bundles with a distinct pith in the centre. The pith and the cortex contain the characteristic spicular cells that are extremely thick-walled and incrusted with crystals of calcium oxalate (Fig. 17.5). These cells in old stem have inner cellulosic walls and outer highly lignified walls. They branch and interlock with each other. In older stems the stele is cup-shaped and consists of two plates of collateral vascular bundles, from which depart leaf traces to the leaves, inflorescence axes and the single tap root. The cortex is parenchymatous but several groups of thick-walled fibres occur scattered in it.

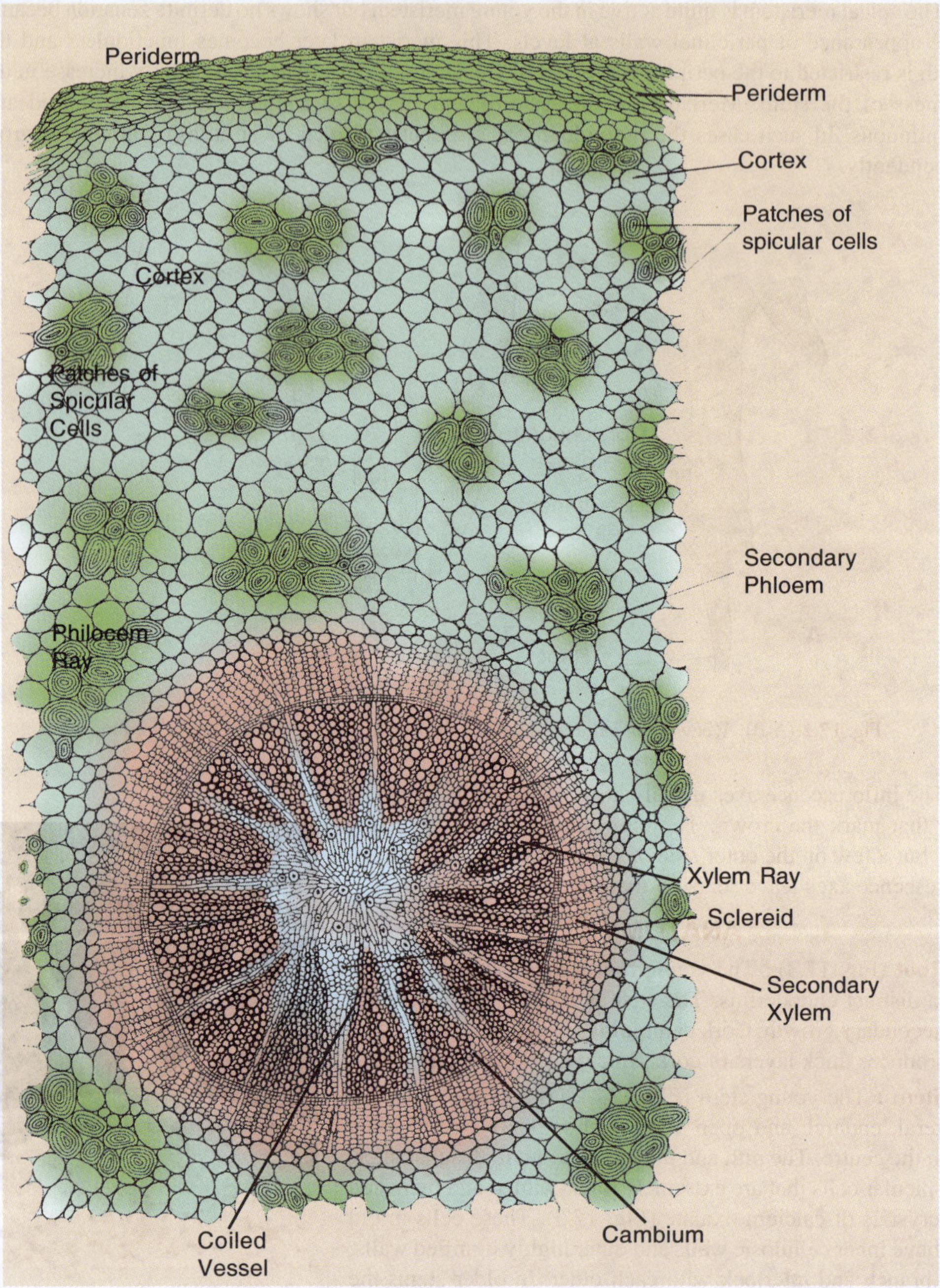

Fig. 17.3. T.S. old root of *Welwitschia bainesii.*

The primary cambium is short lived and becomes functionless after producing a ring of secondary phloem and dxylem. Later successive rings of cambia appear outside the primary ring, in the cortex and form successive incomplete rings of secondary vascular bundles separated by broad medullary (Fig. 17.7) rays. In this respect Welwitschia resembles Gnetum. Cork cambium appears and is subepidermal in position. It cuts off thick layers of cork. The secondary wood is thus polyxylic. The bundles are capped by thick-walled fibres. It is light and not hard. Another peculiarity is the occurrence of isolated, un-connected vessels of upto 12 cells, that are coiled up into worm-like masses (Fig. 17.6).

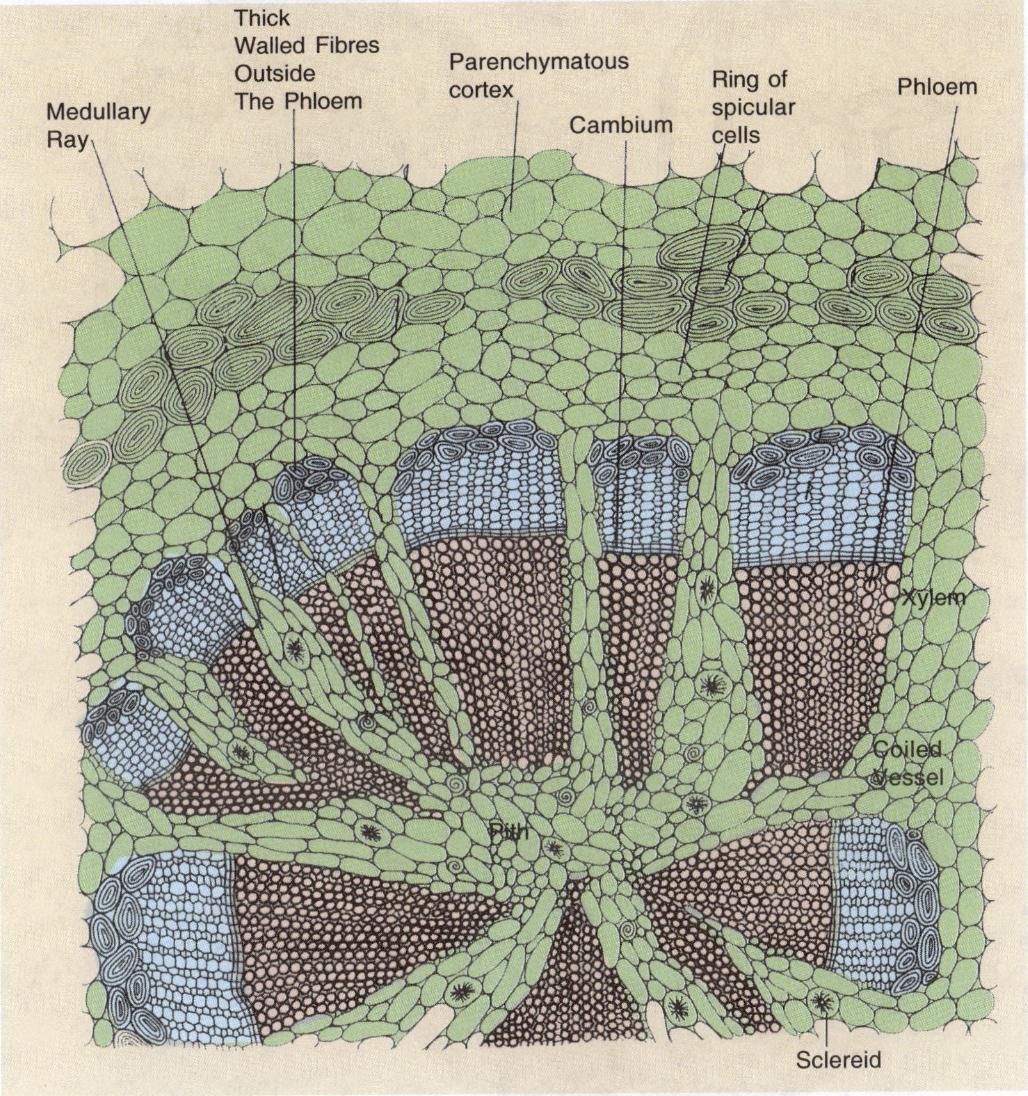

Fig. 17.4. T.S. young stem *Welwitschia bainesii.*

The tracheary elements that develop early have circular bordered pits between the gyres of secondary wall spirals or helix. The vessels are present with usually singly-occurring circular pores on the end walls as well as on lateral walls. The pores may be transversely or obliquely oriented. A few vessel members show pores in pairs and in rare cases a foraminate plate with three pores also occurs. In the later formed xylem elements the walls become thick and possess a complex ramiform system of bordered pits. Xylem parenchyma is quite abundant. Fibre-like cells also occur scattered around the xylem.

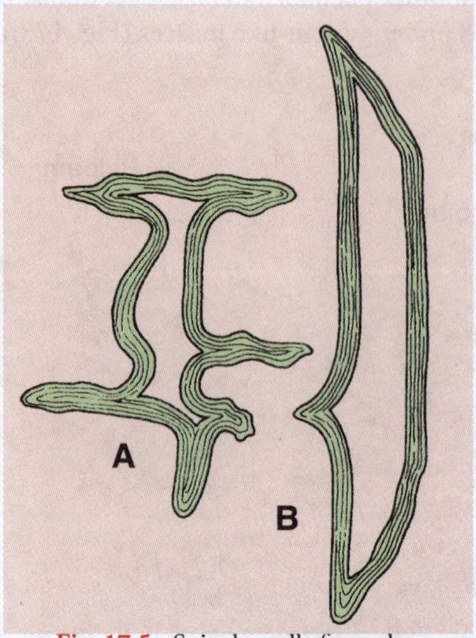

Fig. 17.5. Spicular cells from the stem of *Welwitschia*

Fig. 17.6. Coiled vessel from the stem of *Welwitschia*.

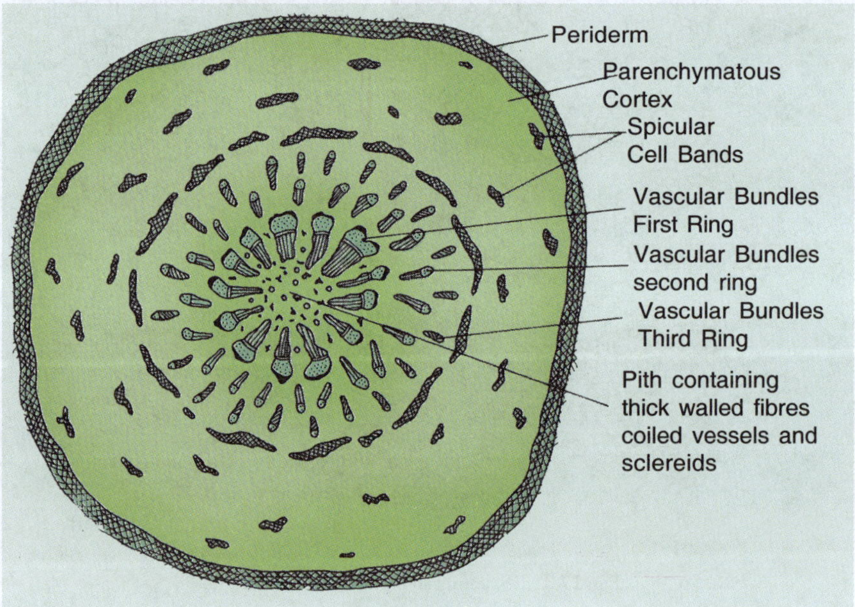

Fig. 17.7. T.S. (diagrammatic) old stem of *Welwitschia* showing secondary growth.

The phloem consists of sieve cells and phloem parenchyma. The sieve cells have sieve areas on the end walls and also sometimes on the side walls. The sieve cells have prominent nuclei.

Leaf (Figure 17.8) : The leaf has a lower and an upper epidermis. The former is covered with a thick cuticle. The stomata are found on both the epidermal layers and are syndetocheilic in development. Below the upper epidermis and above the lower epidermis there are groups of sclerenchyma fibres alternating with palisade tissue that is strongly developed. In between there is a spongy tissue of thin-walled cells. They also contain chloroplasts and enclose small air spaces. Spicular cells are also present in this tissue.

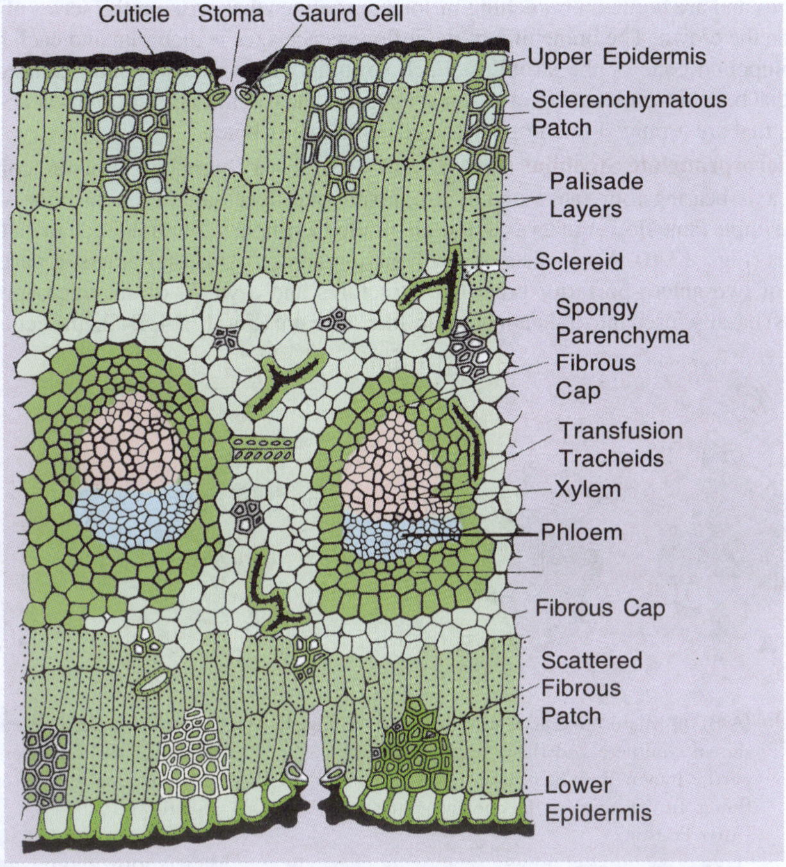

Fig. 17.8. T.S. leaf *Welwitschia*.

The vasculature of the leaf consists of a row of collateral bundles with phloem facing lower epidermis and xylem towards the upper side. Each vascular bundle is surrounded by a transfusion tissue and is capped at either end by groups of thick-walled cells (Fig. 17.8). Small interconnecting bundles (Rhodin, 1953, 1958) are cut lengthwise.

REPRODUCTION

Welwitschia is dioecious. The microsporangiate and megasporangiate strobili are compound

Plate II : Simple male strobilus or flowers of Welwitschia.

structures and are borne on branching inflorescence axes that arise on the series of transverse ridges borne on the crown. The branching of the inflorescence axes is dicharial and each branch terminates in a compound cone or the strobilus. Each strobilus bears opposite and decussate bracts or cone scales that become scarlet red or crimson at maturity. Each compound strobilus bears simple strobili or flowers that are arranged in an opposite and decussate manner.

Microsprangiate Strobilus (Figs. 17.9–17.13). It is a compound structure consisting of a four-angled axis bearing four ranks of broad opposite and decussate bracts or cone scales. Each bract bears a simple male flower in its axil (it is also called a simple strobilus). Each male flower or a simple strobilus (Figs. 17.10, A) consists of a subtending bract, two lateral bracts, a parianth formed by a fusion of two antero-posterior bract-like structures and a whorl of six microsporangiophores (or stamens) basally fused into a sheath of a cup-like structure (Fig. 17.14, B). In the centre there is a sterile

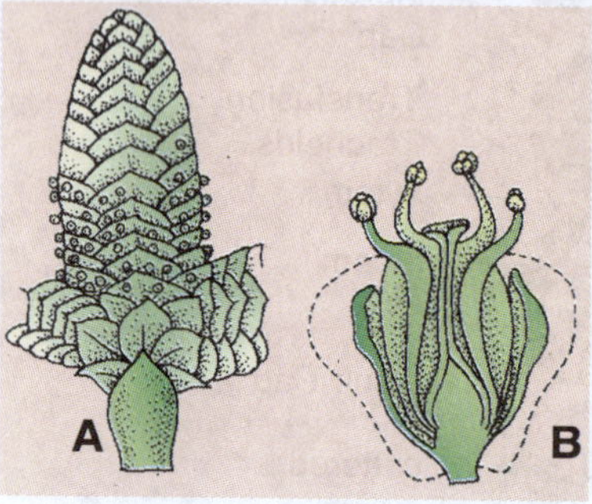

Fig. 17.9. (A-B). A–Microsporangiate strobili, one is shown complete and the other two are partly drawn; B; V.S. of a single male-flower. In the centre is the abortive ovule (After Pearson).

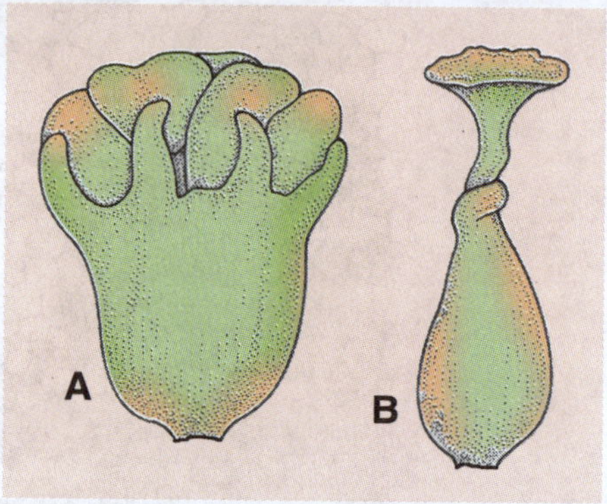

Fig. 17.10. (A-B) A–simple male stobilus or flowers of *Welwitschia* as seen from adaxial side; B–A non-functional ovule from the centre of the male flower.

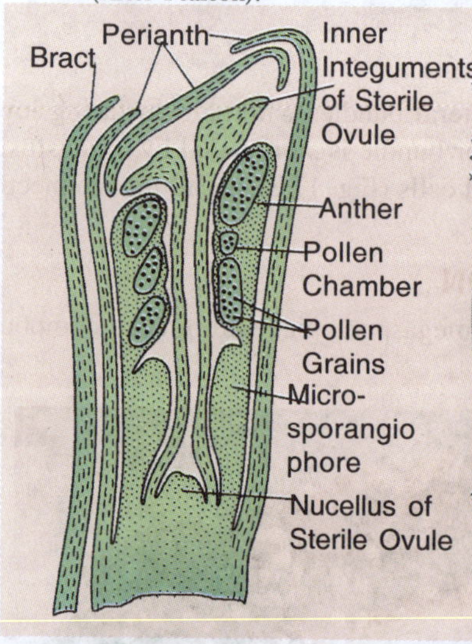

Fig. 17.11. L.S. of a simple male strobilus or flower of *Welwitschia* showing bracts, perianth, stamens and sterile ovule. The stamens have anthers with three lowli. Note the pollen grains also. (After Coulter Chamberlain)

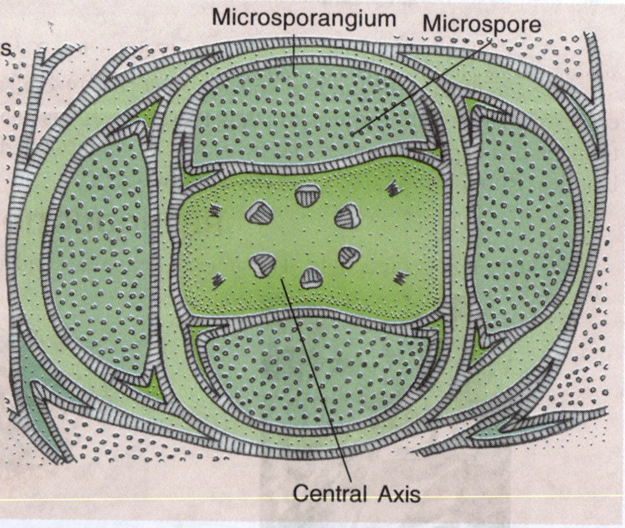

Fig. 17.12. T.S. through a compound male strobilus of *Welwitschia*. Note the central axis and the microsporangia around.

or non-functional ovule that has a small pair of scale-like bracts that appear as slight nipples at its base. The non-functional ovule (Fig. 17.10, B) has a single integument. Each microsporangiophore or the stamen terminates in a synangium made up of three fused sporangia (Fig. 17.10). The sporangia dehisce by means of a vertical slit, their lines of dehiscence radiating from near the point where the septa converge. Two vascular traces innervate the bracts as well as the male flowers.

Each microsporangium (Fig. 17.12) has an outer layer of cells or the epidermis followed by two tapetal layers enclosing large number of microspore mother cells that undergo meiosis to form tetrads of microspores. The young microspore is uni-nucleate and is surrounded by two spore coats, the outer exine and the inner intine. The exine is thick. The spore has no wings.

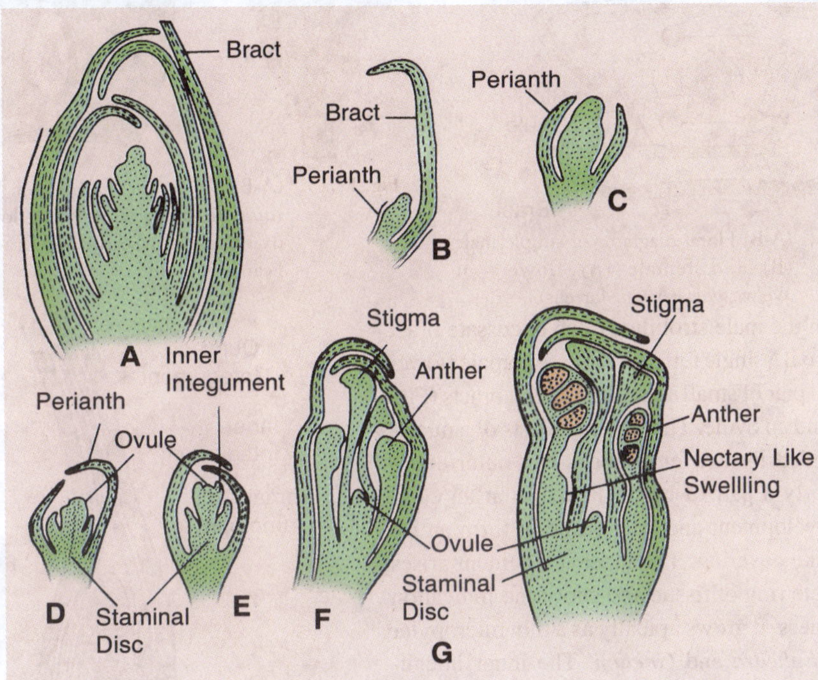

Fig. 17.13. (A–G) *Welwitschia banesii.* A; L.S. through apex or male strobilus; B–G. Stages in development of a simple male strobilus. (After Chamberlain).

Megasporangiate or Ovulate Strobilus (Figs. 17.14 and 17.15) : The megasporangiate strobilus is also a compound structure. It consists of an axis that bears broad decussate bracts. It is somewhat

Plate III : Male strobilus.

Plate IV : L.S. through a male strobilus.

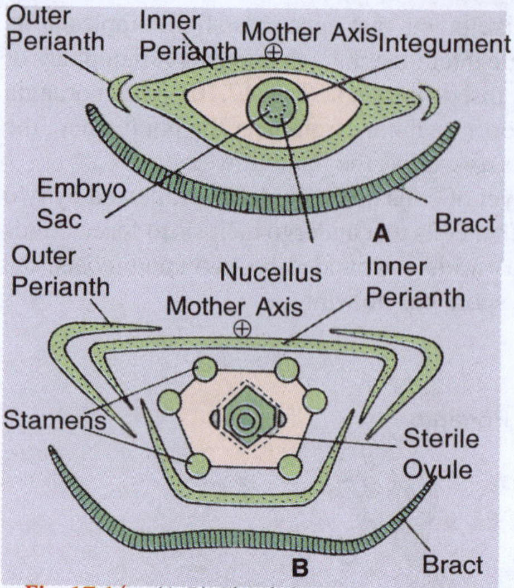

Fig. 17.14. (A-B) Floral diagrams of simple male (B) and female (A) flowers of *Welwitschia*. (After Martens).

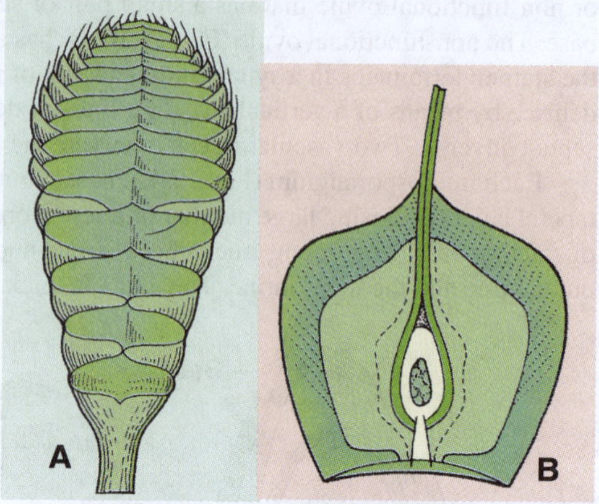

Fig. 17.15. (A-B). A. Female strobilus of *Welwitschia mirabilis*. B–A simple female flower with an ovule and surrounding bracts. (After Pearson).

thicker than the male strobilus. Each decussate bract bears in its axil a single female flower. A female flower consists of a pair of small scale–like lateral bracts (Fig. 17.14, A) and an ovule. The ovule consists of a nucellus surrounded by two envelopes. The outer envelope is clearly a pair of bracts that fuse at an early stage of development and later come to form an integument like covering. The inner integument arises as a complete ring of tissue and shows no indication of bipartedness. It grows apically as a thin micropylar tube as in *Ephedra* and *Gnetum*. The inner integument is fused with (Fig. 17.16) the nucellus in its lower part and is free above. The outer envelope is free (Fig. 17.16). The micropylar tube is lined by an inner epidermal layer of the inner integument. The nucellus contains embedded in it the female gametophyte that becomes cellular before fertilization and lacks archegonia. The cells of the female gametophyte are multinucleate. The nuclei fuse and result in gradate poloidy (Fig. 17.18) as in *Gnetum*. Numerous **embryo sac tubes** arise from the micropylar end of the female gametophyte, after pollination. They meet the pollen tubes during fertilisation. There is no nuclear beak. The pollen chamber is also absent. A slight depression is of course visible at the nucellar apex.

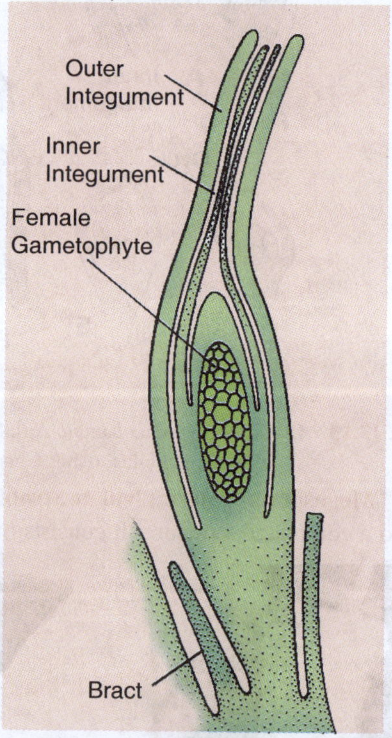

Fig. 17.16. V.S. of ovule of *Welwitschia mirabilis*. (After Pearson).

GAMETOPHYTE

Male Gametophyte (Fig. 17.17) : Microspore is the pioneer structure of the male gametophyte. It is released in a two nucleate condition. These nuclei are the **generative nucleus** and the **tube nucleus**.

There are no prothallial cells in this case. The pollen tube develops when the pollen grain falls on the nucellus. Later the tube nucleus and the generative nucleus migrate into it. The generative nucleus divides into two **male gametes**. According to Sterling (1963) the generative nucleus divides into stalk nucleus and body nucleus. The former degenerates even before pollination. The body nucleus becomes the body cell and migrates into the pollen tube where it divides into two male gametes.

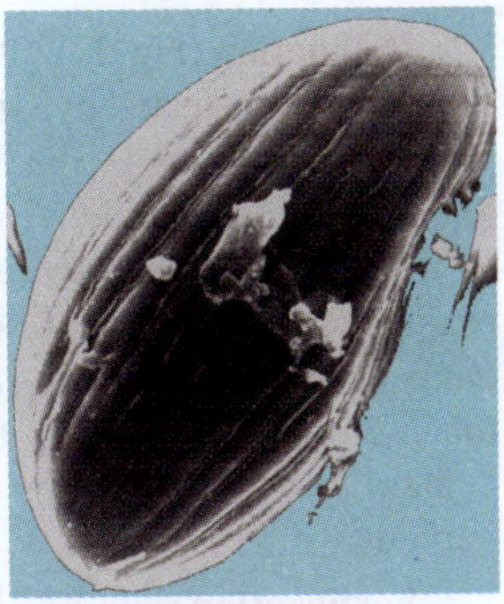

Plate V : Pollen grain of W. mirabili.

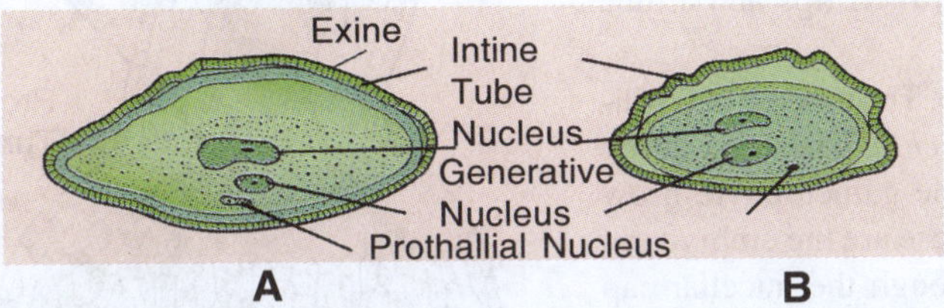

Fig. 17.17. *Welwitschia bainesii.* A–pollen grain at three nucleate stage. B–pollen grain at the time of release from the microsporangium. (After Pearson)

Female Gametophyte (Figs. 17.18 and 17.19). The female gametophyte is tetrasporic like *Gnetum*. A single megaspore mother cell distinguishes quite deep in the nucellus and undergoes meiosis to form four haploid nuclei that lie in its centre. No walls are formed and hence no tetrads of megaspores. Later free nuclear divisions take place to form as many as 1,000 free nuclei (Fig. 17.19, A). Later wall formation takes place to form a cellular female gametophyte. Wall formation is irregular and hence multinucleate cells are formed. Like *Gnetum* the nuclei in these fuse to form polyploid cells of varying degrees. There are no archegonia.

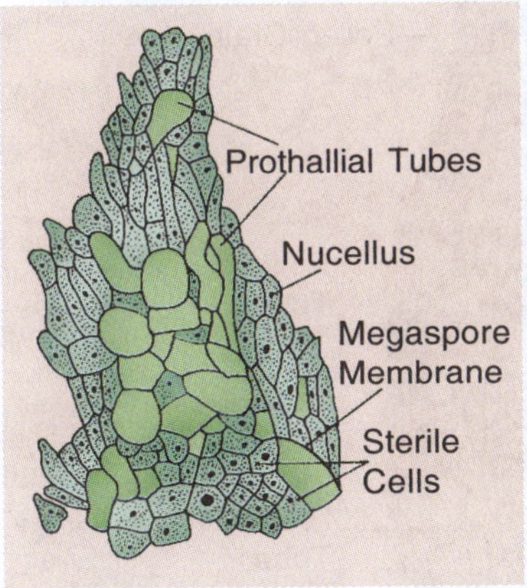

Fig. 17.18. *Welwitschia mirabilis.* Upper portion of female gametophyte showing prothallial tissue.

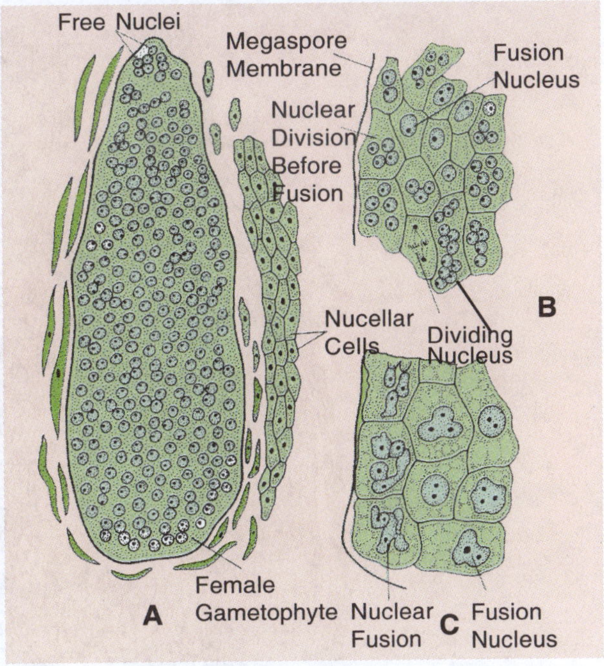

Fig. 17.19. *(A-C) Welwitschia bainesii.* A, it shows free nuclear stage of female gametophyte, B–C, these fig ures illustrate wall formation and gradate polo-idy. The nuclei fuse in cells to form polyploid.

Pollination (Fig. 17.20). It is not certain whether pollination is effected by wind or by insects. micropyle secretes a honey-like fluid into which the pollen grains are embedded. Insects have been seen to visit this drop. Due to evaporation of the fluid the grains are sucked in and lie on the nucellar apex. Here further germination takes place. The pollen tubes grow down through the nucellus. The tubes contain two male gametes at their tips and a tube nucleus.

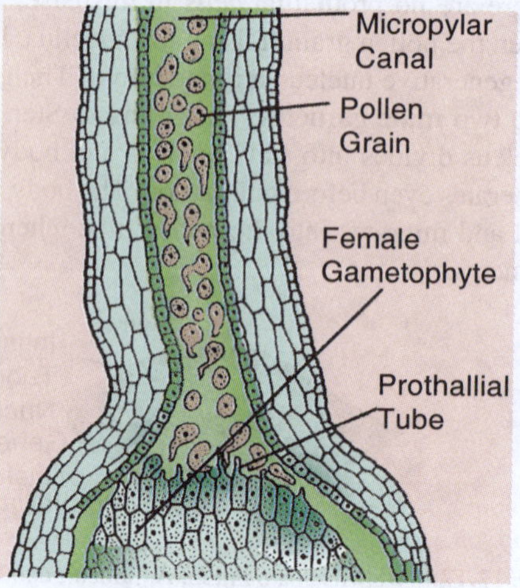

Fig. 17.20. It shows semi-germinated pollen grains in the micropylar canal of the ovule of *Welwitschia bainesii.* (After Martens).

Fertilization (Fig. 17.21). The process of fertilization is peculiar to *Welwitschia*. The cells at the micropylar end of the female-gametophyte grow upwards into small tubes. These are the embryo sac tubes. These tubes grow through the nucellar cap and meet the downward growing pollen tubes. One female nucleus moves to the tip of each embryo sac tube. The wall of contact between the pollen tube and the contact between the pollen tube and the embryo sac tube dissolves and the female nucleus migrates into the pollen tube to fuse with one of the female nuclei to form the **zygote** or the **oospore**. After fertilization the zygote moves down into the female gametophyte, where further development takes place.

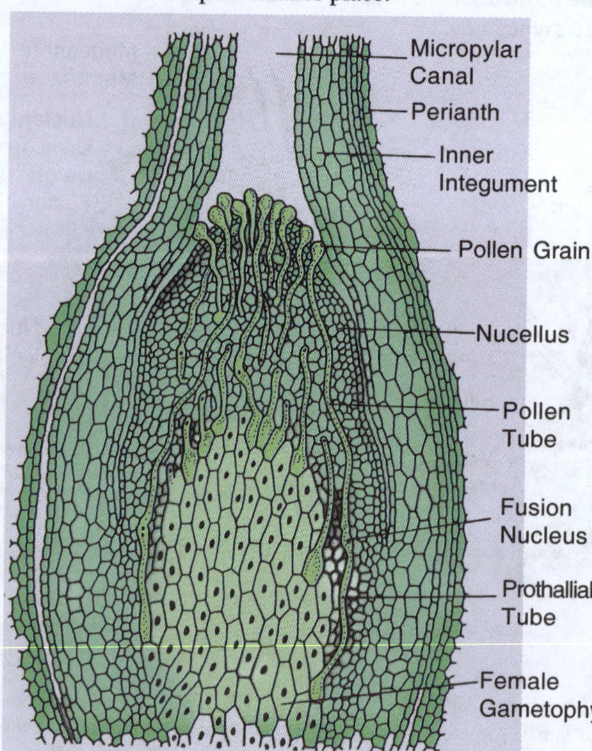

Fig. 17.21. It illustrates the process of fertilisation. Note the fusing prothallial tubes and pollen tube tips. (After Pearson).

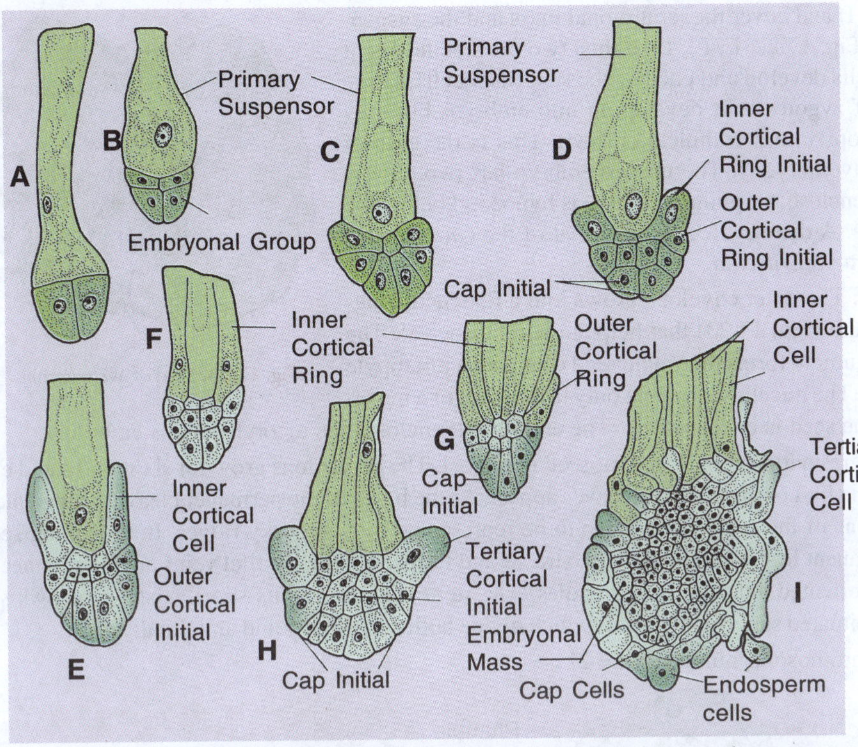

Fig. 17.22. (A-I). Embryology in *Welwitschia mirabilis* (After Pearson). (For explanation see text).

EMBRYOGENY (Fig. 17.22)

The zygote nucleus divides into a two-celled proembryo. The upper cell acts as the primary suspensor cell, whereas the lower cell produces a mass of embryonal cells (Fig. 17.22, A–C). The cells of this mass in the vicinity of the primary suspensor grow into secondary suspensor cells that elongate and push the embryonal mass deep into the female gametophyte. The eight apical cells of the embryonal mass divide to cut off small terminal cells that form the cap cells of the embryo (Fig. 17.22, D). The marginal cells of the embryonal mass act as initials of cortical cells. These cells elongate (Fig.

Plate VI : Seed germination of *W. mirabilis.*

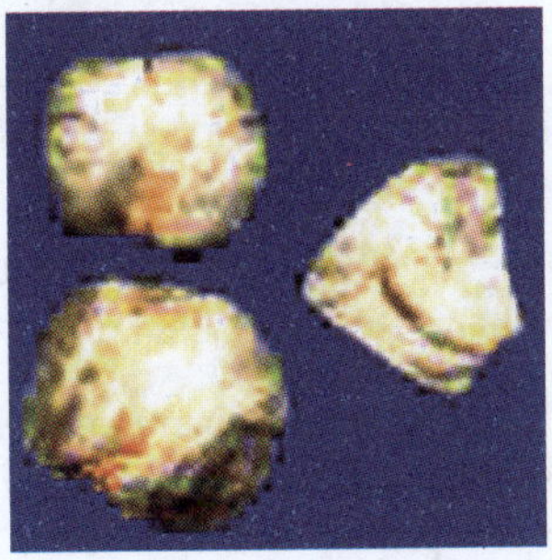

Plate VII : Seeds of *Welwitschia.*

17.22 D–H) and cover the embryonal mass and the suspensor cells (Fig. 17.22, E, F). Like this, two or three layers of cortical cells develop and enclose the suspensor cells. Large number of zygotes start developing into embryos but only one develops into a complete embryo. This is the case of simple polyembryony. The mature embryo has two cotyledons that enclose the plumule. There is hypocotyl below the cotyledons. A distinct 'feeder' like that of the *Gnetum* also develops in *Welwitschia.*

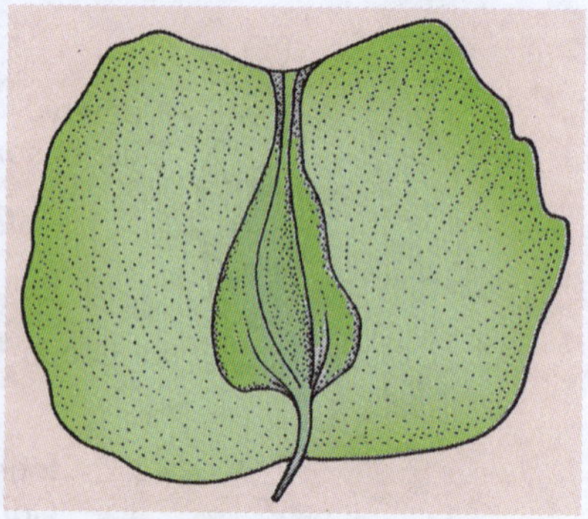

Fig. 17.23. Seed of *Welwitschia.*

Seed : The outer envelope grows into a flattened wing-like structure (Fig. 17.23) that helps in wind dispersal. The inner integument forms the single seed coat with a micropyle at one end. The nucellus is visible only in the form of a nucellar cap. The seed is endospermic. The endosperm encloses the dicotyledonous embryo.

The germination (Fig. 17.24) of seed is epigeal. The cotyledons grow for six months and eventually die. The first or the principal leaves appear. As the bases of the permanent leaves extend laterally, the positions of the cotyledons come to be represented by two corky ridges. In the young seedling each permanent leaf is supplied by two traces and farther out the parallel veins are interconnected by obliquely oriented anastomosing bundles. The stem apex during its short existence shows several medianly situated superficial initials which divide both periclinally and anticlinally.

The chromosome number is $n = 21$

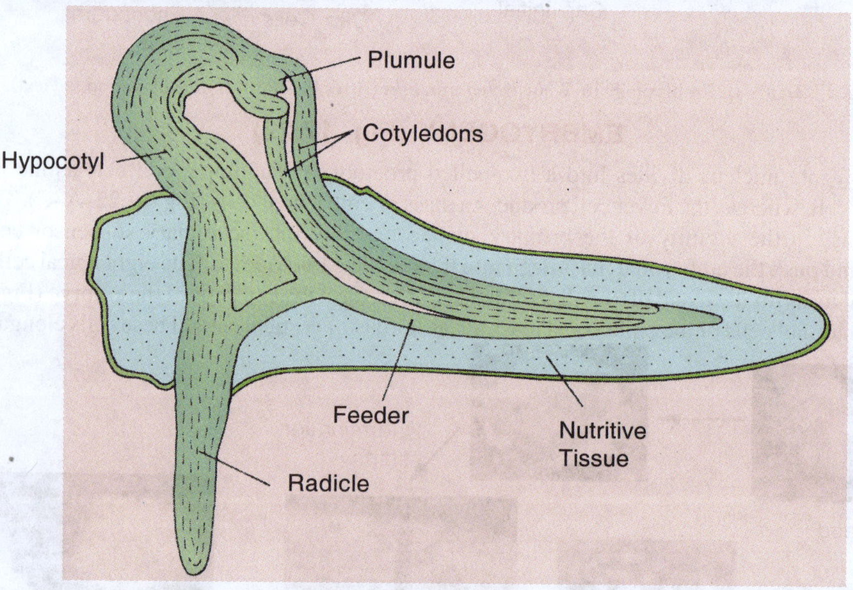

Fig. 17.24. L.S. of germinated seed of *Welwitschia bainesii.*

GENERAL CONCLUSIONS

The three genera of *Gnetopsida, viz., Ephedra, Gnetum* and *Welwitschia* are each highly specialised, both vegetatively and reproductively. Some authors have regarded *Ephedra* as retaining a few primitive features such as : –

1. Its gametophytes become cellular before fertilization.
2. It has archegonia with longer necks.

3. Its cauline style.
4. The gross morphology of its fructifications, which relate well to those of the Cordaitales.

The three genera resemble each other in many respects :

1. Their protoxylem has spiral or annular elements.
2. Transition of protoxylem elements to metaxylem.
3. The structure and development of perforation plate in the vessels.
4. The development of vessels in a special gnetalean manner i.e. by the dissolution of middle lamella and torus of the bordered pits.
5. The structure of the sieve cells and phloem parenchyma. Presence of albuminous cells in the phloem of *Gnetum* and *Ephedra*.
6. The leaves are opposite and decussate.
7. Each has an ovule with an attenuated inner integument that forms a thin micropylar tube and one or two additional integumentsh derived ontogenetically, each from a pair of bracts.
8. The compound nature of the male and female strobili.
9. Broad xylem and phloem rays.
10. The plants are dioecious.
11. The embryos are dicotyledonous.
12. The structure of the microsporangial wall is almost consistent in the three genera.
13. The microspores are wingless.
14. Early formation of unicellular primary suspensors occurs in all the three genera. It is followed by a transverse division.
15. *Gnetum* and *Welwitschia* have both tetrasporic development of female gametophyte. Both lack archegonia. In both, the cells in the female gametophyte show gradate polyploidy.
16. All the three genera possess polyembryony of simple type.
17. All the three genera possess orthotropous ovules.
18. The male flowers in all three genera consist of a stalk bearing synangia composed of 1–8 sporangia. The sporangia usually dehisce by means of apical slits.

The position of the microsporangiate organs in *Gnetum* and that of the inflorescence axis which bears many compound strobili in *Welwitschia* seem anomalous. In each case it can be regarded as an adventitious growth. The kind of meristematic tissue giving rise to these structures in both these genera is homologous. The formation of a linear meristem adjacent to a leaf base in *Welwitschia* forming upward folds of tissue on which the reproductive stalks or in inflorescence axes develop may well have a counterpart in the strobilus of *Gnetum*, where a similar kind of meristem above the bract produces tissues upward on which the microsporangiate simple strobili or male flowers develop. Both of these may further be related to the mode of formation of accessory buds in comparable position in the vegetative shoots of *Gnetum* and *Ephedra*.

Eames (1952) has put forth the view that *Ephedra* has close phylogenetic relationship with the Cordaitales. He was of the opinion that *Gnetum* and *Welwitschia* are not even distantly related to *Ephedra*. Such a conclusion does not seem justified because of certain strong similarities between the three genera. Coulter and Chamberlain (1910) stated, "............ whatever may be the connection of *Ephedra* with other Gymnosperms it cannot be separated from *Welwitschia* and *Gnetum*." This conclusion seems to be sound and it appears to be sufficiently supported by facts unravelled since 1910.

DIFFERENCES BETWEEN GNETUM AND EPHEDRA

The differences between these genera can be tabulated as below :

Ephedra	Gnetum
1. Majority of the species of *Ephedra* are small and bush-like forms. A few are herbaceous.	1. *Gnetum* sporophytes are trees, shrubs or woody climbers.
2. *Ephedra* grows commonly in dry regions or in warm areas. Some species also occur at higher altitudes.	2. *Gnetum* is restricted to humid tropical and subtropical regions.
3. The leaves are small and scale-like.	3. The leaves are foliaceous and green.
4. The venation is parallel.	4. The venation is reticulate-unicostate.
5. The stem is photosynthetic and possesses a well developed palisade layer in cortex.	5. The stem is not photosynthetic. It lacks palisade layer.
6. The sclereids, hair and laticiferous tubes present in the reproductive axes of *Gnetum* are absent in *Ephedra*.	6. Hair, sclereids and laticiferous tubes are abundant in the axes, bracts and collars of the strobili.
7. Anomalous secondary growth by the formation of successive cambial rings is absent in the stem.	7. Secondary growth by the formation of successive cambial rings is characteristic of some species, *e.g. G. ula*.
8. The vessels in *Ephedra* possess a foraminate plate of perforations on their end walls. Simple perforation is rare.	8. In many cases the vessels possess a simple perforation at their end walls.
9. The ovule is terminal on a lateral appendage of the fertile shoot and is not cauline.	9. The ovule is cauline, that is, it terminates the shoot axis.
10. The germination of the pollen grains follows the typical gymnospermous plan *i.e.* the mature microspore consists of a prothallial cell, prothallial nucleus, stalk cell, body cell and a tube nucleus.	10. In *Gnetum* there is only one prothallial cell, a tube nucleus and a generative cell that functions as the body cell and divides into two male gametes. There is no stalk cell. In *Welwitschia* even the prothallial cell is not formed.
11. The embryo sac is monosporic as in other gymnosperms.	11. The embryo sac is tetrasporic like some of the angiosperms.
12. The pollen chamber is deep and well developed as in Cycads and in *Ginkgo*.	12. The pollen chamber is shallow.
13. The tent-pole is as distinct as in *Ginkgo, Taxus*, and Cordaitales.	13. Rudimentary tent-pole is found only in *G. africanum*.
14. The female gametophyte becomes cellular before fertilisation.	14. The female gametophyte becomes completely cellular after fertilisation.
15. The archegonia are present.	15. Archegonia are absent.
16. Zygote undergoes free nuclear division upto the formation of eight nuclei.	16. Zygote divides into two cells, *i.e.* there is no free nuclear division.
17. Embryo lacks 'feeder'.	17. Feeder is present in the embryos of *Gnetum* and *Welwitschia*.

GNETUM AND WELWITSCHIA

The two genera resemble each other in the following respects :–

1. The development of the male gametophyte is similar.
2. The structure of the microsporangium is similar.
3. The ovules in both have their inner integuments produced into a long micropylar tube. The ovules have shallow or no pollen chamber.
4. The development of the female gametophyte is tetrasporic (Martens, 1961, 1962) in both the genera.
5. The cells of female gametophyte are multinucleate. These nuclei fuse together and form polyploid nuclei.
6. The embryo in both the genera develops a distinct feeder or foot.
7. The embryo is dicotyledonous.
8. The zygote does not form free nuclei. Two cells are formed after first division.
9. The xylem possesses vessels and broad medullary rays.
10. Anomalous secondary growth by the formation of successive cambial ring is known in both the genera.
11. Method of pollination is similar.
12. The male and female strobili are compound. The plants are dioecious.

Differences between the two genera are tabulated below :–

Welwitschia	Gnetum
1. The sporophyte is confined in distribution to extremely xerophytic localities in South West Africa where rainfall is sometimes even below 1 inch per year.	1. The plants grow in rain forests and in humid regions of the tropics and subtropics.
2. The plant body is neither a tree nor a shrub. It is like a huge turnip with stem mostly subterranean.	2. The plant body is either a tree, a shrub or a woody climber. The stem is serial or epiterranean.
3. The stem is not differentiated into nodes and internodes.	3. The stem has distinct nodes and internodes. The nodes are swollen.
4. The stem is covered with a thick bark that sometimes becomes longitudinally fissured.	4. The bark is comparatively thin and is not longitudinally fissured.
5. The stem apex above the insertion of leaves is called the crown and possesses a thick rim marked with transverse ridges beset with scars of fallen inflorescence axes. Below the insertion of leaves the stem is called the stock. The stem is unbranched.	5. There is distinction between crown and stock. Transverse ridges are lacking. The stem is extensively branched and exhibits shoot dimorphism.
6. The stem apex has a deep groove.	6. The stem apex has no groove.
7. The tap root is very long and grows deep till it reaches the water table.	7. The tap root is comparatively much less developed.
8. After the cotyledons the plant bears only two opposite leaves that persist	8. The plant bears numerous foliage leaves that fall at intervals and new leaves again

	Welwitschia		Gnetum
	throughout its life. There are no scale leaves. The two principal leaves grow continuously by means of a basal meristem and reach a maximum length of two metres. The leaves split when mature into distinct segments or ribbons. The leaves are thick and leathery.		appear. There are two types of leaves : the scale leaves and foliage leaves. The foliage leaves are small in size and never split. The leaves are herbaceous, thin and glabrous.
9.	The venation is parallel. The parallel veins are joined by transverse connections (Rhodin 1953, 1958).	9.	The venation is reticulate and unicostate.
10.	The male flower is bisexual. In addition to six stamens it bears in the centre an abortive ovule.	10.	The male flowers are unisexual.
11.	The compound strobili are borne terminally on the dichasially branched inflorescence axis.	11.	The compound strobili arise either singly in leaf axils or rarely they are terminal.
12.	There are no male prothallial cells in the male gametophyte.	12.	There is a single male prothallus cell in the male gametophyte.
13.	The female gametophyte forms, 1,000 free nuclei that are evenly distributed throughout the cytoplasm.	13.	The number of free nuclei varies from 256 to 1,500 in various species. They are peripherally arranged. There is a central vacuole.
14.	The female gametophyte becomes completely cellular before fertilisation.	14.	The female gametophyte is partly cellular and partly free nuclear before fertilisation. It becomes completely cellular after fertilisation.
15.	There is no tent-pole.	15.	There is a distinct tent-pole in *G. africanum*.
16.	During fertilisation the cells in the micropylar region of the female gametophyte grow upwards through the nucellus in the form of tubes. These are called the **prothallial tubes**. Some regard these cells as archegonial initials. Each such tube contains a single female nucleus or the egg. These tubes meet with the downwardly growing pollen tubes. Their tips meet and the wall of contact disappears. The female nucleus moves up into the pollen tube and fused with the male gamete to form the zygote or the oospore. The zygote moves down into the female gametophyte and undergoes further development there.	16.	No prothallial tubes are formed in this case. Some free nuclei at the micropylar end get surrounded by cytoplasm and become distinct as eggs. In *G. ula* distinct egg cells are formed at the micropylar end of the female gametophyte. These cells develop at places where the pollen tubes touch the female gametophyte. The pollen tube pierces the membrane of the embryo sac and discharges the male gametes. The nucleus of the larger male gamete fuses with the female nucleus to form the zygote or the oospore.

GNETUM AND CYCADEOIDALES

Resemblances between *Gnetum* and Cycadeoidales have been given sufficient consideration by Thoday (1911) and Berridge (1911). These include :

1. The vascular supply to the peduncle of the cone of *Cycadeoidea wielandii* and the single flower of *Gnetum* is quite similar. The presence of the three concentric rings of bundles at the nodes in both cases has been greatly emphasized by them. Out of these three rings the bundles of the middle ring are inversely oriented.
2. Presence of a cup-shaped layer of fibrous cells at the base of the ovule in *Gnetum africanum* and *Bennettites*.
3. The internal structure of the middle envelope of ovule in *G. africanum* and *Bennettites*. In both the cases this envelope has sub-epidermal alternating bands of sclerenchyma and parenchyma. Next to them is a zone containing palisade layer and a fibrous layer.
4. The micropylar tube in *Gnetum* forms a closing tissue after pollination. This tissue later becomes lignified and forms a central rod-like structure similar to the one present in *Bennettites morierei*.

Thoday opined that these resemblances render it probable that *Gnetum* and *Bennettites* might have a common ancestor.

Gnetum resembles other gymnosperms in possessing the following characters :.

1. Presence of a prothallial cell in the male gametophyte.
2. Occurrence of free nuclear divisions in the female gametophyte.
3. The nature of endosperm.
4. Presence of a small pollen chamber in the nucellus.
5. Presence of pollination drop mechanism during pollination.

GNETUM AND ANGIOSPERMS

For the last so many years resemblances between *Gnetum* and the angiosperms have been greatly emphasized by various authors as portraying a close relationship between the two. Some of them went up to the extent of including *Gnetum* under the angiosperms, whereas others suggested that gnetales and the angiosperms had a common ancestor. Still others believed that the ancestors of angiosperms were like those of *Gnetum* itself. A few of these considerations are outlined below.

Arber and Parkin (1908) put forth the view that both angiosperms and Gnetales originated from a common stock, designated by them as 'Hemiangiosperms'. The two originated from this hypothetical group of plants and in many respects their development continued along parallel lines.

Thompson (1916) regarded that the ancestors of angiosperms very closely resembled with the genus *Gnetum*.

Hagerup (1934) studied the floral development of Gnetales, *Piperaceae* and *Juglandaceae* and came to the conclusion that they closely resemble each other and the differences whatsoever are of minor importance. He postulated the view that Gnetales could be easily included under the angiosperms or at least the *Piperaceae* originated from some gymnospermic plants through the Gnetales.

Anderson (1934), while commenting upon the origin of angiosperms, derived a conclusion from the basic chromosome numbers of the *Magnoliaceae* ($n = 19$), Cycadales ($n = 12$) and Gnetales ($n = 7$). He regarded *Magnoliaceae* as the transitional forms between the gymnosperm and the angiosperms. Recent investigations have shown that the basic chromosome numbers are not $n = 7$ in the Gnetales. This view is only of historical importance.

Gnetum, however, resembles angiosperms in many respects. These are enumerated below:

1. The leaves in *Gnetum* have reticulate venation, that is, an angiospermic character.

2. The presence of vessels in the xylem is also an angiospermic character but the development of vessels in the two is different. McDuffie (1921), however, pointed out that the two types of vessel development occur in some angiosperms belonging to family *Rosaceae, Ranunculaceae* and some families of the order Geraniales. In one species of *Potentialla* both *Gnetum*-type and the usual angiosperm-type of vessel development is present. A close similarity between the gnetalean wood and the angiospermic wood is manifested by the Maule reaction which gives similar results in both the cases (McLean and Evans, 1934).
3. The short apices in *Gnetum* and angiosperms show a distinct tunica and corpus configuration.
4. Some (Lignier and Tison, 1913; Vasil, 1959) regard the inner integument of the gnetalean ovule as comaparable to the ovary that is produced into the long style. The micropylar tube is regarded as the style. It has a nutritive tissue similar to that of the style in many angiosperms. Such an interpretation has been objected to because the micropylar tube is devoid of a vascular supply and if the innermost envelope is regarded as the ovary wall then the ovule becomes entirely devoid of an integument.
5. The tetrasporic development of the female gametophyte.
6. The female gametophyte is only partly cellular before fertilization and becomes completely cellular only after fertilization. Some of the free nuclei act as eggs as there are no archegonia.
7. The zygote does not undergo any free nuclear division.
8. The embryo is dicotyledonous.

The features listed above show that *Gnetum* approaches angiosperms in some respects. These similarities appear to be cases of parallel development and do not portray any close relationship.

In conclusion no words better than Maheshwari and Vasil (1961) can be cited, "The genus *Gnetum* remains largely a phylogenetic puzzle. It is gymnospermous, but possesses some strong angiospermic features."

AFFINITIES OF GNETOPSIDA

Most of the studies about affinities of Gnetopsida were discussed by Eames (1952). According to him *Ephedra* is closer to cordaitales and coniferales rather than *Welwitschia and Gnetum*. Phylogenetically these three taxa showed marked difference both embryologically and morphologically. The ovules and microsporngia of *Ephedra* are appendicular whereas it is cauline in *Welwitschia* and *Gnetum*. Male and female gametophyte also differ in these cases. *Ephedra* has monosporic female gametophyte and male gametophyte is stalked whereas is case of *Welwitschia* and *Gnetum* female gametophyte is tetrasporic and male gametophyte lack stalk cell. Thus *Ephedra* occupies a position separate from *Welwitschia* and *Gnetum*.

Gnetum has the characters of both gymnosperms as well as of angiosperms. According to Muhammad and Sattler, (1982), *Gnetum* shares about 60% characters of angiosperms and 30% with gymnosperms, which shows a closer ancestral relationship of *Gnetum* with angiosperms. Trevisan (1980) has suggested that at least some members of Gnetopsida were present at time of the origin of angiosperms. Ovules of Gnetopsida has multicellular envelopes, unlike other gymnosperms (having a single integument only). This has been considered as integumentary structure homologous with outer integument of an angiosperms (Stewart, 1983). *Gnetum* appears as a typical dicotyledonous angiospermic plant with reticulate venetion in leaves. Both have vessels in their xylem, though the origin of vessels differs in both cases. Vessels develop form tracheids having a large number of bordered pits on the end wall in *Gnetum* where in angiosperms vessels developed from tracheids having narrow scalariform perforations. Tunica and Corpus arrangement of the shoot apex is also reported in *Gnetum*. But presence of central mother cells is a typical character of gymnosperms.

Besides this presence of germinating pollen grains in the micropylar tube of *Gnetum* ovule resembles one of a style of an angiosperm carpel. Tetrasporic megasporogenesis a feature of *Gnetum* is more common in angiosperms, not found in Gymnosperms except *Welwitschia*. Formation of storage tissue or endosperm after fertilization like in angiosperm is also a feature of *Gnetum*. Double fertilization is also reported in *Gnetum gnemon* (Carmichael and Friedman, 1996). Based an these findings Maheshwari (1950) regarded angiosperms to have passed through some stage during the evolution which is now presently shown by *Gnetum*. Maheshwari & Vasil (1961) still consider *Gnetum* a *phylogenetic* puzzle.

QUESTIONS

ESSAY TYPE

1. Give a comparative account of reproduction in *Gnetum, Ephedra* and *Welwitschia*. **(Kanpur M.Sc., 2003, Awadh M.Sc., 2002)**
2. Discuss the reasons for inclusion of *Welwitschia, Ephedra* and *Gnetum* in a single order Gnetales.
3. Discuss the relationships of *Welwitschia, Gnetum* and *Ephedra* with each other and with other groups of plants.
4. Describe the inflorescence and flowers of *Welwitschia*. **(Kanpur M.Sc., 2005)**
5. Describe the megasporogenesis in *Welwitschia*.
6. Give an illustrated account of the male gametophyte in *Welwitschia*.
7. Give a detailed account of the life cycle of *Welwitschia*.
8. Give a brief account of reproduction in *Welwitschia*.

SHORT ANSWER TYPE

9. **Write short notes on :**

(*a*) Secondary growth in *Welwitschia*
(*b*) Flowers of *Welwitschia*
(*c*) Comparison between *Gnetum* and *Welwitschia*
(*d*) Differences between *Ephedra* and *Welwitschia*
(*e*) Wood of *Welwitschia*
(*f*) Flower of *Welwitschia*
(*g*) Habit of *Welwitschia*
(*h*) Seed of *Welwitschia*
(*i*) Germination of seed in *Welwitschia*

10. **Draw a neat labelled diagram of :**

(*a*) T.S. old root of *Welwitschia*
(*b*) T.S. young stem of *Welwitschia*
(*c*) T.S. leaf of *Welwitschia*
(*d*) Flower of *Welwitschia*
(*e*) Floral diagrams of male and female flowers of *Welwitschia*
(*f*) Female strobilus of *Welwitschia*
(*g*) Seed of *Welwitschia*
(*h*) T.S. ovule of *Welwitschia*

11. Give a brief account of secondary growth in *Welwitschia*
12. Briefly describe pollination and fertilisation in *Welwitschia*
13. Describe the ovule of *Welwitschia*

OBJECTIVE TYPE

14. **Select the correct answer :**

(*i*) *Welwitschia* is represented by
 (*a*) one species only
 (*b*) two species
 (*c*) three species
 (*d*) many species

(*ii*) Coiled vessels are characteristic features of
 (*a*) *Gnetum* (*b*) *Ephedra*
 (*c*) *Welwitschia* (*d*) *Citrus*

(*iii*) *Welwitschia* is
 (*a*) Microphyte (*b*) Hydrophyte
 (c) Mesophyte (*d*) Extreme Xerophyte

(*iv*) Venation in the leaves of *Welwitschia* is
 (*a*) Reticulate (*b*) Parallel
 (*c*) Both of the above (*d*) Absent

(*v*) The development of female gametophyte is
 (*a*) monosporic (*b*) tetrasporic
 (*c*) tetrasporic (*d*) none of the above

(*vi*) Winged seeds are present in

(*a*) *Cycas* (*b*) *Gnetum*

(*c*) *Ephedra* (*d*) *Welwitschia*

(*vii*) *Welwitschia* resembles *Gnetum* in

(*a*) Habit (*b*) Stem characters

(*c*) Venation

(*d*) Development of female gametophyte.

(*viii*) Which one of the following characters is not present in *Welwitschia* ?

(*a*) Tetrasporic development of embryo sac

(*b*) Winged seed

(*c*) Parallel venation

(*d*) Stem.

(*ix*) Which one of the following statements is correct ?

(*a*) Both *Gnetum* and *Welwitschia* grow underextreme xerophytic conditions.

(*b*) *Gnetum* and *Welwitschia* have been placed under separate orders.

(*c*) In *Welwitschia* the development of female gametophyte is *tetrasporic*.

(*d*) Both *Gnetum* and *Welwitschia* are giant trees.

CHAPTER 18

General Topics

- Experimental Embryology & Biotechnology
- Female Gametophyte and *Invitro* Gynogeneis
- Embryo-Culture
- World Distribution of Gymnosperms
- Impact of Gymnosperms on Human Society
- Gymnosperms: Areas to be explored.

Sequoia Sempervirens

In this chapter certain topics hither to not described in earlier chapters have been discussed briefly.

I. Experimental Embryology & Biotechnology

Forests in the country today have at least five times more pressure than what they can withstand. Forest wood contributes towards meeting the fuelwood requirements and timber demand for various purposes (FSI, 1996). Gymnosperms are the group of trees which constitute most of the forest tree vegetation and cater to the need of the societies for timber and to some extent their fuelwood demand. With rapid depletion of forest cover not only in our country, but all over the world, there has been a simultaneous loss of forest trees specially gymnosperms which has created special interest in conserving as well as producing gymnosperms through various cultural practices. It is here biotechnological inputs come in the picture (Bonga a 1974; Bonga & Durzan, 1987; Gupta et. al. 1993). Tissue culture practices are being adopted to multiply the plants at commercial level, though in gymnosperms the study is still limited (Table I). It is a technique to regenerate plantlets from unorganised tissues which would make possible the production of haploids and homozygous diploids for tree breeding programmes, induction, isolation and selection of mutants, the manipulation of plant protoplasts; and modification of cells by genetic engineering techniques besides many others like *in vitro* study of host-pathogen interaction and cryopreservation. The establishment of procedure for *in vitro* embryogenesis for gymnosperms is a significant development, because somatic embryos can be reused in large number and can be used for production of desired genotype in large numbers on a regular basis. This can be very much helpful in multiplying various species like *Pinus taeda*, *Picea abies* and *Pseudotsuga menziesii* Gupta et. al., 1993). The gymnosperm biotechnology is mainly used for providing suitable and superior yield of its trees.

Macropropagation through root cuttings, graftings etc. and macropropagation through tissue culture axillary branching, induction of adventitious buds on cultured embryo, excised embryo parts, shoot tips are the two main methods of vegetative propagations in plants including for gymnosperms. Tissue culture is a general term used to describe all techniques in which cells or tissues are grown on specific nutrient medium inaseptic environmental conditions.

It was Ball (1950) who for the first time successfully cultured coniferous tissues *in vitro* by using adventive shoots of buds of *Sequoia sempervirens.* He also maintained the continuous callus using Knop's solution with 3% sucrose and IAA. He found that maximum growth can be obtained in Sucrose (Bal, 1955).

Table I. *List of Gymnosperms regenerated by Tissue culture* technique as per Karnosky. 1981)

Species	*Explant*	*Invstigator/s*
Araucaria cunninghamii	Shoot tip	Haines and de Fossard (1977)
Cryptomeria japonica	Hypocotyl, stem	Isikawa (1974)
Picea abies	Megagametophyte	Bonga (1977)
P. glauca	Hypocotyl	Campbel and Durzan (1976)
P. sitchensis	Embryo, needles, shoot tips	Webb and Street (1977)
Pinus elliottii	Cotyledon	Sommer and Brown (1974)
P. palustris	Cotyledon, embryo	Sommer et al. (1975)
P. pinaster	Cotyledon, hypocotyl	H. David et al. (1978)
P. radiata	Cotyledon, hypocotyl	Reilly and Washer (1977), Smith et al. (1982)
P. rigida *P. sabiniana* *P. virginiana*	Cotyledon	Brown and Sommer (1977)
P. taeda	Cotyledon	Sommer and Brown (1974)
P. strobus	Embryo	Minocha (1980)
Pseudotsuga menziessi	Cotyledon	Cheng and Voqui (1977)
Sequoia sempervirens	Juvenile shoot	Ball (1978)
Thuja plicata	Cotyledon, shoot tip	Coleman and Thorpe (1977)
Tsuga heterophylla	Cotyledon	Cheng (1976)

To understand the nature and intensity of malignancy of cells and biochemical characters of tumour, Reinert and White (1956) cultured normal and abnormal tumorous tissues of *Picea glauca* using mainly cambial regions While experimenting they observed that the behaviour of normal and tumerous tissues differed tumerous tissues require vit B_{12} for growth and multiplication, normal cells do not require it. White and Risser (1964) also made tissue culture studies on *Picea glauca.* They found that maximum growth of callus can be obtained when 20 mg of callus was inoculated on 0.5% agarized substrate at pH 5.5. They formulated a less complex culture medium than those suggested by Reinert and White (1956). Similarly host-parasite relationship has been studied by culturing *Juniperous* infested with *Gymnosporangium* (Constabel, 1957; Maheswari *et. al*; 1967) and on *Pinus monticola* callus infected with *Cronartium ribicola* (Harvey 1967, *a, b*; Harvey and Grasham, 1969 a, b, 1970 a, b) Host-pathogen interactiton at macromolecular level has been studied using *in vitro* callus culture of *pinuslamberbiana* (Dineret. al. 1984) sucrose (Ball, 1955). Constabel (1965) reported that the extent of tanin production in the cultures of *Juniperus communis* depends on growth, the presence of precursors and light. Tanin content decreases with increase in growth rate of the tree. Similarly formation of lignin in the cultures of cambial tissues of *Pinus strobus* was also studied (Hasegawa *et. al.* 1960). It was concluded that shikimic acid and lignin are synthesized from glucose in the isolated cambium tissue and assumed that aromatic nucleus of lignin is formed from sugar *via* shikimic acid.

Callus culture have also been reported from secondary phloem of *Cupressus lusitanica* by Borchert (1967), and mature branch segment of *Pseudotsuga* tissues by Harvey and Grasham (1969b). These *in vitro* cultures were obtained mainly to study morphology, anatomy, nutritional requirements and physiological behaviour of conifers. Konar (1963*a*) grew *Pinus gerardiana* callus as submerged cultures in liquid medium to study the cell behaviour.

In vitro Pollen culture and Male gametophyle

The objective of *in vitro* culture of male and female gametophyte in Gymnosperms was to follow the development of isolated gametophytes *in vitro* and secondly to initiate development of sporophyte from haploid tissues (Rohr, 1987).

Hanging drop technique with 2-20% sugar solution was used for first successful *in vitro* pollen culture of Gymnosperms (Kuhlweini, 1937). Similar technique was later used by Branscheidt (1939) in *Taxus* and *Cupressus* pollen cultures. Here sperm nuclei were formed in about 20 days. Other important plants in pollen culture studies include *Zamia* and *Cupressus* (LaRue, 1954), *Juniperus* (Duhoux, 1973), *Larix, Tsuga* and *Pinus* (HoRh and Rouse, 1970), *Ginkgo* and *Torreya* (*Tulecke*, 1957; Tulecke and Sehgal, 1963), and *Ephedra* (Mehra 1934, 1938). Male cone of *Pinus roxburghii*, at microspore mother cell stage, was cultured by

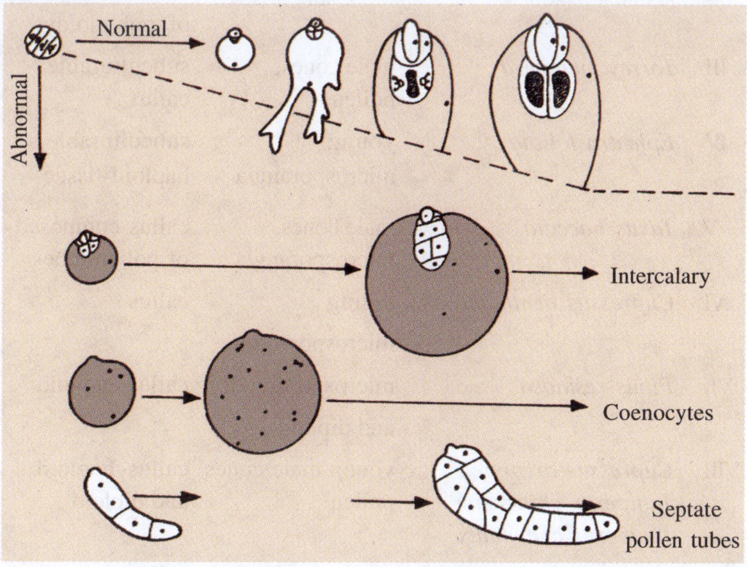

Fig 18.1 Normal and abnormal development of male gametophyte of *Ginkgo biloba* under *in vitro* conditions (a diagrammatic representation). Embryo like structures can be noted (After Tulccke, 1957).

Konar (1963b), in White's medium and nucleic acid. *In vitro* spermatogenesis has been reported from *Encephalartos allensteinii*, where three celled microspores were cultured for eight months. Masses of pollen tubes enlarges from microsporangia within a week after inoculation and ultimately two flagellated microspores were produced per microspores. However, these did not attain maturity. (DeLuca and Sabato, 1979; De Luca *et. al.* 1980).

In vitro androgenesis have been initiated to produce haploid plants of sporophytic form. This has generated considerable interest among geneticist, foresters and horticulturist as it is a tool to produce various genetic combinations which are naturally not possible.

Formation of subculturable callus was first reported by Tulecke (1953, 1957) from the pollen of *Ginkgo biloba* and later by Tulecke and his group (Tulecke and Nickell, 1959,1960; Tulecke *et. al*, 1962, 1965). Several abnormalities have been reported from subculturable callus of *Ginkgo* pollen like formation of septate pollen tubes, division in tube nucleus (containing upto 50 nuclei) and less frequent divisions in prothallial cells, stalk, body and sometimes even in sperm cells to form cluster of cells. (Fig. 18.1) Tulecke and Nickell (1960) have been able to successfully manitain these tissues continuously for several years, without cell differentiation. Subculturable callus has also been reported from other gymnosperms like *Taxus* and *Torreya* (Tulecke & Sehgal, 1963) Callus from micro-gametophyte has also been produced by Rao and Mehta (1969) from *Pinus resinosa* and *Biota orientalis* by Razmologov (1973) from *Cupressus benthansis* and by Duhoux and Norreel (1974) form *Cupressus arizonica* and two species of *Juniperus* viz. *J. chinensis* and *J. communis*.

In vitro culture of male gametophyte amongst Gymnosperms has been summarised in Table II.

Table II. *List of Gymnosperms with in vitro cultures of male gametophyte tissue.*

	Species	Explant	Result	Reference
I	*Ginkgo biloba*	male strobilus or pollen grains	haploid tissue	Tulecke, 1957
II	*Taxus brevifolia*	microsporangia	callus, evidence of polyploidy	Tulecke, 1959
III	*Torreya nucifera*	male cones, pollen	subculturable callus	Tulecke & Sehgal 1963
IV	*Ephedra foliata*	young microsporangia	subculturable haploid tissue	Konar, 1963c
V	*Taxus baccata*	male cones, microsporangia	callus composed of pollen tubes	Rohr, 1973
VI	*Cupressus benthamii*	young microsporangia	callus	Razmologov, 1973
VII	*Pinus resinosa*	microsporophylls and diploid	callus–haploid	Bonga, 1974b
VIII	*Cupressus arizonica* *Juniperus chinensis* *Juniperus communis*	young male cones pollen	callus–haploid and diploid	Duhoux & Norreel, 1974
IX	*Ceratozamia mexicana*	pollen	callus spermatogenesis	De Luca et. al., 1980

Female gametophyte and *in vitro* gynogeneis :

The haploid female gametophyte can be used as an alternate to male gametophyte and being massive in size, is more suitable and amenable than male gametophyte. The earliest report on root formation by female gametophyte was described by Du Chartre (1888) in *Cycas*. There are many reports on root and shoot formation from megagametophyte cultures like that of *Zamia and Cycas* (Coulter and Chrysler, 1904, La Rue, 1954). La Rue (1948 & 1954) also reported embryoids from cultured endosperms of *Zamia*. Later *in vitro* cultures have successfully been obtained in *Picea abies* (Huhtinen, 1972, Steinhauer & Huhtinen, 1978, Simola and Honkanen, 1983), *Taxus baccata* (Zenkteler and Guzowskay, 1970), in various species of *Pinus* (Borchert, 1968; Bonga & Fowler, 1970, Bonga, 1974) and *Encephalartos villosus* (Laliberte *et. al.* 1983). Norstog (1967) and Norstog and Rhamstine (1967) made a detailed study on how deletions, additions and changes in phytohormones (auxins & cytokinins) and nucleic acids affect initiation of callus, adventive embryos, roots and shoots from gametophytic tissues of *Zamia integrifolia* (Fig.).De Luca *et. al.*, (1979) have been able to culture megagametophyte from mature seeds of *Ceratozamia mexicana*, *Cycas revoluta* and *Ecephalartos umbeluziensis* on modified White's media with 2% sucrose. Similarly Konar and Singh (1979) cultured female gametophyte from mature seeds of *Ephedra foliate* and according to him 2-4D stimulates callus formation. However coconut milk was needed for extensive proliferation and formation of subculturable callus.

A significant work has been done on *Ephedra foliata* as Sankhla *et. al.* (1967) first reported callus formation from megagametophyte taken for developing ovule. Later it was reported that haploids can be obtained from mature as well as developing ovule (Konar and Singh, 1979, Singh and Konar, 1981, Bhatnagar and Singh 1984). It has also been reported that explants cultured before archegonia formation do not respond for callus formation. Singh *et. al.* (1981) excised female gametophyte from ovules at the archegonial stage and cultured them on (BM) (MS + 2% sucrose + 10% CM) on based medium. Various combination of BM with 2- 4D, NAA, BAP have been tried for callus growth and rooting, and found that BAP stimulates short bud formation. The meristemoids formed on the endosperm differentiated in roots and shoots both and haploid plantlets have chromosome no. = 7.

Megagametophyte embryogenesis have been reported in plant by various wokers : in *Zamia integrifolia* by Norstog (1965), Norstog and Rhamstine (1967), in *Ceratozamia mexicana* by De Luca *et. al.*, (1979) and in *Larix decidua* by Nagmani and Bonga (1985).

Presently *in vitro* organogenic capacity from megagametophytic tissues has been reported from very few species. A summary of *in vitro* studies on female gametophyte have been given in Table III.

Table III. *List of Gymnosperms with in vitro cultures of female gametophyte*

	Species	Explant	Result	Reference
I	*Zamia floridana*	pre-fertilization stage	buds and root formation	La Rue, 1948, 1950, 1954
II	*Cycas revoluta*	mature megagametophyte	buds and root formation	La Rue, 1950, 1954
III	*Ginkgo biloba*	post-pollination, but unfertilized	callus	Tulecke, 1964
IV	*Zamia integrifolia* *Cycas circinalis*	post-fertilization stage	haploid organ formation	Norstog& Rhamstine, 1967

Table III. *(Contd.)*

	Species	Explant	Result	Reference
V	*Pinus lambertiana*	mature megagametophyte	callus, underwent diploidisation and sometimes became tetraploid	Borchert, 1968
VI	*Taxus baccata*	megagametophyte	callus	Zenkteler & Guzowska, 1970
VII	*Pinus resinosa* *P. nigra, P. mirgo*	megagametophyte	callus, remained haploid	Bonga & Fowler, 1970 Bonga, 1974b
VIII	*Pinus palustris*	megagametophyte	callus	Sommer et. al., 1974
IX	*Picea abies*	mature seed	haploid plantlet, no further growth	Bonga, 1977
X	*Ceratozamia mexicana*	mature megagametophyte	adventive embryos	De Luca et. al., 1979
XI	*Cycas revoluta*	mature megagametophyte	coralloid roots	De Luca & Sabato, 1980
XII	*Microcycas calocoma*	explants from immature seeds	callus and pseudobulbils	Pena et.al., 1982
XIII	*Encephalartos villosus*	infertile megagametophyte	callus	Laliberte et. al., 1983
XIV	*Ephedra foliata*	mature seed	callus, roots and shoots	Konar & Singh, 1979 Singh et. al., 1981 Bhatnagar & Singh, 1984
XV	*Larix decidua*	immature ovules	callus, embryoids plantlets	Nagmani & Bonga, 1985

Embryo Culture :

The earliest reports on embryo culture in Gymnospermes comes from Schmidt (1924) on *Pinus* and from Li and Shen (1934) on *Ginkgo* trees. Here it was shown that excised embryo could be grown into seedlings by substituting megagametophyte with suitable nutrient media. Subsequently embryos of other species like *Pinus*, *Picea* and *Pseudtosuga* were also tried successfully to produce plantlets.

Proembryos of *Ginkgo* and *Pinus nigra* have been successfully cultured for callus formation by Radforth and his associates (Radforth, 1936; Radforth and Pegoraro, 1955; Radforth and Bonga, 1960). However, in various cases proembyos have failed to grow into mature embyos as in case of *Larix* (Sterling, 1949b), *Pinus* and *Keteeleria* (Loo L Wang, 1943) and *Biota orientalis* (Konar and Oberoi, 1965). Part of these failures may be due to non-suitable culture media and presence of uniform accessiblity of nutrients to embryos may be the reason for absence of polarity in the embryo.

It is difficult to culture immature embryo in case of gymnosperms as cotyledons play a very significant role in the nutritional requirements of the embryo placed for culture. Thomas (1970) tried to culture excised young and old embryo of *Pinus sylvestris* and found that growth occur only if they have initiated cotyledons. Young embryos of *Ginkgo* (Wang et. at. 1963, Wang and Chen, 1965) and *Gnetum* (Vasil, 1963) have also been cultured but their growth remains incomplete. According to Bulard (1952), normal growth of embryo occur only when cotyledons and nutrients are well in contact of each other. Cultured embryos show radial symmetry or asymmetry instead of axial symmetry in Gymnosperms. Mostly germinating embryos of gymnosperm develop chlorophyll in the dark but excised embryos on culture medium fail to do so. It is observed that in the nutrient medium specific chlorophyll stimulating molecules were absent/or not involved and this deficiency can be overcome by supplying sucrose, Vit. B, Urea and mixture of amino acids (Engvild, 1964).

Dormancy of embryo has also been studied *in vitro* technique in many gymnosperms (Asakawa, 1959, Lee and Chang, 1963, Le Page Digivry, 1968, 1970, 1973a,b,c). Le Page Degivry and Garello (1973) studied the same in *Taxus baccata.* They found that mature seeds of *Taxus baccata* contain immature and dormant embryo. The presence of abscisic acid (ABA) in the dormant mature seeds inhibits its germination. Therefere, *in vitro* germination of excised seed could be successful if ABA is leached out from the seed. In *Ginkgo*, gibberellic acid promotes *in vitro* germination of excised embryos (Bulard and Le Page-Degivry, 1968).

Minocha (1980) cultured excised embryo of *Pinus strobus* on MS medium supplemented with various growth hormones. He showed how GA_3, 1BA, NAA and T1BA alters growth process of embryo. On BM having 3-6% of sucrose embryos develop proper plantlets however with its lesser concentration (1-2%) root primordia did not grow properly. GA_3 suppressed growth of root primordia, while auxin suppply cause callus formation. Shoots at the tip of cotyledons were formed when T1BA was added to the medium. Various other studies have been done on embryo culture on several modified nutrient media. (Sommer *et. al.*, 1975, Cheng, 1975). With the use of Gresshoff and Doy (1972) medium for embryo culture of *Pinus palustris* for first 4-6 week led to adventitious buds and cotyledons formation. Further these buds when grown on Risser & White (1964) spruce medium led to form plantlets and roots. Reilly and Washer's (1977) medium and Cheng (1975) medium have been tried on embryos of *Pinus ponderosa.*

White's medium with various supplements such as 2-4D, 1AA, CM, CH & YE have been used to culture embryos of *Biota orientalis* by Konav and Singh (1965). Konav & Singh (1980) reported continuous callus growth and regeneration of shoots from hypocotyle cotyledons and callus when obtained from mature embryo of *Pinus wallichiana.*

Some preliminary studies have been done on organogenesis in tissue cultures of *Cedrus deodara* and *Pinus roxburghii* by Bhatnagar et. al. (1983). In *C. dedora* callus could be obtained from mature female gametophytes and mature embryo. Adventitious shoot buds are formed on hypocolyle of embryos cultured on MS + KN (2mg/l) medium. Whereas in case of *P. roxburghii*, the embryos cultured on medium with BAP and NAA, primary needles and shoot buds are differentiated on the cotyledons.

Biotechnology applications in conifers :

Micropropagation techniques are most successful in case of embryo culture of conifers. Konar and Oberoi (1965) were the first to observe adventitious bud formation on cultured conifer embryos, were later this technique was improved and now used for propagation method (Sommer *et. al.*, 1975, Cheng, 1975, 1977, Campbell and Durgan, 1975). Sankla et al. (1967 b) used excised embryos of *Ephedra foliata* to grow callus that subsequently produce plantlets. Similarlly Konar and Singh (1988) reported shoot bud induction and differentiation of young roots from cultured hypocotyle segment and embryos of *Pinus wallichiana.* Callus cultures obtained from hypocotyle differentiated into shoots as well.

Selection of explant is the most important work in micropropagation technique, usually cotyledons and hypocotyles are preferred as explant. Highest potential of organogenesis has been found in germinating seeds and cotyledons. And cotyledous from germinating seeds produce more vigorous buds than those from ungerminated seeds. (Bonga and Durzan, 1987; Mehra-Palta and Thompson, 1988). *In vitro* adventitious bud formation on embryonic explants has been reported in *Pseudotsuga menziesii* (Kirby and Schalk, 1982), *Pinus strobus* (Flinn, *et. al.*, 1988), *P. taeda* (Mott and Amerson, 1981), *P. radiata* (Thorpe, 1988), *Picea abies* (Von Arnold & Hakman, 1988), *P. glauca* and *P. mariana* (Rumary *et. al.*, 1986) and *Abies fraseri* (Saravitz *et. al.*, 1993).

Somatic Embryogenesis :

It has been proved that embryo development is not the monopoly of egg cell and that any cell of embryo sac or any other part of sporophyte may give rise to embryo. The embryogenesis that does not begin with zygote is asexual or somatic. It is a process of development of somatic cells into differentiated plants through the characteristic zygotic stages. Somatic embryogenesis may be of two type i.e. either embryo develop directly from the explant or from callus or cell suspensions. Somatic embryogenesis is influenced by the nature of plant extract, growth regulators and by the physiological state of calli. For example 2-4 D in culture medium is least suitable in some cases for successful embryogenesis. Immature zygotic embryos are the best explants. However, the production of plantlets from somatic embryo is not very encouraging.

A lot of work has been done on somatic embryogenesis in conifers (Toutorus *et. al.*, 1991; Gupta *et. al.*, 1993; Attree and Fowke, 1993). It has been mostly used for rapid and mass production of desired genotype, needed specially for reforestation purposes. "Embryonal suspensor masses" can be obtained in culturing embryos of *Pinus taeda* (Gupta & Durzan, 1987a), *Picea abies* (Hawkman *et. al.*, 1985), *Picea glauca* (Lu & Thorpe, 1987) etc. A detailed account on somatic embryogenesis has been given by Attree & Fowke (1993) specially for gymnosperms. (Table IV)

Production of somatic embryo from seedling cotyledon has been reported by Lelu *et. al.*, (1987) from *Picea abies*. Radicle of mature zygotic embryo was used to obtain the callus in *Pinus lambertiana* (Gupta and Durzan, 1986a). Gupta and Durzan (1987a) used the term "*polyembryogenesis*" to describe the somatic embryo cultures of *Pinus taeda*. A comparative embryogenesis between *Picea glanca* and *Picea abies* was studied by Nagmani *et. al.*, (1987).

Cryptopreservation of somatic embryo-competent cells of *Picea alba* and *Pinus taeda* was also done by Gupta *et. al.*, (1987). He used liquid N_2 (–196°C) to recover viable cultures. Encapsulation of somatic embryos of these species have also been lined by mixing somatic embryos with 1.5% sodium alginate and dropped singly into 50-100 mm solution of Ca $(NO_3)_2$ (Gupta, 1988).

There are various advantages of somatic embryogenesis because

1. A very large number of plantlets of desired species can be produced.
2. Encapsulated somatic embryos with controlled dormancy could help in efficient storage and delivery functions.
3. Somatic embryogenesis can be better utilized in genetic engineering techniques.
4. The cell suspension which are amenable to bioreactor technology can cut down time cost and labour.

Therefore, the development of somatic embryogenesis techniques are an important milestone for rapidly growing culture systems available (Dunstain, 1988). According to (Karnosky, 1980) there are many gymnosperm trees in which the whole plant have been regenerated using tissue culture techniques (see Table I).

These technique have helped to regenerate the degraded forest as well as to meet the ever increasing demande of forest trees.

Table IV. Somatic Embryogenesis in Some Conifers

Species	Explants	Investigator/s
1. *Abies albe*	Immature zygotic embryos	Schuller *et. al.*, (1989)
2. *Larix decidua*	Megagametophytes	Nagmani and Bonga (1985), von Aderkas et. al. (1987) von Aderkas and Bonga (1988)
3. *L. eurolepis*	Immature zygotic embryos	Klimaszewska (1989), von Aderkas et. al. (1990)
4. *Picea abies*	Immature zygotic embryos	Hakman et. al. (1985) Hakman and v Arnold (1985)
	Mature zygotic embryos	Jain et. al. (1988) Verhagen and Wann (1989), von Arnold and Hakman (1986), von Arnold (1987)
	Mature zygotic embryos and 3-7-day old seedlings	Krogstrup (1986) Lelu et. al. (1987)
	Megagametophytes	Simola and Santanen (1990)
5. *P. glauca*	Immature zygotic embryos	Attree et. al. (1989), Hakman and Fowke (1987), Lu and Thorpe (1987)
6. *P. glauca*	Mature zygotic embryos 12-30-day-old seedlings	Tremblay (1990) Attree et. al. (1990)
7. *P. sitchensis*	Immature zygotic embryos	Krogstrup et. al. (1988)
	Mature zygotic embryos	von Arnold and Woodward (1988)
8. *Picea mariana*	Immature zygotic embryos	Hakman and Fowke (1987), Tautorus et. al. (1990)
	Mature zygotic embryos 12-21-day-old seedlings	Tautorus et. al. (1990) Attree et. al. (1990)
9. *Pinus caribaea*	Immature zygotic embryos	Laine and A. David (1990)
10. *P. elliottii*	Immature zygotic embryos	Jain et. al. (1989)
11. *P. lambertiana*	Mature zygotic embryos	Gupta and Durzan (1986a)
12. *P. radiata*	Immature zygotic embryos	Chandler et. al. (1989)
13. *P. strobus*	Immature zygotic embryos	Finer et. al. (1989)
14. *P. taeda*	Immature zygotic embryos	Gupta and Durzan (1987a), Becwar et. al. (1990, 1991)
15. *Pseudotsuga menziesii*	Immature zygotic embryos	Durzan and Gupta (1988)
16. *Sequoia sempervirens*	7-day-old seedlings	Bourgkard and Favre (1988)

II. World Distribution of Gymnosperms

Conifers :

Conifers tend to live in cooler or drier environment. Most exists in the coniferous forests of the northern hemisphere south of arctic tundra and north of the temperate deciduous forests, although there are exceptions (Raven and Johnson, 2002)

Cycads :

Cycads live in tropical and subtropical climate and are generally found between the tropics of cancer and the tropics of capricorn (Norstog and Nicholls, 1997). Some though live in dry forests also (Arens and B.R.S. 2002). Most inhabit parts of Central America, South Africa, Eastern Asia, and Australia (Sperm, 1965). However they can be found on every continent except Europe and Antarctica (Arens and B.R.S., 2002)

Ginkgos :

G. biloba is the only surviving species of the Ginkgos. It is believed to no longer exists in wild. However, they are grown in China and Japan. Also, it is thought to be indigenous to the Zhejiang province in eastern China where it may still inhabit secluded mountain valleys (Kubitzki, 1990).

Gnetophyta :

Gnetophyta exists everywhere in the world except Australia (Raven and Johnson, 2002). Same are found in arid deserts in the United States, Mexico and Africa while other occurs in moist tropical forests in the Amazon, Panama, Australiea, West Africa, India, South China and Malaysia (Sprme, 1965).

III. Impact of Gymnosperms on Human Society

Human Society has since long been influenced by gymnosperms. A brief account follows.

Conifers : Humans rely on conifers for timber, paper, resin and turpentine. Turpentine is made from liquid component of resin and rosin, a solid component of resin, is used in string instruments. Conifers are also valuable in the field of medicine. *Taxol*, a medicine used in the treatment of cancer, is extracted from the bank of the pacific yew (*Taxus brevifolia Nutt*) (Sanganee and Harrison, 2002). Conifers are also used as Christmas trees for decoration during holidays in many Christian households. However timber is the main economic source of conifers to human society (Kubitzki, 1990). Unfortunately destruction of entire forests for this purpose has severely threatened conifers. This rapid deforestation combined with a slow growth rate have endangered conifers. Human Society now is faced with discovering ways to conserve the earths's conifers populations.

Cycads : Cycads are source of food, fibre and gum (Kubitki, 1990). Cycads like other gymnosperms, also have a medical application since they have been used for medicine in the past. Lastly, wreaths can be made out of their fronds. However, like other gymnosperms cycads are also in the danger of extinction as a result of habitat destruction (Kubitzki, 1990). Tourists have been contributing to the problem by taking pieces of cycads as souvenirs, consequently, the conservative of cycads is essential for their survival.

Ginkgos : Because they can withstand air pollution, *Ginkgo* Plants are planted on city streets in many countries (William, 2000). However, a current medical application of *Ginkgo biloba* is the dietary supplement of *Ginkgo biloba* (SEB) extract (Mahady, 2002). Many people take this dietary supplement daily because it is believed to have various positive effects on health. Studies have shown that SGB extract does have some health benefits. It is used to improve the memory and put a stop to memory decline with age (Mahady, 2002). Consequently, it is used to treat dementia-related diseases, such as Alzheimer's disease. It has been shown to be an antioxidant, which reduces harmful free radicals in the body. Lastly, it may well be useful in the treatment of cardiovascular disease and the prevention of strokes. SGB decreases blood clotting, dilates blood vessels, increases blood flow, and regulate blood pressure.

Gnetophytes : Gnetophytes are also used as a source of food. It has also been a source of medicine. The drug Ephedrin was extracted from a member of the genus *Ephedra* in the past and used in the treatment of respiratory problems. Currently Ephedrin is synthetically made (Ravan and Johnson, 2002) but other drugs extracted from Gruetophytes are used as vessel constrictants (Kubitzki, 1990). **Mormon tea** made in the south western United States is also made from the Gnetophyte extract.

IV. GYMNOSPERMS: Areas to be explored.

The Gymnosperms- the naked seed plant are an important group of plants in a sense that most of its species are of high economic value. They are spread all over the world mostly in temperate regions and at higher elevations in the tropics. Coniferous woods are in high demand. Gymnosperms provide a variety of wood used in manufacture of paper, board, timber, building construction furnitures *etc.* at one side, medicines, perfumes some of essential oils on the other side of life we have day to day use of Gymnosperm plants either directly or indirectly. *Ginkgo* is worshiped in China and it is a very familiar tree in most of the Buddhist temples there and in Japan. Gymnosperms play a very important role in all aspects of human life, they are garden tree to provide grace to habitat and to some extent are responsible for fossil fuel in the form of coal. Most of the Gymnosperms are evergreen trees and shrubs forming dominant vegetables in tropical forests. Some of the tallest trees known is plant kingdon belong to Gymnosperms. *Sequoia sempervirens* (Californian redwood) restricted to a narrow coastal belt in California, is probably the tallest living tree reaching to a height of about 112 m and attain a girth of 15m & is as tall as a thirty-six storey building (Laetsch, 1979). *Sequoiadendron giganteum* (big tree) in central California is most massive and is among the oldest tree. The smallest gymnosperms is a cycad i.e. *Zamia pygmaea* whose fronds are only 4 to 5 cm long.

Plate I: *Zamia pygamaea*-smallest known gymnosperm.

It is the wood quality which has made this group economically very importants. Vascular system of the stems are very pronounced which are having open, collateral, endarch vascular bundles arranged in a ring. Secondary growth is very pronounced. An important part of it is that the secondary xylem composed of mainly tracheids lack vessels. Fibre-tracheids are found in late woods, libriform fibres are absent in it. Therefore its wood is referred to as soft wood-a marked difference with dicolytedons having only hardwood.

Plate II : *Ginkgo biloba* *"living fossil"*

Woods are of two types the pycnoxylic and Manoxylic. Manoxylic wood are comparatively soft mainly composed of parenchymatous rays, large pith and cortex. Whereas the commercially important pycnoxylic woods are campact, with small xylem rays, lesser amount of cortex and pith. The manoxylic wood is related to megasporophyllous leaves and radial symmetry of seeds, whereas pycnoxylic wood is related to microphyllous leaves and bilateral symmetry of seeds (Sperm, 1974). As Gymnospermous wood are very important due to its timber quality, it is necessary to have a comprehensive study on seasonal variation on wood anatomy (Sharma, 1994) and effect of external environmental factors on the process of secondary growth. The effect of physiological, structural and genetic factors and the cambial activity may also help us to understand more about wood characters of the plant as a whole (Munting and Willemse, 1987). Little has been studied on secondary phloem, its activity under various external stimuli. Very few studies are available on this (Alfieri and Event, 1973, Denne, 1976).

Demand for quality timber is increasing day by day, it is necessary to have proper techniques and management practices in sericulture so that we are not only able to regenerate the forest trees but are also to meet their demands in society. Micropropagation through cell and tissue culture techniques should be used and encouraged to multiply the production of plantlet of Gymnosperms of economic importance. Not only the molecular genetics can be used specially in conifers as has already been demonstrated that *Agrobacterium* can infect several conifers belonging to Pinaceae. Taxodiaceae and Taxaceae (Smith, 1942). *Agrobacterium* is a naturally occuring bacterium, and can act as vector for transfer and expression of foreign gene. It has got wide host range and has been reported to conifers as well (Decleene and De Ley, 1976, Dunstan, 1988). Though genetic engineering technique is not much in use in Gymnosperms, efforts should be made to get new improved genetic variations and reduce barriers to crossability amongst other species so that wood quality can be improved. Scope of biotechnology input must be increased in Gymnosperms. (Mohammad and Vidavar, 1988; Mohammad et. al. 1989).

Little work has been done on the coralloid roots of *Cycas*. The knowledge about their growth and development is very poor. A detailed study is needed for transitional stages between procoralloid and coralloid stage. Not only the mechanism by which symbionts infect and colonise the coralloid roots, their restriction to a particular layer only needs urgent attention. These studies may provide some significant results which may help us to signify its role, may be in alternative to reliance on chemical fertilizer (Ahern. and Staff, 1994).

There are many plants which can survive in highly air polluted areas, although harmful effects of air pollution are many. Generally air pollutants inhibits seed germination; there is poor quality of growth and photosynthetic efficiency of plant growing in polluted areas decreases. Among Gymnosperms, *Ginkgo biloba* can resist air pollution and can be used for planting on road sides. Renzoni *et. al.* (1990), who has studied effect of air pollution an *Pinus pinea* found that the plant is very sensitive to air pollution, thus can be used as pollution indicator. Air pollution also cause decrease in pollen fertility which may ultimately lead to decrease in seed production (Renzoni, *et. al.* 1990). Such type of studies are still very few among Gymnosperms. Therefore this is the area where work and more research is needed.

Plate III: Trunk of sequoia giganteum the world's largest tree in terms of volume.

Gymnosperms require a great deal of study in the field of pollen biology and reproductive biology. Studies are very meagre in this field. There are a lot of shortcomings when our knowledge is limited specially in physiology of gymnosperms, cellular and biochemical reactions etc. These require immediate attention habit. Latest techniques like TEM, SEM fluorescence and interference microscopy should be employed for studies on cytology and histochemisty, pollen viability, its vectors, dispersal, wall formation in female gametophytes, effect of environmental factors on development of archegonium and antheridium etc. also need more attention if we want to conserve and reforest these natural resources.

Finally, a complete and comprehensive life cycle of many gymnosperms is yet to be studied. Temporal considerations in the reproductive cycle, identification of wild genes in gymnosperms are some of the other areas where knowledge is to be acquired for superior genetic variation in gymnosperms which can be incorporated for tree improvement with the help of biotechnology and tissue culture.

As Gymnosperms are the last group of Archegoniates the detailed study is still confined only to few genera, therefore research priority must be fixed to study about more and more Gymnosperms.

QUESTIONS

ESSAY TYPE

1. Write a short account of Experimental Embryology in Gymnosperms.
2. How is biotechnology used in Gymnosperms.
3. Give a brief account of the development of use of biotechnology in Gymnosperms.
4. Describe world wide distribution of Gymnosperms.

SHORT ANSWER TYPE

5. Write short notes on:
   (*a*) Impact of gymnosperms of Human Society
   (*b*) Areas to be studied in Gymnosperms.
6. What is somatic embryogenesis? How has it been used in conifers?

OBJECTIVE TYPE

7. **Fill in the blanks:**
   (*i*) Biotechnology is used for — of —
   (*ii*) Somatic embryogenesis has been used in ——
   (*iii*) The first to culture —— tissue on artificial culture medium was — ——
   (*iv*) The objective of use of biotechnology is to ———
   (*v*) Gymnoperms have since long affect ———

Bibliography

Abagon, 'M.A. (1938). ***Univ. Philipines nat. and appl. sci. Bull.,*** **6**: 29—58 (Anatomy of Pine needles).

Abraham, A. and Mathews, P.M. (1962). ***Ann. Bot.*** **26**: 261—6 (X and Y chromosomes in Cycas).

Adalkar, P.G. and C.N. Rao, (1963). ***Rec. Geol. Surv. India,*** **92(2)**: 319—323.

Agashe, S.N., (1969). ***Palaeobotanist,*** **17(3)**: 312—316.

Ahmed, P. and N. Ahmed. ***Proc. nat. Inst. Sci. India,*** **28A**: 16—56.

Alfleri, F.J. and R.F. Evert (1968a). ***Planta*** **78**: 93—97 (Albuminous cells in Pinus).

— and — (1968b). ***Amer, Jour. Bot.*** **55**: 518—528 (secondary phloem in Pinus).

Allen, G.S. (1942). ***Jour. Forest,*** **40**: 642—644 (Parthenogensis in Douglass Fir).

— (1943). ***Amer. Jour. Bot.*** **30**: 655—661 (Embryogeny in Pseudotsugataxifolia).

— (1946). ***Ibid (Pseudotsuga taxifolia).***

Alsoi, M.L. and F.J. Alfieri (1972). ***Amer. Jour. Bot.*** **59** (**8**): 818—827 (Secondary phloem ***in Ephedra***).

Andrews, H.N. (1940). ***Ann. Mo. bot. Gdn.*** **27**: 51—118 (wood anatomy of Pteridosperms).

— (1945). ***Ibid.*** **32**: 323—60 (***Schopfiastrum).***

— (1963). ***Science***, **142**: 925—81 (Early seed plants).

— (1973). ***Amer. Jour. Bot.*** **60**: 15 (Abstracts; Heterosporous plant from Middle Devonian).

— and Felix, C.J. (1952). ***Ann. Mo. bot. Gdn.*** **39**: 127—35 (Gametophyte of ***Cardiocarpus***).

Antevs, E. (1914). ***K. svenska Vetensk Akad. Handl.,*** **51**: 1—18 (Peltaspermaceae).

Anderson, E. (1934). ***Nature*** (Lond.) **133**: 462 (orign of Angiosperms).

Apte, V.V., and Kulkarni, L.H. (1953). ***Curr. Sci.*** **22**: 261—262 (Gnetum ula).

Arber, A., (1910). ***Ann. Bot.,*** **24**: 491—509 (Mitrospermum).

— (1946). ***Chronica Bot.,*** **10**: 67—123 (Goethe's Botany).

Arber, E.A.N., (1905). ***Quart. Jour. Geol. Soc. London.*** **61**: 324—38 (Sporangia of Glossopteris).

— and Parkin, J. (1907). ***Jour. Linn. Soc. Bot.,*** **38**: 29—80 (Bennettitales and origin of Angiosperms).

Arnold, C.A. (1948). ***Bot. Gaz.,*** **110**: 2—12 (Classification of Gymnosperms).

— (1953). ***Phytomorph,*** 3: 51—65 (Evolution of Cycads).

Arnold, C.A. (1967). ***Phytomarphology,*** **17(1—4)**: 346—350.

Arnott, H.J. (1959, 1962). ***Amer. Jour. Bot.*** **46:** 405—411; **49:** 658 (Vein anastomoses in ***Ginkgo***).

Asakawa S. (1959). Physiology of embryo dormancy J. Jap. For. Soc., **41:** 389-395.

Attree, S.M. & Fowke, L.C. (1993) Embryogeny of Gymnosperms : Advances in synthetic seed technology of conifers. Plant cell, Tissue and Organ culture, **35 :** 1-15.

Bailey, I.W. (1925). ***Ann. Bot.*** **39:** 587—598 (Tracheary pitting in Gymnosperms).

— (1956). Arnold Arbor., **37:** 269—287 (Nodal anatomy in retrospect).

Baird, A.M, (1955). ***Phytomorph.*** **3:** 258—284 ***(Callitris).***

Banerjee, S.N. (1968). Bot. Magz. (Tokyo) **81:** 11—21 (Pinus).

Banerjee, D. (1968). ***Rev. Palaeobot. Palylnol.*** **6:** 171—176.

— (1973). ***Palaeobotanist***, **20(1):** 1—6.

Banerjee, M, (1969). ***J. Sen Man. Vol.,*** : 359—368.

Banks, H.P. (1968). The early history of land plants. ***Symposium on Evolution and Environment***, E.T. Drake, ed. Yale University Press, New Haven (pp.73—107).

— (1968). The stratigraphic occurrence of early land plants and its bearing on their origin. In ***Proceedings of the international symposium on the Devonian System.*** D. Oswald, ed., Calgary, Alberta, Canada.

Bannan, M.W. (1958). ***New Phytol.*** **57:** 132—134. (Tracheids of ***Thuja Occidentalis***).

Barghoorn, E.S. Jr. (1940). ***Bull. Torrey Bot. Club.,*** **67:** 303—328 (Uniseriate rays in Conifers).

Barnard, P.D. (1959). ***Ann. Bot.*** 23: 285—296 ***(Esperma oxroadense).***

— (1960). Paleantology. **3:** 265—275 ***(Calathospermum fimbriatum).***

Baxter, R.W. (1949). ***Ann. Mo. bot.*** Gdn. 36: 287—352 (Medullosa sp.).

Ball E.A. (1978). Cloning in vitro of ***Sequoia sempervirens***. (Abstr.) p. 163. In : Thorpe T.A. (ed.) frontiers Plant Tissue Culture 1978. Proc. IV Intl. Congr. Plant Tissue & Cell culture calgary (Canada).

Ball, E.(1955). Studies on nutrition of callus cultures of ***sequoia sempervirens.***Ann. Biol., **31:** 281-305.

Ball. E. (1950).Differentiation of callus cultures of ***sequoia sempervirens.Growth***, **14:**295-325.

Bhatnagar S.P. & Moitra Alok. (1996),. ***Gymnosperms***. New Age International (P) Ltd. Publishers, N. Delhi, 470 pp.

Bhatnagar S.P., & Singh M.N. (1984). Organogenesis in the cultured female gametophyte of ***Ephedra foliata***. J. Exp. Bot. **35 :** 268-278.

Bhatnagar. S.P., Singh M.N., Kapur N. (1983), Preliminary investigations on organ differentiation in tissue cultures of ***cedrus deodara*** and ***Pinus roxburghii.*** Indian J. Exp. Biol. **21 :** 524-526.

Biswas, C & Johari B.M (2004). The ***Gymnosperms*** Narosa Publishing House, N. Delhi. 497 pp.

Bonga J.M and Durzan; D.J. (eds). (1987),Cell and Tissue Culture in Forestry Vol. I to III. Martinus Nijhoff, Publ., The Netherlands.

Bonga J.M., & Fowler, D.P. (1970) Growth and differentiation in gametophytes of ***Pinus resinosa*** cultured in vitro. Can. J. Bot. **48 :** 2205-2207.

Bonga, J.M (1974).In vitro culture of microphylls and megagametophyte tissue of Pinus. In Vitro, **9:** 270-277.

Bonga, J.M. (1977). Applications of tissue culture in forestry pp. 93-108. In : Reinert J, Bajaj YPS (eds). Applied and Fundamental Aspects of Plant cell, Tissue and Organ Culture in vitro Can. J. Bot. **48 :** 2205-2207.

Borchert, R. (1968). Spontane Diploidisierung in Gewebekulturen des megagametophyten von ***Pinus Lambertiana*** Z. Pflanzenphysiol. **59 :** 389-392.

Borchert. R, (1967). The cultivation of tissues of cupressus lusitanica in vitro Physiol Plant **20 :** 608–616.

Branscheidt P. (1930) Zur Physiologie der PollenKeimung und ihrer experimentalen Beeinflussung. Planta **11:** 368–456.

Brown C.L. & Sommer H.E. (1977). Bud and root differentiation in conifer cultures. Tappi, **60 :** 72-73

Bulard C. (1952). Culture aseptique de' embryons de ***Ginkgo biloba***. Role des cotyledons dans l'absorption du sucre etla croissance de la tige. C.R. Acad. Sci. Paris 265 D : 1301-1304.

Bullard C, Le page-Degivry M.T. (1968). Quelques pre´cisions sur les conditions d´ obtention d´ une inhibition de la croissance espicotylaire chez. ***Ginkgo biloba*** L. sous l´effect de l´acide gibberellique. CR Acad Sci. Paris, **2660 :** 356 - 359.

Butenko R.G. & Yakovleva S.M. (1962). Controlled organogenesis and regeneration of a whole plant in a culture of non-differentiated plant tissue. zr Akad Nauk USSR Bid. Ser. **2 :** 230-241.

Beck, C.B. (1957). ***Amer. Jour. Bot***. **44:** 350—367 (Tetraxylopteris).

— (1960). ***Brittonia***. **12:** 351—368 (***Archaeopteris*** and ***Callixylon***).

— (1962). ***Amer. Jour***. Bot., **49:** 372—382 (***Archaeopteris***).

— (1967). ***Palaeotographica*** **121:** 1—22 (***Eddya sullivanensis***).

— (1966). Taxon **15**: 337—339 (Origin of Gymnosperms).

— (1970). ***Biological Review*** **45:** 379—400 (Gymnospermous structures).

Benson, M.J. (1914). Trans. ***roy. Soc. Edinb***, **50:** 1—15 (Sphaerostoma).

Berridge, E.M. (1911). ***New Phytol***. **10:** 140—144 (Gnetalean and Bennettitalean seeds).

— (1912). ***Ann. Bot***. (Lond.), **26:** 987—992 (Gnetum).

Bharadwaja, R.C. (1957). ***Jour, Indian Bot. Soc.***, 36: 408—420.

Bharadwaja, D.C., (1952). ***Lloydia*** **15(4):** 234—240.

— D.C. (1953). ***Paloeobotanist***, **2:** 59—70.

Billings, W.D. and Thompson, J.H. (1957). ***Ecology***. **38:** 158—160 (Pinus aristata, over 4,000 years old).

Bierhorst, D.W. (1971). ***Morphology of Vascular plants.*** Macmillan Co., New York.

Biswas, K. (1933). ***J. Indian Bot. Soc.***, **12(1)** : 24—47.

Bose, M.N. (1952). ***Brachyphyllum spiroxylum*** sp. not. from the Rajmahal Hills. India. J. ***Indian bot.*** Soc. **31(4):** 287—296.

— (1953). ***Bucklandia sahnii sp.*** nov. from the Jurassic of the Rajmahal Hills, Bihar. ***Palaeobotanist*** **2:** 41—49.

— (1953a). On some fossil cycadean stems from the Rajmahal Hills, Bihar. ***Palaeobotanist*** 2.

— (1953b). ***Ptilophyllum amarjolensis sp.*** nov. from the Rajmahal Hills, Bihar. ***Proc. nat. Inst. Sci. India.*** **19B(5):** 605—612.

— (1958). Morrisia, a new genus of Cycadophytic fronds from the Rajmahal Hills, Bihar. ***Palaeobotanist,*** **7(1) :** 21—25.

— (1958a). The fossil flora of Jabalpur group 1. ***Ptilophyllum instaticallum*** n. sp. ***Palaeobotanist,*** **7(1):** 26—28.

— (1966). A petrified Bennettitalean flower from the Rajmahal Hills, India. Curr. Sci. 35(22): 569—570.

— (1967). ***Weltrichia singhii*** n. sp. from the Rajmahal Hills, Bihar. ***Curr. Sci***. 36(2): 48—49.

— (1968). A new species of ***Williamsonia*** from the Rajmahal Hills, India. J. Linn. Soc. (Bot). 61: 121—127.

— and J. Hsu (1953). On some coniferous cones, probably of ***Brachyphyllum,*** from the Jurassic of the Rajmahal Hills, Bihar, India. ***Proc. Acad. Sci. India.*** 19(2): 203—209.

— and K.P. Jain (1964). A megastrobilus belonging to the Araucariaceae from the Rajmahal Hills, Bihar, India. ***Paleobotanist*** 12(3): 229—231.

— and M.L. Kasat (1969), The fossil flora of the Jabalpur series. 4. ***Williamsonia seniana*** n. sp. ***Sen. Mem. Volume***: 305—309.

— and (1972). The genus ***Ptilophyllum*** in India. Palaeobotanist 19(2): 115—145.

— and H.K. Maheswari (1973). ***Brachyphyllum sehoraensis,*** a new conifer from Sehora, Narsinghpur District, Madhya Pradesh. ***Geophytology*** 3(2): 121—125.

— and (1973a). Some detached cone-scales belonging to Araucariaceae from the Mesozoic rocks of India. ***Geophytology*** 3(2): 205—214.

— and S.C.D. Sah (1954). On ***Sahnioxylon rajmahalense,*** a new name for **Homoxylon rajmahalense** Sahni and S. ***andrewssi***, a new species of ***Sahnioxylon*** from Amrapara in Rajmahal Hills, Bihar. Palaeobotanist. 3: 1—8.

— and S.C. Srivastava (1971). The genus ***Dicroidium*** from the Triassic of Nidpur, Madhya Pradesh, India. ***Palaeobotanist*** 19(1): 41—51.

— and (1973). ***Nidostrobus*** Gen. nov., a pollen-bearing fructification from the Lower Triassic of Gopad river valley, Nidpur. ***Geophytology***. 2(2): 211—212.

— and (1973a). Some micro and megastrobili from the Lower Triassic of Gopad river valley, Nidpur. ***Geophytology*** 3(1): 69—81.

— and Sukh Dev (1957). A new species of ***Ptilophyllum*** from Bansa, South Rewa Gondwana basin. ***Palaeobotanist***. 6(1): 12-15.

— and (1958). Studies on the fossil flora of the Jabalpur series from the South Rewa Gondwana basin. 1.***Cycadopteris, Nipaniophyllum*** and ***Ginkgoites. Palaeobotanist,*** 7(2): 143—154.

— and (1972). Three new species of Pagiophyllum from Bansa, Madhya Pradesh, India. ***Geophytology,*** 1(2): 116—122.

Bower, F.O. (1882). ***Quart. Jour. Micr. Sci.,*** **22:** 278—298 **(*Gnetum gnemon*).**

Brack, S.D. (1973). ***Amer. Jour. Bot.,*** **60:** 16 (Abstracts) (Additional Studies on ***Dolerotheca*).**

Brongniart, A. (1874). ***Annls. Sci. nat. (Bot.).*** **20:** 234—260 (upper carboniferous seeds).

Brotzman, N.L. and J.T. Schabilion (1972). ***Amer. Jour. Bot.*** (Abstracts) **59:** 659 **(*Cardiocarpus magnicellularis*).**

Buchholtz, J.T. and N.E. Gray (1948-1951). ***Jour. Arnold Arbor.*** **29:** 49—76, 117—151; 32: 82—97 **(*Podocarpus*).**

Calder, M.G. (1938). ***Trans. roy. Soc. Edinb.*** **59:** 309—31 **(*Samaropsis*).**

Camp, W.H. and M.M. Hubbard (1963). ***Amer. Jour. Bot.*** 50: 235—43 **(*Lyginopterid seed*).**

Canright, J.E. (1942). ***Amer. Jour. Bot.*** **59:** 660 **(*Metasequoia*).**

Chamberlain, C.J. (1909). ***Bot. Gaz.*** **47:** 215—236 (***Dioon***).

— (1910). ***Ibid.*** **50:** 415—429 **(*Dioon*).**

— (1916). ***Ibid.*** 61: 352—372 **(*Stangeria paradoxa*).**

Campbell R. & Durzan D. (1976). Vegetative propagation of ***Picea glauca*** by tissue culture. Can J. For Res. **6 :** 240-243.

Campbell, R.A. & Durzan D.J. (1975). Induction of multiple buds and needles in tissue cultures of ***Picea glanca***. Can. J. Bot. **53 :** 1652-1657.

Chang T.Y. (1977). Factors affecting adventitious bud formation of cotyledon culture of Douglas fir. Plant Sci. Lett. **9 :** 179-187.

Cheng T.V. (1975). Adventitious bud formation in culture of Douglas fir (***Pseudotsuga menzisii***): ***mirb***. Plant cell Physiol. **17 :** 1347-1350.

Cheng T.Y. & Voqui T.H. (1977). Regeneration of Douglas fir plantlets through tissue culture. Science **198 :** 306-307.

Cheng T.Y. (1976). Vegetative propagation of western hemlock (***Tsuga heterophyll***a) through tissue culture. Plant cell physiol. **17 :** 1347-1350.

Coleman W.K. & Thorpe T.A. (1977). In vitro culture of western red cedar (***Thuja plicata*** Donn). Plantlet formation. Bot Gaz. **138.** 290-304.

Constabel, F. (1957).Ernahrungsphysiologische und manumetrische Untersuchungen zur Gewebekultur der ***Gymnosporangium*** Gallen von Juniperous Arten.Biol. Abl; **76:** 385- 413.

Constabel. F. (1965) Phenolics in the tissue cultures derived. from ***Juniperos communis L.*** Studies in Tannin synthesis pp 183–190. In : white PR. Grove AR (eds) Intl. Conf. plant tissue culture. Mc Cutchan Publ, Berkely (Cali., USA).

Coulter J.M. & Chrysler M.A. (1904). Regeneration in Zamia. Bot. Gaz. **38 :** 130-139

Chaney, R.W. (1948). Redwoods in China. Nat. ***Hist.*** **57:** 404—445.

Chitaley, S.D. (1949). On a new species of ***Dodoxylon, Dadoxylon eocenum*** sp. nov. from the district of Chhindwara, C.P. India. ***J. Indian bot. Soc.*** **28(4):** 221—237.

Chowdhury, R. (1962). ***Phytomorph.,*** **12:** 313—38 (Conifer embryology).

Christopher, D.C. (1972). ***Ibid.*** **59:** 660 (***Sciadopitys*** like conifer).

Chuang, T.I. and W. Hu (1963). ***Bot. Bull. Acad. Sin.*** (Shanghai) **4:** 10—14 (***Amentotaxis***).

Cook, P.L. (1939). A morphological comparison of two species of ***Thuja.*** Dess. Univ. Illinois, Urbana U.S.A.

Cookson, I.C. (1949). ***Mem. Nat. Mus. Melbourne*** **16:** 117—131.

Coulter, J.M. (1908). ***Bot. Gaz.*** **46:** 43—49 (Embryo of Gnetum gnemon).

Crepet, W.L. and T. Delevoryas (1972). ***Amer. Jour. bot.*** **59(2):** 209—215 (Cycadeoid ovule).

Cross, G.L. (1939). ***Bull. Torrey bot. Cl.*** **66:** 431—452 (***Taxodium distichum***).

— (1941). ***Amer. Jour. Bot.*** **28:** 573—582 (***Cryptomerda japonica***).

David H. Kausa I. & David A. (1978) In vitro studies of adventitious and axillary buds, shoot formation and plantlet regeneration in ***Pinus pinaster*** (Abster.) P. 30. In: Thorpe T.A. (eds). Frontiers of Plant Tissue Culture, 1978, Proc. IV Int. Congr. Plant, Tissue and Cell Culture, Calgary (Canada).

DeLuca P., & Sabato. S. (1979). ***In vitro*** spermatogenesis of ***Encephatortos***. Lehm. ***Caryologia***, **32:** 241–245.

DeLuca P., LaValva, V. & Sabato, S. (1980). Spermatogenesis and tissue formation in cycad pollen grains. ***Caryologia***, **33:** 261–265.

DeLuca, P., Moretti, A. & Sabato. S. (1979). Regeneration in megagametophytes of cycads. Gion. Bot. Il al. **113 :** 129-143.

Duchartre, M.P. (1888). Note sur l'enracinement de l'albumin d'un ***Cycas. Bull. Soc.*** *Fr.* **35 :** 243-256.

Duhoux E., & Norreel, B (1974). Sur Lsolement de Colonies tissulaire d' origin pollinique a parter de cones males du ***Juniperus chinensis L. du J. Cummunis L. et Cupressus arizonica*** G. Cultives in vitro. C.R. hebd. Seanc. Acad. Sci., Paris, **277 :** 2665-2668.

Duhoux, E (1973). Observations infrastructurales des parois au cours de la germination in vitro du pollen du ***Juniperus communis*** (cupressaccaes), ***caryologia***, 25 (suppl.) :191–200.

Dunstan D.I. (1988). Prospects and progress in conifer biotechnology Can. J. For Res. **18 :** 1497-1506.

De Ferrey Y. (1942). ***Bull. Soc. Hist. Nat.*** Toulouse **77:** 3—16 (***Cupressaceae***).

— (1958). ***Ibid.*** **105:** 155—205 (Bibliography of Conifer Literature).

De Lanbenfels, D.J. (1962). Phytomorphology **12:** 296—300.

Delevoryas, T. (1953). ***Amer. Jour. Bot.*** **40:** 144—150 (Cordaitales).

— (1955). Ibid. **42:** 481—488 (***Palaeostachya***).

— (1955). ***Palaeontographica*** **97:** 114—67 (Evolution in Medullosa).

— (1956). ***Contri. Mus. Geol. Univ. Michig.***, **12:** 285—299 (Shoot apex of ***Callistophyton***).

— (1959). ***Amer. Jour. Bot.*** **46:** 657—66 (***Cyadeoidea***).

— (1963). ***Ibid.*** **50:** 45—52 (Flowers of ***Cycadeoidea***).

— (1973). ***Ibid.*** **60:** 16 (Abstract, Gymnosperms from late Triassic).

and Morgan, J. (1954). ***Palaeontographica*** **96:** 12—23 (***Callistophyton***).

De Sloover, J.L. (1963). ***Cellule*** **63:** 331—358 (***Cycadales***).

Dennis, R.L. and D.A. Eggert (1972). ***Amer. Jour. Bot.*** (Abstracts) **59:** 660 (A Medullosan Pollen Organ).

Doak, C.C. (1937). ***Bot. Gaz,*** **98:** 808—815 (***Cupressus***).

Dogra, P.D. (1966). ***Phytomorpholoy.*** 16: 125—141 (Embryogeny of Taxodiaceae).

— (1966). ***Silvae Genet.*** **15:** 11—20 (***Abies pindrow***).

— (1967). ***Studdia Forestalia Seccia*** **45:** 1—97 (Embryogeny in conifers).

Doyle, J. (1945). ***Sci. Proc. R. Dublin Soc.*** **24:** 43—62 (Pollination in conifers).

— (1964). ***Scienti. Proc. Roy. Dublin Soc.*** **26:** 347—373 (Podocarpus).

— (1963). ***Proc. Roy. Irish Acad.*** **62:** 181—216 (Proembryogeny in Pinus).

and A. Kane (1943). ***Scient. and Proc. Roy. Dublin Soc.*** **23:** 57—70 (***Pollination in Tsuga pattoniana and species of Abies and Picea***).

and W.J. Looby (1939). ***Ibid.*** **22:** 127—147 (Embryogeny in Saxegotheca).

and M.O. Leary (1935). ***Ibid.*** **21:** 181—190 (Pollination in ***Pinus***).

and (1935). ***Ibid.*** **21:** 191—204 (***Pollination in Tsuga, Cedrus, Larix,*** etc.).

Dupler, A.W. (1920). ***Bot. Gaz.*** **69:** 492—520 (***Taxus canadensis***).

Duthie, A.V. (1912). Ann. Bot. (Lond). **26:** 593—603 (Anatomy of Gnetum africanum).

Dutt, B.S.M. (1952). ***Sci.*** and ***Cult.*** **18:** 199 (Abnormal cones in ***Gnetum***).

Eames, A.J. (1952). ***Phytomorphology*** **2:** 79—100 (Relationship of Ephedrales).

Edwards, W.N. (1929). ***Ann. Mag. nat. Hist.*** **4:** 385—94 (***Caytonia***).

Eicke Ruth (1957). ***Bot.*** Jb. **77:** 193—217 (Gnetales).

— (1958). Z. Bot. **46:** 5—15 (Phylogeny of Gymnosperms).

Elliot G.C., (1948). ***Proc. Linn. Soc.*** New South Wales **73:** 120—129 (Phaerosphaera hookeriana).

— (1950). ***Ibid.*** 75: 320—333 (P. hookeriana).

— (1951). ***Ibid*. 76:** 36—40 ***(Athrotaxis).***

Emberger, L. (1954). Sv. Bot. ***Tidskr* 48:** 361—367 ***(Ginkgoales).***

Evert, R.F. and F.J. Alfieri (1965). ***Amer. Jour. Bot*.** 52: 1058—1066 (Sieve cells in conifers).

Fagerlind, F. (1938). ***Ark. Bot*.** 29A: 1—15 ***(Ditepalanathus).***

— (1941). ***K. Svenska Vetensk. Akad. Haudl*.** 19 : **(8):** 1—55 (Gametophytes of Gnetum).

— (1946). Ark. Bot. **33A:** 1—57 ***(Gnetum)***.

— (1954). ***Svensk Bot. Tidskr*.** 48: 449—470 (Shoot meristem in ***Gnetum Ephedra*** and other Gymnosperms).

Favre Duchartre, M. (1956). ***Compt. Rend. Acad. Sci*.** (Paris) **243:** 1349—1352 ***(Cephalotaxus).***

— (1957). ***Revue Cytol Biol. Veg*.** 18: 305—343 ***(Cephalotaxus drupacea)***.

— (1960). ***Ibid*. 21:** 329—337 ***(Taxus baccata).***

— (1958). ***Compt. Rend. Acad. Sci*.** (Paris) 246: 979—982 ***(Taxus baccata).***

— (1960). ***Ibid*.** 250: 4435—4437 ***(Araucaria araucana).***

(1964). ***Ibid*. 258:** 661—664 ***(Torreya myristica)***.

Feistmantel, O. (1876). Jurassic (Oolitic) flora of Katch. ***Mem. Geol. Surv. India, Palaeont. Indica***, Ser. II. **2(1):** 1—80.

— (1877). Jurassic (Liassic) flora of the Rajmahal group in the Rajmahal Hills. ***Mem. Geol Surv. India. Palaeont. Indica***, Ser. II. **1(2).**

— (1877a). Jurassic (Liassic) flora of the Rajmahal group from Golapilli near Ellore, South Godavari. ***Mem. Geol. Surv. India, Palaeont. Indica***, Ser. II. **1(3).**

— (1877b). Flora of the Jabalpur group (Upper Gondwanas) in the Sone-Narbada region. ***Mem. Geol. Surv. India. Palaeont. Indica***, Ser. II. **2(2): 1—25.**

— (1879). Upper Gondwana flora of the outliers of the Madras coast. ***Mem. Geol. Surv. India. Palaeont. Indica***. Ser. **2.1(4):** 1—34.

— (1879a). The flora of the Talchir-Karharbari beds. ***Mem. Geol. Surv. India, Palaeont. Indica***, Ser. **12.**

— (1880). The flora of the Damuda and Panchet divisions. ***Mem. Geol. Surv. India. Palaeont*.** Indica. Ser. **12.** 3(2).

— (1882). Fossil flora of the South Rewa Gondwana basin. ***Mem. Geol. Surv. India. Palaeont***, Indica, Ser. 12. **4(2):** 1—52.

Florin, R. (1938-1945). ***Palacontographica* 85B:** 1—729 (Conifers).

— (1931). ***K. Sv. Vet.—akad. Handle*. 10:** 1—588 (Coniferales and Cordaitales).

— (1940). The Tertiary fossil conifers of South Chile and their phytogeographical significance, with a review of the fossil conifers of southern lands. ***K. svenska Vetensakad. Handl.* 19(2):** 1—107.

— (1948). ***Bot. Gaz*., 110:** 31—39 (Relationships of ***Taxaceae).***

— (1948). *Acta Harti Bergiani* **14:** 385—395 (*Nothotaxus*).

— (1951). Evolution in Cordaites and conifers. *Acta. Hort. Berg.* **15:** 285—388.

— (1951). *Ibid.* **15:** 285—388 (Evolution in Cordaitales and Coniferales).

— (1952). *Bot. Notiser* **105:** 1—29 (*Metasequoia*).

— (1954). *Biol. Rev.* **19:** 367—389 (Female reproductive organs of conifers and taxads).

— (1955). The Systematics of Gymnosperms. In: A Century of Progress in the Natural Sciences 1853-1953, San Francisco. pp.323—403.

— (1958). *Acta. Horti Bergiani* **17:** 403—411 (Podocarpaceae).

— (1963). *Ibid.* **20:** 121—312 (Distribution of conifers and taxads).

E´tude en culture in vitro de la dormance embryonnaire chez *Taxus baccata* L. Biol. Plant, **15 :** 264-269.

Engvild, K.C. (1964). Growth and Chlorophyll formation of dark-grown pine embryos on different media. Physiol Plant **17 :** 866-874.

Flinn, B.S. Web. DT. & Newcomb, W (1988). The role of cell clusters and promeristemoids in determination and competence for caulogenesis by *Pinus strobus* cotyledons in vitro. Can. J. Bot. **66 :** 1556-1565.

Foster, A.S. (1938). *Bull. Torrey Bot. Club* **65:** 531—556 (Shoot apex of *Ginkgo*).

Frenguelli J. (1943). *Rev. Mus. La Planta*, **2:** 225—342 (Peltaspermaceae and Corystospermaceae).

— (1944). *Ibid.* **2:** 1—30 (Corystospermaceae from Argentina).

Fry, W.L. (1955). *Bull. Torrey Bot. Club*, **82:** 486—490 (*Cordaitanthus*).

FSI, 1996: Forest Survey of India.

Ganju, P.N. (1943). On a new species of *Psygmophyllum* from the Lower Gondwana beds of Kashmir. *J. Indian Bot. Soc.* **22(2-4):** 201—207.

— (1946). On a collection of Jurassic plants from the Rajmahal Hills, Bihar. *J. Indian bot.* Soc. *Iyengar Comm. Volume*: 51—96.

— (1947). On *Beaniopsis rajmahalensis* Gen. et sp. nov. a new type of gymnosperm female fructification from the Jurassic of Bihar. *Proc. Ind. Aad. Sci.* **25B(5):** 95—104.

George, L. (1931). *C.R. Acad. Sci. Paris*, **193:** 1451—1453 (Gnetales).

— (1932). *Ibid.* **194:** 1090—1092 (Gnetales).

Gibbs, R.D. (1957). The Maule reaction, and the relationships between woody plants. In: *The Physiology of Forest trees* (Ed. K.U. Thimann). The Ronald Press Company, New York.

Gifford, E.M. (1943). *Bull Torrey bot. club*. **70:** 15—25 (Shoot apex of *Ephedra*).

Goldring, W. (1924). *Bull. N.Y. St. Mus.*, **251:** 50—72 (*Cospermatopteris*).

Gonzalves, E. and R.R. Ladwa (1951). *Proc. 38th Ind. Sci. Cong.* (Bangalore) **1951:** 12 (*Gnetum*)..

Gopal, V. and J. Jacob (1955). Geology of the Upper Gondwanas in the Ramnad District, Madras, *Rec. Geo. Surv. India*, 84(4).

Gordon, W.T. (1912). ***Trans. roy. soc. Edinb*. 48:** 813—825 ***(Rhetinangium).***

— (1935). ***Ibid*. 58:** 279—311 ***(Pitys).***

— (1938). ***Ibid*. 59:** 351—70 ***(Tetrastichia).***

Gordon, W.T., (1941). ***Ibid*. 60:** 427—61 (Salpingostoma).

Gray, N.E. (1953-62). ***J. Arnold Arbor*.,** 34: 67—76, 163—75; 36: 199—206; 37: 160—72; 39: 424—77; **41:** 36—39; 43: 67—79 ***(Podocarpus).***

Greguss, P. (1955). Identification of living gymnosperms on the basis of xylotomy. Budapest.

Gunckel, J.E. and R.H. Wetmore (1946). ***Amer. Jour. Bot*., 33:** 532—543 (Nodal anatomy of ***Ginkgo*).**

— K.V. Thimann and R.H. Wetmore (1949). ***Ibid*. 36:** 309—318 (Long and short shoots in ***Ginkgo biloba*).**

Gupta, K.M. (1934). On the wood anatomy and theoretical significance of homoxylous angiosperms. ***J. Indian bot. Soc*. 13(1):** 71—100.

— (1943). A new species of Williamsonia (W. sahnii) from the Rajmahal Hills. ***J. Indian Bot. Soc*. 22(2-4):** 191—200.

— (1954). Notes on some Jurassic plants from the Rajmahal Hills, Bihar, India. ***Palaeobotanist*, 3:** 18—25.

— (1958). ***Williamsonia*** fructifications from the Jurassic of Rajmahal Hills—their preservation and plan of construction. ***J. Palaeont. Soc. India* 3:** 230—232.

— (1971). Investigations on the Jurassic flora of the Rajmahal Hills, India. On the structure and affinities of ***Sewardioxylon sahnii*** Gen. et sp. nov. A petrified cycadean wood from the Rajmahal Hills, India, ***Palaeontographica,* 131B:** 160—166.

— and B.D. Sharma (1960—64). Investigations of the Jurassic flora of the Rajmahal Hills, India.— 1.On the Bennettitalean Genus ***Dictyozamites*,** with description of ***D. sahnii*** sp. nov., ***Jour. Palaeont. Soc. India*,** V-IX: 21—28.

Gresshoff PM. & Doy CH (1972). Development and differentiation of haploid ***Lycopersicon esculentum*** (tomato). Planta **107 :** 161 - 170.

Gupta P.K. & Durzan D.J. (1986). Somatic polyembryogenesis of protoplasts from sugar pine (***Pinus lambertiana***). Plant cell Rep. **5 :** 147-151.

Gupta P.K. & Durzan D.J. (1987). Biotechnology of somatic polyembryogenesis and plantlet regeneration of loblolly pine. Biotechnology, **5 :** 147-159.

Gupta P.K. (1988). Advances in Biotechnology of conifers. Curr. Sci., **57 :** 629-637.

Gupta P.K., Durzan D.J. Finkle B.J. (1987). Somatic polyembryogenesis in embryogenic cell masses of ***Picea abies*** (Norway spruce) and ***Pinus taeda*** (lobolly pine) after thawing from liquid nitrogen. Can J. For Res. **17 :** 1130-1134.

Gupta P.K., Pullman G., Timmis R., Kreitinger M.,Carlson W.C.,Grob J, and Welty, E.(1993). Forestry in 21st century .The Biotechnology of somatic embryogenesis. Biotechnology **11:** 454-459.

Hagerup, O. (1934), ***K. danske Vidensk. Selsk, Biol. Medd*. 11(4):** 1—83 (Gnetales).

— (1938). ***Ibid.*** **14(4):** 1—34 (Origin of Angiosperms through Gnetales).

— (1939). ***Ibid.*** 15: 1—39 (Gnetales and Coniferae).

Haining, H.I. (1920). ***Bot. Gaz.*** **70:** 436—445 ***(Gnetum).***

Halket, A.C. (1930). ***Ann. Bot.*** **44:** 865—905 (Rootlets of Amyelon).

Hall, J.W. (1954). ***Boy. Gaz.*** **115 :** 346—60 ***(Stephanospermum).***

Halle, T.G. (1929). ***K. Svensk Vetensk Akad. Handl.*** **6:** 1—24 (Permian seed plants of China).

— (1935). ***Ibid.*** **12:** 1—103 (Male organs of Pteridosperms).

— (1937). ***C.r. Congr. Avanc. Etudes Stratigr. carb. Heerlen,*** **1935:** 227—35 (Evolution of seeds and male organs in Pteridosperms).

Harris, T.M. (1932). ***Medd. Greenland,*** **85:** 1—112 (Mesozoic Pteridosperms and bennettitales).

— (1932). ***Ibid.*** **85:** 1—133 (Caytoniaceae and Bennettitales).

— (1933). ***New Phytol.*** **32:** 97—114 (Caytoniaceae).

— (1940). ***Ann. Mag. Nat. Hist.,*** **6:** 249—65 ***(Sagenopteris).***

— (1941). ***Ann. Bot.*** **5:** 47—58 ***(Caytonanthus).***

— (1941). ***Phil. Trans.*** B. **231:** 75—98 (Fossil Cycads).

— (1943). ***Ann. Mag. nat. Hist.*** **10:** 505—522 ***(Anomozamites).***

— (1944). ***Phil. Trans.*** B. **231:** 313—28 ***(Williamsoniella).***

— (1951). ***Ibid.*** **235:** 483—508 ***(Czekanowskia*** **and** ***Leptostrobus).***

— (1951). ***Phytomorph,*** 1: 29—39 (Caytoniaceae).

— (1961). ***Palaeontology,*** **4:** 313—23 (Fossil cycads).

— (1962). ***Trans. ray. Soc. N. Z. (Geol.)*** 1: 17—27 ***(Cornoconites).***

Holden, H.S. (1954). ***Ann. Bot.*** **18:** 407—415 ***(Physostoma stellatum).***

Holden, R. (1917). On the anatomy of two Palaeozoic stems from India. ***Ann. Bot.*** **31:** 315—326.

Holloway, J.T. (1937). ***Trans. roy. Soc. N.Z.*** **67:** 149—165 ***(Phyllocladus alpinus).***

Hooker, J.D. (1862). ***Trans. Linn. Soc. Lond.*** (Bot)., **24:** 1—48 (Welwitschia).

Hoskins, J.H. and A.T. Cross (1946). Am. Middl. Nat. **36:** 207—250, 331—61 (Pachytesta).

Hsu, J. and M.N. Bose (1952). Further information on ***Homoxylon rajmahalense Sahni. J. Indian Bot. Soc.*** **31(1-2):** 1—12.

Hu, H.H. and W.C. Cheng (1948). ***Bull. F. Memoir. Insti. Biol.*** **1:** 153—161 ***(Metasequoia glyptostroboides).***

Hutchinson, A.H. (1914). ***Bot. Gaz.*** **57:** 148—153 ***(Abies).***

Haines R.J. & de Fossard R.A. (1977). Propagation of hoop pine (***Araucaria cunninghamii Ait***) Acta. Hort. **78 :** 297-302.

Hakman I, Fowke, F.C. , Von Arnold S. Eriksson, T. (1985). The development of somatic embryos in tissue cultures initiated from immature embryos of ***Picea abies*** (Norwayspruce). Plant Sci. **38 :** 53-59.

Harvey, A.E & Grasham J.L. (1970b). Growth of ***Cronartium ribicola*** in the absence of physical contact with its host. ***Can.J. Bot***., **48:** 71-73.

Harvey, A.E. & Grasham J.L. (1970b). Innoculation of western pine tissue cultures with basidiospores of ***Cronartium ribicola***. Can J. Bot., **48 :** 1909-1311.

Harvey, A.E. (1967 b). Axenic cultures of western white Pine cambial explants infected with blister rust. Phytopathology, **57:** 547-549.

Harvey, A.E. (1967a). Tissue culture of ***Pinus monticola*** on a chemically defined medium. ***Can. J. Bot***. **45 :** 1783-1787.

Harvey, A.E. and Grasham, J.L. (1969a). Effect of Plant growth regulators on ***Cronartium ribicola*** in tissue cultures of ***Pinus monticola. Phytopathology***, **59 :** 1029-1039.

Harvey., AE & Graham J.L. (1969b). Procedures and media for obtaining tissue cultures of 12 conifer species. Can J. Bot. **47:** 547–549.

Hasegawa M, Higuchi T, Ishikawa H. (1960). Formation of lignin in tissue culture of ***Pinus strobus.*** Plant Cell Physiology **1:** 173–182.

Ho, R.H & Rouse G.E. (1970). Pollen germination of ***Larix sibirica*** in vitro ***can J. Bot.*** **48:** 213–215.

Huhtinen, O. (1972). Production and use of haploids in breeding conifers. IUFRO Genetics-SABRAO Joint Symposium, Tokyo D-3 **(1) :** 1-8.

Iner, A.M ; Mott, R.L. & Amerson, H.V. (1984). Cultured cells of white pine show genetic resistance to axenic blister rust hyphae. ***Science*** **224:** 407–408.

------------- (1973B) Influence de1´ acide abcissique sur le development des embryons de ***Taxus baccata*** L. Cuturs in vitro Z. Pflenzenphysiol., **70:** 406-413.

------------ (1973C) Intervention O´un inhibiteur lie dans la dormance embryonnaire de ***Taxus baccata*** L. C.R. Acad. Sci. Paris **277 D:** 177-180.

Isikawa H (1974). In vitro formation of adventitious buds and roots on the hypocotyl of ***Cryptomeria japonica***. Bot. Mag. (Tokyo), **87 :** 73-77.

Jacob, K. (1951). ***Dictyozamites bagjoriensis*** sp. nov. from the Mesozoic of the Rajmahal Hills, with notes on the distribution of the genus. Proc. Nat. Inst. Sci. India. **17(1):** 7—13.

— and C. Jacob (1954). Cuticular study of Indian Ptilophyllum from Kutch and Jubbulpore. ***Palaeont. Indica*. 33(1):** 1—34.

Jain, K.P. (1964). ***Fascisvarioxylon mehtae*** Gen. et sp. nov., A new petrified cycadean wood from the Rajmahal Hills, Bihar, India. ***Paleeobotanist*, 11(3):** 138—143.

— (1965). A new species of ***Mesembrioxylon, M. rajmahalense*** from the Rajmahal Hills, Bihar. ***Palaeobotanist*. 13:** 154—156.

Jager, L. (1899). ***L. Flora*, 86:** 241—288 **(*Taxus baccata*).**

Jain, K.P. (1972). Taxonomic observations on the genus ***Sewardioxylon*** Gupta—A junior synonym of ***Fascisvarioxylon*** Jain. ***Palaeobotanist*, 19(3):** 251—252.

Jennings, J.R.J. (1973). ***Amer. Jour. Bot.*** (Abstracts) **60:** 17 (Relationships of Genus Rhodea).

Johnson, M.A. (1950). Bull. Torrey bot. Club., **77:** 354—67 (stem apex of Gnetum).

— (1951). ***Phytomorph.* 1:** 188-204 (Stem apices of Gymnosperms).

Johri, B.M. (1936). ***Proc. Indian Acad. Sci.*** B. **3:** 246—257 (Cedrus deodara).

Johari B.M. (2000). ***Botany in India.*** INSA Publication, N. Delhi.

Kaur, D. (1958). Proc. Indian Sci. Congr. **45:** 291—292 ***(Cephalotaxus drugpacea var. pedunculata).***

Kanis, A. and Karstens, W.K.H. (1963). ***Acta bot. Neerl.*, 12:** 281—286 (Amphistomatic leaves in ***Ginkgo).***

Kendall, W. (1947). On five species of ***Brachyphyllum*** from the Jurassic of Yorkshire and Wiltshire. ***Ann. Mag. Nat. Hist.*** (Ser. 11), **14:** 225—251.

— (1948). On six species of ***Pagiophyllum*** from the Jurassic of Yorkshire and Southern England. Ann. Mag. Nat. Hist. (Ser. 12). **1:** 73—108.

Kern, E.M. and H.N. Andrews (1946). Ann. Mo. bot. Gdn., **33:** 291—306 ***(Rhabdospermum*** and ***Kamarospermum).***

Kershaw, E.M. (1912). ***Ann. Bot.* 26:** 625—646 (ovule of ***Bowenia spectabilis).***

Khan, R. (1943). ***Proc. Nat. Acad. Sci. India.* 13:** 357—75 (Embryology of ***Ephedra foliata).***

— (1940). ***Curr. Science*** (Bangalore) **9:** 323—324 (Double fertilisation in ***Ephedra foliata).***

Khoshoo, T.N. (1961). ***Silvae. Geneti.* 10:** 1—32 (Chromosome numbers in Gymnosperms).

—(1962). ***Proc. Summer Sch. Bot. Darjeeling,*** 119—35 (Karyohypes of Gymnosperms).

— (1957). ***Curr. Sci.*, 26:** 118—120 (Meiosis in ***Cephalotaxus drupaeca* var. *Pedunculata).***

— (1957). ***Cytologia*, 22:** 80—89 (Cytology of conifers III).

Kidston, R. and D.T. Gwynne-Vaughan (1912). ***Trans. Roy, Soc. Edinb.*, 48:** 263—271 ***(Stenomyelon).***

— and W.H. Lang (1923). ***Ibid.* 53:** 409—417 ***(Palaeoptys).***

Koidzumi, G. (1942). ***Acta. Phytotax. Geobot.*** Kyoto **11:** 135—136 (Amentotaxaceae).

Konar, R.N. (1960). ***Phytomorphology*, 10:** 305-319 (Embryology of ***Pinus roxburghii*** and ***P. wallichiana).***

— (1962). ***Ibid.* 12:** 120—195 (Male cones in ***Fitzroya cuppressoides*** and ***Pilgerodendron*** uniferum).

— and S.K. Banerjee (1963). ***Ibid.*, 13:** 321—338 ***(Cupressus funiberis).***

— and Y.P. Oberoi (1968). ***Beitr. Biol. Pflan.* 45:** 329—376 ***(Podocarpus gracilior).***

— and S. Ramchandani (1958). ***Phytomorphology*, 8:** 328—346 ***(Pinus wallichiana).***

— and Y.P. Oberoi (1969). ***Bot. Rev.* 35:** 89—116 (Living conifers and Taxads—A Review).

Krausel, R. (1943). Palaeontographica B **87:** 59—93 (Sphenobaiera).

— (1949). ***Ibid.*** B **89:** 35—82 ***(Westersheimia*** and ***Williamsonia wettstenii).***

— (1948). ***Senckenbergiana*. 29:** 141—9 ***(Sturriella)*.**

— and K.P. Jain (1964). New fossil coniferous woods from the Rajmahal Hills, Bihar, India. ***Palaeobotanist*, 12(1):** 59—67.

P.K. Maithy and H.K. Maheswari (1961). Gymnospermous woods with primary structures from Gondwana rocks. A review. ***Palaeobotanist*. 10(2):** 97—107.

Krishnan, M.S. (1960). ***Geology of India and Burma. Madras*.**

Karnosky, D.F. (1981). Potential for forest tree improvement via tissue culture. Bioscience **31 :** 114-120.

Kirby E.G. & Schalk M.E. (1982) Surface structural analysis of cultured cotyledons of Douglas fir. Can J. Bot., **60 :** 2729-2733.

Konar R.N. & Singh M.N. (1980). Induction of shoot buds from tissue culture of ***Pinus wallichiana*.** Z. Pflenzenphysiol. **99 :** 173-177.

Konar, R N. & Singh M.N. (1979). Production of plantlets from the female gametophytes of ***Ephedra*** foliata. Bioss Zlanzenphysiol. **95 :** 87-90.

Konar, R.N. & Oberoi. Y.P. (1965). In vitro development of embryoids on the cotyledons of ***Biota orientalis. Phytomorphology* 15 :** 137-140.

Konar, R.N. (1963b). A haploid tissue from pollen of ***Ephedra foliata*** Bioss., ***Phytomorphology*, 13:** 170–173.

Konar. R.N. & Oberoi, Y.P. (1965). ***In vitro*** development of embryoids on cotyledons of ***Biota orientalis***. Phytomorphology, **19 :** 122-133.

Konar. R.N. (1963a). Anatomical studies on Indian pines with special reference to ***Pinus roxburghii*** sarg. Phytomorphology, **13:** 388– 402.

Kuhlwein, H (1937). Zur Physiologie der Pollenkeimung insbesonders der Frage nach dem Befruchtung- sverzug bei Gymnospermen. Beih Bot Zbl. **57:** 37.

Lacey, W.S. (1953). ***Ann. Bot*. 17:** 579—596 (***Eristophyton*** and ***Endoxylon*).**

Lam, H.J. (1952). ***Annls. Biol*., 28:** 57—88 (Classification of vascular plants).

(1950). ***Sv. Bot. Tdskr*. 44:** 517—534 (Stachyspory and Phyllospory).

Land, W.J.G. (1904). ***Bot. Gaz*. 38:** 1—18 ***(Ephedra trifurcata)*.**

— (1907). ***Ibid*. 44:** 273-292 ***(Ephedra)*.**

Lee, C.L. (1954). ***Amer. Jour*.** Bot. **41:** 545—9 (Sex chromosomes in Ginkgo).

— (1955). ***Bot. Gaz*., 117:** 79—100 (Fertilisation in ***Ginkgo*).**

Leisman, C.A. (1972). ***Amer. Jour. Bot*. 59:** 661 (Abstracts; ***Linopteris*).**

Lee Goc. M.J. (1914). ***Ann. Bot*., 28:** 183—93 (Centripetal xylem in Cycad leaves).

Lele, K.M. (1953). On the occurrence of ***Pterophyllum*** in the Parsora beds, South Rewa, India. ***Nature*,** 172: 1195.

— (1955). Plant fossils from Parsora in the Rewa Gondwana Basin, India. ***Palaeobotanist*, 4:** 23—34.

— (1961). Studies in the Indian Middle Gondwana flora. 1.On ***Dicroidium*** from South Rewa Gondwana basin. Palaeobotanist, **10(1-2):** 48—69.

— (1961a). Studies in the Indian Middle Gondwana flora. 2.Plant fossils from the South Rewa Gondwana basin. Palaeobotanist, **10(1-2):** 69—83.

— (1962). Studies in the Indian Middle Gondwana flora. 3.Platyspermic seeds and megaspore impressions from the South Rewa Gondwana basin. Palaeobotanist, **11(1-2):** 13—18.

— (1964). The problem of Middle Gondwana in India. ***XXII Intern. Geol. Congr.*** Pt. IX, Sect. 9: 181—202.

— (1966). Studies in the Talchir flora of India—4.Quest for the early traces and subsequent development of the Glossopteris flora in the Talchir stage. ***Symposium on Floristics and Stratigraphy of Gondwana Land,*** Lucknow : 85—97.

— and P.K. Maithy, Studies in the ***Glossopteris*** flora of India - 15. Revision of the epidermal structure of ***Noeggerathiopsis*** Feistmantel. ***Palaeobotanist,*** **12(1):** 7—17.

Lee, H.L. (1952). Jour. Arnold Arbor. **38:** 192—198 (Amentotaxus).

Long, A.G. (1960—1961). Trans. roy. Soc. Edinb. **64:** 29—44, 201—215, 261—280, 281—295, 401—419 (Lower Carboniferous seeds).

— (1963). ***Ibid.*** **65:** 211—224 (Fronds associated with ***Pitys***).

— (1964). ***Ibid.*** **65:** 435—437 (***Stenomyclon***).

Looby, W.J., and J. Doyle (1937). ***Scient. Proc. roy. Dublin Soc.,*** **21:** 457—76 (***Sequoia***). La Rue CD (1948). Regeneration in the megagametophyte of ***Zamia Floridiana. Bull Torrey Bot. Club***, **75 :** 597-603.

Laliberte. S. Bertrand C.e Veith. H. (1983). Callogenesis and degree of differentiation in endosperm cultures of ***Encephalartos villosus***. Rev. Can. Biol. Exp; **42 :** 7-12.

LaRue, CD (1954). Studies on growth and regeneration in gametophytes and sporophylls of gymnosperms. Brookhaven Natl Lab Symp. **6:** 187–208.

Le page-Degivry. M.T. & Garello, G. (1973). La dormance embryonnaire Chez ***Taxus bacata*** : Influence de la composition du milieu liquide sur l' induction de la germination. Physiol. Plant. **29 :** 204-207.

Le Page–Digivry, M.T. (1970). Acid abcissique et dormance chez les embryons de ***Taxus baccata*** L. C.R. Acad. Sci. Paris. **271D :** 482-484

Lee C.L. & Chang, H. (1963). Comparative studies on the mature embryo of ***Pinus koraiensis*** grown in vitro. Scientia Sinica, **12 :** 1149-1159.

Lelu M.A., Boulay M., Arnaud Y. (1987). Obtention de cals embryoge´ne´s a partier de cotyledons de ***Picea abies*** (L.) Karst. Pre´ leves sur de jeunes plantees g´ee´s de 3'a 7 Jours apre's germination C.R. Acad Sci. Ser III Sci. Vie **305 :** 105-181.

Li. T. & Shen, T. (1934). The effect of "Pantothenic acid" on growth of yeast and on growth of the radicle of ***Ginkgo*** embryos in artificial media. Peking Nat. ***Tsing Hua Univ. Sci. Rep.*** **B2 :** 53-60.

Loo, S.W. & Wang F.H. (1943). The culture of young conifer embryos in vitro. ***Science*** : **98 :** 544

Lu, CY. Thorpe, T.A. (1987). Somatic embryogenesis and plantlet regeneration in cultured immature embryos of ***Picea glauca***. J. Plant Physiol. **128 :** 297-302.

— and — (1939). ***Ibid.*, 22:** 95—117 **(*Saxegothaea*).**

— and — (1940). ***Ibid.* 22:** 241—255 **(*Callitris*).**

— and — (1942). ***Ibid.* 23:** 35—54 **(*Sequoia*)**

— and — (1944). ***Ibid.* 23:** 227—237 **(*Podocarpus andinus*).**

— and — (1944). ***Ibid.* 23:** 257—270 **(*P. andinus*).**

Lotsy, J. — (1899). ***Annl. Jard. Bot. Buitenz* 1:** 46—114 **(*Gnetum gnemon*).**

Loze, J.C. (1965). ***Rev. Cotol. Biol. Vegetale* 28:** 211—256 **(*Taxus baccata*).**

Magdefrau, K. (1931). ***Beih. Bot. Cbl.* 48:** 119—140.

— (1932). ***Ibid.*, 49:** 706—718.

— (1966). ***Die Geschichte der Pflanzen.*** Mannheim.

Maheswari, H.K. (1964). Studies on the ***Glossopteris*** flora of India — 24. On two new species of fossil woods from the Raniganj stages of Raniganj coal fields, Bengal. Palaeobotanist, **13(2):** 148—152.

— (1967). Studies on the ***Glossopteris*** flora of India — 28. On some fossil woods from the Raniganj stage of the Raniganj coal field, Bengal. ***Palaeobotanist* 15(3):** 243—257.

Maheswari, P. & Vimla Vasil, 1961. Botanical Monograph No. I ***Gnetum. C.S.I.R.*** New Delhi.

Maheswari, P. (1935). ***Proc. Indian Acad. Sci.*, 1:** 586—606 (Gametophytes of Ephedra).

— and V. Vasil (1961). ***Gnetum.*** Botanical Monograph No. I. Council of Sceintific and Industrial Research, New Delhi.

— and (1961). ***Ann. Bot. (n-s)* 25:** 313—319 **(*Gnetum*).**

— and M. Sanwal (1963). ***Memoris Indian Bot. Soc.* 4:** 103—119 (Archegonia in gymnosperms).

— and H. Singh (1967). Biol. Rev., **42:** 88—130 (Female gametophytes of gymnosperms).

Mamay, S.H. (1954). ***Prof. Pap. U.S. Geol. Surv.* 254D:** 81—95 **(*Tyliosperma*).**

Marchant, C.J. (1968). ***Chromosoma*, 24:** 100—134 (Chromosome patterns in Cycads).

Marsden, M.P.F. and T.A. Steaves (1955). ***J. Arnold Arbor.*,** 36: 241—258 (Nodal anatomy in ***Ephedra*).**

Martens, P. (1959, 1961, 1962). ***La Cellulae*,** 60: 171—286; **62:** 7—91; 63: 309—329 **(*Welwitschia*).**

— and L. Waterkeyn (1963). ***Phytomorphology*, 13:** 359—363 **(*Welwitschia*).**, L. Waterkeyn and M. Huyskens (1967). ***Ibid.* 17:** 114—118 (Microspores in ***Pinus sylvestris*).**

Martin, P.C., (1950). ***Proc. Paris Acad. Sci.*, 24:** 65—112 (Comparison of ***Biota*** and ***Thuja*).**

Masters, M.T. (1891). ***J. Linn. Soc. Bot.*, 27:** 226—232 (coniferae).

Mathews, A.C., (1939). ***Jour. Elisha Mitschell Scient. Soc.*, 55:** 7—62 (Juniperus virginiana).

Matten, L.C. (1968). ***Amer. Jour. Bot.* 55:** 773—782 (***Actinoxylon banksii***).

Mc. Williams. J.R. (1958). ***Bot. Gaz.*, 120:** 109—117 (Pollination in Pinus).

— and F. Mergen (1958). ***Ibid.*, 119:** 246—249 (cytology of fertilisation in Pinus).

Mehra, P.N. (1938). ***Proc. Indian Acad. Sci.* 8 : (3):** 218—30 (***Ephedra***).

— (1946). ***Proc. Nat. Acad. Sci. India*, 16:** 259—286 (***Ephedra***).

— (1947). ***Jour. Indian Bot.Soc.*** M.O.P. Iyengar Commemoration Volume, pp.121—132 (Diploid pollen in Ephedra).

— (1950). ***Proc. Nat. Acad. Sci. India.*, 16:** 259—286 (Ephedra).

— and T. N. Khoshoo (1956). ***Jour. Genet.*, 54:** 165—185 (Cytology of conifers).

— and R.K. Malhotra (1947). ***Proc. Nat. Acad. Sci. India***, **17B:** 129—153 (Embryogeny of Cupressus sempervirens).

— and M.K. Sirkar (1949). ***Proc. Nat. Inst. Sci. India.*, 15B:** 15—23 (Cupressus funiberis).

Mehta, K.R. and J.D. Sud (1953). ***Paleobot.* 2:** 51—54 (ginkgoalean impressions in Rajmahal Hills, India).

Melville, R. (1962). ***Kew Bull.* 17:** 1—50 (A new theory of Angiosperm flower).

Mettre, V. (1953). A male flower of Pentoxyleae, with remarks on the structure of the female cones of the group. ***Palaeobotanist*, 2:** 75—84.

Millay, M.A. and T.N. Taylor (1972). ***Amer. Jour. Bot.*** (Abstracts) **59:** 661 (Early evolution of saccate pollen).

Miller, C.N. and Colemann, R.R., (1972). ***Amer. Jour. Bot.*** (Abstracts) **59:** 66(2) (Cones of Pinaceae from Late Cretaceous).

Mittre, V. (1956). Studies on the fossil flora of Nipania, Rajmahal series, India. -Bennettitales. ***Palaeobotanist.* 5(2):** 95—99.

— (1956a). ***Masculostrobus sahnii*** sp. nov. A petrified conifer male cone from the Jurassic of Rajmahal Hills, Bihar. ***Grana Palynol.*** (N.S.) **1(2) :** 99—107.

— (1957). Studies on the fossil flora of Nipania (Rajmahal series) India—Pentoxyleae. ***Palaeobotanist*, 6(1):** 31—46.

— (1957a). Studies on the fossil flora of Nipania (Rajmahal series), Bihar-Coniferales. ***Palaeobotanist*, 6(2):** 82—112.

— (1963). The ice ages and the evolutionary history of the Indian Gymnosperms. ***J. Indian bot. Soc.*** XLII(2) : 301—308.

Mukherjee, S.M., M. Banerjee and J. Sen (1966). Further Glossopteridean fructifications from India. ***Palaeontographica* 117B:** 99—113.

Millay, M. A. and T. N. Taylor (1972). ***Amer. Jour. Bot.*** (Abstracts) **59 :** 661 (Early evolution of saccate pollen).

Miller, C. N., and Colemann, R. R., (1972) ***Amer. Jour. Bot.*** (Abstracts) **59 :** 662 (Cones of ***Pinaceae from*** Late Cretaceous)

Moseley, M.F. (1943). ***Lloyava* 6:** 109—132 (Windringtonia cupressoides).

Moule, A.C. (1944). ***Jour. Roy. Hort. Soc.* 69:** 166 (The name Ginkgo).

Mullay, B.N. (1941). ***Proc. Indian Sci. Congr.,* 28:** Part III, p.158 (***Ephedra foliata***).

— and N.C. Nair (1952). ***Biol. Sci. Studies, Univ. Rajputana*** 1—20 (***Ephedra foliata***).

Maheshwari, R., Hildebrandt, A.C. and Allen P.J. (1967). Factors affecting the growth of rust fungi on host tissue cultures. Bot. Gaz, **128:** 153-159.

Mehra P.N. (1934). The germination of pollen grains in artificial cultures in ***Ephedra foliata*** and ***E. gerardiana*** and a study of the number and morphology of their chromosomes. ***Curr. Sci.***, **3:**11–14.

Mehra Palta A.& Thompson D.G. (1988). Invitro propagation of conifers. In nalentine, F.A. (eds) : ***Forest and crop Biotechnology*** : Progress and Prospects, pp. 12-34. Springer Verlag, Berlin.

Minocha, S.C. (1980). Callus and adventitious shoot formation in excised embryos of white pine *(Pinus strobus*). Can. J., Bot. **58 :** 366-370.

Minocha, Sc. (1980). Callus and adventitious shoot formation in excised embryos of white pine (***Pinus strobus***). Can. J. Bot. **58 :** 366-370.

Mott. R.L. & Amerson H.V. (1981). A tissue culture process for the clonal production of lobloly pine plantlets. ***North Carolina Agr. Res. Tech. Bul***. **271 :** 1-14.

Nakai, T., (1938). ***Bull. For. Soc. Korea***, **19:** 1—29 (***Austrotaxaceae***).

Namboodiri, K.K. and C.B. Beck (1968a). ***Amer. Jour. Bot.* 55:** 447—457 (Primary vascular system of conifers I).

— and (1968b). ***Ibid.,* 55:** 458—463 (Primary vascular system of conifers II).

— and (1968c). ***Ibid.,* 55:** 464—472 (Primary vascular system of conifers III).

Nathorst, A.G. (1909). ***K. Svensk. Vetensk, Akad. Handl.***, **45:** 1—37 (***Williamsonia and Wiellandiella).***

Negi., V. and Madhulata (1957). ***Phytomorphology***, **7:** 230—236 (***Gnetum***).

Negi, V., (1958). ***Ph. D. thesis, Univ. Delhi (Gnetum ula).***

Nagmani R, Becwar M.R., Wann SR (1987). Single cell origin and development of somatic embryos in ***Picea abies*** (L.) Karst. (Norway Spruce) and ***P. glauca*** (Moench.) voss (white spruce). Plant cell Rep. **6:** 157-159.

Nagmani, R. & Bonga J.M. (1985). Embryogenesis in subcultured callus of *Larix decidua. Can. J. For. Res*., **15 :** 1088-1091.

Norstog, K. & Rhamstine. E. (1967). Isolation and culture of haploid and diploid cycad tissue. Phytomorphology, **17 :** 374-381.

Norstog, K. (1965) Induction of apogamy in megagametophyte of ***Zamia integrifolia.*** Am. J. Bot, **52:** 993-999.

Norstog, K. (1967). Fine structure of the spermatozoids of Zamia with special reference to the flagellar apparatus, Am. J. Bot. **54 :** 831-840.

Oliver, F.W. (1903). ***Ann. Bot.,* 17:** 451—76 (Hyposperm in seeds).

— (1904). ***Trans. Linn. Soc. Lond.*** (Bot)., **6:** 361—400 (***Stephanospermum***).

— (1909). ***Ann. Bot.*** **23:** 73—116 (***Physostoma***).

— and E.J. Salisbury (1911). ***Ibid.*** **15:** 1—50 (***Conostoma and Gnetopsis***).

— and D.H. Scott (1904). ***Phil. Trans***, **B197:** 193-247 (***Lagenostoma***).

Oinuma, T. (1943). ***Bot. Mag.*** (Tokyo), **56:** 285—291 (***Torreya***).

Osborn, T.G.B. (1960). ***Aust. Jour. Bot.*** **8:** 243—255 (***Podocarpus falcatus***).

Page, V.M. (1972). Amer. Jour. Bot. 59: 662 (Abstracts; A new gymnosperm from upper Cretaceous).

Pant, D.D. (1958). ***Bull. British Mus. Nat. Hist. (Geology),*** **3:** 127—175 (***Glossopteris*** flora of Tanganyika).

— (1962). In ***Proc. Summ. Sch. of Bot.*** Darjeeling, pp.302—319 (***Glossopteris***).

and D.D. Nautiyal. ***Palaeontographica,*** **B107:** 41—64 (Seeds and pollen in Glossopteridaceae).

— (1957). The classification of gymnospermous plants. ***Palaeobotanist,*** **6(2):** 65—69.

— and Nautiyal (1967). On the structure of ***Buriadia heterophylla*** (Feistmantel) Seward and Sahni and its fructification. ***Phil. Trans. Roy. Soc. London***, Ser. **B.252:** 27—48.

— and G.K. Srivastava, 1968. On the cuticular structure of ***Araucaria(Araucarites)*** cutchensis ***(Feistmantel)*** comb. Nov. from the Jabalpur series, India. J. Linn. Soc. (Bot.) 61: 201—206.

and B. Mehra (1964). ***Jour. Linn. Soc. Bot.,*** **58:** 491—7 (Stomata in Cycads and ***Ginkgo***).

Pearson, H.H.W., (1906). ***Phil. Trans. Royal. Soc. London.,*** **B.198:** 265—304.

— (1912). ***Ann. Bot.*** (London), **26:** 603—620 (***Gnetum***).

— (1914). ***Trans. Linn.***, Soc. London, II 8: 311—330 (***Gnetum***).

— (1915a). ***Jour. Linn.,*** So. Bot., **43:** 55—56 (***Gnetum***).

— (1915b). ***Ann. Boluls Herb.*** **1 :** 152—173 (***Gnetum***).

— (1917). ***Trans. Royal. Soc. S. Africa***, **6:** 69—87 (***Gnetum***).

(1929). ***Gnetales***. Cambridge.

— and M.R.H. Thompson (1917). ***Trans. Roy. Soc. S. Africa*** **6:** 231—267 (***Gnetum***).

Pettitt, J.M. (1966). ***Jour. Linn. Soc. Bot.,*** **59:** 253—263 (Megaspore membrane in some gymnosperm ovules).

— and C.B. Beck (1967). Science **156:** 1727—1728 (seed from Upper Devonian).

— and (1968). ***Contr. Mus. Paleontol. Univ. Michig.,*** **22:** 139—154 (***Archaeosperma arnoldi***).

Phillips, T.L. and H.N. Andrews (1963). ***Ann. Mo. bot. Gdn.,*** **50:** 29—51 (***Sutcliffa*** in U.S.A.).

Pijl, Van Der. L. (1952). ***K. Nederl. Akad. Wentenschaffen-Amsterdam*** **55:** 574—586 (***Gnetum***).

— (1953). ***Ann. Bogor.*** **1:** 77—99 (***Gnetum***).

Pilger, R. (1926). Gymnospermae. In A. Engler and K. Prantl, ***Die Naturlischen Pflanzenfamilien*** **13. *Band***. Wilhelm Engelmann Verlag, Leipzig.

Plumstead, E.P., (1952, 1956 and 1958). ***Trans. geol. Soc. S. Africa***, 55: 281—328; 59: 211—236; **61:** 51—75, 82—94 (***Glossopteridaceae***).

Prakash, U. (1956). Studies in the Deccan Intertrappean flora—On a petrified ovuliferous cone from Mangaon cherts in the Deccan. ***Palaeobotanist***, 5(2): 91—94.

— (1961). Further observations on a petrified ovuliferous cone (***Mohgaostrobus sahnii*** Gen. et sp. nov.) from Mohgaon Cherts in the Deccan. ***Palaeobotanist***, **10(1-2):** 1—4.

Quinn, C.J. (1964). ***Phytomorphology***, **14:** 342—351 (***Podocarpus***).

— (1965). ***Ibid.***, **15:** 37—45 (***Dacrydium taxifolium***).

— (1966). ***Ibid.***, **16:** 81—91 (***Dacrydium bidwillii***).

Radforth, N.W. (1937). ***Trans. Roy. Soc. Edinb.***, **59:** 385—396 (***Dactylothea plumosa and Seftenbergia ophiodermatica***).

Raizada, M.B. and K.C. Sahni, (1960). Living Indian Gymnosperms. Pt. I (Cycadales, Ginkgoales and Coniferales). ***Indian For. Rec.*** **5(2):** 1—150.

Ramanujan, C.G.K. (1953). On two new species of ***Mesembrioxylon*** from the vicinity of Pondicherry, South India. ***Palaeobotanist***, **2:** 101—106.

— (1954). On some silicified woods from near Pondicherry, South India. Palaeobotanist, **3:** 40—50.

— (1957). Cone-scales of ***Araucarites cutchensis*** from the Rajmahal Hills, Bihar. Curr. Sci. **26:** 87—88.

— (1960). Silicified woods from the Tertiary of South India. ***Palaeontographica***. **106B:** 99—140.

Rao, A.R. (1943). ***Nipaniostrobus***, A genus of ***Dacrydium***-like seed bearing cones, and other silicified plants from the Rajmahal series. ***Proc. Nat. Acad. Sci. India,*** **13(2):** 113—133.

— (1943a). The structure and affinities of ***Taeniopteris spatulata*** McCl. Proc. ***Nat. Acad. Sci. India*** **13(6):** 333—355.

— (1946). ***Nipanioruha granthia*** Gen. et sp. nov., a new petrified coniferous shoot from the Rajmahal Hills, Bihar. ***J. Indian bot. Soc. Iyengar Comm. Volume***: 389—397.

— (1949). The megastrobilus of ***Nipanioruha granthia. Curr. Sci.*** **18:** 447—448.

— (1963). The Podocarpaceae in India. ***Mem. Indian Bot. Soc.*** **4:** 150—157.

— (1963a). ***Stachyotaxus sampathkumarani*** sp. nov. from Onthea in the Rajmahal Hills, ***Bihar. Palaeobotanist***, **12(3):** 217—219.

— and M.N. Bose, Further observations on ***Ontheodendron florini*** Sahni and Rao. ***Palaeobotanist***, 7(1): 29-31.

— and (1972). ***Podostrobus*** Gen. nov. A petrified podocarpaceous male cone from the Rajmahal Hills, India. ***Palaeobotanist***, 19(1): 83-85.

— and Vimla Achutan (1968). Further contribution to our knowledge of ***Ptilophyllum. Palaeobotanist***, **16(3):** 249-257.

Radforth, N.W. & Bonga J.M. (1960). Differentiation induced as season advances in embryo-gametophyte complex of ***Pinus Nigra*** Var. ***austriaca*** using IAA. Nature. **185 :** 332.

Radforth, N.W. & Pegoraro, L. (1955). Assessment of early differentiation in Pinus proembryos transplanted to *in vitro* conditions. ***Trans. Roy. Soc***. canada, **49 :** 69-82.

Radforth, N.W. (1936). The Development invitro of the proembryo of *Ginkgo*. Trans. Roy. Con. Inst. Toronto, **21 :** 87-95.

Rao, N.M. & Mehta, A.R., (1969). Callus tissue from pollen of ***Thuja orientali Indian J. Exp. Biol***. **7 :** 132 :

Razmologov., UP. (1973). Tissue culture from the generative cell of the pollen grain of ***Cupressus*** sps. *Bull Torrey Bot. Club*, **100 :** 18-22.

Rao, C.N. and K. Jacob, (1957). On a new species of ***Taeniopteris*** from the Upper Gondwanas of the Rajmahal Hills, Bihar. ***Rec. geol. Surv. India*, 84(4).**

Rao, H.S. 1935. On a sphaerosiderite containing a new species of ***Dadoxylon* (*D. parbeliense*)** from Lower Gondwana coal measures of India. ***Rec. geol. Surv. India***, **59:** 174—183.

Rastogi, B. (1964). Morphology, anatomy and embryology of ***Agathis brownsi*** **:** M.Sc. Thesis, Delhi Univ., Delhi.

Reilly K.&. Washer J. (1977). Vegetative propagation of radiata pine by tissue culture : Plantlet formation from embryonic tissue. NZ J. for Sci. **7 :** 199-206.

Reinert, J. and White, P.R. (1956). The cultivation in vitro of tumor tissues and normal tissues of Picea glaca.Physio.Plant, **17:** 620-635.

Risser PG. & White P.R. (1964). Nutritional requirements of spruce tumor cells in vitro. Physiol. Plant. **17:** 620-635.

Rohr. R. (1987). Haploids (Gymnosperms): In : Bong a, J.M. & Durzan, D.J. (eds). ***Cell and Tissue Culture in Forestry***. Vol. II 230–246. Mortinus Nishoff Publ., Netherland.

Rumary C., Patel K.R. & Thorpe T.A. (1986). Plantlet formation in black and white spruce II. Histological analysis of adventitious shoot formation in vitro. Can. J. Bot. **64 :** 997-1002.

Sah, S.C.D. (1953). On some species of ***Ginkgoites*** from the Jurassic of Rajmahal Hills, Bihar, ***Palaeobotanist*, 2:** 55—58.

Read, C.B. (1936). ***Prof. Pap.*** U.S. ***geo. surv*., 185:** 149—161 **(*Diichnia*).**

— (1937). ***Ibid*., 186:** 81—104 **(*Stenomyelon and Calamopitys*).**

Rhodin, R.J. (1953). ***Amer. Jour. Bot*. 40:** 280—85, 371—78 **(*Welwitschia*).**

— (1958). ***Ibid*., 45:** 90—103 **(*Welwitschia*).**

Roy Chowdhury, C. (1961). ***Phytomorphology*, 11:** 283—304 (***Cedrus deodara***).

— (1962). ***Ibid*., 12:** 313—338 (Embryogeny of conifers).

— and S.K. Dutta (1968). Palynostratigraphy of the Tertiary sedimentary formations of Assam. 2.Stratigraphic significance of spores and pollen in the Tertiary succession of Assam. ***Palaeobotanist*, 16(2):** 173—199.

— and K.P. Jain (1964). Some fossil woods from the Jurassic of Rajmahal Hills, Bihar, India. ***Palaeobotanist*. 12(2):** 169—180.

Sahni, B. (1928). Revisions of Indian fossil plants. Pt.1. Coniferales (a.impressions and incrustations).

Palaeont. Indica (N.S.), 11: 1—49.

Sankhla N., Sankhla D. & Chatterjee U.N. (1967). ***In vitro*** induction of proliferation in female gametophyte tissue of ***Ephedra foliata*** Boiss. ***Naturwioss***., **54 :** 203.

Saravitz C.H. Blazich, F.A. & Amerson, H.V. (1993). Histology of in vitro adventitious bud development on cotyledons and hypocotyl of Fraser fir. J. Amer. Soc. Hort. Sci. **118 :** 163-167.

Schmidt., A. (1924). Uber die chlorophyll bild in Konifer embryo. *Botan. Arch*., **5 :** 260-282.

Simola, L.K. & Honkanen, J. (1983). Organogenesis and fins structure in Megagametophyte callus lines of ***Picea abies***. Physiol. Plant, **59 :** 557-561.

Singh M.N. & Konar R.N. (1981). ***In vitro*** production of haploid roots and shoots from female gametophyte of ***Ephedra foliata***. Beitr. Biol. Pflanzen. **55 :** 169-177.

Singh M.N., Konar, R.N. & Bhatnagar S.P. (1981). Haploid Plantlet formation from female gametophyte of ***Ephedra foliata*** Boiss in vitro. Ann. Bot. **48 :** 215-220.

Smith D.R, Horgan K.J., Aitken-Christie J. (1982). Micropropagation of ***Pinus radiata*** for afforestation pp. 723-724. In : Fujiwara A. (ed.) Plant Tissue and Cell Culture, Maruzen, Tokyo.

Sommer H.E. & Brown C.L. (1974). Plantlet formation in pine tissue cultures Am. J. Bot. **61 :** 11.

Sommer H.E., Brown C.L. & Kormanik P.P. (1975). Differentiation of Plantlets in longleaf pine (***Pinus Palustris Mill***.) tissue cultures ***in vitro***. Bot-Gaz. 136-200.

Steinhauer, A. & Huhtinen, O. (1978), Haploid tissue culture in Forest trees. Callus and shoot regeneration from megagametophyte cultures of ***Picea abies***. IV Intr. Cangr. Pl. Tissue and cell culture, Calgary, Canada.

Sterling C. (1949). Embryonic differentiation in ***Taxus Cupsideta*** Bull. Torrey ***Bot. Cl***. **76 :** 116-133.

— (1931). Revisions of Indian fossil plants. Pt. 2. Coniferales (b.petrifactions). Palaeont. Indica (N.S.), **11:** 51—124.

— (1932). ***Homoxylon rajmahalense*** Gen. et sp. nov., a fossil angiospermous wood devoid of vessels from the Rajmahal Hills, Bihar. ***Palaeont. Indica*** (N.S.) 20(2): 1—19. (1932a) A petrified Williamsonia (W. sewardiana sp. nov.) from the Rajmahal Hills, India, Palaeont. Indica (N.S.), **20(3):** 1—19.

— (1948). The Pentoxyleae: A new group of Jurassic Gymnosperms from the Rajmahal Hills of India. ***Bot. Gaz***. 110(1): 47—80.

— and A.R. Rao (1931). On some Jurassic plants from Rajmahal Hills. ***Proc. Asiat. Soc. Bengal,*** (N.S.) 27**(2):** 183—208.

Sahni, K.C. (1953). Botanical Exploration in the great Nicobar Island. ***Indian For.*** **79(1):** 3—16.

Saksena, S.D. (1952). Correlation of the Gondwanas based on the evidence of plant fossils. ***Agra Univ. Jour. Res.*** 1 (Sci.): 1—13.

— (1955). On two new species of Samaropsis from the South Rewa Gondwana basin, Central India. ***Palaeobotanist,*** **4:** 73—76.

Sahni, B., (1920). ***Ann. Bot.,*** **34:** 117—133 (Taxales).

— (1932). ***Mem. geol. surv. India,*** **20:** 1—19 **(*Williamsonia Sewardiana*).**

— (1948). *Bot. Gaz.*, **110:** 47—80 (***Pentoxylales***).

Sexston, W.T. (1934). ***Ann. Bot.*, 48:** 411—427 (***Austrotaxus***).

— (1936). ***Ibid.*, 50:** 519—522 (***Taxus baccata***).

Schopf, J.M. (1949). ***J. Palaeontol.* 22:** 681—724 (***Dolerotheca***).

Scheckler, S.E. and H.P. Banks (1972). ***Amer. Jour. Bot***. (Abstracts) **59:** 663 (Cladoxylales).

and (1972). ***Ibid.* 59:** 663 (Aneurophylales).

Scott, D.H. (1897). ***Ann. Bot.*, 11:** 399—419 (Centripetal xylem in cycad cones).

— (1906). ***Trans. Linn. Soc. Lond. (Bot.)*, 7:** 45—68 (***Sutcliffa***).

and A.J. Massllu (1907). ***Ann. Bot. Lond.* 21:** 89—134 (***Trigonocarpus***).

Seelinger, I. (1954). ***Flora*, 141:** 114—162 (***Ephedra fragalis var. campylopoda***).

Sen, I. (1955). ***Bot. Notiser*, 108:** 244—252 (***Fertile Glossopteridaceae***).

— (1958). ***Ibid.* 111:** 436—448 (***Vertebraria***).

Seward, A.C. (1903). ***Quart. Jour. Geol. Soc. Lond.*, 59:** 217—233 (***Dictyozamites***).

— (1919). Fossil plants Vol. III and IV. ***Cambridge.***

— (1938). ***Sci. Progress*, 127:** 420—440 (***Ginkgo***).

— and B. Sahni (1920). Indian Gondwana Plants. A. revision ***Palaentol. India*** (N.S.J.) **7(1):** 1—41.

Shadle, G.L. and B.M. Slidd (1972). ***Amer. Jour. Bot***. (Abstracts) **59:** 663 (***Heterangium***).

Shapiro, S., (1951). ***Amer. Jour. Bot.*, 38**: 47—53 (Stomata on the nucellus of ***Zamia***).

Shimamura, T. (1937). ***Int. Jour. Cytol.* (*Tokyo*)**, 416—23 (Sperms of ***Ginkgo***).

— (1956). ***Bot. Mag.*** (Tokyo), **69:** 524—529 (***Pinus thunbergii***).

Singh, H., (1961). ***Phytomorphology*, 11:** 153—197 (***Cephalotaxus drupacea***).

— (1964). ***Ibid.***, 14: 527—529 (***C. drupacea***).

— and, Maheshwari (1962). ***Ibid.*, 12(4):** 361—372 (***Ephedra***).

—and J. Chatterjee (1963). ***Ibid.*, 13:** 429—445 (***Cryptomeria japonica***).

— and Y.P. Oberoi (1962). ***Ibid.*, 12:** 373—393 (***Biota orientalis***).

Shah, S., Gopal Singh and Gururaja (1973). Observations on the Post-Triassic Gondwana sequence of India. ***Palaeobotanist*, 20(2):** 221—237.

Sharma, B.D. (1967). A review of the genus ***Ptilophyllum*** of Morris, with description of two new species from Amarjola in the Rajmahal Hills. ***Palaeontographica*, 120B:** 139—150.

— (1967a). Investigations on the Jurassic flora of Rajmahal Hills, India. 4.On a new species of Indian ***Bucklandia*, *B.guptai*** with remarks on B.Sahnii Bose. ***Ameghiniana*, 4(2):** 35—46.

— (1968). Investigations on the Jurassic flora of Rajmahal Hills, India. 5. Epidermal studies on the bracts in two new species of ***Williamsonia, W. guptai and W. amarjolense. Acta Botanica Hungary*. 14:** 373—383.

— (1969). ***Bucklandia dichotoma*** sp. nov. from the Middle Jurassic of Rajmahal Hills, India. ***Ameghiniana*** **6(4):** 303—308.

— (1969a). Further observations on ***Williamsonia santalensis*** Sitholey and Bose with the description of a new species. ***Palaeontographica,*** **125B:** 93—103.

— (1969b). ***Guptioxylon amarjolense*** Gen. et sp. nov. from Amarjola in the Rajmahal Hills, India. ***Palaeontographica,*** **126B:** 145—153.

— (1969c). On Pentoxyleae remains from Amarjola in the Rajmahal Hills, India. ***Ameghiniana*** **6(1):** 50—56.

— (1970). On the structure of the seeds of ***Williamsonia*** collected from the Middle Jurassic rocks of Amarjola in the Rajmahal Hills, India. ***Ann. Bot.*** 34: 1071—1078.

— (1971). Further observations on ***Sewardioxylon sahnii*** Gupta collected from the Middle Jurassic of Rajmahal Hills, India. ***Acta Bot. Neerl.*** **20(5):** 475—480.

— (1972). ***Guptioxylon endocentrica*** sp. nov. du Jurassique D' Amarjola dans les Rajmahal Hills (Inde). ***Bull. Soc. Linn. Lyon.*** **41(6):** 114-119.

— (1973). Further observations on ***Pentoxylon sahnii*** Srivastava from the Jurassic of Amarjola in the Rajmahal Hills, India. ***Palaeobotanist*** **20(2):** 216-220.

Shukla, V.B. (1938). On a new species of ***Dadoxylon, D. deccani*** sp. nov. from the Deccan Intertrappean series. ***J. Indian bot. Soc.*** **17(5-6):** 335—362.

— (1944). ***Dadoxylon resinosum*** sp. nov. from the Chhindware district of the Central Provinces. ***J. Indian bot. Soc.*** **23(3):** 83—90.

Singh, G. (1956). ***Araucarites nipaniensis*** sp. nov. a female araucarian cone-scale from the Rajmahal series. ***Palaeobotanist,*** **5(2):** 54—65.

Singhai, L.C. (1958). On a new species of ***Dadoxylon, D. shuklai*** sp. nov. from the Deccan Intertrappean beds of Chindwara district. ***J. Palaeont. Soc. India,*** **3:** 136—141.

Sitholey, R.V. (1954). The Mesozoic and Tertiary floras of India—A review. ***Palaeobotanist,*** **3:** 55—69.

— (1963). Gymnosperms of India - 1. Fossil forms. ***Bull. Nat. Bot. Gard. No.*** **86:** 1—78.

— and M.N. Bose (1953). ***Williamsonia santalensis*** sp. nov. a male fructification from the Rajmahal series, with remarks on the structure of ***Ontheanthus polyandra*** Ganju. ***Palaeobotanist*** **2:** 29—39.

— and (1953). ***Palaeobotanist,*** **2:** 29—39 (***Williamsonia santalensis***).

— and (1971). ***Wiltrichia santalensis*** (Sitholey and Bose) and other Bennettitalean male fructifications from India. ***Palaeontographica***, **131B:** 151—159.

Slidd, B.M. and T.L. Phillips (1972). ***Ameri. Jour. Bot***. (Abstracts) 59: 664 (***Schopfiastrum decussatum***).

Smith, D.L. (1964). ***Biol. Rev.*** **39:** 137—59 (Evolution of the ovule).

— (1959). ***Ann. Bot.*** (n.s.) **23:** 477—491 (Geminitheca scotica).

— (1964). ***New Phytol.*** **63:** 418—421.

Sporne, K.R. (1965). ***The Morphology of Gymnosperms***. London, pp.216.

Spruce, R. (1908). Notes of a botanist on the Amazon and Andez, London.

Srivastava, B.P. (1946). Silicified plant remains from the Rajmahal series of India. ***Proc. Nat. Acad. Sci. India***, **15(6):** 185—211.

Srivastava, P.N. (1956). Studies in Glossopteris flora of India-4. ***Glossopteris, Gangamopteris*** and ***Palaeovittaria*** from the Raniganj coal field. Palaeobatoanist **5(1):** 1—45.

Srivastava, S.C. (1969). Two new species of ***Glossopteris*** from the Triassic of Nidpur, Madhya Pradesh, India. ***Sen. Mem. Volume:*** 299—303.

— (1971). Some gymnospermic remains from the Triassic of Nidpur, Sidhi district, Madhya Pradesh. ***Palaeobotanist***, **18(3):** 280—296.

Steeves, M.W., and E.S. Barghoorn 91959). ***J. Arnold Arbor***., **40:** 221—255 (Pollen of ***Ephedra***).

Sterling, C. (1948a). ***Bull. Torrey Bot***. Club., **75:** 147—165 (Gametophytes of ***Taxus cuspidata***).

— (1948b). ***Ibid***., **75:** 469—485 (Proembryo in T. ***cuspidata***).

— (1948c). ***Bot. Gai***., **109:** 531—534 (***Tsuga canadensis***).

— (1949). ***Amer***. Jour. Bot. **36:** 461—471 (***Metasequoia***).

— (1963). ***Biol***. Rev., **38 :** 167—203 (Male gametophytes in gymnosperms).

Stewart, W.N., and T. Delevoryas (1956). ***Bot. Rev.,*** **22:** 45—80 (Medullosan pteridosperms).

Stopes, M.C., (1905). ***Ann. Bot.*** **19:** 561—66 (Double nature of Cycad integuments).

Sugihara, Y., (1938). ***Sci. Rep. Tokoku Imp. Univ., Biol. Ser***. 4, **13:** 9—14 (Embryogeny of ***Chamaecyparis pisifera***).

— (1939). ***Ibid.***, **14:** 291—303 (Embryology of ***Thujopsis***).

— (1941). ***Ibid***., **16:** 187—192 (Embryology of ***Cunninghamia***).

— (1941b). ***Ibid***., **16:** 291—295 (Embryology of ***Taiwania***).

Sugihara, Y. (1943). ***Ibid***., **17:** 215—221 (Embryology of Keteleeria sp.).

— (1946). ***Bot. Mag***. (Tokyo) **59:** 61—67 (***Amentotaxus argotaenia***).

— (1947). ***Ibid***., **60:** 47—52 (Embryogeny of ***Cryptomeria japonica***).

— (1947b). ***Ibid***., **60:** 53—57 (Embryogeny of ***Cunninghamia sp***.).

— (1947c). ***Ibid***., **60:** 58—62 (Embryogeny of ***Abies firma***).

— (1947d). ***Ibid***., **60:** 45—46 (male gametes of Cephalotaxus drupacea).

— (1956a). ***Sci. Rep. Tohoku Univ***. (Biol). **22:** 1—4 (Embryology of Cupressus sempervirens).

— (1956b). Bot. Mag. (Tokyo) **69:** 439—441 (Embryology of Cupressus funiberis).

— (1956c). Saito Ho-on Kai Mus. Res., Bull. No. 25 (Torreya fruiticosa).

Surange, K.R. (1966). Distribution of Glossopteris flora in the Lower Gondwana formations of India.

Symposium on Floristics and Stratigraphy of Gondwana Land, Lucknow : 55—68.

— and P.N. Srivastva (1956). ***Palaeobotanist***, 5: 46—9 (Glossopteridaceae).

— and S. Chandra (1973). ***Denkania indica Gen***. et sp. nov. a Glossopteridean fructification from the Lower Gondwana of India. Palaeobotanist, **20(2):** 264—268.

and K.M. Lele, (1956). Studies in the ***Glossopteris*** flora of India-3. Plant fossil from Talchir Needle shales from Girdih coal field. ***Paleobotanist***, **4 :** 153—157.

— and (1973a). Partha, a new type of female fructification from the Lower Gondwana of India. ***Palaeobotanist*** **20(3):** 356—360.

— and K.M. Lele (1956a). Studies in the ***Glossopteris*** flora of India-6. Plant fossils from Talchir beds of South Rewa Gondwana basin. ***Palaeobotanist*** **5:** 82—90.

— and H.K. Maheswari (1970). Some male and female fructifications of Glossopteridales from India. ***Palaeontographica*** **129B:** 178—191.

— and P.K. Maithy (1962). Studies in the ***Glossopteris*** flora of India-14. Two new fossil woods from the Lower Gondwanas of India. ***Palaeobotanist*** **11:** 96—102.

and S.C.D. Sah (1956). Studies in the ***Glossopteris*** flora of India-7. Dadoxylon jhariense sp. nov. from the Jharia coal field, India. ***Palaeobotanist*** **5(2):** 100—103.

and P. Singh (1951). ***Walkomiella indica***, a new conifer from the Lower Gondwanas of India. ***J. Indian bot***. Soc. **30(1-4):** 143—147.

— and (1952). The seeds of ***Walkomiella indica*** — a conifer from the Lower Gondwanas of India. ***Curr. Sci***. **21:** 40—41.

— and (1953). A female dwarf shoot of ***Walkomiella indica*** — a conifer from the Lower Gondwanas of India. ***Palaeobotanist***, **2:** 5—8.

— and P.N. Srivastava (1956). Studies in the ***Glossopateris*** flora of India-5. Generic status of Glossopateris and Palaeovittaria. ***Palaeobotanist***, **5(1):** 46—49.

Suryanarayana, K. (1953). ***Mesembrioxylon tirumangalense*** a new species from the Sripermatur group near Madras. J. Indian bot. Soc. **32(4):** 159—164.

— (1954). Fossil plants from the Jurassic rocks of the Madras coast, India. ***Palaeobotanist*** 3: 87—90.

— (1955). ***Dadoxylon rajmahalense*** Sahni from the coastal Gondwanas of India. ***Palaeobotanist***, 4: 89—90.

Takeda, H., (1913). ***Ann. Bot***. (Lond.) **27:** 365—366 (***Gnetum gnemon***).

— (1913). ***Ibid***., **27:** 347—57 (***Gnetum and Welwitschia***).

— (1913). ***Ibid.***, **27:** 547—552 (***Gnetum and Welwitschia***).

Tahara, M., (1937). ***Cytologia*** (Fujii), 14—19 (***Sciadopitys verticillata***).

— (1940a). ***Sci. Rep. Tohoku Univ***. (***Biol***.) **15:** 19—28 (***Sciadopitys***).

— (1940b). ***Ibid***., **15:** 419—420 (Embryogeny of ***Torreya nucifera***).

— (1941). ***Ibid.***, **16:** 91—98 (***Podocarpus***).

Takao, A., (1959). ***Bot. Mag***. (Tokyo) **72:** 289—297 (***Pinus thunbergii***).

Tang. S.H. (1948a). ***Bot. Bull. Acad. Sincia*** (Shanghai) **2:** 13—18 (Embryogeny in ***Juniperus chinensis***).

— (1948b). ***Ibid.***, **2:** 269—275 (Embryogeny in ***Torrey grandis***).

Taylor, T.N. (1969). ***Science***, **164:** 294—295 (Fossil Cycads).

— (1973). ***Amer. Jour. Bot.*** (***Abstract***) **60:** 19 (Multicellular microgametophytes in Paleozoic and Mesozoic seed plants).

Thoday, M.G. and E.M. Berridge, (1912). Ann. Bot., **26:** 953—985 (***Ephedra,*** flowers and inflorescence).

Thomas, H.H. (1913). ***Quart. J. geol. soc. Lond.***, **69:** 223—51 (***Williamsonia***).

— (1915). ***Phil. Trans.***, **B207:** 113—48 (***Williamsoniella***).

— (1925). ***Ibid***., **B213 :** 299—363 (***Caytonianceae***).

— (1925). ***Ibid***., **B222:** 193—265 (Mesozoic pteridosperms).

— (1947). ***Adv. Sci***., **4:** 243—54 (History of plant form).

— (1952). ***Palaeobotanist,*** **1:** 435—438 (***Glossopteris***).

— (1955). ***Phytomorph***.,. **5:** 177—85 (Mesozoic Pteridosperms).

— (1958). ***Bull. Brit. Mus. nat. Hist***. (***Geol.***) **3:** 179—189 (***Lidgettonia***).

— and N. Bancroft (1913). ***Trans. Linn. Soc. Lond***. (***Bot***.) **8:** 155—204 (Cuticles of Cycas and Bennettitales).

— and T.M. Harris (1960). ***Senckenbergiaama***, **41:** 139—161 (Cycads).

Thompson, W.P., (1919). ***Bot. Gaz.***, **68:** 451—459 (Phloem of Gnetum).

— (1912). ***Ann. Bot***. **26:** 1077—1104 (***Ephedra***, Anatomy).

— (1916). ***Amer. Jour***. ***Bot***. 3: 135—184 (Affinities of ***Gnetum***).

— (1918). ***Bot. Gaz.,*** **65 :** 83—90 (Vessels in Gnetum and angiosperms).

Townrow, J.A. (1956). ***Avh. Norske***. ***Vidensk. Akad. Oslo,*** 1—28 (***Lepidopteris***).

— (1957). ***Trans. geol. soc. S. Africa***, **60:** 21—56 (***Dicroidium***).

— (1960). ***Palaentology***, **3:** 333—61 (Peltaspermaceae).

— (1962). ***Bull. Brit***. ***Mus. nat. Hist***. (Geol.) **6:** 289—320.

Traverse, A. (1950). ***Amer. Jour. Bot.***, **37:** 318—325 (***Mesoxylon***).

Turrill, W.B. (1959). In ***Vistas in Botany***, ed. W.B. Turrill. London (Gymnospermae).

Tulecke, W., (1957). ***Amer. Jour. Bot.***, **44:** 602—608 (Pollen of ***Ginkgo***).

------------- (1938) The germination of pollen grains in artificial cultures in ***Ephedra foliata.*** Bioss and ***E. gerardiana*** wall Proc. Ind. Acad. Sci., **8:** 218–230.

------------- (1957) The pollen of *Ginkgo biloba* : in vitro culture and tissue formation. *Am. J. Bot*. **44:** 602-608.

Thomas. M.J. (1970). Premieres recherches sur les besoins Mutrifits des embryons isoles du Pinus silvestries L. Embryons indifferencies et en voie de differenciation. C.R. Acad. Sci., Paris, **270 :** 1120-1123.

Thorpe T.A. (1988). Physiology of bud induction in conifers In vitro. In : J. Hanover & D. Keathley (eds.) *Genetic Manipulation in Woody plants* Plenum Press, NY.

Tulecke, W & Sehgal, N. (1963). Cell proliferation from the pollen of *Torreya nucifera* Contr. Boyce Thompson Inst., **23:** 33–46.

Tulecke, W, Taggart, R. & Colavito, L. (1965). Continuous cultures of higher plant cells in liquid media. Contr. Boyce Thompson Inst. **23** :33-46

Tulecke, W. & Nickell, L.G. (1960). Methods, problems and results of growing plants under submersed conditions. trans. *N.Y. Acad. Sci. Sec.* III. **22 :** 196-206.

Tulecke, W. (1953). A tissue derived from pollen of *Ginkgo biloba*. *Science* (Wash.) **117:** 599–600.

Tulecke, W., Weinstein, L.H., Rutner, A. & Laurencot Jr., H.J. (1962). Biochemical physiological studies of tissue cultures and the plant parts from which they were derived, II, *Ginkgo biloba* cont. *Boyce Thompson* Inst. **21 :** 291-302

Varma, C.P., (1954). On two new species of *Mesembrioxylon* from the Cretaceous rocks of Trichinopoly district, Madras. *Palaeobotanist*, **3:** 97—102.

(1963). *Glossopteris* fructifications from Chintalpudi sandstone, S. India. *Curr. Sci*. 32: 75—77.

Vasil, Vimla (1959). *Phytomorphology*, **9:** 167—214 (*Gnetum ula*).

— and R. K. Sahni (1964). *Ibid*., **14:** 369—384 (*Taxodium mucronatum*).

Vazart, B., (1958). *Rend. Acad. Sci*. (*Paris*). **247:** 952—955.

Vishnu-Mittre (1953). *Palaeobotanist*, **2:** 75—84 (Pentoxylales).

— (1957). *Ibid*., **6:** 31—46 (Pentoxylales).

Voth, P.D., (1934). *Bot. Gaz*., **96:** 298—313 (*Ephedra*).

Van Arnold S., Hakman. (1986). Effects of sucrose on imiation of embryogenic callus from mature zygotic embryos of *Picea abies* (L.) Karst. (Norway spruce). J. Plant. Physiol. **122 :** 251-265.

Vasil, V. (1963). In vitro culture of embryos of *Gnetum ula* Brong, pp278-280. In : Maheshwari P. & Rangaswamy. N.S. (eds) Symp. Plant tissue and organ culture. Intl. Soc. Plant Morphologists; Univ. Delhi.

Wang, F., Chen, T.K. & Lee, S.C. (1963). Experimental studies of young embryos of Ginkgo. I. The effect of the bee royal jelly on the embryo growth. *Acta Botanica Sinica*. **13 :** 217-222.

Wang. F. & Chen T.K. (1965). Experimental Studies on young Ginkgo embryos. The effect of coconut milk on the embryos cultured in vitro. *Acta Botanica Sinica*. **13 :** 224-231.

Wadia, D.N. (1957). *Geology of India*. London.

Walton, J. (1926). *Phil. Trans*. **B215:** 201—224 (Carboniferous fern like fronds).

— (1931). ***Ibid.***, **B219:** 347—379 (Carboniferous fern like fronds).

— (1949). ***Trans. roy. soc. Edinb.***, **51:** 719—728 (***Calathospermum***).

— (1953). ***Adv. Sci.***, 10: 223—230 (Evolution of pteridosperms ovules).

— (1957). ***Trans. roy. soc***. Edinb., **63:** 333—40 (Protopitys).

— and J.A.R. Wilson, (1932). ***Proc. roy. soc. Edinb***. 52: 200—207 (***Vertebraria***).

Waterkeyen, L. (1954). ***Cellule***, **56:** 105—46 (***Gnetum africanum***).

— (1959). ***Cellule***, **60**: 7—78 (***G. africanum***).

— (1960). ***Cellule***, **61**: 81—95 (***G. africanum***).

Wang, F.H. (1947). ***Bot. Bull. Acad. Sinica*** (Shanghai), 1: 133—140 (***Keteleeria***).

— (1948a). ***Amer. Jour. Bot.***, 35: 21—27 (***Keteleeria***).

— (1948b). ***Bot. Bull. Acad. Sinica*** (***Shanghai)*** 2: 1—12 (***Glyptostrobus***).

— (1950). Ibid. 3: 141—145 (***Podocarpus nagi***).

Wesley, A. (1963). In Advances in ***Botanical Research***, Vol. I. ed. R.D. Preston. London and New York. (The status of some fossil plants).

Whiteside, K.L. (1972). ***Amer. Jour. Bot.***, 59: 665 (Cordaitean stem).

Wilde, M.H., (1944). ***Ann. Bot.***, 8: 1—41 (A new interpretation of coniferous cones—Podocarpaceae).

Wieland, G.R. (1906). American fossil cycads. ***Carnegie Inst. Publ. No***. 34.

Wilson, CW. (1942). ***Amer. Jour. Bot***. 29: 759—764.

Williamson, W.C., (1870). ***Trans. Linn. Soc. Lond***. (***Bot.***), 26: 663—74 (***Williamsonia gigas***).

Webb K.J. & Street H.E. (1977). Morphogenesis in vitro of ***Pinus*** and ***Picea***. Acta Hort. 78 : 259-269.

White, P.R. 1967; Risser, P.G.(1964). Some basic parameters in the cultivation of spruce tissue. Physiol. Plant, **17:** 600-619.

Yamamoto, Y., (1932). ***Icon. Plant. formos., Supp***l., 5: 7—11 (***Amentotaxaceae***).

Ziegler, H. (1959). Planta 52: 587—599.

Zimmermann, W. (1933). ***Palaeobiologia***, 5: 321—348 (***Williamsoniella***).

— (1952). ***Palaeobototanist***, 1: 456—470 (The main result of Telome theory).

Zenkteler, M.A. & Guzowska, I (1970). Cytological studies on regenerating mature female gametophyte of ***Taxus baccata*** and mature endosperm of ***Tilia*** platyphyllus scop. in vitro cultures. ***Acta Soc. Bot. Pol.*** **39 :** 161-173.

REFERENCE BOOKS

Andrews, H.N. (1961). ***Studies in Paleobotany***. New York and London.

Arber, A., (1950). ***The Natural Philosophy of Plant Form***. Cambridge.

Arnold, C.A., (1947). ***An Introduction to Palaeobotany***. New York and London.

Bierhorst. D.W. (1971). ***Morphology of Vascular Plants***. New York and London.

Bowen, R.N.C (1958). ***The Exploration of Time***. **London.**

Chamberlain, C.J. (1934). ***Gymnosperms—Structure and Evolution***. Chicago.

Coulter, J.M. and Chamberlain, C.J. (1917). ***Morphology of Gymnosperms.*** Chicago.

Dallimore, W. and Jackson, A.B. (1948). ***A Handbook of Coniferae including Ginkgoaceae***. London.

Delevoryas. T. (1962). ***Morphology and Evolution of Fossil Plants***. New York and London.

Foster, A.S. and Gifford, E.M. (1959). ***Comparative Morphology of Vascular Plants***. San Francisco.

Gausoen, H. (1950). ***Les gymnosperms actuelles et Fossiles***. Tailouse.

Greguss, P. (1955). ***Identification of Living Gymnosperms on the Basis of Xylotomy***. Budapest.

Guttenberg, Herman Ven (1961). ***Grundzuge Der Histogenose Hoherer Pflanzen II. Die Gymnospermen***. Berlin.

Napp-Zinn, Klaus (1966). ***Anatomic des Blattes I. Blattananatomic der Gyymnospermen***, Berlin.

Maheshwari, P., and Vasil, Vimla (1961). ***Gnetum*** Delhi.

Pant, D.D. and Mehra, B. (1962). ***Cycas***. Allahabad, India.

Pearson, H.H.W. (1929). ***Gnetales***. Cmpbridge.

Scott. D.H. (1929). ***Studies in Fossil botany . Vol. 2. Spermatophyta***. London.

Seward, A.C. (1898—1919). ***Fossil Plants for Students of Botany and Geology***. 4 volumes. Cambridge.

Walton, J. (1953). ***An Introduction to the Study of Fossil Plants***. London.

Wieland, (1906 and 1916). ***American Fossil Cycads***. 2 volumes. Washington.

Zimmermann, W. (1959). ***Die phylogenie der Pflanzen***. Stuttgart.

Index

I

J

L

M